Metallocenes

Volume 1

Edited by
A. Togni and R. L. Halterman

⊛ WILEY-VCH

Metallocenes

Volume 1

Synthesis • Reactivity • Applications

Edited by
Antonio Togni and Ronald L. Halterman

⊛ WILEY-VCH

Weinheim · New York · Chichester · Brisbane · Singapore · Toronto

Professor Dr. Antonio Togni
Laboratorium für Anorganische Chemie
Eidgenössische Technische Hochschule
Universitätstrasse 6
CH-8092 Zürich
Switzerland

Professor Dr. Ronald L. Halterman
Department of Chemistry and Biochemistry
University of Oklahoma
620 Parrington Oval
Norman, OK 73019
USA

Library of Congress Card No. applied for

A catalogue record for this book is available from the British Library.

Deutsche Bibliothek – CIP-Einheitsaufnahme

Metallocences : synthesis reactivity applications / ed. by Antonio Togni and Ronald L. Halterman.
– Weinheim ; New York ; Chichester ; Brisbane ; Singapore ; Toronto : Wiley-VCH
    ISBN 3-527-29539-9

Vol. 1 (1998)

Vol. 2 (1998)

© WILEY-VCH Verlag GmbH, D-69469 Weinheim (Federal Republic of Germany), 1998

Printed on acid-free and chlorine-free paper.

Composition: Johanna Boy, D-93179 Brennberg
Printing: Strauss Offsetdruck, D-69509 Mörlenbach
Bookbinding: Großbuchbinderei Osswald&Co., D-67433 Neustadt

Preface

The cyclopentadienyl ligand (Cp') is ubiquitous in the organometallic chemistry of the transition elements. It is hard to imagine what this research field would be without this ligand. Soon after the discovery of ferrocene in 1951, the chemistry of so-called metallocenes, bis(cyclopentadienyl) metal derivatives, has been developed for most d-elements. Thus, strictly speaking a metallocene is a "sandwich" compound, being expressed by the formula $[M(\eta^5C_5R_5)_2]$. However, it is currently accepted to designate as metallocenes also compounds, e.g., containing further ligands, the so-called bent metallocenes such as Cp_2WH_2, or complexes in which one of the Cp fragments has been replaced by an anionic ligand fulfilling similar coordination-chemical purposes.

In this monograph a dozen of internationally recognized experts in the field provide up-to-date and critical overviews concerning metallocenes and their applications. However, the book is not intended to be a comprehensive and systematic treatise of a compound class representing roughly three quarter of the elements. Therefore, the topic of each single chapter has been selected according to its importance, as reflected by current reasearch activities. Part 1 is focussed on synthesis, properties, and reactivity, whereas Part 2 concentrates on applications, in particular homogeneous catalysis and materials science.

Metallocenes containing main-group elements have long been considered as the "exotic" representatives of this class of compounds. Jutzi and Burford (Chapter 1) demonstrate the wealth of structural and coordination-chemical features of general theoretical interest of derivatives of this kind that have been reported in recent years. Organolanthanide chemistry is to be identified with the use of cyclopentadienyl ligands, much more than for the transition metals. Chapter 2 (Edelmann) covers mainly synthetic aspects related to the lanthanocenes whose importance also in catalytic chemistry is steadily increasing. The related scandocenes and yttrocenes have received much attention in the last decade because of their role in C-H activation reactions and in catalytic processes involving olefins. Bercaw and Chirik review this area in Chapter 3. Titanocenes (Beckhaus, Chapter 4) and Zirconocenes (Negishi and Montchamp, Chapter 5) are nowaday extremely important compounds with an enormous application potential. Thus, from a variety of stereoselective organic transformations to modern Ziegler-Natta olefin polymerization, these two metallocenes have contributed most significantly to the progress of modern synthetic chemistry. For the organometallic chemist, an important class of metallocenes is the one containing the elements of groups 5 and 6 (Royo and Ryan, Chapter 6). This because of the multifaceted chemistry these compounds display, and the wealth of derivatives and structural types that is not to be found for other metallocenes. Further, Okuda and Eberle discuss in Chapter 7 the chemistry of "half-sandwich" complexes, as these are recognized as metallocene analogues. Since titanocene and zirconocene dichlorides are fundamental starting materials for use in organic synthesis, both as reagents and catalysts, Haltermann describes in Chapter 8 recent developments in the synthesis of chiral derivatives for applications in stereoselective reactions.

Part 2 of the book starts with a review of what probably is to be seen as the hottest topic in metallocene chemistry, i.e., their application as catalysts for olefin polymerization. This most exciting area is reviewed by Janiak in Chapter 9. Hoveyda and Morken in Chapter 10 present an overview of organic synthesis utilizing enantiomerically pure titanocene and zirconocene derivatives, most prominently C_2-symmetric *ansa*-metallocenes. Ferrocene is the "oldest" prototype metallocene but research in the field of ferrocene chemistry is still very active. This has been reviewed in the companion monograph of this book: *Ferrocenes. Homogeneous Catalysis. Organic Synthesis. Materials Science* (A. Togni, T. Hayashi, Eds., VCH, 1995). Therefore, the only aspect that has been taken into consideration here is the application of new chiral ferrocenyl ligands in asymmetric catalysis (Togni, Chapter 11). This particular area has recently seen the first large scale industrial application of a metallocene ligand. Metallocene fragments, and most prominently ferrocenes, have also been incorporated into polymeric structures. This kind of materials is of particular appeal because of the interesting electrical, magnetic, and optical properties they exhibit, differentiating them from other, more common polymers. This interdisciplinary area is reviewed in the final Chapter 12 by Manners and coworkers.

We anticipate that the monograph will be of utility to many members of the chemical community, from graduate students to practitioners in industry, as well as from organic to inorganic chemists. It is also our hope that it will further stimulate an already extremely active research area.

We gratefully acknowledge the excellent work done by all authors, as well as VCH-Wiley for the collaboration at all stages of this book project.

ETH Zürich,   Antonio Togni
Norman, Oklahoma   Ronald Halterman
April 1998

Contents

2   Lanthanocenes

F. T. Edelmann

5   Zirconocenes

E. Negishi and J.-L. Montchamp

6   Group 5 and Group 6 Metallocenes

P. Royo and E. Ryan

7   Half-Sandwich Complexes as Metallocene Analogs

J. Okuda and T. Eberle

8   Synthesis of Chiral Titanocene and Zirconocene Dichlorides

R. L. Halterman

Volume 2 Applications

10   Chiral Titanocenes and Zirconocenes in Synthesis

A. H. Hoveyda and J. P. Morken

11   New Chiral Ferrocenyl Ligands for Asymmetric Catalysis

A. Togni

12   Metallocene-Based Polymers

T. J. Peckham, P. Gómez-Elipe and I. Manners

List of Contributors

Rüdiger Beckhaus
Department of Inorganic Chemistry
Aachen Technical University
D-52056 Aachen
Germany

John E. Bercaw
Arnold and Mabel Beckman Laboratories
of Chemical Synthesis
California Institute of Technology
Pasadena, CA 91125
USA

Neil Burford
Department of Chemistry
Dalhousie University
Halifax
Nova Scotia B3H 4J3
Canada

Paul J. Chirik
Arnold and Mabel Beckman Laboratories
of Chemical Synthesis
California Institute of Technology
Pasadena, CA 91125
USA

Thomas Eberle
Departmet of Inorganic Chemistry
and Analytical Chemistry
Johannes-Gutenberg-University Mainz
Johann-Joachim-Becher-Weg 24
D-55099 Mainz
Germany

Frank T. Edelmann
Department of Chemistry
Otto-von-Guericke-University Magdeburg
Universitätsplatz 2
D-39106 Magdeburg
Germany

Paloma Gómez-Elipe
Department of Chemistry
University of Toronto
80 St. George Street
Toronto, ON M5S 3H6
Canada

Ronald L. Halterman
Department of Chemistry and Biochemistry
University of Oklahoma
Norman, OK 73019
USA

Amir H. Hoveyda
Department of Chemistry
Merkert Chemistry Center
Boston College
Chestnut Hill, Massachusetts
USA

Christoph Janiak
Department of Inorganic and
Analytical Chemistry
Freiburg University
Albertstraße 21
D-79104 Freiburg
Germany

Peter Jutzi
Department of Chemistry
University of Bielefeld
Universitätsstraße 25
D-33615 Bielefeld
Germany

Ian Manners
Department of Chemistry
University of Toronto
80 St. George Street
Toronto, ON M5S 3H6
Canada

Jean-Luc Montchamp
Department of Chemistry
Purdue University
West Lafayette, Indiana 47907
USA

James P. Morken
Department of Chemistry
University of North Carolina
Chapel Hill
North Carolina
USA

Ei-ichi Negishi
Department of Chemistry
Purdue University
West Lafayette, Indiana 47907
USA

Jun Okuda
Department of Inorganic Chemistry
and Analytical Chemistry
Johannes-Gutenberg-University Mainz
Johann-Joachim-Becher-Weg 24
D-55099 Mainz
Germany

Timothy J. Peckham
Department of Chemistry
University of Toronto
80 St. George Street
Toronto, ON M5S 3H6
Canada

Pascual Royo
Department of Inorganic Chemistry
Alcalá University
28871 Alcalá de Henares
Spain

Evelyn Ryan
Department of Inorganic Chemistry
Alcalá University
28871 Alcalá de Henares
Spain

Antonio Togni
Laboratory of Inorganic Chemistry
Swiss Federal Institute of Technology
ETH-Zentrum
CH-8092 Zürich
Switzerland

Volume 1
Synthesis and Reactivity

1 Main Group Metallocenes

Peter Jutzi and Neil Burford

1.1 Abstract

Cyclopentadienyl complexes of the elements of groups 1, 2, 13, 14 and 15 are classified and compared in terms of their synthesis, characteristic properties and structural types, which include a variety of sandwich and half sandwich complexes as well as dimeric, oligomeric and polymeric arrangements. The variety of structure and bonding features reveal exciting new directions in the coordination chemistry of the main group elements, and the factors governing the structures of multicyclopentadienyl complexes are discussed.

1.2 Introduction

The metallocene icon of organometallic chemistry originally refers to bis(cyclopentadienyl)metal complexes ('sandwich'), but wider usage is now accepted to include cyclopentadienyl complexes ('half sandwich') and multicyclopentadienyl complexes ('multidecker sandwich') as well as complexes with additional substitution at the metal center [1]. Nevertheless, the metallo- prefix is restrictive when one considers the more recent development of complexes involving nonmetallic elements such as boron, silicon or arsenic. In fact, metallocene-like complexes are now known for many elements in the periodic table.

This chapter provides an overview of the extensive series of compounds in which a main group element (Groups 1, 2, 13, 14 and 15) is π-bound to one or more cyclopentadienyl ligands. Synthetic procedures, characterization features, structural types and bonding are classified in general terms and then classic examples of complexes involving elements from each of the periodic groups are examined in some detail [2]. The discussion is restricted to complexes involving a 6π-electron cyclopentadienyl ligand, however, it is important to recognize that an extensive chemistry is also developing for main group complexes involving cyclopentadienyl ligands with extended π-frameworks [3], as well as heterocyclopentadienyls [4], and other isolobal analogs of cyclopentadienyl such as arene [3, 5], carbollyl [3], and borazolyl [6].

Current assessments of main group metallocene chemistry are delayed by about two decades in comparison to the established chemistry of transition metallocenes, and most conclusions derive from solid state and gas phase structural data. While these compounds have features in common with transition metal analogs, a new avenue of organometallic chemistry is defined by structural novelty and synthetic utility. Most complexes are air sensitive materials and involve standard formal oxidation states, but the cyclopentadienyl ligand imparts unusual stability to the lower oxidation state for the elements of group 13 (+1) and group 14 (+2).

Main group elements interact with cyclopentadienyl ligands in a variety of fashions [3, 7, 8] which we specify structurally in terms of the position of the element with respect to the pentagonal cylinder defined by the cyclopentadienyl ring. For compounds in which the element resides inside the cylinder, the element may be closer to one carbon center of the cyclic frame, which is referred to as a monohapto interaction, and is abbreviated as η^1. If the element is equidistant from more than one carbon center of the cyclic frame, the complexes are described as having hapticities of η^2, η^3, η^4 (rare) or η^5, respectively. Each of these types of complex ($\eta^1-\eta^5$) are classified here as metallocenes. When the element resides outside the cyclopentadienyl cylinder, the carbon atom to which it is closest usually exhibits significant distortion from trigonal planarity. In addition, the bond lengths and angles within the cyclopentadienyl ring are nonequivalent and tend towards those of cyclopentadiene [9]. Such compounds are considered to involve sp^3 hybridization at the element bound carbon center and are classified as σ-complexes.

Di- and tri-hapto complexes are sometimes described as involving peripheral coordination [10, 11], which imposes a substantial distortion on the cyclopentadienyl plane for transition metal analogs and a localized double bond, as illustrated in **1**. In contrast, main group metallocenes typically exhibit relatively minor perturbation of the C–C bonds within the cyclopentadienyl framework and the planarity is retained as the hapticity is apparently decreased, as depicted in **2**. Moreover, the energetic distinctions between σ- and multi-hapto structures are small for main group elements [12]. This accounts for routine observation of haptotropic shifts and consequential fluxionality [13], as well as the observation of a single ^{13}C NMR resonance for the carbon atoms of the cyclopentadienyl cyclic frame in the solid state spectra of complexes that are shown to be less than pentahapto coordinated on the basis of X-ray crystallographic studies [14]. Most convincing are the recent experimental and theoretical studies of tricyclopentadienylaluminum derivatives, which reveal negligible energy differences (4–8 kJ · mol^{-1}) between hapticities [15]. Clear distinction between σ and π interaction is not possible for many cyclopentadienyl derivatives of the main group elements, and there are uncertainties in the classification of hapticity.

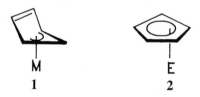

    M   E

    1   **2**

The definition of a bonding model for cyclopentadienyl derivatives is further complicated by the degree of ionic character in the ligand–element interaction. A much wider range of electronegativities is available in the main group than for the transition metal elements. Therefore, while most p-block metallocenes are considered to involve covalent interactions between the element and the cyclopentadienyl ligand, the alkali and heavier alkaline earth metallocenes are viewed as principally ionic, and yet both have structural features related to the transition metal analogs. This diversity in bonding types has resulted in the evolution of two compound nomenclatures. The more ionic compounds are usually referred to as 'element cyclopentadienides' with the formulation ECp_n, while the covalent compounds are usually referred to as 'cyclopentadienylelement' with the formulation Cp_nE. We retain both types of nomenclature and formulation in this chapter but do not necessarily infer a type of bonding, as most compounds exhibit both ionic and covalent characteristics.

The development of main group metallocenes has been facilitated by the use of substituted cyclopentadienyl ligands. The most common of these is pentamethylcyclopentadienyl (abbreviated as Cp*), which often gives complexes greater thermal stability than the parent cyclopentadienyl analogs [16, 17]. The range of other derivatized ligands are abbreviated here as $Cp^{nSUBSTITUTION}$, where the superscript indicates the number (n = 1–5) and type of the substituents (M = methyl, I = isopropyl, B = tertiary butyl, Ph = phenyl, Bz = benzyl, D = dimethylsilyl, T = trimethylsilyl, P = diphenylphosphino; $Cp^* = Cp^{5M}$; a number of unique and mixed substitution patterns are defined when referred to in the text; Cp' is used as a general abbreviation for Cp and substituted derivatives).

The bulky nature of many Cp' ligands can also impose interesting distortion on the complex [18, 19, 20, 21], but less obvious is the fact that they are good leaving groups, allowing for substitution reactions with element–carbon bond cleavage in many situations [16, 22]. The introduction of a side chain functionality (Cp^N, Cp^O, N = dimethylaminoethyl, O = methoxyethyl) with the potential for auxiliary coordination to the element center also offers a new dimension to the development of metallocenes with diversified chemistry and novel molecular structure [23, 24].

1.3 Synthetic Procedures

Consistent with general synthetic principles of organometallic chemistry, the formation of main group element–Cp' bonds can be most conveniently achieved using metathesis reactions. For this reason, the alkali metal cyclopentadienides, magnesium dicyclopentadienides and thallium cyclopentadienides are important in terms of their role as synthetic sources of Cp' ligands. This and other generally applicable methods, as well as specific reactions (used for particular complexes), are classified below with key examples, and synthetic procedures for particular complexes can be found in the references given in the discussion of those complexes.

1.3.1 Reactions Involving Cyclopentadienes

$$2E + 2Cp'H \rightarrow 2ECp' + H_2 \text{ or } E + 2Cp'H \rightarrow ECp'_2 + H_2 \tag{1-1}$$
$$ER + Cp'H \rightarrow ECp' + RH \tag{1-2}$$
$$EH + Cp'H \rightarrow ECp' + H_2 \tag{1-3}$$
$$ENH_2 + Cp'H \rightarrow ECp' + NH_3 \tag{1-4}$$

Cyclopentadiene carries a relatively acidic proton and reacts smoothly with a number of the more reactive main group elements Eq. (1-1), such as sodium [25] and magnesium [26], and certain metallocenes can be prepared by co-condensation of the cyclopentadiene with the element [27, 28, 29]. LiCp is most conveniently prepared by reaction of a lithium alkyl (usually Li^nBu) with cyclopentadiene Eq. (1-2) [30]. This procedure has been extrapolated to other Cp' derivatives [31, 32], and applies to other organometallic reagents such as $MgMe_2$ [33, 34, 35] and Al_2Me_6 [36]. Sodium or potassium cyclopentadienides can be obtained from reaction of the element hydride with cyclopentadiene Eq. (1-3), or reaction of the element amide with the cyclopentadiene in liquid ammonia Eq. (1-4) [32]. A similar amide displacement reaction is used for calcocene [37] and barocene [8] derivatives, although some of these complexes are formed in donor solvents and are consequently isolated with these solvents bound as auxiliary ligands. Cp'Tl derivatives are routinely prepared from the corresponding cyclopentadiene Eq. (1-2) and thallium hydroxide or thallium salts (usually TlOEt) in basic aqueous solution [38, 39].

1.3.2 Salt Metathesis Reactions

$$nMCp' + EX_n \rightarrow Cp'_nE + nMX \text{ (M = Li, Na)} \tag{1-5}$$
$$\text{or } n/2MgCp'_2 + EX_n \rightarrow Cp'_nE + n/2MgX_m$$

The most generally applied procedure for the preparation of metallocenes involving the elements of groups 2, 13, 14 and 15, is the appropriate stoichiometric combination of an element halide (X) or polyhalide with an alkali metal cyclopentadienide salt (usually lithium or sodium) or a magnesium dicyclopentadienide salt [3, 18, 40]. Complexes of aluminum and gallium involving the element in less familiar low oxidation states are obtained from metastable halides 'AlCl' [41] and 'GaCl' [42, 43], which are formed *in situ*. In comparison, InCl [44] is readily available as a starting material, and a variety of thallium(I) salts can be used [45].

1.3.3 Disproportionation, Comproportionation and Decomposition

$$2Cp'EX \rightarrow Cp'_2E + EX_2 \text{ and } 2Cp'EX \leftarrow Cp'_2E + EX_2 \qquad (1\text{-}6)$$

Magnesocene and its methylated derivatives can be prepared by thermal decomposition of Cp'MgBr [46, 47, 48, 49], and the opposite comproportionation reaction can be used to obtain substituted derivatives of half sandwich complexes for magnesium [50], germanium [51] and tin [52, 53]. Reductive elimination of Cp ligands from Cp_3In [54] and of phenyl groups from Cp_2PbPh_2 [55] is also thermally induced.

1.3.4 Halide Ion Abstraction

$$Cp'_nEX + A \rightarrow [Cp'_nE][AX] \qquad (1\text{-}7)$$

Lewis acids such as aluminum halides (A) are well known as effective halide ion abstraction agents, thereby generating a cationic system from a formerly covalent halide. The procedure is responsible for the formation of half sandwich complexes (n = 1) of boron [56, 57], germanium [3] and tin [58, 59, 60, 61] and the sandwich complexes (n = 2) of arsenic [62], antimony [63], and bismuth [64].

1.3.5 Substitution onto the Cp Ligand Frame

$$ECp_n + AR \rightarrow Cp^RECp_{n-1} + AH \quad \text{or} \quad ECp_n + nAR \rightarrow ECp^R_{n-1} + nAH \qquad (1\text{-}8)$$

Thallium and tin metallocenes have a relatively high stability, which allows for modification of the Cp ligand with retention of the complex. The procedure has contributed to the preparation of a particularly extensive series of stannocenes, including Cp' ligands bearing silyl [65], stannyl [66], phosphine [67], or phosphenium [68] moieties, and is responsible for the formation of the novel $Cp^{tricyanovinyl}Tl$ [69, 70].

1.3.6 Reduction

$$Cp'_nEX_2 + M \rightarrow Cp'_nE + MX_2 \qquad (1\text{-}9)$$

Reductive dehalogenation can be achieved for silicon, germanium and tin in $Cp*_2EX_2$ [71, 72], for aluminum in $[Cp*AlCl_2]_2$ [73] and can be thermally promoted for Cp_3In [54] and Cp_2PbPh_2 [55] (see above).

1.3.7 Addition of Cp Anions

$$Cp'E + [Cation][Cp'] \rightarrow [Cation][Cp'_2E] \tag{1-10}$$

Additional cyclopentadienide anions can be introduced onto lithium, sodium, cesium, thallium, tin and lead cyclopentadienyl compounds to give anionic multicyclopentadienyl complexes. This is achieved by reaction of the neutral metallocene with sources of nucleophilic cyclopentadienide in salts with complex cations such as tetraphenylphosphonium [74, 75] and trisdimethylaminosulfonium [76].

1.4 Characterization Features

Characteristic infrared and Raman spectroscopic features were pivotal in the early identification and assignment of main group metallocenes, however, as with many areas of modern inorganic chemistry, X-ray crystallography has played the leading role in the discovery and development of these compounds. Consequently, the majority of the discussion in this chapter focuses on structural features in the solid state, and the implications for bonding and reactivity. Nevertheless, there are a number of important spectroscopic and physical properties to make note of.

The most enlightening spectroscopic feature of main group metallocenes is the distinctive element NMR chemical shift (for elements with NMR active nuclei), examples of which are listed in Table 1-1 in comparison with typical δ values

Table 1-1. Examples of element chemical shifts (ppm) for metallocene derivatives

Nucleus	Metallocene	Ref.	δ	δ [ER$_n$]	Ref.
^{7}Li	LiCp^{3T}	[82]	−11	0 to −0.7 [LiMe]	[83]
^{6}Li	[Li(12-crown-4)$_2$][Li^{iso}Cp$_2$]	[78]	−13	−2 to +2 [LiR]	[78]
^{27}Al	Cp*Al ({Cp*Al}$_4$)	[84]	−150 (−80)	153 [AlMe$_3$]	[85]
^{27}Al	[Cp*$_2$Al][Cp*AlCl$_3$]	[86]	−114	153 [AlMe$_3$]	[85]
^{25}Mg	Cp*Mg·OEt$_2$	[85]	−34	99 [MgEt$_2$]	[85]
^{25}Mg	Cp*$_2$Mg	[85]	−78	99 [MgEt$_2$]	[85]
^{29}Si	Cp*$_2$Si	[87]	−398	0 [SiMe$_4$]	[88]
^{71}Ga	Cp*Ga	[43]	−653	730 [GaMe$_3$]	[89]
^{119}Sn	[Cp*Sn][BF$_4$]	[81]	−2247	0 [SnMe$_4$]	[88]
^{119}Sn	Cp*$_2$Sn	[81]	−2129	0 [SnMe$_4$]	[88]
^{207}Pb	[Cp*Pb][BF$_4$]	[81]	−5041	0 [PbMe$_4$]	[88]
^{207}Pb	Cp*$_2$Pb	[81]	−4390	0 [PbMe$_4$]	[88]

(ppm) for alkyl derivatives of those elements. The dramatic upfield shift is general-ly rationalized in terms of ring current phenomena [77, 78, 79]. Observation of coupling between Cp' nuclei and NMR active elements can be interpreted as an in-dication of covalent character for the E–Cp' interaction and the J_{E-H} coupling for derivatives of tin and lead represent appropriate examples. The simulated 31 line multiplet for $Cp*_2Sn$ and $Cp*_2Pb$ in the ^{119}Sn and ^{207}Pb NMR spectra, respectively, are observed (Sn δ, –2129 ppm; $^3J_{Sn-H}$, 3.7 Hz [72, 80]; Pb δ, –4384 ppm; $^3J_{Pb-H}$, 11.2 Hz) [81] as 16 lines for intensity reasons, as shown for the stannocene in Fig. 1-1.

Melting point trends for the sandwich metallocenes of groups 2 and 14, summa-rized in Table 1-2, have been analyzed in detail [21]. The parent sandwich metal-locenes and symmetrically substituted derivatives (*e.g.* decamethyl) typically have a relatively high melting point in comparison to derivatives with single or asym-metric substitution patterns around the Cp' ligand. The observations are interpreted

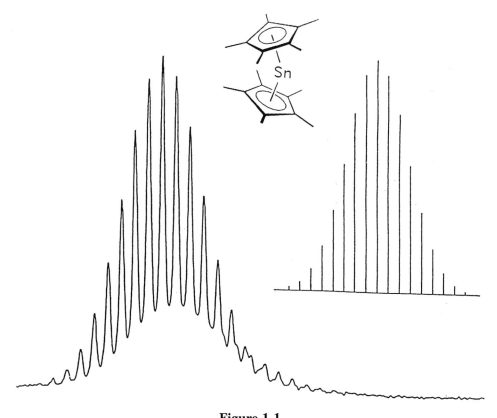

Figure 1-1
^{119}Sn NMR signal for $Cp*_2Sn$ and comparison with the simulated signal

in terms of the disruption of crystal packing patterns with the introduction of partial substitution which is restored as the substitution is completed [21]. Similar trends are evident for volatility and solubility, although mediated by the increase in molecular weight [19]. For example, a decrease in sublimation temperature is observed as the degree of Cp' substitution is increased, consistent with a decrease in the magnitude of intermolecular interactions.

Table 1-2. Melting points (°C) for selected sandwich metallocene complexes, Cp'$_2$E, for elements of groups 2 and 14 (taken from Ref. [21])

	Cp	Cp^B	Cp^{2T}	Cp^{3I}	Cp^{4I}	Cp*
Mg	176	<25	80–83	92–100	229–231	289–292
Ca	>265	301–304	189	40–45	196–200	207–210
Sr	>360	375–380	150	<25	151–153	216–218
Ba	>420	320	222	92–94	149–150	265–268
Ge	78	–	72	–	–	90–94
Sn	105	<25	solid	18	162–163	100–121
Pb	140	–	–	–	–	110–105

1.5 Structural Types

The structural motifs exhibited in main group metallocene chemistry are varied in terms of the position of the element with respect to the Cp' ligand, the number of Cp' ligands associated with a element center and their relative orientation, the involvement of auxiliary ligands (other than Cp' derivatives) with the element, and the association of the complexes into dimers, oligomers or polymers. The general cyclic fluxionality of the Cp' ligand implies that all five of the carbon atoms in the Cp' frame are involved in the Cp'–element π-interaction, even for systems exhibiting low hapticity. This is apparently frozen out in the solid state so that the nonequivalent element–carbon distances obtained from X-ray crystallographic data often illustrate a closer interaction to some framework carbon atoms than others. This can be defined by the similarity or difference between the Cp'$_{centroid}$–E distance and the Cp'$_{plane}$–E distance (**3**).

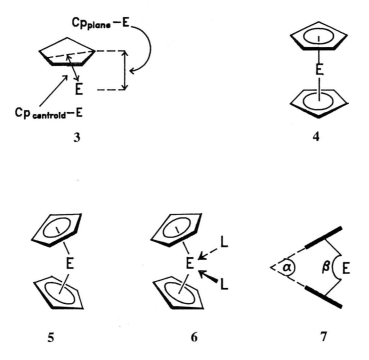

3

4

5

6

7

1.5.1 Sandwich Complexes

Sandwich structures represent the prototypical metallocenes and, as implied by the term, they involve two Cp' ligands, one on either side of the element. Within each periodic group of elements, there are examples of sandwich complexes with the two ligand planes parallel **4** (usually staggered D_{5d}, but eclipsed D_{5h} conformations are often close in energy), as well as examples of nonparallel **5** (often referred to as 'bent' sandwich structures) complexes. The presence of one (*e.g.* $Cp^{2T}_2Ca\cdot THF$) [27] or more (*e.g.* $Cp^B_2Sr\cdot THF_2$) [90] auxiliary ligands is usually observed with a nonparallel arrangement of Cp' ligands **6**, but are not necessarily responsible for the nonparallel structure. Most sandwich complexes involve an equivalent interaction of the element with each Cp' ligand, although the unusual dimeric structure of $LiCp^{5Bz}$ involves a rare example of asymmetric coordination of two Cp' ligands to one of the lithium centers [91].

For the purpose of structural comparison between all sandwich complexes (**7**) we define the angle α between the Cp' ligand planes ($\alpha = 0°$ in complexes of type **4**), as well as the Cp'$_{centroid}$–E–Cp'$_{centroid}$ angle β ($\beta = 180°$ in complexes of type **4**). The anionic sandwich complexes of the alkali metals, $[LiCp_2]^-$ [74] and $[NaCp_2]^-$ [75], represent examples of D_{5d} arrangements **4**, as do magnesocene [92] and the aluminocenium cation $[Cp^*_2Al]^+$ [86]. Cp^*_2Si adopts two conformations

($\alpha = 0°$ and 25°) in the same crystal [87],but most parent and methylated complexes for elements of groups 2, 13, 14 and 15 are typically nonparallel **5**, with α angles as large as 47° in Cp_2Sn [93]. However, these angles decrease as the size and number of substituents around the Cp' frame is increased so that Cp^{5Ph}_2Sn has a parallel structure **4** [94]. The ions $[Cp_2Tl]^-$ [95], $Cp'_2As]^+$ [62], and $[Cp^*_2Sb]^+$ [63] adopt nonparallel sandwich structures **5**, however, bulkier substituents (*e.g.* $[Cp^{3B}_2Sb]^+$ [63] and $[Cp^{3B}_2Bi]^+$) [64] again impose a parallel planar arrangement **4**.

1.5.2 Half Sandwich Complexes

A wider range of element environments are accessible for complexes involving a single π-coordinated Cp' ligand, summarized in the general structural representations **8–12**, in which Y corresponds to a nonbonding electron pair, a covalent bond (substituent R), or a dative covalent bond (auxiliary ligand). Univalent group 13 elements engage cyclopentadienyl derivatives in monomeric half sandwich 'Milking stool' type structures **8** (Y = lone pair) in the gas phase (*e.g.* Cp*Al) [96], and the isovalent cations $[Cp^*E]^+$ (E = group 14) have been characterized in the solid state (*e.g.* $[Cp^*Sn][BF_4]$) [61]. Molecular half sandwich complexes are readily obtained for most elements with the presence of an auxiliary substituent(s) or ligand(s), depending on the valence and the size of the element. The most extensive series of structural type **8** are known for lithium (*e.g.* $Cp^{3T}Li·THF$) [97] and beryllium (*e.g.* CpBeCl) [98], and cationic systems are accessible for boron (*e.g.* $[Cp^*BBr][AlBr_4]$) [99]. The presence of two auxiliary electron pairs **9** can be achieved for lithium with a bidentate ligand (*e.g.* $Cp^TLi·TMEDA$) [100]. However, for elements of group 14, reduction of the Cp' hapticity gives structures of type **10** (*e.g.* Cp*GeCl, Y = lone pair and Cl [101]; $[Cp^*Sn·pyridine][BF_4]$, Y = lone pair and pyridine) [102]. 'Piano stool' structures **11** are evident in the unique examples of $[CpMg·PMTEDA]^+$ containing a tridentate ligand [95] and the inverted sandwich cation $[TAS–Cp–TAS]^+$ (TAS = trisdimethylaminosulfonium) [76], as well as in the aluminum dimer $\{Cp^*AlCl_2\}_2$ [103]. However, the hapticity is reduced, as shown by **12**, in alkylated derivatives, such as $\{Cp^*Al(Cl)Me\}_2$ [104].

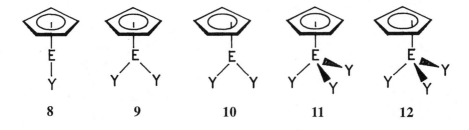

8   9   10   11   12

1.5.3 μ–π–Cyclopentadienyl Molecular Complexes

Association of element centers by π-bound μ-Cp' ligands is a general feature of polymeric structures (see below), but is also evident in a number of molecular anionic systems, representing neutral metallocenes bound together by an additional cyclopentadienide anion. The general formula $[E_2Cp_3]^-$ is observed as a nonparallel multidecker sandwich **13** for cesium [105] and thallium [106], and plumbocene sandwich units, Cp_2Pb, are associated in the unusual anions $[Cp_5Pb_2]^-$ **14** and $[Cp_9Pb_4]^-$ **15** [107]. Each lead center accommodates three cyclopentadienyl ligands, as in the tris(cyclopentadienyl) anions **16**, which are known for both tin [108] and lead [109] in the salts $[Mg(THF)_6][Cp_3E]_2$.

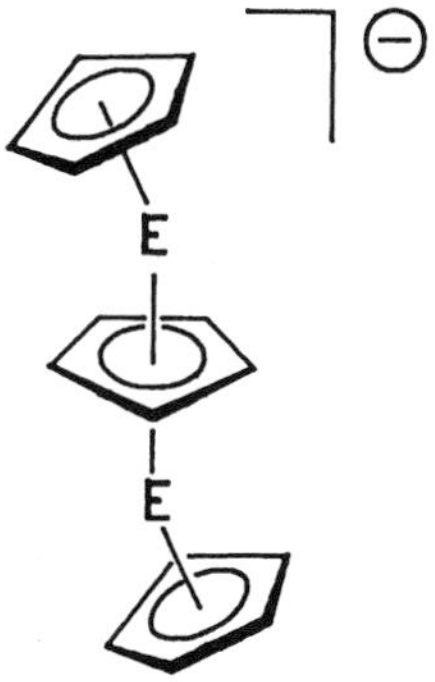

13

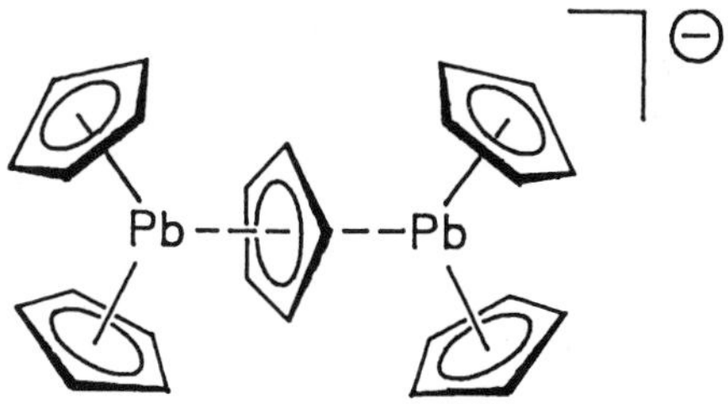

14

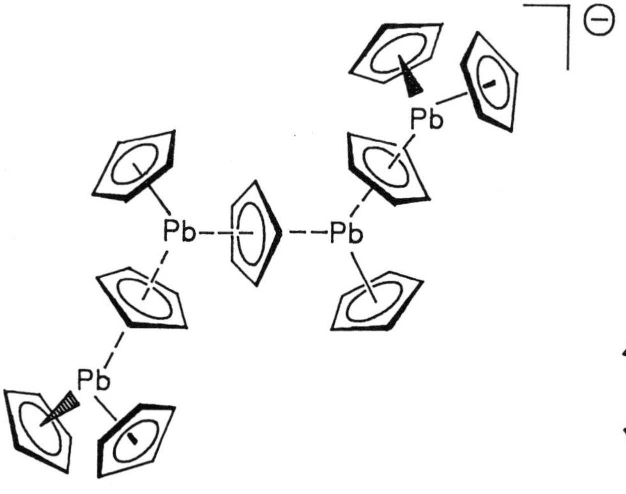

15

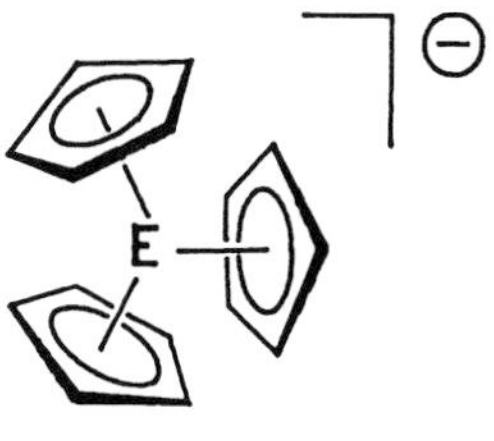

16

1.5.4 μ–π–Cyclopentadienyl Super-Sandwich Polymers

The half sandwich formulation for alkali metal Cp' derivatives leads to polymeric arrays in the solid state involving alternating element and μ–π–Cp' ligands. The planes of the Cp' ligands may be parallel or almost parallel (referred to as super-sandwich), as in the case of LiCp and NaCp **17**, or nonparallel, as in the case of KCp **18** [110]. The branched polymeric structure observed in the solid state for the sandwich metallocene Cp$_2$Pb **19** [111] reveals a tricoordinate environment for each lead center composed of two μ–π–Cp ligands (dashed lines) and one terminal Cp ligand (full line), although a slight distortion from pentahapticity is apparent. The solid state structure of BaCp*$_2$ **20** [112] is at first glance related, but the monomers are associated by means of relatively weak Ba–methyl interactions.

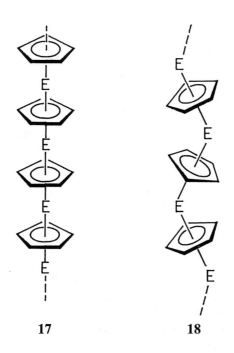

17   **18**

19   **20**

1.5.5 Dimers

Unique examples of dimeric arrangements **21** are observed in the solid state for Cp^{5Bz}In [113] and Cp^{5Bz}Tl [114] (one of two crystalline forms), whilst more familiar dimeric structures involving halide bridges with terminal half sandwich Cp' complexes **22** (X = halogen; Y = halogen or donor) are adopted by compounds of magnesium [115] calcium [116], aluminum [117], and bismuth [118].

21   **22**

1.5.6 Clusters

With the influence of certain types of Cp' ligands, complexes involving monovalent elements of group 13 (Cp'E), which are monomeric in the gas phase, adopt highly symmetric cluster arrangements in the solid state. A tetrahedral structure is observed for Cp*Al **23** [41] while Cp*Ga [41a] and Cp*In [119] adopt octahedral hexameric structures **24** (methyl groups have been omitted for clarity). Cp^{2T}Tl forms a hexameric cluster structure, but involves only μ–π–Cp interactions between element centers **25** [120].

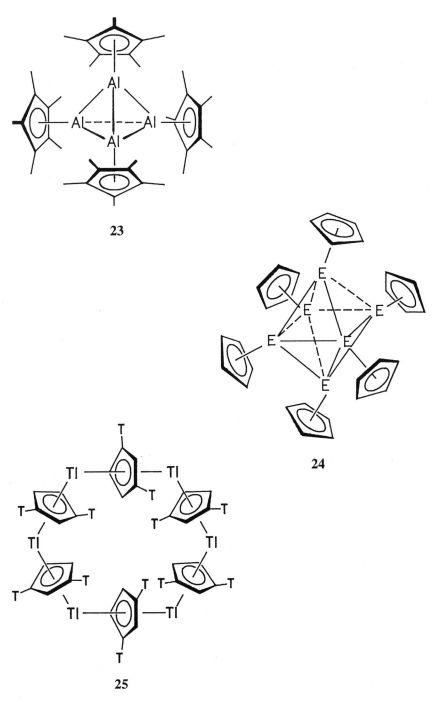

23

24

25

1.6   Cp-Element Orbital Interactions

Main group elements encompass a much wider range of electronegativities and hardness than the transition series so that the Cp'–E bond varies from interactions which are almost purely ionic for elements from group 1, to predominantly covalent for elements from groups 14 and 15. Theoretical studies of LiCp at various levels of sophistication promote an ionic classification and describe a structure with C_{5v} symmetry [78, 121, 122, 123]. Nevertheless, some degree of covalent character is apparent in all main group metallocenes, and orbital interactions likely dominate for metallocenes of the p block elements.

In this context, the decrease in $Cp'_{centroid}$–E distances observed for germanium and tin metallocenes within the series Cp'_2E, CpECl and CpE$^+$(Ge, 220, 211, 199 pm; Sn, 240, 230, 229 pm) is somewhat consistent with the 'octet rule'. Although the differences in these distances are small and may be attributed to a number of factors, including the variety of hapticities, the general trends can be rationalized with the Lewis drawings **26, 27** and **28**, illustrating the adjustment in Cp'E bond order.

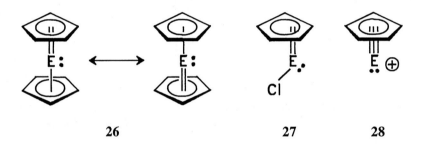

| **26** | **27** | **28** |

1.6.1   Orbital Interaction Model for Half Sandwich Complexes

A general qualitative molecular orbital energy correlation diagram for half sandwich complexes with C_{5v} symmetry is given in Fig. 1-2. The most effective stabilization occurs between the fully bonding (lowest energy a_1) occupied π-orbital of the Cp ligand and the s and p(axis) orbitals of E. A secondary interaction involves the two occupied π-orbitals (e_1) of Cp with the degenerate p orbitals of E (perpendicular to the E–$Cp_{centroid}$ axis).

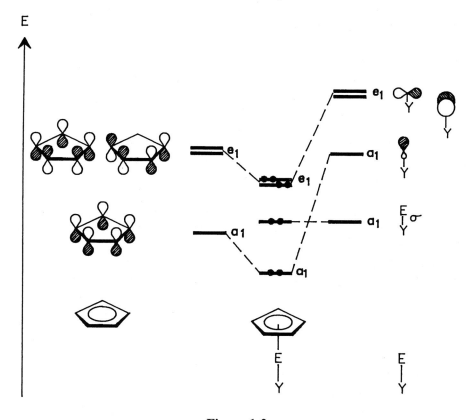

Figure 1-2
General molecular orbital energy diagram for main group half sandwich complexes
with C_{5v} symmetry, showing only occupied molecular orbitals

In Fig. 1-2, Y is a coordinate bond if one invokes the model for the elements of group 1, and a covalent bond for elements of group 2 involved in neutral complexes. Cationic complexes of group 13 elements involve a covalent bond (*e.g.* [CpB-Br]+), while neutral group 13 (*e.g.* CpAl) complexes and cationic complexes of group 14 (*e.g.* [CpSn]+) carry a nonbonding electron pair for Y. The validity of this model is demonstrated by calculated MO energies for beryllium complexes being in close agreement with measured vertical ionization energies obtained from PE spectra [3, 124] and calculated NMR chemical shifts for aluminum and gallium complexes in good agreement with experimental data [125].

Most of the compounds for which this model is invoked are characteristic of *nido* clusters with a pentagonal pyramidal framework [126] when the Cp–E interaction is considered in isolation, but the three occupied bonding MOs [123] (a_1 and e_1) can also be described in terms of a 'three dimensional aromaticity' model [122]. Inclusion of the σ-pair on E nicely fulfills the octet model as depicted in **28**, and recognition of the lower hapticity generally observed for CpECl complexes is also consistent in **27** if the Cp ligand is considered as contributing four valence electrons rather than six.

1.6.2 Orbital Interaction Model for Sandwich Complexes

Covalent bonding in sandwich complexes can be modeled qualitatively by the general molecular orbital schemes for parallel planar (D_{5d}) and bent (C_{2v}) structures containing 14 valence electrons, shown in Fig. 1-3. The stabilization of three electron pairs (D_{5d}: $1a_{1g}$, $1e_{1u}$; C_{2v}: $1a_1$, $2b_1$, $1b_2$) is primarily responsible for the Cp–element bond in both structures. The absence of an element orbital of a_2 symmetry in the C_{2v} structure and element orbitals of e_{1g} symmetry in the D_{5d} structure distinguishes the models from those for transition metallocenes, imposing ligand based nonbonding orbitals. While the 'octet rule' is inappropriate for systems considered as coordination complexes, it generally provides useful simple models to appreciate the bonding in compounds of p block elements and is applied consistently for half sandwich complexes in **27** and **28**. A Lewis model for the sandwich structures is less obvious, but the six electrons which are unaccounted for in the resonance drawings **26** for the 14 valence electron system correspond to three nonbonding ligand based molecular orbitals, and the $2a_{1g}$ orbital (D_{5d}) and $2a_1$ orbital (C_{2v}) correspond to the respective element based 'lone pair'.

  Most discussions of structural preferences for 14 electron sandwich metallocenes focus attention on the energy and occupation of the $2a_{1g}$ orbital in a D_{5d} structure, which correlates with the 2_{a1} orbital in a C_{2v} structure. Structural adjustment from D_{5d} to C_{2v} symmetry effects a stabilization of the $2a_{1g}$ orbital due to loss of partial antibonding character on transformation into the more nonbonding $2a_1$ orbital (C_{2v}). The observed structures of neutral Cp_2E (E = group 14) and cationic $[Cp_2E]^+$ (E = pnictogen) sandwich metallocenes are consistent with this model, which can be understood in more simple terms using VSEPR arguments, implicating the $2a_1$ orbital (C_{2v}) as the so-called 'lone pair'. However, the energy of the $2a_{1g}$ orbital in the D_{5d} structure is also sensitive to the separation of the Cp ligands [127]. For a hypothetical planar (D_{5d}) germanocene, the $2a_{1g}$ orbital is calculated to be above the $1e_{1g}$ level at small Cp–Cp separations (400 pm), it decreases in energy with increasing Cp–Cp separation, and is below the $1e_{1g}$ level for Cp–Cp separations in excess of 500 pm, consistent with the antibonding character of the $2a_{1g}$ orbital. Therefore, observed sandwich metallocene structures are likely the result of a number of factors rather than solely, the most obvious, electronic control. In fact, the electronic imposition or consequences of the nonparallel structure may be minimal. Experimental support for the deep energy of the formal lone pair ($2a_{1g}$ or $2a_1$) on the element center is provided in the photoelectron spectroscopic data for Cp^*_2E (E = Si, Ge, Sn, Pb), which are compared in Table 1-3 [128]. Although the ordering of the orbital energy levels for the heavier elements (Ge, Sn, Pb) is as

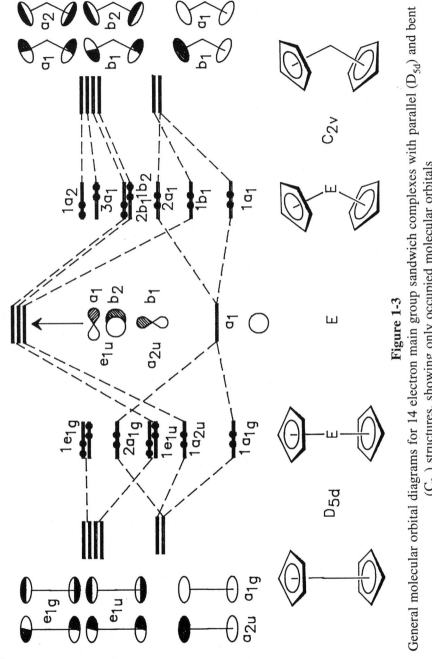

Figure 1-3

General molecular orbital diagrams for 14 electron main group sandwich complexes with parallel (D_{5d}) and bent (C_{2v}) structures, showing only occupied molecular orbitals

shown in Fig. 1-3, the 2_{a1} orbital is anomalously high in relative energy (7.50 eV) in the silicocene, consistent with the relatively close proximity of the Cp* ligands and the more effective overlap of the 3p valence orbitals of silicon in comparison with the heavier elements.

Table 1-3. Orbital energies (eV) for derivatives of Cp'_2E (E = Si, Ge, Sn, Pb) from photoelectron spectroscopic and theoretical data

Orbital	$Cp*_2Si$	$Cp*_2Ge$	$Cp*_2Sn$	$Cp*_2Pb$
1_{a2}	6.70	6.60	6.60	6.33
3_{a1}	6.96	6.75	6.60	6.88
2_{b1}	8.06	7.91	7.64	7.38
1_{b2}	8.30	8.05	7.64	7.38
2_{a1}	**7.50**	8.36	8.40	8.93

The wide variety of theoretical molecular orbital studies on 14 electron systems has been comprehensively reviewed [19, 20], concluding that although nonparallel (bent) structures represent global minima, this can only be recognized with higher order theory, as very small energy differences (<10 kJ $\cdot$ mol^{-1}) distinguish parallel from nonparallel structures.

Although metallocenes of elements such as the alkali metals are predominantly ionic, modification of Fig. 1-3 involving destabilization and vacancy of the $2a_{1g}$ orbital provides a model for 12 electron systems consistent with the observed structures for anionic (*i.e.* [ECp$_2$]$^-$, E = alkali metal), neutral (*i.e.* [Cp'$_2$Mg]) and cationic (*i.e.* [Cp*$_2$Al]$^+$) sandwich metallocenes. However, observation of nonparallel structures for the neutral 12 electron Cp$_2$E metallocenes involving the heavier alkaline earth elements (E = Ca, Sr, Ba) are inconsistent with this model. As a result, these complexes have attracted intense theoretical molecular orbital (extensively reviewed in Refs 19 and 21) and molecular mechanics [129, 130] studies. An overview assessment of the factors governing the structure of multicyclopentadienyl complexes is presented toward the end of this chapter.

The absence of d orbitals impacts both the electronic structure and the reactivity of main group metallocenes. Electron density remains at the ligands in nonbonding molecular orbitals, which directs the majority of reactivity. This contrasts the chemistry of transition metallocenes where the frontier orbitals are more metal based.

1.7 Metallocenes of the Alkali Metals

LiCp [121] as well as the sandwich anion [LiCp$_2$]$^-$ [131] have both been termed the 'simplest metallocenes'. Although the colorless microcrystalline solid formulated as LiCp was first prepared in 1901 [132], conclusive structural characterization was delayed due to the fact that it is essentially insoluble in nonpolar solvents. The structure of the sandwich anion was also only recently determined.

1.7.1 Monomers and Polymers

The solubility of alkali metallocenes of the general formula ECp' is enhanced by the introduction of trimethylsilyl or tert-butyl substituents onto the cyclopentadienyl frame, and hexane solutions of these complexes can be prepared *in situ*. How-

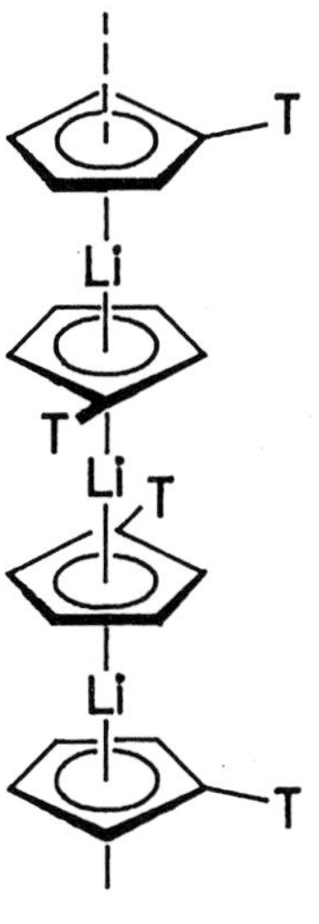

29

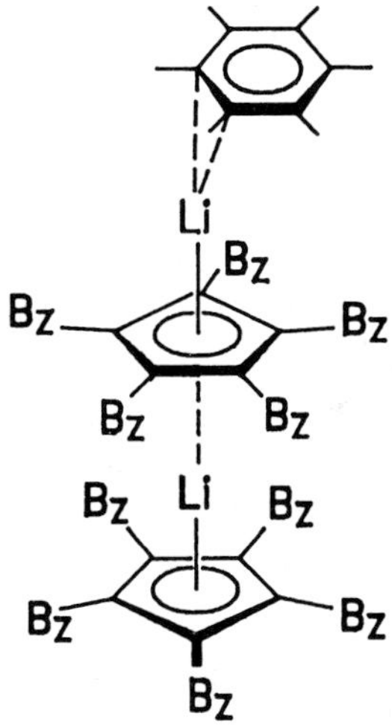

30

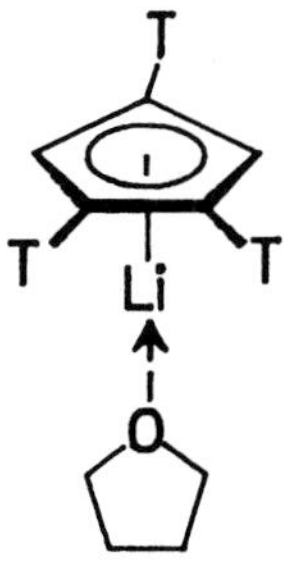

31

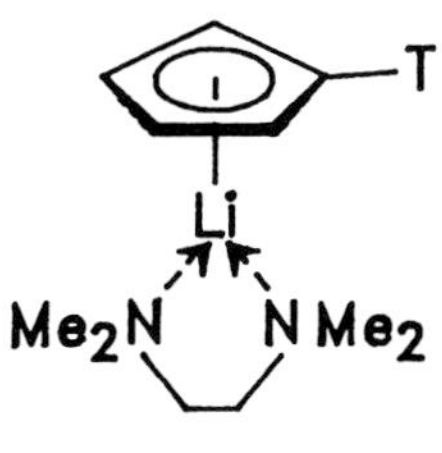

32

ever, when concentrated, a sudden and irreversible increase in viscosity occurs with precipitation of an amorphous powder which is subsequently insoluble in hexane [97], and crystalline solids are inaccessible [133]. Nevertheless, the mono-silylated derivative LiCp^T was unexpectedly isolated as a crystalline by-product of the reaction between [(η^5Cp^T)Y(O^{t}Bu)]$_2$ and LiCH$_2$SiMe$_3$ [134]. It crystallizes as a 'super-sandwich' of stacked half sandwiches, with the Cp^T ligands staggered and the positioning of the SiMe$_3$ groups describing a screw trace along the stack, as depicted in **29**. A contrasting dimeric structure **30** is observed for the more heavily substituted salt LiCp^{5Bz}, where a benzene solvate molecule essentially fragments the polymer [91]. This prevention of coordinative polymerization is complete with more effective donors, such as nitrogen or oxygen, which enhance the solvation of LiCp' derivatives, and numerous derivatives (*e.g.* **31** [97] and **32** [100]) have been comprehensively characterized.

The dilithium cationic complex [Cp^M(Li·TMEDA)$_2$]$^+$ (Cp$^M_{centroid}$–Li, 200 pm) **33** and the related mixed metal complexes Cp[(SiMe$_3$)$_2$N]Sn(μ-Cp)Li·PMDETA **34** (Cp$_{centroid}$–Li, 225 pm) and CpTl(μ-Cp)Li·TMEDA **35** (Cp$_{centroid}$–Li, 223 pm) [95] involve μ–π–Cp interactions, which are substantially weaker than those in the super-sandwich structure of LiCp^T (Cp$^T_{centroid}$–Li 196 and 198 pm), but represent models of the 'inverted sandwich' moiety.

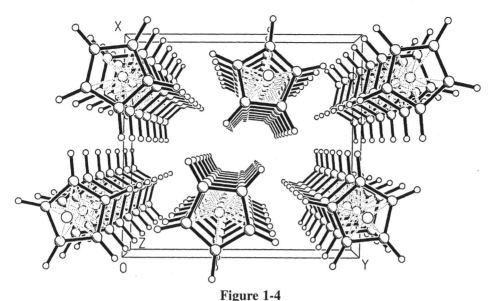

Figure 1-4

The solid state structure of LiCp viewed down the super-sandwich strands

Nearly 100 years after the initial discovery of a parent alkali metal cyclopentadienide salt [132] the structure was ultimately confirmed for LiCp, NaCp and KCp
by high resolution powder X-ray diffraction. A super-sandwich structure of type **17**
is observed for LiCp and NaCp, with eclipsed and almost parallel–planar Cp rings
(LiCp, β = 176°; NaCp, β = 178°) [110] as shown in Fig. 1-4. In contrast, KCp
adopts a 'zig-zag' pattern of type **18** with a dramatically smaller β angle (138°),
and auxiliary interactions to the potassium center from two Cp ligands of neighboring polymer chains [110]. The presence of a silyl substituent in KCp^T (cf. LiCp^T)
imposes a slightly larger β angle at potassium (151°) [135]. Nevertheless, the
availability of additional coordination sites made possible by the nonparallel arrangement of the Cp^T ligands essentially modifies the relative stability of the polymeric array with respect to the influence of auxiliary ligands.

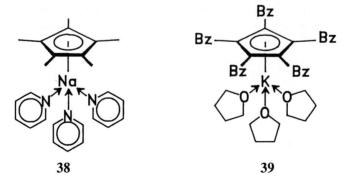

36   **37**

In contrast to derivatives of LiCp, derivatives of NaCp and KCp accommodate donor solvents (auxiliary ligands) with retention of the super-sandwich. The structure of Cp*K·(pyridine)$_2$ **36** (β = 138°) [136] can be considered a modification of KCp (β = 138°) and KCp^T (β = 151°), in which the interchain Cp'–K contacts are replaced by the two auxiliary ligand interactions, but with minimal or no impact on the structure of the polymer. The auxiliary coordination in CpNa·TMEDA **37** (β = 128 and 119°) [137] effects a dramatic distortion of the polymeric backbone, relative to the base free NaCp polymer (β = 178°). Nevertheless, NaCp' and KCp' polymers can be disrupted by the introduction of excess donors. For example, NaCp* dissolves in pyridine to give the monomeric piano stool structure **38** [136] which is related to **39** [138].

38   **39**

1.7.2 Lithocene and Sodocene Anions

The exothermicity for the association of alkali metal ions with cyclopentadienide is calculated [75, 139] to be substantially greater for lithium than for sodium:

$[Li]^+_{(g)} + [Cp]^-_{(g)} \rightarrow LiCp_{(g)}$ $\Delta H = -703$ kJ $\cdot$ mol^{-1} [75] (-761 kJ.mol^{-1}) [139] (1-11)

$[Na]^+_{(g)} + [Cp]^-_{(g)} \rightarrow NaCp_{(g)}$ $\Delta H = -531$ kJ $\cdot$ mol^{-1} [75] (-607 kJ.mol^{-1}) [139] (1-12)

However, formation of the sandwich anions from the corresponding half sandwiches is more exothermic for sodium:

$LiCp_{(g)} + [Cp]^-_{(g)} \rightarrow [LiCp_2]^-_{(g)}$ $\Delta H = -163$ kJ $\cdot$ mol^{-1} [75]   (1-13)

$NaCp_{(g)} + [Cp]^-_{(g)} \rightarrow [NaCp_2]^-_{(g)}$ $\Delta H = -184$ kJ $\cdot$ mol^{-1} [75]   (1-14)

This is due to stronger repulsive interactions between the [Cp]$^-$ anions, which are closer in lithocene, and outweighs the advantage of the stronger [Cp]$^-$–[Li]$^+$ interaction. Consistent with the discussion of polymer–monomer relationships above, donor solvents such as THF dramatically influence coordination of the second Cp ligand, and an equilibrium is envisaged:

$CpE \cdot L_{n\ (solv)} + [Cp]^-_{\ (solv)} \rightleftharpoons [ECp_2]^-_{\ (solv)} + nL_{(solv)}$ (1-15)

Conductivity measurements on liquid ammonia solutions [140], as well as IR [141] and NMR [78] spectroscopic data on THF solutions, first indicated the existence of lithocene anions. However, the use of complex cations enabled isolation and structural confirmation of [Li(12-crown-4)$_2$][Li^{iso}Cp$_2$] (containing anion **40**) [142] and [Ph$_2$PMe$_2$][LiCp$^B_2$] [143], as well as tetraphenylphosphonium [74] and trisdimethylaminosulfonium [76] salts of the parent anion **41**. The anions are parallel planar and almost perfectly staggered with consistent Cp'$_{centroid}$–Li distances (200 pm). The introduction of excess NaCp is necessary to effect a shift of the equilibrium in favor of the sodocene **42** salt [PPh$_4$][NaCp$_2$] [74], which has a crystal structure that is remarkably similar to that of the analogous lithocene salt [74]. The equilibrium can also be modified by tethering the Cp ligands together to give the *ansa*-sodocene compound **43** [74], which has a bent sandwich structure, opening a coordination site for an auxiliary THF molecule. In cesium chemistry, the use of the complex cation [PPh$_4$]$^+$ also facilitates the formation of the multidecker anion [Cs$_2$Cp$_3$]$^-$ of the type **13** [105].

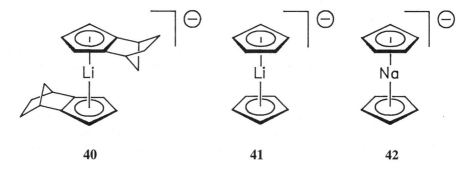

40   41   42

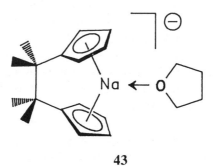

43

1.8   Metallocenes of the Alkaline Earth Metals

Comparisons are frequently made between transition metallocenes and those of the
alkaline earth metals in light of their preferred oxidation state of +2 [144]. The variety of structural arrangements observed for alkaline earth metallocenes include
the typical parallel sandwich for magnesium, the surprising nonparallel sandwich
for heavier elements, and the rare 'slip–sandwich' structure of beryllocene.

44   45

1.8.1 Beryllocene

The solid state structure of beryllocene has been determined a number of times [10, 145, 146] due to the unique asymmetric arrangement of the parallel planar Cp ligands shown in **44**, which is also observed in the gas phase [147]. A model **45** involving a σ-bonded ligand has been proposed on the basis of theoretical studies, invoking disruption of the parallel relationship of the ligands [148]. However, structure **44** accounts for the microwave spectroscopic data [149], the large dipole moment ($\mu_{25\,^\circ C}$ in benzene, 2.46 D; in cyclohexane, 2.24 D) [150], and the fluxionality of the complex in solution, which is clearly demonstrated by a sharp singlet observed in the ^1H NMR spectrum at $-135°C$ [151]. The estimated rate of 10^{10} s^{-1} at 300K corresponds to an activation energy of 5.2 kJ $\cdot$ mol^{-1} [152] and the process is proposed to occur *via* two isomerization mechanisms with barriers of 5 ('gear wheel' having a η^5/η^5 transition state) and 8 kJ $\cdot$ mol^{-1} ('molecular inversion' having a η^3/η^3 transition state), respectively.

Consistent with the fluxional behavior in solution, the solid state structure of beryllocene is disordered with the beryllium atom in two sites of 50% occupancy so that each Cp ligand is engaged in both η^5 and η^1 bonding, as illustrated in **44**. The beryllium atom is 150.5 pm from one Cp plane and 151.3 pm from the other [146], and the structural parameters, Raman data [153], and PE data [154] imply that it is bound to an sp^2 (η^1) hybridized carbon center with minor perturbation of the π-bonding in the ligand. A similar structure has been observed for the isovalent Cp*$_2$Zn [155].

The preference for the η^5/η^1 structure of Cp$_2$Be can be understood when one considers the Cp$_{plane}$–Cp$_{plane}$ distance for the hypothetical D$_{5d}$ structure (302 pm), which is substantially shorter than the Cp$_{plane}$–Cp$_{plane}$ distance in ferrocene (332 pm) [156]. The term 'steric oversaturation' has been used to account for systems in which no more than one ligand of a particular type can be accommodated by an element center [18]. The phenomenon can be applied to Cp$_2$Be, as attempts to prepare the substituted derivative Cp*$_2$Be have been unsuccessful giving instead the monosubstituted half sandwich compound Cp*BeCl [157].

1.8.2 Alkaline Earth Sandwich Metallocenes

The numerous derivatives of sandwich complexes for the heavier alkaline earth metals have been extensively reviewed [19–21,40], as well as their potential utility as MOCVD sources [158]. Magnesocene adopts a typical staggered, parallel sandwich arrangement **46** in the solid state [92] but is eclipsed in the gas phase [159]. Although a similar structure is observed for Cp'$_2$Mg in the gas phase [47, 48] a slight distortion is imposed in the solid state by the introduction of silyl substituents in Cp$^{3T}_2$Mg ($\alpha = 8°$) [35]. Calcocene [160], strontocene, and barocene [161] were amongst the first metallocenes to be identified. Cp$_2$Ca was found to adopt a

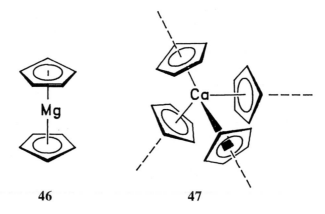

46   **47**

surprising nonparallel polymeric structure in the solid state ($\alpha = 57°$), in which each calcium atom engages four Cp rings with varying hapticity, and each ring adopts a μ–π–coordination with two calcium atoms **47** [162]. It can be considered in terms of optimal packing of two cyclopentadienide anions with Ca^{2+}, nevertheless, the sandwich metallocene molecular ion is observed in the mass spectrum. Moreover, a nonparallel structure is generally observed for the heavier alkaline earth sandwich metallocenes, even in the gas phase (Cp^*_2Ca [48, 49], Cp^*_2Sr [163], Cp^*_2Ba [163]).

Table 1-4. Comparison of interplane angles (α, °), $Cp'_{centroid}$–E–$Cp'_{centroid}$ angles (β, °) and $Cp'_{centroid}$–E distances (pm) for selected alkaline earth sandwich metallocenes

	α	β	$Cp'_{centroid}$–E	Ref.
Cp^*_2Ca	35	147	235	[164]
Cp^*_2Ba	48	131	274	[112]
Cp^{4I}_2Ca	11	162	235	[165]
Cp^{4I}_2Ba	17	154	268	[165]
$Cp^{2T}_2Ca \cdot THF$	46	135	240	[27]
$Cp^{2T}_2Sr \cdot THF$	48	134	255	[27]
$Cp^B_2Ca \cdot (THF)_2$	48	133	246	[90]
$Cp^B_2Sr \cdot (THF)_2$	48	133	260	[90]
$Cp^{3I}_2Ba \cdot (THF)_2$	56	132	279	[166]

In contrast to Cp_2Ca, most sandwich derivatives are monomeric in the solid state (Cp^*_2Ba exhibits weak Ba–Me intermolecular interactions) [112], and the conformation of the Cp' ligands seems dependent upon the degree of substitution around the Cp' frame. Structural features for representative examples are compared in Table 1-4 and show smaller α angles and larger β angles for complexes possessing more sterically imposing substituents on the Cp' ring. Derivatives such as Cp^{4I}_2Ca are described as encapsulating the element center [165], which is responsible for the air stability of this complex in contrast to Cp^*_2Ca, and likely prevents formation of adducts. In addition, the solubility and volatility of these heavily substituted metallocenes is enhanced and DNMR studies indicate hindered rotation of the Cp' rings. Although the derivatives listed carry different substituents, the trends indicate that introduction of auxiliary ligands has only a minor effect on the α angle, and is essentially independent of the element, the number of ligands (one or two) or the Cp'$_{centroid}$–E distance. A decrease in the β angle describes a withdrawal of the element center from the Cp'$_{centroid}$–Cp'$_{centroid}$ vector relative to complexes without auxiliary ligands.

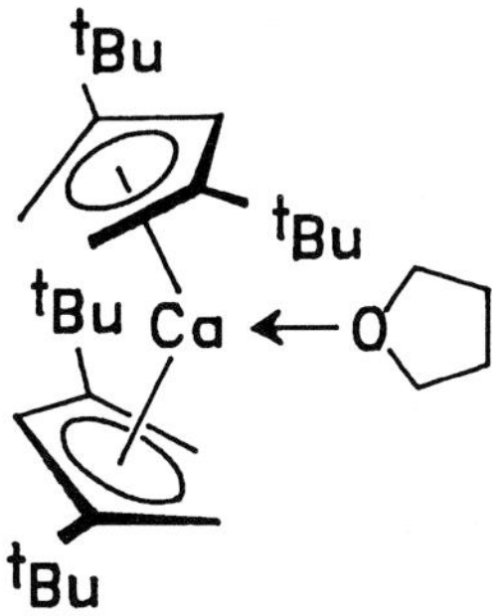

48

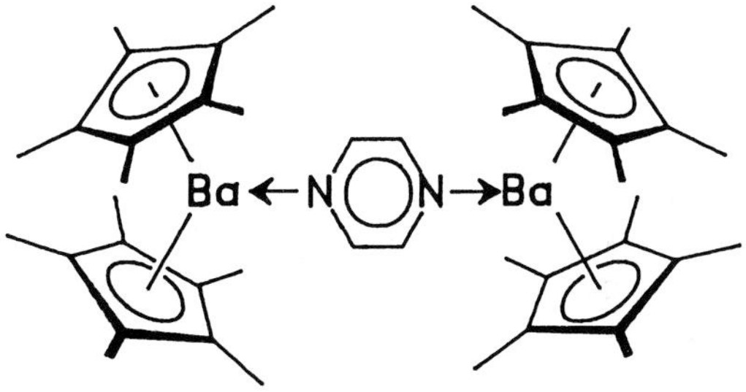

49

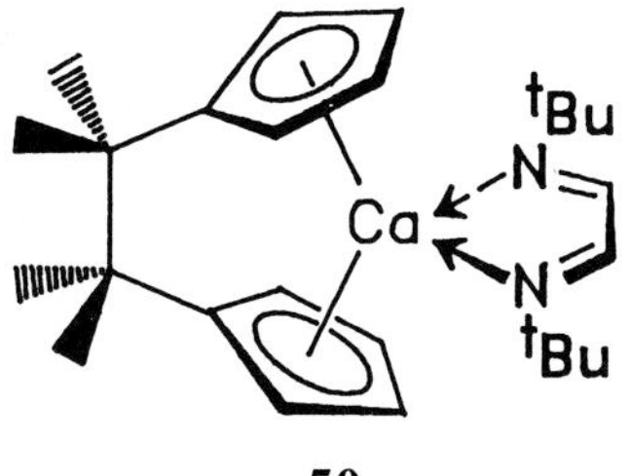

50

The 'open-calcocene' **48** is a derivative of these complexes [167] and there are rare examples of complexes involving bifunctional ligands as a link between two metallocene units **49** [168] as well as a chelate **50** [169]. Several derivatives of calcium, strontium and barium metallocenes carry donor functionalized side-chain cyclopentadienyl ligands [23, 170] including the first optically active calcocene **51** [171]. Organic ligands exhibit more flexible coordination, for example, the solid state structure of the bistrimethylsilylbutadiyne complex of Cp*$_2$Ca contains two slightly different molecules, both with a butadiyne molecule wedged between the Cp* rings of the metallocene **52** [172]. In one, the diyne is symmetrically disposed about the Cp*$_{centroid}$–E–Cp*$_{centroid}$ plane and in the other the ligand is slipped to the side by about 80 pm, possibly due to crystal packing forces. The Lewis acidity of the calcium center in Cp*$_2$Ca can be further exploited to form a number of mixed metal complexes, such as [Cp*$_2$Ca·MeAlMe$_2$THF]$_2$ [173] and Cp*$_2$Ca·ClHfCp'$_2$Cl [174].

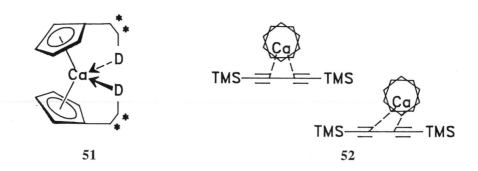

51   52

1.8.3 Half Sandwich Metallocenes of the Group 2 Elements

Half sandwich beryllium complexes of the general type **53** are monomeric in solution, in the gas phase and in the solid state, and all exhibit pentahapto coordination with similar Cp'$_{centroid}$–Be distances (*e.g.* CpBeCl$_{(g)}$ 148 pm [98]; CpBeCl$_{(s)}$ 145 pm [175]; CpBeBr$_{(g)}$ 153 pm [176]; CpBeMe$_{(g)}$ 150 pm [177]; CpBeCCH$_{(g)}$ 149 pm [176]; Cp*BePBu$_{2(s)}$ 148 pm [178]) despite the varying degrees of π-interaction possible between the auxiliary ligand and beryllium. This comparison extends to include the borane complexes (*e.g.* CpBeB$_5$H$_8$ solid 147 pm) [179], and beryllocene (Cp$_{plane}$–Be 151 pm) [146].

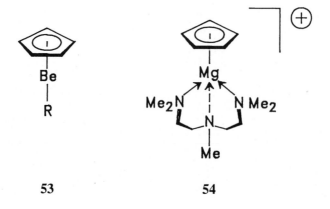

53   54

Synthesis of half sandwich complexes for the heavier alkaline earth metals is complicated by disproportionation equilibria, but the use of donor solvents and bulky Cp' units impose a kinetic stability. Monomeric structures are observed for complexes with 'monodentate' (*e.g.* Cp^{4I}CaN(SiMe$_3$)$_2$·THF) [180], bidentate (*e.g.* CpMgBr·TMEDA [181], Cp^{3T}MgBr·TMEDA) [35], and tridentate (*e.g.* **54**) [95] ligands, while dimeric arrangements are bound by halide bridges (*e.g.* {Cp*MgCl·OEt$_2$}$_2$ [115], {Cp*CaI·THF$_2$}$_2$ [116], {Cp^{4I}CaI·THF}$_2$) [182] or amino bridges (*e.g.* {Cp^{4ME}Ca[SiMe$_2$–CH$_2$CH$_2$SiMe$_2$N]}$_2$) [180], and have the general structure **22**.

1.9 Metallocenes of the Group 13 Elements

The elements of group 13 provide the most structurally diverse series of metallocene complexes, including examples of monomeric half sandwich molecules, various arrangements of dimeric half sandwich complexes, oligomeric clusters, polymeric chains analogous to those observed for the alkali metals, and rare examples of sandwich complexes. A recent review of the syntheses, structures and chemistry of aluminum(I) and gallium(I) compounds includes derivatives of the general formula Cp'E and their oligomers [125]. A comprehensive discussion of the organometallic chemistry of subvalent thallium is also available, which is dominated by examples of cyclopentadienyl derivatives [183].

1.9.1 Half Sandwich Monomers, Dimers, Oligomers and Polymers

A monomeric half sandwich η^5-structure **8** (Y = lone pair) is consistently observed in the gas phase for derivatives of the general formula Cp'E, including Cp*Al [96], Cp*Ga [184], CpIn [185], Cp^{M}In [186], Cp'In [119], CpTl [187], and Cp*Tl [188]. Cp*Ga is stable enough to be distilled at ambient pressure (b.p. 150°C) and only decomposes under MOCVD conditions at 600°C.[14] NMR spectroscopic data indicate that Cp'Ga derivatives also have monomeric structures in solution [43].

Half sandwich Cp'E complexes of indium and thallium with relatively minimal substitution around the Cp' ligand typically form 'zig-zag' polymeric structures of type **18** in the solid state, which are analogous to those observed for alkali metallocenes. The β angles are reasonably consistent (*e.g.* CpIn,128° [186]; Cp^{4M}In,134° [119]; CpTl, 137° [189]; Cp^{4M}Tl, 134°) [190], and the respective Cp'$_{centroid}$–E distances (*e.g.* CpIn, 269 and 273 pm; Cp^{4M}In, 250 and 281 pm; CpTl, 319 pm; Cp^{4M}Tl, 268 and 271 pm) are considerably larger than in representative gas phase monomers (*e.g.* Cp*In, 229 pm [119]; Cp*Tl, 237 pm [188]). Monomeric solid state structures are observed for the phosphine substituted derivatives Cp^{4MP}In [191] and Cp^{4MP}Tl [192].

Nucleophilic addition of the cyclopentadienide anion to CpTl gives either the unique anionic thallate sandwich metallocene of structural type **5** in [CpMg·PMDETA][Cp$_2$Tl] [95], or the multidecker sandwich [Cp$_3$Tl$_2$]$^-$ [106], which has structure **13**, analogous to [Cs$_2$Cp$_3$]$^-$ [105]. However, unlike the cesium derivative, the nonparallel conformations of these sandwich derivatives are consistent with the presence of a lone pair at the thallium center(s), the environment of the element being isovalent with neutral sandwich metallocenes for elements of group 14 and cationic sandwich metallocenes for elements of group 15.

The heavily substituted Cp^{5Bz}Tl adopts two crystalline forms depending upon the rate of crystallization. Rapid precipitation effects an alignment of the monomeric units in a super-sandwich polymeric array **17** [193], whilst a dimeric structure **21** is formed upon slow crystallization [114] in which the thallium centers are adjacent and the molecular structure is described as a *trans* bent geometry. The dimer presumably represents the thermodynamically favored arrangement, and is isostructural with the indium complex {Cp^{5Bz}In}$_2$, which has an identical element–element distance (363 pm) [113]. The possibility of In–In and Tl–Tl bonds has been theoretically assessed at various levels and there is much debate about the strength of such interactions [194, 195]. Although there is indication of element–element interactions, they are weak and the {Cp$^{5Bz}_2$E}$_2$ dimers likely involve both element–element and ligand–ligand attractive interactions [195].

1.9.2 Clusters

Cp*Al sublimes at 140°C as the monomer, but in the solid state it forms a unique tetrahedron of aluminum atoms with the apices capped by η^5-bound Cp* ligands, as shown in **23** [41]. The Al–Al bond is shorter (277 pm) than that in aluminum metal (9 pm longer) and is consistent with that calculated for the parent compound {CpAl}$_4$ (279 pm) [196], but is longer than that calculated for {AlX}$_4$ tetrahedra (X = H, Si^{t}Bu$_3$, SiH$_3$, Cl, F; 262–264 pm) [196, 197].

While Cp*Tl is polymeric in the solid state [198], the indium analog Cp*In adopts an unusual hexameric structure with Cp* ligands surrounding an octahedral element cluster **24** [119, 199]. Cp*Ga [41a] is surprisingly isostructural with Cp*In including identical cell volumes. The Ga–Ga distances (407–417 pm, cf. atomic radius 130 pm) are longer than the In–In distances in {Cp*In}$_6$ (394–396 pm, cf. atomic radius 155 pm) to accommodate the differences in Cp*$_{centroid}$–E distances (Ga, 208 pm; In 230 pm). Such comparisons indicate that, in contrast to {Cp*Al}$_4$, the ligand–ligand interactions define the cluster structure, with element–element interactions of low significance. Consistently, the gallium metallocene cluster is colorless, whilst the indium metallocene cluster is yellow implicating more effective element–element interactions made possible by the larger atomic size. Interestingly, the cluster structures **24** are distorted from O$_h$ to S$_6$

symmetry by the slight off-setting of the $Cp*_{centroid}$–normal from the center of the octahedron.

1.9.3 Metallocenes Involving Group 13 Elements in Oxidation State III

The availability of vacant valence orbitals on high oxidation state group 13 centers might be expected to support π-coordination of Cp' ligands, however, examples are relatively rare. While CpAlMe$_2$ is monomeric in the gas phase and is interpreted in terms of a η^2-interaction [200], alkyl derivatives of CpAR$_2$ [201] are typically polymeric in the solid state with bridging Cp ligands [202], and CpGaMe$_2$ has a similar structure [203]. Haloaluminum complexes involving cyclopentadienyl ligands are dimeric bis-halogen bridged structures of the type **22**, with varying hapticity for the Cp' ligand [103, 104, 117]. Dimerization can be prevented with the presence of a side chain donor functionality on the cyclopentadienyl ligand (Cp^N), which imposes a chelate interaction in place of the bridge, but most derivatives are best considered as σ-complexes [23, 24, 204]. The delicate energetic distinction between σ- and π-coordination is illustrated in the typically σ-coordinated derivatives of Cp'$_3$Al, by the unique $\pi(\eta^5)$, σ, σ structure **55** observed in the solid state for Cp$^{3M}_3$Al [15].

55   56

While the neutral Cp' substituted boranes are also σ-bonded compounds [205], cationic complexes of the form [Cp'BR]$^+$ have definitive half sandwich structures comparable to the isoelectronic Cp'BeR complexes **53**. ^{11}B, ^{1}H, and ^{13}C NMR spectral data are unambiguously interpreted in terms of an η^5-interaction, and this is confirmed in a crystallographic study on [Cp*BBr][AlBr$_4$] [99].

  Cationic sandwich metallocenes of general formula [Cp'$_2$E]$^+$ have only been identified for boron and aluminum. Cp*$_2$BCl (^{11}B, –74 ppm) reacts with BCl$_3$ to give the tetrachloroborate salt of [Cp*$_2$B]$^+$ as a microcrystalline precipitate (^{11}B, +44 ppm) [57]. ^{1}H NMR spectroscopic studies show the cation to be structurally analogous to the isoelectronic slip–sandwich Cp$_2$Be complex **44**. The decamethyla-luminocenium cation **56** has been identified in the salts [Cp*$_2$Al][Cp'AlCl$_3$] [86] and [Cp*$_2$Al][LiCp$^{5Bz}_2$] [91], and the crystal structures confirm the staggered parallel planar D$_{5d}$ structure. Although salts of the parent cation [Cp$_2$Al]$^+$ have not yet been isolated, it has been shown to be an effective olefin polymerization initiator [206].

1.9.4 Heteroatom Clusters

Reaction of {Cp*Al}$_4$ with elemental selenium or tellurium gives heterocubanes of the general formula {Cp*AlZ}$_4$ (Z = Se or Te) [73], where the tetrahedral cluster of aluminum atoms is retained. The higher oxidation state for aluminum effects a significant shortening of the Cp*$_{centroid}$–Al distances (Se: 195 pm; Te: 197 pm) relative to the starting material {Cp'Al}$_4$ (202 pm) [41], but the gallium analogs [207, 208] involve σ-coordination of the Cp* ligands. Reaction of {Cp*Al}$_4$ with elemental phosphorus (P$_4$) results in cleavage of all Al–Al and P–P bonds and formation of an unusual {(Cp*Al)$_6$P$_4$} cage, which can be described as two face-sharing heterocubanes with opposite corners missing [209]. Related clusters {(Cp*AlF)$_4$(SiPh$_2$)$_2$} and {(Cp*Al)$_3$Sb$_2$} are also accessible [210], as well as {NiCp$_2$(Cp*Al)$_2$}, which shows the Cp*Al monomer as a bridging ligand [211].

1.10 Metallocenes of the Group 14 Elements

1.10.1 Group 14 Sandwich Complexes

The group 14 sandwiches represent the most extensive series of isolated main group metallocenes. Silicocene derivatives are less accessible than germanocenes, stannocenes and plumbocenes, perhaps due to the difficulties associated with the stabilization of a divalent silicon center, and Cp*$_2$Si is the only isolated derivative [212]. It is most conveniently prepared by the quantitative reduction of dibromobis(pentamethylcyclopentadienyl)silane using potassium anthracenide [87], and melts without decomposition. Two geometrical isomers are observed in the solid state, one with the Cp* ligand planes parallel **57a** and the other **57b** with α = 25° (β = 155°). The solid state CP-MAS ^{29}Si NMR spectrum contains two signals, –403.2 ppm (nonparallel) and –423.4 ppm (parallel), and the solution shift (–398 ppm) indicates that the nonparallel isomer (α = 25°) predominates [80].

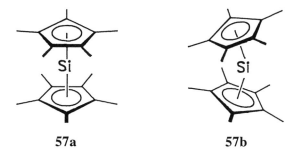

57a   **57b**

Sandwich complexes of group 14 elements are generally monomeric in all phases and in solution, with the parent plumbocene as the only exception, composed of a zig-zag chain of lead atoms linked by π-coordinated μ-Cp ligands **19**, and with each lead atom bearing a third terminal Cp ligand [111]. The larger atomic radius of the heavier elements allows for greater structural flexibility, so that interplane angles (α) for Cp*$_2$Ge$_{(g)}$ [101], Cp*$_2$Sn$_{(s)}$ [58], and Cp*$_2$Pb$_{(s)}$ [93], listed for comparison with other derivatives in Table 1-5, are significantly larger than observed for Cp*$_2$Si. Nevertheless, α is sensitive to steric encumbrances, and is smaller for complexes bearing more heavily substituted Cp' derivatives.

Table 1-5. Comparison of interplane angles (α, °) for examples of group 14 sandwich metallocenes (averages where necessary)

	Ge		Sn		Pb	
	α	Ref.	α	Ref.	α	Ref.
Cp$_2$E	50	[213]	46	[93]	60[a]	[111]
Cp'$_2$E$_{(g)}$	34[b]	[127]	50[b]	[127]	45[c]	[214]
Cp*$_2$E	22[c]	[101]	36	[58]	37	[93]
Cp$^{5Bz}_2$E	31	[215, 216]	33	[216]	33	[216]
Cp$^{4M(SiB2M)}_2$E [d]	–	–	0	[217]	0	[218]

[a] Polymeric
[b] Cp$^M_2$E
[c] Gas phase
[d] SiB2M = di^{t}butylmethylsilyl

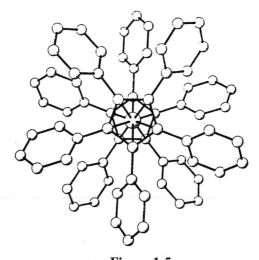

Figure 1-5

Decaphenylstannocene viewed perpendicular to the plane of the Cp^{5Ph} ring

In some cases, the substitution pattern is so sterically demanding that a parallel planar structure is observed, as for Cp$^{5I}_2$Sn [219] and Cp$^{5Ph}_2$Sn (Fig. 1-5) [94]. However, the structural impact of the substituents is often difficult to predict. For example, Cp$^{4I}_2$Sn adopts a nonparallel structure ($\alpha = 28°$; Cp$^{4I}_{centroid}$–Sn 242 pm) [220] while the apparently less sterically loaded Cp$^{3I}_2$Pb is parallel [20], despite having a larger Cp'$_{centroid}$–E distance (247 pm). Cp$^{4M(SiB2M)}_2$Sn [217] and Cp$^{4M(SiB2M)}_2$Pb [218] also exhibit parallel structures with less bulky substitution, which has led to the conclusion that the lone pair at the element center is stereochemically inactive. The asymmetric complex Cp^{5Ph}SnCp is noticeably nonparallel ($\alpha = 43°$) [221] consistent with the release of steric strain in comparison with Cp$^{5Ph}_2$Sn. The structural flexibility of sandwiches involving heavier elements is further demonstrated by the introduction of auxiliary chelate ligands onto Cp$_2$Pb, for which **58** is an example [222] (see also tricyclopentadienyl complexes below).

58   **59**

60 **61**

1.10.2 Half Sandwich Complexes of Group 14 Elements

[CpGe]$^+$ and [CpSn]$^+$ were first observed by mass spectrometry, but use of the Cp*
ligand enabled the isolation of salts containing [Cp*Ge]$^+$ [58, 223], [Cp*Sn]$^+$ [58,
60, 61], and [Cp*Pb]$^+$ cations [81]. [Cp*Pb][BF$_4$] adopts a dimeric structure in the
solid state bound by bridging anions [81], while the crystal structure of
[Cp*Sn][BF$_4$] confirms the presence of isolated ions **59** involving pentahapto interaction [61]. Cations of the type [Cp'Sn]$^+$ are also observed as units of more complicated structures including the solid state structure of Cp$_2$Sn·BF$_3$ [224] and the
novel cluster [Cp$^{4M(SiB2M)}{}_6$Sn$_9$Cl$_{12}$] [225].

The nucleophilicity of [Cp*E]$^+$ cations is demonstrated by the formation of coordination complexes with neutral ligands such as pyridine to give representatives
of 'slip–sandwich' structures (*e.g.* **60**) [102]. Similar structural features are observed for the isoelectronic neutral half sandwich derivatives of general formula
Cp'EX (E = Sn [52, 53, 226], Pb [227]; X = Cl, Br, I, CH$_3$COO). The solid state
structure of CpSnCl involves a short Sn–Cl bond as well as long Sn–Cl contacts,
responsible for a coordination polymeric structure, illustrated in **61** [53, 226].
However, the absence of tin coupling in the ^{1}H and ^{13}C NMR spectra implies disproportionation in solution [228], and Cp*SnCl is described as monomeric in the
solid state [226]. Molecular units are observed in the solid state structures of **62**
[51] as well as in the chelated derivative **63** [230], but the same 'slip–sandwich'
interaction is evident and is also apparent for Cp*GeCl in the gas phase [101]. Finally, one of the Cp ligands of Cp$_2$Sn can be substituted by reaction with
LiN=C(NMe$_2$)$_2$ to give the unusual cyclic dimeric structure **64** [231] involving
pentahapto coordination.

62   63   64

1.10.3 Tricyclopentadienyl Complexes

Tin and lead exhibit the unique ability to accommodate three π-bound Cp ligands in molecular species. The fundamental complexes $[Cp_3E]^-$ are obtained by nucleophilic addition of $[Cp]^-$ (NaCp or LiCp, or Cp_2Mg) to Cp_2Sn or Cp_2Pb, and isolation of the salt $[Mg(THF)_6][Cp_3Sn]_2$ reveals an nonassociated (cation contact free) anion [108] with a 'Paddle-Wheel' structure **16**. The geometry for tin can be described as a shallow pyramid with symmetrical η^3 attachment of the three Cp ligands. Although the same anion is present in the $[Na \cdot PMDETA]^+$ salt [232], one of the Cp ligands behaves as a bridge (μ–π–Cp) between the metals, which distorts the anion. The $[Li \cdot PMDETA]^+$ salt **34** of the aminostannyl anion $[Cp_2SnN(SiMe_3)_2]^-$ [233] also shows the presence of both terminal and μ–π–Cp ligands.

Reaction of $[Cp]^-$ with Cp_2Pb also accounts for the two chain-like multinuclear anions $[Cp_5Pb_2]^-$ **14** and $[Cp_9Pb_4]^-$ **15**, in the same $[Li(12\text{-crown-}4)_2]^+$ salt [107]. Although, at first glance, their structures are related to the solid state structure of Cp_2Pb **19**, they exhibit a disruption of the polymer planarity featured in the neutral parent metallocene. Nevertheless, similar variations in the $Cp_{centroid}$–Pb distances (terminal $Cp_{centroid}$–Pb, 251–262 pm, are significantly shorter than those involving the μ-Cp ligands, 273–292 pm) prompt models involving the Cp_2Pb unit, so that **14** is viewed as two Cp_2Pb molecules bound by a $[\mu$-Cp$]^-$ and **15** is viewed as **14** bearing two additional Cp_2Pb terminal units [107].

1.10.4 Reaction Chemistry

The extensive series of reactions already established for the metallocenes of group 14 can be summarized in five classifications.

1) Cp' ligand exchange is generally observed with the trend Ge<Sn<Pb, and has been synthetically exploited to diversify stannocene and germanocene derivatives [52, 234, 235].

2) Oxidative addition is a typical reaction for Cp*$_2$Si leading to compounds of the type Cp*$_2$Si(Y)X, or intermediates of the type Cp*$_2$Si=Y [236, 237]. Metallocenes of germanium and tin are also oxidized to give the corresponding E(IV) product [229, 238].

3) Nucleophilic substitution occurs with reagents such as alkyllithiums, which readily transform Cp*GeR compounds into R$_2$Ge compounds [239] and Cp*$_2$Sn into R$_2$Sn [235], and can be envisaged to involve oxidative addition and reductive elimination steps.

4) Nucleophilic addition or the demonstration of Lewis acidity is a general theme for metallocenes generating tricyclopentadienyl complexes (Sn and Pb) and complexes with a variety of auxiliary ligands.

5) Homolytic bond cleavage occurs on thermal treatment of group 14 decamethyl metallocenes to give the element, tetramethylfulvene and Cp*H, under MOCVD conditions. The temperature of decomposition (Cp*$_2$Si, ~550°C; Cp*$_2$Ge, ~300°C; Cp*$_2$Sn, ~200°C; Cp*$_2$Pb, ~200°C) is lower for the heavier congeners. Cp*$_2$Pb even decomposes in sunlight [240].

1.11 Metallocenes of the Pnictogens

While X-ray crystallographic studies of Cp$_3$Sb [241] and Cp$_3$Bi [242] clearly show a σ-interaction, π-coordination is evident in compounds with fewer cyclopentadienyl ligands (Cp'$_2$ECl or Cp'ECl$_2$). The solid state structure of Cp$^{3B}_2$BiCl shows an almost parallel ($\alpha = 18°$) sandwich arrangement **65**, but with the bismuth center withdrawn from the Cp$^{3B}_{centroid}$–Cp$^{3B}_{centroid}$ axis ($\beta = 145°$) [118].

65   **66**

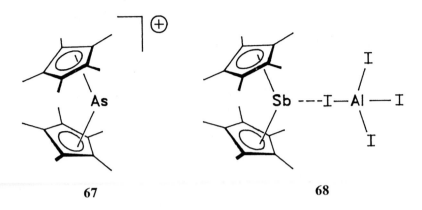

67   68

Cationic sandwich metallocenes exhibit varying degrees of distortion from a parallel planar arrangement. Consistent with observations for the more extensive series of neutral group 14 derivatives, heavily substituted Cp' ligands enforce a close to parallel planar structure as in the case of $[Cp^{3B}_2Sb][AlCl_4]$ ($\alpha = 12°$) [63] and $[Cp^{3B}_2Bi][AlCl_4]$ ($\alpha = 10°$) [64] (cation shown in **66**, E = Sb or Bi), whilst the structures of $[Cp^*_2As][BF_4]$ ($\alpha = 36°$) [62] (cation shown in **67**) and $[Cp^*_2Sb][AlI_4]$ ($\alpha = 36°$) [63] **68** are substantially nonparallel and the element is displaced from the centroid normal, with the latter displaying a significant interaction between antimony and iodine of the anion. Theoretical studies of the hypothetical parent phosphorus complex $[Cp_2P]^+$ present a surprising interplane angle ($\alpha = 111°$) for the lowest energy structure and postulate η^2/η^1 coordination [243]. However, a nonbonding lone pair MO of stereochemical significance could not be identified.

The monosubstituted cyclopentadienyl complexes $CpSbCl_2$ [244] and $CpBiCl_2$ [245] are classified as η^3 and $\eta^3-\eta^2$, respectively. While $CpSbCl_2$ adopts a chiral polymeric arrangement due to long range intermolecular Cp–element π-interactions, chlorine bridges between bismuth centers are responsible for the polymeric solid state structure of $CpBiCl_2$. Thermal stability imposed by the Cp^{4I} ligand enables the isolation and structural characterization of an homologous series, $Cp^{4I}ECl_2$ with E = P, As, Sb and Bi, allowing for useful comparisons [246]. The solubility in common aprotic solvents decreases from P to Bi, and the color varies from yellow (P) to red–orange (Bi). Fluxional behavior observed in the ^1H NMR spectra can be frozen out for P and As, but not for Sb or Bi. In the solid state, the position of the pnictogen center shows a smooth progression of the element from outside the pentagonal cylinder in the case of phosphorus, to a η^3 bonded arrangement in the case of bismuth, as illustrated in **69**. The halogen bridged dimer structure **70** and the unusual cluster **71** [118] are consistent with other areas of bismuth coordination chemistry.

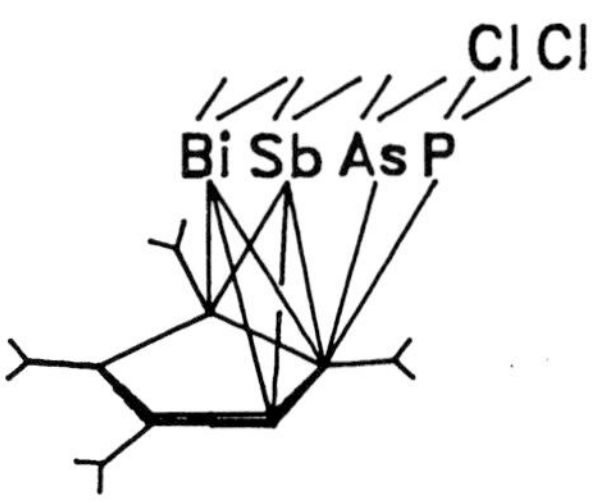

69

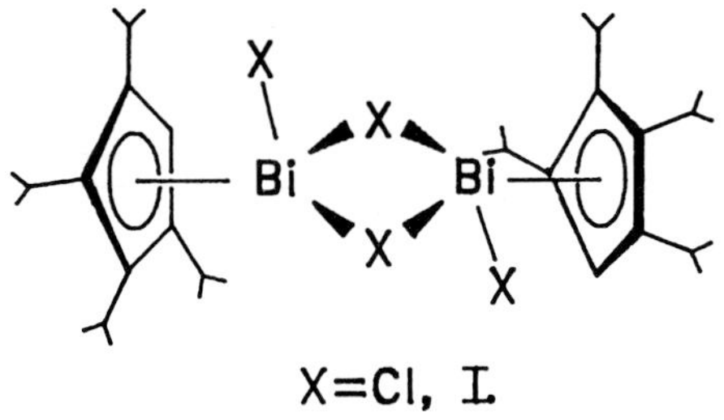

X=Cl, I

70

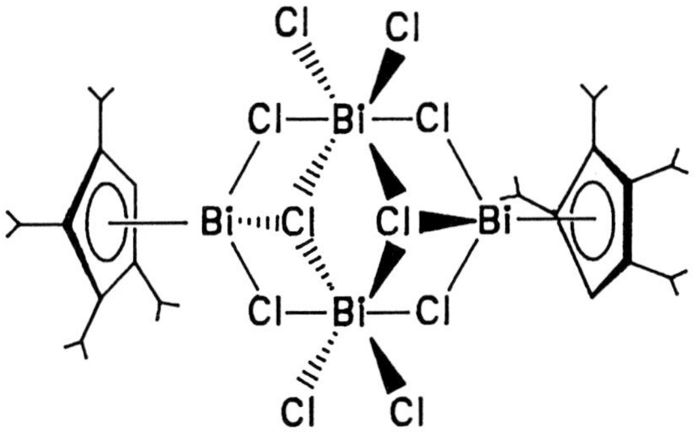

71

The only structurally confirmed example of a phosphorus metallocene was obtained by chloride ion abstraction from Cp*P(Cl)NH^{t}Bu with either AlCl$_3$ or AgSO$_3$CF$_3$ [247]. The aluminate salt of the phosphanylium cation **72** (phosphenium) shows a definitive η^2 interaction in the crystal structure, but NMR spectroscopic studies on complexes of this type confirm fluxionality in solution down to −78°C [248].

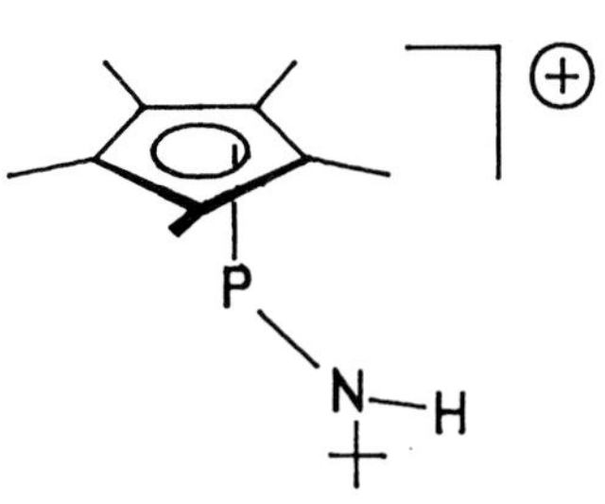

72

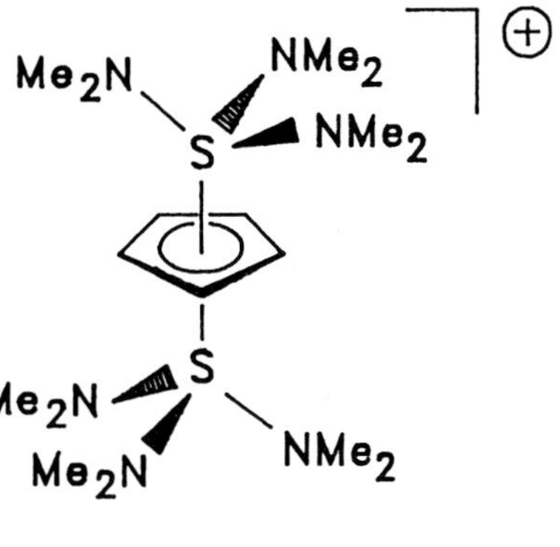

73

Cyclopentadienyl derivatives of sulfur are definitively σ-complexes, with $Cp*_2S$ as the classic representative [249]. Nevertheless, the solid state structure of the ionic [TAS–Cp–TAS][Cp] [76], containing cation **73**, reveals a distinct interaction between the two sulfur centers and the centroid of the cyclopentadienide anion, which may be considered the first example of a π-interaction with sulfur (TAS = trisdimethylaminosulfonium).

1.12   Factors Governing the Structure of Multicyclopentadienyl Complexes

The interplane angles (α) for the Cp' ligands in molecular sandwich complexes and polymeric super-sandwich structures range from 0°, for elements such as Li, Na, Mg and Al, to as much as 50° in Cp_2Ge [213]. Experimental observations can be rationalized in covalent terms on the basis of the valence electron count, with 12 electron systems typically adopting parallel structures and 14 electron systems having nonparallel arrangements, implicating a stereochemical role for the two extra electrons, as discussed earlier. However, the inability of electronic models to rationalize the nonparallel structures observed for 12 electron sandwich metallocenes of the heavier group 2 elements (without steric loading) despite the absence of such a nonbonding pair [21], highlights the importance of other factors such as the degree of ionic character in the Cp'–E interaction, crystal packing forces and interligand nonbonded interactions.

Molecular orbital calculations in general reveal small energy differences between parallel and nonparallel structures. For example, the linearization of Cp_2Ba ($\beta = 147°$) is calculated at 1.5 kJ.mol^{-1} [250], and the nonparallel structures for the decamethylmetallocenes of Ca, Sr and Ba are 2 to 3.5 kJ · mol^{-1} more stable than the corresponding parallel structures [163]. While estimates of such energy differences for the group 14 sandwich metallocenes span a wide range, the values obtained from higher level calculations are generally small (>10 kJ · mol^{-1}) and are comparable with those of the alkaline earth derivatives [19, 20], as are the values for the isoelectronic $[Cp_2Tl]^-$ [95]. Therefore, noncovalent factors which are normally of minor structural consequence are expected to impose considerable conformational control. In this context, molecular mechanics force field calculations [129, 130] reveal that nonbonded attractive interactions between the Cp' ligands are significant in alkaline earth sandwich metallocenes. Crystal packing forces influence the structures to some extent, but are considered less significant [130]. It is interesting to note that a number of complexes of the general formula $Cp*_2E$ exhibit a consistent inter-ligand (methyl–methyl) closest contact distance of around 410 pm [48, 129], which is essentially the Van der Waal's distance for two methyl groups (400 pm) [251], although the closest inter-ligand distance in the nonparallel conformer of $Cp*_2Si$ is as low as 380 pm. It is also instructive to consider inter-ligand distance trends for all sandwich complexes. For this purpose, Table 1-6 lists typical $Cp'_{centroid}$–E distances in sandwich metallocenes or super-sandwich polymeric arrays, representing half the $Cp'_{centroid}$–$Cp'_{centroid}$ distance and therefore the inter-ligand distance for parallel parent complexes Cp_2E. The introduction of substituents around the Cp' frame (such as methyl) will obviously result in the closest inter-ligand distance being less than the $Cp'_{centroid}$–$Cp'_{centroid}$ distance, as would any distortion from parallel geometry.

Table 1-6. Typical $Cp'_{centroid}$–E distances (pm) in sandwich metallocenes or super-sandwich polymeric arrays

Group 1		Group 2		Group 13		Group 14		Group 15	
Li	199	Be	151	B	–	C	–	N	–
Na	236	Mg	199	Al	184	Si	211	P	–
K	281	Ca	235	Ga	–	Ge	225	As	215
Cs	310	Sr	260	In	255	Sn	240	Sb	230
Fr	–	Ba	270	Tl	260	Pb	250	Bi	250

The $Cp'_{centroid}$–$Cp'_{centroid}$ distances are relatively large in comparison to transition metallocenes (*e.g.* ferrocene 332 pm) [156] as would be expected on the basis of a comparison of the orbital overlap model (due to the absence of d orbitals in the main group). In fact, most $Cp'_{centroid}$–$Cp'_{centroid}$ distances for sandwich main group metallocenes exceed the Van der Waal's contact for two aromatic systems (340 pm) [251]. An increase in $Cp'_{centroid}$–E distance down each group, as well as a general decrease across each period, are consistent with the trends in atomic size and effective nuclear charge. The nonsterically loaded parallel sandwich metallocenes of lithium, magnesium and aluminum are predictably the shortest, and a slightly larger separation in $Cp*_2Si$ allows for both a parallel and nonparallel ($\alpha = 25°$) form in the same crystal structure [87], illustrating the minimal energy difference between the two conformations. The very short $Cp'_{centroid}$–Be distance prevents beryllocene from adopting a typical sandwich arrangement and a distortion to 'slip–sandwich' is observed with one of the ligands η^1-bound [146]. For heavier elements, the $Cp'_{centroid}$–$Cp'_{centroid}$ distances are substantially greater than the typical inter-ligand distance (410 pm) [48, 129], so that distortion from the parallel conformation is possible until this distance is achieved, or slightly less as in nonparallel $Cp*_2Si$.

Realization of the close structural similarity for identically substituted sandwich metallocenes of elements from groups 2 and 14 (the structure of Cp^{4I}_2Ca can be superimposed on Cp^{4I}_2Sn, and $Cp*_2Ca$ on $Cp*_2Sn$) has also prompted the conclusion that the steric factors play a more dominant role in the control of molecular structure than electronic factors [21]. These arguments can be extended to rationalize other metallocene structures. At first glance, the dimers $\{Cp^{5Bz}E\}_2$ **21** of indium [113] and thallium [114] are defined by element–element interactions, however, the similarity of the bond lengths and angles in these isostructural complexes implicate the importance of inter-ligand interactions, and a comparison of structural features for the $\{Cp^*E\}_6$ octahedral clusters of gallium [41a] and indium [119] draws similar conclusions.

1.13 Conclusions

Metallocene chemistry is no longer the sole preserve of the d- and f-block elements, as an extensive series of main group metallocene complexes have now been identified and comprehensively examined. Most elements of groups 1, 2, 13, 14 and 15 engage Cp' ligands in a π-interaction, and there are rare examples which may be considered metallocenes of sulfur. Although there are obvious structural similarities with transition metallocenes, a wider structural diversity is made possible by a relatively weaker Cp'–E bond, the nature of which ranges from mainly ionic to highly covalent. As a result, although there are rare examples of catalytic behavior analogous to transition metallocenes, the chemistry of the main group metallocenes is highlighted by their role as synthetic reagents, as precursors for the MOCVD process, and a novel reactivity in general. This stems from the coordinatively flexible Cp'–E interaction and the ability of Cp' to behave as a leaving group, allowing for an extensive substitution chemistry. Further diversification involving isolobal analogs of Cp' has already begun and represents the future of main group metallocene chemistry.

1.14 Acknowledgments

We thank the Alexander von Humboldt Foundation for a fellowship (NB), Ms. M. Welschof–Hildebrandt for preparing the drawings, Ms. B. Neumann for assistance with analysis of the structural data and a number of colleagues for informing us of recent results and providing preprints of their work.

References

[1]   Elschenbroich, C; Salzer, A. *'Organometallics'* VCH Verlagsgesellschaft; Weinheim, 1989.
[2]   Comprehensive account: Jutzi, P.; Burford, N. *Chem. Rev.* to be submitted.
[3]   Jutzi, P. *Adv. Organomet. Chem.* **1986**, *26*, 217-295.
[4]   For examples of recent developments, see, West, R.; Sohn, H.; Powell, D. R.; Müller, T.; Apeloig, Y. *Angew. Chem.* **1996**, *108*, 1095-1097; *Angew. Chem. Int. Ed. Engl.* **1996**, *35*, 1002-1004. Freeman, W. P.; Tilley, T. D.; Liable-Sands, L. M.; Rheingold, A. L. *J. Am. Chem. Soc.* **1996**, *118*, 10457-10468. Freeman, W. P.; Tilley, T. D.; Yap, G. P. A.; Rheingold, A. L. *Angew. Chem.* **1996**, *108*, 960-961. Hong, J. -H.; Pan, Y.; Boudjouk, P. *Angew. Chem.* **1996**, *108*, 213-215; *Angew. Chem. Int. Ed. Engl.* **1996**, *35*, 186-188.
[5]   Schmidbaur, H. *Angew. Chem.* **1985**, *97*, 893-904; *Angew. Chem. Int. Ed. Engl.* **1985**, *24*, 893-904.
[6]   Parkin, G. E. *Adv. Inorg. Chem.* **1995**, *42*, 291-393. Reger, D. L. *Coord. Chem. Rev.* **1996**, *147*, 571-595.
[7]   Jutzi, P. *J. Organomet. Chem.* **1990**, *400*, 1-17.
[8]   Jutzi, P. *Pure & Appl. Chem.* **1989**, *61*, 1731-1736.
[9]   Daminani, D.; Ferretti, L.; Gallinella, E. *Chem. Phys. Lett.* **1976**, *37*, 265-269.
[10] Beattie, J.; Nugent, K. W. *Inorg. Chim. Acta* **1992**, *198-200*, 309-318.
[11] Cradock, S.; Duncan, W. *J. Chem. Soc., Faraday Trans. II* **1978**, *74*, 194-202.
[12] Anh, N. T.; Elian, M.; Hoffmann, R. *J. Am. Chem. Soc.* **1978**, *100*, 110-116.
[13] Jutzi, P. *Chem. Rev.* **1986**, *86*, 983-996.
[14] Jutzi, P.; Reumann, G. unpublished results.
[15] Fisher, J. D.; Budzelaar, P. H. M.; Shapiro, P. J.; Staples, R. J.; Yap, G. P. A.; Rheingold, A. L. *Organometallics* **1997**, *16*, 871-879.
[16] Jutzi, P. *Comments Inorg. Chem.* **1987**, *6*, 123-144.
[17] King, R. B.; Bisnette, M. B. *J. Organomet. Chem.* **1967**, *8*, 287-294. Gassmann, P. G.; Macomber, D. W.; Hershberger, J. W. *Organometallics*, **1983**, *2*, 1470-1472.
[18] Janiak, C.; Schumann, H. *Adv. Organomet. Chem.* **1991**, *33*, 291-393.
[19] Hanusa, T. P. *Chem. Rev.* **1993**, *93*, 1023-1036.
[20] Hays, M. L.; Hanusa, T. P. *Adv. Organomet. Chem.* **1996**, *40*, 117-170.
[21] Burkey, D. J.; Hanusa, T. P. *Comments Inorg. Chem.* **1995**, *17*, 41-77.
[22] Jutzi, P. *'The Chemistry of Organo Silicon Compounds Part 3'*, Rappoport, Z. ed. Wiley, 1997, in press.
[23] Jutzi, P.; Dahlhaus, J. *Phosphorus, Sulfur and Silicon.* **1994**, *87*, 73-82. Jutzi, P.; Siemeling, U. *J. Organomet. Chem.* **1995**, *500*, 175-185.
[24] Jutzi, P.; Dahlhaus, J. *Coord. Chem. Rev.* **1994**, *137*, 179-199.
[25] King, R. B.; Stone, F. G. A. *Inorg. Synth.* **1963**, *7*, 99-115.
[26] Barber, W. A. *J. Inorg. Nucl. Chem.* **1957**, *4*, 373-374.
[27] Engelhardt, L. M.; Junk, P. C.; Raston, C. L.; White, A. H. *J. Chem. Soc., Chem. Commun.* **1988**, 1500-1501.
[28] Kuz'yants, G. M. *Izv. Akad. Nauk SSSR, Ser. Khim.* **1976**, *25*, 1785-1786.
[29] Tacke, M. *Organometallics* **1994**, *13*, 4124-4125.
[30] Wagner, B.O.; Ebel, H. F. *Tetrahedron*, **1970**, *26*, 5155-5167.
[31] Abel, E. W.; Dunster, M. O. *J. Organomet. Chem.* **1971**, *33*, 161-167.
[32] Kohl, F. X.; Kanne, D.; Jutzi, P. in *'Organometallic Syntheses'* Vol. 3, p.381-383, King, R. B.; Eisch, J. J. eds. Elsevier, Amsterdam, 1986.
[33] Parris, G. E.; Ashby, E. C. *J. Organomet. Chem.* **1974**, *72*, 1-10.
[34] Strohmeier, W.; Landsfeld, H.; Gernert, F.; Langhäuser, W. *Z. Anorg. Allg. Chem.* **1960**, *307*, 120-122.
[35] Morley, C. P.; Jutzi, P.; Krüger, C.; Wallis, J. M. *Organometallics* **1987**, *6*, 1084-1090.
[36] Kroll, W. R.; Hudson, Jr., B. E. *J. Organomet. Chem.* **1971**, *28*, 205-210. Kroll, W. R.; Naegele, W. *J. Chem. Soc., Chem. Commun.* **1969**, 246-247.
[37] Jutzi, P.; Leffers, W.; Müller, G.; Huber, B. *Chem. Ber.* **1989**, *122*, 879-884.

[38] Kurosova, H. Thallium in *'Comprehensive Organometallic Chemistry'* Wilkinson, G.; Stone, F. G. A.; Abel, E. W. eds. Vol. 1, p. 749, Pergamon, 1982. Anderson, G. K.; Cross, R. J.; Phillips, I. G. *J. Chem. Soc., Chem. Commun.* **1978**, 709. Nielson, A. J.; Rickard, C. E. F.; Smith, J. M. *Inorg. Synth.* **1986**, *24*, 97. Nielson, A. J.; Rickard, C. E. F.; Smith, J. M. *Inorg. Synth.* **1990**, *28*, 315-316.

[39] Fritz, H. P.; Köhler, F. H. *J. Organomet. Chem.* **1971**, *30*, 177-185. Koridze, A. A.; Ogorodnikova, N. A.; Petrovsky, P. V. *J. Organomet. Chem.* **1978**, *157*, 145-151. Rausch, M. D.; Edwards, B. H.; Rogers, R. D.; Atwood, J. L. *J. Am. Chem. Soc.* **1983**, *105*, 3882-3886. Conway, B. G.; Rausch, M. D. *Organometallics* **1985**, *4*, 688-693. Spink, W. C.; Rausch, M. D. *J. Organomet. Chem.* **1986**, *308*, C1-C4. Schumann, H.; Janiak, C.; Khani, H. *J. Organomet. Chem.* **1987**, *330*, 347-355. Arthurs, M.; Al-Daffaee, H. K.; Haslop, J.; Kubal, G.; Pearson, M. D.; Thatcher, P.; Curzon, E. *J. Chem. Soc., Dalton Trans.* **1987**, 2615-2619. Singh, P.; Rausch, M. D.; Bitterwolf, T. E. *J. Organomet. Chem.* **1988**, *352*, 273-282. Morcos, D.; Tikkanen, W. *J. Organomet. Chem.* **1989**, *371*, 15-18. Rausch, M. D.; Spink, W. C. Atwood, J. L.; Baskar, A. J.; Bott, S. G. *Organometallics* **1989**, *8*, 2627-2631. Hughes, R. P.; Lomprey, J. R. *Inorg. Chim. Acta* **1995**, *240*, 653-656.

[40] Hanusa, T. P. *Polyhedron* **1990**, *9*, 1345-1352.

[41] Dohmeier, C.; Robl, C.; Tacke, M.; Schnöckel, H. *Angew. Chem.* **1991**, *103*, 594-595; *Angew. Chem. Int. Ed. Engl.* **1991**, *30*, 564-565; Loos, D.; Baum, E.; Ecker, A.; Schnöckel, H.; Downs, A.J. *Angew. Chem.* **1997**, *109*, 894; *Angew. Chem. Int. Ed. Engl.* **1997**, *36*, 860.

[42] Loos, D.; Schnöckel, H.; Gauss, J.; Schneider, U. *Angew. Chem.* **1992**, *104*, 1376-1378; *Angew. Chem. Int. Ed. Engl.* **1992**, *31*, 1362-1364.

[43] Loos, D.; Schnöckel, H. *J. Organomet. Chem.* **1993**, *463*, 37-40.

[44] Peppe C.; Tuck, D. G.; Victoriano, L. *J. Chem. Soc., Dalton Trans.* **1981**, 2592-2582. Jutzi, P.; Leffers, W.; Müller, G. *J. Organomet. Chem.* **1987**, *334*, C24-C25. Schumann, H.; Lentz, A. *Z. Naturforsch.* **1994**, *49B*, 1717-1724.

[45] Jutzi, P.; Leffers, W. *J. Chem. Soc., Chem. Commun.* **1985**, 1735-1736. Rausch, M. D.; Lewison, J. F.; Hart, W. P. *J. Organomet. Chem.* **1988**, *358*, 161-168. Jones, S. S.; Rausch, M. D.; Bitterwolf, T. E. *J. Organomet. Chem.* **1990**, *396*, 279-287. Schumann, H.; Kucht, H.; Kucht, A. *Z. Naturforsch.* **1992**, *47B*, 1281-1289. Schumann, H.; Kucht, A.; Kucht, H.; Lentz, A., Esser, L. *Z. Naturforsch.* **1993**, *48B*, 297-312.

[46] Cotton, F. A.; Wilkinson, G.; Birmingham, J. M. *J. Inorg. Nucl. Chem.* **1956**, *2*, 95-113. Wilkinson, G.; Cotton, F. A. *Chem. Ind.* **1954**, *11*, 307. Fischer, E. O.; Hafner, W. *Z. Naturforsch.* **1954**, *B9*, 503-504.

[47] Robbins, J. L.; Edelstein, N.; Spencer, B.; Smart, J. C. *J. Am. Chem. Soc.* **1982**, *104*, 1882-1893.

[48] Andersen, R. A.; Blom, R.; Boncella, J. M.; Burns, J. C.; Volden, H. V. *Acta Chem. Scand.* **1987**, *A41*, 24-35.

[49] Andersen, R. A.; Boncella, J. M.; Burns, J. C.; Blom, R.; Haaland, A.; Volden, H. V. *J. Organomet. Chem.* **1986**, *312*, C49-C52.

[50] House, H. O.; Latham, R. A.; Whitesides, G. M. *J. Org. Chem.* **1967**, *32*, 2481-2496.

[51] Jutzi, P.; Hampel, B.; Hursthouse, M. B.; Howes, A. J. *Organometallics* **1986**, *5*, 1944-1948.

[52] Bos, K. D.; Bulten, E. J.; Noltes, J. G. *J. Organomet. Chem.* **1972**, *39*, C52-C54.

[53] Bos, K. D.; Bulten, E. J.; Noltes, J. G.; Spek, A. L. *J. J. Organomet. Chem.* **1975**, *99*, 71-77.

[54] Poland, J. S.; Tuck, D. G. *J. Organomet. Chem.* **1972**, *42*, 307-314.

[55] Gaffney, C.; Harrison, P. G. *J. Chem. Soc., Dalton Trans.* **1982**, 1055-1060.

[56] Jutzi, P.; Seufert, A.; Buchner, W. *Chem. Ber.* **1979**, *112*, 2488-2493.

[57] Jutzi, P.; Seufert, A. *J. Organomet. Chem.* **1978**, *161*, C5-C7.

[58] Jutzi, P.; Kohl, F.; Hofmann, P.; Krüger, C.; Tsay, Y. -H. *Chem. Ber.* **1980**, *113*, 757-769.

[59] Jutzi, P.; Dickbreder, R. *J. Organomet. Chem.* **1989**, *373*, 301-306.

[60] Kohl, F. X.; Jutzi, P. *Chem. Ber.* **1981**, *114*, 488-494.

[61] Jutzi, P.; Kohl, F.; Krüger, C. *Angew. Chem.* **1979**, *91*, 81-82; *Angew. Chem. Int. Ed. Engl.* **1979**, *18*, 59-60.

[62] Jutzi, P.; Wippermann, T. Krüger, C.; Kraus, H. -J. *Angew. Chem.* **1983**, *95*, 244-245; *Angew. Chem. Int. Ed. Engl.* **1983**, *22*, 250.

[63] Sitzmann, H.; Ehleiter, Y.; Wolmershäuser, G.; Ecker, A.; Üffing, C.; Schnöckel, H. *J. Organomet. Chem.* **1997**, *527*, 209-213.

[64] Sitzmann, H.; Wolmershäuser, G. *Z. Naturforsch.* **1997**, *52b*, 398.

[65] Cowley, A. H.; Jutzi, P.; Kohl, F. X.; Lasch, J. G.; Norman, N. C.; Schlüter, E. *Angew. Chem.* **1984**, *96*, 603-604; *Angew. Chem. Int. Ed. Engl.* **1984**, *23*, 616-617.

[66] Bulten, E. J.; Budding, H. A. *J. Organomet. Chem.* **1978**, *157*, C3-C4.

[67] Cowley, A. H.; Lasch, J. G.; Norman, N. C.; Stewart, C. A.; Wright, T. C. *Organometallics* **1983**, *2*, 1691-1692.

[68] Cowley, A. H.; Kemp. R. A.; Stewart, C. A. *J. Am. Chem. Soc.* **1982**, *104*, 3239-3240.

[69] Freeman, M. B.; Sneddon, L. G.; Huffman, J. C. *J. Am. Chem. Soc.* **1977**, *99*, 5194-5196.

[70] Freeman, M. B.; Sneddon, L. G. *Inorg. Chem.* **1980**, *19*, 1125-1132.

[71] Jutzi, P.; Hielscher, B. *J. Organomet. Chem.* **1985**, *291*, C25-C27.

[72] Jutzi, P.; Hielscher, B. *Organometallics* **1986**, *5*, 1201-1204

[73] Schulz, S.; Roesky, H. W.; Koch, H. J.; Sheldrick, G. M.; Stalke, D.; Kuhn, A. *Angew. Chem.* **1993**, *105*, 1828-1830; *Angew. Chem. Int. Ed. Engl.* **1993**, *32*, 1729-1731.

[74] Harder, S.; Prosenc, M. H.; *Angew. Chem.* **1994**, *106*, 1830-1832; *Angew. Chem. Int. Ed. Engl.* **1994**, *33*, 1744-1746.

[75] Harder, S.; Prosenc, M. H.; Rief, U. *Organometallics* **1996**, *15*, 118-122.

[76] Wessel, J.; Lork, E.; Mews, R. *Angew. Chem.* **1995**, *107*, 2565-2567; *Angew. Chem. Int. Ed. Engl.* **1995**, *34*, 2376-2377.

[77] Exner, M. M.; Waack, R.; Steiner, E. C. *J. Am. Chem. Soc.* **1973**, *95*, 7009-7018.

[78] Paquette, L. A.; Bauer, W.; Sivik, M. R.; Buhl, M.; Feigel, M.; Schleyer, P. v. R. *J. Am. Chem. Soc.* **1990**, *112*, 8776-8788.

[79] Schleyer, P. v. R.; Jiao, H. *Pure & Appl. Chem.* **1996**, *68*, 209-218.

[80] Wrackmeyer, B.; Sebald, A.; Merwin, L. H. *Magn. Reson. Chem.* **1991**, *29*, 260-263.

[81] Jutzi, P.; Dickbreder, R.; Nöth, H. *Chem. Ber.* **1989**, *122*, 865-870.

[82] Jutzi, P.; Schlüter, E.; Pohl, S.; Saak, W. *Chem. Ber.* **1985**, *118*, 1959-1967.

[83] Scherr, P. A.; Hogan, R. J.; Oliver, J. P. *J. Am. Chem. Soc.* **1974**, *96*, 6055-6059.

[84] Gauss, J.; Schneider, U.; Ahlrichs, R.; Dohmeier, C.; Schnöckel, H. *J. Am. Chem. Soc.* **1993**, *115*, 2402-2408.

[85] Benn, R.; Rufinska, A. *Angew. Chem.* **1986**, *98*, 851-871; *Angew. Chem. Int. Ed. Engl.* **1986**, *25*, 861-881.

[86] Dohmeier, C.; Schnöckel, H.; Robl, C.; Schneider, U.; Ahlrichs, R. *Angew. Chem.* **1993**, *105*, 1714-1716; *Angew. Chem. Int. Ed. Engl.* **1993**, *32*, 1655-1657.

[87] Jutzi, P.; Holtmann, U.; Kanne, D.; Krüger, C.; Blom, R.; Gleiter, R.; Hyla-Kryspin, I. *Chem. Ber.* **1989**, *122*, 1629-1639.

[88] Harris, R. K.; Mann, B. E. '*NMR and the Periodic Table*', Academic Press, London, 1978.

[89] Cerny, Z.; Machacek, J.; Fusek, J.; Kriz, O.; Casensky, B.; Tuck, D. G. *J. Organomet. Chem.* **1993**, *456*, 25-30.

[90] Gardiner, M. G.; Raston, C. L.; Kennard, C. H. L. *Organometallics* **1991**, *10*, 3680-3686.

[91] Dohmeier, C.; Baum, E.; Ecker, A.; Köppe, R.; Schnöckel, H. *Organometallics* **1996**, *15*, 4702-4706.

[92] Bünder, W.; Weiss, E. *J. Organomet. Chem.* **1975**, *92*, 1-6.

[93] Atwood, J. L.; Hunter, W. E.; Cowley, A. H.; Jones, R. A.; Stewart, C. A. *J. Chem. Soc., Chem. Commun.* **1981**, 925-927.

[94] Heeg, M. J.; Janiak, C.; Zuckerman, J. J. *J. Am. Chem. Soc.* **1984**, *106*, 4259-4261.

[95] Armstrong, D. R.; Herbst-Irmer, R.; Kuhn, A.; Moncrieff, D.; Paver, M. A.; Russell, C. A.; Stalke, D.; Steiner, A.; Wright, D. S.; *Angew. Chem.* **1993**, *105*, 1807-1809; *Angew. Chem. Int. Ed. Engl.* **1993**, *32*, 1774-1776.

[96] Haaland, A.; Martinsen, K. -G.; Shlykov, S. A.; Volden, H. V.; Dohmeier, C.; Schnöckel, H. *Organometallics*, **1995**, *14*, 3116-3119.

[97] Jutzi, P.; Leffers, W.; Pohl, S.; Saak, W. *Chem. Ber.* **1989**, *122*, 1449-1456.

[98] Drew, D. A.; Haaland, A. *Acta Chem. Scand.* **1972**, *26*, 3351-3356.

[99] Dohmeier, C.; Köppe, R.; Robl, C.; Schnöckel, H. *J. Organomet. Chem.* **1995**, *487*, 127-130.

[100] Lappert, M. F.; Singh, A.; Engelhardt, L. M.; White, A. H. *J. Organomet. Chem.* **1984**, *262*, 271-278.

[101] Fernholt, L.; Haaland, A.; Jutzi, P.; Kohl, F. X.; Seip, R. *Acta. Chem. Scand.* **1984**, *A38*, 211-216.

[102] Kohl, F. X.; Schlüter, E.; Jutzi, P.; Krüger, C.; Wolmershäuser, G.; Hofmann, P.; Stauffert, P. *Chem. Ber.* **1984**, *117*, 1178-1193.

[103] Koch, H. -J.; Schulz, S.; Roesky, H. W.; Noltemeyer, M.; Schmidt, H. -G.; Heine, A.; Herbst-Irmer, R.; Stalke, D.; Sheldrick, G. M. *Chem. Ber,* **1992**, *125*, 1107-1109.

[104] Schonberg, P. R.; Paine, R. T.; Campana, C. F. *J. Am. Chem. Soc.* **1979**, *101*, 7726-7728.

[105] Harder, S.; Prosenc, M. H. *Angew. Chem.* **1996**, *108*, 101-103; *Angew. Chem. Int. Ed. Engl.* **1996**, *35*, 97-99.

[106] Armstrong, D. R.; Edwards, A. J.; Moncrieff, D.; Paver, M. A.; Raithby, P. R.; Rennie, M. -A.; Russell, C. A.; Wright, D. S. *J. Chem. Soc., Chem. Commun.* **1995**, 927-928.

[107] Duer, M. J.; Page, N. A.; Paver, M. A.; Raithby, P. R.; Rennie, M. -A.; Russell, C. A.; Stourton, C.; Steiner, A.; Wright, D. S. *J. Chem. Soc., Chem. Commun.* **1995**, 1141-1142.

[108] Edwards, A. J.; Paver, M. A.; Raithby, P. R.; Russell, C. A.; Stalke, D.; Steiner, A.; Wright. D. S. *J. Chem. Soc., Dalton Trans.* **1993**, 1465-1466.

[109] Paver, M. A.; Russell, C. A.; Wright, D. S. *Angew. Chem.* **1995**, *107*, 1679-1688; *Angew. Chem. Int. Ed. Engl.* **1995**, *34*, 1545-1554.

[110] Dinnebier, R. E.; Behrens, U.; Olbrich, F. *Organometallics* **1997**, *16*, 3855-3858.

[111] Panattoni, C.; Bombieri, G.; Croatto, U. *Acta Crystallogr.* **1966**, *21*, 823-826.

[112] Williams, R. A.; Hanusa, T. P.; Huffman, J. C. *J. Chem. Soc., Chem. Commun.* **1988**, 1045-1046.

[113] Schumann, H.; Janiak, C.; Görlitz, F.; Loebel, J.; Dietrich, A. *J. Organomet. Chem.* **1989**, *363*, 243-251.

[114] Schumann, H.; Janiak, C.; Pickardt, J.; Börner, U. *Angew. Chem.* **1987**, *99*, 788-789; *Angew. Chem. Int. Ed. Engl.* **1987**, *26*, 789-790.

[115] Dohmeier, C.; Loos, D.; Robl, C.; Schnöckel, H.; Fenske, D. *J. Organomet. Chem.* **1993**, *448*, 5-8.

[116] McCormick, M. J.; Sockwell, S. C.; Davies, C. E. H.; Hanusa, T. P.; Huffman, J. C. *Organometallics* **1989**, *8*, 2044-2049.

[117] Schonberg, P. R.; Paine, R. T.; Campana, C. F.; Duesler, E. N. *Organometallics* **1982**, *1*, 799-807.

[118] Sitzmann, H.; Wolmershäuser, G. *Chem. Ber.* **1994**, *127*, 1335-1342.

[119] Beachley, Jr., O. T.; Blom, R.; Churchill, M. R.; Faegri, Jr., K.; Fettinger, J. C.; Pazik, J. C.; Victoriano, L. *Organometallics* **1989**, *8*, 346-356.

[120] Harvey, S.; Raston, C. L.; Skelton, B. W.; White, A. H.; Lappert, M. F.; Srivastava, G. *J. Organomet. Chem.* **1987**, *328*, C1-C6.

[121] Alexandratos, S.; Streitwieser, A.; Schaefer, H. F. *J. Am. Chem. Soc.* **1976**, *78*, 7959-7962. Watermann, K. C.; Streitwieser, A. *J. Am. Chem. Soc.* **1984**, *106*, 3138-3140.

[122] Jemmis, E. D.; Schleyer, P. v. R. *J. Am. Chem. Soc.* **1982**, *104*, 4781-4788.

[123] Lattman, M.; Cowley, A. H. *Inorg. Chem.* **1984**, *23*, 241-247.

[124] Böhm, M. C.; Gleiter, R.; Morgan, G. L.; Lusztyk, J.; Starowieyski, K. B. *J. Organomet. Chem.* **1980**, *194*, 257-268.

[125] Dohmeier, C.; Loos, D.; Schnöckel, H. *Angew. Chem.* **1996**, *108*, 141-161; *Angew. Chem. Int. Ed. Engl.* **1996**, *35*, 129-149.

[126] Wade, K. *Adv. Inorg. Radiochem.* **1976**, *18*, 1-66.

[127] Almlof, J.; Fernholt, L.; Faegri, Jr., K.; Haaland, A.; Schilling, B. E. R.; Seip, R.; Taugbol, K. *Acta Chem. Scand.* **1983**, *A37*, 131-140.

[128] Jutzi, P. in *'Frontiers of Organogermanium, Tin and Lead Chemistry'*, Lukevics, E.; Ignatovich, I. eds. Latvian Institute of Organic Synthesis 1993.

[129] Hollis, T. K.; Burdett, J. K.; Bosnich, B. *Organometallics* **1993**, *12*, 3385-3386. Bosnich, B. *Chem. Soc. Rev.* **1994**, *23*, 387-395.

[130] Timofeeva, T. V.; Lii, J. -H.; Allinger, N. L. *J. Am. Chem. Soc.* **1995**, *117*, 7452-7459.

[131] Stalke, D. *Angew. Chem.* **1994**, *106*, 2256-2259; *Angew. Chem. Int. Ed. Engl.* **1994**, *33*, 2168-2171.

[132] Thiele, J. *Ber. Dtsch. Chem. Fes.* **1901**, *34*, 68.

[133] Leffers, W. Thesis, University of Bielefeld, 1988.

[134] Evans, W. J.; Boyle, T. J.; Ziller, J. W. *Organometallics* **1992**, *11*, 3903-3907.

[135] Jutzi, P.; Leffers, W.; Hampel, B.; Pohl, S.; Saak, W. *Angew. Chem.* **1987**, *99*, 563-564; *Angew. Chem. Int. Ed. Engl.* **1987**, *26*, 583-584.

[136] Rabe, G.; Roesky, H. W.; Stalke, D.; Pauer, F.; Sheldrick, G. M. *J. Organomet. Chem.* **1991**, *403*, 11-19.

[137] Aoyagi, T.; Shearer, H. M. M.; Wade, K.; Whitehead, G. *J. Organomet. Chem.* **1979**, *175*, 21-31.

[138] Lorberth, J.; Shin, S. -H.; Wocadlo, S.; Massa, W. *Angew. Chem.* **1989**, *101*, 793-794; *Angew. Chem. Int. Ed. Engl.* **1989**, *28*, 735-736.

[139] Lambert, C.; Schleyer, P. v. R. *Angew. Chem.* **1994**, *106*, 1187-1199; *Angew. Chem. Int. Ed. Engl.* **1994**, *33*, 1129-1140.

[140] Strohmeier, W.; Landsfeld, H.; Gernert, F. *Z. Electrochem.* **1962**, *66*, 823-827.

[141] Ford, W. T. *J. Organomet. Chem.* **1971**, *32*, 27-33.

[142] Zaegel, F.; Gallucci, J. C.; Meunier, P.; Gautheron, B.; Sivik, M. R.; Paquette, L. A. *J. Am. Chem. Soc.* **1994**, *116*, 6466-6467.

[143] Wong, W. -K.; Zhang, L.; Wong, W. -T.; Xue, F.; Mak, T. C. W. *Polyhedron*, **1996**, *15*, 4593-4597.

[144] Faegri, K.; Almlöf, J.; Lüthi, H. P. *J. Organomet. Chem.* **1983**, *249*, 303-313. Evans, S.; Green, M. L. H.; Jewitt, B.; Orchard, A. F.; Pygall, C. F. *J. Chem. Soc., Faraday Trans. 2* **1972**, *68*, 1847-1864. Cauletti, C.; Green, J. C.; Kelly, M. R.; van Tilborg, J.; Robbins, J.; Smart, J. *J. Electro. Spectr. Rel. Phenom.* **1980**, *19*, 327.

[145] Wong, C. H.; Lee, T. Y.; Chao, K. J.; Lee, S. *Acta Crystallogr.* **1972**, *28b*, 1662-1665. Wong, C. H.; Lee, T. Y.; Lee, T. J.; Chang, T. W.; Liu, C. S. *Inorg. Nucl. Chem. Lett.* **1973**, *9*, 667-673.

[146] Nugent, K. W.; Beattie, J. K.; Hambley, T. W.; Snow, M. R. *Aust. J. Chem.* **1984**, *37*, 1601-1606.

[147] Nugent, K. W.; Beattie, J. K. *Inorg. Chem.* **1988**, *27*, 4269-4273. Almenningen, A.; Haaland, A.; Lusztyk, J. *J. Organomet. Chem.* **1979**, *170*, 271-284. Haaland, A. *Acta Chem. Scand.* **1968**, *22*, 3030-3032.

[148] Marynick, D. S. *J. Am. Chem. Soc.* **1977**, *99*, 1436-1441. Chiu, N. -S.; Schäfer, L. *J. Am. Chem. Soc.* **1978**, *100*, 2604-2607. Jemmis, E. D.; Alexandratos, S.; Schleyer, P. v. R.; Streitwieser, A.; Schaefer, H.F. *J. Am. Chem. Soc.* **1978**, *100*, 5695-5700. Gleiter, R.; Böhm, M. C.; Haaland, A.; Johansen, R.; Lusztyk, J. *J. Organomet. Chem.* **1979**, *170*, 285-292.

[149] Pratten, S. J.; Cooper, M. K.; Aroney, M. J.; Filipczuk, S. W. *J. Chem. Soc., Dalton Trans.* **1985**, 1761-1765.

[150] Fischer, E. O.; Schreiner, S. *Chem. Ber.* **1959**, *92*, 938-948.

[151] Wong, C.; Wang, S. *Inorg. Nucl. Chem. Lett.* **1975**, *11*, 677-678.

[152] Nugent, K. W.; Beattie, J. K.; Field, L. D. *J. Phys. Chem.* **1989**, *93*, 5371-5377. Nugent, K. W.; Beattie, J. K. *J. Chem. Soc., Chem. Commun.* **1986**, 186-187.

[153] Lusztyk, J.; Starowieyski, K. B. *J. Organomet. Chem.* **1979**, *170*, 293-297.

[154] Gleiter, R.; Böhm, M. C.; Haaland, A.; Johansen, R.; Lusztyk, J. *J. Organomet. Chem.* **1979**, *170*, 271-284.

[155] Blom, R.; Haaland, A.; Weidlein, J. *J. Chem. Soc., Chem. Commun.* **1985**, 266-267.

[156] Cp₂Fe: Seiler, P.; Dunitz, J. D. *Acta Crystallogr.* **1979**, *B35*, 2020-2032. Cp'₂Fe: Freyberg, D. P.; Robbins, J. L.; Raymond, K. N.; Smart, J. C. *J. Am. Chem. Soc.* **1979**, *101*, 892-897.

[157] Burns, C. J.; Andersen, R. A. *J. Organomet. Chem.* **1987**, *325*, 31-37.

[158] Wojtczak, W. A.; Fleig, P. F.; Hampden-Smith, M. J. *Adv. Organomet. Chem.* **1996**, *40*, 215-340.

[159] Haaland, A.; Lusztyk, J.; Brunvoll, J.; Starowieyscki, K. Z. *J. Organomet. Chem.* **1975**, *85*, 279-285.

[160] Ziegler, K.; Froitzheim-Kuhlhorn, H.; Hafner, H. *Chem. Ber.* **1956**, *89*, 434-443.

[161] Fischer, E. O.; Stölzle, G. *Chem. Ber.* **1961**, *94*, 2187-2193.

[162] Zerger, R.; Stucky, G. *J. Organomet. Chem.* **1974**, *80*, 7-17.

[163] Andersen, R. A.; Blom, R.; Burns, C. J.; Volden, H. V. *J. Chem. Soc., Chem. Commun.* **1987**, 768-769. Blom, R.; Faegri Jr., K.; Volden, H. V. *Organometallics*, **1990**, *9*, 372-379.

[164] Williams, R. A.; Hanusa, T. P.; Huffman, J. C. *Organometallics*, **1990**, *9*, 1128-1134.

[165] Williams, R. A.; Tesh, K. F.; Hanusa, T. P. *J. Am. Chem. Soc.* **1991**, *113*, 4843-4851.

[166] Burkey, D. J.; Williams, R. A.; Hanusa, T. P. *Organometallics*, **1993**, *12*, 1331-1337.

[167] Overby, J. S.; Hanusa, T. P. *Angew. Chem.* **1994**, *106*, 2300-2302; *Angew. Chem. Int. Ed. Engl.* **1994**, *33*, 2191-2193.

[168] Williams, R. A.; Hanusa, T. P.; Huffman, J. C. *J. Organomet. Chem.* **1992**, *429*, 143-152.

[169] Rieckhoff, M.; Pieper, U.; Stalke, D.; Edelmann, F. T. *Angew. Chem.* **1993**, *105*, 1102-1104; *Angew. Chem. Int. Ed. Engl.* **1993**, *32*, 1079-1081.

[170] Jutzi, P.; Dahlhaus, J.; Kristen, M. O. *J. Organomet. Chem.* **1993**, *450*, C1-C3. Rees, W. S. Jr.; Lay, U. W.; Dippel, K. A. *J. Organomet. Chem.* **1994**, *483*, 27-31. Hays, M. L.; Hanusa, T. P.; Nile, T. A. *J. Organomet. Chem.* **1996**, *514*, 73-79.

[171] Molander, J. A.; Schumann, H.; Rosenthal, E. C. E.; Demtsdunk, J. *Organometallics* **1996**, *15*, 3817-3824.

[172] Williams, R. A.; Hanusa, T. P.; Huffman, J. C. *J. Am. Chem. Soc.* **1990**, *112*, 2454-2455.

[173] Tanner, P. S.; Williams, R. A.; Hanusa, T. P. *Inorg. Chem.* **1993**, *32*, 2234-2235.

[174] Sockwell, S. C.; Tanner, P. S.; Hanusa, T. P. *Organometallics* **1992**, *11*, 2634-2638.

[175] Goddard, R.; Akhtar, J.; Starowieyski, K. B. *J. Organomet. Chem.* **1985**, *282*, 149-154.

[176] Haaland, A.; Novak, D. P. *Acta Chem. Scand.* **1974**, *A28*, 153-156.

[177] Drew, D. A.; Haaland, A. *Acta Chem. Scand.* **1972**, *26*, 3079-3084.

[178] Atwood, J. L.; Bott, S. G.; Jones, R. A.; Koschmieder, S. U. *J. Chem. Soc., Chem. Commun.* **1990**, 692-693.

[179] Gaines, D. F.; Coleson, K. M.; Calabrese, J. C. *J. Am. Chem. Soc.* **1979**, *101*, 3979-3980.

[180] Sockwell, S. C.; Hanusa, T. P.; Huffman, J. C. *J. Am. Chem. Soc.* **1992**, *114*, 3393-3399.

[181] Johnson, C.; Toney, J.; Stucky, G. D. *J. Organomet. Chem.* **1972**, *40*, C11-C13.

[182] Burkey, D. J.; Alexander, E. K.; Hanusa, T. P. *Organometallics* **1994**, *13*, 2773-2786.

[183] Janiak, C. *Coord. Chem. Rev.* **1997**, *163*, 107-216.

[184] Haaland, A.; Martinsen, K. -G.; Volden, H. V.; Loos, D.; Schnöckel, H. *Acta Chem. Scand.* **1994**, *48*, 172-174.

[185] Shibata, S.; Bartell, L. S.; Gavin, Jr., R. M. *J. Chem. Phys.* **1964**, *41*, 717-722.

[186] Beachley, Jr., O. T.; Pazik, J. C.; Glassman, T. E.; Churchill, M. R.; Fettinger, J. C.; Blom, R. *Organometallics* **1988**, *7*, 1051-1059.

[187] Tyler, J. K.; Cox, A. P.; Sheridan, J. *Nature [London]*, **1959**, *183*, 1182-1183.

[188] Blom, R.; Werner, H.; Wolf, J. *J. Organomet. Chem.* **1988**, *354*, 293-299.

[189] Berar, J. F.; Calvarin, G.; Pommier, C.; Weigel, D. *J. Appl. Crystallogr.* **1975**, *8*, 386-387.

[190] Schumann, H.; Kucht, H.; Kucht, A.; Görlitz, F. H.; Dietrich, A. *Z. Naturforsch.* **1992**, *47b*, 1241-1248.

[191] Schumann, H.; Ghodsi, T.; Esser, L. *Acta Crystallogr.* **1992**, *C48*, 618-620.

[192] Schumann, H.; Ghodsi, T.; Esser, L. Hahn, E. *Chem. Ber.* **1993**, *126*, 591-594.

[193] Schumann, H.; Janiak, C.; Khan, M. A.; Zuckerman, J. J. *J. Organomet. Chem.* **1988**, *354*, 7-13.

[194] Budzelaar, P. H. M.; Boersma, J. *Recl. Trav. Chim. Pays-Bas* **1990**, *109*, 187-189. Schwerdtfeger, P. *Inorg. Chem.* **1991**, *30*, 1660-1663. Treboux, G.; Barthelat, J. -C. *J. Am. Chem. Soc.* **1993**, *115*, 4870-4878.

[195] Janiak, C.; Hoffmann, R. *J. Am. Chem. Soc.* **1990**, *112*, 5924-5946.

[196] Ahlrichs, R.; Ehrig, M.; Horn, H. *Chem. Phys. Lett*, **1991**, *183*, 227-233.

[197] Schneider, U.; Ahlrichs, R.; Horn, H.; Schäfer, A. *Angew. Chem.* **1992**, *104*, 327-329; *Angew. Chem. Int. Ed. Engl.* **1992**, *31*, 353-355.

[198] Werner, H.; Otto, H.; Kraus, H. J. *J. Organomet. Chem.* **1986**, *315*, C57-C60.

[199] Beachley, Jr., O. T.; Churchill, M. R.; Fettinger, J. C.; Pazik, J. C.; Victoriano, L. *J. Am. Chem. Soc.* **1986**, *108*, 4666-4668.

[200] Drew, D. A.; Haaland, A. *Acta Chem. Scand.* **1973**, *27*, 3735-3745.

[201] Haaland, A.; Weidlein, J. *J. Organomet. Chem.* **1972**, *40*, 29-33. Stadelhofer, J.; Weidlein, J.; Haaland, A. *J. Organomet. Chem.* **1975**, *84*, C1-C4. Stadelhofer, J.; Weidlein, J.; Fischer, P. *J. Organomet. Chem.* **1976**, *116*, 55-63.

[202] Tecle, B.; Corfield, W. R.; Oliver, J. P. *Inorg. Chem.* **1982**, *21*, 458-461.

[203] Mertz, K.; Zettler, F.; Hausen, H. D.; Weidlein, J. *J. Organomet. Chem.* **1976**, *122*, 159-170.

[204] Jutzi, P.; Dahlhaus, J.; Kristen, M. O. *J. Organomet. Chem.* **1993**, *450*, C1-C3. Jutzi, P.; Dahlhaus, J.; Bangel, M. *J. Organomet. Chem.* **1993**, *460*, C13-C15. Jutzi, P.; Dahlhaus, J.; Neumann, B.; Stammler, H. -G. *Organometallics* **1996**, *15*, 747-752.

[205] Einstein, F. W. B.; Gilbert, M. M.; Tuck, D. G. *Inorg. Chem.* **1972**, *11*, 2832-2836. Beachley, Jr., O. T.; Getman, T. D.; Kirss, R. U.; Hallock, R. B.; Hunter, W. E.; Atwood, J. L. *Organometallics* **1985**, *4*, 751-754. Beachley, Jr., O. T.; Hallock, R. B.; Zhang, H. M.; Atwood, J. L. *Organometallics* **1985**, *4*, 1675-1680. Schumann, H.; Nickel, S.; Weimann, R. *J. Organomet. Chem.* **1994**, *468*, 43-47.

[206] Bochmann, M.; Dawson, D. M. *Angew. Chem.* **1996**, *108*, 2371-2372; *Angew. Chem. Int. Ed. Engl.* **1996**, *35*, 2226-2228.

[207] Schulz, S.; Gillan, E. G.; Ross, J. L.; Rogers, L. M.; Rogers, R. D.; Barron, A. R. *Organometallics* **1996**, *15*, 4880-4883.

[208] Schulz, S.; Andruh, M.; Pape, T.; Heinze, T.; Roesky, H. W.; Häming, L.; Kuhn, A.; Herbst-Irmer, R. *Organometallics* **1994**, *13*, 4004-4007.

[209] Dohmeier, C.; Schnöckel, H.; Robl, C.; Schneider, U.; Ahlrichs, R. *Angew. Chem.* **1994**, *106*, 225-227; *Angew. Chem. Int. Ed. Engl.* **1994**, *33*, 199-200.

[210] Schulz, S.; Schoop, T.; Roesky, H. W.; Häming, L.; Steiner, A.; Herbst-Irmer, R. *Angew. Chem.* **1995**, *107*, 1015-1016; *Angew. Chem. Int. Ed. Engl.* **1995**, *34*, 919-920.

[211] Dohmeier, C.; Krautscheid, H.; Schnöckel, H. *Angew. Chem.* **1994**, *106*, 2570-2571; *Angew. Chem. Int. Ed. Engl.* **1994**, *33*, 2482-2483.

[212] Jutzi, P.; Kanne, D.; Krüger, C. *Angew. Chem.* **1986**, *98*, 163; *Angew. Chem. Int. Ed. Engl.* **1986**, *25*, 164

[213] Grenz, M.; Hahn, E.; du Mont, W. -W.; Pickardt, J. *Angew. Chem.* **1984**, *96*, 69-70; *Angew. Chem. Int. Ed. Engl.* **1984**, *23*, 61-62.

[214] Almenningen, A.; Haaland,, A.; Motzfeldt, T. *J. Organomet. Chem.* **1967**, *7*, 97-104.

[215] Schumann, H.; Janiak, C.; Hahn, E.; Loebel, J.; Zuckerman, J. J. *Angew. Chem.* **1985**, *97*, 765; *Angew. Chem. Int. Ed. Engl.* **1985**, *24*, 773.

[216] Schumann, H.; Janiak, C.; Hahn, E.; Kolax, C.; Loebel, J.; Rausch, M. D.; Zuckerman, J. J.; Heeg, M. J. *Chem. Ber.* **1986**, *119*, 2656-2667.

[217] Constantine, S. P.; Hitchcock, P. B.; Lawless, G. A.; De Lima, G. M. *J. Chem. Soc., Chem. Commun.* **1996**, 1101-1102.

[218] Constantine, S. P.; Hitchcock, P. B.; Lawless, G. A. *Organometallics* **1996**, *15*, 3905-3906.

[219] Sitzmann, H.; Boese, R.; Stellberg, P. *Z. Anorg. Allg. Chem.* **1996**, *622*, 751-755.

[220] Burkey, D. J.; Hanusa, T. P. *Organometallics* **1995**, *14*, 11-13.

[221] Heeg, M. J.; Herber, R. H.; Janiak, C.; Zuckerman, J. J.; Schumann, H.; Manders, W. F. *J. Organomet. Chem.* **1988**, *346*, 321-332.

[222] Beswick, M. A.; Cromhout, N. L.; Harmer, C. N.; Raithby, P. R.; Russell, C. A.; Smith, J. S. B.; Steiner, A.; Wright, D. S. *J. Chem. Soc., Chem. Commun.* **1996**, 1977-1978.

[223] Jutzi, P.; Kohl, F. X.; Schlüter, E.; Hursthouse, M. B.; Walker, N. P. C. *J. Organomet. Chem.* **1984**, *271*, 393-402.

[224] Dory, T. S.; Zuckerman, J. J.; Barnes, C. L. *J. Organomet. Chem.* **1985**, *281*, C1-C7.

[225] Constantine, S. P.; De Lima, G. M.; Hitchcock, P. B.; Keates, J. M.; Lawless, G. A. *J. Chem. Soc., Chem. Commun.* **1996**, 2337-2338.

[226] Constantine, S. P.; De Lima, G. M.; Hitchcock, P. B.; Keates, J. M.; Lawless, G. A.; Marziano, I. *Organometallics* **1997**, *16*, 793-795.

[227] Holliday, A. K.; Makin, P. H.; Puddephatt, R. J. *J. Chem. Soc., Dalton Trans.* **1976**, 435-438.

[228] Williams, R. E. *Adv. Inorg. Radiochem.* **1976**, *18*, 67-142. Onak, T. in *'Comprehensive Organometallic Chemistry'*, Wilkinson, G.; Stone, F. G. A.; Abel, E. W. eds. Pergamon, Oxford, 1982.

[229] Harrison, P. G.; Zuckerman, J. J. *J. Am. Chem. Soc.* **1969**, *91*, 6885-6886.

[230] Jutzi, P.; Schmidt, H.; Neumann, B.; Stammler, H. -G. *J. Organomet. Chem.* **1995**, *499*, 7-10.

[231] Edwards, A. J.; Paver, M. A.; Raithby, P. R.; Rennie, M. -A.; Russell, C. A.; Wright. D. S. *J. Chem. Soc., Dalton Trans.* **1995**, 1587-1591. Stalke, D.; Paver, M. A.; Wright, D. S. *Angew. Chem.* **1993**, *105*, 445-446; *Angew. Chem. Int. Ed. Engl.* **1993**, *32*, 428-429.

[232] Davidson, M. G.; Stalke, D.; Wright, D. S. *Angew. Chem.* **1992**, *104*, 1265-1267; *Angew. Chem. Int. Ed. Engl.* **1992**, *31*, 1226-1227.

[233] Paver, M. A.; Russell, C. A.; Stalke, D.; Wright. D. S. *J. Chem. Soc., Chem. Commun.* **1993**, 1349-1351.

[234] Harris, D. H.; Lappert, M. F. *J. Chem. Soc., Chem. Commun.* **1974**, 895-896.

[235] Jutzi, P.; Hielscher, B. *Organometallics* **1986**, *5*, 2511-2514.

[236] Jutzi, P.; Möhrke, A.; Müller, A.; Bögge, H. *Angew. Chem.* **1989**, *101*, 1527-1528; *Angew. Chem. Int. Ed. Engl.* **1989**, *28*, 1518-1519.

[237] Jutzi, P. *'Frontiers of Organosilicon Chemistry'*, Bassindale, A. R.; Gaspar, P. eds. Royal Society of Chemistry, 1991.

[238] Jutzi, P.; Kohl, F. X. *J. Organomet. Chem.* **1979**, *164*, 141-152. Jutzi, P.; Hampel, B. *Organometallics* **1986**, *5*, 730-734.

[239] Jutzi, P.; Becker, A.; Stammler, H. -G.; Neumann, B. *Organometallics* **1991**, *10*, 1647-1648.

[240] Jutzi, P.; Dahlhaus, J. unpublished results.

[241] Birkhahn, M.; Krommes, P.; Massa, W.; Lorberth, J. *J. Organomet. Chem.* **1981**, *208*, 161-167.

[242] Lorberth, J.; Massa, W.; Wocadlo, S.; Sarraje, I.; Shin, S.-H.; Li, X.-W. *J. Organomet. Chem.* **1995**, *485*, 149-152.

[243] Lee, T. J.; Schaefer, III, H. F.; Magnusson, E. A. *J. Am. Chem. Soc.* **1985**, *107*, 7239-7243.

[244] Frank, W. *J. Organomet. Chem.* **1991**, *406*, 331-341.

[245] Frank, W. *J. Organomet. Chem.* **1990**, *386*, 177-186.

[246] Ehleiter, Y.; Wolmershäuser, G.; Sitzmann, H.; Boese, R. *Z. Anorg. Allg. Chem.* **1996**, *622*, 923-930.

[247] Gudat, D.; Nieger, M.; Niecke, E. *J. Chem. Soc., Dalton Trans.* **1989**, 693-700.

[248] Baxter, S. G.; Cowley, A. H.; Mehrotra, S. K. *J. Am. Chem. Soc.* **1981**, *103*, 5572-5573. Cowley, A. H.; Mehrotra, S. K. *J. Am. Chem. Soc.* **1983**, *105*, 2074-2075.

[249] Bard, A. J.; Cowley, A. H.; Leland, J. K.; Thomas, G. J. N.; Norman, N. C.; Jutzi, P.; Morley, C. P.; Schlüter, E. *J. Chem. Soc., Dalton Trans.* **1985**, 1303-1307.

[250] Kaupp, M.; Schleyer, P. v. R.; Dolg, M.; Stoll, H. *J. Am. Chem. Soc.* **1992**, *114*, 8202-8208.

[251] Pauling, L. *'The Nature of the Chemical Bond'*, 3rd ed., Cornell University Press, New York 1960, p 261.

2 Lanthanocenes

Frank T. Edelmann

2.1 Introduction

The beginnings of organolanthanide chemistry date back to the year 1954, when Birmingham and Wilkinson described the tris(cyclopentadienyl)lanthanide complexes, $Ln(C_5H_5)_3$ [1]. Thus the chemistry of lanthanocenes is not much younger than that of other metallocenes of the *d*-transition metals. However, the development of this area of organometallic chemistry was relatively slow due to the intrinsic instability of organolanthanide compounds towards moisture and oxygen. The highly oxophilic character of the 4*f*-elements requires rigorous exclusion of traces of air and moisture during preparation and characterization of organometallic compounds of these metals. In addition, it turned out that simple alkyl and aryl derivatives of the type LnR_3 could not be isolated due to severe steric unsaturation. With today's more sophisticated experimental and analytical techniques combined with the design of new ligand sets organolanthanides and -actinides are routinely synthesized and handled and their unique properties are easily studied. Historically, the early discovery of $Ln(C_5H_5)_3$ was followed by a period of relative stagnation which lasted about two decades. The past twenty years, however, have witnessed an enormous and exciting development of organolanthanide chemistry [2–7]. This spectacular growth in research activities is mainly because many organo-*f*-element complexes exhibit unique structural and physical properties and many of them are highly active in various catalytic processes. In fact, the catalytic activity of certain organolanthanide hydrocarbyls and hydrides is often dramatically higher than that of comparable *d*-transition metal catalysts. Applications of lanthanocenes in homogeneous catalysis have been compiled in a recent review article [8]. Thus the present review will focus mainly on synthetic aspects of lanthanocene chemistry.

2.2 Lanthanide(II) Metallocenes

Lanthanocene(II) complexes are among the oldest known rare earth organometallics. Like alkali metals, several rare earth elements can be dissolved in liquid ammonia. $Eu(C_5H_5)_2$ was first obtained by reacting the blue solution of Eu metal in liquid ammonia with cyclopentadiene [9]. The base-free material was obtained by heating the initially produced ammonia adduct to 200°C under high vacuum followed by vacuum sublimation of the thermally highly stable product at 400°C. The behavior of $Eu(C_5H_5)_2$ nicely illustrates the typical properties of most organolanthanide complexes: The compound is thermally extremely stable while it is rapidly decomposed in the presence of traces of oxygen. $Yb(C_5H_5)_2$ too has been prepared by adopting the same ammonia route [2–7].

$$\text{Ln} + 3C_5H_6 \xrightarrow[\text{2. subl.}]{\text{1. NH}_3(l)} \text{Ln}(C_5H_5)_2 + C_5H_8 \qquad (2\text{-}1)$$
$$\text{Ln = Eu, Yb}$$

Subsequently several other synthetic routes leading to lanthanocene(II) complexes have been developed. They are normally carried out in THF solution and initially yield the THF adducts of $Ln(C_5H_5)_2$. An elegant transmetallation reaction starts with the metal powders of Sm, Eu, or Yb which are treated with $Hg(C_5H_5)_2$. Similarly, Yb metal reacts with TlC_5H_5 in DME solution to give $(C_5H_5)_2Yb(DME)$ (Eqs. (2-2) and (2-3)) [10].

$$\text{Ln} + Hg(C_5H_5)_2 \xrightarrow{\text{THF}} (C_5H_5)_2Ln(THF)_n + Hg \qquad (2\text{-}2)$$
$$\text{Ln = Sm, Eu, Yb}$$

$$Yb + 2TlC_5H_5 \xrightarrow{\text{DME}} (C_5H_5)_2Yb(DME) + 2Tl \qquad (2\text{-}3)$$

In the solid-state structures of $(C_5H_5)_2Yb(DME)$ [11], $(C_5H_5)_2Yb(THF)_2$ [12, 13], and $(MeC_5H_4)_2Yb(DME)$ [14] the central Yb atom is tetrahedrally surrounded by two cyclopentadienyl ligands and two oxygen atoms of the ether ligands. Perhaps the most straightforward access to $Sm(C_5H_5)_2$ was reported by Kagan et al. and involves the reaction of SmI_2 with two equivalents of sodium cyclopentadienide (Eq. (2-4)) [15–17].

$$SmI_2 + 2Na(C_5H_5) \xrightarrow{\text{THF}} Sm(C_5H_5)_2 + 2KI \qquad (2\text{-}4)$$

Yet another possible synthesis of $(C_5H_5)_2Ln(II)$ derivatives is the reduction of various lanthanocene(III) complexes [2–7, 13, 18, 19]. Either sodium, sodium naphthalide, potassium cyclooctadienide, or metallic Yb have been used to reduce $(C_5H_5)_2YbCl(THF)$ to the divalent oxidation state:

$$(C_5H_5)_2YbCl(THF) + Na \xrightarrow{\text{THF}} (C_5H_5)_2Yb(THF)_n + NaCl \qquad (2\text{-}5)$$

$$3(C_5H_5)_2YbCl(THF) + Yb \xrightarrow{\text{THF}} 3(C_5H_5)_2Yb(THF)_n + YbCl_3 \qquad (2\text{-}6)$$

Tris(cyclopentadienyl)lanthanides can also be used as precursors for lanthanide(II) metallocenes. Purple $(C_5H_5)_2Sm(THF)_n$ was obtained by reductive cleavage of $Sm(C_5H_5)_3$ with potassium naphthalide [2–7]. Deacon et al. have studied the redox chemistry of $Yb(C_5H_5)_3$ and found that this compound can be reduced by metallic ytterbium to give $Yb(C_5H_5)_2$. In turn, $Yb(C_5H_5)_2$ can be oxidized back to $Yb(C_5H_5)_3$ by treatment with TlC_5H_5 [15]. Reduction of $Yb(C_5H_5)_3$ with sodium naphthalide was originally reported to give $Yb(C_5H_5)_2$ (after sublimation at 400°C). A recent reinvestigation of this reaction revealed that the sublimed green material was polymeric $[NaYb(C_5H_5)_3]_n$ [20] instead. Thus the reaction pathway has to be reformulated according to Eq. (2-7).

$$Yb(C_5H_5)_3 + Na \longrightarrow NaYb(C_5H_5)_3 \qquad (2\text{-}7)$$

In this unusual 'ate' complex of divalent ytterbium the different metal atoms are bridged by $\mu\text{-}\eta^5\text{:}\eta^5$-cyclopentadienyl ligands to give a polymeric network. Somewhat surprising is the fact that the compound can be sublimed at very high temperatures (ca. 400 °C) without any sign of decomposition. Retention of sodium cyclopentadienide is an explanation for the various color changes associated with $Yb(C_5H_5)_2$ in solution and in the solid state. For solid $Yb(C_5H_5)_2$ a green and a red modification had been described in the early literature [2–7]. It has now become clear that the green color is associated with $[NaYb(C_5H_5)_3]_n$ while the red material is genuine $Yb(C_5H_5)_2$. $Yb(C_5H_5)_2$ itself is also polymeric in the solid state [20].

Ring-substituted ligands such as t-butylcyclopentadienyl or 1, 3-di-t-butylcyclopentadienyl have been successfully employed in the preparation of highly reactive lanthanide(II) organometallics. The resulting products are often more soluble in hydrocarbon solvents than the parent cyclopentadienyl derivatives. Simple bis-substituted lanthanide(II) complexes containing these ligands have been prepared by metathetical reactions between the lanthanide diiodides and the corresponding sodium or potassium cyclopentadienides. Monomeric $(^tBuC_5H_4)_2Yb(THF)_2$ was isolated as the disolvate and characterized by X-ray crystallography [21]. Monomeric $(^tBuC_5H_4)_2Sm(DME)$ has been prepared analogously [22, 23]. In the case of divalent samarium the use of the sterically more demanding 1,3-di-t-butylcyclopentadienyl ligand leads to the formation of the mono-THF adduct $(^tBu_2C_5H_3)_2Sm(THF)$ [25]. Trimethylsilyl-substituted cyclopentadienyl ligands are equally useful in the preparation of soluble, highly reactive organolanthanide(II) complexes. For example, dimeric $[(Me_3SiC_5H_4)_2Yb(\mu\text{-}Cl)]_2$ can be reduced by sodium amalgam in THF to give the bis(THF) solvate of $Yb(Me_3SiC_5H_4)_2$. The THF ligands can be replaced by TMEDA, and simple heating of the THF adduct yields the green, base-free Yb(II) derivative (Scheme 2-1) [25].

$$[(Me_3SiC_5H_4)_2Yb(\mu\text{-}Cl)]_2 \quad + \quad 2\ Na/Hg$$

$$\xrightarrow[\text{- Hg}]{\text{THF}} \quad (Me_3SiC_5H_4)_2Yb(THF)_2 \quad + \quad 2\ NaCl$$

$$\xrightarrow{\Delta} \quad Yb(Me_3SiC_5H_4)_2 \quad + \quad 2\ THF$$

Scheme 2-1. Preparation of $Yb(Me_3SiC_5H_4)_2$.

$SmI_2(THF)_2$ reacts with two equivalents of $K[(Me_3Si)_2C_5H_3]$ to give the monosolvate $[(Me_3Si)_2C_5H_3]_2Sm(THF)$ which can be desolvated to afford base-free $Sm[(Me_3Si)_2C_5H_3]_2$ [26]. $[(Me_3Si)_2C_5H_3]_2Yb(THF)$ was made analogously and structurally characterized [27]. The bent metallocenes $Ln[(Me_3Si)_2C_5H_3]_2$ (Ln = Eu, Yb) have been structurally investigated. Both compounds are polymeric and display unusual intermolecular contacts between the metal centers and γ-methyl groups of neighboring $(Me_3Si)_2C_5H_3$ ligands. The europium derivative exhibits an unprecedented structure with a cyclopentadienyl ring bridging $\eta^3:\eta^5$ between two nonequivalent europium atoms [28].

Various O- or N-donor-substituted cyclopentadienyl ligands have been shown to be very useful in the stabilization of divalent lanthanide metallocenes. A typical example is the methoxyethylcyclopentadienyl ligand, which forms solvated and unsolvated complexes with divalent Sm and Yb (Eq. (2-8)) [29]:

$$LnI_2 + 2\ K(MeOCH_2CH_2C_5H_4) \xrightarrow{\text{THF}} (MeOCH_2CH_2C_5H_4)_2Ln + 2\ KI \quad (2\text{-}8)$$
$$Ln = Sm,\ Yb$$

Equally useful properties are exhibited by the 1,1'-(3-oxa-pentamethylene)dicyclopentadienyl ligand, which allowed the preparation of the structurally characterized Yb(II) complex $[O(CH_2CH_2C_5H_4)_2]Yb(DME)$ [30–32]. Intramolecular chelate stabilization has also been achieved with the amino-functionalized ligand $C_5Me_4CH_2CH_2NMe_2$. Treatment of SmI_2 with two equivalents of $Li(C_5Me_4CH_2CH_2NMe_2)$ produced dark green, unsolvated $Sm(C_5Me_4CH_2CH_2NMe_2)_2$ (**1**) [33, 34].

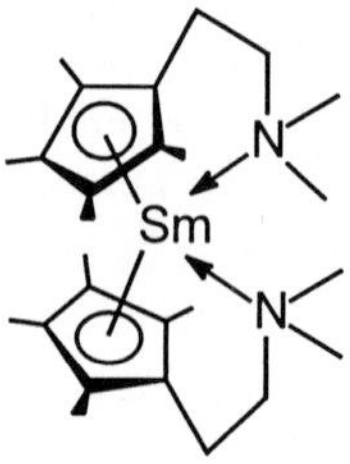

1

An interesting novel aspect of divalent lanthanocene chemistry is the successful synthesis of a 'metalloligand'. The phosphinoytterbocenes $(C_5H_4PPh_2)_2Yb(THF)$ and $(C_5H_4PPh_2)_2Yb(DME)$ have been prepared using different synthetic approaches (Eqs. (2-9)–(2-11)) [35].

$$Yb(C_6F_5)_2 + 2C_5H_5PPh_2 \xrightarrow{\text{THF}} (C_5H_4PPh_2)_2Yb(THF) + 2C_6F_5H \qquad (2\text{-}9)$$

$$Yb + Hg(C_6F_5)_2 + 2C_5H_5PPh_2 \xrightarrow{\text{THF}} (C_5H_4PPh_2)_2Yb(THF) + Hg + 2C_6F_5H \quad (2\text{-}10)$$

$$Yb + 2TlC_5H_4PPh_2 \xrightarrow{\text{DME}} (C_5H_4PPh_2)_2Yb(DME) + 2Tl \qquad (2\text{-}11)$$

Divalent lanthanide metallocenes are generally highly reactive. Especially the Sm(II) and Yb(II) derivatives are powerful one-electron reducing agents. They easily undergo redox reactions and have frequently been used as precursors to prepare bis(cyclopentadienyl)lanthanide(III) complexes. For example, $(C_5H_5)_2Yb(DME)$ can be treated with various Hg, Tl, Cu, or Ag salts to give the corresponding trivalent $(C_5H_5)_2YbX$ derivatives. The main advantage of such oxidative transformations is that the products can be isolated free of alkali metal halides, which tend to be readily incorporated in the products (Eqs. (2-12)) [36, 37].

$$(C_5H_5)_2Yb(DME) + MX \xrightarrow[-DME]{} (C_5H_5)_2YbX + M \qquad (2\text{-}12)$$

$$M = \tfrac{1}{2}\ Hg,\ Tl,\ Cu,\ Ag$$
$$X =\quad Cl,\ Br,\ I,\ C{\equiv}CPh,\ C_6F_5,\ O_2CMe,$$
$$O_2CC_6F_5,\ O_2CC_5H_4N,$$
$$(MeCO)_2CH,\ (PhCO)_2CH$$

Decamethyllanthanocenes(II) of the divalent lanthanides form a truly remarkable class of organolanthanide complexes. The successful preparation and characterization of unsolvated decamethylsamarocene by Evans *et al.* can be considered a landmark in organo-*f*-element chemistry. This compound is perhaps the most reactive organometallic compound ever made and has given rise to a large number of unusual reactions and unexpected products. The most straightforward synthesis involves the reaction of SmI_2 with two equivalents of $K(C_5Me_5)$ (Eq. (2-13)) [38]. Unsolvated $Sm(C_5Me_5)_2$ can be obtained by desolvation and sublimation of the THF adduct under high vacuum [39]. $Eu(C_5Me_5)_2$ is prepared analogously from $EuI_2(THF)_2$ and $K(C_5Me_5)$. In this case the resulting THF solvate is more difficult to desolvate than the samarium analog and pure $Eu(C_5Me_5)_2$ was obtained only after repeated vacuum sublimation [40]. Base-free $Yb(C_5Me_5)_2$ is accessible analogously.

$$\text{SmI}_2(\text{THF})_2 + 2\text{K}(\text{C}_5\text{Me}_5) \xrightarrow{\text{THF}} (\text{C}_5\text{Me}_5)_2\text{Sm}(\text{THF})_2 + 2\text{KI}$$

$$\downarrow$$

$$\text{Sm}(\text{C}_5\text{Me}_5)_2 \qquad\qquad (2\text{-}13)$$

Ether adducts of the type $(\text{C}_5\text{Me}_5)_2\text{LnL}$ (L = Et_2O, THF) can also be obtained by reacting LnCl_2 with two equivalents of $\text{Na}(\text{C}_5\text{Me}_5)$ in THF or diethyl ether (Eq. (2-14)) [41]. In the case of europium EuCl_3 may be employed as starting material because reduction to the divalent oxidation state takes place during the course of the reaction with three equivalents of $\text{Na}(\text{C}_5\text{Me}_5)$ (Eq. (2-15)) [40].

$$\text{LnCl}_2 + 2\text{Na}(\text{C}_5\text{Me}_5) \xrightarrow{+\,\text{L}} (\text{C}_5\text{Me}_5)_2\text{LnL} + 2\text{NaCl} \qquad (2\text{-}14)$$
$$\text{Ln} = \text{Sm, Eu, Yb}$$

$$\text{EuCl}_3 + 3\text{Na}(\text{C}_5\text{Me}_5) \xrightarrow[\text{2. Et}_2\text{O}]{\text{1. THF}} (\text{C}_5\text{Me}_5)_2\text{Eu}(\text{Et}_2\text{O})(\text{THF}) + 3\text{NaCl} + \tfrac{1}{2}(\text{C}_5\text{Me}_5)_2 \quad (2\text{-}15)$$

$(\text{C}_5\text{Me}_5)_2\text{Sm}(\text{THF})_2$ (**2**) is isolated as a purple solid, while $\text{Sm}(\text{C}_5\text{Me}_5)_2$ (**3**) forms dark green crystals. Both complexes are highly air-sensitive, turning yellow in the presence of traces of oxygen. The molecular and crystal structure of $(\text{C}_5\text{Me}_5)_2\text{Sm}(\text{THF})_2$ was determined by X-ray diffraction. The molecule adopts a typical bent metallocene structure with a *pseudo*-tetrahedral coordination geometry around the divalent samarium [2–7, 39].

2   **3**

The unusual molecular structure of $\text{Sm}(\text{C}_5\text{Me}_5)_2$ consists of bent metallocene molecules with a centroid–Sm–centroid angle of 140.1°. The shortest separation between two monomeric units is 3.22(1) Å [42]. Base-free $\text{Eu}(\text{C}_5\text{Me}_5)_2$ and $\text{Yb}(\text{C}_5\text{Me}_5)_2$ have also been structurally characterized by X-ray analyses [42, 43], showing once again a bent metallocene geometry similar to that in the samarium analog or in $\text{M}(\text{C}_5\text{Me}_5)_2$ (M = Ca, Ba) [44, 45]. Weak *van der Waals* attractive forces between C_5Me_5 rings are made responsible for the bent structures in the solid state [46, 47].

Very few examples of π-adducts of lanthanocenes(II) have been described. In the case of π-bonded alkynes it was possible to isolate the adduct $(C_5Me_5)_2Yb(\eta^2\text{-}MeC{\equiv}CMe)$, made from $Yb(C_5Me_5)_2$ and the free alkyne ligand [48]. The structural data indicate a weak coordination of the alkyne ligand and little or negligible π-backbonding ($Yb\text{--}C(C_5Me_5)$ 2.659(9), $Yb\text{--}C(\eta^2)$ 2.85(1) Å). The same is true for the first organolanthanide olefin complex. Bimetallic $(C_5Me_5)_2Yb(\mu\text{-}C_2H_4)Pt(PPh_3)_2$ was prepared by reacting $Yb(C_5Me_5)_2$ and $(C_2H_4)Pt(PPh_3)_2$ (Eq. (2-16), $Yb\text{--}C(C_5Me_5)$ 2.67(2), $Yb\text{--}C(\eta^2)$ 2.781(6) Å) [49].

$$Yb(C_5Me_5)_2 + (C_2H_4)Pt(PPh_3)_2 \longrightarrow (C_5Me_5)_2Yb(\mu\text{-}C_2H_4)Pt(PPh_3)_2 \quad (2\text{-}16)$$

In the platinum complex $(C_2H_4)Pt(PPh_3)_2$ the coordinated ethylene is more electron-rich than the free olefin and can thus interact as a Lewis base with the Lewis acid $Yb(C_5Me_5)_2$ [49]. A truly remarkable compound is the substituted methane complex $(C_5Me_5)_2Yb(\mu\text{-}Me)Be(C_5Me_5)$, which was prepared according to Eq. (2-17) and structurally characterized. The most unusual structural feature is the nearly linear Yb–Me–Be unit (177.2°) [50].

$$Yb(C_5Me_5)_2 + MeBe(C_5Me_5) \longrightarrow (C_5Me_5)_2Yb(\mu\text{-}Me)Be(C_5Me_5) \quad (2\text{-}17)$$

Several examples of *ansa*-lanthanocenes(II) have been reported in the literature. For example, the trimethylene-bridged bis(cyclopentadienyl) ligand has been successfully employed in the preparation of divalent lanthanide *ansa*-metallocenes. Treatment of $LnCl_2$ with the disodium salt of the ligand in THF solution afforded the solvated species $[C_5H_4(CH_2)_3C_5H_4]Ln(THF)_2$ (Ln = Sm, Yb). The ytterbium(II) derivative was structurally characterized (Yb–C 2.67(2)–2.72(2) Å, ring centroid–Yb–ring centroid 127(1)°) [51]. Several complexes containing the tetramethylethylene-bridged dicyclopentadienyl ligand have also been reported. Direct reaction of activated ($HgCl_2$) Ln metal powders with 6,6-dimethylfulvene in THF afforded the tetramethylethylene bridged lanthanide(II) metallocenes $[Me_4C_2(C_5H_4)_2]Ln$ (Ln = Sm, Yb; Eq. (2-18)) [52].

$$\text{Ln = Yb, Sm}$$

2.3 Lanthanide(III) Metallocenes

2.3.1 Lanthanide(III) Metallocenes Containing Ln–Group 17 Element Bonds

Lanthanocene fluorides are not generally available starting directly from LnF_3 due to the insolubility of anhydrous lanthanide trifluorides. Thus the normal metathetical reactions cannot be applied to these systems, although a few fluorides have been synthesized by other routes. For example, $Yb(MeC_5H_4)_2$ reacts with perfluoroolefins under defluorination to give $(MeC_5H_4)_2YbF(THF)$ [53]. In contrast to the less-investigated fluorides, lanthanocenes of the type $(C_5H_5)_2LnCl$ play a key role as intermediates in the synthesis of other organolanthanide complexes. The synthetic routes leading to lanthanocene chlorides are summarized in Eqs. (2-19)–(2-24) [2–7].

$$LnCl_3 + 2MC_5H_5 \longrightarrow (C_5H_5)_2LnCl + 2MCl \qquad (2\text{-}19)$$
$$M = Na, Tl$$

$$ScCl_3 + Mg(C_5H_5)_2 \longrightarrow (C_5H_5)_2ScCl + MgCl_2 \qquad (2\text{-}20)$$

$$Ln(C_5H_5)_3 + NH_4Cl \longrightarrow (C_5H_5)_2LnCl + C_5H_6 + NH_3 \qquad (2\text{-}21)$$

$$Ln(C_5H_5)_3 + HCl \longrightarrow (C_5H_5)_2LnCl + C_5H_6 \qquad (2\text{-}22)$$

$$\overset{NaC_5H_5}{(C_5H_9C_5H_4)ErCl_2(THF)_n \longrightarrow (C_5H_5)_2ErCl(THF)} \qquad (2\text{-}23)$$
$$(C_5H_9C_5H_4 = \text{cyclopentylcyclopentadienyl})$$

$$LnCl_3 + 2Ln(C_5H_5)_3 \longrightarrow 3(C_5H_5)_2LnCl \qquad (2\text{-}24)$$

Due to coordinative unsaturation $(C_5H_5)_2LnCl$ derivatives are often not available for the lighter lanthanide elements La–Nd. In general, the compounds $(C_5H_5)_2LnCl$ are unstable as monomers. In benzene solution as well as in the solid state, these compounds form chloro-bridged dimers of the type $[(C_5H_5)_2Ln(\mu\text{-}Cl)]_2$ (**4**).

4

Treatment with THF readily cleaves the chloride bridges in $[(C_5H_5)_2Ln(\mu-Cl)]_2$ under formation of the monomeric 1:1 adducts $(C_5H_5)_2LnCl(THF)$ [2–7]. The molecular and crystal structure of $(C_5H_5)_2LuCl(THF)$ was determined by X-ray crystallography [54, 55]. The molecule exhibits a *pseudo*-tetrahedral coordination geometry. In addition, the THF-solvated dimers $[(C_5H_5)_2Ln(\mu-Cl)(THF)]_2$ (Ln = Nd, Er, Yb) were structurally characterized [56–58]. Bis(cyclopentadienyl)lanthanide hydrocarbyls can also serve as starting materials for complexes with classical co-ligands. For example, the heterobimetallic lanthanocene chloride complex $(C_5H_5)_2Yb(\mu-Cl)_2AlMe_2$ was synthesized by reacting dimeric $[(C_5H_5)_2Yb(\mu-Me)]_2$ with $MeAlCl_2$. The reaction proceeds *via* cleavage of the ytterbium–carbon bonds [2–7].

Mono- and disubstituted cyclopentadienyl ligands have been extensively employed in the preparation of lanthanocene halides. The chloro-bridged dimers $[(MeC_5H_4)_2Yb(\mu-Cl)]_2$ [59] and $[(MeC_5H_4)_2Sm(\mu-Cl)(THF)]_2$ have been structurally characterized [60]. *t*-Butyl-substituted cyclopentadienyl ligands are also increasingly utilized in organolanthanide chemistry. Dimeric chloro-bridged molecular structures were established by X-ray crystallography for the *t*-butylcyclopentadienyl lanthanide chlorides $[(^tBuC_5H_4)_2Ln(\mu-Cl)]_2$ (Ln = Pr, Gd, Sm, Lu) [61, 62] and $[(^tBu_2C_5H_3)_2Ln(\mu-Cl)]_2$ (Ln = Ce, Nd, Lu) [63, 64].

Disubstituted lanthanide(III) halides containing the trimethylsilylcyclopentadienyl ligand include the base-free dimers $[(Me_3SiC_5H_4)_2Yb(\mu-X)]_2$ (X = Cl, I) [25] as well as the 'ate' complex precursors $(Me_3SiC_5H_4)_2Y(\mu-Cl)_2Li(L)$ (L = DME, 1/2 THF) [65], $(Me_3SiC_5H_4)_2Yb(\mu-Cl)_2Li(Et_2O)_2$ [66], and $(Me_3SiC_5H_4)_2Lu(\mu-Cl)_2Li(TMEDA)$ [66]. Especially useful for the synthesis of lanthanocene halides (besides C_5Me_5) is the very bulky 1,3-bis(trimethylsilyl)cyclopentadienyl anion. With this ligand two series of disubstituted complexes have been prepared for nearly all elements of the lanthanide series. Treatment of lanthanide trichlorides with two equivalents of lithium 1,3-bis(trimethylsilyl)cyclopentadienide initially affords the solvated lithium chloride adducts $[(Me_3Si)_2C_5H_3]_2Ln(\mu-Cl)Li(THF)_2$ which upon heating above 140°C yield the unsolvated dimers $[\{(Me_3Si)_2C_5H_3\}_2Ln(\mu-Cl)]_2$ (Scheme 2-2) [2–7].

LnCl$_3$ + 2 Li[(Me$_3$Si)$_2$C$_5$H$_3$]

$$\xrightarrow[- LiCl]{THF} [(Me_3Si)_2C_5H_3]_2Ln(\mu-Cl)Li(THF)_2$$

$$\xrightarrow[- 2\ LiCl,\ -\ 4\ THF]{\Delta} [\{(Me_3Si)_2C_5H_3\}_2Ln(\mu-Cl)]_2$$

Scheme 2-2. Preparation of $[\{(Me_3Si)_2C_5H_3\}_2Ln(\mu-Cl)]_2$.

Dimeric $[\{MeC_5H_2(CH_2)_5\}_2Nd(\mu\text{-}Cl)]_2$ contains fused-ring bicyclic cyclopentadie-nyl ligands (**5**) [67].

5

The use of cyclopentadienyl ligands containing 'pendant-arm' donor functions, which are able to coordinatively saturate the lanthanide ion, provides an interesting route to base-free lanthanocene halides. A typical example of such a ligand is the 3-methoxypropylcyclopentadienyl anion, $C_5H_4(CH_2)_3OMe^-$. The halide bridged dimers $[\{C_5H_4(CH_2)_3OMe\}_2Ln(\mu\text{-}Cl)]_2$ have been prepared by reacting anhydrous lanthanide trichlorides with the sodium salt of the ligand in a 1:2 molar ratio (Eq. (2-25)) [68, 69].

$$2LnCl_3 + 4Na[C_5H_4(CH_2)_3OMe] \tag{2-25}$$

$$\xrightarrow{\text{THF}} [\{C_5H_4(CH_2)_3OMe\}_2Ln(\mu\text{-}Cl)]_2 + 4NaCl$$
$$Ln = Y, La, Pr, Nd, Gd, Ho, Er, Yb$$

The molecular and crystal structure of the La derivative has been determined [68]. The $C_5H_4(CH_2)_3OMe^-$ anions act as chelating ligands with the dangling methoxy functions being coordinated to the central La atoms. Together with the bridging chlorine ligands this coordination mode results in a formal coordination number of 9 around the metal atoms [70]. Various lanthanocene chlorides containing the relat-ed 2-methoxyethylcyclopentadienyl ligand have been synthesized. The lanthanum derivative $[\{C_5H_4(CH_2)_2OMe\}_2La(\mu\text{-}Cl)]_2$ is a dimer with the formal coordination number 10 around lanthanum. In contrast, the structurally characterized complexes $[C_5H_4(CH_2)_3OMe]_2LnCl$ (Ln = Dy, Yb) are unsolvated monomers in which the for-mal coordination number is 9 [71].

A related tetramethyldisiloxane-bridged ligand provided the lanthanocene chlo-rides $[O(Me_2SiC_5H_4)_2]PrCl(THF)_n$ (n = 1, 2) and $[\{O(Me_2SiC_5H_4)_2\}Yb(\mu\text{-}Cl)]_2$. In this case the oxygen atoms in the bridge did not participate in the coordination to ytterbium [72]. Nitrogen-chelation was found in the chloro-bridged dimers of the type [{MeN(CH$_2$CH$_2$C$_5$H$_4$)$_2$}Ln(μ-Cl)]$_2$ (6, Ln = Y, Nd, Sm, Yb) and [{C$_5$H$_3$N(CH$_2$C$_5$H$_4$)$_2$-2,6}Ln(μ-Cl)]$_2$ (Ln = Y, Pr, Nd, Sm, Dy, Er, Yb, Lu) which have been made by using amine-linked bis(cyclopentadienyl) ligands [73, 74].

6

The use of the donor-substituted (dimethylaminoethyl)cyclopentadienyl ligand has recently allowed the synthesis and structural characterization of the first solvent-free, monomeric lanthanocene chloride. (C$_5$H$_4$CH$_2$CH$_2$NMe$_2$)$_2$NdCl **7** was prepared by reacting NdCl$_3$ with two equivalents of K(C$_5$H$_4$CH$_2$CH$_2$NMe$_2$) followed by sublimation [75].

7

Especially in the area of lanthanocene(III) complexes the use of the bulky penta-methylcyclopentadienyl ligand has had a strong influence on the development of organolanthanide chemistry. This particular ligand provides enhanced stability and better solubility to many such complexes and its use was crucial for the successful synthesis of many highly reactive and catalytically active lanthanocene derivatives.

Decamethyllanthanocene fluorides are not generally available *via* the normal metathetical routes because of the insolubility of the LnF$_3$ precursors. However, defluorination of perfluoroolefins by the divalent organolanthanides (C$_5$Me$_5$)$_2$Ln(L) (Ln = Sm, Eu, Yb; L = Et$_2$O, THF) has been found to be an elegant way to make the corresponding lanthanide(III) fluorides. The compounds (C$_5$Me$_5$)$_2$LnF(L) (Ln = Sm, Eu, Yb; L = Et$_2$O, THF) have been prepared and fully characterized. Crystal structure determinations of (C$_5$Me$_5$)$_2$YbF(Et$_2$O) and (C$_5$Me$_5$)$_2$YbF(THF) revealed the first terminal Ln–F bonds (Yb–F 2.015(4) and 2.026(2), respectively) [53].

Dimeric mixed-valence $(\mu\text{-F})[Yb(C_5Me_5)_2]_2$ has been prepared and structurally characterized. The compound contains an asymmetric $Yb^{2+}\text{-F-}Yb^{3+}$ bridge [76].

Complexes containing a $(C_5Me_5)_2LnX$ (X = Cl, Br, I) unit are easily available by reacting anhydrous lanthanide trihalides with two equivalents of a C_5Me_5 transfer reagent. These reactions are generally carried out in a coordinating solvent such as THF or DME. Normally a monomeric $(C_5Me_5)_2LnX$ unit is coordinatively unsaturated. As a consequence, $(C_5Me_5)_2LnX$ complexes show a strong tendency to either form adducts with coordinating solvent molecules or to retain one equivalent of alkali metal halide and form 'ate' complexes. At least five different types of decamethyllanthanocene halides have been characterized: 1. A monomeric $(C_5Me_5)_2LnCl$ complex is stable only in the case of the small scandium(3+) ion. 2. THF adducts of the type $(C_5Me_5)_2LnX(THF)$ have been prepared and structurally characterized for several lanthanide elements. 3. In most cases $(C_5Me_5)_2LnX$ derivatives are initially isolated as 'ate' complexes of the type $(C_5Me_5)_2Ln(\mu\text{-X})_2ML_2$ (X = Cl, Br, I; M = Li, Na, K; L = THF, Et$_2$O, 1/2DME, 1/2TMEDA). In these materials one equivalent of alkali metal halide is incorporated to increase the formal coordination number around the lanthanide atom. Lanthanide and alkali metal are bridged by two halide ligands to give a central four-membered $Ln(\mu\text{-X})_2M$ ring. Formally these compounds could also be described as 'ate' complexes of the type $[ML_2][(C_5Me_5)_2LnX_2]$. One or two solvent molecules are needed to complete the coordination sphere around the alkali metal ion. 4. Unsolvated species normally exist as symmetrical halide-bridged dimers of the type $[(C_5Me_5)_2Ln(\mu\text{-X})]_2$. 5. The compound $(C_5Me_5)_2Y(\mu\text{-Cl})Y(C_5Me_5)_2Cl$ is a rare example of an asymmetrical dimer, while $[(C_5Me_5)_2Sm(\mu\text{-Cl})]_3$ is trimeric in the solid state.

A two-step procedure was reported for the preparation of $(C_5Me_5)_2YbCl(THF)$. Unsolvated $(C_5Me_5)_2Yb(\mu\text{-Cl})_2Li$ reacts with aluminum trichloride to give the blue bimetallic complex $(C_5Me_5)_2Yb(\mu\text{-Cl})_2AlCl_2$. Subsequent treatment of the intermediate tetrachloroaluminate with THF results in elimination of $AlCl_3$ and formation of $(C_5Me_5)_2YbCl(THF)$ [2-7]. An elegant way of preparing salt-free THF adducts of the type $(C_5Me_5)_2LnX(THF)$ (Ln = Sm, Eu, Yb; X = Cl, I) is the controlled oxidation of lanthanocenes(II) using alkyl or aryl halides [77, 78]. All $(C_5Me_5)_2LnX(THF)$ derivatives were found to be isostructural exhibiting a *pseudo*-tetrahedral geometry around the metal atom.

Synthesis and characterization of 'ate' complexes of the type $(C_5Me_5)_2Ln(\mu\text{-X})_2ML_2$ (X = Cl, Br, I; M = Li, Na, K; L = THF, Et$_2$O, 1/2DME) have been described in detail in a number of publications. A typical reaction sequence is shown in Eq. ((2-26)).

$$LnCl_3 + 2Li(C_5Me_5) \xrightarrow{\quad THF \quad} (C_5Me_5)_2Ln(\mu\text{-Cl})_2Li(THF)_2 + LiCl \qquad (2\text{-}26)$$

Most of these reactions have been carried out with the lanthanide trichlorides *(cf.* **8**), while the corresponding bromides and iodides are much less investigated. $Li(C_5Me_5)$, $Na(C_5Me_5)$, and $K(C_5Me_5)$ have been used as pentamethylcyclopentadienyl transfer reagents, and suitable co-ligands are diethyl ether, THF, and DME or TMEDA. Structurally characterized complexes include $[(C_5Me_5)_2Ce(\mu\text{-}Cl)_2K(THF)]_n$ [79], $(C_5Me_5)_2Pr(\mu\text{-}Cl)_2Na(DME)_2$ [80], and $(C_5Me_5)_2Yb(\mu\text{-}Cl)_2Li(Et_2O)_2$ [2–7].

Ring-bridged bifunctional cyclopentadienyl ligands play an important role in the preparation of lanthanocene derivatives. It was found that complexes containing a $(C_5H_5)_2LnCl$ unit are unstable for the early lanthanide elements La–Nd. This problem can be circumvented with the use of bridged cyclopentadienyl ligands. Several ligand systems have been developed, in which the cyclopentadienyl rings are connected through simple hydrocarbon chains. More sophisticated ligands have donor atoms such as oxygen or nitrogen incorporated in the bridge. These allow the isolation of unsolvated lanthanocene halide derivatives. The first ligand of this type which has been employed in organolanthanide chemistry was the dianion of 1,1'-trimethylene-bis(cyclopentadiene), $^-C_5H_4(CH_2)_3C_5H_4^-$. The disodium salt of this ligand was reacted with anhydrous lanthanide trichlorides to give exclusively *ansa*-lanthanocenes of the type $[C_5H_4(CH_2)_3C_5H_4]LnCl(THF)$ (Eq. (2-27)). The formation of stable products with the lighter lanthanide elements was in marked contrast to the corresponding reactions with NaC_5H_5 [81–83].

$$LnCl_3 + Na_2[C_5H_4(CH_2)_3C_5H_4] \qquad\qquad (2\text{-}27)$$

$$\xrightarrow{\text{THF}} [C_5H_4(CH_2)_3C_5H_4]LnCl(THF) + 2NaCl$$
$$Ln = La, Ce, Pr, Nd, Gd, Dy, Ho, Er, Yb, Lu$$

In several recent papers Marks *et al.* reported the synthesis, structural systematics, absolute configurations, and structural interconversions of a series of C1-symmetric organolanthanide chloro, alkyl, and amide complexes based on chiral chelating $Me_2Si(C_5Me_4)(C_5H_3R^*)^{2-}$ ligands (= $[Me_2SiCp''(R^*Cp)]^{2-}$), where R^* = (+)-neomenthyl, (–)-menthyl, and (–)-phenylmenthyl [84–86]. These ligands are accessible in three steps starting from known chiral cyclopentadienes (Scheme 2-3). The synthetic procedure involves metallation of the chiral cyclopentadienes followed by condensation with $Me_4C_5HSiMe_2Cl$ and subsequent *in situ* lithiation to give the corresponding dianions of the chelating ligands.

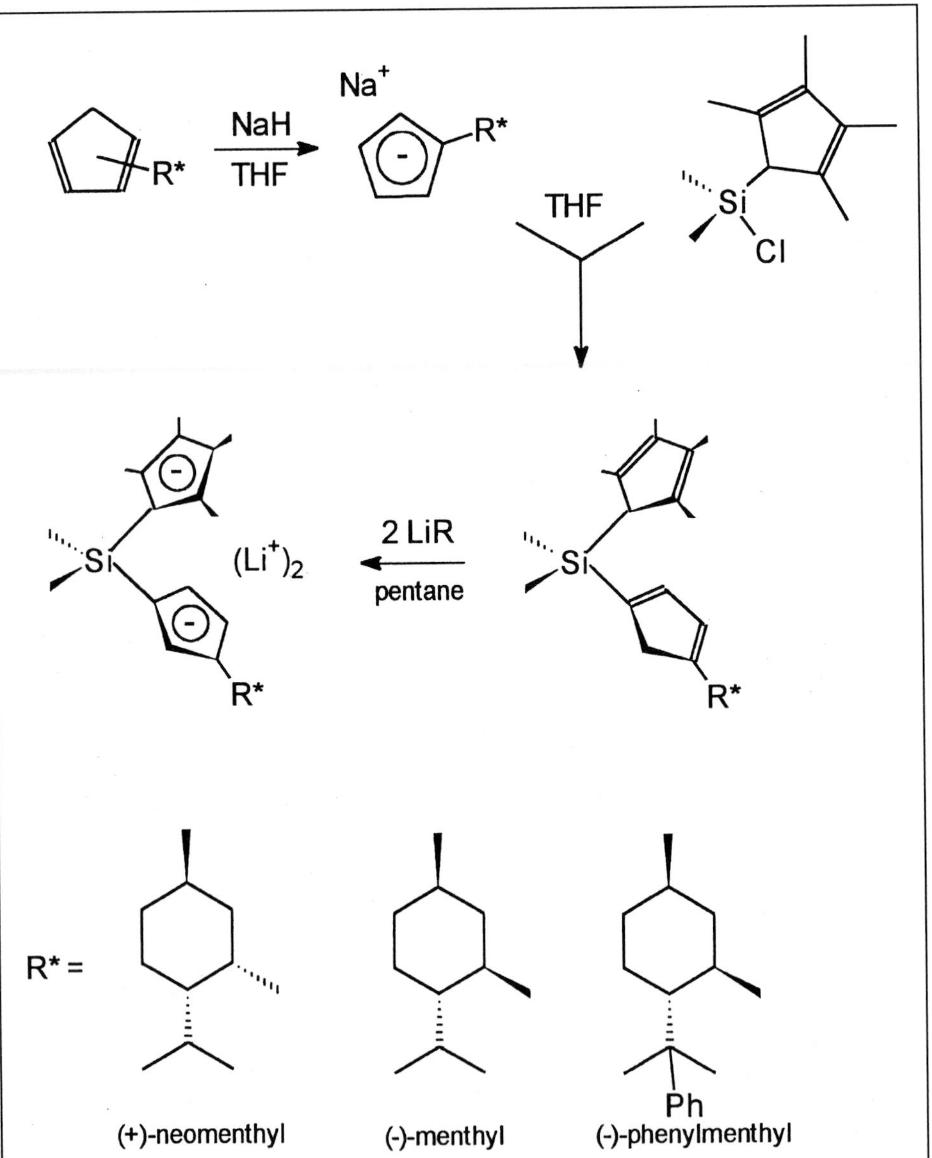

Scheme 2-3. Synthesis of chiral chelating cyclopentadienyl ligands.

The lanthanide chloro precursors can be prepared in high yield by transmetallation of the dilithiated ligands with anhydrous lanthanide trichlorides followed by work-up with diethyl ether. Alternatively, the mono-DME adducts can be prepared (Eqs. (2-28) and (2-29)) [86].

$Li_2[Me_2SiCp''(R^*Cp)] + LnCl_3$ (2-28)

$$\xrightarrow[\text{2. Et}_2\text{O}]{\text{1. THF}} [Me_2SiCp''(R^*Cp)]Ln(\mu\text{-Cl})_2Li(Et_2O)_2 + LiCl$$

$R^* = $ (+)-neomenthyl, (−)-menthyl,
(−)-phenylmenthyl
Ln = Y, La, Nd, Sm, Lu

$Li_2[Me_2SiCp''(R^*Cp)] + LnCl_3$ (2-29)

$$\xrightarrow[\text{2. DME}]{\text{1. THF}} [Me_2SiCp''(R^*Cp)]Ln(\mu\text{-Cl})_2Li(DME) + LiCl$$

$R^* = $ (+)-neomenthyl, (−)-menthyl
Ln = Sm, Lu

As established by NMR spectroscopy and circular dichroism, the diastereomerical-ly pure chloro complexes can be epimerized in the presence of donor solvents to afford mixtures of *R*- and *S*-configurational isomers. The isomer ratio depends on solvent, R*, and lanthanide ion. Importantly, selective epimerization allowed en-richment in either antipode with diastereomerically pure complexes obtained in a single recrystallization step. This was a crucial prerequisite for the preparation of σ-alkyl or amide precatalysts of desired configuration. The molecular structure of $[Me_2SiCp''\{(+)\text{-neomenthylCp}\}]Lu(\mu\text{-Cl})_2Li(Et_2O)_2$ has been determined by single-crystal X-ray analysis [86].

2.3.2 Lanthanide(III) Metallocenes Containing Ln–Group 16 Element Bonds

Various lanthanocenes containing Ln–O, Ln–S, Ln–Se, and Ln–Te bonds have been reported in the literature, with oxygen-bonded species forming by far the largest class of compounds. In recent years several examples of μ-hydroxo- and μ-oxo-bridged organolanthanide complexes of the type $[Cp_2Ln(\mu\text{-OH})]_2$ and (μ-O)$[LnCp_2]_2$ have been prepared and structurally characterized. They are generally made by carefully controlled hydrolysis of the corresponding Cp_2LnX precursors (X = halide, cyclopentadienyl). The μ-oxo bridged ytterbium complex (μ-O)$[Yb(MeC_5H_4)_2]_2$ exhibits a linear central Yb–O–Yb unit (Yb–O 2.915(1), Yb–C 2.667 Å) [87].

Various dimeric alkoxides of the type $[(C_5H_5)_2Ln(\mu\text{-}OR)]_2$ (and in some cases monomeric THF solvates $(C_5H_5)_2Ln(OR)(THF)_n$) have been prepared by treatment of the chloro complexes with sodium alkoxides [2–7, 88] or by protonation of $Ln(C_5H_5)_3$ with stoichiometric amounts of alcohols [89–92]. The latter method has been used to prepare a series of bis(cyclopentadienyl)ytterbium alkoxides containing chiral alkoxide ligands such as (–)-mentholate [92, 93]. A typical example of a monomeric species is $(C_5H_5)_2Nd(OC_6H_3Ph_2\text{-}2,6)(THF)_2$ [89]. The alkoxide complexes $(C_5H_5)_2Yb(THF)(OR_F)$ [Ln = Nd, Sm, Yb; R_F = 2,4,6-tris(trifluoromethyl)phenyl, $\text{-}C_6H_2(CF_3)_3\text{-}2,4,6$] were obtained by reaction of the appropriate $(C_5H_5)_3Ln$ with 2,4,6-tris(trifluoromethyl)phenol (R_FOH) (Eq. (2-30)) [94]. The thiolate complex $(C_5H_5)_2Yb(THF)(SR_F)$ was prepared analogously from $(C_5H_5)_3Yb$ and R_FSH. According to X-ray diffraction studies the $(C_5H_5)_2Yb(THF)(SR_F)$ is a monomer, in which the ytterbium atom is coordinated by two C_5H_5-rings, one sulfur of the $SC_6H_2(CF_3)_3\text{-}2,4,6$ ligand and one oxygen of solvating THF molecule. Thus the ytterbium has a *pseudo*-tetrahedral arrangement and the formal coordination number 8.

Ln = Nd, Sm, Yb   (2-30)

Lanthanocene(III) carboxylates are also dimeric in the solid state. They can be obtained by reacting $(C_5H_5)_2LnCl(THF)$ with various sodium carboxylates in THF solution [2–7, 36]:

$2\ (C_5H_5)_2LnCl(THF)\ +\ 2\ NaO_2CR$

Ln = Sc, Sm, Gd, Tb, Dy, Ho, Er, Tm, Yb, Lu
R = H, Me, CH_2Cl, CCl_3, Et, ^nBu, ^tBu, Ph,
$o\text{-}C_6H_4R'$, C_6F_5, C_5H_4N (2-31)

In some cases anhydrous lanthanide triflates have been found to be better precursors for lanthanocenes than the previously used lanthanide trichlorides [95]. Dimeric structures have been reported for lanthanocene triflates, which have been prepared according to Eq. (2-32) $[(C_5H_5)_2Yb(\mu\text{-}O_3SCF_3)]_2$ was crystallographically characterized [96, 97].

$$Ln(O_3SCF_3)_3 + 4Na(C_5H_5) \tag{2-32}$$

$$\xrightarrow{\text{THF}} [(C_5H_5)_2Ln(\mu\text{-}O_3SCF_3)]_2 + 4NaO_3SCF_3$$
$$Ln = Sc, Yb, Lu$$

An unusual complex containing a tetradentate oxygen ligand was formed by insertion of carbon monoxide into the reactive Lu–C bond in $(C_5H_5)_2Lu^tBu(THF)$. The initial product in this reaction is the yellow acyl complex $(C_5H_5)_2Lu[C(O)^tBu]$ which can be described by two resonance forms. Further CO insertion yields the final product *via* postulated ketene and carbene intermediates (Scheme 2-4) [2–7].

A large number of $(C_5Me_5)_2Ln$ derivatives containing additional Group 16 ligands have been prepared and characterized. Many of them result from transformations of the corresponding divalent samarium or ytterbium precursors. A typical reaction product is easily observed when $(C_5Me_5)_2Sm(THF)_2$ decomposes in the presence of traces of oxygen. The purple color of the samarium(II) complex then changes to orange–yellow, indicating the formation of $(\mu\text{-}O)[Sm(C_5Me_5)_2]_2$. This oxygen-bridged dimer is also commonly the final product when $(C_5Me_5)_2Sm(THF)_2$ is reacted with various oxygen-containing substrates such as nitrosobenzene, pyridine-*N*-oxide *etc* [98]. The binuclear complex $(\mu\text{-}O)[Sm(C_5Me_5)_2]_2$ has been structurally characterized. As imposed crystallographically, the compound contains an exactly linear Sm–O–Sm bridge. The two $(C_5Me_5)_2Sm$ units are in a perpendicular arrangement relative to each other [99]. Few alkoxide and aryloxide complexes of the type $(C_5Me_5)_2LnOR$ have been reported. An unsolvated samarium complex was prepared by reacting $(C_5Me_5)_2Sm(THF)_2$ with 2,3,5,6-tetramethylphenol in toluene solution (Eq. (2-33)) [100]:

$$(C_5Me_5)_2Sm(THF)_2 + HOC_6HMe_4 \xrightarrow[-2THF]{} (C_5Me_5)_2Sm(OC_6HMe_4) + \tfrac{1}{2}H_2 \tag{2-33}$$

A reaction sequence leading to an unusual alkoxide derivative was performed by reacting $(C_5Me_5)_2Sm(THF)_2$ first with diphenylacetylene and subsequently treating the initial product with carbon monoxide [101]. Formal addition of two CO molecules to diphenylacetylene resulted in the formation of an indenoindene ring system. In the final product two $(C_5Me_5)_2Sm$ units were coordinated only to oxygen atoms of the new ligand showing again the strong preference of lanthanides for hard donor ligands (Scheme 2-5) [101].

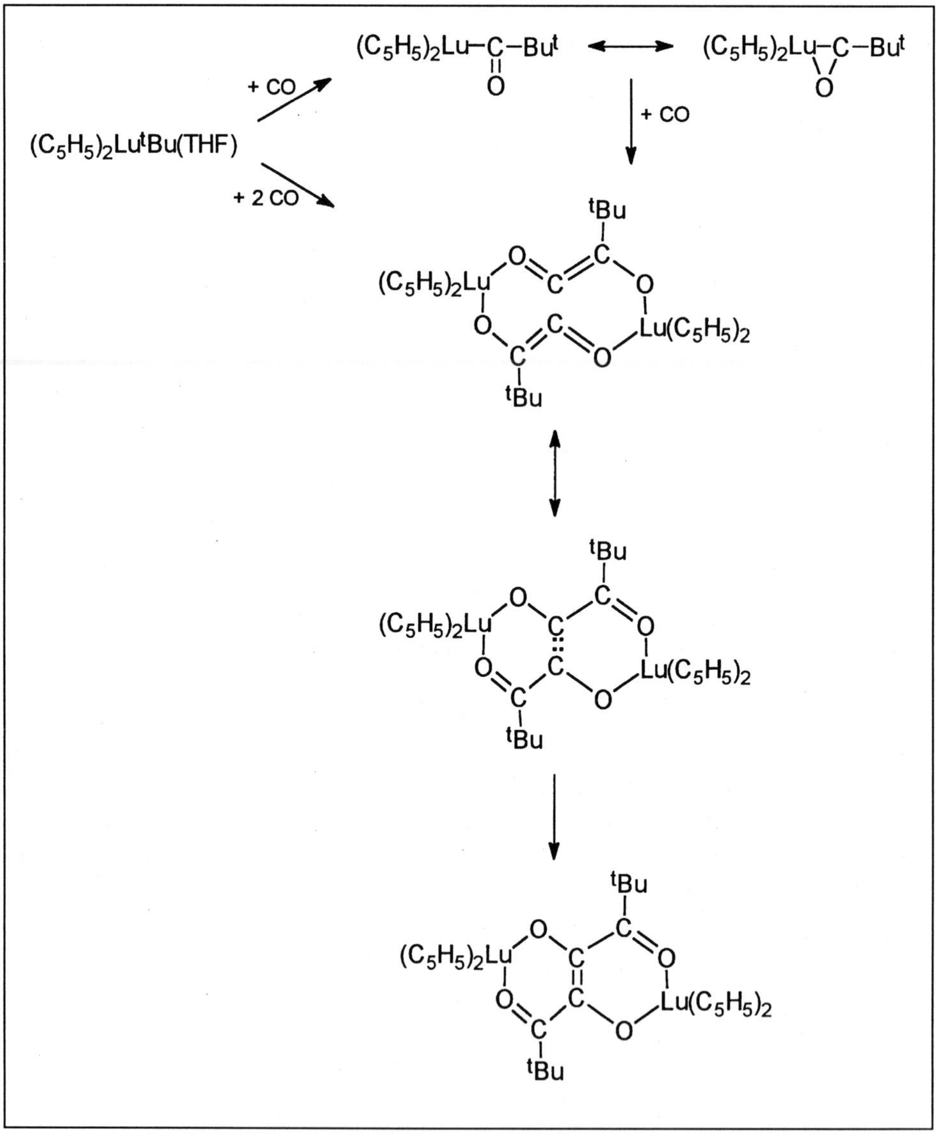

Scheme 2-4. Reactions of $(C_5H_5)_2Lu^tBu(THF)$ with carbon monoxide.

One of the most unusual organolanthanide complexes results from the reaction of $(C_5Me_5)_2Sm(THF)_2$ with carbon monoxide alone. Treatment of the samarium(II) precursor with CO under pressure gave a black crystalline material which was shown by X-ray structure analysis to be the tetranuclear samarium(III) complex $[(C_5Me_5)_4Sm_2(\mu\text{-}O_2CCCO)(THF)]_2$ [102]. The result of this unexpected reaction is a trimerization of carbon monoxide to give a ketenecarboxylate dianion,

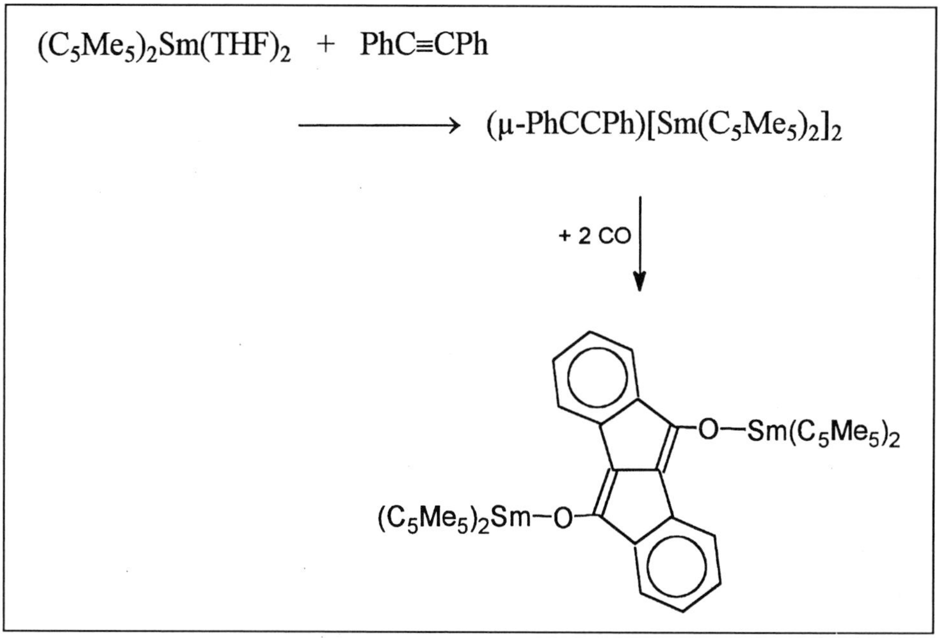

$(C_5Me_5)_2Sm(THF)_2$ + $PhC{\equiv}CPh$

$$\longrightarrow \quad (\mu\text{-PhCCPh})[Sm(C_5Me_5)_2]_2$$

$+ 2\ CO$

(C_5Me_5)_2Sm—O—
—O—Sm(C_5Me_5)_2

Scheme 2-5. Sequential reaction of $(C_5Me_5)_2Sm(THF)_2$ with $PhC{\equiv}CPh$ and CO.

O_2CCCO^{2-}. In the product, two of these ketenecarboxylate dianions act as bridging ligands between four $(C_5Me_5)_2Sm$ units.

   $(C_5Me_5)_2Ln$ derivatives containing sulfur ligands can be prepared in a conventional manner from the corresponding chloro complexes by salt-elimination reactions [103]:

$$(C_5Me_5)_2LnCl(THF) + NaS_2CNEt_2 \xrightarrow{\ THF\ } (C_5Me_5)_2Ln(S_2CNEt_2) + NaCl \quad (2\text{-}34)$$
$$Ln = Nd,\ Yb$$

Protolysis of hydrocarbyls by free alcohols or thiols also leads to decamethyllanthanocene alkoxides and thiolates. For example, methane is eliminated upon treatment of $(C_5Me_5)_2LuMe$ with ethanol to afford the monomeric alkoxide $(C_5Me_5)_2LuOEt$ [104]. One of the first organolanthanide complexes containing bridging thiolate ligands was prepared according to Eq. (2-35) [105].

$$(C_5Me_5)_2Lu(\mu\text{-Me})_2Li(THF)_2 + 2^tBuSH \qquad\qquad (2\text{-}35)$$

$$\longrightarrow (C_5Me_5)_2Lu(\mu\text{-S}^tBu)_2Li(THF)_2 + 2CH_4$$

Other complexes in which $(C_5Me_5)_2Ln$ units are bonded to organic ligands *via* Group 16 atoms have been obtained starting from the divalent lanthanide precursors $(C_5Me_5)_2Sm(THF)_2$ or $(C_5Me_5)_2Yb(Et_2O)$ *via* reductive cleavage of E–E bonds (E = S, Se, Te) [106]:

$$2(C_5Me_5)_2Sm(THF)_2 + [Me_2NC(S)S]_2 \qquad\qquad (2\text{-}36)$$

$$\xrightarrow{\hspace{3cm}} 2(C_5Me_5)_2Sm(S_2CNMe_2) + 4THF$$

$$2(C_5Me_5)_2Sm(THF)_2 + REER \qquad\qquad (2\text{-}37)$$

$$\xrightarrow{\hspace{3cm}} 2(C_5Me_5)_2Sm(ER)(THF) + 2THF$$
$$E = S, Se, Te$$
$$R = Mes, C_6H_2(CF_3)_3\text{-}2,4,6$$

Mono- and dinuclear $(C_5Me_5)_2Yb$ complexes of the type $(C_5Me_5)_2Yb(ER)L$ (E = O, S, Se, Te; L = Lewis base), $(\mu\text{-}E)[Yb(C_5Me_5)_2]_2$ (E = S, Se, Te) and $(\mu\text{-}Te_2)[Yb(C_5Me_5)_2]_2$ have been investigated by Andersen *et al.* [107–109]. Monomeric species are readily formed upon treatment of $(C_5Me_5)_2Yb(Et_2O)$ with dichalcogenides. The ytterbium(II) precursor also reacts with elemental sulfur, selenium, or tellurium to afford the $\mu\text{-}E$ or $\mu\text{-}E_2$-bridged complexes. The monomeric ytterbium(III) chlacogenolates $(C_5Me_5)_2Yb(SPh)(NH_3)$ (Yb–C 2.64(2), Yb–N 2.428(5), Yb–S 2.674(5) Å) and $(C_5Me_5)_2Yb(TePh)(NH_3)$ (Yb–C 2.63(3), Yb–N 2.50(1), Yb–Te 3.039(1) Å) were crystallographically characterized [108, 109]. In the binuclear complex $(\mu\text{-}Te_2)[Yb(C_5Me_5)_2]_2$ the two $(C_5Me_5)_2Yb$ fragments are bridged by a side-on coordinated $\mu\text{-}Te_2$ unit (Yb–C 2.63(2), Yb–Te 3.156(4) Å) [109]. Related scandium tellurides have been prepared either from $(C_5Me_5)_2ScR$ or $(C_5Me_5)_2ScD$. Elemental tellurium slowly inserts into the Sc–C bond of the scandium hydrocarbyls to afford the corresponding tellurolates. More rapid Te insertion can be achieved when $^nBu_3PTe$ is used as a soluble tellurium source. The tellurolates decompose upon heating in solution to afford the Te-bridged binuclear complex $(\mu\text{-}Te)[Sc(C_5Me_5)_2]_2$ (Scheme 2-6). Photochemically induced reversion of the thermal reaction was observed upon UV irradiation of a mixture of $(\mu\text{-}Te)[Sc(C_5Me_5)_2]_2$ and TeR_2 [110].

$$(C_5Me_5)_2ScR \ + \ Te \ or \ ^nBu_3PTe \ \longrightarrow \ (C_5Me_5)_2ScTeR$$
$$R = CH_2SiMe_3, CH_2Ph$$

$$\xrightarrow[-\ TeR_2]{\Delta} \ (\mu\text{-}Te)[Sc(C_5Me_5)_2]_2$$

Scheme 2-6. Formation of $(\mu\text{-}Te)[Sc(C_5Me_5)_2]_2$.

2.3.3 Lanthanide(III) Metallocenes Containing Ln–Group 15 Element Bonds

Basically two different synthetic routes have been employed to prepare lantha-nocene complexes with Ln–N bonds. One of them is the metathesis of $(C_5H_5)_2LnCl$ derivatives with metallated nitrogen compounds and the other one involves the pro-tolysis of lanthanide–carbon bonds by N–H functions. A typical example for the met-athetical preparation is the synthesis of the lanthanide pyrrolyl complexes $(C_5H_5)_2Ln(NC_4H_4)$ (Ln = Sm, Dy, Yb, Lu) and $(C_5H_5)_2Lu(NC_4H_2Me_2)(THF)$ from $(C_5H_5)_2LnCl(THF)$ and the corresponding sodium pyrrolyls. The dimethylpyrrolyl derivative has been established by X-ray crystallography (Lu–N 2.289(4) Å) [111, 112].

Both oxygen and nitrogen coordination has been observed in the compound bis[μ-acetoneoximato-bis(cyclopentadienyl)gadolinium], $[(C_5H_5)_2Gd(\mu\text{-}\eta^2\text{-ONCMe}_2)]_2$, which was prepared by reacting $Gd(C_5H_5)_3$ with acetone oxime in THF solution (Eq. (2-38)). An X-ray crystal structure analysis revealed an interesting structure with the N–O fragment of the acetone oximato ligand acting as both a bridging and side-on donating group. This bonding situation represents a novel coordination mode of an oximato ligand to a metal [113].

$$2Gd(C_5H_5)_3 + 2HON=CMe_2 \qquad\qquad (2\text{-}38)$$

$$\xrightarrow{\text{THF}} [(C_5H_5)_2Gd(\mu\text{-}\eta^2\text{-ONCMe}_2)]_2 + 2C_5H_6$$

Among all Group 15 elements nitrogen donor ligands are best suited for coordina-tion to lanthanides. In accordance with the hard Lewis acid character of the lan-thanide(3+) cations, the number of stable organolanthanide complexes containing Ln–P, Ln–As, Ln–Sb, and Ln–Bi bonds is quite small. The first example of an 'or-ganolanthanide-phosphine' was reported by Schumann *et al.* The phosphide bridged lutetium complex $(C_5H_5)_2Lu(\mu\text{-PPh}_2)_2Li(TMEDA)(toluene)_{0.5}$ was charac-terized by an X-ray analysis (Lu–P 2.782(1) and 2.813(2) Å) [114]. $(C_5H_5)_2Lu(\mu\text{-AsPh}_2)_2Li(TMEDA)$ was prepared analogously (Lu–C 2.59(1), Lu–As 2.88(1) Å) [114–116]. Other lanthanocene phosphides have been prepared according to Eq. (2-39) by treatment of halide precursors with lithium diorganophosphides [2–7, 117].

$$(C_5H_5)_2LnCl(THF) + LiPRR' \qquad\qquad (2\text{-}39)$$

$$\xrightarrow{\text{THF}} \quad (C_5H_5)_2LnPRR'(THF) + LiCl$$
$$Ln = Tb, Ho, Er, Tm, Yb, Lu$$
$$R = R' = {}^tBu, c\text{-}C_6H_{11}, Ph$$
$$R = {}^tBu, R' = Ph$$

The lutetium hydrocarbyl $(C_5H_5)_2Lu(CH_2SiMe_3)(THF)$ has been used to prepare compounds containing Lu–P and Lu–As bonds. Treatment of the precursor with diphenylphosphine or diphenylarsine resulted in the formation of $(C_5H_5)_2LuPPh_2(THF)$ and $(C_5H_5)_2LuAsPh_2(THF)$, respectively (Eq. (2-40)). The preparation is very clean because volatile tetramethylsilane is the only by-product [118].

$$(C_5H_5)_2Lu(CH_2SiMe_3)(THF) + HEPh_2 \qquad\qquad (2\text{-}40)$$

$$\longrightarrow \quad (C_5H_5)_2LuEPh_2(THF) + SiMe_4$$
$$E = P, As$$

Reactions of $(C_5Me_5)_2Sm(THF)_2$ with Group 15 compounds are often quite unusual and in several cases further transformations of the organic ligands in the coordination sphere of samarium can be carried out. A typical example for such a reaction sequence is the Sm-mediated insertion of carbon monoxide into the N=N double bond of azobenzene [119, 120]. Treatment of $(C_5Me_5)_2Sm(THF)_2$ with PhN=NPh in toluene (molar ratio 2:1) initially affords the binuclear complex $(\mu\text{-}\eta^1{:}\eta^1\text{-PhNNPh})[Sm(C_5Me_5)_2]_2$ in which the two decamethylsamarocene units are in a *trans* arrangement. The samarium azobenzene complex was subsequently treated with carbon monoxide, which resulted in a double insertion of CO into the nitrogen–nitrogen bond to give the binuclear complex $[\mu\text{-}\eta^4\text{-}$(PhN)OCCO(NPh)][Sm(C_5Me_5)_2]_2$ (Scheme 2-7) [119–121].

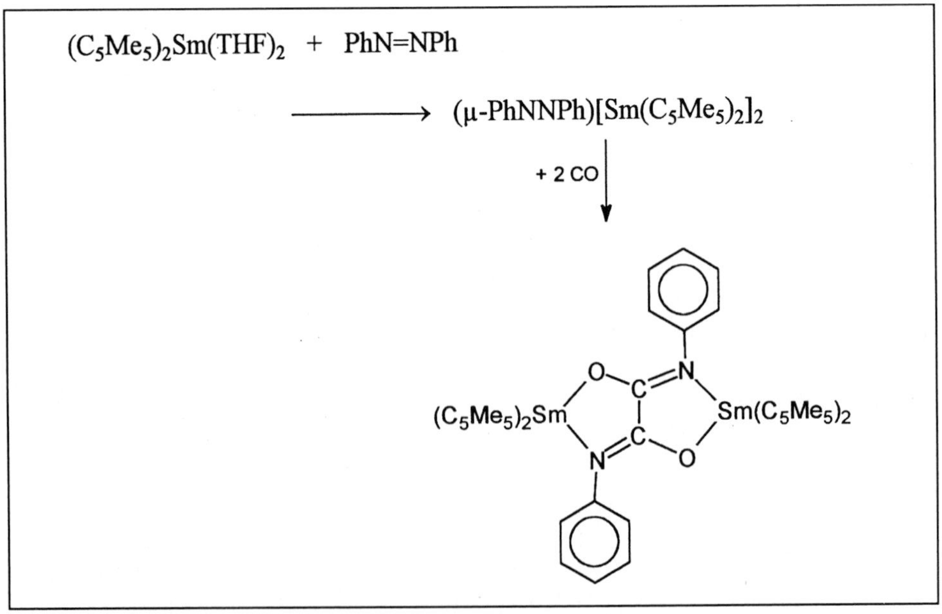

Scheme 2-7. Sequential reaction of $(C_5Me_5)_2Sm(THF)_2$ with PhN=NPh and CO.

The use of the bulky bis(trimethylsilyl)amide ligands allows the synthesis of unsolvated organolanthanide amides of the type $(C_5Me_5)_2LnN(SiMe_3)_2$. The molecular structure of $(C_5Me_5)_2SmN(SiMe_3)_2$ has been determined [2–7, 122, 123].

$$(C_5Me_5)_2Ln(\mu\text{-}Cl)_2Li(Et_2O)_2 + MN(SiMe_3)_2 \qquad (2\text{-}41)$$

$$\longrightarrow (C_5Me_5)_2LnN(SiMe_3)_2 + MCl + LiCl$$
$$Ln = Y, Ce, Nd, Sm$$
$$M = Li, Na, K$$

Similar amidation of the chiral chloro complexes $[Me_2SiCp''(R*Cp)]Ln(\mu\text{-}Cl)_2Li(Et_2O)_2$ affords the corresponding chiral amides in high yields (Eq. (2-42)) [86].

$$[Me_2SiCp''(R*Cp)]Ln(\mu\text{-}Cl)_2Li(Et_2O)_2 + MN(SiMe_3)_2 \qquad (2\text{-}42)$$

$$\longrightarrow [Me_2SiCp''(R*Cp)]LnN(SiMe_3)_2 + LiCl + MCl$$
$$R* = (+)\text{-neomenthyl}, (-)\text{-menthyl}, (-)\text{-phenylmenthyl}$$
$$M = Na, K; Ln = Y, La, Nd, Sm, Lu$$

Single crystal structure analyses have been carried out on the complexes S-$[Me_2SiCp''\{(+)\text{-neomenthylCp}\}]SmN(SiMe_3)_2$, S-$[Me_2SiCp''\{(-)\text{-menthylCp}\}]SmN(SiMe_3)_2$, and R-$[Me_2SiCp''\{(-)\text{-menthylCp}\}]YN(SiMe_3)_2$. These amide complexes are configurationally stable in toluene at 60 °C for many hours [86]..

Adducts of $(C_5Me_5)_2Sm$ with bipyridine and diazadienes (Eq. (2-43)) are readily formed by treatment of $(C_5Me_5)_2Sm(THF)_2$ with the corresponding nitrogen donor ligands [124, 125]:

$$(C_5Me_5)_2Sm(THF)_2 + RN=CH\text{–}CH=NR \qquad (2\text{-}43)$$

$$\longrightarrow (C_5Me_5)_2Sm(\eta^2\text{-}RN=CH\text{–}CH=NR) + 2\ THF$$
$$R = {}^iPr, {}^tBu, c\text{-}C_6H_{11}, p\text{-}C_6H_4Me$$

The N-t-butyl derivative $(C_5Me_5)_2Sm(\eta^2\text{-}{}^tBuN=CH\text{–}CH=N{}^tBu)$ has been structurally characterized (Sm–C 2.773(7), Sm–N 2.480(5) and 2.489(5) Å) [125]. Structural as well as the spectroscopic data are consistent with the presence of Sm(3+) complexes of the bipyridine or diazadiene radical anions. The reaction of $(C_5Me_5)_2La(\mu\text{-}Cl)_2K(DME)_2$ with one equivalent of $Na_2(DAD)$ [DAD = (Ph)N=C(Ph)C(Ph)=N(Ph)] in the presence of DAD or with two equivalents of Na(DAD) gave the ionic complex $[Na(DAD)][(C_5Me_5)_2La(DAD)]$ (Scheme 2-8) [126]. The interaction of $(C_5Me_5)_2Y(\mu\text{-}Cl)_2Li(OEt_2)_2$ with Na(DAD) [DAD = (p-MeC_6H_4)N=C(Ph)C(Ph)=N(C_6H_4Me-p)] led to $(C_5Me_5)_2Y(DAD)$, which was structurally characterized. The DAD ligand in the complex is reduced to the radical anion. In the structure the yttrium atom has pseudotetrahedral arrangement formed by the two C_5Me_5 groups and two N atoms of the cis-1,4-diazadiene. The Y–(C_5Me_5-ring) and Y–N distances are 2.408(4) and 2.362(6) Å, respectively. The metallacycle Y(N–C–C–N) is planar.

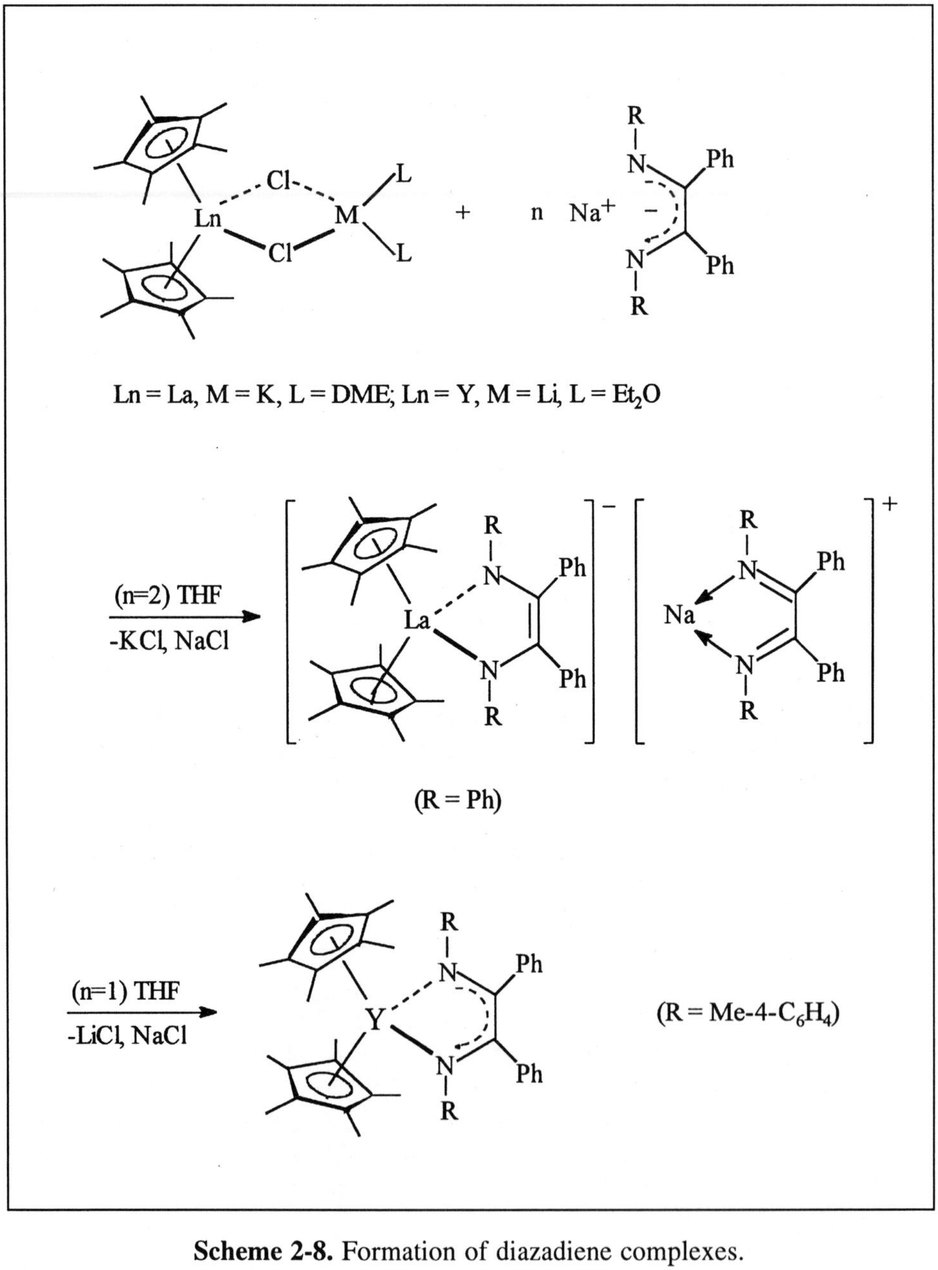

Scheme 2-8. Formation of diazadiene complexes.

Dianions of certain nitrogen heterocycles have been found to act as anti-aromatic bridging ligands between decamethyllanthanocene units. The first complexes of this type were obtained by *in situ* preparation of the dianion followed by treatment with $(C_5Me_5)_2La(\mu\text{-}Cl)K(DME)_2$. Eq. (2-44) shows the example of 2,3-dimethylquinoxaline [127].

$$(2\text{-}44)$$

$(\mu\text{-}\eta^2{:}\eta^2\text{-}N_2)[Sm(C_5Me_5)_2]$ **9**, the first dinitrogen complex of an *f*-element, was prepared by slow evaporation of a dark green toluene solution of $Sm(C_5Me_5)_2$ in a nitrogen atmosphere inside a dry-box. The resulting red–brown crystals were structurally characterized. In the unusual binuclear complex the dinitrogen ligand is side-on coordinated to two $Sm(C_5Me_5)_2$ units. The central $Sm(\mu\text{-}\eta^2{:}\eta^2\text{-}N_2)Sm$ unit is planar (Sm–C 2.73(2), Sm–N 2.36(1) Å) [128].

9

Thus far no polyphosphide complex derived from $(C_5Me_5)_2Sm(THF)_2$ has been reported, although some interesting results have been obtained with antimony and bismuth. $Sb(^nBu)_3$ reacts with unsolvated $Sm(C_5Me_5)_2$ to give the structurally characterized trinuclear product $(\mu-\eta^2:\eta^2:\eta^1-Sb_3)[Sm(C_5Me_5)_2]_3(THF)$ which can be described as a derivative of the triantimony Zintl ion (Sm–C 2.72(2), Sm–Bi 3.29(2) Å) [129]. The analogous reaction with $BiPh_3$ illustrates the fact that it is often difficult if not impossible to predict the outcome of reactions involving $(C_5Me_5)_2Sm(THF)_2$ or $Sm(C_5Me_5)_2$. In this case the bismuth analog of the above-mentioned dinitrogen complex is formed among other products (Eq. (2-45)). $(\mu-\eta^2:\eta^2-Bi_2)[Sm(C_5Me_5)_2]_2$ was structurally characterized [130].

$$4Sm(C_5Me_5)_2 + 2BiPh_3 \qquad\qquad\qquad (2\text{-}45)$$

$$\longrightarrow (\mu-\eta^2:\eta^2-Bi_2)[Sm(C_5Me_5)_2]_2 + PhPh + 2(C_5Me_5)_2SmPh$$

More recent studies have revealed that $Sm(C_5Me_5)_2$ (2 equivalents) reacts with Ph_2EEPh_2 to produce $(C_5Me_5)_2SmEPh_2$ (E = P, As), while treatment of $(C_5Me_5)_2Sm(THF)_2$ with Ph_2EEPh_2 affords the THF adducts $(C_5Me_5)_2SmEPh_2(THF)$. The mixed-valence dinuclear complexes $(C_5Me_5)_2Sm(\mu-EPPh_2)Sm(C_5Me_5)_2$ were prepared by reacting 4 equivalents of $Sm(C_5Me_5)_2$ with Ph_2EEPh_2 (E = P, As) [131].

2.3.4 Lanthanide(III) Metallocenes Containing Ln–Group 14 Element Bonds

Lanthanocene hydrocarbyls are normally prepared from the corresponding $(C_5H_5)_2LnCl$ precursors. A generally applicable synthetic route involves reactions of $(C_5H_5)_2LnCl$ derivatives with equimolar amounts of lithium alkyls or aryls in THF solution (Eq. (2-46) [2–7, 132, 133]. Alternative but not so widely used preparations involve treatment of $Ln(C_5H_5)_3$ with LiR or oxidation of lanthanide(II) metallocenes with HgR_2 [134–137].

$$(C_5H_5)_2LnCl(THF) + LiR \xrightarrow[-78°C]{THF} (C_5H_5)_2LnR(THF) + LiCl \qquad (2\text{-}46)$$

Ln = Y, Nd, Sm, Dy, Er, Tm, Yb, Lu
R = Me, Et, ^iPr, ^nBu, ^tBu, $CH_2{}^tBu$,
CH_2SiMe_3, CH_2Ph, Ph, $C_6H_4Me\text{-}p$,
$C_6H_4Cl\text{-}p$, CH_2PMe_2

$(C_5H_5)_2LuCH_2SiMe_3(THF)$ has been structurally characterized by X-ray diffraction [88]. The coordination geometry in this complex is *pseudo*-tetrahedral. A similar structure was found for $(C_5H_5)_2YbMe(THF)$ (Yb–C(σ) 2.36(1)Å) [138]. Monomeric lanthanocene hydrocarbyls in which the lanthanide atoms and lithium are bridged by methyl groups are obtained when $(C_5H_5)_2LnCl$ precursors are treated with two equivalents of methyllithium. The resulting products are hydrocarbyl analogs of the well-known lithium chloride adducts $(C_5H_5)_2Ln(\mu\text{-}Cl)_2Li(THF)_2$ [139, 140].

$$(C_5H_5)_2LnCl(THF) + 2MeLi \qquad\qquad (2\text{-}47)$$

$$\xrightarrow{\text{THF}} (C_5H_5)_2Ln(\mu\text{-}Me)_2Li(THF)_2 + LiCl \quad (2\text{-}47)$$
$$Ln = Er, Lu$$

Unsolvated $(C_5H_5)_2LnR$ species have been successfully prepared by a two-step procedure. In the first step, dimeric lanthanocene chlorides are reacted with lithium tetraalkylaluminates, $LiAlR_4$, to give the bimetallic complexes $(C_5H_5)_2Ln(\mu\text{-}R)_2AlR_2$ (Eq. (2-48)) [2–7]. From these intermediates the trialkylaluminum can be abstracted by addition of pyridine, which forms an insoluble 1:1 adduct of the type $R_3AlNC_5H_5$. This reaction provides the unsolvated $(C_5H_5)_2LnR$ derivatives in the form of alkyl-bridged dimers (Eq. (2-49)) [2–7]. In some cases simple metathetical reactions also lead to the μ-Me bridged dimers (Eq. (2-50)) [141].

$$[(C_5H_5)_2Ln(\mu\text{-}Cl)]_2 + 2LiAlR_4 \xrightarrow{\text{toluene}} 2(C_5H_5)_2Ln(\mu\text{-}R)_2AlR_2 + 2\,LiCl \quad (2\text{-}48)$$
$$R = Me; Ln = Sc, Y, Gd, Dy, Ho, Er, Tm, Yb$$
$$R = Et; Ln = Sc, Y, Ho$$

$$2(C_5H_5)_2Ln(\mu\text{-}Me)_2AlMe_2 + 2C_5H_5N \qquad\qquad (2\text{-}49)$$

$$\xrightarrow{\text{toluene}} [(C_5H_5)_2Ln(\mu\text{-}Me)]_2 + Me_3AlNC_5H_5$$

$$[(^tBuC_5H_4)_2Nd(\mu\text{-}Cl)]_2 + 2MeLi \qquad\qquad (2\text{-}50)$$

$$\xrightarrow{\hspace{3cm}} [(^tBuC_5H_4)_2Nd(\mu\text{-}Me)]_2 + 2LiCl$$

Lanthanocene derivatives containing neutral phosphoylides as ligands have been prepared by displacement of THF by the ylide. Halides as well as hydrocarbyls can be used as precursors in these ligand exchange reactions [2–7, 142, 143].

$$(C_5H_5)_2LuCl(THF) + Ph_3P=CH_2 \qquad\qquad (2\text{-}51)$$

$$\xrightarrow{\hspace{3cm}} (C_5H_5)_2LuCl(CH_2PPh_3) + THF$$

$(C_5H_5)_2LuR(THF) + R'_3P=CHR''$ (2-52)

$$\longrightarrow (C_5H_5)_2LuR(CHR''PR'_3) + THF$$
$$R = {}^tBu, R' = Ph, R'' = H$$
$$R = CH_2SiMe_3, R' = Ph, R'' = H$$
$$R = {}^tBu, R' = Me, R'' = SiMe_3$$

Intramolecular metallation of a phenyl ring was observed when the phosphoylide adduct $(C_5H_5)_2LuCl(CH_2PPh_3)$ was treated with strong bases such as MeLi or sodium hydride (Scheme 2-9) [2–7, 143].

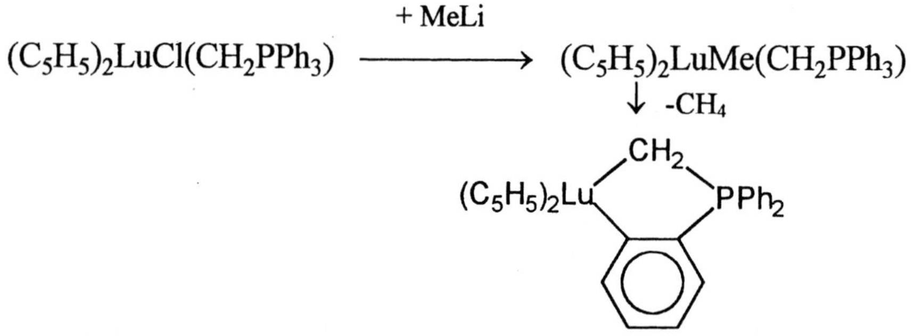

Scheme 2-9. Reaction of $(C_5H_5)_2LuCl(CH_2PPh_3)$ with MeLi.

Effective stabilization of lanthanide–carbon bonds can also be affected by chelating anionic phosphoylide complexes of the type $R_2P(CH_2)_2^-$. Several lanthanocene derivatives of these ligands have been reported. The cyclopentadienyl complexes $(C_5H_5)_2Ln[(CH_2)_2PR_2]$ are monomeric in benzene solution. Their preparation is easily achieved by reacting lanthanocene chlorides with lithiated phosphoylides [2–7, 98, 144].

$(C_5H_5)_2LuCl(THF) + Li(CH_2)_2P^tBu_2$ (2-53)

$$\xrightarrow{\text{THF}} (C_5H_5)_2Lu[(CH_2)_2P^tBu_2] + LiCl$$

Only a very small number of heavier homologs of the lanthanocene hydrocarbyls (*i.e.* silyls, germyls *etc.*) have been studied. The anionic silyl complexes $[Li(DME)_3][(C_5H_5)_2Ln(SiMe_3)_2]$ were prepared by Schumann *et al.* according to Eq. (2-54) and the lutetium derivative was structurally characterized (Lu–Si 2.888(2) Å) [114, 145–147].

$(C_5H_5)_2LnCl + 2LiSiMe_3$ (2-54)

$$\xrightarrow{\text{DME}} [Li(DME)_3][(C_5H_5)_2Ln(SiMe_3)_2] + LiCl$$
$$Ln = Sm, Dy, Ho, Er, Tm, Lu$$

Organolanthanide complexes containing Ln–Ge and Ln–Sn bonds are also accessible. These were obtained by reactions of lanthanocene chloride precursors with MPh_3^- (Eq. (2-55), M = Ge, Sn) [2–7]. A similar synthetic route was adapted to prepare a series of monomeric bis(cyclopentadienyl)scandium silyl and germyl complexes (Eq. (2-56)) [148].

$$(C_5H_5)_2LnCl(THF) + LiMPh_3 \xrightarrow{\text{THF}} (C_5H_5)_2LnMPh_3 + LiCl \qquad (2\text{-}55)$$

$[(C_5H_5)_2Sc(\mu\text{-Cl})]_2 + 2(THF)_3LiER_3$ (2-56)

$$\xrightarrow{} 2(C_5H_5)_2Sc(ER_3)(THF) + 2LiCl + 2THF$$
$$ER_3 = Si(SiMe_3)_3, Si(SiMe_3)_2Ph,$$
$$Si(^tBu)Ph_2, SiPh_3, Ge(SiMe_3)_3$$

$(C_5H_5)_2Sc[Si(SiMe_3)_3](THF)$ was structurally characterized by X-ray crystallography (Sc–Si 2.863(2) Å). Like organolanthanide hydrocarbyls, the scandium silyls and germyls polymerize ethylene and undergo facile migratory insertion reactions with CO and xylylisocyanide [148].

The lanthanocene hydrocarbyls are highly reactive towards a variety of reagents. They easily undergo hydrogenolysis to form the corresponding organolanthanide hydrides. Facile insertion reactions are observed with carbon monoxide and isonitriles. A typical example is the CO insertion into the Lu–C bond in $(C_5H_5)_2Lu^tBu$. The initial product can be described by a resonance between an acyl complex and the mesomeric alkoxycarbene structure [2–7]:

$$(C_5H_5)_2Lu^tBu(THF)$$

$$+ CO \downarrow$$

$$(C_5H_5)_2Lu\text{–}\underset{\overset{\|}{O}}{C}\text{–}Bu^t \longleftrightarrow (C_5H_5)_2Lu\underset{\overset{\diagup}{O}}{\diagdown}C\text{–}Bu^t \qquad (2\text{-}57)$$

Especially important as catalytically active species and precursors for catalytically active organolanthanide hydrides are organolanthanide hydrocarbyls of the type $(C_5Me_5)_2LnR$ [2–7, 80, 149, 150]. They are readily prepared from 'ate' complexes of the type $(C_5Me_5)_2Ln(\mu\text{-}Cl)_2ML_2$ (M = alkali metal, L = Et_2O, THF, 1/2DME) by treatment with alkyllithium reagents. In the case of small alkyl groups such as methyl the products are again 'ate' complexes in which the lanthanide and alkali metal are bridged by the hydrocarbyl ligands. The stepwise formation of such complexes is outlined in Scheme 2-10:

$(C_5Me_5)_2Ln(\mu\text{-}Cl)_2Li(THF)_2$ + MeLi

$$\xrightarrow[-\text{ LiCl}]{} (C_5Me_5)_2Ln(\mu\text{-}Cl)(\mu\text{-}Me)Li(THF)_2$$

$$\xrightarrow[-\text{ LiCl}]{+\text{ MeLi}} (C_5Me_5)_2Ln(\mu\text{-}Me)_2Li(THF)_2$$
$$Ln = La, Pr, Yb, Lu$$

Scheme 2-10. Stepwise formation of $(C_5Me_5)_2Ln(\mu\text{-}Me)_2Li(THF)_2$.

Halide-free $(C_5Me_5)_2Sm$ hydrocarbyls can be obtained by the reaction of $[(C_5Me_5)_2Sm(THF)_2][BPh_4]$ with lithium alkyls according to Eq. (2-58) [151]:

$$[(C_5Me_5)_2Sm(THF)_2][BPh_4] + LiR \qquad\qquad (2\text{-}58)$$

$$\xrightarrow{\quad\quad} (C_5Me_5)_2SmR(THF) + LiBPh_4$$
$$Ln = Me, Ph$$

The THF adducts $(C_5Me_5)_2LnMe(THF)$ (Ln = Sc, Y, Sm, Yb, Lu) have been isolated and all are thermally stable. Structurally characterized solvated organolanthanide methyl complexes are $(C_5Me_5)_2YMe(THF)$ (Y–C(Me) 2.55(2) Å) [152], $(C_5Me_5)_2SmMe(THF)$ (Sm–C(Me) 2.48(1) Å) [153], $(C_5Me_5)_2YMe(THF)$ and $(C_5Me_5)_2YbMe(Et_2O)$ [154]. Monomeric organoscandium hydrocarbyls have been prepared in an analogous manner from $(C_5Me_5)_2ScCl$ and organolithium reagents (Eq. (2-59)) [155].

$$(C_5Me_5)_2ScCl + LiR \xrightarrow{\quad\quad} (C_5Me_5)_2ScR + LiCl \qquad (2\text{-}59)$$
$$R = Me, Ph, C_6H_4Me(o, m, p), CH_2Ph$$

In some cases solvent-free hydrocarbyls have been prepared by adding stoichiometric amounts of olefins to the corresponding hydrides (Eq. (2-60)) [155].

$$1/n\ [(C_5Me_5)_2ScH]_n + H_2C=CHR \xrightarrow{\quad\quad} (C_5Me_5)_2ScCH_2CH_2R \qquad (2\text{-}60)$$
$$R = H, Me$$

The synthesis of the base-free, highly reactive lanthanide methyls $(C_5Me_5)_2ScMe$ [155] and $[(C_5Me_5)_2LnMe]_2$ (Ln = Y, Lu) was a major achievement in organo-*f*-element chemistry [2–7, 104, 156]. Base-free $[(C_5Me_5)_2YbMe]_2$ was recently prepared starting from $Yb(C_5Me_5)_2$ and methylcopper. It forms an asymmetrically μ-Me bridged dimer in the solid state [157]. The monomeric 14-electron species $(C_5Me_5)_2ScMe$ was structurally characterized. Unexpectedly, α-agostic C–H–Sc interaction does not play any significant role in this compound [155]. A dimeric structure was found by X-ray crystallography for $[(C_5Me_5)_2LuMe]_2$. One methyl group acts as a bridge between the two lutetium atoms, while the second methyl ligand is terminal. The methyl complexes are active catalysts for olefin polymerizations, and they activate C–H bonds. Hydrogenolysis of $(C_5Me_5)_2ScMe$ and $[(C_5Me_5)_2LnMe]_2$ (Ln = Y, Lu) yields the corresponding hydrides, $[(C_5Me_5)_2Sc(μ-H)]_n$ and $[(C_5Me_5)_2Ln(μ-H)]_2$ (Ln = Y, Lu) [104, 155]. All three methyl derivatives activate the C–H bonds of a range of hydrocarbons including $^{13}CH_4$, arenes, styrene, and alkynes [145, 146, 156, 158]. Facile elimination of methane is observed when the σ-methyl complexes are treated with benzene, tetramethylsilane, pyridine or methylenetriphenylphosphorane (Scheme 2-11) [104, 156].

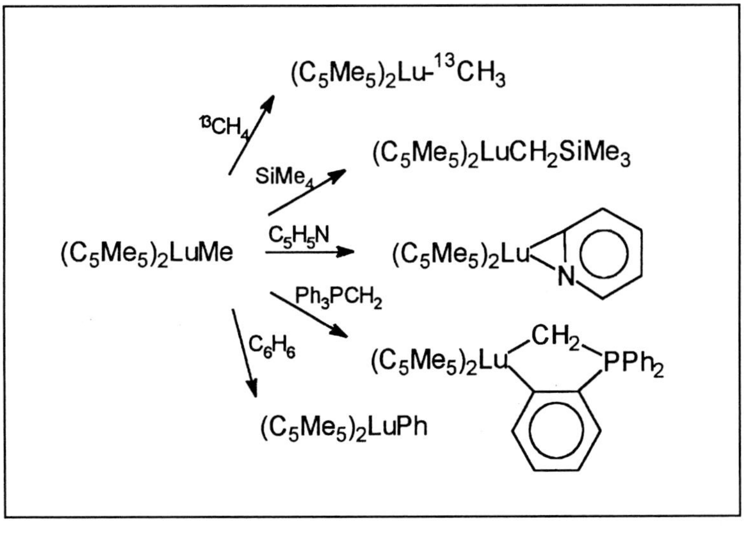

Scheme 2-11. Reactivity of $(C_5Me_5)_2LuMe$ (dimer).

C–H activation of $^{13}CH_4$ by $(C_5Me_5)_2LnMe$ species in cyclohexane solution has been studied by a kinetic analysis. The relative rates of this reaction are in the order Y (250) > Lu (5) > Sc (1). A four-center transition state was formulated as an intermediate. Scheme 2-12 shows the proposed scenario. H–H and C–H activation reactions involving $(C_5Me_5)_2LnH$ and $(C_5Me_5)_2LnR$ derivatives have also been the subject of theoretical studies [159, 160].

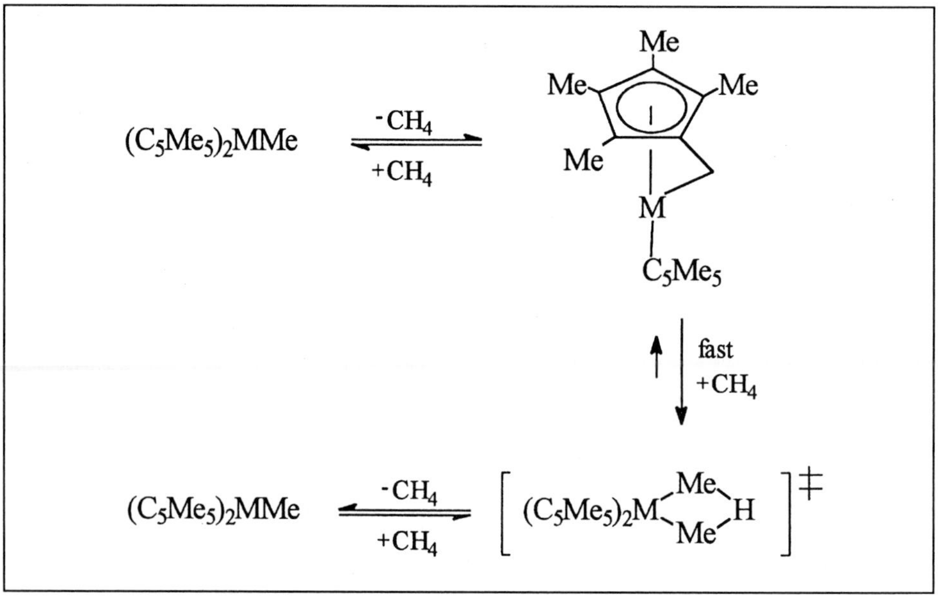

Scheme 2-12. C–H activation of methane by $(C_5Me_5)_2MMe$ (M = Sc, Y, Ln).

For $(C_5Me_5)_2ScR$ the relative reactivity toward various types of C–H bonds has been studied in detail (Eq. (2-61)). The reaction rates have been found to depend strongly on the s character of the reacting bonds, *i.e.* it follows the order R, R' = H > sp C > sp^2 C > sp^3 C [154, 161].

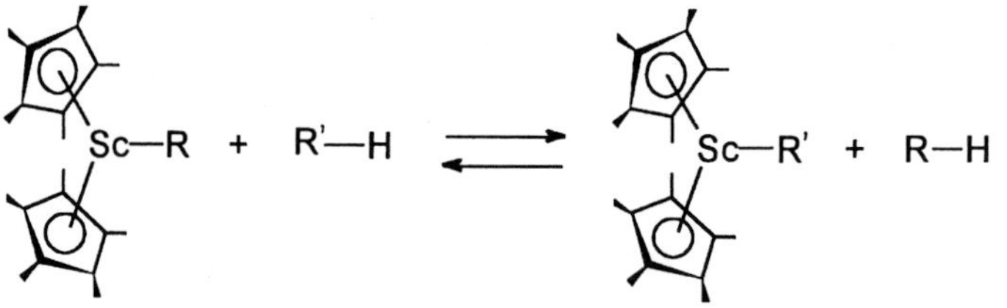

$$Sc—R + R'—H \rightleftharpoons Sc—R' + R—H \qquad (2\text{-}61)$$

$(C_5Me_5)_2ScMe$ reacts with substituted olefins (*e.g.* propylene) at the terminal sp^2 C–H bond to afford the corresponding vinylic C–H activation products (Eq. (2-62)).

$$(C_5Me_5)_2ScMe + H_2C=CHR \longrightarrow (C_5Me_5)_2ScCH=CHR + CH_4 \quad (2\text{-}62)$$

Decamethyllanthanocene hydrocarbyls **10** containing the bulky bis(trimethylsilyl)methyl ligand are of great importance as precatalysts in organo-lanthanide-catalyzed processes. They also appear to be the most suitable precursors for the preparation of the dimeric hydrides $[(C_5Me_5)_2Ln(\mu\text{-}H)]_2$. Due to the increased steric bulk of the $CH(SiMe_3)_2$ ligand, such compounds can be isolated completely free of coordinating solvents and alkali metal halides (Eq. (2-63)) [122, 162–167].

$$(C_5Me_5)_2Ln(\mu\text{-}Cl)_2Li(Et_2O)_2 + LiCH(SiMe_3)_2 \tag{2-63}$$

$$\xrightarrow{} (C_5Me_5)_2LnCH(SiMe_3)_2 + 2LiCl + 2Et_2O$$
$$Ln = Y, La, Ce, Nd, Sm, Lu$$

10

An interesting synthetic difference was reported for the corresponding organocerium hydrocarbyl $(C_5Me_5)_2CeCH(SiMe_3)_2$ [163, 168]. The original preparation called for treatment of the 'ate' complex $(C_5Me_5)_2Ce(\mu\text{-}Cl)_2Li(THF)_2$ with $LiCH(SiMe_3)_2$. However, the yields were low because the LiCl by-product (two equivalents) readily adds to the formed $(C_5Me_5)_2CeCH(SiMe_3)_2$. A more reliable preparation of $(C_5Me_5)_2CeCH(SiMe_3)_2$ involves the reaction of sublimed $[(C_5Me_5)_2CeCl]_n$ with the organolithium reagent (Eq. (2-64)):

$$1/n\ [(C_5Me_5)_2CeCl]_n + LiCH(SiMe_3)_2 \tag{2-64}$$

$$\xrightarrow{} (C_5Me_5)_2CeCH(SiMe_3)_2 + LiCl$$

An X-ray structure determination of $(C_5Me_5)_2YCH(SiMe_3)_2$ revealed α-C–H–Y and γ-Me–Y agostic interactions. Both types of secondary interactions relieve the electronic unsaturation in the formally 14-electron complex [122]. The molecular structures of both $(C_5Me_5)_2CeCH(SiMe_3)_2$ and $(C_5Me_5)_2NdCH(SiMe_3)_2$ [165] have also been determined by X-ray diffraction. Both complexes exhibit an agostic α-C–H–Ln interaction as well as a secondary β-Si–Me–Ln interaction.

Bifunctional ligands in which tetramethylcyclopentadienyl groups are connected by Me_2Si or Me_2Ge [169, 170] units have been demonstrated to be very useful in lanthanocene chemistry. Such compounds, especially the hydrocarbyl and hydride derivatives, show interesting catalytic behavior. They can be regarded as 'tied-back' analogs of the $(C_5Me_5)_2Ln$ derivatives. Halides, hydrocarbyls, and hydrides containing these ligands can be synthesized by the methods established in pentamethylcyclopentadienyl chemistry (Scheme 2-13) [171].

LnCl$_3$ + Li$_2$[Me$_2$Si(C$_5$Me$_4$)$_2$]

$$\xrightarrow[- \text{LiCl}]{\text{Et}_2\text{O}} [\text{Me}_2\text{Si}(\text{C}_5\text{Me}_4)_2]\text{Ln}(\mu\text{-Cl})_2\text{Li}(\text{Et}_2\text{O})_2$$

$$\xrightarrow[- 2 \text{LiCl}]{+ \text{LiCH}(\text{SiMe}_3)_2} [\text{Me}_2\text{Si}(\text{C}_5\text{Me}_4)_2]\text{LnCH}(\text{SiMe}_3)_2$$
$$\text{Ln} = \text{Nd,Sm,Lu}$$

Scheme 2-13. Synthesis of [Me$_2$Si(C$_5$Me$_4$)$_2$]LnCH(SiMe$_3$)$_2$.

The 'tied-back' organolanthanide hydrocarbyls [Me$_2$Si(C$_5$Me$_4$)$_2$]NdCH(SiMe$_3$)$_2$ [164], [Me$_2$Si(C$_5$Me$_4$)(C$_5$H$_4$)]LuCH(SiMe$_3$)$_2$ [172], [Me$_2$Ge(C$_5$Me$_4$)$_2$]HoCH(SiMe$_3$)$_2$ [169] and [Me$_2$Si(C$_5$Me$_4$){C$_5$H$_3$CH$_2$CH$_2$P(^{t}Bu)$_2$}]ScCH(SiMe$_3$)$_2$ [173] as well as the halide precursors [Me$_2$Ge(C$_5$Me$_4$)$_2$]Ln(μ-Cl)$_2$Li(THF)$_2$ (Ln = Sm, Lu) [169] and [Me$_2$Si{C$_5$H$_2$(^{t}Bu-4)(SiMe$_3$-2}$_2$]Y(μ-Cl)$_2$Li(THF)$_2$ [174] have been structurally characterized, with all hydrocarbyls exhibiting a characteristic γ-C–H–Ln agostic interaction [175].

Alkylation of the chloro precursors [Me$_2$SiCp''(R*Cp)]Ln(μ-Cl)$_2$Li(Et$_2$O)$_2$ affords the corresponding chiral σ-alkyl complexes (Eq. (2-65)) [86]:

[Me$_2$SiCp''(R*Cp)]Ln(μ-Cl)$_2$Li(Et$_2$O)$_2$ + LiCH(SiMe$_3$)$_2$ (2-65)

$$\xrightarrow{\hspace{2cm}}$$ [Me$_2$SiCp''(R*Cp)]LnCH(SiMe$_3$)$_2$ + 2LiCl
R* = (+)-neomenthyl, (–)-menthyl, (–)-phenylmenthyl
Ln = Y, La, Nd, Sm, Lu

Single crystal structure analyses have been carried out on the complexes (R/S)-[Me$_2$SiCp''{(+)-neomenthylCp}]YCH(SiMe$_3$)$_2$, R-[Me$_2$SiCp''{(–)-menthylCp}]SmCH(SiMe$_3$)$_2$, and R-[Me$_2$SiCp''{(–)-menthylCp}]YCH(SiMe$_3$)$_2$. These σ-alkyl complexes are configurationally stable in toluene at 60°C for many hours. Rapid hydrogenolysis of the hydrocarbyl complexes at ambient temperature affords the corresponding hydrides with retention of the configuration [86].

C–H activation takes place upon treatment of the decamethyllanthanocene hydrocarbyls (C$_5$Me$_5$)$_2$LnCH(SiMe$_3$)$_2$ (Ln = La, Ce) with acetonitrile. In this case the binuclear μ-cyanomethyl complexes [(C$_5$Me$_5$)$_2$LnCH$_2$CN]$_2$ have been isolated. The La derivative was structurally characterized (La–C(σ) 2.748(4), La–N 2.537(4) Å) [176]. Monomeric (C$_5$Me$_5$)$_2$ScR complexes also metallate pyridine to give the structurally characterized compound (C$_5$Me$_5$)$_2$Sc(C, N-η^2-C$_5$H$_4$N) [155]. An extensive derivative chemistry based on the decamethylyttrocene hydrocarbyls $(C_5Me_5)_2YCH(SiMe_3)_2$ and $(C_5Me_5)_2YMe(THF)$ has been developed by Teuben *et al* [177, 178]. Typical reactions of $(C_5Me_5)_2YCH(SiMe_3)_2$ are summarized in Scheme 2-14. As expected, facile cleavage of the Y–C bond occurs upon treatment of the hydrocarbyl with protic reagents such as alcohols or terminal alkynes [177, 179]. Other characteristic reactions include insertions of CO_2 or ^tBuCN into the yttrium–carbon bond.

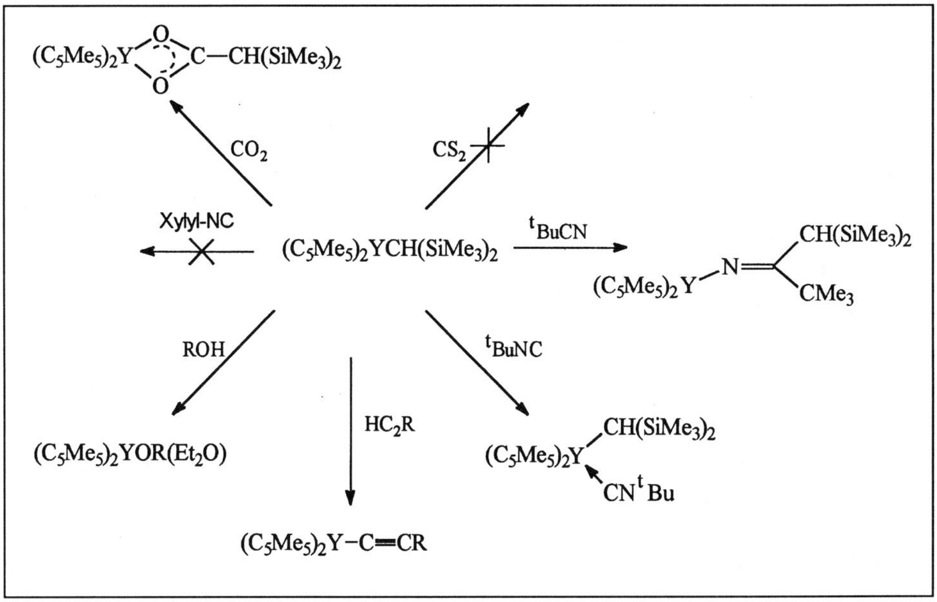

Scheme 2-14. Derivative chemistry of $(C_5Me_5)_2YCH(SiMe_3)_2$.

The lanthanocene hydrocarbyls $(C_5Me_5)_2LnCH(SiMe_3)_2$ (Ln = La, Ce) display a differentiated reactivity toward ketones [180]. It was proposed that the differences in reactivity are thermodynamic and not kinetic in origin. Both early lanthanide hydrocarbyls did not react with the sterically hindered di-*t*-butylketone. Hydrogen transfer and formation of the lanthanide aldolates $(C_5Me_5)_2LnOCMe_2CH_2C(=O)Me$ (Ln = La, Ce) was observed when the hydrocarbyls were treated with acetone (Eq. (2-66)). In contrast, addition of diethylketone to $(C_5Me_5)_2LnCH(SiMe_3)_2$ did not result in C–C coupling but afforded the enolate–ketone adducts $(C_5Me_5)_2Ln[OC(Et)=C(H)Me](O=CEt_2)$ [180].

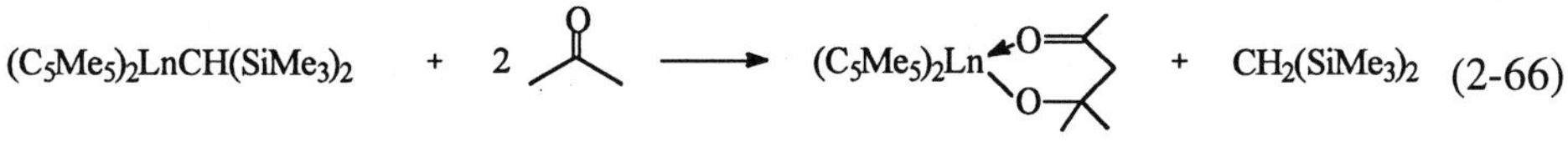

(2-66)

Interesting bimetallic methyl and ethyl species have been prepared by reacting lanthanocenes with alkyl aluminum or gallium reagents. The bimetallic hydrocarbyls $(C_5Me_5)_2Ln(\mu\text{-Me})_2MMe_2$ (Ln = Y, Lu; M = Al, Ga) are formed when the dimeric $(C_5Me_5)_2LnMe$ species are reacted with MMe_3 [153, 181, 182]. In solution there is an equilibrium between the monomeric complexes and cyclic dimers. The latter contain eight-membered ring systems with methyl bridges between the different metal atoms (Eq. (2-67)). Similar bimetallic alkyls can also be prepared by using organolanthanide(II) precursors. For example, excess Al_2Me_6 reacts with $(C_5Me_5)_2Sm(THF)_2$ to produce the cyclic dimer $(C_5Me_5)_2Sm[(\mu\text{-Me})_2Al(\mu\text{-Me})_2]_2Sm(C_5Me_5)_2$ [153]. Two derivatives of this type, $(C_5Me_5)_2Ln[(\mu\text{-Me})_2Al(\mu\text{-Me})_2]_2Ln(C_5Me_5)_2$ (Ln = Y, Sm) have been structurally characterized [153, 182].

$$(2\text{-}67)$$

A mononuclear bimetallic compound with bridging μ-Et ligands can be isolated, when unsolvated $Sm(C_5Me_5)_2$ is reacted with an excess of $AlEt_3$ (Eq. (2-68)). An X-ray crystal structure determination of $(C_5Me_5)_2Sm(\mu\text{-Et})_2AlEt_2$ revealed a monomeric complex with highly asymmetrical ethyl bridges between samarium and aluminum ($Sm–C(C_5Me_5)$ 2.712(2), $Sm–C(CH_2)$ 2.662(4) Å) [183].

$$3Sm(C_5Me_5)_2 + 4AlEt_3 \ 3 \ \longrightarrow \ (C_5Me_5)_2Sm(\mu\text{-Et})_2AlEt_2 + Al \qquad (2\text{-}68)$$

Yet another coordination mode of a μ-Et group was found in the structurally characterized bimetallic ytterbium complex $(C_5Me_5)_2Yb(\mu\text{-Et})AlEt_2(THF)$ (11) (Eq. (2-69)) [184]. In this case ytterbium and aluminum are bridged by a single ethyl group but both carbon atoms are coordinated to the ytterbium atom ($Yb–C(C_5Me_5)$ 2.68(2), $Yb–C(1)$ 2.85(2), $Yb–C(2)$ 2.94(2) Å).

$$(C_5Me_5)_2Yb(THF) + AlEt_3 \ \longrightarrow \ (C_5Me_5)_2Yb(\mu\text{-Et})AlEt_2(THF) \qquad (2\text{-}69)$$

11

A major achievement in recent organolanthanide chemistry was certainly the successful synthesis of the first rare earth carbene complexes [185–187]. Treatment of either $(C_5Me_4Et)_2Yb(THF)$ or $(C_5Me_5)_2Sm(THF)$ with stable carbenes of the Arduengo-type (imidazol-2-ylidenes) afforded the mono and bis(carbene) complexes **12** and **13**. Crystal structure determinations revealed that the carbene-metal distances in theses derivatives are far longer than in related *s*-, *p*-, and *d*-block metal complexes.

R = Me, ^{i}Pr

12 **13**

The first lanthanocene butadiene complex was synthesized by reacting $(C_5Me_5)_2La(\mu\text{-Cl})_2K(DME)$ with 'magnesium butadiene', $Mg(C_4H_6)(THF)_2$. The product, binuclear $(C_5Me_5)_2(THF)La(\mu\text{-}\eta^1,\eta^3\text{-CH}_2CHCHCH_2)La(C_5Me_5)_2$ was structurally characterized (La–C(η^1) 2.633(4), La–C(η^3) 2.73(3) Å) [188]. A mononuclear η^2-butadienyl complex, $(C_5Me_5)_2Sm[\eta^2\text{-}^tBuCH=CC(^tBu)=CH_2]$ was isolated from a reaction of $[(C_5Me_5)_2Sm(\mu\text{-H})]_2$ with the terminal alkyne $^tBuC\equiv CH$ and structurally characterized. The ligand can be formally regarded as a 1,3-di-*t*-butyl-1,3-butadiene metallated in the 2-position [123]. Various other hydrocarbyl complexes containing a $(C_5Me_5)_2Sm$ unit have been prepared from decamethylsamarocene or its THF adduct $(C_5Me_5)_2Sm(THF)_2$. For example, treatment of $(C_5Me_5)_2Sm(THF)_2$ with diphenylmercury (Eq. (2-70)) gave the structurally characterized σ-phenyl complex (Sm–C(σ) 2.511(8) Å). $(C_5Me_5)_2SmPh(THF)$ was also formed in a C–H activation reaction between $(C_5Me_5)_2SmMe(THF)$ and benzene [153].

$$2(C_5Me_5)_2Sm(THF)_2 + HgPh_2 \tag{2-70}$$

$$\longrightarrow 2(C_5Me_5)_2SmPh(THF) + Hg + 2THF$$

Highly reactive decamethylsamarocene also reacts with a variety of hydrocarbons to give samarium(III) hydrocarbyls. A typical example is the reaction of unsolvated $(C_5Me_5)_2Sm$ with styrene, which affords $(\mu\text{-}\eta^2\text{:}\eta^4\text{-CH}_2CHPh)[Sm(C_5Me_5)_2]_2$. Similarly, *trans*-stilbene yields the binuclear complex $(\mu\text{-}\eta^2\text{:}\eta^4\text{-}$ PhCHCHPh)[Sm(C$_5$Me$_5$)$_2$]$_2$, while *cis*-Stilbene is isomerized to the *trans* form in the presence of (C$_5$Me$_5$)$_2$Sm(THF)$_2$ [189]. Aliphatic olefins generally react with Sm(C$_5$Me$_5$)$_2$ under formation of samarium(III) η^3-allyl complexes. One equivalent of olefin is hydrogenated during the course of the reaction (Eq. (2-71)). The same complexes are accessible *via* addition of olefins to the dimeric hydride [(C$_5$Me$_5$)$_2$Sm(μ-H)]$_2$ [189].

$$Sm(C_5Me_5)_2 + RCH=CH_2 \qquad\qquad (2\text{-}71)$$
$$R = Me, Et, CH_2Ph$$

$$\xrightarrow{\hspace{2cm}} (C_5Me_5)_2Sm(\eta^3\text{-}CH_2CHCHR') + CH_3CH_2CH_2R'$$
$$R' = H, Me, Ph$$

1,3-Butadiene is dimerized in the presence of (C$_5$Me$_5$)$_2$Sm to form the bis(η^3-allyl) complex $(\mu\text{-}\eta^3\text{:}\eta^3\text{-}CH_2CHCHCH_2CH_2CHCHCH_2)$[Sm(C$_5Me_5$)$_2$]$_2$. In an extension of this work, a detailed study on the reactivity of (C$_5$Me$_5$)$_2$Sm with polycyclic aromatic hydrocarbons has been carried out. Decamethylsamarocene reacts with hydrocarbons which have reduction potentials more positive than –2.22 V *vs*. SCE to form a series of binuclear complexes in high yield. The hydrocarbons included in this investigation were anthracene, 9-methylanthracene, pyrene, 2,3-benzanthracene, acenaphthylene, and azulene. In all cases the (C$_5$Me$_5$)$_2$Sm fragments are coordinated to η^3-allyl units [190].

Diphenylacetylene was found to add two equivalents of (C$_5$Me$_5$)$_2$Sm(THF)$_2$ to give a black compound formulated as (μ-PhC=CPh)[Sm(C$_5$Me$_5$)$_2$]$_2$ [191]. This material should be regarded as a samarium(III) complex of the dianion of *trans*-stilbene rather than a samarium(II) adduct of neutral PhC≡CPh. In accordance with this description of the bonding the complex yields pure *trans*-stilbene upon hydrolysis. (μ-PhC=CPh)[Sm(C$_5$Me$_5$)$_2$]$_2$ is a useful starting material for the preparation of the highly reactive hydride species [(C$_5$Me$_5$)$_2$Sm(μ-H)]$_2$, which is obtained in high yield by simple hydrogenolysis with molecular hydrogen under normal pressure [192]. Structurally different acetylides have been prepared by reacting (C$_5$Me$_5$)$_2$Ln(Et$_2$O) (Ln = Eu, Yb) with phenylacetylene. With (C$_5$Me$_5$)$_2$Eu(Et$_2$O) binuclear [(C$_5$Me$_5$)$_2$Eu(μ-C≡CPh)(THF)$_2$]$_2$ was formed, while the ytterbium(II) complex yielded trinuclear Yb[(μ-C≡CPh)$_2${Yb(C$_5$Me$_5$)$_2$}$_2$]$_2$ [193]. Monomeric acetylides of the type (C$_5$Me$_5$)$_2$Ln(C≡CR)(THF) (R = ^{t}Bu, Ph) were prepared by reacting the cationic species [(C$_5$Me$_5$)$_2$Sm(THF)$_2$][BPh$_4$] with KC≡CR or by protonation of (C$_5$Me$_5$)$_2$LnN(SiMe$_3$)$_2$ (Ln = Ce, Nd, Sm) with phenylacetylene [109, 151]. Acetylide bridging between yttrium and lithium was found by X-ray analysis in the organoyttrium acetylide (C$_5$Me$_5$)$_2$Y(μ-C≡C^{t}Bu)$_2$Li(THF), which was made straightforwardly from (C$_5$Me$_5$)$_2$YCl(THF) and LiC≡C^{t}Bu [194]. Mixed lanthanide potassium acetylides were prepared similarly according to Eq. (2-72) [123].

$(C_5Me_5)_2Ln(\mu\text{-}Cl)_2K(THF)_2 + 2KC\equiv CPh$ \hfill (2-72)

$$\xrightarrow{} 1/n\ [(C_5Me_5)_2Ln(\mu\text{-}C\equiv CPh)_2K]_n + 2KCl$$
$$Ln = Ce,\ Nd,\ Sm$$

Several other preparative routes leading to $(\mu\text{-}\eta^2{:}\eta^2\text{-}PhC_4Ph)[Sm(C_5Me_5)_2]_2$ *via* coupling of PhC≡C units have been reported. The $\mu\text{-}\eta^2{:}\eta^2$-diacetylene complex was formed during thermolysis of $(C_5Me_5)_2Sm(C\equiv CPh)(THF)$ as well as by react-ing $(C_5Me_5)_2SmCH(SiMe_3)_2$ or $[(C_5Me_5)_2Sm(\mu\text{-}H)]_2$ with phenylacetylene. Later it has been demonstrated that acetylide carbon–carbon coupling in the coordination sphere of lanthanocenes is a very common reactivity pattern for unsolvated acetyl-ides of the type $[(C_5Me_5)_2Ln(\mu\text{-}C\equiv CR)]_n$ (Ln = La, Ce, Sm; R = Me, ^tBu, Ph) [123, 195, 196]. Reactions of $Sm(C_5Me_5)_2$ with terminal alkynes in the absence of THF may even take a different course. The novel trienediyl complexes $(\mu\text{-}\eta^2{:}\eta^2\text{-}RCH_2CH_2C=C=C=CCH_2CH_2R)[Sm(C_5Me_5)_2]_2$ (R = ^iPr, Ph) have been isolated from reactions of $Sm(C_5Me_5)_2$ with $HC\equiv CCH_2CH_2R$ (R = ^iPr, Ph) [123].

Phosphaalkynes have only rarely been employed in organo-*f*-element chemistry. An unprecedented reductive dimerization of a stable phosphaalkyne occurred when $(C_5Me_5)_2Sm(THF)_2$ was treated with one equivalent of $^tBuC\equiv P$ (Eq. (2-73)). The resulting binuclear samarium complex was structurally characterized (Sm–C 2.557(6), Sm–P 2.949(2) Å). The structural data suggest, that the new ligand is the dianion of a 2,3-diphosphabutadiene. Formally the reductive dimerization of a phosphaalkyne closely resembles the above-mentioned dimerization of acetylide units in the coordination sphere of lanthanocenes.

$2(C_5Me_5)_2Sm(THF)_2 + 2^tBuC\equiv P$ \hfill (2-73)

$$\xrightarrow{} (\mu\text{-}^tBuC=PP=C^tBu)[Sm(C_5Me_5)_2]_2 + 4THF$$

Anionic chelating phosphoylide ligands have frequently been proven to be versatile ligands in lanthanocene chemistry. The combination of these ligands with decame-thyllanthanocene fragments results in thermally highly stable rare earth hydrocar-byl complexes. The first complex of this type was synthesized according to Eq. (2-74) [2–7].

$(C_5Me_5)_2Lu(\mu\text{-}Cl)_2Na(Et_2O)_2 + Li(CH_2)_2PMe_2$ \hfill (2-74)

$$\xrightarrow{} (C_5Me_5)_2Lu[(CH_2)_2PMe_2] + LiCl + NaCl$$

LiCl adducts of $(C_5Me_5)_2Ln[(CH_2)_2PRR']$ (Ln = Nd, Sm; R = Me, R' = Ph; R = R' = Me, ^tBu, Ph) have been prepared analogously from the corresponding $(C_5Me_5)_2Ln(\mu\text{-}Cl)_2Li(Et_2O)_2$ precursors. The molecular structure of $(C_5Me_5)_2Lu[(CH_2)_2PMe_2]$ was established by X-ray diffraction. It shows a typical bent lanthanocene(III) derivative with a *pseudo*-tetrahedral coordination around the central Lu atom [198].

Even the unusual (trimethylsilyl)diazomethyl substituent has been attached to decamethyllanthanocenes (Eq. (2-75)). The products are thermally quite sensitive [199].

$$(C_5Me_5)_2LnCl(THF) + LiC(=N_2)SiMe_3 \tag{2-75}$$

$$\longrightarrow (C_5Me_5)_2Ln[C(=N_2)SiMe_3](THF) + LiCl$$
$$Ln = Y, Yb, Lu$$

Recently the first $(C_5Me_5)_2Ln$ derivatives containing Ln–Si bonds have been prepared and characterized. Treatment of $(C_5Me_5)_2LnCH(SiMe_3)_2$ with $H_2Si(SiMe_3)_2$ in the absence of solvent afforded the neutral silyl complexes $(C_5Me_5)_2LnSiH(SiMe_3)_2$ (Ln = Y, Sm, Nd) (Eq. (2-76)). An X-ray crystal structure analysis of the Sm derivative revealed the presence of a dimer stabilized by intermolecular Sm•••Me–Si contacts [200, 201]. The complexes $(C_5Me_5)_2LnSiH_2Ph$ (Ln = Y, Nd) were postulated to be intermediates in the organolanthanide catalyzed dehydrogenation of $PhSiH_3$ and the hydrosilylation of olefins [202, 203].

$$(C_5Me_5)_2LnCH(SiMe_3)_2 + H_2Si(SiMe_3)_2 \tag{2-76}$$

$$\longrightarrow (C_5Me_5)_2LnSiH(SiMe_3)_2 + CH_2(SiMe_3)_2$$

2.3.5 Lanthanide(III) Metallocenes Containing Ln–Hydrogen Bonds

Lanthanocene hydrides play an important role in organolanthanide chemistry as they are highly reactive and offer a great potential in homogeneous catalysis. Theoretical studies on the bonding in hydride-bridged $(C_5H_5)_2Ln$ complexes have been carried out by Hoffmann *et al.* [204]. Subsequently several of these hydrides have been synthesized and structurally characterized. Structural types include dimeric and trimeric complexes of the types $[(C_5H_5)_2Ln(\mu-H)]_2$ and $[(C_5H_5)_2Ln(\mu-H)]_3$, the trimer anions $[(\mu_3-H)\{(C_5H_5)_2Ln(\mu-H)\}_3]^-$ as well as heterobimetallic hydrides, including tetrahydroborates and tetrahydroaluminates [2–7].

Two main synthetic routes leading to lanthanocene hydrides have been developed. One of them is hydrogenolysis of suitable $(C_5H_5)_2Ln$ hydrocarbyl precursors [2–7, 133] and the other one involves metathetical reactions of the corresponding halide precursors with metal hydrides, including $NaBH_4$ and $LiAlH_4$ [205]. THF solvates of the dimeric hydrides are obtained when the hydrogenolysis of lanthanocene hydrocarbyls is carried out in the presence of THF [2–7, 133, 206, 207].

$2(C_5H_5)_2LnR(THF) + 2H_2$ (2-77)

$$\xrightarrow{\text{toluene}} [(C_5H_5)_2Ln(\mu\text{-H})(THF)]_2 + 2RH \quad (2\text{-}77)$$

R = Me, ^{t}Bu, Ph; Ln = Y, Er, Lu

R = ^{t}Bu, $CH_2{}^t$Bu, CH_2SiMe_3; Ln = Lu

The molecular and crystal structures of $[(MeOCH_2CH_2C_5H_4)_2Y(\mu\text{-H})]_2$ [208], $[(MeC_5H_4)_2Ln(\mu\text{-H})(THF)]_2$ (Ln = Y, Er) [133, 209], and $[(C_5H_5)_2Lu(\mu\text{-H})(THF)]_2$ [210, 211] have been determined. The complexes are dimerized *via* two hydride ligands and the coordination sphere around each lanthanide atom is completed by an additional THF ligand of the pending methoxy function. Unsolvated dimeric lanthanocene hydrides have been prepared with the use of bulky cyclopentadienyl ligands such as *t*-butylcyclopentadienyl or 1,3-di-*t*-butylcyclopentadienyl. Structurally characterized examples are $[(^tBuC_5H_4)_2Lu(\mu\text{-H})]_2$ [212] and $[(^tBu_2C_5H_3)_2Ln(\mu\text{-}H)]_2$ (Ln = Ce, Sm) [212, 213].

The hydride-bridged dimers have been shown to be highly reactive. For example, $[(MeC_5H_4)_2Er(\mu\text{-H})(THF)]_2$ reacts with ^{t}BuNC to give the insertion product $[(MeC_5H_4)_2Er(\mu\text{-HC}=N^tBu)]_2$ [191]. Similar treatment of $[(C_5H_5)_2Ln(\mu\text{-H})]_2$ with ^{t}BuNC afforded the dimeric insertion products $[(C_5H_5)_2Ln(\mu\text{-HC}=N^tBu)]_2$ (Ln = Y, Er) with bridging formimidoyl ligands [191, 214]. $[(C_5H_5)_2Y(\mu\text{-H})(THF)]_2$ and $[(MeC_5H_4)_2Y(\mu\text{-H})(THF)]_2$ have also been found to react with alkenes, alkynes, allenes, nitriles, and pyridine to afford the new compounds $(RC_5H_4)_2Y(Et)(THF)$, $(RC_5H_4)_2Y(^nPr)(THF)$, $(RC_5H_4)_2Y(\eta^3\text{-allyl})(THF)$, $[(RC_5H_4)_2Y(\mu\text{-C}\equiv C^tBu)]_2$, $(RC_5H_4)_2Y[C(R')=CHR'](THF)$ (R' = Et, Ph), $[(RC_5H_4)_2Y(\mu\text{-NCH}^tBu)]_2$, $[(RC_5H_4)_2Y(\mu\text{-H})(NC_5H_5)]_2$, and $(RC_5H_4)_2Y(NC_5H_6)(NC_5H_5)$ (R = H, Me) [206]. The dimeric molecular structure of $[(RC_5H_4)_2Y(\mu\text{-NCH}^tBu)]_2$ was determined by X-ray methods. Scheme 2-15 summarizes the reactivity of the bis(cyclopentadienyl)yttrium hydrides $[(RC_5H_4)_2Y(\mu\text{-H})(THF)]_2$ (R = H, Me) towards various substrates.

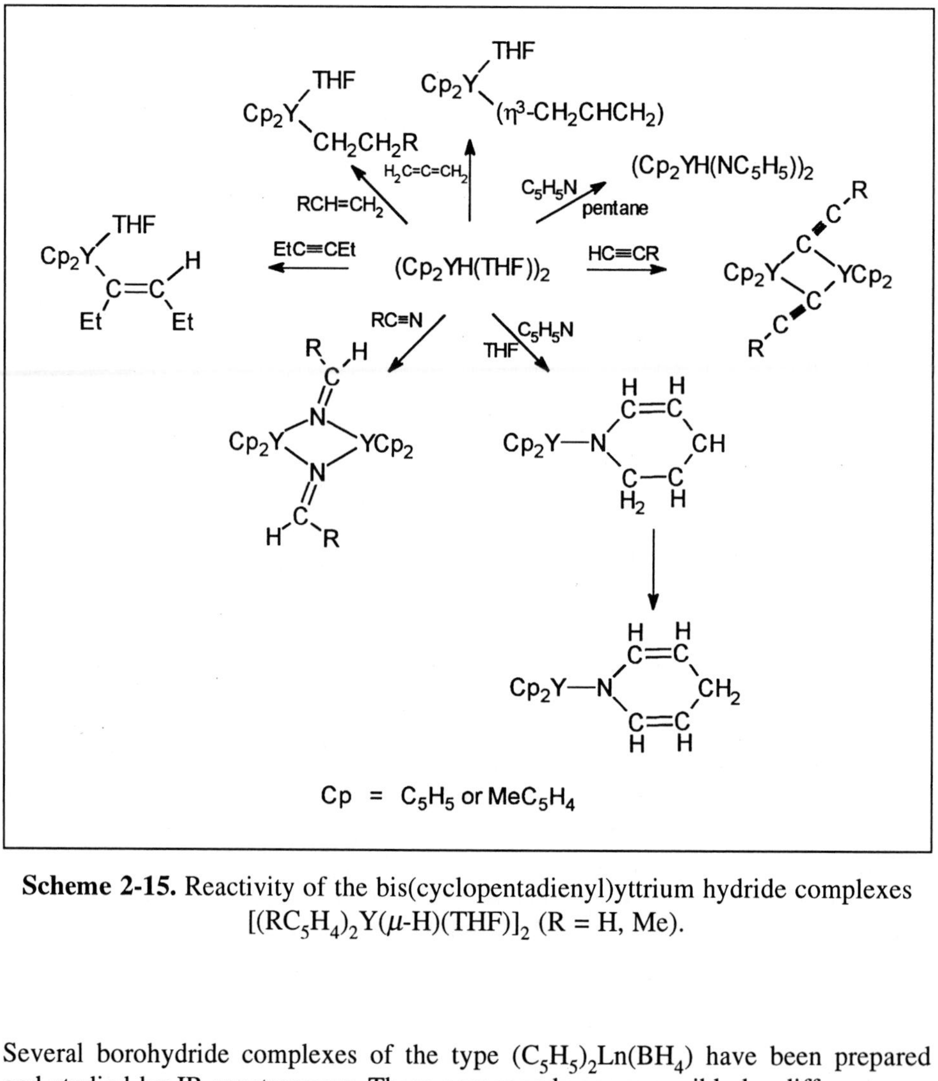

Scheme 2-15. Reactivity of the bis(cyclopentadienyl)yttrium hydride complexes $[(RC_5H_4)_2Y(\mu\text{-H})(THF)]_2$ (R = H, Me).

Several borohydride complexes of the type $(C_5H_5)_2Ln(BH_4)$ have been prepared and studied by IR spectroscopy. These compounds are accessible by different synthetic routes (Eqs. (2-78)–(2-80)) [2–7, 69, 88, 215–217].

$$(C_5H_5)_2LnCl(THF) + NaBH_4 \tag{2-78}$$

$$\xrightarrow{\text{THF}} (C_5H_5)_2Ln(BH_4)(THF) + NaCl$$
$$Ln = Sm, Er, Yb, Lu$$

$$[(Me_3Si)_2C_5H_3]_2LnCl(THF) + NaBH_4 \qquad\qquad (2\text{-}79)$$

$$\xrightarrow{\quad THF \quad} [(Me_3Si)_2C_5H_3]_2Ln(BH_4)(THF) + NaCl$$
$$Ln = Sc, Y, La, Pr, Nd, Sm, Yb$$

$$Sc(BH_4)_3(THF)_2 + 2\,Na(RC_5H_4) \qquad\qquad (2\text{-}80)$$

$$\xrightarrow{\quad THF \quad} (RC_5H_4)_2Sc(BH_4) + 2\,NaBH_4$$
$$R = H, Me$$

In the case of $[(Me_3Si)_2C_5H_3]_2Ln(BH_4)(THF)$ (Ln = Y, Yb) the IR data indicated a bidentate chelating coordination of the borohydride ligand. This was later confirmed by an X-ray structural analysis of unsolvated $[(Me_3Si)_2C_5H_3]_2Sc(BH_4)$ [216]. According to the IR spectra, the coordination of the BH_4^- ligand is tridentate in all other complexes of this type. Crystal structure determinations have also been reported for $(MeOCH_2CH_2C_5H_4)_2Ln(BH_4)$ (Ln = Y, Yb) [69].

Lanthanocene hydrides of the type $(C_5Me_5)_2LnH$ are important as rare earth-based homogeneous catalysts. Thus this group of compounds has been thoroughly investigated and elaborate methods for their synthesis have been designed [104, 122, 162, 163, 179, 192, 218]. A monomeric $(C_5Me_5)_2LnH$ fragment appears to be coordinatively unsaturated. As a consequence, these hydride species are usually isolated as dimers or solvated species. The decamethyllanthanocene hydrides **14** are generally prepared by hydrogenolysis of the corresponding hydrocarbyls. Unsolvated hydrocarbyls containing the bulky $-CH(SiMe_3)_2$ ligand have been shown to be especially useful for the preparation of hydrides. A typical synthesis starting from the THF adduct of yttrium trichloride is outlined below (Eqs. (2-81)–(2-83)) [219].

$$YCl_3(THF)_3 + 2Na(C_5Me_5) \xrightarrow{\quad THF \quad} (C_5Me_5)_2YCl(THF) + NaCl \qquad (2\text{-}81)$$

$$(C_5Me_5)_2YCl(THF) + LiCH(SiMe_3)_2 \qquad\qquad (2\text{-}82)$$

$$\xrightarrow{\quad Et_2O \quad} (C_5Me_5)_2YCH(SiMe_3)_2 + LiCl + THF$$

$$2(C_5Me_5)_2YCH(SiMe_3)_2 + H_2 \qquad\qquad (2\text{-}83)$$

$$\xrightarrow{\quad\quad} [(C_5Me_5)_2Y(\mu\text{-}H)]_2 + 2CH_2(SiMe_3)_2$$

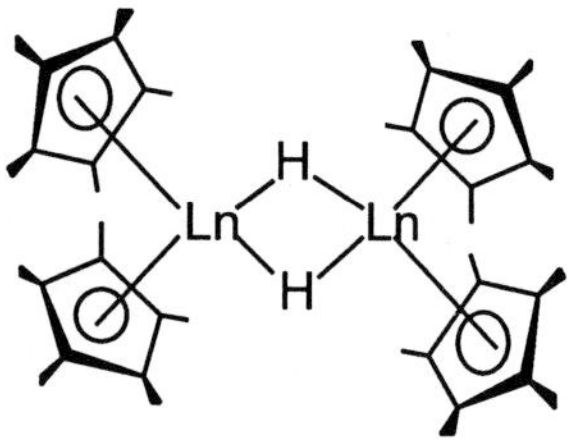

14

The dimeric lanthanocene hydrides $[(C_5Me_5)_2Ln(\mu\text{-}H)]_2$ (**14**, Ln = Ce, Nd, Sm) have been prepared analogously by hydrogenation of $(C_5Me_5)_2LnCH(SiMe_3)_2$ [151, 162, 165]. Unsolvated $[(C_5Me_5)_2Lu(\mu\text{-}H)]_2$ was prepared by hydrogenolysis of base-free $[(C_5Me_5)_2LuMe]_2$ in hexane solution. For this complex ^1H NMR studies revealed an equilibrium at room temperature between the dimeric form and the monomer $(C_5Me_5)_2LuH$ (Scheme 2-15) [2–7, 104].

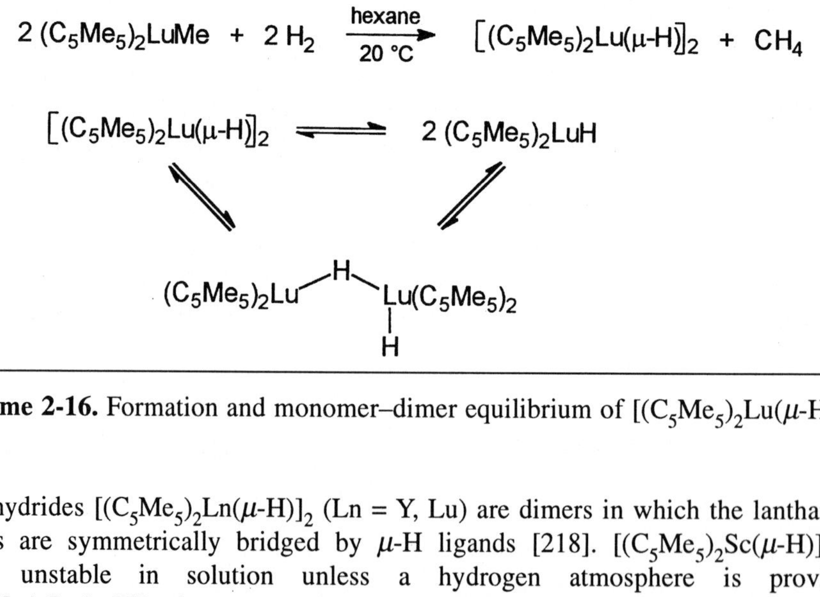

Scheme 2-16. Formation and monomer–dimer equilibrium of $[(C_5Me_5)_2Lu(\mu\text{-}H)]_2$.

The hydrides $[(C_5Me_5)_2Ln(\mu\text{-}H)]_2$ (Ln = Y, Lu) are dimers in which the lanthanide atoms are symmetrically bridged by μ-H ligands [218]. $[(C_5Me_5)_2Sc(\mu\text{-}H)]_n$ is quite unstable in solution unless a hydrogen atmosphere is provided $[(C_5Me_5)_2Lu(\mu\text{-}H)]_2$ decomposes in diethyl ether solution to give $(C_5Me_5)_2LuOEt$ and ethane. Generally more stable are the THF solvates $(C_5Me_5)_2LnH(THF)$ (Sc, Y, Lu). Reactions leading to hydride species have also been designed for lanthanide *ansa*-metallocenes. Treatment of the appropriate halide precursors with alkyllithium or aryllithium afforded a series of new lanthanide hydrocarbyls containing the ring-bridged cyclopentadienyl ligands. Subsequent hydrogenolysis produced the dimeric hydrides (Scheme 2-17) [82, 83, 220]:

$[C_5H_4(CH_2)_3C_5H_4]LnCl(THF)$ + RLi

$$\longrightarrow [C_5H_4(CH_2)_3C_5H_4]Ln(R)(THF) + LiCl$$
Ln = Y,La,Pr,Dy,Er,Lu
R = $^tBu,CH_2^tBu,Ph,p-C_6H_4Me$

$$\xrightarrow{H_2} [\{C_5H_4(CH_2)_3C_5H_4\}Ln(\mu\text{-}H)(THF)]_2 + RH$$
Ln = Y,Dy,Er,Lu

Scheme 2-17. Formation of $[\{C_5H_4(CH_2)_3C_5H_4\}Ln(\mu\text{-}H)(THF)]_2$.

Similar reaction sequences have been reported for the corresponding 1,1'-pentame-thylene-bis(cyclopentadienyl) complexes, *e.g.* $[C_5H_4(CH_2)_5C_5H_4]LnCl(THF)$ (Ln = Y, Sm, Gd, Dy, Er, Lu) [221]. Hydride complexes of the type $[(C_5Me_5)_2Ln(\mu\text{-}H)]_2$ are exceedingly reactive. A most astonishing reaction is the formation of (μ-$OSiMe_2OSiMe_2O)[(C_5Me_5)_2Sm(THF)]_2$. This compound is formed when a THF solution of $[(C_5Me_5)_2Sm(\mu\text{-}H)]_2$ comes into contact with high-vacuum grease! An alternative preparation involves treatment of the dimeric hydride with hexamethyl-cyclotrisiloxane [222]. The dimeric hydrides have also been found to polymerize olefins and to activate small molecules such as carbon monoxide as well as sp^2 and sp^3 C–H bonds [2–7, 104, 156]. $[(C_5Me_5)_2Sc(\mu\text{-}H)]_n$ adds ethylene or propylene to afford the alkyls $(C_5Me_5)_2ScEt$ and $(C_5Me_5)_2Sc(^nPr)$, respectively [155]. Treatment of $[(C_5Me_5)_2Sc(\mu\text{-}H)]_n$ with allene produced the addition product $(C_5Me_5)_2Sc(\eta^3\text{-}C_3H_5)$, while pyridine was metallated to give $(C_5Me_5)_2Sc(\eta^2\text{-}C_5H_4N)$. The methyl groups of the C_5Me_5 ligands in $[(C_5Me_5)_2Sc(\mu\text{-}D)]_n$ have been reported to undergo intramolecular H/D exchange [155]. Treatment of other hydrides of the type $[(C_5Me_5)_2Ln(\mu\text{-}H)]_2$ (Ln = La, Nd, Sm, Lu) with α-olefins normally results in the formation of η^3-allyl complexes. As shown in Scheme 2-18 their formation in-volves insertion of the olefin into the Ln–H bond followed by hydrogen abstraction.

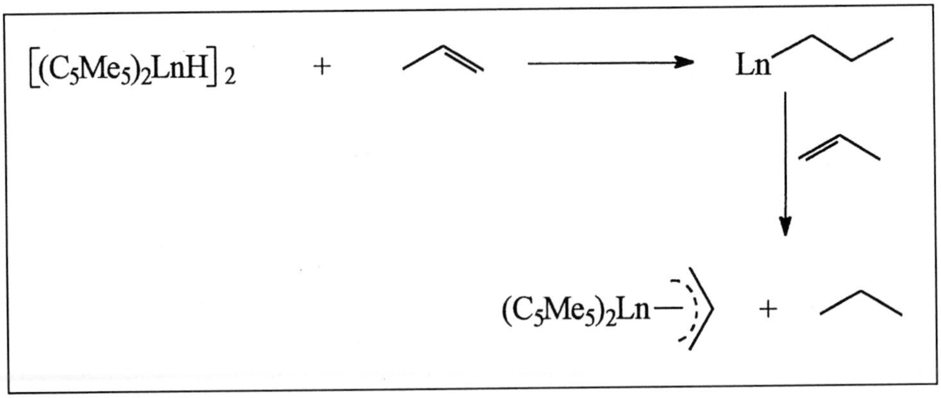

$$[(C_5Me_5)_2LnH]_2 \quad + \quad \diagup\!\!\diagdown\!\!\diagup \quad \longrightarrow \quad Ln\diagdown\!\!\diagup\!\!\diagdown\!\!\diagup$$

$$(C_5Me_5)_2Ln\!-\!\!\langle \quad + \quad \diagup\!\!\diagdown\!\!\diagup$$

Scheme 2-18. Formation of η^3-allyl complexes from organolanthanide hydrides and α-olefins.

Various C–H activation reactions have also been reported for $[(C_5Me_5)_2Y(\mu\text{-}H)]_2$. Metallation of benzene and toluene leads to the σ-aryl compounds $(C_5Me_5)_2YPh$ and $(C_5Me_5)_2YCH_2Ph$, respectively. With substituted aromatic molecules RX (X = OMe, SMe, NMe$_2$, CH$_2$NMe$_2$, PMe$_2$, PPh$_2$=CH$_2$, F, Cl, Br), *ortho*-metallation is the dominant reaction. The product resulting from *ortho*-metallation of Ph$_3$P=CH$_2$, $(C_5Me_5)_2Y(o\text{-}C_6H_4PPh_2CH_2)$, was structurally characterized. An especially remarkable case of C–H activation is the reaction of $(C_5Me_5)_2LuH$ with tetramethylsilane, which yields $(C_5Me_5)_2LuCH_2SiMe_3$ [104]:

$$(C_5Me_5)_2LuH + SiMe_4 \longrightarrow (C_5Me_5)_2LuCH_2SiMe_3 + H_2 \qquad (2\text{-}84)$$

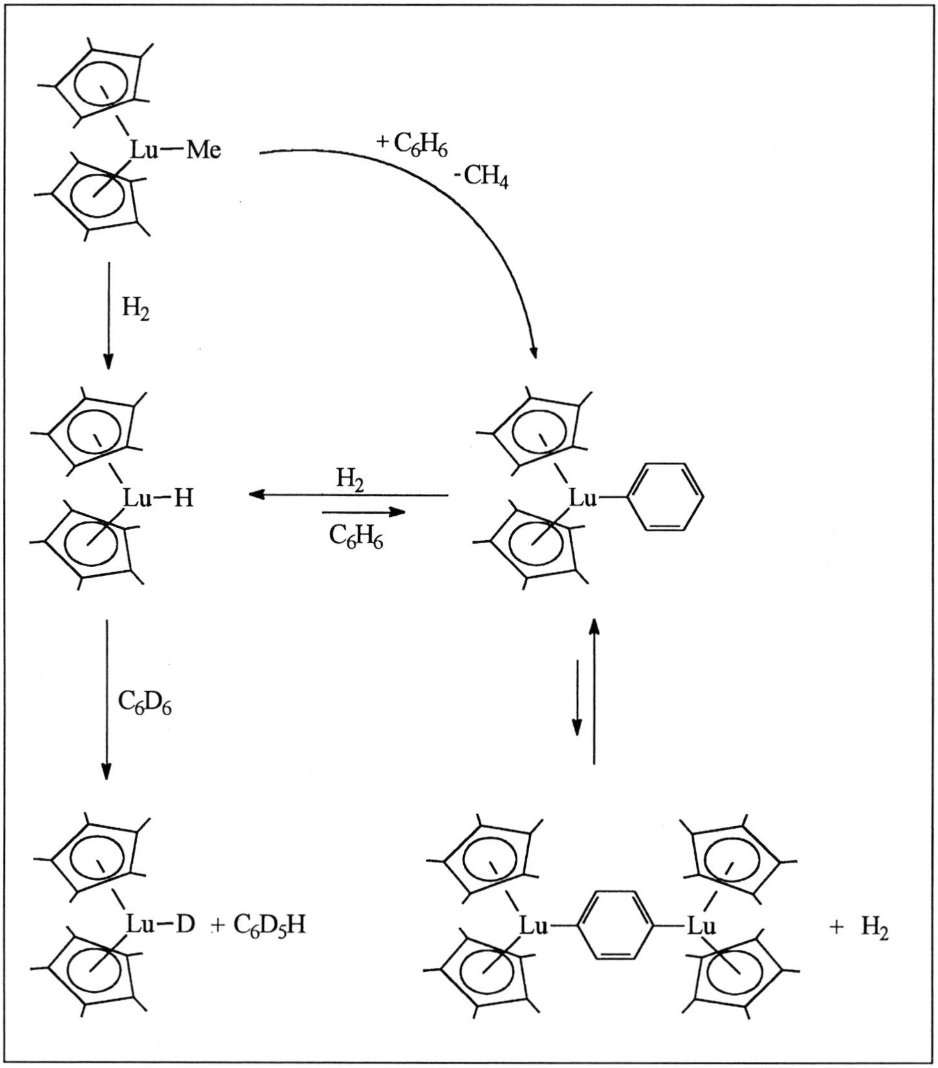

Scheme 2-19. C–H activation reactions of $(C_5Me_5)_2LuH$ (dimer).

The C–H bonds in benzene are also readily attacked by $(C_5Me_5)_2LuH$ (dimer). In the first step the σ-phenyl complex $(C_5Me_5)_2LuPh$ and hydrogen are formed. $(C_5Me_5)_2LuPh$ can further react with $(C_5Me_5)_2LuH$ yielding a phenylene bridged binuclear complex (Scheme 2-19) [13, 104]:

Activation of carbon monoxide by $[(C_5Me_5)_2Sm(\mu\text{-}H)]_2$ was investigated by Evans *et al.* [223]. $[(C_5Me_5)_2Sm(\mu\text{-}H)]_2$ reacts with CO in aromatic hydrocarbon solvents to give *cis*- and *trans*-$(\mu\text{-}OCH{=}CHO)[(C_5Me_5)_2Sm(OPPh_3)]_2$ after recrystallization in the presence of triphenylphosphine oxide as auxiliary ligand. The *cis-*

isomer was found to isomerize at room temperature to give the *trans*-complex. The *cis*-isomer was structurally characterized by X-ray diffraction. Facile addition of Ln–H bonds has also been reported for metal-coordinated carbon monoxide. This reaction was investigated for $(C_5Me_5)_2ScH$ [161]:

$$(C_5Me_5)_2ScH + (C_5H_5)M(CO)_2 \qquad\qquad (2\text{-}85)$$

$$\xrightarrow{\hspace{2cm}} (C_5Me_5)_2ScOC(H)=M(C_5H_5)(CO)$$
$$M = Co,\ Rh$$

Novel $(C_5Me_5)_2Ln$ borohydride derivatives have been prepared by addition of $[Mes_2BH]_2$ to dimeric decamethyllanthanocene hydrides (Eq. (2-86)). Thermolyses of these complexes did not result in isolatable boryl complexes of the type $(C_5Me_5)_2LnBMes_2$ [200].

$$[(C_5Me_5)_2Ln(\mu\text{-}H)]_2 + [Mes_2BH]_2 \xrightarrow{\hspace{2cm}} 2(C_5Me_5)_2Ln(\mu\text{-}H)_2BMes_2 \quad (2\text{-}86)$$

The monomeric 'tied-back' scandocene hydride $[Me_2Si(C_5H_3{}^tBu)_2]ScH$ was synthesized in order to study its reactions with unsaturated hydrocarbons (Scheme 2-20). This hydride reacts with butadiene and several nonconjugated dienes to give η^3-allyl complexes of the type $[Me_2Si(C_5H_3{}^tBu)_2]Sc(\eta^3\text{-}allyl)$. The same complexes were obtained from reactions of the hydride with methylenecyclopropane or substituted conjugated dienes [224].

2.4   Conclusion and Perspective

During the past 15 years the knowledge about lanthanocenes and their chemistry has increased enormously. A large body of synthetic and structural data has been accumulated and various members of this large class of rare earth complexes are now readily available. A large number of unusual molecular structures have been discovered. Especially impressive is the very high reactivity of certain lanthanide(II) metallocenes and lanthanide(III) derivatives containing σ-alkyl and hydride ligands. This includes facile C–H activation reactions with substrates such as benzene, tetramethylsilane, or even methane, which are usually regarded as inert under the given conditions. A major contribution to the rapid development of lanthanocene chemistry can be attributed to the favorable properties of the pentamethylcyclopentadienyl ligand. The coordination environment in a $Ln(C_5Me_5)_2$ unit allows the preparation of thermally stable though highly reactive organolanthanide hydrocarbyls and hydrides such as $(C_5Me_5)_2LnCH(SiMe_3)_2$ and $[(C_5Me_5)_2Ln(\mu\text{-}H)]_2$. More recent work has also been focused on the preparation of *ansa*-lanthanocenes as well as lanthanocenes containing chiral cyclopentadienyl ligands. The

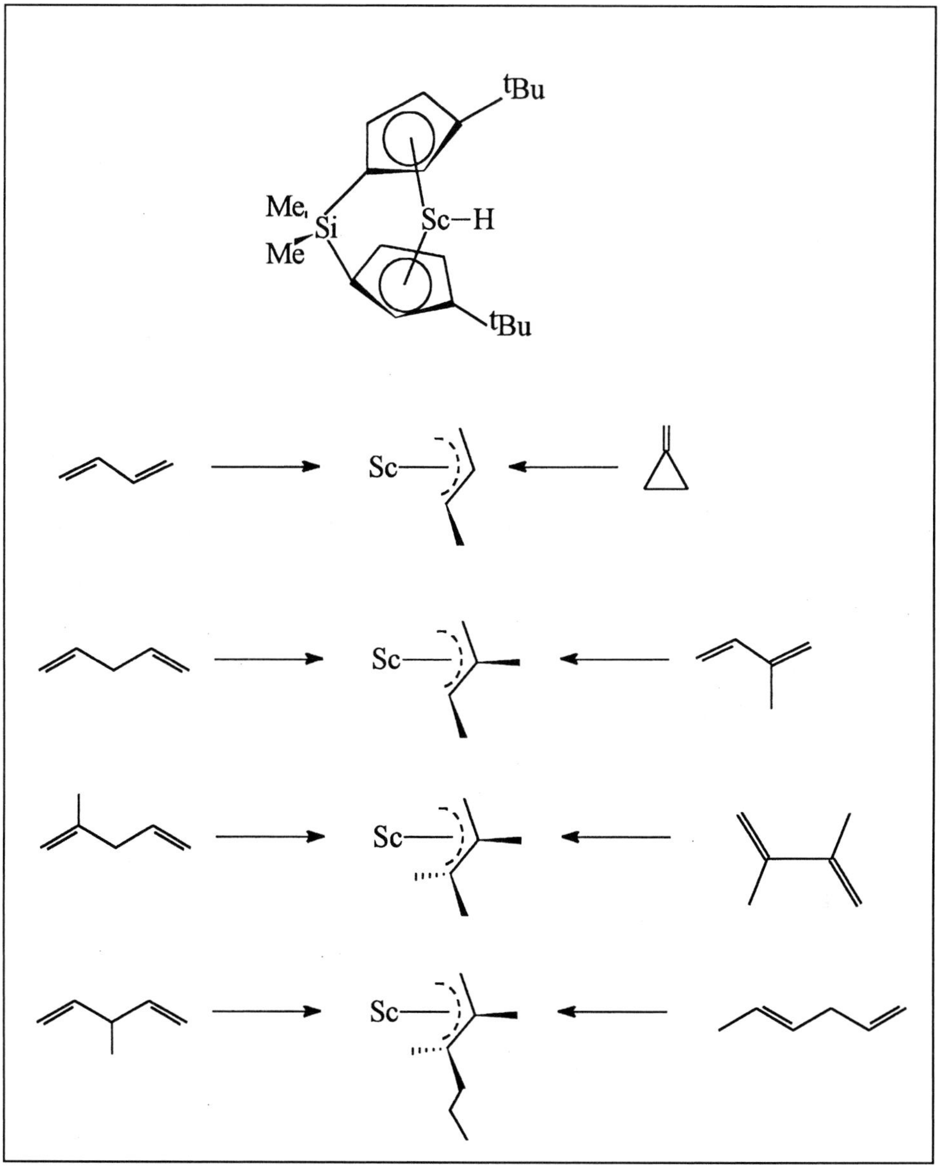

Scheme 2-20. Formation of scandium η^3-allyl complexes.

driving force for much of this work is the expectation that these lanthanocenes may exhibit useful catalytic activities. In fact, the catalytic activity of lanthanocene hydrocarbyls of the type $(C_5Me_5)_2LnCH(SiMe_3)_2$ *e.g.* in olefin hydrogenation reactions is exceedingly high and surpasses by far the activity of commonly used *d*-transition metal hydrogenation catalysts. However, the high air-sensitivity of such lanthanocene hydrocarbyls has thus far prevented them from coming into practical, *i.e.* industrial use [225]. Similar considerations can be made for the use of lanthanide organometallics in materials science [226, 227]. Thus future research in this area is expected to continue producing unusual molecular structures and unexpected reaction patterns. However, it will be the great challenge in the up-coming years to put lanthanocenes into practical uses, especially in the area of industrial catalysis. Great efforts have to be made to overcome the practical problems associated with the typical high moisture-sensitivity of organolanthanide complexes. This will be mainly a question of sophisticated ligand design. Imagining a water-soluble organolanthanide catalyst is certainly asking too much at the present state, but it should be possible to design highly reactive lanthanocene catalysts which can be handled under the conditions of industrial processes.

2.5 Acknowledgments

Our own contributions to organolanthanide chemistry at Göttingen and Magdeburg have been generously supported by the Deutsche Forschungsgemeinschaft, the Fonds der Chemischen Industrie and the Otto-von-Guericke-Universität Magdeburg. I am also indebted to Professor Herbert W. Roesky for continuing support.

References

[1]   G. Wilkinson and J. M. Birmingham, *J. Am. Chem. Soc.*, **1954**, *76*, 6210.
[2]   H. Schumann, *Angew. Chem.*, **1984**, *96*, 475; *Angew. Chem. Int. Ed. Engl.*, **1984**, *23*, 474.
[3]   W. J. Evans, *Polyhedron*, **1987**, *6*, 803.
[4]   R. D. Köhn, G. Kociok-Köhn, H. Schumann, *Scandium, Yttrium & the Lanthanides: Organometallic Chemistry* in: *Encyclopedia of Inorganic Cemistry* (Ed. R. B. King), Wiley, **1994**.
[5]   C. J. Schaverien, *Adv. Organomet. Chem.*, **1994**, *36*, 283.
[6]   H. Schumann, J.A. Meese-Marktscheffel and L. Esser, *Chem. Rev.*, **1995**, *95*, 865.
[7]   F. T. Edelmann in: *Comprehensive Organometallic Chemistry II* (Eds. E. W. Abel, F. G. A. Stone, G. Wilkinson), Pergamon Press, **1995**.
[8]   F. T. Edelmann, *Top. Curr. Chem.*, **1996**, *179*, 247.
[9]   E. O. Fischer and H. Fischer, *Angew. Chem.*, **1964**, *76*, 52; *Angew. Chem. Int. Ed. Engl.*, **1964**, *3*, 132.
[10]   G. B. Deacon, P. I. MacKinnon, T. W. Hambley and J. C. Taylor, *J. Organomet. Chem.*, **1983**, *259*, 91.
[11]   A. Hammel, W. Schwarz and J. Weidlein, *J. Organomet. Chem.*, **1989**, *378*, 347.
[12]   J. Jin, S. Jin and W. Chen, *J. Organomet. Chem.*, **1991**, *412*, 71.
[13]   X. Jusong, G. Wei, Z. Jin and W. Chen, *Zhongguo Xitu* Xuebao, **1992**, *3*, 203.
[14]   T. Jiang, Q. Shen, Y. Lin and S. Jin, *J. Organomet. Chem.*, **1993**, *450*, 121.
[15]   G. B. Deacon, G. N. Pain and T. D. Tuong, *Polyhedron*, **1985**, *4*, 1149.
[16]   G. Z. Suleimanov and I. P. Beletskaya, *Metalloorg. Khim.*, **1989**, *2*, 704.
[17]   J. L. Namy, P. Girard, H. B. Kagan and P. Caro, *Nouv. J. Chem.*, **1981**, *5*, 479.

[18] G. Z. Suleimanov, T. K. Kurbanov, Y. A. Nuriev, L. F. Rybakova and I. P. Beletskaya, *Dokl. Akad. Nauk SSSR*, **1982**, *265*, 896; *Proc. Acad. Sci. USSR*, **1982**, *265*, 254.

[19] G. Z. Suleimanov, L. F. Rybakova, Y. A. Nuriev, T. K. Kurbanov and I. P. Beletskaya, *J. Organomet. Chem.*, **1982**, *235*, C19.

[20] C. Apostolidis, G.B. Deacon, E. Dornberger, F.T. Edelmann, B. Kanellakopulos, P. MacKinnon and D. Stalke, *J. Chem. Soc., Chem. Commun.*, **1997**, 1047.

[21] V. D. Makhaev, Y. B. Zvedov, N. G. Chernorukov and V. I. Berestenko, *Vysokchist. Veshchestva*, **1990**, 210.

[22] A. L. Wayda, *J. Organomet. Chem.*, **1989**, *361*, 73.

[23] Q. Shen, D. Zheng, L. Lin and Y. Lin, *J. Organomet. Chem.*, **1990**, *391*, 307.

[24] V. K. Bel'skii, Y. K. Gun'ko, B. M. Bulychev, A. I. Sizov and G. L. Soloveichik, *J. Organomet. Chem.*, **1990**, *390*, 35.

[25] M. F. Lappert, P. I. W. Yarrow, J. L. Atwood and R. Shakir, *J. Chem. Soc., Chem. Commun.*, **1980**, 987.

[26] W. J. Evans, R. A. Keyer and J. W. Ziller, *J. Organomet. Chem.*, **1990**, *394*, 87.

[27] R. D. Rogers, *J. Organomet. Chem.*, **1996**, *512*, 97.

[28] P. B. Hitchcock, J. A. K. Howard, M. F. Lappert and S. Prashar, *J. Organomet. Chem.*, **1992**, *437*, 177.

[29] D. Deng, C. Qian, F. Song, Z. Wang, G. Wu and P. Zheng, *J. Organomet. Chem.*, **1992**, *443*, 79.

[30] C. Qian, C. Ye, H. Lu, Y. Li, J. Zhou and Y. Ge, *J. Organomet. Chem.*, **1983**, *247*, 161.

[31] C. Qian and D. Zhu, *J. Organomet. Chem.*, **1993**, *445*, 75.

[32] C. Qian, C. Zhu, Y. Lin and Y. Xing, *J. Organomet. Chem.*, **1996**, *507*, 41.

[33] P. Jutzi, J. Dahlhaus and M. O. Kristen, *J. Organomet. Chem.*, **1993**, *450*, C1.

[34] P. Jutzi and J. Dahlhaus, *Phosphorus, Sulfur, and Silicon*, **1994**, *87*, 73.

[35] G. B. Deacon, C. M. Forsyth, W. C. Patalinghug, A. H. White, A. Dietrich and H. Schumann, *Aust. J. Chem.*, **1992**, *45*, 567.

[36] G. B. Deacon, G. D. Fallon, P. I. MacKinnon, R. H. Newnham, G. N. Pain, T. D. Tuong and D. L. Wilkinson, *J. Organomet. Chem.*, **1986**, *277*, C21.

[37] G. B. Deacon, G. D. Fallon and C. M. Forsyth, *J. Organomet. Chem.*, **1993**, *462*, 183.

[38] W. J. Evans, *Inorg. Chim. Acta*, **1987**, *139*, 169.

[39] W. J. Evans, J. W. Grate, H. W. Choi, I. Bloom, W. E. Hunter and J. L. Atwood, *J. Am. Chem. Soc.*, **1985**, *107*, 941.

[40] W. J. Evans, L. A. Hughes and T. P. Hanusa, *Organometallics*, **1986**, *5*, 1285.

[41] P. L. Watson, *J. Chem. Soc., Chem. Commun.*, **1980**, 652.

[42] W. J. Evans, L. A. Hughes and T. P. Hanusa, *J. Am. Chem. Soc.*, **1984**, *106*, 4270.

[43] R. A. Andersen, J. M. Boncella, C. J. Burns, J. C. Green, D. Hohl and N. Rösch, *J. Chem. Soc., Chem. Commun.*, **1986**, 405.

[44] R. A. Williams, T. P. Hanusa and J. C. Huffman, *Organometallics*, **1990**, *9*, 1128.

[45] R. A. Williams, T. P. Hanusa and J. C. Huffman, *J. Chem. Soc., Chem. Commun.*, **1988**, 1045.

[46] T. K. Hollis, J. K. Burdett and B. Bosnich, *Organometallics*, **1993**, *12*, 3385.

[47] M. Kaupp, P. R. von Schleyer, M. Dolg and H. Stoll, *J. Am. Chem. Soc.*, **1992**, *114*, 8202.

[48] C. J. Burns and R. A. Andersen, *J. Am. Chem. Soc.*, **1987**, *109*, 941.

[49] C. J. Burns and R. A. Andersen, *J. Am. Chem. Soc.*, **1987**, *109*, 915.

[50] C. J. Burns and R. A. Andersen, *J. Am. Chem. Soc.*, **1987**, *109*, 5853.

[51] S. J. Swamy, J. Loebel and H. Schumann, *J. Organomet. Chem.*, **1989**, *379*, 51.

[52] A. Recknagel and F. T. Edelmann, *Angew. Chem.*, **1991**, *103*, 720; *Angew. Chem., Int. Ed. Engl.*, **1991**, *30*, 693.

[53] P. L. Watson, T. H. Tulip and I. Williams, *Organometallics*, **1990**, *9*, 1999.

[54] C. Qian, D. Deng, Z. Zhang and C. Ni, *Youji Huaxue*, **1985**, 403.

[55] C. Ni, Z. Zhang, D. Deng and C. Qian, *J. Organomet. Chem.*, **1986**, *306*, 209.

[56] Z. Lin, Y. Liu and W. Chen, *Sci. Sin., Ser. B*, **1988**, *30*, 1136.

[57] H. Yasuda, H. Yamamoto, K. Yokota and A. Nakamura, *Chem. Lett.*, **1989**, 1309.

[58] J. Jizhu, Z. Jin and W. Chen, *Jiegou Huaxue*, **1992**, *11*, 204.

[59]   W. J. Evans, R. A. Keyer and J. W. Ziller, *J. Organomet. Chem.*, **1993**, *450*, 115.

[60]   E. C. Baker, L. D. Brown and K. N. Raymond, *Inorg. Chem.*, **1975**, *14*, 1376.

[61]   V. K. Bel'skii, S. Y. Knyazhanskii, Y. K. Gun'ko, B. M. Bulychev and G. L. Soloveichik, *Metalloorg. Khim.*, **1991**, *4*, 1135.

[62]   S. Song, Q. Shen, S. Jin, J. Guan and Y. Lin, *Polyhedron*, **1992**, *12*, 2857.

[63]   V. K. Bel'skii, S. Y. Knyazhanskii, B. M. Bulychev and G. L. Soloveichik, *Metalloorg. Khim.*, **1989**, *2*, 567.

[64]   A. Recknagel, F. Knösel, H. Gornitzka, M. Noltemeyer and F. T. Edelmann, *J. Organomet. Chem.*, **1991**, *417*, 363.

[65]   J. L. Atwood, W. E. Hunter, R. D. Rogers, J. Holton, J. McMeeking, R. Pearce and M. F. Lappert, *J. Chem. Soc., Chem. Commun.*, **1978**, 140.

[66]   P. L. Watson, J. F. Whitney and R. L. Harlow, *Inorg. Chem.*, **1981**, *20*, 3271.

[67]   W. A. Herrmann, R. Anwander, H. Riepl, W. Scherer and C. R. Whitaker, *Organometallics*, **1993**, *12*, 4342.

[68]   D. Deng, C. Qian, G. Wu and P. Zheng, *J. Chem. Soc., Chem. Commun.*, **1990**, 880.

[69]   D. Deng, X. Zheng, C. Qian, S. Jin and Y. Lin, *Huaxue Xuebao*, **1992**, *10*, 1024.

[70]   D. Deng, B. Li and C. Qian, *Polyhedron*, **1990**, *9*, 1435.

[71]   C. Qian, B. Wang, D. Deng, J. Hu, J. Chen, G. Wu and P. Zheng, *Inorg. Chem.*, **1994**, *33*, 3382.

[72]   J. Gräper, R. D. Fischer and G. Paolucci, *J. Organomet. Chem.*, **1994**, *471*, 87.

[73]   C. Qian and D. Zhu, *J. Organomet. Chem.*, **1993**, *445*, 79.

[74]   G. Paolucci, R. D'Ippolito, C. Ye, C. Qian, J. Gräper and R. D. Fischer, *J. Organomet. Chem.*, **1994**, *471*, 97.

[75]   W. A. Herrmann, R. Anwander, F. C. Munck and W. Scherer, *Chem. Ber.*, **1993**, *126*, 319.

[76]   C. J. Burns and R. A. Andersen, *J. Chem. Soc., Chem. Commun.*, **1989**, 136.

[77]   R. G. Finke, S. R. Keenan, D. A. Schiraldi and P. L. Watson, *Organometallics*, **1987**, *6*, 1356.

[78]   R. G. Finke, S. R. Keenan and P. L. Watson, *Organometallics*, **1989**, *8*, 263.

[79]   W. J. Evans, J. M. Olofson, H. Zhang and J. L. Atwood, *Organometallics*, **1988**, *7*, 629.

[80]   H. Schumann, I. Albrecht, J. Loebel, E. Hahn, M. B. Hossain and D. van der Helm, *Organometallics*, **1986**, *5*, 1296.

[81]   C. Qian, C. Ye, H. Lu, Y. Li and Y. Huang, *J. Organomet. Chem.*, **1984**, *263*, 333.

[82]   C. Qian, C. Ye and Y. Li, *J. Organomet. Chem.*, **1986**, *302*, 171.

[83]   C. Qian, C. Ye and Y. Li, *Youji Huaxue*, **1986**, 130.

[84]   V. P. Conticello, L. Brard, M. A. Giardello, Y. Tsuji, M. Sabat, C. L. Stern and T. J. Marks, *J. Am. Chem. Soc.*, **1992**, *114*, 2761.

[85]   M. R. Gagné, L. Brard, V. P. Conticello, M. A. Giardello, C. L. Stern and T. J. Marks, *Organometallics*, **1992**, *11*, 2003.

[86]   M. A. Giardello, V. P. Conticello, L. Brard, M. Sabat, A. L. Rheingold, C. L. Stern and T. J. Marks, *J. Am. Chem. Soc.* **1994**, *116*, 10212.

[87]   M. Adam, G. Massarweh and R. D. Fischer, *J. Organomet. Chem.*, **1991**, *405*, C33.

[88]   H. Schumann, W. Genthe, N. Bruncks and J. Pickardt, *Organometallics*, **1982**, *1*, 1194.

[89]   G. B. Deacon, S. Nickel and E. R. T. Tiekink, *J. Organomet. Chem.*, **1991**, *409*, C1.

[90]   Z. Wu, Z. Xu, X. You, X. Zhou and Z. Jin, *Polyhedron*, **1992**, *12*, 2673.

[91]   Z. Wu, Z. Xu, X. You, X. Zhou, Y. Xing and Z. Jin, *J. Chem. Soc., Chem. Commun.*, **1993**, 1494.

[92]   J. Stehr and R. D. Fischer, *J. Organomet. Chem.*, **1993**, *459*, 79.

[93]   G. Massarweh and R. D. Fischer, *J. Organomet. Chem.*, **1993**, *444*, 67.

[94]   P. Poremba, M. Noltemeyer, H.-G. Schmidt and F.T. Edelmann, *J. Organomet. Chem.* **1995**, *501*, 315.

[95]   H. Schumann, J. A. Meese-Marktscheffel and A. Dietrich, *J. Organomet. Chem.*, **1989**, *377*, C5.

[96]   J. Stehr and R. D. Fischer, *J. Organomet. Chem.*, **1992**, *430*, C1.

[97]   H. Schumann, J. A. Meese-Marktscheffel, A. Dietrich and F. H. Görlitz, *J. Organomet. Chem.*, **1992**, *430*, 299.

[98] W. J. Evans, D. K. Drummond, L. A. Hughes, H. Zhang and J. L. Atwood, *Polyhedron*, **1988**, 7, 1693.

[99] W. J. Evans, J. W. Grate, I. Bloom, W. E. Hunter and J. L. Atwood, *J. Am. Chem. Soc.*, **1985**, 107, 405.

[100] W. J. Evans, T. P. Hanusa and K. R. Levan, *Inorg. Chim. Acta*, **1984**, 110, 191.

[101] W. J. Evans, L. A. Hughes, D. K. Drummond, H. Zhang and J. L. Atwood, *J. Am. Chem. Soc.*, **1986**, 108, 1722.

[102] W. J. Evans, J. W. Grate, L. A. Hughes, H. Zhang and J. L. Atwood, *J. Am. Chem. Soc.*, **1985**, 107, 3728.

[103] T. D. Tilley, R. A. Andersen, A. Zalkin and D. H. Templeton, *Inorg. Chem.*, **1982**, 21, 2644.

[104] P. L. Watson, *J. Chem. Soc., Chem. Commun.*, **1983**, 276.

[105] H. Schumann, I. Albrecht, M. Gallagher, E. Hahn, C. Muchmore and J. Pickardt, *J. Organomet. Chem.*, **1988**, 349, 103.

[106] A. Recknagel, M. Noltemeyer, D. Stalke, U. Pieper, H.-G. Schmidt and F. T. Edelmann, *J. Organomet. Chem.*, **1991**, 411, 347.

[107] D. J. Berg, C. J. Burns, R. A. Andersen and A. Zalkin, *Organometallics*, **1989**, 8, 1865.

[108] D. J. Berg, R. A. Andersen and A. Zalkin, *Organometallics*, **1988**, 7, 1858.

[109] A. Zalkin and D. J. Berg, *Acta Cryst.*, **1988**, C44, 1488.

[110] W. E. Piers, *J. Chem. Soc., Chem. Commun.*, **1994**, 309.

[111] Y. Yu, S. Wang, H. Ma and Z. Ye, *Polyhedron*, **1992**, 12, 265.

[112] H. Schumann, P. R. Lee and A. Dietrich, *Chem. Ber.*, **1990**, 123, 1331.

[113] Z. Wu, X. Zhou, W. Zhang, Z. Xu, X. You and X. Huang, *J. Chem. Soc., Chem. Commun.*, **1994**, 813.

[114] H. Schumann, I. Albrecht, M. Gallagher, E. Hahn, C. Janiak, C. Kolax, J. Loebel, S. Nickel and E. Palamidis, *Polyhedron*, **1988**, 7, 2307.

[115] H. Schumann, E. Palamidis, G. Schmid and R. Boese, *Angew. Chem.*, **1986**, 98, 726; *Angew. Chem., Int. Ed. Engl.*, **1986**, 25, 718.

[116] H. Schumann, E. Palamidis, J. Loebel and J. Pickardt, *Organometallics*, **1988**, 7, 1008.

[117] W. J. Evans, I. Bloom, W. E. Hunter and J. L. Atwood, *Organometallics*, **1983**, 2, 709.

[118] H. Schumann, E. Palamidis and J. Loebel, *J. Organomet. Chem.*, **1990**, 384, C49.

[119] W. J. Evans, D. K. Drummond, S. G. Bott and J. L. Atwood, *Organometallics*, **1986**, 5, 2389.

[120] W. J. Evans, J. W. Grate, K. R. Levan, I. Bloom, T. T. Peterson, R. J. Doedens, H. Zhang and J. L. Atwood, *Inorg. Chem.*, **1986**, 25, 3614.

[121] W. J. Evans, D. K. Drummond, L. R. Chamberlain, R. J. Doedens, S. G. Bott, H. Zhang and J. L. Atwood, *J. Am. Chem. Soc.*, **1988**, 110, 4983.

[122] K. H. den Haan, J. L. De Boer, J. H. Teuben, A. L. Spek, B. Kojic-Prodic, G. R. Hays and R. Huis, *Organometallics*, **1986**, 5, 1726.

[123] W. J. Evans, R. A. Keyer and J. W. Ziller, *Organometallics*, **1993**, 12, 2618.

[124] W. J. Evans and D. K. Drummond, *J. Am. Chem. Soc.*, **1989**, 111, 3329.

[125] A. Recknagel, M. Noltemeyer and F. T. Edelmann, *J. Organomet. Chem.*, **1991**, 410, 53.

[126] A. Scholz, K.-H. Thiele, J. Scholz and R. Weimann, *J. Organomet. Chem.*, **1995**, 501, 195.

[127] J. Scholz, A. Scholz, R. Weimann, C. Janiak and H. Schumann, *Angew. Chem.*, **1994**, 106, 1220; *Angew. Chem., Int. Ed. Engl.*, **1994**, 33, 1171.

[128] W. J. Evans, T. A. Ulibarri and J. W. Ziller, *J. Am. Chem. Soc.*, **1988**, 110, 6877.

[129] W. J. Evans, S. L. Gonzales and J. W. Ziller, *J. Chem. Soc., Chem. Commun.*, **1992**, 1138.

[130] W. J. Evans, S. L. Gonzales and J. W. Ziller, *J. Am. Chem. Soc.*, **1991**, 113, 9880.

[131] W. J. Evans, J. T. Leman, J. W. Ziller and S. I. Khan, *Inorg. Chem.*, **1996**, 35, 4283.

[132] C. Qian, C. Ye, H. Lu, Y. Li, J. Zhou, Y. Ge and M. Tsutsui, *J. Organomet. Chem.*, **1983**, 247, 161.

[133] W. J. Evans, J. H. Meadows, A. L. Wayda, W. E. Hunter and J. L. Atwood, *J. Am. Chem. Soc.*, **1982**, 104, 2008.

[134] H. Schumann and G. Jeske, *Angew. Chem.*, **1985**, 97, 208; *Angew. Chem., Int. Ed. Engl.*, **1985**, 24, 255.

[135] H. Schumann and G. Jeske, *Z. Naturforsch.*, **1985**, *B40*, 1490.

[136] H. Schumann, F. W. Reier and E. Palamidis, *J. Organomet. Chem.*, **1985**, *297*, C30.

[137] G. B. Deacon and D. L. Wilkinson, *Inorg. Chim. Acta*, **1988**, 142, 155.

[138] W. J. Evans, R. Dominguez and T. P. Hanusa, *Organometallics*, **1986**, *5*, 263.

[139] H. Schumann, H. Lauke, E. Hahn, M. J. Heeg and D. Van der Helm, *Organometallics*, **1985**, *4*, 321.

[140] H. Schumann, F. W. Reier and E. Hahn, *Z. Naturforsch.*, **1985**, *B40*, 1289.

[141] Q. Shen, Y. Cheng and Y. Lin, *J. Organomet. Chem.*, **1991**, *419*, 293.

[142] H. Schumann, F. W. Reier and M. Dettlaff, *J. Organomet. Chem.*, **1983**, *255*, 305.

[143] H. Schumann and F. W. Reier, *Inorg. Chim. Acta*, **1984**, *95*, 43.

[144] V. I. Bregadze, N. A. Kovalchuk, N. N. Godovikov, G. Z. Suleimanov and I. P. Beletskaya, *J. Organomet. Chem.*, **1983**, *241*, C13.

[145] H. Schumann, S. Nickel, E. Hahn and M. J. Heeg, *Organometallics*, **1985**, *4*, 800.

[146] H. Schumann, J. A. Meese-Marktscheffel and F. E. Hahn, *J. Organomet. Chem.*, **1990**, *390*, 301.

[147] H. Schumann, S. Nickel, J. Loebel and J. Pickardt, *Organometallics*, **1988**, *7*, 2004.

[148] B. K. Campion, R. H. Heyn and T. D. Tilley, *Organometallics*, **1993**, *12*, 2584.

[149] H. Schumann, I. Albrecht, J. Pickardt and E. Hahn, *J. Organomet. Chem.*, **1984**, *276*, C5.

[150] M. L. H. Green, A. K. Hughes, N. A. Popham, A. H. H. Stephens and L.-L. Wong, *J. Chem. Soc., Dalton Trans.*, **1992**, 3077.

[151] W. J. Evans, T. A. Ulibarri, L. R. Chamberlain, J. W. Ziller and D. Alvarez, *Organometallics*, **1990**, *9*, 2124.

[152] K. H. den Haan, J. L. de Boer, J. H. Teuben, W. J. J. Smeets and A. L. Spek, *J. Organomet. Chem.*, **1987**, *327*, 31.

[153] W. J. Evans, L. R. Chamberlain, T. A. Ulibarri and J. W. Ziller, *J. Am. Chem. Soc.*, **1988**, *110*, 6423.

[154] P. L. Watson and T. Herskovitz, *ACS Symp. Ser (Initiation Polym.)*, **1983**, *212*, 459.

[155] M. E. Thompson, S. M. Baxter, A. R. Bulls, B. J. Burger, M. C. Nolan, B. D. Santarsiero, W. P. Schaefer and J. E. Bercaw, *J. Am. Chem. Soc.*, **1987**, *109*, 203.

[156] P. L. Watson, *J. Am. Chem. Soc.*, **1983**, *105*, 6491.

[157] P. T. Matsunaga, *Energy Res. Abstr.*, **1992**, *17(6)*.

[158] G. M. Forsyth, S. P. Nolan and T. J. Marks, *Organometallics*, **1991**, *10*, 2543.

[159] T. Ziegler, E. Folga and A. Berces, *J. Am. Chem. Soc.*, **1992**, *115*, 636.

[160] E. Folga, T. Ziegler and L. Fan, *New. J. Chem.*, **1992**, *15*, 741.

[161] W. E. Piers, P. J. Shapiro, E. E. Bunel and J. E. Bercaw, *Synlett*, **1990**, 74.

[162] P. J. Fagan, J. M. Manriquez, T. J. Marks, V. W. Day, S. H. Vollmer and C. S. Day, *J. Am. Chem. Soc.*, **1980**, *102*, 5393.

[163] H. J. Heeres, J. Renkema, M. Booij, A. Meetsma and J. H. Teuben, *Organometallics*, **1988**, *7*, 2495.

[164] G. Jeske, L. E. Schock, P. N. Swepston, H. Schumann and T. J. Marks, *J. Am. Chem. Soc.*, **1985**, *107*, 8103.

[165] H. Mauermann, P. N. Swepston and T. J. Marks, *Organometallics*, **1985**, *4*, 200.

[166] J. Renkema and J. H. Teuben, *Recl. Trav. Chim. Pays-Bas*, **1986**, *105*, 241.

[167] G. Jeske, H. Lauke, H. Mauermann, P. N. Swepston, H. Schumann and T. J. Marks, *J. Am. Chem. Soc.*, **1985**, *107*, 8091.

[168] J. Renkema and J. H. Teuben, *Recl. Trav. Chim. Pays-Bas*, **1988**, *105*, 241.

[169] C. M. Fendrick, L. D. Schertz, V. W. Day and T. J. Marks, *Organometallics*, **1988**, *7*, 1828.

[170] H. Schumann, L. Esser, J. Loebel, A. Dietrich, D. Van der Helm and X. Ji, *Organometallics*, **1991**, *10*, 2585.

[171] H. Schumann, J. Loebel, J. Pickardt, C. Qian and Z. Xie, *Organometallics*, **1991**, *10*, 215.

[172] D. Stern, M. Sabat and T. J. Marks, *J. Am. Chem. Soc.*, **1990**, *112*, 9558.

[173] W. P. Schaefer, R. D. Köhn and J. E. Bercaw, *Acta Cryst.*, **1992**, *C48*, 251.

[174] R. E. Marsh, W. P. Schaefer, E. B. Coughlin and J. E. Bercaw, *Acta Cryst.*, **1992**, *C48*, 1773.

[175] N. Koga and K. Morokuma, *J. Am. Chem. Soc.*, **1988**, *110*, 108.

[176] H. J. Heeres, A. Meetsma and J. H. Teuben, *Angew. Chem.*, **1990**, *102*, 449; *Angew. Chem., Int. Ed. Engl.*, **1990**, *29*, 420.

[177] J. H. Teuben, in *"Fundamental and Technological Aspects of Organo-f-Element Chemistry"*, NATO ASI Ser., ed. T. J. Marks and I. L. Fragalà, D. Reidel, Boston, **1985**, *155*, 195.

[178] K. H. den Haan, G. A. Luinstra, A. Meetsma and J. H. Teuben, *Organometallics*, **1987**, *6*, 1509.

[179] K. H. den Haan, Y. Wielstra and J. H. Teuben, *Organometallics*, **1987**, *6*, 2053.

[180] H. J. Heeres, M. Maters and J. H. Teuben, *Organometallics*, **1992**, *11*, 350.

[181] P. L. Watson and G. W. Parshall, *Acc. Chem. Res.*, **1985**, *18*, 51.

[182] M. A. Busch, R. Harlow and P. L. Watson, *Inorg. Chim. Acta*, **1987**, *140*, 15.

[183] W. J. Evans, L. R. Chamberlain and J. W. Ziller, *J. Am. Chem. Soc.*, **1987**, *109*, 7209.

[184] H. Yamamoto, H. Yasuda, K. Yokota, A. Nakamura, Y. Kai and N. Kasai, *Chem. Lett.*, **1988**, 1963.

[185] H. Schumann, M. Glanz, J. Winterfeld, H. Hemling, N. Kuhn and T. Kratz, *Angew. Chem.*, **1994**, *106*, 1829; *Angew. Chem., Int. Ed. Engl.*, **1994**,

[186] A. J. Arduengo, III, M. Tamm, S. J. McLain, J. C. Calabrese, F. Davidson and W. J. Marshall, *J. Am. Chem. Soc.*, **1994**, *116*, 7927.

[187] W. A. Herrmann, F. C. Munck, G. R. J. Artus, O. Runte and R. Anwander, *Organometallics*, **1997**, *16*, 682.

[188] A. Scholz, A. Smola, J. Scholz, J. Loebel, H. Schumann and K.-H. Thiele, *Angew. Chem.*, **1991**, *103*, 444; *Angew. Chem., Int. Ed. Engl.*, **1991**, *30*, 435.

[189] W. J. Evans, T. A. Ulibarri and J. W. Ziller, *J. Am. Chem. Soc.*, **1990**, *112*, 219.

[190] W. J. Evans, S. L. Gonzales and J. W. Ziller, *J. Am. Chem. Soc.*, **1994**, *116*, 2600.

[191] W. J. Evans, J. H. Meadows, W. E. Hunter and J. L. Atwood, *Organometallics*, **1983**, *2*, 1252.

[192] W. J. Evans, I. Bloom, W. E. Hunter and J. L. Atwood, *J. Am. Chem. Soc.*, **1983**, *105*, 1401.

[193] J. M. Boncella, T. D. Tilley and R. A. Andersen, *J. Chem. Soc., Chem. Commun.*, **1984**, 710.

[194] W. J. Evans, D. K. Drummond, T. P. Hanusa and J. M. Olofson, *J. Organomet. Chem.*, **1989**, *376*, 311.

[195] H. J. Heeres, J. Nijhoff, J. H. Teuben and R. D. Rogers, *Organometallics*, **1991**, *12*, 2609.

[196] C. M. Forsyth, S. P. Nolan, C. S. Stern, T. J. Marks and A. L. Rheingold, *Organometallics*, **1993**, *12*, 3619.

[197] A. Recknagel, D. Stalke, H. W. Roesky and F. T. Edelmann, *Angew. Chem.*, **1989**, *101*, 496; *Angew. Chem., Int. Ed. Engl.*, **1989**, *28*, 445.

[198] W.-K. Wong, H. Chen and F.-L. Chow, *Polyhedron*, **1990**, *9*, 875.

[199] H. Siebald, M. Dartiguenave and Y. Dartiguenave, *J. Organomet. Chem.*, **1992**, *438*, 83.

[200] N. S. Radu and T. D. Tilley, *Phosphorus, Sulfur, and Silicon*, **1994**, *87*, 209.

[201] N. S. Radu, T. D. Tilley and A. L. Rheingold, *J. Am. Chem. Soc.*, **1992**, 114, 8293.

[202] T. Kobayashi, T. Sakakura, T. Hayashi, M. Yumura and M. Tanaka, *Chem. Lett.*, **1992**, 1158

[203] T. Sakakura, H.-J. Lautenschlager and M. Tanaka, *J. Chem. Soc., Chem. Commun.*, **1991**, 40.

[204] J. V. Ortiz and R. Hoffmann, *Inorg. Chem.*, **1985**, *24*, 2095.

[205] E. N. Zavadovskaya, O. K. Sharaev, G. K. Borisov, Y. P. Yampolskii, E. I. Tinyakova and B. A. Dolgoplosk, *Dokl. Akad. Nauk SSSR*, **1985**, *284*, 143.

[206] W. J. Evans, J. H. Meadows, W. E. Hunter and J. L. Atwood, *J. Am. Chem. Soc.*, **1984**, *106*, 1291.

[207] I. P. Beletskaya, A. Z. Voskoboinikov and G. K.-I. Magomedov, *Metalloorg. Khim.*, **1989**, *2*, 810.

[208] D. Deng, Y. Jiang, C. Qian, G. Wu and P. Zheng, *J. Organomet. Chem.*, **1994**, *470*, 99.

[209] W. J. Evans, D. K. Drummond, T. P. Hanusa and R. J. Doedens, *Organometallics*, **1987**, *6*, 2279.

[210] H. Schumann, W. Genthe, E. Hahn, M. B. Hossain and D. Van der Helm, *J. Organomet. Chem.*, **1986**, *299*, 67.

[211] C. Qian, D. Deng, C. Ni and Z. Zhang, *Inorg. Chim. Acta*, **1988**, *146*, 129.

[212] S. Y. Knyazhanskii, V. K. Bel'skii, B. M. Bulychev and G. L. Soloveichik, *Metalloorg. Khim.*, **1989**, *2*, 570.

[213] Y. K. Gun'ko, B. M. Bulychev, G. L. Soloveichik and V. K. Bel'skii, *J. Organomet. Chem.*, **1992**, *424*, 289.

[214] W. J. Evans, T. P. Hanusa, J. H. Meadows, W. E. Hunter and J. L. Atwood, *Organometallics*, **1987**, *6*, 295.

[215] T. J. Marks and G. W. Grynkewich, *Inorg. Chem.*, **1976**, *15*, 1302.

[216] M. F. Lappert, A. Singh, J. L. Atwood and W. E. Hunter, *J. Chem. Soc., Chem. Commun.*, **1983**, 206.

[217] V. D. Makhaev and A. P. Borisov, *Koord. Khim.*, **1992**, 466.

[218] M. Booij, B.-J. Deelman, R. Duchateau, D. S. Postma, A. Meetsma and J. H. Teuben, *Organometallics*, **1993**, *12*, 3531.

[219] K. H. den Haan and J. H. Teuben, *Recl. Trav. Chim. Pays-Bas*, **1984**, *103*, 333.

[220] C. Ye, C. Qian and X. Yang, *J. Organomet. Chem.*, **1991**, *407*, 329.

[221] C. Qian, Z. Xie and Y. Huang, *Inorg. Chim. Acta*, **1987**, *139*, 195.

[222] W. J. Evans, T. A. Ulibarri and J. W. Ziller, Organometallics, **1991**, *10*, 134.

[223] W. J. Evans, J. W. Grate and R. J. Doedens, *J. Am. Chem. Soc.*, **1985**, *107*, 1671.

[224] E. Bunel, B. J. Burger and J. E. Bercaw, *J. Am. Chem. Soc.*, **1988**, *110*, 976.

[225] B. Cornils and W. A. Herrmann, "*Applied Homogeneous Catalysis with Organometallic Compounds*", VCH, Weinheim, **1996**.

[226] G. B. Deacon, P. MacKinnon, R. S. Dickson, G. N. Pain and B. O. West, *Appl. Organomet. Chem.*, **1990**, *4*, 439.

[227] Y. K. Gun'ko and F. T. Edelmann, *Comments Inorg. Chem.*, **1997**, *19*, 153.

3 Group 3 Metallocenes

Paul J. Chirik and John E. Bercaw

3.1 Introduction

Cyclopentadienyl is perhaps the most important and widespread ancillary ligand in organometallic chemistry. Cyclopentadienyl ligands stabilize a wide variety of organometallic complexes with group 3 metals (Sc, Y, La), as they do for most other transition metals. The focus of this review is to present a general synopsis of the chemistry of group 3 complexes based on the bis(cyclopentadienyl) (or substituted cyclopentadienyl) framework. It concentrates primarily on the chemistry of scandocene and yttrocene compounds. Lanthanum metallocenes have thus far received less attention, and thus lanthanocenes will be mentioned only briefly. Traditionally, group 3 metallocene chemistry has been reviewed with organolanthanide chemistry [1], however, elements beyond La will not be considered here.

The definition of a metallocene for the purpose of this review is a complex containing two η^5-coordinated cyclopentadienyl or substituted cyclopentadienyl rings. The scope of this definition is expanded in some cases to include monocyclopentadienyl complexes having alkoxide or amide ligands, in cases where such compounds undergo chemistry similar to that of the bis(cyclopentadienyl) systems. Group 3 metallocene chlorides $[Cp_2MCl]_x$ ($Cp = (\eta^5\text{-}C_5H_5)$, M = Sc, Y) were prepared by Wilkinson and Birmingham in the mid-1950s, and apart from some important studies by Lappert, Wailes, Manzer and Atwood in the 1970s, most of the chemistry of the group 3 metallocenes has been developed during the past 10 years. This review will focus primarily on these recent developments, especially the use of group 3 metallocene complexes as reagents for activation of the C–H bonds of hydrocarbons, for polymerization of olefins, and more recently, as reagents for organic synthesis. The first sections concern methods for preparing group 3 metallocene complexes and summarize the general reactivity trends observed for these systems. Special attention is devoted to alkyls and hydrides, since these are most often encountered in catalytic applications, most notably polymerization of α–olefins. Later sections review reactions of the group 3 metallocene compounds with acetylenes, with other small molecules and the applications of some of this chemistry to organic synthesis. The final two brief sections are devoted to some studies of the photophysics of group 3 metallocene compounds and to ligands that are designed as alternatives to the cyclopentadienyl ligand array.

3.2 Synthesis of Group 3 Metallocenes

3.2.1 Scandium

Although the organometallic chemistry of first row transition elements (Ti through Cu) has been extensively investigated over the past three decades, the first member of the series, scandium, has received much less attention. The slow development of organoscandium chemistry may be traced to the perception that organometallic compounds of scandium differ little from those of its group 13 counterparts, being plagued by insolubility due to their tendency to form polymeric materials supported by 3–center, 2- or 4-electron (M–(μ-X)–M) bonds (X = halide, hydrocarbyl, *etc.*) [2]. However, the employment of the cyclopentadienyl ligand (and its substituted derivatives) has provided an opportunity to the prepare soluble, well-defined organoscandium compounds. A veritable renaissance in its organometallic chemistry has followed.

   The first report of a scandocene complex was communicated by Wilkinson and Birmingham [3] with the synthesis of Cp_3Sc from the reaction of anhydrous $ScCl_3$ and NaCp. This material is thermally robust, but is polymeric and insoluble and thus a rather intractable material [4]. Scandocene chloride, $(Cp_2ScCl)_2$, prepared from scandium trichloride and $MgCp_2$, may be isolated as a green solid [5]. Ebulliometric molecular weight measurements are indicative of a chloro-bridged dimeric structure, which has been confirmed by X-ray crystallography [6]. The syntheses of $[Cp_2Sc(O_2CCH_3)]_2$, $Cp_2Sc(\eta^3$-allyl), and $[Cp_2Sc(C\equiv CPh)]_2$ have also been described [5]. Manzer [7] reports a more convenient preparation for $Cp_2ScCl(THF)$ from the reaction of two equivalents of TlCp with anhydrous $ScCl_3$ in THF solution (Eq. (3-1)).

$$ScCl_3 + 2\ TlCp \xrightarrow{\ THF\ } \underset{X}{\overset{}{Sc}}\text{–THF} \xrightarrow{\ Li(CH_2)PPh_2\ } Sc\overset{-CH_2}{\underset{CH_2}{<}}\overset{+}{PPh_2} \quad (1) \quad (3\text{-}1)$$

Alkylation with $Li(CH_2)_2PPh_2$ affords the scandium chelate complex, $Cp_2Sc(\mu_2\text{-}CH_2)_2PPh_2$; reaction with $LiC_6H_4\text{-}o\text{-}CH_2NMe_2$ yields $Cp_2Sc(\eta^2\text{-}C_6H_4\text{-}o\text{-}CH_2NMe_2)$ [8,9]. The product obtained from scandium fluoride is quite different. Reaction of ScF_3 with Cp_2Mg results in a mixture of Cp_3Sc and 'Cp_2ScF' [10,11]. The same mixture also results from treatment of the trifluoride with NaCp [11]. X-ray crystallographic studies show that 'Cp_2ScF' is actually a trimer with bridging fluoride ligands.

   Lappert [12] reported the first example of a scandocene having substituted cyclopentadienyl ligands with the preparations of $[(\eta^5\text{-}C_5H_3(SiMe_3)_2)_2ScCl]_2$ and $\{(\eta^5\text{-}C_5H_3(SiMe_3)_2\}_2Sc(\mu_2\text{-}H)_2BH_2$. Despite the presence of two bulky trimethylsilyl substituents, the resulting scandocene chlorides still undergo dimerization. However, use of the sterically very demanding pentamethylcyclopentadienyl ligand (Cp*) affords monomeric scandium chlorides and alkyls. Thus, reaction of $ScCl_3 \cdot (THF)_3$ with LiCp* produces the monomeric permethylscandocene chloride, $Cp*_2ScCl$ [13]. Addition of the appropriate lithium reagent, LiR, results in the formation of the corresponding hydrocarbyl derivative, $Cp*_2ScR$ (R = CH_3, C_6H_5, $CH_2C_6H_5$). Hydrogenolysis of the Sc–C bonds of these derivatives produces $Cp*_2ScH$, which is stable only at low temperatures or under a dihydrogen atmosphere. Hydrogenation of the alkyls in a mixture of hexane and THF affords the more stable THF adduct, $Cp*_2ScH(THF)$ (Eq. (3-2)) [14].

$$ScCl_3 \cdot (THF)_3 \xrightarrow{2\ LiCp*} Sc\!-\!Cl \xrightarrow{LiR} Sc\!-\!R \xrightarrow{H_2} Sc\!-\!H \quad (3\text{-}2)$$

Since the reactivity of $Cp*_2ScR$ compounds is limited by the sterically demanding Cp* ligands, two strategies have been devised to relieve the excess crowding at scandium: (1) the synthesis of mixed ring scandocenes where one of the Cp* rings is replaced by a less bulky cyclopentadienyl ligand and (2) synthesis of *ansa* scandocenes, where the two cyclopentadienyl rings are 'tied back' by a dimethylsilylene linker. For the former strategy, the challenge is the synthesis of the requisite mono(pentamethylcyclopentadienyl) scandium dichloride precursor, since the reaction of one equivalent of Cp*Li with $ScCl_3$ results in the formation of ill-defined products [15]. Despite these difficulties, high yield routes to $[Cp*ScCl_2]_x$ have been developed from the reaction of $Sc(acac)_3$ and Cp*MgCl·THF, producing Cp*Sc(acac), which is then converted to the dichloride through reaction with a stoichiometric amount of $AlCl_3$ (Eq. (3-3)) [16].

$$Sc(acac)_3 + Cp*MgCl \longrightarrow \underset{O\ \ O\ \ O\ \ O}{Sc} \xrightarrow{AlCl_3} [Cp*ScCl_2]_x \quad (3\text{-}3)$$

Mixed-ring scandocene chlorides can be obtained from the reaction of $[Cp*ScCl_2]_x$ and LiCp (Cp = cyclopentadienyl or substituted cyclopentadienyl). Conversion of the mixed-ring chloride to alkyls and hydrides can be achieved through similar synthetic procedures as with the permethylscandocene system, except that the addition of PMe_3 is used to stabilize the resulting alkyl complexes (Scheme 3-1) [16]. Mixed ring scandocene hydrides have also been prepared; however, these compounds readily undergo bimolecular decomposition with release of H_2 to give the metallated dimeric compound $Cp*(\eta^5\text{-}C_5H_5)Sc(\mu_2\text{-}\eta^5,\eta^1\text{-}C_5H_4)(\mu_2\text{-}H)ScCp*$[17].

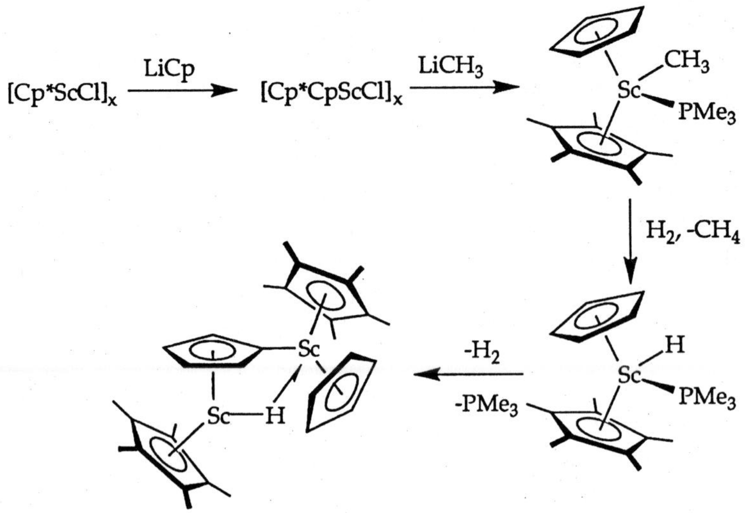

$$[Cp*ScCl]_x \xrightarrow{\text{LiCp}} [Cp*CpScCl]_x \xrightarrow{\text{LiCH}_3}$$

Scheme 3-1

The second approach to preparing more reactive scandocene complexes is the use of 'tied-back' cyclopentadienyl rings, thus producing an *ansa*-scandocene. A variety of linked cyclopentadienyl ligands has been synthesized [18] and coordinated to scandium, as shown in Scheme 3-2.

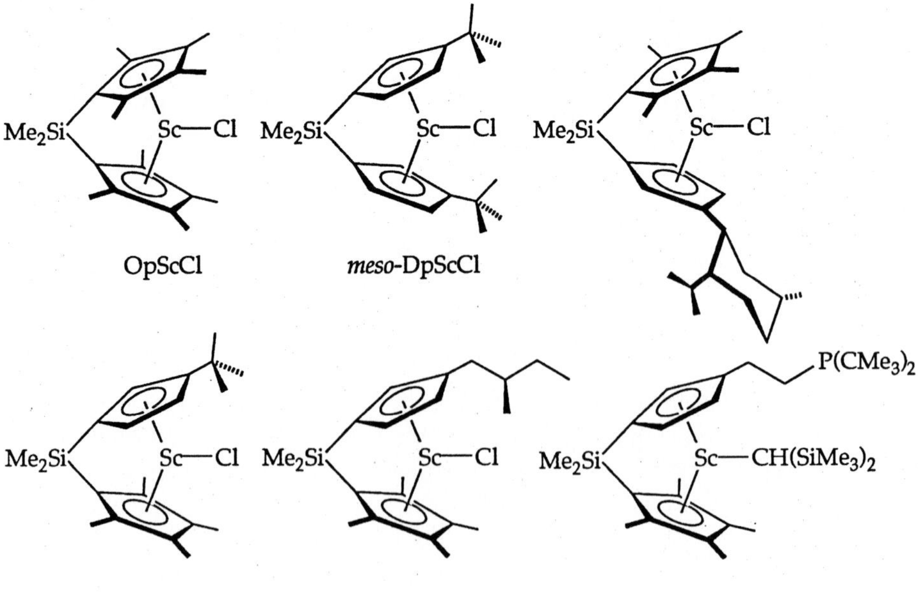

Scheme 3-2

OpScCl and *meso*-DpScCl are prepared via reactions of equimolar mixtures of the dilithium salt of the respective linked bis(cyclopentadienides) and $ScCl_3 \cdot 3THF$ in refluxing toluene. For the OpScCl system, one equivalent of coordinated lithium chloride is inseparable from the scandocene under normal workup. However, *meso*-DpScCl is an oligomeric (probably dimeric) material that is free of lithium chloride and Lewis bases. Reaction of bulky alkylating agents such as $LiCH(SiMe_3)_2$ with the scandocene chlorides results in the formation of the corresponding bis(trimethylsilyl)methyl complex. Treatment with dihydrogen results in the formation of the scandium hydride species. 'OpScH' is most conveniently isolated as a PMe_3 adduct as identified by X-ray diffraction [18], whereas *meso*-DpSc-H forms a stable dimeric species.

Linked cyclopentadienyl–amide ligands have also been developed for use as ancillary ligands for scandium. These ligands were devised as a result of the striking effect on the observed reactivity of the scandium center as the steric properties on the ligand framework are varied. These Cp-amido systems render the metal more electron-deficient and less sterically encumbered than their bis(cyclopentadienyl) counterparts. Reaction of $ScCl_3(THF)_3$ with $Li_2\{(C_5Me_4)Me_2Si(NCMe_3)\}$ results in (Cp*SiNR)ScCl (Cp*SiNR = $\{(\eta^5\text{-}C_5Me_4)SiMe_2(\eta^1\text{-}NCMe_3)\}$, as an adduct with LiCl and THF. Removal of THF generates a scandium derivative that can be alkylated with $LiCH_2(SiMe_3)_2$ to form the alkyl derivative [19]. Hydrogenation in the presence of PMe_3 forms the double hydrogen bridged dimer of C_2 symmetry (Scheme 3-3).

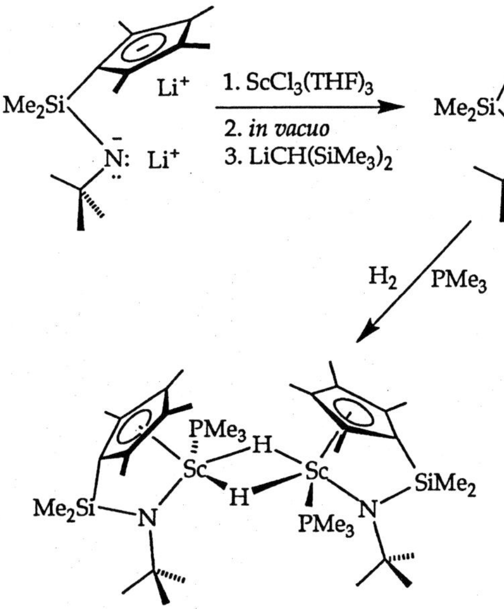

Scheme 3-3

Piers *et. al.* [20] have developed a more direct route to scandium alkyls that does not require ligand deprotonation or preparation of the requisite scandium chloride species. *In situ* generation of $Sc(CH_2SiMe_3)_3 \cdot 2THF$ from $ScCl_3 \cdot (THF)_3$ followed by treatment with $CpH^{NMe}SiN(H)R$ results in the formation of $[(Cp^{NMe}SiNR)Sc(CH_2SiMe_3)]$ in 55% yield.

$$ScCl_3 \cdot (THF)_3 \xrightarrow[\text{2. } CpH^{NMe}SiN(H)(R)]{\text{1. } 3\, Me_3SiCH_2Li,\, 0\,°C} \qquad (3\text{-}4)$$

Hydrogenolysis of the scandium alkyl results in the formation of two doubly-hydride bridged species. One of the products was characterized by X-ray crystallography and identified as a C_i symmetric diastereomeric scandium hydride dimer. Of the three other diastereomeric products that are possible, the identity of the other product was tentatively assigned as the '1R-*cis*-1'-R' isomer based on steric arguments and literature precedent.

3.2.2 Yttrium and Lanthanum

Cp_3Y, the first cyclopentadienyl derivative of yttrium, was first prepared by Birmingham and Wilkinson from the reaction of YCl_3 and NaCp [21]. The crystal structure of the THF adduct, $Cp_3Y(THF)$ has since been reported [22]. The lanthanum complex can be synthesized in a similar manner and has been identified as a 'zigzag' polymeric material in which each metal ion is coordinated in an η^5 fashion to three Cp ligands and η^2-coordinated to a fourth Cp ring [23]. Substitution of one methyl group on the cyclopentadienyl ligand in $La(C_5H_4Me)_3$ reduces the polymeric nature to a tetramer, in which each metal is η^5-coordinated to three Cp anions and η^1-coordinated to a fourth bridging cyclopentadienyl [24].

Since the sterically encumbered *tris*(cyclopentadienyl) complexes do not offer much opportunity for catalytic activity, more attention has been devoted to the preparation of bis(cyclopentadienyl) complexes of yttrium. Alkyl and hydride derivatives have received the most attention, due to their function as homogeneous catalysts. Reaction of $[Cp_2Y(\mu_2\text{-}Cl)]_2$ with $LiCH_2SiMe_3$ in dimethoxyethane (DME)/dioxane mixture forms $[Cp_2Y(CH_2SiMe_3)_2]_2Li_2(DME)_2(dioxane)$ [25]. This dimeric alkyl, characterized by X-ray crystallography, reacts with dihydrogen slowly to produce the trimeric yttrium hydride, $[Cp_2Y(\mu_2\text{-}H)]_3(\mu_3\text{-}H)][Li(DME)_2]$. It was discovered early on that the rate of the hydrogenolysis of an yttrium–carbon bond is highly dependent on the specific alkyl complex involved in the reaction [26]. The cyclopentadienyl complexes $(RCp)_2Y(tert\text{-}C_4H_9)(THF)$ (R = H, CH_3), synthesized from the reaction of $[(RCp)_2Y(\mu_2\text{-}Cl)]_2$ with $tert\text{-}C_4H_9Li$ in THF [27], react rapidly with H_2 at room temperature and atmospheric pressure, while the dimeric methyl species take weeks for the reaction to reach completion [25]. In both cases the crystallographically characterized yttrocene hydride, $[(RCp)_2Y(THF)(\mu_2\text{-}H)]_2$, is generated.

Reaction of $[Cp_2Y(\mu_2\text{-}H)(THF)]_2$ with LiH, CH_3Li, and *tert*-C_4H_9Li has been investigated and a variety of products identified [28]. For example, reaction of $[Cp_2Y(\mu_2\text{-}H)(THF)]_2$ with *tert*-C_4H_9Li results in the trimeric hydride, $[Cp_2Y(\mu_2\text{-}H)]_3(\mu_3\text{-}H)][Li(THF)_4]$ in 75% yield, with $Cp_2Y(tert\text{-}C_4H_9)(THF)$ and $Cp_3Y(THF)$ identified as byproducts. The trimer has been characterized by ^1H NMR spectroscopy, with diagnostic resonances being observed for the yttrium-coupled hydride signals. The mechanism for trimer formation is believed to be *in situ* generation of $[Cp_2YH_2]^-$ which then reacts with another equivalent of the original hydride dimer to form the trimeric hydride. In support of this proposal, reaction of $[(MeCp)_2ZrH_2]_x$ with $[(MeCp)_2Y(\mu_2\text{-}H)(THF)]_2$ forms the bimetallic trimer. Reaction with CH_3Li does not produce the trimeric hydride, but rather results in an oil identified as $Cp_2Y(CH_3)_2Li(THF)$ as well as other unidentified oligomeric products.

Unlike for the organoscandium analogs, use of the sterically demanding pentamethylcyclopentadienyl ligand generally does not produce monomeric yttrocenes free of coordinated salts or Lewis bases. Reaction of YCl_3 with 2 equivalents of KCp^* in THF results in the formation of $Cp^*_2YCl_2K(THF)_2$ [29]. Heating of this compound at 285°C under vacuum allows for isolation of the asymmetric, solvent-free dimer, $Cp^*_2Y(\mu_2\text{-}Cl)Y(Cl)Cp^*_2$ (Eq. (3-5)).

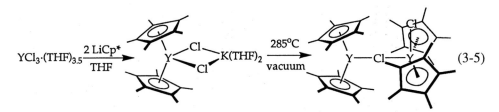

A byproduct during the synthesis of $Cp^*_2YCl_2K(THF)_2$ is $Cp^*_2YCl(THF)$ [30], which has been identified by X-ray crystallography. Reaction of $Cp^*_2YCl(THF)$ with $LiOCMe_3$ in refluxing toluene yields a mixture of products. Both reaction products, $[Cp^*_2Y(\mu_2\text{-}Cl)_2Li(THF)_2]$ and $[Cp^*_2YCl(\mu_2\text{-}Cl)Li(THF)_3]$, cocrystallize from the reaction mixture and are found in the same unit cell [31]. Teuben and co-workers [32] have prepared monomeric, base-free derivatives of permethylyttrocene by using bulky alkyl and amide groups. Reaction of $Cp^*_2YCl(THF)$ with either $NaN(SiMe_3)_2$ or $LiCH(SiMe_3)_2$ results in the crystallographically characterized $Cp^*_2YN(SiMe_3)_2$ and $Cp^*_2YCH(SiMe_3)_2$ compounds. The monomeric alkyl, being an electron-deficient, 14-electron species displays a δ agostic C–H interaction, whereas the isoelectronic amide contains a short Y–N bond, suggesting double bond character [32].

Attempts to prepare monomeric yttrium hydrocarbyls of smaller alkyl groups is plagued by preferential yttrate formation with salts such as LiCl and MgX_2 and co-ordination of ethereal solvents (Et_2O, THF) [33]. For example, reaction of $Cp^*_2Y(\mu_2\text{-Cl})_2Li\cdot2OEt_2$ with two equivalents of methyllithium at $-80°C$ results in the bridged dimethyl yttrate, $Cp^*_2Y(\mu_2\text{-CH}_3)_2Li\cdot OEt_2$, while the same reaction in THF solution produces $Cp^*_2Y(CH_3)(THF)$, whose structure has been confirmed by NMR spectroscopy, elemental analysis and X-ray diffraction (Scheme 3-4) [34]. Coupling of the methyl group to a single ^{89}Y (spin 1/2, 100% abundant) nucleus is observed in both ^1H and ^{13}C NMR spectra, indicative of a monomer in solution as well.

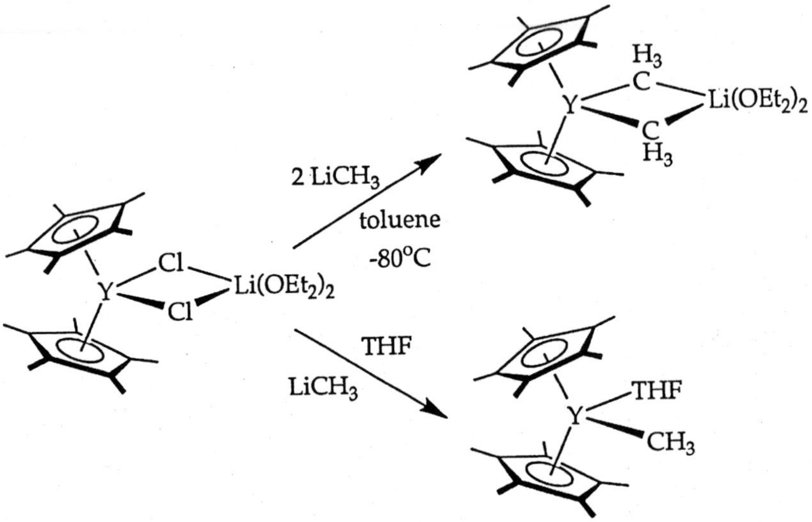

Scheme 3-4

Attempts to prepare salt- and solvent-free allyls using the chelating nature of this ligand led instead to $Cp^*_2Y(\eta^1\text{-CH}_2CMe=CH_2)\cdot[MgCl_2\cdot2THF]$, from reaction of $Cp^*_2YCl(THF)$ with 2-methyl-2-propenylmagnesium chloride. Apparently, the strong tendency for yttrates to form effectively blocks η^3 coordination of the allyl [33].

Hydrogenation of the monomeric, $Cp^*_2YCH(SiMe_3)_2$ in pentane results in the hydride dimer, $[Cp^*_2Y(\mu_2\text{-H})]_2$ [35,36]. Unlike the analogous scandium system which is unstable unless a dihydrogen atmosphere is maintained, the yttrium hy-dride can be isolated in the solid state, presumably due to its dimeric nature. Ther-molysis of the dimer results in the formation of the red fulvene-bridged dimer, identified as $Cp^*_2Y(\eta^1,\eta^5\text{-CH}_2C_5Me_4)(\mu_2\text{-H})YCp^*$ on the basis of NMR evidence [37].

In view of the common use of alkyl aluminum co-catalysts in Ziegler–Natta α-olefin polymerization systems, a variety of tetraalkylaluminates of yttrocene have been prepared. Reaction of $[Cp_2YCl]_2$ and $Li[AlR_4]$ ($R = CH_3$, CH_3CH_2) in toluene results in the formation of $Cp_2YR_2AlR_2$; the ethyl derivative has been structurally characterized by single crystal X-ray diffraction [38].

$$(3\text{-}6)$$

$$(R = CH_3, CH_2CH_3)$$

NMR data indicate the complexes are fluxional, but the pseudorotation can be frozen out at $-40°C$. Similarly, the alane adducts, $[Cp_2YCl]_2 \cdot AlH_3 \cdot OEt_2$ [39] and $[Cp_2YCl]_2 \cdot AlH_3 \cdot NEt_3$ [40] have also been prepared from $[Cp_2YCl]_2$ and $MAlH_4$ (M = Li or Na) in the presence of diethyl ether or triethylamine [41]. The structures of these compounds have been confirmed by X-ray crystallography [41,42,43].

Symmetrical, linearly-bridged dimers of $[Cp*_2Y(\mu_2\text{-}CH_3)_2Al(CH_3)_2]_2$ and $[Cp*_2Y(\mu_2\text{-}CH_3)_2Ga(CH_3)_2]_2$ have been reported by Watson [44]. These products are formed from the reaction of solvent-free $Cp*_2Y(CH_3)_2Li$ with $M'(CH_3)_3$ (M' = Al, Ga). NMR spectra indicate two species in solution, which were shown to be in equilibrium (Eq. (3-7)).

$$(3\text{-}7)$$

Under normal concentrations and at room temperature the monomer is the dominant species in solution. Mixing $Cp*_2Y(\mu_2\text{-}CH_3)_2Al(CH_3)_2$ and $Cp*_2Y(\mu_2\text{-}CH_3)_2Ga(CH_3)_2$ results in the formation of a random distribution of symmetric and asymmetric dimers. The correlation of relative bond strengths, analogous to those for a bimolecular transition state, have been established through kinetic studies which show that $Y-(\mu_2\text{-}CH_3)-Al$ bonds are stronger than $Y-(\mu_2\text{-}CH_3)-Ga$. The data do not, however, correlate well with the release of steric strain in the four membered ring as the driving force for the dimerization. The free energies of dimerization are approximately zero with the enthalpy of the reaction favoring the formation of dimers with linear methyl-bridge bonds over monomers with bent methyl-bridge bonds. From these data, it can be concluded that the methyl group is a ligand with a high propensity for hypercoordination with the electropositive metal center.

Ethyltetramethylcyclopentadienyl (η^5-C_5Me_4Et) derivatives of yttrium have also been prepared and their reactivity explored. Reaction of yttrium trichloride and two equivalents of Na[C_5Me_4Et] in THF forms [(η^5-C_5Me_4Et)$_2$YCl(THF)] [45]. Conversion to monomeric amide or alkyl derivatives is accomplished by addition of NaN(SiMe$_3$)$_2$ or LiCH(SiMe$_3$)$_2$ in toluene at 0°C. It is apparent from the crystal structure of the amide derivative that the amide ligand is moved from the center of the metallocene wedge by a (μ-CH_3) bridging yttrium and silicon.

$$YCl_3 \cdot (THF)_{3.5} \xrightarrow{2\ Li(C_5Me_4Et)} \quad Y\!-\!Cl \xrightarrow{LiCH(SiMe_3)_2} \quad Y\!-\!CH(SiMe_3)_2 \quad (3\text{-}8)$$

Other alkyl substituted cyclopentadienyl ligands have been synthesized and coordinated to yttrium using analogous procedures. Reaction of YCl$_3$ with 2 equivalents of K(1,3-$C_5H_3Me_2$) in THF results in the formation of (η^5-1,3-$C_5H_3Me_2$)$_2$Y(THF)Cl, which can be converted to [(η^5-1,3-$C_5H_3Me_2$)$_2$YCl]$_2$ by stirring in toluene [46]. The methyl complex, prepared by the reaction of (η^5-1,3-$C_5H_3Me_2$)$_2$Y(THF)Cl and 2 equivalents of methyllithium, is a dimer, [(η^5-1,3-$C_5H_3Me_2$)$_2$Y(μ_2-CH_3)]$_2$. Hydrogenolysis in THF/hexane results in an yttrium hydride species, whose structure is dependent on how the product is isolated. Removal of solvent results in the formation of the trimeric, [(η^5-1,3-$C_5H_3Me_2$)$_2$Y(μ_2-H)]$_3$, but recrystallization of this compound from THF yields the dimeric species [(η^5-1,3-$C_5H_3Me_2$)$_2$Y(THF)(μ_2-H)]$_2$. It was concluded that dimethyl substitution has only a small effect on the yttrium chemistry compared to (η^5-C_5H_5) or (η^5-C_5H_4Me) derivatives [46].

Since synthesis of group 3 metallocenes is often plagued by facile coordination of unwanted solvent, halide or alkyl groups, incorporation of Lewis basic sites into the cyclopentadienyl ligand framework has been pursued in an attempt to prepare 'base-free' metallocenes. Treatment of YCl$_3$(THF)$_{3.5}$ with two equivalents of $C_5H_4CH_2CH_2OMe$·Li(TMEDA) in toluene at −80°C results in the formation of [(η^5-$C_5H_4CH_2CH_2OMe$)$_2$Y(μ_2-Cl)]$_2$ [47]. Conversion to the aluminohydride or borohydride adduct, [(η^5,η^1-$C_5H_4CH_2CH_2OMe$)$_2$Y]EH$_4$ (E = Al, B) can be accomplished through addition of either LiAlH$_4$ or NaBH$_4$ to the chloride dimer. The crystal structure of the borohydride complex indicates that both ether functions of the cyclopentadienyl ligand are coordinated to the yttrium center. Treatment of the aluminohydride complex with excess NEt$_3$ results in the hydride dimer, [(η^5,η^1-$C_5H_4CH_2CH_2OMe$)$_2$Y(μ_2-H)]$_2$. This conversion can be carried out *in situ* from the chloride derivative, resulting in the formation of the hydride dimer in moderate yield [47]. Hydride dimers have also been prepared by reactions of [(η^5,η^1-

$C_5H_4CH_2CH_2OMe)_2M(\mu_2\text{-}Cl)]_2$ (M = Y, La) with NaH in THF [48]. *Tris*(2-meth-oxyethylcyclopentadienyl) sodium reacts with MCl_3 (Ln = La, Y) to yield $[M(\eta^5,\eta^1\text{-}C_5H_4CH_2CH_2OMe)_3]$ [49]. X-ray crystallography reveals that the complexes have two of the oxygens of the cyclopentadienyl rings coordinated, while the third oxygen is not coordinated.

Diphenylphosphino-substituted cyclopentadienyl yttrium complexes have also been prepared [50]. Reaction of lithium diphenylphosphinocyclopentadienide with $YCl_3\cdot(THF)_{3.5}$ results in the formation of three species: $(\eta^5\text{-}Ph_2PC_5H_4)_2Y(\mu_2\text{-}Cl)_2Li(THF)_2\cdot0.5LiCl$, $\{(\eta^5\text{-}Ph_2PC_5H_4)_2Y(\mu_2\text{-}Cl)\}_2$ and $[((\eta^5\text{-}Ph_2PC_5H_4)_2YCl(THF)]$. Reaction of the chloro-bridged dimer with $[Mo(CO)_6]$ in refluxing toluene illustrates the ability of the phosphines to ligate to another metal center, forming $[\{CO)_4Mo\{(Ph_2PC_5H_4)_2Y(\mu_2\text{-}Cl)\}\}_2]$. Incorporation of a phosphorus into the cyclopentadienyl ring has also been accomplished. The yttrocene $(\eta^5\text{-}C_4Me_4P)_2YCl(THF)$ has been prepared from the reaction of $YCl_3\cdot(THF)_{3.5}$ and $Li(PC_4Me_4)$ (Eq. (3-9)) [51].

$$YCl_3\cdot(THF)_{3.5} \xrightarrow{} \qquad\qquad (3\text{-}9)$$

Another strategy for incorporation of heteroatoms into the cyclopentadienyl ligand array has been through the use of a heteroatom-containing cyclopentadienyl linking group. Ether-oxygen bridged cyclopentadienyls such as 1,1'-(3-oxa-pentamethylene)dicyclopentadienyl have been reported. Reaction of the disodium salt with either YCl_3 or $LaCl_3$ produces the unsolvated monomeric chloride containing an intramolecular metal-oxygen dative bond [52]. These compounds in the presence of NaH reduce alkenes such as 1-hexene to the corresponding alkane. Shumann et al. [53] reported that the resulting chloride complex can be converted to the dimethylpyrazolyl derivative by reaction of (3,5-dimethylpyrazolyl) sodium in THF (Scheme 3-5).

Replacement of the oxygen with a carbon linker in 1,1'-pentamethylenedicyclopentadiene forms only THF solvated complexes when coordinated to yttrium [54]. The pyrazol complex reacts with water to produce the μ-hydroxide dimer, whose structure has been characterized crystallographically. A shorter carbon linker of 3 atoms has also been employed in the reaction of 1,1'-trimethylenedicyclopentadienyl sodium with $LaCl_3$ or YCl_3 forming the bridged, THF-adduct, $[(\eta^5\text{-}C_5H_4)_2(CH_2)_3]MCl\cdot THF$ [55]. Addition of aryllithium (Aryl = C_6H_5; p-MeC_6H_4) or alkyllithium (alkyl = CMe_3, $(CH_3)_3CCH_2$) reagents to the chloride–THF adduct produce THF-solvated alkyl or aryl complexes.

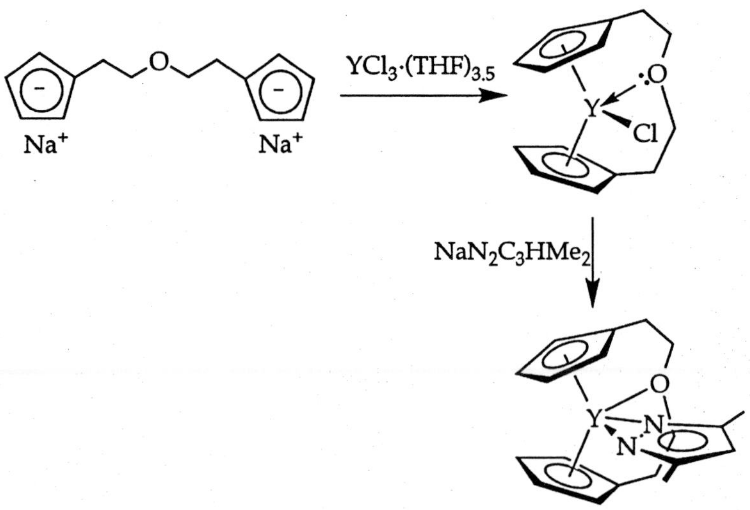

Scheme 3.5

Furan-linked cyclopentadienyls have also been prepared and coordinated to yttrium [48]. In these chloro-bridged dimeric compounds, the furan rings coordinate to the yttrium centers.

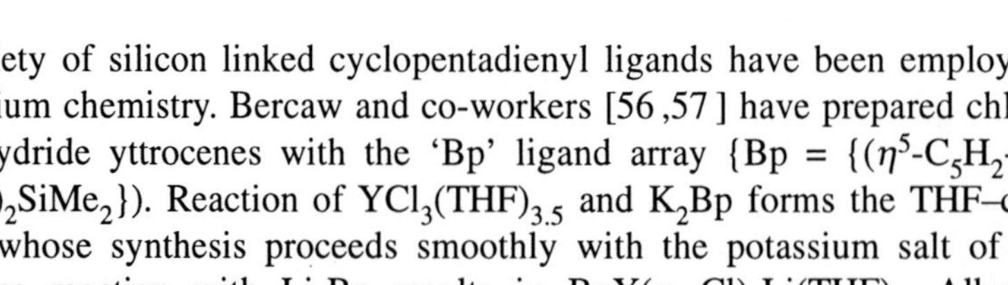

$$(3\text{-}10)$$

A variety of silicon linked cyclopentadienyl ligands have been employed in orga-noyttrium chemistry. Bercaw and co-workers [56,57] have prepared chloride, alkyl and hydride yttrocenes with the 'Bp' ligand array {Bp = {(η^5-C$_5$H$_2$-2-SiMe$_3$-4-CMe$_3$)$_2$SiMe$_2$}). Reaction of YCl$_3$(THF)$_{3.5}$ and K$_2$Bp forms the THF–chloride ad-duct, whose synthesis proceeds smoothly with the potassium salt of the ligand, whereas reaction with Li$_2$Bp results in BpY(μ_2-Cl)$_2$Li(THF)$_2$. Alkylation with LiCH(SiMe$_3$)$_2$ yields BpY{CH(SiMe$_3$)$_2$}, which can be hydrogenated to form the hydride dimer (Scheme 3-6).

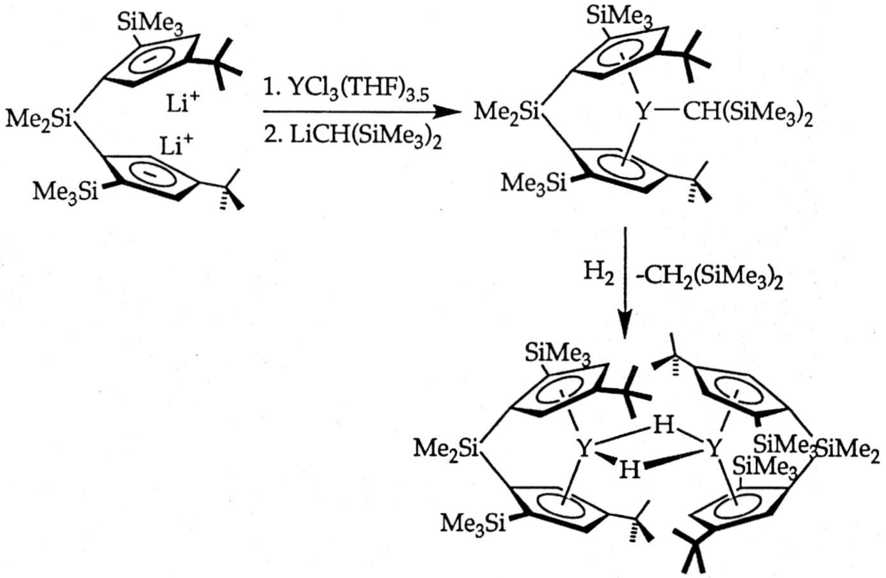

Scheme 3-6)

Piers [58] has prepared silicon linked cyclopentadienyl-amide yttrocenes via amine elimination from $Y[N(SiMe)_2]_3$, to allow for direct isolation of the yttrium amide such that metallation to the yttrium chloride species can be achieved without ligand deprotonation. Reaction of $Y[N(SiMe_3)_2]_3$ and $Cp^*HSiNHCMe_3$ occurs at 110°C in toluene solution, producing $(Cp^*SiNR)YN(SiMe_3)_2$ in 40% yield. Unfortunately, conversion of the yttrium-amide to the yttrium-alkyl using standard alkylating agents such as $LiCH_2SiMe_3$ has not been successful, presumably due to the thermal instability of the resulting alkyl complex.

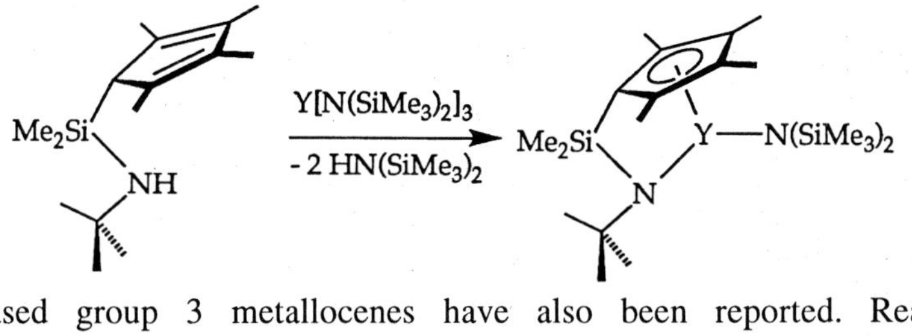

(3-11)

Indenyl-based group 3 metallocenes have also been reported. Reaction of $YCl_3(THF)_{3.5}$ with two equivalents of Na^+Ind^- (Ind = indenyl) results in a 2:1 mixture of $(\eta^5-C_9H_7)YCl_2(THF)_n$ and $(\eta^5-C_9H_7)_2YCl(THF)_m$ [59]. These compounds have proven difficult to isolate, requiring several recrystallizations to obtain pure material. On the other hand, $\{\eta^5-C_9H_4-2,4,7-(CH_3)_3\}_2MX_2Li(THF)_2$ may be prepared by treatment of 2,4,7-trimethylindenide with $MX_3(THF)_n$ (M = Y, X = Cl; M = La, X = Br). These compounds can be cleanly converted to alkyls and amides by reaction with $LiCH(SiMe_3)_2$ or $LiN(SiMe_3)_2$, respectively.

3.3 Group 3 Metallocenes for the Activation of Hydrocarbons

The activation of the C–H bonds of saturated and unsaturated hydrocarbons has been actively pursued in organometallic chemistry [60]. By virtue of their strong Lewis acidity and high electrophilicity, 14-electron, group 3 metallocene alkyl and hydride complexes are attractive candidates for activating these types of generally unreactive bonds. A variety of scandocenes and yttrocenes have been employed for hydrocarbon activation and considerable mechanistic information has been gained from the studies of their reactivity.

Watson [61] has reported the reactivity of $Cp^*_2YCH_3$ towards the unactivated sp^3 bonds of methane, representing the first well-characterized example of methane activation by a homogeneous organometallic complex. Exchange between $^{13}CH_4$ and the yttrium methyl group is observed upon heating the reaction to 70°C in cyclohexane-d_{12}.

$$Y{-\!}CH_3 \ + \ ^{13}CH_4 \ \rightleftharpoons \ Y{-\!}^{13}CH_3 \ + \ CH_4 \qquad (3\text{-}12)$$

Other hydrocarbons such as ethane and propane react, but the corresponding organometallic products decompose via β-hydrogen elimination.

Similarly, the monomeric permethylscandocene Cp^*_2ScR (R = hydride, alkyl, alkenyl, alkynyl, aryl) complexes have been studied for their reactivity towards the C–H bonds of saturated and unsaturated hydrocarbons [62]. The hydride derivative, Cp^*_2ScH, catalyzes H/D exchange reactions for a variety of C–H bonds. For reaction of the aromatic C–H bonds of benzene with Cp^*_2Sc-H, an upper limit of 100 seconds has been established as the half life of the reaction at 25°C. For toluene, the rate of exchange for aromatic positions is 60–70 times faster than exchange into the methyl group, and the ratio for aryl exchange remains approximately constant at 39:41:20 for *ortho:meta:para* hydrogens. Deuterium exchange into the C–H bonds of the Cp* ligand is also observed by heating Cp^*_2Sc-D in benzene-d_6 under a D_2 atmosphere. Methane and tetramethylsilane also undergo H/D exchange with the kinetic products having one deuterium incorporated, *e.g.* CH_3D. Hydrogen/deuterium exchange is also observed for 2° C–H bonds (*e. g.* for cyclopentane), but the rate is much slower. The overall reactivity of R–H bonds with Cp^*_2Sc–R' bonds is as follows: R = R' = H >> R = H, R' = alkyl >> R–H = sp C–H R' = alkyl > R–H = sp² C–H, R' = alkyl > R–H = sp C–H, R' = alkyl > R–H = sp² C–H R' = alkyl > R–H = sp³ C–H, R' = alkyl.

From these studies, the mechanistic features of scandocene reactivity toward the C–H bonds of hydrocarbons has been established. A concerted rather than a step-wise process is indicated. The rate of the reaction increases with the amount of s character in the reacting C–H σ bonds of the substrate. A four-center transition state is consistent with these findings.

$$Cp^*_2Sc\text{-}R + R'\text{-}H \;\rightleftharpoons\; \left[Cp^*_2Sc \begin{smallmatrix} R' \\ H \\ R \end{smallmatrix} \right]^{\ddagger} \;\rightleftharpoons\; Cp^*_2Sc\text{-}R' + R\text{-}H \qquad (3\text{-}13)$$

A nondirectional "s" orbital provides better overlap and hence more bonding and stability in the transition state [62]. This picture has been supported theoretically by Goddard and Steigerwald [63], who concluded that the d orbitals of scandium are ideally suit-ed to provide good overlap with all three hydrogen atoms in the transition state. The geometry of the transition state was found to be 'kite shaped' rather than square, fa-voring positioning of the H atom in the β (central) position relative to scandium. Therefore, the mechanistic picture for these concerted σ-bond metathesis reactions is one where the incoming H–H or C–H bond approaches the vacant $1a_1$ orbital of the scandocene alkyl, forms the aforementioned transition state, from which the newly formed H–H or C–H bond departs from the opposite side of the $1a_1$ orbital [62].

The relative bond dissociation energies (BDE) have been obtained for Sc–aryl and Sc–alkyl by the equilibration of the metallated complex, $Cp^*(\eta^5,\eta^1\text{-}C_5Me_4CH_2CH_2CH_3)Sc\text{-}C_6H_5$ with $Cp^*(\eta^5,\eta^1\text{-}C_5Me_4CH_2CH_2CH_2)Sc$ and C_6H_6 [64]. The $Sc\text{-}C_6H_5$ BDE was estimated as 16.6(3) kcal·mol^{-1} stronger than the Sc–$CH_2CH_2CH_2C_5Me_4$ bond strength. In combination with other experiments [64], a lower limit was placed on scandium alkynyl bonds, such that BDE(Sc–alkynyl) – BDE(Sc–aryl) ≥ 29(5) kcal·mol^{-1}. It was found that the early transition metal relative M–R BDEs correspond to the analogous relative R–H BDEs (M–alkynyl > M–aryl > M–alkyl), although the M–R BDEs increase more rapidly with increasing s character than their R–H counterparts. The polarity of Sc–aryl bonds has been probed through a study of the metallation reactions of two different substituted benzyl groups. A small dependence of K_{eq} on the nature of X or Y suggests that the Sc–aryl bond is essentially covalent with a small ionic contribution (Eq. (3-14)) [64].

$$(3\text{-}14)$$

$(X, Y = H, CH_3, NMe_2, CF_3)$

Teuben has found that $Cp*_2Y(H)(THF)$, like its scandium counterparts, catalyzes H/D exchange between sp^3 C–H bonds and sp^2 C–D bonds [34]. The half life for H/D scrambling between benzene-d_6 and $CH_2(SiMe_3)_2$ is about 27 h. It was also found that the methyl group in toluene is deuterated in an exchange reaction with benzene-d_6. However, in contrast to the permethylscandocene system, scrambling of the deuterium into the Cp* ligands is not observed under these particular conditions (*vide infra*). The yttrium alkyl complex $Cp*_2YCH(SiMe_3)_2$ reacts with mesitylene at elevated temperature to yield monomeric $Cp*_2Y-CH_2(3,5-(CH_3)_2C_6H_3)$ (Eq. (3-15)).

$$\text{(3-15)}$$

With toluene, a mixture of benzyl, *ortho*, *meta* and *para*-tolylyttrocenes has been identified, indicating that C–H activation may involve both sp^2- and sp^3-hybridized C–H bonds. No reaction with methane was observed up to 100°C.

The dimeric permethylyttrocene hydride, $[Cp*_2Y(\mu_2-H)]_2$ undergoes thermolysis in the presence of *n*-octane, cyclohexane or benzene to preferentially activate its pentamethylcyclopentadienyl C–H bonds, yielding the fulvene complex, $Cp*_2Y(\mu_2-H)(\mu_2-\eta^1,\eta^5-CH_2C_5Me_4)YCp*$ [65]. However, in deuterated aromatic solvent, H/D scrambling between solvent and yttrocene ligand C–H positions is observed, similar to the permethylscandocene system. Rapid removal of H_2 gas allows for isolation of $Cp*_2Y-C_6H_5$ and $Cp*_2Y-CH_2C_6H_5$, which are believed to be formed by σ-bond metathesis. For substituted aromatics, C_6H_5X ($X = NMe_2$, OCH_3, SCH_3, CH_2NMe_2, PMe_2, F, Cl, Br), *ortho*-metallation predominates [65].

3.4 Group 3 Metallocenes as Homogeneous Olefin Polymerization Catalysts

It is now generally accepted that the active catalysts in group 4 metallocene-based Ziegler–Natta olefin polymerization systems are cationic, d^0 alkyls, either as free 14-electron cations, weakly solvated or weakly ion paired with a large counter-anion [66]. The neutral, 14 electron, group 3 metallocene alkyl and hydride derivatives offer an opportunity to investigate the mechanisms of chain initiation, propagation and transfer without the complication of ill-defined cocatalysts such as methylalumoxane (MAO) and the possible complications of ion pairing effects. As a result, detailed studies into the rates, enantioselectivities and α agostic assistance for olefin insertion have been performed with group 3 metallocenes, lending considerable insight as to the essential features of a useful Ziegler–Natta olefin polymerization catalyst.

Yttrocene complexes were identified as single component homogeneous olefin polymerization catalysts as early as the mid-1970s. The methyl-bridged dimeric yttrocene complexes, $[(\eta^5\text{-}C_5H_4R)_2Y(\mu_2\text{-}CH_3)]_2$ (R = H, CH$_3$, SiMe$_3$), are active catalysts for ethylene polymerization [67]. Although the dimer is the only organometallic species observed during polymerization, monomeric $(\eta^5\text{-}C_5H_4R)_2YCH_3$ is believed to be the active species. Rapid scrambling of bridging alkyl groups between yttrium centers is observed when $[(\eta^5\text{-}C_5H_4R)_2Y(\mu_2\text{-}CH_3)]_2$ and $[(\eta^5\text{-}C_5H_4R)_2Y(\mu_2\text{-}CH_2CH_2CH_2CH_3)]_2$ are mixed, in accord with a monomer/dimer equilibrium.

Evans and coworkers have reported the reaction of α-olefins (ethylene, propylene) with $[(\eta^5\text{-}C_5H_4R)_2YH(THF)]_2$ (R = H, CH$_3$), producing the corresponding alkyl complexes [68]. In the presence of excess ethylene, formation of the dimeric, $\mu\text{-}n$-butyl complex is observed, arising from olefin insertion into the yttrium–alkyl bond. Propylene undergoes 1,2-addition of the Y–H moiety to yield the less hindered primary alkyl product. Linked, mixed-ring yttrocene dimeric hydrides, $[\{(\eta^5\text{-}C_5Me_4)(\eta^5\text{-}C_5H_4)SiMe_2\}Y(\mu_2\text{-}H)]_2$ have been prepared where the Cp ligand array spans two metal centers [69]. Although ethylene is polymerized, reaction with α-olefins slowly yields the μ-hydrido, μ-alkyl complex, but no polymer or further insertion products are observed.

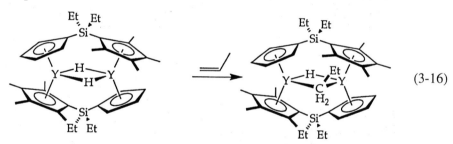

$$(3\text{-}16)$$

Permethylscandocene complexes Cp*$_2$ScR (R = H, alkyl) insert ethylene rapidly at low temperature (–80°C) without competition from chain termination via β-hydrogen elimination; hence, living ethylene polymerization may be achieved [70]. The kinetics of ethylene insertion have been measured. The relative rates of ethylene insertion are reflected in part by the strength of the Sc–C bond, where stronger bonds result in slower rates. The order of decreasing rate of olefin insertion is Sc–H >> Sc–CH$_2$CH$_2$CH$_3$ ≥ Sc–CH$_2$(CH$_2$)$_n$CH$_3$ (n ≥ 2) > Sc–CH$_3$ > Sc–CH$_2$CH$_3$ > Sc–C$_6$H$_5$ [70]. The increased rate of insertion for the hydride derivative is likely due to greater orbital overlap (nondirectional 1s valence orbital) in the transition state, while the greater rate for the propyl species relative to the methyl is presumably due to a weaker Sc–CH$_2$CH$_2$CH$_3$ relative to Sc–CH$_3$ [70,71]. The anomalously sluggish rate of ethylene insertion into the scandium–ethyl bond is a result of β-agostic interaction, which further stabilizes the ground state. Such an interaction is not observed for the scandium–propyl complex, presumably since such an interaction is sterically unfavorable, since the methyl group of the β-carbon would be directed toward one of the (η^5-C$_5$Me$_5$) rings (Eq. 3-17).

$$(3\text{-}17)$$

The transition states for both β-hydrogen elimination, and hence also for ethylene insertion have been probed by systematically varying the substituent at the β-carbon. For the scandium phenylethyl series a linear free energy correlation is found [70], indicating that the transition state for these processes is quite polar with the electropositive scandium center abstracting hydride (and by extension alkyl (R$^{\delta -}$) for insertion/β-alkyl elimination) of the positively charged β–carbon atom.

The linked cyclopentadienyl complexes [DpScH]$_2$ and OpSc(H)(PMe$_3$) rapidly polymerize ethylene, whereas reaction with higher α-olefins (propylene, 1-butene) yield head-to-tail dimers [72,73]. Addition of either *cis*- or *trans*-2-butene to either (DpScH)$_2$ or OpSc(H)(PMe$_3$) results in isomerization about the carbon–carbon bond as well as the formation of head-to-tail dimer [74]. Moreover, α,ω-dienes are cleanly cyclized to the corresponding methylenecycloalkane [72]. Five, six, seven, eight or nine-membered rings can be synthesized from the appropriate α,ω-diene in 85–99% yields. For the smaller methylenecycloalkanes, methylenecyclobutane and methylenecyclopropane, the reverse reactions, exothermic ring openings to 1,4-pentadiene and butadiene, respectively, are catalyzed (Scheme 3-7).

The structure of the transition state for chain propagation of scandocene and yttrocene catalysts has been investigated by examining the possibility of an α agostic C–H interaction which assists olefin insertion, thus supporting a 'modified Green–Rooney' mechanism [75]. [DpScH]$_2$, OpSc(H)(PMe$_3$) and [(Cp*SiNR)Sc(PMe$_3$)]$_2$(μ_2-H$_2$) are active catalysts for hydrocyclization of 1,5-hexadiene to methylcyclopentane. By employing 'deuterium isotopic perturbation of stereochemistry', originally developed by Grubbs [76], this mechanistic probe has been adapted for the scandium systems. Addition of an achiral α,ω-diene, such as *trans, trans*-1,6-d_2-1,5-hexadiene to achiral OpScH yields precisely a 50:50 mixture of R- and S-6-hexenyl-1-d_1-6-scandium complexes. Due to ring strain there should be a preference for the *cis* ring fusion in the transition state for olefin insertion. Face selection for the insertion of the pendant olefin then depends on whether H or D occupies the α agostic position. The expected preference for H to occupy the bridging position leads to an expected excess of the R,R (*trans*) (and S,S (*trans*)) over the R,S (*cis*) (and S,R *cis*)) products, if an α agostic interaction assists olefin insertion (see Scheme 3-9, below).

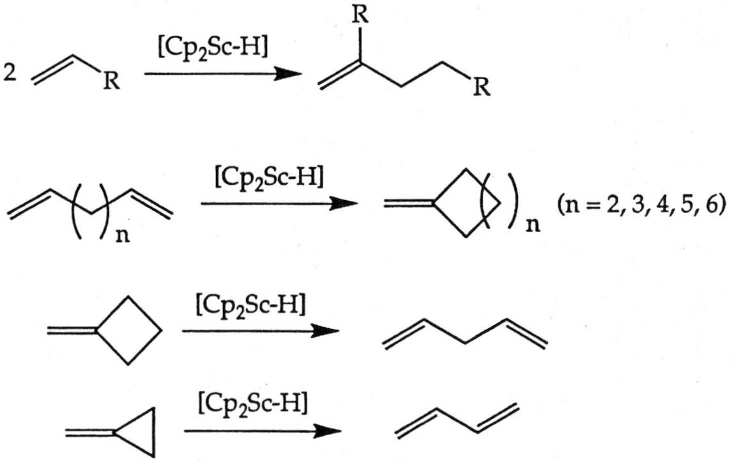

(Cp$_2$Sc-H = DpScH or OpSc(H)PMe$_3$)

Scheme 3-7

Analysis of the product mixture by ^{2}H{^{1}H} NMR yields an average *trans:cis* ratio of 1.226(12):1 at 25°C [77]. Similarly, isotopic perturbation of stereochemistry is also observed for the hydrocyclization of deuterated 1,5-hexadienes with other scandium catalysts [78]. [DpScH]$_2$ hydrocyclizes *trans, trans*-1,6-d_2-1,5-hexadiene with a *trans:cis* ratio of 1.203(7):1, while [(Cp*SiNR)Sc(PMe$_3$)(μ_2-H$_2$)]$_2$ produces d_2-methylcyclopentanes with a *trans:cis* ratio of 1.209(15):1. These results provide evidence for the 'modified Rooney–Green' pathway for chain propagation with these Ziegler–Natta systems. Moreover, these results provide an explanation for the requirement that the active bis(cyclopentadienyl)metal catalysts be 14-electron derivatives with two vacant orbitals: one to accommodate the incoming olefin, another for the α agostic interaction [66,76] (Scheme 3-8).

The mechanism of olefin insertion and chain transfer processes by both β-hydride and β-methyl elimination for Ziegler–Natta polymerizations has been examined with OpSc(H)(PMe$_3$) [79]. Reaction of isobutene with OpSc(H)(PMe$_3$) forms OpSc(CH$_3$)(PMe$_3$) along with isobutene, 2-methylpentane, isobutane, 2-methyl-1-pentene, propane, and *n*-pentane. The organic products are formed from a series of reactions involving olefin insertion, β-methyl elimination and β-hydride elimination. It was established that β-H elimination proceeds faster than β-CH$_3$ elimination, which proceeds until only 2-methyl-1-alkenes and the predominant organometallic product, OpSc(CH$_3$)(PMe$_3$) remain. Alkane products result from σ-bond metathesis reactions of the C–H bonds of PMe$_3$ and the Sc–C bonds of scandium alkyls. However, β-ethyl elimination is not observed for the related 2-ethylbutyl derivative, OpSc{CH$_2$CH(C$_2$H$_5$)CH$_2$CH$_3$}(PMe$_3$), which is obtained from reaction of 2-ethyl-1-butene with OpSc(H)(PMe$_3$).

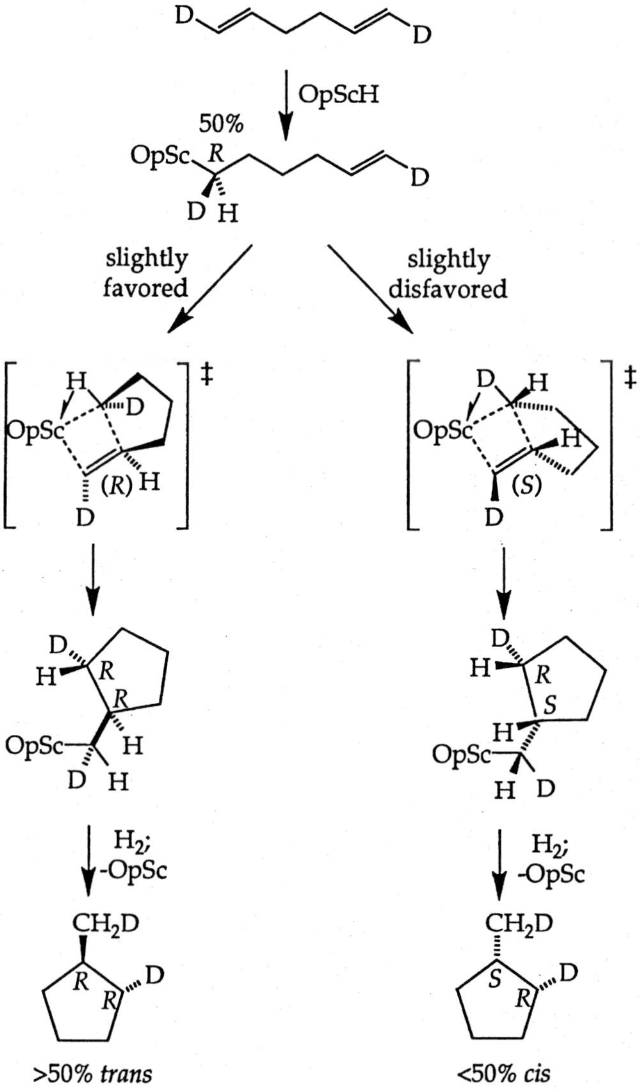

Scheme 3-8

Linked cyclopentadienyl-amido scandium systems have proven to be active catalysts for olefin polymerization. Treatment of [(Cp*SiNR)(PMe₃)Sc(μ₂-H)]₂ with two equivalents of propylene at low temperature yields the phosphine-free alkyl, [(Cp*SiNR)Sc(μ-CH₂CH₂CH₃)]₂, which is an active catalyst for the polymerization of ethylene and α-olefins [80,81]. Kinetic analysis of 1-pentene polymerization in addition to ¹³C and ³¹P NMR analysis reveals that the active catalyst for this system is the 12-electron scandium alkyl [(Cp*SiNR)ScCH₂CHRP] (**P** = growing polymer chain). Although polymerization has been achieved, the rates of polymerization and the molecular weights of the polymers are rather low: polypentene (25¡C, neat 1-pentene): M_n = 6,000; PDI = 1.5; polypropylene (25°C, 25 vol. % in toluene): M_n = 9,600; PDI = 1.8; polypropylene (–9°C, 25 vol% toluene): M_n = 16,500; PDI = 1.7. ^{13}C NMR analysis of the resulting polymers indicates a predominantly atactic structure.

Isospecific polymerization of α-olefins has been achieved with the single component, [*rac*-Me$_2$Si(2-SiMe$_3$-4-CMe$_3$C$_5$H$_2$)Y-R] ('BpY-R') catalysts [82]. The ligand array is designed such that, upon coordination, only the desired racemic metallocene is formed thereby eliminating tedious separation of the unwanted meso isomer. Inspection of the X-ray crystal structure [83] of *rac*-BpYCl$_2$Li(THF)$_2$ confirms expectations that unfavorable steric interactions between the SiMe$_3$ groups in the narrow portion of the Cp–M–Cp wedge are avoided only for the racemic isomer (Scheme 3-9).

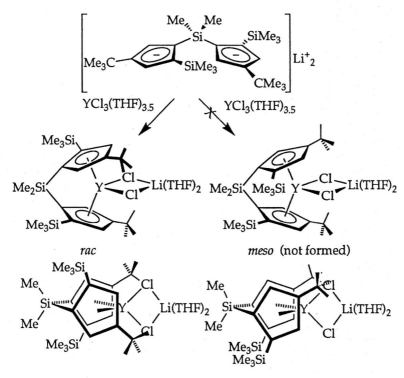

Scheme 3-9

Ethylene is rapidly polymerized by BpYCH(SiMe$_3$)$_2$ producing moderately high molecular weight polyethylene (M_n = 720,000, PDI = 2.0, T_m = 137°C). Solutions of [BpY(μ_2-H)]$_2$ polymerize α-olefins slowly to form moderately high molecular weight polymers. Polymerization is slow, presumably due to inactivity of the 16-electron hydride dimer; thus, slow dissociation into the active monomers dictates low activity. For propylene polymerization, ^{13}C NMR pentad analysis reveals that the *mmmm* resonance accounts for >99% of the methyl resonances, while chain end analysis by ^1H NMR shows <0.1% 2,1 mis-insertions and both vinyl and vinylidene end groups, indicative of both β hydrogen and β methyl elimination for chain termination. Similarly, polymers generated from 1-butene, 1-pentene and 1-hexene are also highly isotactic and contain very few regiomistakes.

Yasuda [84] has modified the Bp ligand array by replacement of the *tert*-butyl groups with $Si(CH_3)_2(CMe_3)$ groups and has prepared the corresponding $\{CH(SiMe_3)_2\}$ and hydride dimers. Polymerization of 1-pentene and 1-hexene produced polymers with $M_n = 1 \times 10^4$ and high degrees of isotacticity.

Although coordination of the Bp ligand avoids formation of the *meso* isomer, the *racemo* isomer naturally contains both enantiomers. Resolution of enantiomers is necessary to allow for isolation of an enantiopure metallocene. A chiral auxiliary attached to the silylene linking group with groups extending toward the [SiMe$_3$] groups at the 2-positions of the Cp rings forces the formation of only one enantiomer. Accordingly, a new ligand, $\{[C_5H_2-2-SiMe_3-4-CMe_3)_2Si(OC_{10}H_6C_{10}H_6O)\}^{2-}$ [(BnBp)$^{2-}$], has been prepared for this purpose and coordinated to yttrium [85]. Use of optically pure ligand yields optically pure metallocene, where the (*R*)-(+)-1,1'-bi-2-naphthol directs the formation of the (*S*)-yttrocene and (*S*)-(-)-1,1'-bi-2-naphthol directs formation of the (*R*)-yttrocene (Scheme 3-10).

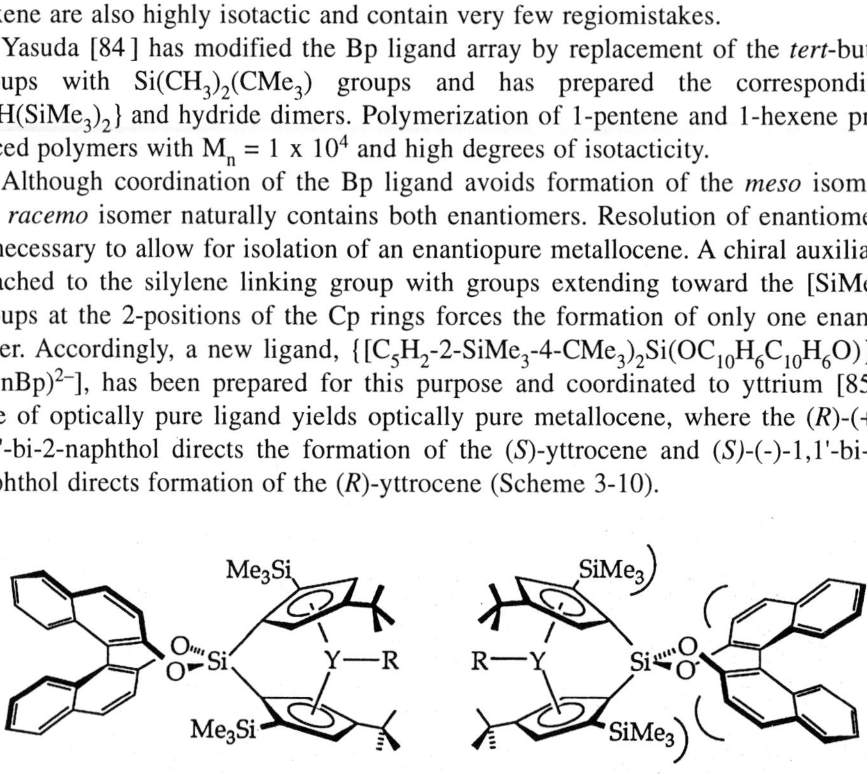

R,S-[BnBp]Y-R
(*preferred*)   R,R-[BnBp]Y-R
                                 (*not preferred*)

Scheme 3-10

The alkyl derivative, (BnBp)YCH(SiMe$_3$)$_2$, is an active catalyst for both olefin polymerization and hydrogenation [86]. Hydrogenolysis of (BnBp)YSiCH(SiMe$_3$)$_2$ cleanly yields the hydride dimer [(BnBp)Y(μ_2-H)]$_2$ which can be isolated as a white solid. Immediately upon addition of H$_2$ to the alkyl complex the formation of *two different* yttrium hydride species is apparent. From NMR investigations, it was found that initially both heterochiral and homochiral hydride dimers are formed, then over time the heterochiral dimers are converted to the more thermodynamically stable homochiral form [85]. The same transformation occurs in the racemic [BpY(μ_2-H)]$_2$ system as well. Significantly, no monomeric hydride species (no doublet in ^{1}H NMR spectrum) is observed, suggesting that monomeric hydrides rapidly dimerize to form much more stable dimers. This behavior greatly reduces the activity of these systems for olefin hydrogenation and polymerization.

Development of a resolved, enantiopure group 3 metallocene offers an unprecedented opportunity to determine the stereoselectivity of additions to carbon–carbon double bonds. Optically active deutero-1-pentenes have been synthesized and used to evaluate the stereoselectivity of Y–H and Y–pentyl additions to optically pure (BnBp)Y–R complexes [87]. Insertion into metal–pentyl bonds proceeds with a much higher facial selectivity ($\geq$40:1) than does insertion into metal–hydride bonds (2:1). Moreover, insertions into metal hydride and metal–*n*-pentyl bonds proceed with opposite enantioselectivities.

Cyclopentadienyl–alkoxide yttrocene systems have also been employed as olefin polymerization catalysts. Terminal alkenes (ethylene, propylene, 1-butene) react with [Cp*(OAr)Y(μ_2-H)]$_2$ (Ar = C$_6$H$_3$(CMe$_3$)$_2$) to produce the μ_2-*n*-alkyl species *trans*-[{Cp*(OAr)Y}$_2$(μ_2-H)(μ_2-CH$_2$CH$_2$R)] [88]. Similar alkyls are produced with longer chain alkyls (*e. g.* from 1-hexene), but no insertion or isomerization occurs with internal olefins (*i. e. trans*-3-hexene). For the polymerization of ethylene, it is proposed that the active catalyst is the undetectable, monomeric [Cp*(OAr)YX] (X = H, CH$_2$R).

More recently, group 3 metallocene catalysts have been investigated for their reactivity towards cyclic functionalized olefins. The catalytic oligomerization of 2-cyclopentene-1-ones has also been investigated with the group III hydrides, [Cp*$_2$La(μ_2-H)]$_2$ and [Cp*$_2$Y(μ_2-H)]$_2$ [89]. Isolation of oligomers reveals that propagation occurs via 1,4 addition, a conclusion also supported by the structure of the isolated trimer. Approximately equimolar amounts of *cis* and *trans* isomers are formed, indicating that the process is not stereoselective. Attempts to isolate cyclic intermediates have not been successful, therefore mechanistic details of this process have not yet been elucidated.

3.5 Reactions of Group 3 Metallocenes with Acetylenes

Catalytic reactions of group 3 metallocenes with alkynes have received much less attention as compared with their reactions with olefins, although several well documented examples have been reported. Reaction of acetylene with Cp*$_2$ScR (R =H, alkyl, aryl, alkenyl, alkynyl, amide) at low temperature (-78°C) yields R–H and Cp*$_2$Sc–C$\equiv$CH (Eq. (3-18)) [90].

(R =H, alkyl, aryl, alkenyl, alkynyl, amide)

The scandocene acetylide derivative then reacts with excess acetylene to produce polyacetylene. The acetylide also cleanly decomposes to give $Cp*_2Sc-C\equiv C-ScCp*_2$, a compound that is cleanly generated by addition of 0.5 equivalents of acetylene to $Cp*_2ScR$.

Permethylscandocene alkyls promote the rapid dimerization of terminal alkynes selectively to 2,4-disubstituted 1-buten-3-ynes [62]. Likewise, Teuben and coworkers report that $Cp*_2Y-CH(SiMe_3)_2$ also serves as a precatalyst for selective head-to-tail dimerization of aliphatic alkynes (Eq. (3-19)) [91].

With the permethylyttrocene system mixtures are obtained for phenylacetylene and (trimethylsilyl)acetylene, indicating that both steric and electronic effects are significant in determining the stereochemical outcome of the reaction. For both systems rapid σ-bond metathesis between the metal alkyl (or alkynyl) and the sp-hybridized C–H bonds of the acetylenes appears to occur to the exclusion of insertion, since higher oligomers are not detected.

Evans reports that reaction of $[(C_5H_4R)_2Y(\mu_2-H)(THF)]_2$ (R = H, CH_3) with terminal alkynes does not involve insertion into the Y–H bond to yield an alkenyl derivative, but rather σ-bond metathesis at the terminal C–H bond of the acetylene occurs to yield the alkynyl derivative and dihydrogen [92]. This reaction is taken as evidence for the hydridic nature of this class of metal–hydrogen bonds. Reaction of $[(C_5H_4R)_2Y(\mu_2-H)(THF)]_2$ (R = H, CH_3) with internal alkynes does however proceed to the corresponding alkenyl complexes, where the addition of the Y–H bond is exclusively cis.

The yttrocene alkyls, $Cp*_2YCH(SiMe_3)_2$ and $Cp*_2Y(CH_3)(THF)$, react with a variety of terminal acetylenes in Et_2O to yield $Cp*_2Y-C\equiv CR \cdot solv$ (solv = Et_2O; THF) [93]. Complexation of solvent is believed to be a result of the relatively small, rodlike nature of the acetylide ligand, allowing sufficient space around the metal for solvent coordination. In the absence of ethereal solvents, dimerization of the acetylene is observed at a rate of 5,400 turnovers/hr at 20°C.

Cyclodimerization of disubstituted alkynes has been reported [94]. The lanthanum alkyl, $Cp*_2LaCH(SiMe_3)_2$, is an active catalyst for the cyclodimerization of 2-alkynes of the form, $CH_3-C\equiv C-R$ ($R = CH_3$, CH_2CH_3, $CH_2CH_2CH_3$), to 1,2-disubstituted, 3-alkylidenecyclobutenes (Eq. (3-20)).

$$2\ CH_3\text{-}C\equiv\text{C-R} \xrightarrow{[Cp*_2\text{La-CH(SiMe}_3)_2]} \quad (3\text{-}20)$$

$$(R = CH_3,\ CH_2CH_3,\ CH_2CH_2CH_3)$$

NMR investigations indicate that the first step in the reaction is the activation of the C–H bond in the acetylene methyl group, yielding $Cp*_2La-CH_2C\equiv CCH_3$. The yttrium analog is unreactive under these conditions, but $Cp*_2YCH_2C\equiv CCH_3$ has been prepared from the reaction of 2-butyne with $[Cp*_2Y(\mu_2\text{-}H)]_2$.

3.6 Reactions of Group 3 Metallocenes with Group 6 Compounds

The combination of well-defined organometallic complexes with the chemistry of heavier group 16 elements (Se, Te) offers the opportunity to develop new procedures for the synthesis of binary materials. It is thus of considerable interest to isolate intermediates in the process of going from the molecular level to the bulk material. Group 3 metallocene alkyl complexes are very promising reagents in this regard.

Reaction of the permethylscandocene alkyl, $Cp*_2ScCH_2SiMe_3$, with elemental tellurium results in the formation of $Cp*_2ScTeCH_2SiMe_3$ [95,96]. This compound can be prepared by reaction with $TeP(CH_2CH_2CH_2CH_3)_3$, but complete removal of $P(CH_2CH_2CH_2CH_3)_3$ is difficult and hence, the elemental tellurium route is preferred. Insertions of tellurium into other scandium alkyls ($R = CH_3$, CH_2Ph, $CH_2CH_2CMe_3$) has also been observed. Heating of the scandium tellurate leads to extrusion of TeR_2 and formation of the tellurium-bridged dimer, $Cp*_2Sc$-Te-ScCp*_2$. Interestingly, the telluride dimer can be photochemically reverted to the scandium alkyltellurate by addition of TeR_2. The corresponding selenides have also been prepared by similar methodology [96].

Through a series of detailed experiments [96], the general mechanistic features of these systems have been explored. Retention of configuration is observed for the thermal extrusion reaction, suggesting a concerted reaction pathway [97]. Photoaddition of TeR$_2$ to Cp*$_2$Sc-Te-ScCp*$_2$ likely occurs through an excited state with a weaker Sc–Te bond that allows the dimer to bend and become susceptible to attack by TeR$_2$. The trapping agent employed in the photoreaction of the Cp*$_2$Sc-Te-ScCp*$_2$ process behaves as a Lewis base, since no reaction occurs in the presence of H-Sn(CH$_2$CH$_2$CH$_2$CH$_3$)$_3$ or Me$_3$SnSnMe$_3$ [96]. Reaction of the excited state of Cp*$_2$Sc-Te-ScCp*$_2$ with TeR$_2$ results in formation of one equivalent of the scandocenetellurate in addition to the Cp*$_2$ScTe·/R· radical pair, which is evidenced by the cyclization in the case of R = 5-hexenyl alkyl.

Other scandocenes behave similarly to the permethylscandocene system. Reaction of elemental tellurium with DpScCH$_2$SiMe$_3$ results in the formation of the scandium tellurate DpScTeR [98]. The dimeric μ_2-telluride DpSc-Te-ScDp can be synthesized from either extrusion of TeR$_2$ from DpSc-TeR or from direct reaction of [DpSc(μ_2-H)]$_2$ with 0.5 equivalents of TeP(CH$_2$CH$_2$CH$_2$CH$_3$)$_3$. Addition of phosphine to the telluride dimer results in the formation of Dp(PMe$_3$)Sc-Te-Sc(PMe$_3$)Dp, which has been characterized crystallographically, revealing the linearity of the Sc–Te–Sc linkage [98].

In addition to organoscandium–tellurium complexes, several selenide and sulfide derivatives of yttrium have also been reported [99]. Reaction of [(η^5-C$_5$H$_4$CMe$_3$)$_2$Y(μ_2-CH$_3$)]$_2$ with two equivalents of either RSSR (R = C$_6$H$_5$, CMe$_3$, CH$_3$CH$_2$CH$_2$CH$_2$, CH$_2$Ph) or PhSeSePh produce dimeric yttrium organosulfides or organoselenides [(η^5-C$_5$H$_4$CMe$_3$)$_2$Y(μ_2-ER)]$_2$ (E = S, Se). The reactions are quantitative and several of these compounds have been characterized crystallographically.

3.7 Reactions of Group 3 Metallocenes with Other Small Molecules

Group 3 metallocenes are coordinatively unsaturated and hence are voracious Lewis acids. As a result, extremely strong bonds between the metal center and hard ligand atoms such as O, N, F and Cl are expected. Thus, the prospect for catalytic reactions with small molecules having heteroatoms are poor. Nonetheless, the reactivity of group 3 metallocenes with heteroatom-containing substrates has been extensively developed and several examples of catalytic chemistry have been reported.

The addition of Sc–H and Sc–C bonds to carbon monoxide, nitriles and carbon dioxide have been examined using permethylscandocene derivatives. The chemistry observed in these systems closely parallels the reactivity patterns of group 4 permethylmetallocenes [100]. Insertion of CO and CO$_2$ into Sc–aryl bonds occurs readily (Scheme 3-11) [13].

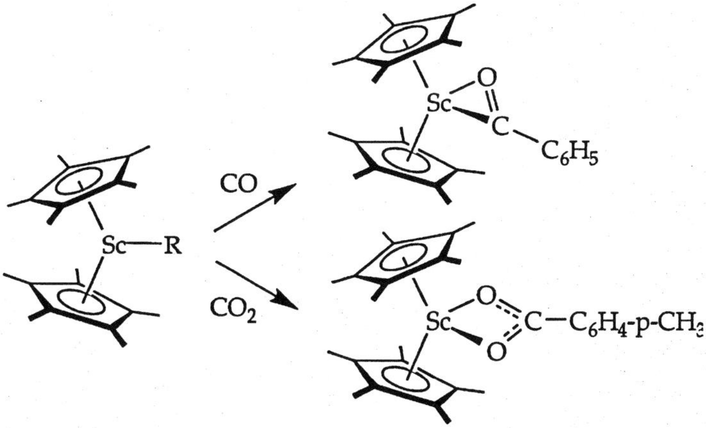

Scheme 3-11

The permethylscandocene hydride, Cp*$_2$Sc-H, undergoes facile addition of its Sc–H bond to the carbonyl ligands of transition metal carbonyls to yield scandoxy-carbenes (*e. g.* Cp(CO)M=CHO-ScCp*$_2$ (M = Co, Rh) and Cp$_2$M=CHO-ScCp*$_2$(M = Mo, W) [100]. This reaction is highly dependent on steric factors. Bulkier sub-stituents such as those of Cp*$_2$Sc-CH$_2$CH$_2$C$_6$H$_5$, Cp*$_2$Sc-N(CH$_3$)$_2$ and Cp*$_2$Sc-C$_6$H$_5$ do not react with Cp$_2$M–C...O (M= Mo, W). However, the less sterically en-cumbered CpM(CO)$_2$ system, (M = Co, Rh) produces the scandoxycarbene product in good yield.

Insertion of nitriles into Sc–R bonds of Cp*$_2$Sc-R (R = H, CH$_3$) proceeds smoothly to form the alkylideneamido complex, which undergoes further additions of the nitrile (Scheme 3-12) [101].

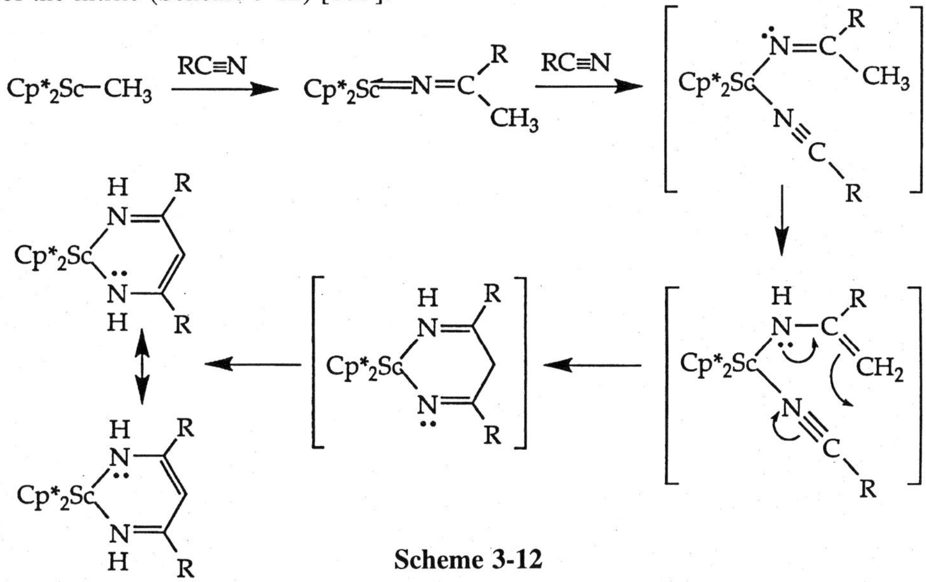

Scheme 3-12

Permethylscandocene hydride also catalyzes the hydrogenation of *tert*-butyl cyanide, yielding neopentylamine (Eq. (3-21)).

$$Me_3CC\equiv N + 2 H_2 \xrightarrow{[Cp^*_2ScH]} Me_3CCH_2NH_2 \qquad (3-21)$$

Reaction of one equivalent of hydrazine with $Cp^*_2ScCH_3$ yields the hydrazido(1–) complex, $Cp^*_2ScN(H)NH_2$ [102]. The hydrazido complex reacts with acetonitrile to form scandocyclic $Cp^*_2ScN(H)CCH_3NNH_2$. The mechanism of the cyclization reaction, explored through the use of ^{15}N labeled acetonitrile, likely proceeds through insertion of the acetonitrile into the Sc–N bond of $Cp^*_2ScNHNH_2$, followed by tautomerization to form the observed product (Scheme 3-13).

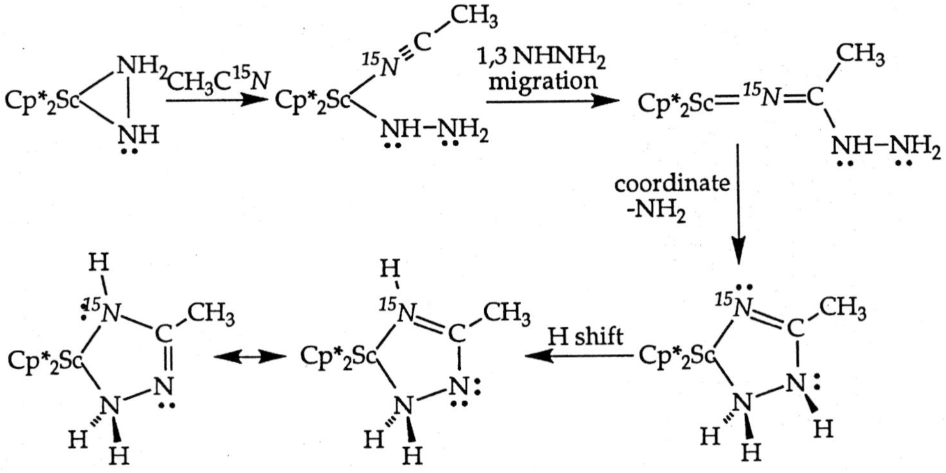

Scheme 3-13

Tilley and co-workers [103] have reported the synthesis of silyl derivatives of scandium from reaction of $[Cp_2ScCl]_2$ and lithium silyls [103]. These react with CO_2, rapidly forming dimerization products (Eq. (3-22)).

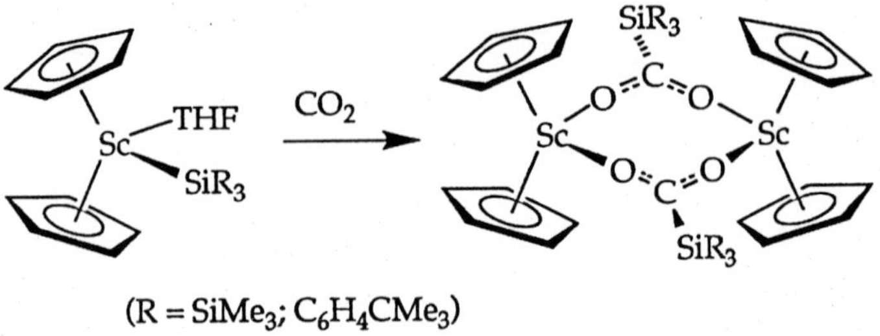

$(R = SiMe_3; C_6H_4CMe_3)$

The dimer can be independently prepared from reaction of $Cp_2Sc(SiR_3)(THF)$ and $HO_2CSi(SiMe_3)_3$ [104]. Recrystallization of the scandium carboxylate leads to extrusion of CO, forming the siloxide, $[Cp_2ScOSi(SiMe_3)_3]_x$. However, upon heating of the dimer, $\{Cp_2Sc[\mu\text{-}O_2CSi(SiMe_3)_3]\}_2$, decomposition to the siloxide is not observed, and no structure has definitely been assigned. Reaction of the scandocene silyls with CO and isocyanides have also been reported [105]. Carbonylation of $Cp_2Sc[Si(SiMe_3)_3](THF)$ forms the ene–diolate, methyltetrahydrofuran adduct (Eq. (3-23)).

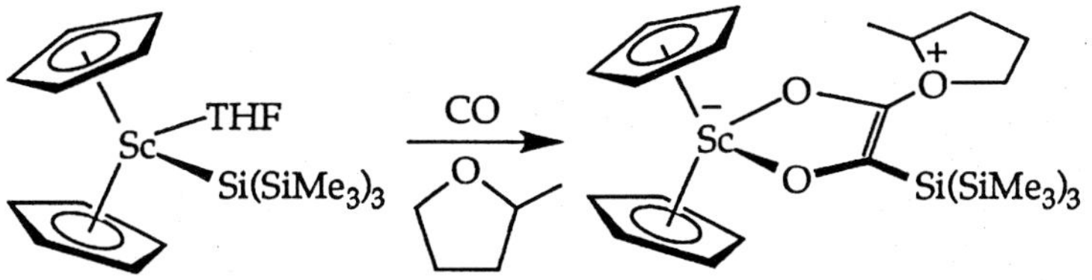

$$(3\text{-}23)$$

The isoelectronic isocyanate, $C\equiv NC_6H_3Me_2$, reacts with $Cp_2Sc[Si(SiMe_3)_3](THF)$ to yield the η^2-iminosilylacyl, which undergoes further addition of $C\equiv NC_6H_3Me_2$ ultimately to yield the bicyclic structure 1:

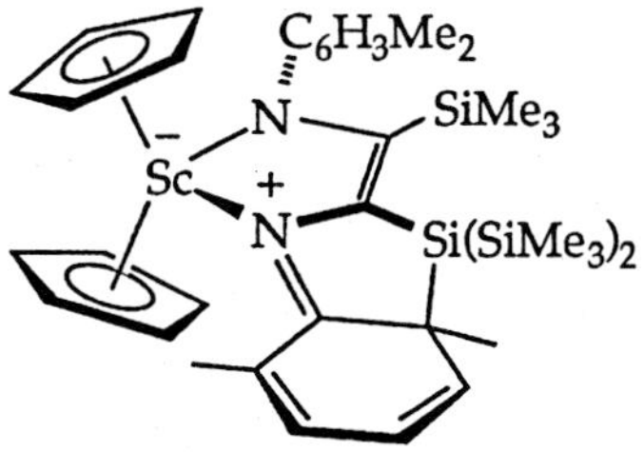

1

The mechanism for this rearrangement is postulated to involve a silylene intermediate, although attempts to isolate or characterize this species have been unsuccessful.

A variety of yttrocene complexes has been studied for their reactivity with pyridines. Reaction of the yttrium alkyls $Cp*_2YCH(SiMe_3)_2$ and $Cp*_2Y(CH_3)(THF)$ with pyridine results in metallation at the α position [106]. The mechanism for the reaction is proposed to involve precoordination of the pyridine, followed by hydrogen transfer from the *ortho* position to the alkyl group by σ-bond metathesis reaction. Coordination of THF to the pyridyl complexes is also observed, forming the 18-electron $Cp*_2Y(\eta^2\text{-}NC_5H_4)(THF)$ adduct. Permethylyttrocene hydride, $[Cp*_2Y(\mu_2\text{-}H)]_2$, also reacts with pyridine selectively at the *ortho* C–H position to yield $Cp*_2Y(\eta^2\text{-}NC_5H_4)$ [107]. The pyridyl complex reacts with H_2 to give the hydride addition product, $Cp*_2Y(NC_5H_6)$. Addition of ethylene and propylene to the pyridyl complexes yields the monoinsertion yttrocyclic products, while acetylenes undergo C–H activation, forming yttrium acetylides. Reaction with carbon monoxide produces $[Cp*_2Y]_2\{\mu_2\text{-}\eta^2\text{:}\eta^2\text{-OC(2-NC}_6H_4)_2\}$, which has been identified by X-ray crystallography (Scheme 3-14).

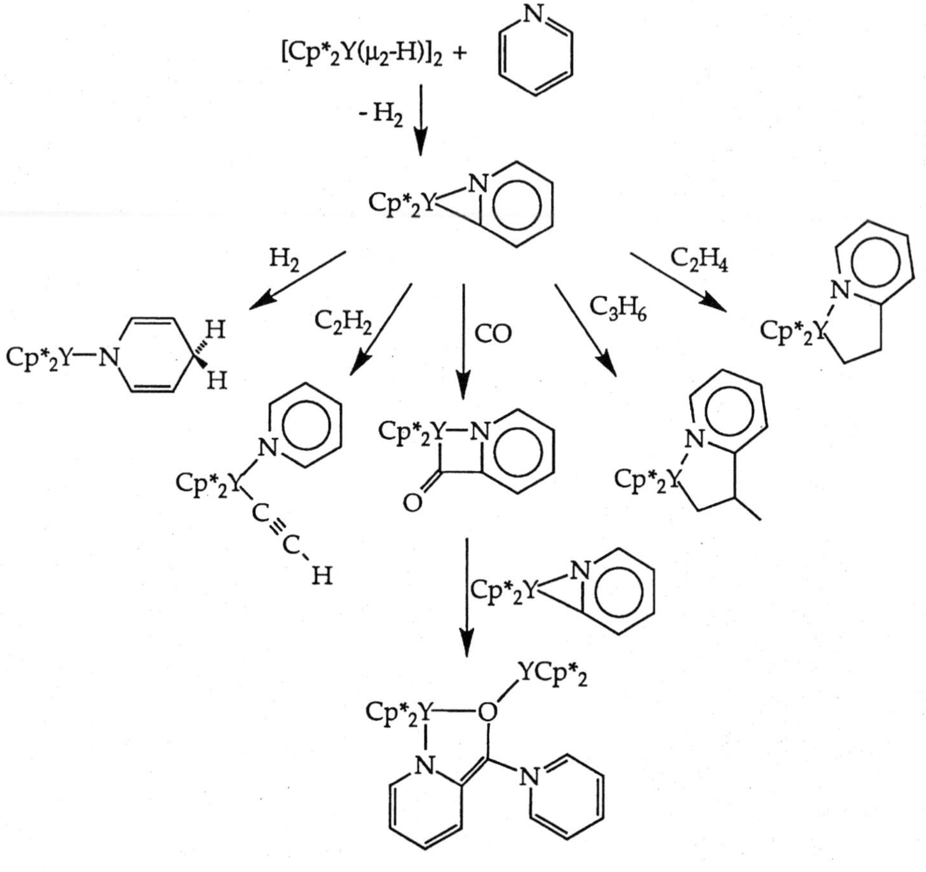

Scheme 3-14

In the presence of excess ethylene, pyridine can be converted to 2-ethylpyridine using $Cp*_2Y(\eta^2\text{-NC}_5H_4)$ as a catalyst. Trace amounts of 2-*n*-butylpyridine and 2-*n*-hexylpyridine are also formed, indicating that multiple insertions are possible [107].

The yttrium hydrides, $[(\eta^5\text{-C}_5H_4R)_2Y(\mu_2\text{-H})(THF)]_2$ (R = H, CH$_3$), react with *tert*-butyl isocyanide in THF to yield $[(\eta^5\text{-C}_5H_4R)_2Y(\mu_2\text{-CHNCMe}_3)]_2$ (Eq. (3-24)) [108].

$$[Cp^*_2YH(THF)]_2 + 2\,C\equiv NCMe_3 \longrightarrow \qquad (3\text{-}24)$$

Spectroscopic and crystallographic data [109] suggest that the structure of the complex is a formimidoyl dimer where nitrogen atoms donate a lone pair of electrons to the yttrium center. Reaction of the chloro-bridged dimer, $[Cp_2Y(\mu_2\text{-}Cl)]_2$, with lithium N,N-dimethylbenzylamine produces $Cp_2Y(C_6H_4\text{-}2\text{-}CH_2NMe_2)$, which has been characterized crystallographically [110]. The crystal structure reveals the yttrium center is σ-bound to the *ortho* carbon and the nitrogen atoms of the N,N-dimethylbenzylamine ligand (Eq. (3-25)).

$$[Cp^*_2Y(\mu_2\text{-}Cl)]_2 + \qquad \longrightarrow \qquad (3\text{-}25)$$

Reaction of $[Cp_2Y(\mu_2\text{-}H)(THF)]_2$ with CH_2O or CH_3OH forms the dicyclopentadienylyttrium methoxide complex [111]. This compound also can be prepared directly from reaction of $[Cp_2Y(\mu_2\text{-}H)(THF)]_2$ with $KOCH_3$ at low temperature ($-78°C$). At higher temperatures the polymetallic oxide, $Cp_5Y_5(\mu\text{-}OCH_3)_4(\mu_3\text{-}OCH_3)_4(\mu_5\text{-}O)$, is isolated from the reaction mixture. However, reaction of $[Cp_2Y(\mu_2\text{-}H)(THF)]_2$ with $NaOCH_3$ forms $[Cp_2Y(\mu\text{-}OCH_3)]_2$ in 80–95% yield [112]. Further investigation into the $KOCH_3$ reaction reveals that $[Cp_2Y(\mu_2\text{-}OCH_3)]_2$ can be formed from $[Cp_2Y(\mu_2\text{-}H)(THF)]_2$ when the $KOCH_3$ is prepared from reaction of $KN(SiMe_3)_2$ with dry methanol. A byproduct of this reaction is the trimetallic compound, $\{[Cp_2Y(\mu_2\text{-}OCH_3)]_2[Cp_2Y](\mu_3\text{-}O)\}$, whose formation is strongly dependent on the amount of $KOCH_3$ present. The pentametallic complex is also isolated from this reaction, and its formation is dependent on the amount of solvated CH_3OH.

The yttrium polyhydride, $[[Cp_2Y(\mu_2\text{-}H)]_3(\mu_3\text{-}H)[Li(THF)_4]$ [113] reacts with one equivalent of CH_3OH to form $[[Cp_2Y(\mu_2\text{-}OCH_3)][Cp_2Y(\mu_2\text{-}H)]_2(\mu_3\text{-}H)][Li(THF)_4]$ [114]. Reaction with two equivalents of CH_3OH forms $[[Cp_2Y(\mu_2\text{-}OCH_3)]_2[Cp_2Y(\mu_2\text{-}H)](\mu_3\text{-}H)][Li(THF)_4]$, which can also be formed from reaction of $[[Cp_2Y(\mu_2\text{-}OCH_3)][Cp_2Y(\mu_2\text{-}H)]_2(\mu_3\text{-}H)][Li(THF)_4]$ with one equivalent of CH_3OH. It is noteworthy that in all cases, the central μ_3-hydride core does not react.

Yttrium–enolate complexes, $[(\eta^5\text{-}C_5H_4R)_2Y(\mu_2\text{-}OCH=CH_2)]_2$, have been synthesized from the reaction of $[(\eta^5\text{-}C_5H_4R)_2Y(\mu_2\text{-}Cl)]_2$ (R = H, CH_3) and

LiOCH=CH$_2$ [115]. Crystallographic studies reveal that the complex is a dimer with bridging enolate oxygen atoms spanning the two yttrium centers. Thermolysis of a sealed NMR tube containing the yttrium alkyl, Cp$_2$Y(CH$_2$SiMe$_3$)(THF), in C$_6$D$_6$ forms the yttrium enolate, [Cp$_2$Y(μ_2-OCH=CH$_2$)]$_2$, generating an equivalent of SiMe$_4$. However, thermolysis of the methyl derivatives, (η^5-C$_5$H$_4$R)$_2$Y(CH$_3$)(THF), under the same conditions forms (η^5-C$_5$H$_4$R)$_3$Y with no evidence for the yttrium enolates. The difference in reactivity is attributed to steric factors, where more crowded complexes have a higher propensity to undergo thermolysis with coordinated THF to form the corresponding enolate complexes.

Teuben [106] has reported the reactivity of Cp*$_2$YCH(SiMe$_3$)$_2$ and Cp*$_2$Y(CH$_3$)(THF) toward a variety of complexes with 'active' hydrogen containing molecules. Reaction with HCl produces [Cp*Y(μ_2-Cl)]$_2$ and Cp*$_2$Y(Cl)(THF), while reaction with acetylacetone (acacH) yields Cp*$_2$Y(acac). Reaction of Cp*$_2$YCH(SiMe$_3$)$_2$ and Cp*$_2$Y(CH$_3$)(THF) with alcohols forms the yttrocene alkoxides, Cp*$_2$Y(OR)(Et$_2$O). The mechanism for this reaction likely involves protonolysis of the Y–C bonds. Similarly, [Cp*$_2$Y(μ_2-H)]$_2$ reacts with 2 equivalents of furan in pentane/THF to form Cp*$_2$Y(2-C$_4$H$_3$O)(THF), which has been characterized crystallographically (Eq. (3-26)) [116].

((Equation 3-26 here))   (3-26)

Cp$_3$Y(THF) reacts with NaOH in THF to form NaCp and Cp$_2$Y(OH)(THF) [117]. Addition of hexachlorocyclotriphosphazene (NPCl$_2$)$_x$ (80%, x = 3; 20%, x = 4) to remove free NaCp from the reaction mixture allows for efficient isolation of Cp$_2$Y(OH)(THF). The bridging hydroxide dimer, [Cp$_2$Y(μ_2-OH)]$_2$(C$_6$H$_5$C...CC$_6$H$_5$), is prepared by hydrolysis of Cp$_2$Y(CMe$_3$)(THF) in the presence of diphenylacetylene. It has been characterized crystallographically.

Mixed metal yttrium–rhenium polyhydride complexes have been prepared from reaction of Cp$_2$Y(THF)(CH$_3$) and ReH$_7$(PPh$_3$)$_2$ or ReH$_5$(PMe$_2$Ph)$_3$, eliminating CH$_4$ forming Cp$_2$Y(THF)Re(H)$_6$(PPh$_3$)$_2$ and Cp$_2$Y(THF)Re(H)$_4$(PMe$_2$Ph)$_3$, respectively [118]. p$_2$Y(THF)Re(H)$_6$(PPh$_3$)$_2$ is unreactive toward a CO, CO$_2$ and PhC≡CPh, but is active for oligomerization of ethylene, propylene and styrene. Trimetallic complexes have been prepared from the reaction of Cp$_2$Y(THF)(CH$_3$) and Re$_2$H$_8$(PMe$_2$Ph)$_4$ yielding Cp$_2$Y(THF)H$_7$Re$_2$(PMe$_2$Ph)$_4$ [118].

3.8 Applications of Group 3 Metallocenes in Organic Synthesis

Although group 3 metallocenes have proven to be efficient catalysts for a variety of organic transformations, their application toward organic synthesis has remained somewhat limited. Attempts to employ group 3 metallocenes as selective hydrogenation catalysts have been explored by Molander. Using 1–mol%

Cp*$_2$Y(CH$_3$)(THF) as a catalyst, a variety of diolefins are hydrogenated to the corresponding alkane in yields of 70% to 99% [119]. Disubstituted alkenes such as limonene, 5-methylenebicyclo[2.2.1]hept-2-ene, or bicycloheptenes cannot be reduced. As a result, the yttrocene catalyst offers the opportunity to selectively hydrogenate terminal olefins in the presence of other types of substituted olefins. Interestingly, ether groups are tolerated by the reaction. Allylic ethers which are rapidly hydrogenated using late metal systems remain intact. For α,ω-dienes, in which two olefins are differentiated by allylic substitution on one of the alkenes, good selectivity is observed. Cp*$_2$Y(CH$_3$)(THF) also catalyzes the hydrocyclization of α,ω-dienes diastereoselectively with excellent yields [120]. In addition to ethers, acetal and dithioacetal functionalities are tolerated by the catalyst. Hydrogenation and isomerization of 1-hexene with a variety of yttrocene chloride complexes and NaH has also been reported [121].

Chiral yttrocenes, such as the chiral yttrocene hydride precursor *rac*-Me$_2$Si[η^5-C$_5$H$_3$SiMe$_3$][η^5-C$_5$H$_3$)(−)-menthyl]YCH(SiMe$_3$)$_2$, catalyze the asymmetric hydrogenation of unfunctionalized olefins. Olefins such as α-ethylstyrene and 2-ethyl-1-hexene are hydrogenated with enantiomeric excess of 25% and <1% respectively [122]. Reductive cyclization of 1,5-dienes is also observed with this system, with olefins such as 1,5-hexadiene undergoing quantitative cyclization. Regio- and stereoselective hydroamination/cyclization of amino olefins such as 1-aminopent-4-ene, 2-amino-hex-5-ene, and dimethyl-1-aminohex-5-ene have been reported [123] with C_1 symmetric yttrocene catalyst precursors such as *rac*-Me$_2$Si[η^5-C$_5$H$_3$SiMe$_3$][η^5-C$_5$H$_3$)R']YCH(SiMe$_3$)$_2$ (Eq. (3-27)) [124]. Enantiomeric excess as high as 69% has been observed in these systems.

$$(3\text{-}27)$$

[chiral yttrocene-R] = [*rac*-Me$_2$Si[η^5-C$_5$H$_3$SiMe$_3$][η^5-C$_5$H$_3$)(-)-menthyl]YCH(SiMe$_3$)$_2$], *etc*

Organoyttrium catalysts are also efficient for the hydrosilation of olefins. Cp*$_2$YCH(SiMe$_3$)$_2$ is activated by reaction with PhSiH$_3$, ultimately generating an yttrium hydride that catalyzes the hydrosilylation of a variety of substituted and unsubstituted olefins [125]. A combination of the cyclooligomerization process and hydrosilation methodology results in the cyclization and silation of 3-phenyl-1,5-hexadiene in the presence of 5 mol percent of Cp*$_2$YCH(SiMe$_3$)$_2$ [126]. Improved results are obtained when Cp*$_2$Y(CH$_3$)(THF) is substituted for the yttrium alkyl (Eq. (3-28)).

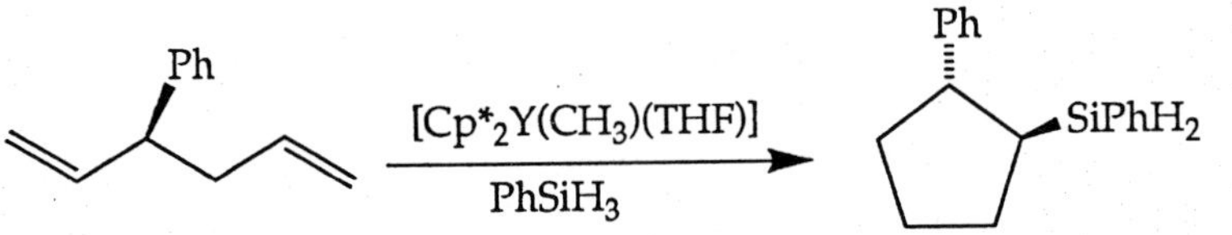

(3-28)

The yttrocene systems result in high regio- and stereoselectivity in addition to tolerance of certain functional groups such as ethers, thioacetals and tertiary amines. The phenylsilane products of the reaction can conveniently be converted to alcohols. This methodology has been applied in the total synthesis of (±)-epilupidine, in which the key step is the selective preparation of a polycyclic ring system through the use of an yttrocene catalyzed cyclization/silylation (Scheme 3-15) [127].

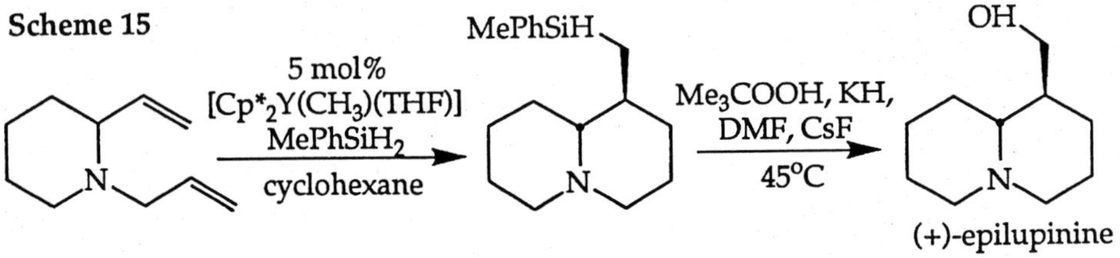

Scheme 15

Scheme 3-15

Organolanthanocenes have been employed in a variety of organic transformations as well. Cp*$_2$LaR complexes (R = H, CH(SiMe$_3$)$_2$) are active catalyst precursors for the hydroboration of terminal olefins in the presence of catecholborane [128]. The reaction is useful for both terminal or substituted olefins. The regioselectively is greater than 98% for the anti-Markovnikov product (Eq. (3-29)).

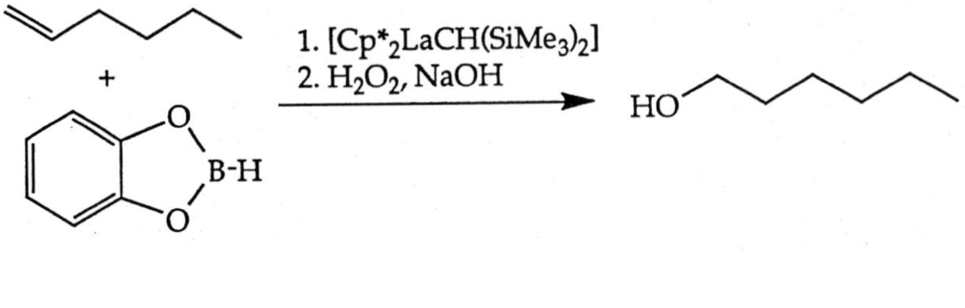

(3-29)

3.9 Photophysics of Group 3 Metallocenes

Inorganic and organometallic complexes that emit at room temperature have been studied extensively in hopes of understanding the fundamental properties of their excited states. Group 3 metallocenes offer a new type of luminescent complex, since classical inorganic photochemistry has focused mainly on second or third row complexes with heterocyclic amine ligands. For Cp^*_2ScCl, three UV–vis absorptions are present in solution and a fourth transition can be observed in the excitation spectrum [129]. From studies of the excitation spectrum, the lowest energy state is taken to be the emissive state. The luminescent lifetimes for permethylscandocene chloride in the solid state at 298K are 2.0 μs, 15.0 μs and 6.5 μs. The long lifetimes are believed to be a result of phosphorescence which arises from an emission from the triplet excited state. The observed transitions in the optical spectrum are assigned to the excitations from the p-Cl nonbonding HOMO to the $1a_1$, $1b_2$ and the $2a_1$ orbitals on scandium.

In addition to the photophysical properties of permethylscandocenes, calculated molecular orbital diagrams of some of the compounds have also been reported [130]. Calculations indicate that the long-lived emissions of these complexes are due to LMCT from the Cp* ligands to the scandium center. It appears that the lone pairs of the halogen are required for the emission to occur, since no emission is observed for either $Cp^*_2ScCH_3$ or $Cp^*_2ScCH_2Ph$, although the molecular orbital diagrams are similar to the chloride complex.

In addition to calculations for observed emission and visible spectra, Group 3 metallocenes have also been the subject of other theoretical calculations. The general features of the bonding in metallocene complexes have been presented [131], but details specific to scandium and yttrium have since appeared. Schleyer [132] has performed *ab initio* calculations for the hypothetical $[Cp_2M]^+$ (M = Sc, La) and found that these species prefer bent geometries due to a small σ-contribution, even though that represents only a small part of the overall covalent bonding. The insertion of ethylene into Cp_2ScCH_3 has been studied through the use of nonlocal density functional calculations [133]. The geometries of Cp_2ScR (R = H, CH_3, SiH_3) have also been calculated by *ab initio* methods [134]. It has been found that the scandium systems prefer planar configurations (*i. e.* the hydride or alkyl substituents prefer the central position in the equatorial plane of the bent metallocene), ultimately attributable to the $(4s)^2(3d)^1$ ground state of the metal. It was hypothesized that this planarity would prohibit the syndiotactic polymerization of α-olefins by catalysts having a group 3 metal with the syndiospecific ligand system, $[Me_2C(\eta^5-C_5H_4)(\eta^5-fluorenyl)]$.

3.10 Alternatives of Cyclopentadienyl Ligands

Since the reactivity of group 3 metallocenes is profoundly influenced by the ancillary ligand array, several attempts have been made to employ ligands alternative to the cyclopentadienyl environment [135]. One approach has been replacement of one of the cyclopentadienyl moieties with 'hard' oxygen or nitrogen-containing ligands. Such examples of these complexes are the 'Cp*SiNR' metallocenes discussed previously. Other approaches have used dicarboranes as cyclopentadienyl substitutes. Although the detailed chemistry of these types of complexes is beyond the scope of this report, a brief overview of a few representative cyclopentadienyl replacements is presented. Alkoxide ligands in conjunction with the cyclopentadienyl ligand have been successful in allowing for the preparation and isolation group 3 metallocene analogs. Reaction of $Y[OC_6H_3(CMe_3)_2]_3$ with KCp* in toluene results in the formation of $Cp*Y[OC_6H_3(CMe_3)_2]_2$ in 70–80% yield [136]. An alkyl derivative has been prepared by reaction of $Cp*Y[OC_6H_3(CMe_3)_2]_2$ with $LiCH(SiMe_3)_2$, which can be hydrogenated under 10 bar of H_2 to give the hydride dimer, $[Cp*Y(OC_6H_3(CMe_3)_2(\mu_2\text{-}H)]_2$ (Scheme 3-16).

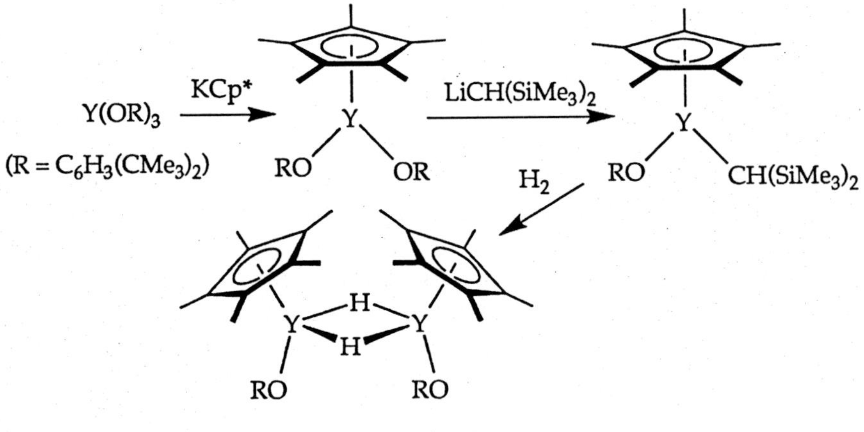

Scheme 3-16

Reactivity studies indicate that the isomer with a *cis* arrangement of Cp* ligands is the only one formed, and that this species is an active catalyst for the polymerization of ethylene. The bridging alkoxide dimers, $[CpY(\mu_2\text{-}OCMe_3)(OCMe_3)]_2$ (Cp = Cp*, Cp, $(\eta^5\text{-}C_5H_4Me)$, $[\eta^5\text{-}C_5H_3(SiMe_3)_2]$) have been prepared from the corresponding NaCp and $Y_3(OCMe_3)_7Cl_2$ [137]. The indenyl complex, $[(\eta^5\text{-}C_9H_7)Y(\mu_2\text{-}OR)(OR)]_2$ (R = CMe_3), has also been synthesized. Attempts at alkylation with $LiCH(SiMe_3)_2$ have resulted in loss of LiCp and formation of bridged alkoxide species.

Teuben and coworkers have prepared $Cp*/[PhC(NSiMe_3)_2]$ yttrium alkyls. Reaction of $YCl_3(THF)_{3.5}$ with $Cp*K$ followed by reaction with $[PhC(NSiMe_3)_2]LiOEt_2$ in THF yields the chloride dimer [138]. Alkylation of the chloride dimer with traditional alkylating agents such as $ME(SiMe_3)_2$ (M = Li, Na; E = CH, N) resulted in formation of complex mixtures; however, a well-defined methyl dimer, $Cp*[PhC(NSiMe_3)_2]Y(\mu_2\text{-}CH_3)_2Li\cdot TMEDA$, has been isolated from the reaction of the chloride dimer with CH_3Li. The methyl dimer is subject to protolysis under mild conditions by terminal acetylenes and alcohols. Reaction of $M[N(SiMe_3)_2]$ (M = Y, La) with $Cp*H$ afforded a mixture of both mono and bis-$Cp*$ complexes [139]; however employing bulky chelating alkyl groups such as in the case of $Y(o\text{-}C_6H_4CH_2NMe_2)_3$ allows for the isolation of the monocyclopentadienyl complex.

A variety of monocyclopentadienyl lanthanum complexes has been prepared. By employing larger iodides to assist in stabilization, $Cp*LaI_2(THF)_3$ has been prepared from $LaI_3(THF)_3$ and $KCp*$ [140]. The thermally unstable alkyl complex, $Cp*La[CH(SiMe_3)_2](THF)$ can be prepared by addition of two equivalents of $KCH(SiMe_3)_2$, while the coordinated THF can be removed by reaction with Me_3SiI [141]. Reaction of the THF-free bis-alkyl with $[PhNMe_2H][BPh_4]$ yields the first cationic lanthanum alkyl complex [142].

Preparation of some mixed pentamethylcylopentadienyl/dicarbollide derivatives of scandium has also been described [143]. Reactions of $[Cp*ScCl_2]_x$ with $Na_2[C_2B_9H_{11}]$ or $[Cp*ScMe_2]_x$ with $C_2B_9H_{13}$, followed by treatment with THF yield $Cp*(C_2B_9H_{11})Sc(THF)_3$. Alkylation of $Cp*(C_2B_9H_{11})Sc(THF)_3$ with $LiCH(SiMe_3)_2$ yields $Cp*(C_2B_9H_{11}ScCH(SiMe_3)_2Li(THF)_3$ and $\{[Cp*(C_2B_9H_{11})ScCH(SiMe_3)_2]_2Li\}\cdot Li(THF)_3$, which has been characterized structurally. This alkyl derivative reacts slowly with H_2 to yield $[Cp*(C_2B_9H_{11})ScH]_2[Li(THF)_n]_2$, a surprisingly unreactive scandium hydride dimer. Recrystallization from toluene affords a crystalline form with less coordinated THF, $[Cp*(C_2B_9H_{11})ScH]_2[Li(THF)]_2\cdot 3/2(C_6H_5CH_3)$, whose structure reveals that the two anionic $[Cp*(C_2B_9H_{11})ScH]^-$ fragments are held together by reciprocal B–H dative bonding from the dicarbollide ligand to the electron deficient scandium.

3.11 Concluding Remarks

Since the initial reports of Cp_3M complexes over 40 years ago, there has been explosive growth in group 3 metallocene chemistry. An enormous number of both scandocene and yttrocene complexes have been prepared and their reaction chemistry has been applied in a variety of areas. Although many group 3 metallocenes have been prepared, their synthesis is by no means trivial. Most syntheses of the metallocene derivatives rely on salt metathesis reactions in a coordinating solvent such as THF or diethyl ether. This methodology coupled with the extreme Lewis acidity of the metal center often leads to unwanted coordination of solvent or salt molecules which can interfere with further synthesis or catalytic activity. Despite these limitations, a large number of useful group 3 metallocenes have been synthesized. Such complexes are active catalysts for olefin hydrogenation, silylation, hydroboration, hydroamination, cyclization and polymerization. Incorporation of the properly substituted cyclopentadienyl framework has lead to the development of chiral metallocenes which are successful in catalyzing some of the aforementioned reactions stereospecifically. Group 3 metallocenes have been particularly effective in elucidating the mechanistic features in olefin polymerization. Essential polymerization steps, including olefin insertion and β elimination have been probed with a variety of scandocenes and yttrocenes, revealing the key features of a Ziegler–Natta catalyst.

Group 3 metallocenes also exhibit high reactivity toward the C–H bonds of hydrocarbons. Once again, the relatively clean nature of many of these reactions has allowed investigations of the fundamental processes, notably σ-bond metathesis. Additionally, scandocenes and yttrocenes react with a variety of small molecules that include carbon monoxide, carbon dioxide, nitriles and amines. Despite the strong bonds formed between the metal center and the heteroatoms of these substrates, some catalytic transformations have been developed.

The catalytic transformations promoted by yttrocenes are now beginning to be applied in organic synthesis. A variety of stereo- and regiospecific reactions have been developed and are being used to prepare natural products. Notwithstanding the synthetic challenges associated with these complexes, particularly their intolerance to heteroatom-containing substrates and their inherent air and water sensitivity, a wealth of chemistry has been uncovered.

References

[1]   a) Evans, W.J. *Adv. Organomet. Chem.* **1985**, *24*, 131. b) Shaverien, C.J. *Adv. Organomet. Chem.* **1994**, *36*, 283. c) Rogers, R.D.; Rogers, L.M. *J. Organomet. Chem.* **1992**, *443*, 83. d) Evans, W.J. *Polyhedron,* **1987**, *6*, 803. e) Nolan, S.P.; Stern, D.; Hedden, D.; Marks, T.J. *ACS Symp. Ser.* **1990**, *428*, 159. f) Burns, C.J.; Burstein, B.E. *Comments Inorg. Chem.* **1989**, *9*, 61. g) Evans, W.J.; Foster, S.E. *J. Organomet. Chem.* **1992**, *433*, 79.

[2]   a) Lappert, M.F.; Holton, J.; Ballard, D.G.H.; Pearce, R.; Atwood, J.L.; Hunter, W.E. *J. Chem. Soc. Dalton Trans.* **1979**, 49. b) Lappert, M.F.; Holton, J.; Ballard, D.G.H.; Pearce, R.; Atwood, J.L.; Hunter, W.E. *J. Chem. Soc. Dalton Trans.* **1979**, 54.

[3]   Wilkinson, G.; Birmingham, J.M. *J. Am. Chem. Soc.* **1954**, *76*, 6210.

[4]   Hart, F.A.; Massey, A.G.; Saran, M.S. *J. Organomet. Chem.* **1970**, *21*, 147.

[5]   Coutts, R.S.P.; Wailes, P.C. *J. Organomet. Chem.* **1970**, *25*, 117.

[6]   Atwood, J.L.; Smith, K.D. *J. Chem. Soc. Dalton Trans.* **1973**, 2487.

[7]   Manzer, L.E. *J. Organomet. Chem.* **1976**, *110*, 291.

[8]   Manzer, L.E. *Inorg. Chem.* **1976**, *15*, 2567.

[9] Manzer, L.E. *J. Am. Chem. Soc.* **1978**, *100*, 8068.

[10] Reid, A.; Wailes, P.C. *Inorg. Chem.* **1996**, *5*, 1213.

[11] Bottomly, F.; Paez, D.E.; White, P.S. *J. Organomet. Chem.* **1985**, *291*, 35.

[12] (a) Lappert, M.F.; Singh, A.; Atwood, J.L.; Hunter, W.E. *J. Chem. Soc. Chem. Comm.* **1983**, 206. (b) Lappert, M.F.; Singh, A.; Atwood, J.L.; Hunter, W.E. *J. Chem. Soc. Chem. Comm.* **1983**, 69.

[13] Thompson, M.E.; Bercaw, J.E. *Pure Appl. Chem.* **1984**, *56*, 1.

[14] Although Cp*₂ScH(THF) is moderately stable in solution in the absence of H_2, it does decompose in the solid state after several days at room temperature.

[15] Watson, P.L.; Herskovitz, T. *ACS Symposium Series* **1983**, *212*, 459.

[16] Piers, W.E.; Shapiro, P.J.; Bunel, E.E.; Bercaw, J.E. *Synlett* **1990**, 74.

[17] Schaefer, W.P.; Köhn, R.D.; Bercaw, J.E. *Acta Crystallogr.* **1992**, *C48*, 251.

[18] Bunel, E.E., Ph. D Thesis, California Institute of Technology, Pasadena, CA (**1989**).

[19] Shapiro, P.J.; Bunel, E.; Schaefer, W.P.; Bercaw, J.E. *Organometallics* **1990**, *9*, 867.

[20] Mu, Y.; Piers, W.P.; MacQuarrie, D.C.; Zawaorotko, M.J.; Young, V.G. *Organometallics* **1996**, *15*, 2720.

[21] Birmingham, J.M.; Wilkinson, G. *J. Am. Chem. Soc.* **1956**, *78*, 62.

[22] Rogers, R.D.; Atwood, J.L.; Emad, A.; Sikora, D.J.; Rausch, M.D.; *J. Organomet. Chem.* **1981**, *216*, 383.

[23] Eggers, S.H.; Kopf, J.; Fischer, R.D. *Organometallics* **1986**, *5*, 383.

[24] Xie, Z.; Hahn, F.E.; Qian, C. *J. Organomet. Chem.* **1991**, *414*, C12.

[25] Evans, W.J.; Dominguez, R.; Levan, K.R.; Doedens, R.J. *Organometallics* **1985**, *4*, 1836.

[26] Evans, W.J.; Meadows, J.H.; Wayda, A.L.; Hunter, W.E.; Atwood, J.L. *J. Am. Chem. Soc.* **1982**, *104*, 2008.

[27] Maginn, R.E.; Manastyrskyj, S.; Dubeck, J. *J. Am. Chem. Soc.* **1963**, *85*, 672.

[28] Evans, W.J.; Meadows, J.H.; Hanusa, T.P. *J. Am. Chem. Soc.* **1984**, *106*, 4454.

[29] Evans, W.J.; Peterson, T.T; Rausch, M.D.; Hunter, W.E.; Zhang, H.; Atwood, J.L. *Organometallics* **1985**, *4*, 554.

[30] Evans, W.J.; Grate, J.W.; Levan, K.R.; Bloom, I.; Peterson, T.T.; Doedens, R.J.; Zhang, H.; Atwood, J.L. *Inorg. Chem.* **1986**, *25*, 3614.

[31] Evans, W.J.; Boyle, T.J.; Ziller, J.W. *Inorg. Chem.* **1992**, *31*, 1120.

[32] den Haan, K.H.; de Boer, J.L.; Teuben, J.H.; Spek, A.L.; Kojic-Prodic, B.; Hays, G.R.; Huis, R. *Organometallics* **1986**, *5*, 1726.

[33] den Haan, K.H.; Wielstra, Y.; Eshius, J.J.W.; Teuben, J.H. *J. Organomet. Chem.* **1987**, *323*, 181.

[34] den Haan, K.H.; De Boer, J.L.; Teuben, J.H.; Smeets, W. J. J. J.; Spek, A.L. *J. Organomet. Chem.* **1987**, *327*, 31.

[35] den Haan, K.H.; Teuben, J.H. *Recl. Trav. Chim. Pays-Bas* **1984**, *103*, 333.

[36] den Haan, K.H.; Wielstra, Y.; Teuben, J.H. *Organometallics* **1987**, *6*, 2053.

[37] den Haan, K. H.; Teuben, J.H. *J. Chem. Soc. Chem. Comm.* **1986**, 682.

[38] Holton, J.; Lappert, M.F.; Scollary, G.R.; Ballard, D.G.H.; Pearce, R.; Atwood, J.L.; Hunter, W.E. *J. Chem. Soc. Chem. Com.* **1976**, 425.

[39] Lobkovskii, E.B.; Soloveichik, G.L.; Erofeev, A.B.; Bulychev, B.M.; Bel'skii, V. *J. Organomet. Chem.* **1982**, *235*, 151.

[40] Lobkovsky, E.B.; Soloveichik, G.L.; Bulychev, B.M.; Erofeev, A.B.; Gusev, A.I.; Kirillova, N.I. *J. Organomet. Chem.* **1983**, *254*, 167.

[41] Belsky, V.K.; Erofeev, A.B.; Bulychev, B.M.; Soloveichik, G.L. *J. Organomet. Chem.* **1984**, *265*, 123.

[42] Erofeev, A.B.; Bulychev, B.M.; Bel'skii, V.K.; Soloveichik, G.L. *J. Organomet. Chem.* **1987**, *335*, 189.

[43] Erofeev, A.B.; Bulychev, B.M.; Bel'skii, V.K.; Soloveichik, G.L. *J. Organomet. Chem.* **1987**, *335*, 189.

[44] Busch, M.A.; Harlow, R.; Watson, P.L. *Inorg. Chim. Acta* **1987**, *140*, 15.

[45] Schumann, H.; Rosenthal, E.C.E.; Kociok-Kohn, G.; Molander, G.A.; Winterfeld, J. *J. Organomet. Chem.* **1995**, *496*, 233.

[46] Evans, W.J.; Drummond, D.K.; Hanusa, T.P.; Doedens, R.J. *Organometallics* **1987**, *6*, 2279.

[47] Laske, D.A.; Duchateau, R.; Teuben, J.H.; Spek, A.L. *J. Organomet. Chem.* **1993**, *462*, 149.

[48] Deng, D.; Jiang, Y.; Qian, C.; Wu, G.; Zheng, P. *J. Organomet. Chem.* **1994**, *470*, 99.

[49] Qian, C.; Wang, B.; Deng, D.; Sun, J.; Hahn, F.E.; Chen, J.; Zheng, P. *J. Chem. Soc. Dalton Trans.* **1996**, 955.

[50] Broussier, R.; Delmas, G.; Perron, P.; Gautheron, B.; Peterson, J.L. *J. Organomet. Chem.* **1996**, *511*, 185.

[51] Nief, F.; Mathey. F. *J. Chem. Soc. Chem. Comm.* **1989**, 800.

[52] Qian, Changtao;Xie, Z..; Huang, Y. *J. Organomet. Chem.* **1987**, *323*, 285.

[53] Schumann, H.; Loebel, J.; Pickardt, J.; Qian, C.; Xie, Z. *Organometallics* **1991**, *10*, 215.

[54] Qian, C.; Xie, Z. Huang, Y. *Inorg. Chim. Acta* **1988**, *139*, 195.

[55] Qian, C.; Ye, C.; Li, Y. *J. Organomet. Chem.* **1986**, *302*, 171.

[56] Coughlin, E.B.; Bercaw, J.E. *J. Am. Chem. Soc.* **1992**, *114*, 7606.

[57] Chacon, S.T.; Coughlin, E.B.; Henling, L.M.; Bercaw, J.E. *J. Organomet. Chem.* **1995**, *497*, 171.

[58] Mu, Y.; Piers, W.E.; MacDonald, M.A.; Zaworotko, M.J. *Can. J. Chem.* **1995**, *73*, 2233.

[59] Kretschmer, W.P. *Verslagen Werkgroep Organometaalchemie en Homogene Katalyse* **1996**, *9*, 67.

[60] For general reviews see: (a) Crabtree; R.H. *Chem. Rev.* **1985**, *85*, 245. (b) Tanaka, M.; Sakakura, T. In *Homogeneous Transition Metal Catalyzed Reactions*; Moser, W.R., Slocum, D.W. Eds.; American Chemical Society: Washington, DC, **1992**; p. 181. (c) Jones, W.D. In *Selective Hydrocarbon Activation*; Davies, J.A., Watson, P.L., Liebman, J.F.; Greenberg, A., Eds.; VCH Publishers: New York, **1990**; p. 113.

[61] Watson, P.L. *J. Am. Chem. Soc.* **1983**, *105*, 6491.

[62] Thompson, M.E.; Baxter, S.M.; Bulls, A.R.; Burger, B.J.; Nolan, M.C.; Santarsiero, B.D.; Schaefer, W.P.; Bercaw, J.E. *J. Am. Chem. Soc.* **1987**, *109*, 203.

[63] Steigerwald, M.L.; Goddard, W.A. *J. Am. Chem. Soc.* **1984**, *106*, 308.

[64] Bulls, A.R.; Bercaw, J.E.; Manriquez, J.M.; Thompson, M.E. *Polyhedron*, **1988**, *7*, 1409.

[65] Booij, M.; Deelman, B.J.; Duchateau, R.; Postma, D.S.; Meetsma, A.; Teuben, J.H. *Organometallics* **1993**, *12*, 3531.

[66] Burger, B.J.; Cotter, W.D.; Coughlin, E.B.; Chacon, S.T.; Hajela, S.H.; Herzog, T.A.; Kohn, R.; Mitchell, J.; Piers, W.E.; Shapiro, P.J.; Bercaw, J.E. In *Ziegler Catalysts*; Fink, G., Mülhaupt, R.; Brintzinger, H.H. Eds.; Springer-Verlag: Berlin, **1995**; pp 318.

[67] Ballard, D.G.H.; Courtis, A.; Holton, J.; McMeeking, J.; Pearce, R. *J. Chem. Soc. Chem. Com.* **1978**, 994.

[68] Evans, W.J.; Meadows, J.H.; Hunter, W.E.; Atwood, J.L. *J. Am. Chem. Soc.* **1984** *106*, 1291.

[69] Stern, D.; Sabat, M.; Marks, T.J. *J.Am. Chem. Soc.* **1990**, *112*, 9558.

[70] Burger, B.J.; Thompson, M.E.; Cotter, W.D.; Bercaw, J.E. *J. Am. Chem. Soc.* **1991**, *112*, 1566.

[71] (a) Bryndza, H.E.; Fong, L.K.; Paciello, R.A.; Tam, W.; Bercaw, J.E. *J. Am. Chem. Soc.* **1987**, *107*, 1444. (b) Bryndza, H.E.; Domaille, P.J.; Fong; L.K.; Paciello, R.A.; Tam, W.; Bercaw, J.E. *Polyhedron* **1988**, *7*, 1441.

[72] Bunel, E.E.; Burger, B.J.; Bercaw, J.E. *J. Am. Chem. Soc.* **1988**, *110*, 976.

[73] Bercaw, J.E. *Pure & Appl. Chem.* **1990**, *62*, 1151.

[74] Chirik, P.J.; Bercaw, J.E. unpublished results.

[75] (a) Brookhart, M.; Green, M.L.H. *J. Organomet. Chem.* **1983**, *250*, 395. (b) Brookhart, M.; Green, M.L.H.; Wong, L. *Prog. Inorg. Chem.* **1988**, *36*, 1. (c) Grubbs, R.H.; Coates, G. W. *Acc. Chem. Res.* **1996**, *29*, 85.

[76] Clawson, L.; Soto, J.; Buchwald, S.L.; Steigerwald, M.L.; Grubbs, R.H.*J. Am. Chem. Soc.* **1985**, *107*, 3377.

[77] Piers, W.E.; Bercaw, J.E. *J. Am. Chem. Soc.* **1990**, *112*, 9406.

[78] Herzog, T.A. *Ph. D. Thesis California Institute of Technology* **(1996)**.

[79]   Hajela, S.; Bercaw, J.E. *Organometallics* **1994**, *13*, 1147.

[80]   Shapiro, P.J.; Bunel, E.E.; Schaefer, W.P.; Bercaw, J.E. *Organometallics* **1990**, *9*, 867.

[81]   Shapiro, P.J.; Cotter, W.D.; Schaefer, W.D.; Labinger, J.A.; Bercaw, J.E. *J. Am. Chem. Soc.* **1994**, *116*, 4623.

[82]   Coughlin, E.B.; Bercaw, J.E. *J. Am. Chem. Soc.* **1992**, *114*, 7607.

[83]   Marsh, R.E.; Schaefer, W.P.; Coughlin, E.B.; Bercaw, J.E. *Acta Cryst.* **1992**, *C48*, 1773.

[84]   Yasuda, H,.; Ihara, E. *Tetrahedron* **1995**, *51*, 4563.

[85]   Mitchell, J.P.; Hajela, S.; Brookhart, S.K.; Hardcastle, K. I.; Henling, L.M.; Bercaw, J.E. *J. Am. Chem. Soc.* **1996**, *118*, 1045.

[86]   Under an atmosphere of H_2, neat 1-pentene is readily polymerized to afford a polymer of relatively narrow molecular weight distribution and modest molecular weight (PDI = 1.44; M_n = 119,000; *mmmm* > 95%).

[87]   Gilchrist, J.H.; Bercaw, J.E. *J. Am. Chem. Soc.* **1996**, *118*, 12021.

[88]   Schaverien, C.J. *J. Chem Soc. Chem. Comm.* **1992**, 11.

[89]   Meijboom, R. *Verslagen Werkgroep Organometaalchemie en Homogene Katalyse* **1996**, *9*, 139.

[90]   St. Clair, M.; Schaefer, W.P.; Bercaw, J.E. *Organometallics* **1991**, *10*, 525.

[91]   Heeres, H.J.; Teuben, J.H. *Organometallics* **1991**, *10*, 1980.

[92]   Evans, W.J.; Meadows, J.H.; Hunter, W.E.; Atwood, J.L. *J. Am. Chem. Soc.* **1984** , *106*, 1291.

[93]   den Haan, K.H.; Wielstra, Y.; Teuben, J.H. *Organometallics* **1987**, *6*, 2053.

[94]   Heeres, H.J.; Heeres, A.; Teuben, J.H. *Organometallics* **1990**, *10*, 1508.

[95]   Piers, W.E.; MacGillivray, L.R.; Zaworotko, M. *Organometallics* **1993**, *12*, 4723.

[96]   Piers, W.E.; Parks, D.J.; MacGillivray, L.R.; Zaworotko, M.J. *Organometallics* **1994**, *13*, 4547.

[97]   Piers, W.E. *J. Chem. Soc. Chem. Com.* **1994**, 309.

[98]   Piers, W.E.; Ferguson, G.; Gallagher, J.F. *Inorg. Chem.* **1994**, 3784.

[99]   Beletskaya, I.P.; Voskoboynikov, A.Z.; Shestakova, A.K.; Yanovsky, A.I.; Fukin, G.K.; Zacharov, L.V.; Struchkov, Y.T.; Schumann, H. *J. Organomet. Chem.* **1994**, *468*, 121.

[100] St. Clair, M.A.; Santarsiero, B.D.; Bercaw, J.E. *Organometallics* **1989**, *8*, 17.

[101] Bercaw, J.E.; Davies, D.L.; Wolczanski, P.T. *Organometallics* **1986**, *5*, 443.

[102] Shapiro, P.J.; Henling, L.M.; Marsh, R.E.; Bercaw, J.E. *Inorg. Chem.* **1990**, *29*, 4560.

[103] Campion, B.K.; Heyn, R.H.; Tilley, T.D. *Inorg. Chem.* **1990**, *29*, 4356.

[104] Brook, A.G.; Yau, l. *J. Organomet. Chem.* **1984**, *271*, 9.

[105] Campion, B.K.; Heyn, R.; Tilley, T.D. *J . Am. Chem. Soc.* **1990**, *112*, 2011.

[106] den Haan, K. H.; Wielstra, Y.; Teuben, J.H. *Organometallics* **1987**, *6*, 2053.

[107] Deelman, B.J.; Stevels, W.M.; Teuben, J.H.; Lakin, M.T.; Spek, A.L. *Organometallics* **1994**, *13*, 3881.

[108] Evans, W.J.; Meadows, J.H.; Hunter, W.E.; Atwood, J.L. *Organometallics* **1983**, *2*, 1252.

[109] Evans, W.J.; Hanusa, T.P.; Meadows, J.H.; Hunter, W.E.; Atwood, J.L. *Organometallics* **1987**, *6*, 295.

[110] Rausch, M.D.; Foust, D.F.; Rogers, R.D.; Atwood, J.L. *J. Organomet. Chem.* **1984**, *265*, 241.

[111] vans, W.J.; Sollberger, M.S. *J. Am. Chem. Soc.* **1986**, *108*, 6095.

[112] Evans, W.J.; Sollberger, M.S.; Shreeve, J.L.; Olofson, J.M.; Hain, J.H.; Ziller, J.W. *Inorg. Chem.* **1992**, *31*, 2492.

[113] Evans, W.J.; Meadows, J.H.; Hanusa, T.P. *J. Am. Chem. Soc.* **1984**, *106*, 4454.

[114] Evans, W.J.; Sollberger, M.S.; Khan, S.I.; Bau, R. *J. Am. Chem. Soc.* **1988**, *110*, 439.

[115] Evans, W.J.; Dominguez, R.; Hanusa, T. P. *Organometallics* **1986**, *5*, 1291.

[116] Ringelberg, S.N. *Verslagen Werkgroep Organometaalchemie en Homogene Katalyse*, **1996**, *9*, 79.

[117] Evans, W.J.; Hozbor, M.A.; Bott, S.G.; Robinson, G.H.; Atwood, J.L. *Inorg. Chem.* **1988**, *27*, 1990.

[118] Alvarez, D.; Caulton, K.G.; Evans, W.J.; Ziller, J.W. *Inorg. Chem.* **1992**, *31*, 5500.

[119] Molander, G.A.; Hoberg, J.O. *J. Org. Chem.* **1992**, *57*, 3266.

[120]   Molander, G.A.; Hoberg, J.O. *J. Am. Chem. Soc.* **1992**, *114*, 3123.

[121]   Qian, C.T.; Zie, Z.; Huang, Y. *Chinese Sci. Bull.* **1989**, *34*, 1106.

[122]   Haar, C.M.; Stern, C.L.; Marks, T.J. *Organometallics* **1996**, *15*, 1765.

[123]   Giardello, M.A.; Conticello, V.P.; Brard, L.; Gagne, M.R.; Marks, T.J. *J. Am. Chem. Soc.* **1994**, *116*, 10241.

[124]   R' = (1*S*,2*S*,5*R*)-*trans*-5-methyl-*cis*-2-(2-propyl)cyclohexyl((+)-neomenthyl),
        (1*R*, 2*S*,5*R*)-*cis*-5-methyl-*trans*-2-(2-propyl)-cyclohexyl((-)-menthyl)
        and (1*R*,2*S*,5*R*)-*cis*-5-methyl-*trans*-2-(2-phenyl-2-propyl)-cyclohexyl((-)-phenylmethyl); (E = N, CH).

[125]   Molander, G.A.; Julius, M. *J. Org. Chem.* **1992**, *57*, 6347.

[126]   Molander, G.A.; Nichols, P.J. *J. Am. Chem. Soc.* **1995**, *117*, 4415.

[127]   Molander, G.A.; Nichols, P.J. *J. Org. Chem.* **1996**, *61*, 6040.

[128]   Harrison, K.N.; Marks, T.J. *J. Am. Chem. Soc.* **1992**, *112*, 9220.

[129]   Pfennig, B.W.; Thompson, M.E.; Bocarsly, A.B. *J. Am. Chem. Soc.* **1989**, *111*, 8947.

[130]   Pfennig, B.W.; Thompson, M.E.; Bocarsly, A.B. *Organometallics* **1993**, *12*, 649.

[131]   Lauher, J.W.; Hoffmann, R. *J. Am. Chem Soc.* **1976**, *98*, 1729.

[132]   Kaupp, M.; Charkin, O.P.; Schleyer, P. v. R. *Organometallics* **1992**, *11*, 2765.

[133]   Woo, T.K.; Fan, L.; Ziegler, T. *Organometallics* **1994**, *13*, 432.

[134]   Bierwagen, E.P.; Bercaw, J.E.; Goddard, W.A. *J. Am. Chem. Soc.* **1994**, *116*, 1481.

[135]   For *bis*(N,N'-bistrimethylsilyl)benzamidinato)yttrium complexes see the following:
        a) Duchateau, R.; Wee, C.T. van; Meetsma, A.; Teuben, J.H. *J. Am. Chem. Soc.* **1993**, *115*, 4931.
        b) Bijpost, E.A.; Duchateau, R.; Teuben, J.H. *J. Mol. Cat.* **1995**, *95*, 121. For yttracarborane
        complexes: Oki, A.R.; Zhang, H.; Hosmane, N.S. *Organometallics* **1991**, *10*, 3964.
        For scandium benzamidinate complexes see: Hagadorn, J.R.; Arnold, J. *Organometallics* **1996**,
        15, 984.

[136]   Schaverien, C.J.; Frijins, J.H.G.; Heeres, H.J.; van den Hende, J.R., Teuben, J.H.; Spek, A. *J. Chem. Soc.* **1991**, 642.

[137]   Evans, W.J.; Boyle, T.J.; Ziller, J.W. *Organometallics* **1993**, *12*, 3998.

[138]   Duchateau, R.; Meetsma, A.; Teuben, J.H. *Organometallics* **1996**, *15*, 1656.

[139]   Booij, M.; Kiers, N.H.; Heeres, H.J.; Teuben, J.H. *J. Organomet. Chem.* **1989**, *364*, 79.

[140]   Hazin, P.N.; Huffman, J.C.; Bruno, J.W. *Organometallics* **1987**, *6*, 23.

[141]   van der Heijen, H.; Schaverien, C.J.; Orpen, A.G. *Organometallics* **1989**, *8*, 255.

[142]   Schaverien, C.J. *Organometallics* **1992**, *11*, 3476.

[143]   Marsh, R.E.; Schaefer, W.P.; Bazan, G.C.; Bercaw, J.E. *Acta. Cryst.*, C48, 1416.

4   Titanocenes

Rüdiger Beckhaus

4.1   Introduction

The bright green, so-called *'titanocene'*, obtained from the reaction of $TiCl_2$ with CpNa and first reported in 1956 by A. K. Fischer and G. Wilkinson [1], marked the discovery of one of the most extensively utilized complex systems in organotransition metal research. In this chapter, complexes exhibiting a $[(\eta^5\text{-}Cp)_2Ti]$ unit or derivatives with comparable substitution pattern (Cp', Cp*) will be discussed with the exclusion of complexes containing only one Cp ligand. Typical η^5-Cp type ligands are summarized in Scheme 4-1. Free titanocene $[Cp_2Ti]$, as a formally 14 electron species, is obviously not isolable. However, a wide range of derivatives with additional neutral or anionic ligands exists, characterized by the typical *bent-metallocene* structure and chemical reactions which take place on the more open side of the electronically flexible and sterically compact $[Cp_2Ti]$ fragment.

The chemistry of titanium is characterized by its multiple oxidation states. For such titanocenes the available oxidation states are normally +2, +3, +4 (Scheme 4-1). In this chapter, the organometallic properties of titanocenes will be discussed from the point of view of coordination chemistry. During the last few years a wide range of titanocene derivatives exhibiting different substitution patterns have been synthesized and characterized. This article focuses on selected topics in the synthesis and reactivity pattern of titanocene derivatives. For further details, excellent and extensive summaries already exist [2–7].

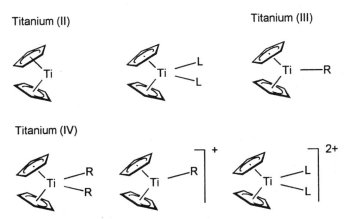

Scheme 4-1. General ligand and structure types of titanocene complexes.

Typical η^5-Ligands

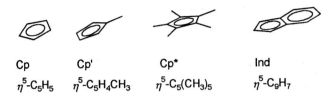

Cp	Cp'	Cp*	Ind
η^5-C$_5$H$_5$	η^5-C$_5$H$_4$CH$_3$	η^5-C$_5$(CH$_3$)$_5$	η^5-C$_9$H$_7$

4.1.1 Bonding Modes of Cyclopentadienyl Ligands in Titanium Complexes

At least 10 distinct bonding modes for the C$_5$H$_5$ ligand have been structurally char-acterized to date [8]. Of these, the η^5-C$_5$H$_5$, η^3-C$_5$H$_5$, and η^1-C$_5$H$_5$ bonding modes are of particular interest due to their interconversion chemistry (*ring slippage*) [8] and the consequent implications for the design of catalytic systems. In 1971 Cotton reported the variable-temperature ^{1}H NMR studies of [(η^5-C$_5$H$_5$)$_2$Ti(η^1-C$_5$H$_5$)$_2$] (**1**), for which he was able to observe η^1 to η^5 cyclopentadienyl interconversions [9, 10].

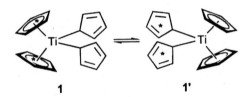

1   **1'**

Although no intermediates were observed, Cotton later proposed an [(η^1-C$_5$H$_5$)(η^4-C$_5$H$_5$)$_2$(η^5-C$_5$H$_5$)Ti] intermediate for this η^1 to η^5 interconversion [8].

4.1.2 Synthesis of Titanocene Complexes

Generally, Cp ligands are introduced by using the salt metathesis reaction. The common derivatives *e.g.* [Cp$_2$TiCl$_2$] are available by this method (Scheme 4-2, path **a**) [11], but, the preparation of the permethylated [Cp*$_2$TiCl$_2$] starting from TiCl$_4$ and Cp*Li failed. The only product of this reaction is the monocyclopentadi-enyl derivative [Cp*TiCl$_3$] [12]. Higher yields, up to 65% of purple–brown [Cp*$_2$TiCl$_2$], were obtained when TiCl$_3$ was treated with Cp*Na or Cp*Li in dimethoxyethane solution in a molar ratio of 1:3, and then oxidized by treating with hydrochloric acid [13, 14].

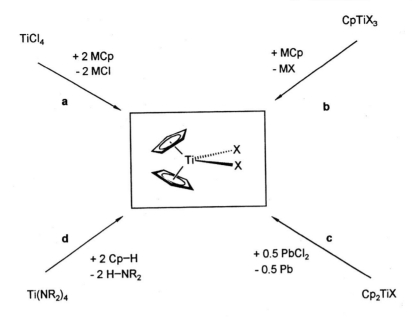

Scheme 4-2. Possible synthetic pathways **a–d** for titanocene complexes.

The use of cyclopentadiene-substituted tin compounds to prepare monocyclopenta-dienyl transition metal complexes, first reported by Abel and co-workers [15], is now an established method for the synthesis of [CpTiCl$_3$] and related compounds [16, 17]. Selective and high-yielding preparations of mixed ring complexes [Ti(η^5-C$_5$H$_5$)(η^5-C$_9$H$_7$)Cl$_2$] and [Ti(η^5-C$_5$H$_5$)(η^5-C$_5$H$_4$Bu^t)Cl$_2$] are reported by McCamley and co-workers using the corresponding tin reagents (path b) [18]. The reaction between Cp*SiMe$_3$ and group 4 element tetrahalides MX$_4$ (M = Ti, X = Cl, Br, I; M = Zr, Hf, X = Cl) gives the corresponding [(η^5-C$_5$Me$_5$)MX$_3$] derivatives in near quantitative yields in a one-step procedure without the need for further purification [19]. Tetravalent monocyclopentadienyl complexes are useful starting materials for lower valent titanium complexes like [CpTiCl$_2$], from which mixed [CpCp^{R}TiCl] complexes become available by halide exchange reaction [20]. The oxidation of this type of complex using PbCl$_2$ as an oxidizing agent is very useful for the prep-aration of mixed L$_2$TiCl$_2$ complexes or substituted derivatives exhibiting function-alized side arms (path c) [20]. Lead dichloride is a convenient reagent to cleanly oxidize a wide variety of tervalent [Cp*$_2$TiR] complexes to their tetravalent ana-logs. Monobromide complexes can be obtained by the same procedure using PbBr$_2$ [21]. The oxidation proceeds smoothly at room temperature, but also takes place at a reasonable rate at temperatures as low as –30°C.

Amine elimination reactions from titanium amides and Cp-H, first reported by Chandra and Lappert [22] for the synthesis of [CpTi(NMe$_2$)$_3$] from [Ti(NMe$_2$)$_4$] and Cp-H (72%) are now a well developed method for the preparation of a wide range of other metallocene derivatives (path **d**) [23–25]. However, lower reactivity is observed in titanium complexes than in the corresponding zirconium complexes and only monocyclopentadienyl type complexes are normally formed [26]. The reactivity of the titanium amides can be increased by using the less crowded titanium azetidide complexes instead of the [Ti(NMe$_2$)$_4$], as shown for the formation of **2** [23].

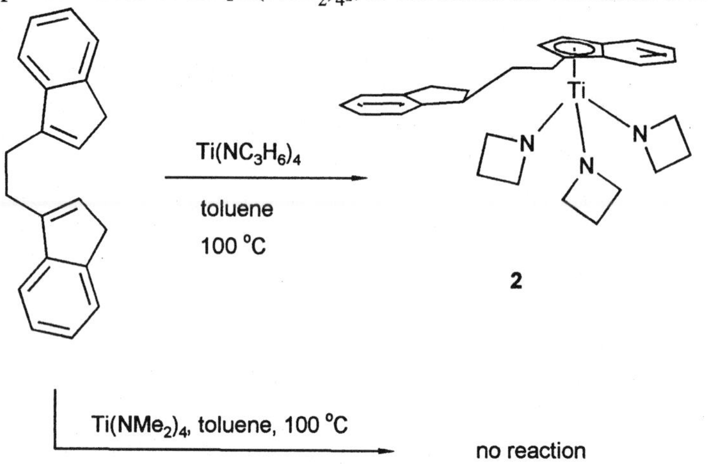

Scheme 4-3. Amine elimination reaction for titanium complexes.

However, mixtures of TiCl$_4$, Cp-H and HNEt$_2$ can be used for the preparation of [Cp$_2$TiCl$_2$] without using the highly sensitive NaCp [27]. By ligand exchange reactions with the corresponding chlorides, fluorides become available in reactions with NaF, whereas bromides and iodides can be prepared using the corresponding boron halides (BBr$_3$, BI$_3$) [28, 29].

Table 4-1. Selected titanocene complexes, syntheses and properties

Compound	Properties	Synthesis	Ref.
[Cp$_2$TiCl$_2$]	bright-red crystals mp 287-291 °C, ^{1}H NMR (CDCl$_3$) δ 6.60 (s)	TiCl$_4$ + 2 CpNa, 90%	[11]
[Cp*$_2$TiCl$_2$]	dark purple crystals from CHCl$_3$, long needles from CCl$_4$, mp 273 °C,	TiCl$_3$ + Cp*Li, oxidation, + HCl, 65 %	[13]
[Cp*$_2$TiF$_2$]	orange crystals, mp 208-210 °C, ^{1}H NMR (CDCl$_3$) δ 1.93 (s)	halide-exchange using NaF, 41%	[28]
[Cp*CpTiF$_2$]	bright orange crystals, mp 153-155 °C, ^{1}H NMR (CDCl$_3$) δ 6.16 (5H, t J_{HF} = 1.34 Hz) 1.99 (15H, s)	halide-exchange using NaF, 25%	[28]
[Cp*$_2$TiBr$_2$]	black crystals, mp 260-262 °C, ^{1}H NMR (CDCl$_3$) δ 2.09 (s)	halide-exchange using BBr$_3$, 54 %	[28]
[Cp*CpTiBr$_2$]	black crystals, mp 250-222 °C, ^{1}H NMR (CDCl$_3$) δ 6.37 (5H, s) 2.12 (15H, s)	halide-exchange using BBr$_3$, 58%	[28]

4.1.3 Sources of the Titanocene Fragment as a Versatile Building Block

In many applications the bent titanocene fragment of the 'Cp$_2$Ti' type is useful in coordinating new unsaturated ligands or in finding new types of C–C bond as well as other coupling reactions. This means it becomes necessary to use titanocene-generating complexes, because a Cp$_2$Ti fragment does not exist without further stabilizing ligands (Scheme 4-4). In practice, ligand exchange reactions have to take place between selected titanocene precursors and substrates to give the required products. Depending on the reaction conditions, the substrates and the final products, different types of starting material are used. Carbonyls and phosphines can be substituted by stronger π-acceptor ligands, but in many cases, only one ligand can be substituted at a time or the leaving ligand itself is involved in unwanted subsequent reactions. The generation of the titanocene fragments from alkyl derivatives is possible in photochemical reactions, but side reactions sometimes occur under these reaction conditions.

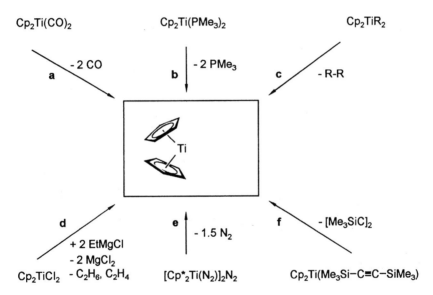

Scheme 4-4. Sources for the generation of titanocene building blocks.

In organic syntheses, path **d** is very useful [30]. By this path the reductive cyclization of ester-containing enynes can be achieved in order to synthesize bicyclic cyclopentenones and iminocyclopentenes. Nitrogen complexes of titanocenes can also be used as titanocene sources. But due to the high sensitivity of the nitrogen complexes to air and moisture, the handling of these complexes requires much experience. The bistrimethylsilyl acetylene complexes are excellent reagents in generating Cp$_2$Ti as well as Cp*$_2$Ti building blocks [31]. Generally, the reactivity of these complexes is characterized by the liberation of the inert silylalkyne, which does not react further, as carbon monoxide or phosphines frequently do. In some cases, the olefin complex [Cp*$_2$Ti(η^2-C$_2$H$_4$)] can also act as a titanocene generating complex. Generally, Cp* complexes are of higher stability and more soluble than the equivalent nonmethylated Cp complexes. Cp*$_2$Ti complexes are mostly monomers, whereas the normal Cp$_2$Ti complexes react to form dimers or oligomers. As a result of this, some Cp* complexes are available for which the equivalent Cp systems are unknown. Additionally, there are some reaction paths, which are quite different, depending on whether Cp or Cp* complexes are used. Some examples are given in Table 4-2.

Table 4-2. Selected complexes and reactions courses of Cp$_2$Ti *vs* Cp*$_2$Ti complexes

Complexes	Cp$_2$Ti	Ref.	Cp*$_2$Ti	Ref.
L$_2$TiN$_2$ complexes	in solution, or additional ligands [Cp$_2$Ti(PMe$_3$)]$_2$(μ-N$_2$)	[4, 32]	[Cp*$_2$Ti(N$_2$)]$_2$N$_2$	[33]
L$_2$Ti(η^2-C$_2$H$_4$) complexes	in solution	[34]	Cp*$_2$Ti(η^2-C$_2$H$_4$)	[35]
Reactions				
L$_2$Ti(CH=CH$_2$)	reductive elimination → Cp$_2$Ti + C$_4$H$_6$	[36]	α-H elimination → [Cp*$_2$Ti=C=CH$_2$] + H$_2$C=CH$_2$	[37-39]
L$_2$Ti(R)OH (R: CH=CH$_2$)	condensation reaction → Ti-O-Ti species	[40]	monomeric complexes, no condensation	[41]
"L$_2$Ti" + ROH	redox reaction → (Cp$_2$TiOR)$_2$ + H$_2$		oxidative addition → Cp*$_2$TiH(OMe)	[42]

4.2 Electronic Properties of Titanocenes

Normal bis(η^5-cyclopentadienyl) transition metal complexes such as ferrocene are highly symmetric molecules with parallel cyclopentadiene rings. In the bent bis(η^5-cyclopentadienyl)transition metal complexes the rings are not parallel *i.e.* the angle between the normals to the planes of the cyclopentadienyl ligands is less than 180°, and there are other ligands present at the metal center. A bent Cp$_2$M fragment has C_{2v} symmetry if the Cp ligands have an eclipsed geometry, but only C_s symmetry if the rings are staggered. Generally, the chemistry of the bent metallocene fragment is determined by the three low-lying frontier orbitals 1a$_1$, b$_2$ and 2a$_1$ [43]. All three orbitals have significant density in the mirror plane of the metallocene fragment. The b$_2$ orbital is chiefly dyz in character. The two a$_1$ orbitals each contain some contribution from the s and pz orbitals as well as dx^2–y^2 and

dz^2 contributions. The $1a_1$ orbital can be described as similar to a dy^2 orbital. This leads to the well established bent metallocene geometry, for which the electronic constraints are well documented both through their extensive derivative chemistry and through detailed quantum chemical calculations [44]. For a *bent*-metallocene fragment, such as Cp_2Ti, the ancillary ligands interact with the three low lying orbitals which project in the so-called 'equatorial binding plane' or 'binding wedge' of the metallocene fragment. The orientation of the ligands will then be dominated by the interactions of their p or π-symmetry orbitals with the π-symmetry combination in this plane [45, 46].

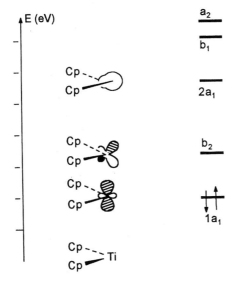

Fig. 4-1. Valence orbitals of bent titanocenes.

4.2.1 Titanocene Complexes in Low Oxidation States

$[Cp^*_2TiX]$ complexes (X: monodentate, one-electron ligand, *e.g.* halide, amide, alkoxide) appear to be ideal for the study of metal-to-ligand π-bonding. They are monomeric, unlike $[Cp_2TiX]_2$ in which the electrons are coupled. The trivalent decamethyltitanocenes have an empty b_2 orbital available for π-bonding, unlike $[Cp_2MX_2]$ in which the b_2-orbital is used for σ-bonding. From the electronic and the EPR spectra of a series of d^1-bent metallocene compounds Cp^*_2TiX the order of decreasing π-donor ability was found to be $N(Me)H \approx NH_2 \approx OMe > OPh \approx F > NMePh > Cl > Br > I > H$ [47]. Among the known alkyls, the π-donor ability was $Et > Me > n\text{-}Pr \approx CH_2CMe_3 > CH_2C_6H_5$ which is rationalized in part by a β-agostic interaction in the case of Et.

This observation is in accordance with the general observation, that the reactivity of bent titanocene complexes is determined by the electron deficient character of the metal center in combination with the orientation of the acceptor orbitals in the equatorial plane of the metallocene fragment. This can also be illustrated by the orientation of the substituents on nitrogen atoms in $[Cp^*_2TiNR^1R^2]$ complexes. If there is no steric hindrance, the substituents on the nitrogen atom ($R^1 = R^2 = H$ [48]; $R^1 = H$, $R^2 = Me$ [47]) are rotated out of the equatorial plane of the metallocene to maximize overlap between the nitrogen lone pair (located in the pz orbital) and the lowest unoccupied molecular orbital (LUMO) of the metal. This effect is also found in the case of $[Cp^*_2Hf(H)(NHMe)]$ [49]. If larger substituents are present, as in $[Cp^*_2TiN(CH_3)(Ph)]$, the methyl and the phenyl group are now located in the equatorial plane, and consequently a $d\pi$–$p\pi$ interaction is no longer possible [50].

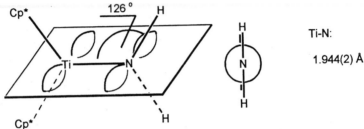

E. Brady, et. al. 1994

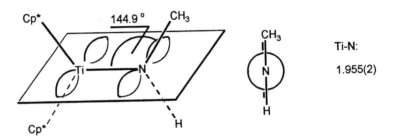

R. A. Andersen, et. al. 1996

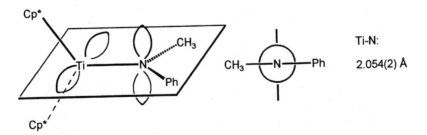

J. Feldman, et. al. 1991

Scheme 4-5. Schematic drawing of structures of $[Cp^*_2TiNR_2]$ complexes, Ti–N distances depends on the orientation of the lone pair on nitrogen atom.

A similar orientation of the substituents can be observed in carbene complexes and their cumulogous derivatives. The character of the metal fragment MO b_2 determines that in the structure of [$Cp_2Ti=C=CH_2$] only the $C=CH_2$ rotamer shown allows a Ti–C_a π-bond ('back bonding') [38]. In accordance with this, model calculations on [$Cp_2Ti=C=CH_2$] yield a rotational barrier of 134 kJ mol^{-1} [38]. For asymmetrically substituted vinylidenetitanium compounds it is possible to show a potential rotation around the Ti=C bond in solution [51]. For the titanaallenylidene, no structural dynamics are observed, and NMR data indicate that all C atoms of the cumulene lie in a mirror plane of the molecule [52].

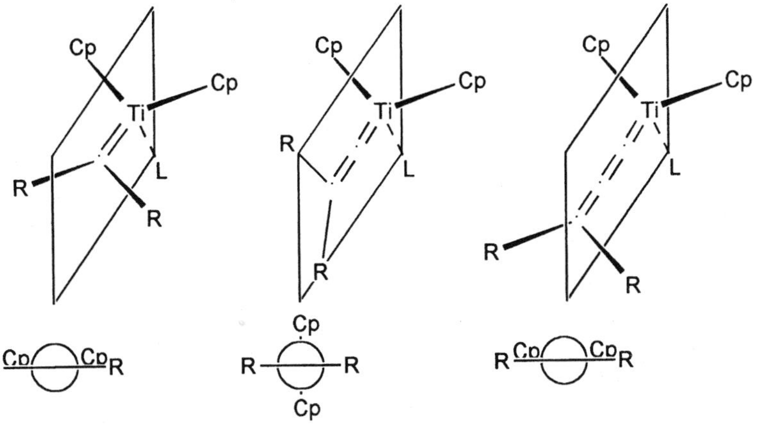

Scheme 4-6. Geometries of titanocenemethylene, -vinylidene, and -allenylidene complexes.

4.2.2   Substituents on the Cp Rings

Since its first general use as a ligand [12], the pentamethylcyclopentadienyl moiety has attracted considerable attention and widespread use. It is commonly known that replacement of the cyclopentadienyl group (Cp) by the pentamethylcyclopenta- dienyl ligand (Cp*) in certain bis- and mono(cyclopentadienyl) complexes of transition metals results in significant changes in chemical reactivity, stability, sensitivity to oxidation, and many other properties. ESCA studies of a series of titanocenes, zirconocenes, and hafnocenes have shown that substitution of methyls for hydrogens on the cyclopentadienyl ligand results in a significant electronic effect, as reflected by the binding energies of the inner-shell electrons of the metal. The substitution of two pentamethylcyclopentadienyl groups for cyclopentadienyl ligands is approximately equivalent to a one-electron reduction of the complexed metal [53]. This replacement (Cp→Cp*) in metallocene derivatives leads to increased stability of complexes owing to the inductive and hyperconjugation effects of the methyl group and to the increased steric shielding of the coordination space about the central metal atom. The inductive and hyperconjugative effects influence the chemical and physical properties of the complexes by increasing the electron density at the metal atom, and decreasing the ionization potential of the Cp ligand [54, 55]. In the titanocene dihalide series the substitution of Cp with Cp* lowers the oxidation potential by 0.5–0.6 V and decreases the binding energies of the Ti(2p) inner shell electrons by 0.8 eV [28, 53, 56]. The ^{49}Ti NMR chemical shifts in Cp* derivatives appear well downfield from where they are in the equivalent Cp complexes, thus following the inverse halogen-dependence found for titanocene dihalides [28]. The methyl substituents have a substantial effect on the energy of valence electrons in titanocene dihalides and cyclopentadienyltitanium trihalides as has been observed by photoelectron spectroscopy [57]. An excellent linear correlation of binding energy (Ti2p) with the number of methyl substituents (1–10) was found, indicating that the electronic influence of methyl groups on the titanium center is additive [58]. Furthermore, the trialkylsilyl group was found to be slightly more electron-donating than methyl [59].

4.3 Ligands in the Coordination Sphere of Titanocenes

4.3.1 Olefins

Starting from $[Cp_2Ti(PMe_3)_2]$ solutions of $[Cp_2Ti(\eta^2\text{-}C_2H_4)]$ can be prepared [34]. The same complex is generated from $[Cp_2TiCl_2]$ with magnesium/1,2-dibromethane in the presence of ethylene [60]. However, a well characterized olefin complex has been prepared by reduction of $[Cp*_2TiCl_2]$ with sodium amalgam in toluene at 25°C in the presence of ethylene. The resulting olefin complex is isolated in 80% yield in the form of green crystals. The bonding of the ethylene ligand is characterized by extensive backbonding. The C–C bond (1.438(5) Å) is long compared with that of free ethylene (1.337 Å) and the hydrogen atoms are bent away from the metal so that the CH_2 plane forms an angle of 35° with the C–C axis [35, 61]. The metallacyclopropane form was found to be lower in energy than the π-complex form for $[Cl_2Ti(C_2H_4)]$ complexes, using wave function descriptions [62]. A wide range of substitution and metal centered cycloaddition reactions can be carried out using the $[Cp*_2Ti(\eta^2\text{-}C_2H_4)]$ complex [35, 63, 64].

4.3.2 Acetylenes

The first titanocene–acetylene complexes were obtained by partial replacement of carbon monoxide [65, 66] or a trialkylphosphine [67–69] by diphenylacetylene or per(pentafluorophenyl)acetylene [70]. The complexes of dialkylacetylenes and diphenylacetylene were prepared either by the displacement of ethylene from

[Cp*$_2$Ti(η^2-C$_2$H$_4$)] [35] or by acetylene coordination to further titanocene generating systems (L$_2$TiCl$_2$/Mg/THF) L: C$_5$H$_x$Me$_{5-x}$ (x = 0–5) [71–74, 31]. As excellent π-acceptors, alkynes can act as a four- as well as a two-electron ligand. While the shift in the $\nu_{C\equiv C}$ frequency on coordination of the alkyne proves to be rather insensitive to the coordination mode of the alkyne, the ^{13}C NMR chemical shift is apparently a better guide. As a general rule $\delta_{C\equiv C}$ values of 100–150 are taken to indicate that alkyne acts as two-electron donor, for example when coordinated to nickel complexes, whereas alkynes as four-electron ligands give $\delta_{C\equiv C}$ values of 190–270, when coordinated for example to titanium. In agreement with this differing π-acceptor behavior, the chemical shift differences between free and coordinated disubstituted alkynes C$_2$R$_2$ are much higher for early transition metals ($\Delta\delta_{C\equiv C}$ = 106–140) than for nickel ($\Delta\delta_{C\equiv C}$ = 38–55). The complex [Cp$_2$Ti(η^2-PhC≡CPh)] has been isolated by reduction of the [Cp$_2$TiCl$_2$] with magnesium in the presence of the alkyne as a brown microcrystalline solid which decompose at 117–119°C. The extensive backbonding contribution to the alkyne in this complexes is confirmed by the low ^{13}C chemical shift ($\delta_{C\equiv C}$ = 196) as well as the low $\nu_{C\equiv C}$ IR frequency of 1712 cm^{-1}, which has led to the consideration of a metallacyclopropenylium resonance structure.

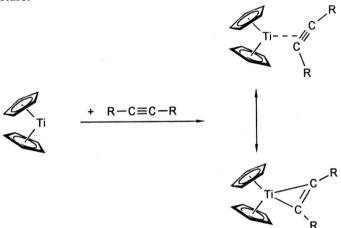

Scheme 4-7. Titanocene alkyne complexes.

The dialkylacetylene and diphenylacetylene complexes are highly reactive towards acetylenes to give titanacyclopentadiene derivatives and low yields of catalytically formed hexasubstituted benzenes. Thus, only the [L$_2$Ti(η^2-PhC≡CPh)] (L: Cp, Cp*) complexes were isolated and characterized by spectroscopic and chemical means [35, 72]. In contrast complexes with mono(trimethylsilyl)acetylene ligands RC≡CSiMe$_3$ (R=Ph, ^{t}Bu,^{n}Bu) as well as the bis(trimethylsilyl)acetylene complexes [Cp$_2$Ti(η^2-Me$_3$SiC≡CSiMe$_3$)] [73] and [Cp*$_2$Ti(η^2-Me$_3$SiC≡CSiMe$_3$)] [74] are more stable and were obtained in high yields. The X-ray crystal data were published for the latter complex and both complexes were characterized by spectroscopic data [75]. These two complexes do not form titanacyclopentadiene complexes and do not undergo addition reactions. Instead the silylalkyne $Me_3SiC\equiv CSiMe_3$ is displaced by more strongly coordinating reagents [31].

Table 4-3. Selected analytical data of acetylene complexes $[L_2Ti(\eta^2\text{-}RC\equiv CR')]$
(L = Cp, Cp*)

	R = R' = SiMe₃	R = SiMe₃, R' = Ph	R = R' = Ph
R-C≡C-R			
$\nu(C\equiv C)$ [cm⁻¹]	2107	2160	2223
$\delta(C_2)$	114.0	92.5; 104.4	90.1
distance (C≡C) [Å] [76a]	1.215 (2)	-	1.193 (5)
angle (R-C≡C) [°]	179.4 (1) [76b]	-	178.2 (4) [76c]
[Cp₂Ti(η²-R'≡CR)]	golden crystals, mp 80-81 °C [73]	brown crystals, mp 98-99 °C [75]	brown crystals, mp 117-119 °C [72]
$\nu(C\equiv C)$ [cm⁻¹]	1687	1686	1713
$\Delta\nu(C\equiv C)$ [cm⁻¹]	420	474	510
$\delta(C_2)$	244.9	213.0; 219.6	196.5
$\Delta\delta(C_2)$*	130.9	120.5; 115.2	106.4
distance (C≡C) [Å]	-	1.289(4) 1.279(4)	-
angle (R-C≡C) [°]	-	148.2(2); 140.8(3) 151.9(2); 141.0(3)	-
[Cp*₂Ti(η²-RC≡CR)]	golden crystals, mp 191-192 °C [74, 75]	dark red crystals, mp 135 °C [75]	brown crystals [77]
$\nu(C\equiv C)$ [cm⁻¹]	1598 / 1563	1625	1647
$\Delta\nu(C\equiv C)$ [cm⁻¹]*	527	535	576
$\delta(C_2)$	248.5	213.2; 224.9	200.9
$\Delta\delta(C_2)$*	134.5	120.7; 120.5	110.8
distance (C≡C) [Å]	1.309(4)	1.308(3)	-
angle (R-C≡C) [°]	134.8(3) 136.8(3)	140.5(2) 138.1(2)	-

$\Delta\nu(C\equiv C) = |\nu(C\equiv C)_{acetylene} - \nu(C\equiv C)_{complex}|$, $\Delta\delta(C_2) = |\delta(C_2)_{acetylene} - \delta(C_2)_{complex}|$
*⁾ two independent molecules.

The silylalkyne complexes [Cp$_2$Ti(η^2-Me$_3$SiC≡CSiMe$_3$)] and [Cp*$_2$Ti(η^2-Me$_3$SiC≡CSiMe$_3$)] are suitable reagents for the generation of [Cp$_2$Ti] and [Cp*$_2$Ti] species because they are (a) simple prepared (b) thermally stable at ambient temperature and (c) the coordinated Me$_3$SiC≡CSiMe$_3$ does not undergo addition reactions with an excess of substituted acetylenes.

4.3.3 Carbonyls

The synthesis and reactivity of carbonyl complexes of titanium has been reviewed extensively [2, 79], and detailed syntheses of [Cp$_2$Ti(CO)$_2$] and [Cp*$_2$Ti(CO)$_2$] have been described [80]. The mixed-ligand complex [CpCp*Ti(CO)$_2$] is similarly prepared by reduction of the corresponding titanocene dichloride with zinc powder. The electronic structure of [Cp$_2$Ti(CO)$_2$] has been researched by gas-phase He^I and He^{II} photoelectron spectroscopy and X$_a$ molecular orbital calculations. Of high diagnostic value are the ν_{CO} frequencies of some carbonyl derivatives as summarized in Table 4-4.

Table 4-4. Infrared spectroscopic data for bis(cyclopentadienyl)titanium carbonyl complexes

Compound	ν_{CO} [cm^{-1}]	Measurement	^{1}H NMR data δ	Ref.
[Cp$_2$Ti(CO)$_2$]	1977, 1899	*n*-hexane	4.62	[80]
[Cp*$_2$Ti(CO)$_2$]	1940, 1858	*n*-hexane	1.67	[80]
[CpCp*Ti(CO)$_2$]	1956, 1875	*n*-hexane	-	[67]
[Cp$_2$Ti(CO)PF$_3$]	1932	*n*-hexane	4.75 (J_{HP} = 3.5 Hz)	[70]
[Cp$_2$Ti(CO)PMe$_3$]	1864	*n*-hexane	4.62 (J_{HP} = 2.6 Hz)	[81]

Kinetic studies were performed for CO substitution reactions of [(η^5-L)$_2$Ti(CO)$_2$] derivatives, where (η^5-L) = Cp, Cp* or Ind [82]. Nucleophiles used for these reactions include PMe$_2$Ph, PMePh$_2$, PPh$_3$, P(OEt)$_3$, P(*n*-Bu)$_3$, and CO. The reaction rates of the titanium compound were found to be first order in the substrate and zero order in the entering nucleophile at nucleophile concentrations which gave the limiting reaction rate. The results indicate a dissociative mechanism. Activation parameters for these reactions are also in agreement with a dissociative process. In contrast, reaction rates of the zirconium and hafnium compounds are also first order in both substrate and entering nucleophile, indicating an associative mechanism. This mechanistic difference may be attributed to steric considerations caused by the smaller size of titanium [82].

4.3.4 Phosphines

$[Cp_2Ti(PMe_3)_2]$ is prepared in high yield by reduction of $[Cp_2TiCl_2]$ with magnesium in THF in the presence of PMe_3 [83–85]. It has proved to be a very versatile starting material for a wide range of reactions. In such reactions the PMe_3 ligands are easily displaced under milder conditions than from $[Cp_2Ti(CO)_2]$. Both ligand exchange and oxidative addition reactions are common.

4.3.5 Hydrides

The first structurally characterized monomeric bis(cyclopentadienyl)titanium(III) hydride, $[(PhC_5Me_4)_2TiH]$, was synthesized by hydrogenolysis of $[(PhC_5Me_4)_2TiMe]$. This hydride was found to be a monomeric bent sandwich complex by X-ray diffraction methods, and it can be concluded that the $[Cp*_2TiH]$ analog possesses a similar molecular structure by comparison of the spectroscopic and reactivity data [86]. A red–brown solution of the hydride is formed by exposing a green pentane solution of $[(PhC_5Me_4)_2TiMe]$ to 1 atm H_2 at room temperature, from which red–brown crystals can be separated. The Cp–Ti–Cp angle (Cp–Centroid) is found to be very large (150.84°), which apparently occurs because the hydride ligand is very small. The Ti–H distance (1.768(15) Å) is comparable to other monomeric hydrides.

Table 4-5. Selected structural data of titanium hydrides

Compound	Ti-H [Å]	Ref.
$[(PhC_5Me_4)_2TiH]$	1.768(15)	[86]
$[Cp*(C_5Me_4CH_2(C_5H_3MeN)TiH]$	1.70(4)	[87]
$[CpTi(CO)_2H]$	1.75(7)	[88]
$[CpTi(dmpe)H]$	1.96(6)	[89]
$[\mu\text{-}\eta^5{:}\eta^5\text{-}C_{10}H_8][(\eta^5\text{-}C_5H_5)Ti(\mu\text{-}H)]_2$	1.73 (av)	[90]
$[\mu\text{-}C_5H(CH_3)_2(CH_2)_2][(C_5HMe_4)Ti(\mu\text{-}H)]_2$	1.81	[91a]
$rac\text{-}[(C_2H_4(\eta^5\text{-tetrahydroindenyl})_2)Ti(\mu\text{-}H)]_2$	1.90	[92]

As expected, oxidation of $[(PhC_5Me_4)_2TiH]$ with $PbCl_2$ yields $[(PhC_5Me_4)_2Ti(H)Cl]$ (δ_{Ti-H} = 5.16), while the reaction of $[Cp*_2TiH]$ gives $[Cp*_2Ti(H)Cl]$ [86].

4.3.6 Titanium Ligand Multiple Bonds

4.3.6.1 Carbene Complexes

The chemistry of transition-metal alkylidenes and metallacycles has attracted much recent attention and has resulted in a number of useful synthetic transformations [39, 93]. Titanocene carbene complexes can be synthesized *via* routes starting from the Tebbe reagent (**b**), [94–97] from titanacyclobutanes [98–100] (**c**) and from alkyl derivatives (**a**) [101–115], which act as sources of the parent titanium methylidene derivatives (Scheme 4-8) [116, 117]. These carbene complexes normally act as intermediates in subsequent reactions [118], but also a few examples of titanium group metal carbene complexes have been isolated and characterized [119–121]. The variety of reactions available for the formation of carbene complexes of electron-deficient transition metals indicates a high synthetic potential. Preparative applications of carbene complexes generated by path **a**, as opposed to those obtained by path **b**, frequently give higher yields and selectivities, and a wider choice of reaction media and substituents on the carbene ligand as well as more tolerant functional groups, when used in stoichiometric and catalytic reactions.

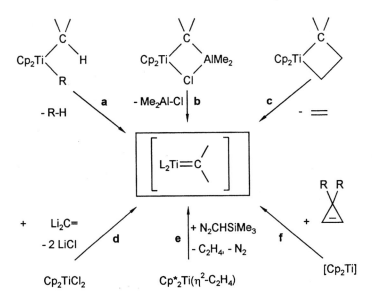

Scheme 4-8. Possible synthetic pathways (**a–f**) for complexes with Ti–C double bonds.

Even though α-H eliminations have been well-established since the discovery of Schrock carbenes, this method (**a**) for the generation of [L_2Ti=CR$_2$] complexes has only recently been successfully applied to electron-deficient transition metal complexes. This is in contrast to the extensive use of dimetallic species, of which for example the Tebbe reagent (path **b**) has assumed a prominent role in organic chemistry. Also, dititanacyclobutane a bismethylene complex reacts rapidly and quantitatively with PMe_3 to yield $[Cp_2Ti=CH_2(PMe_3)]$ [117]. The cleavage of bis-μ-methylene complexes fails, however, for other metal combinations, such as Ti–Zr, Zr–Zr, or Ti–Si [117, 122–125].

Metallacyclobutanes derived from the Tebbe reagent have also enhanced our knowledge of carbenoid transition metal compounds (path **c**). The use of titanacyclobutanes as the starting material has the advantage over the Tebbe reagent that additional Lewis acidic components such as $[Me_2AlCl·L]$ do not have to be separated off after the reaction. For the synthesis of titanacyclobutanes, procedures can be employed that were developed by Grubbs, using the Tebbe reagent [51, 126–138], by Bickelhaupt [139], preferably using 1,3-di-Grignard species, or by Stryker [140], employing allyl complexes. Titanacyclobutanes can also be used to generate substituted [131] or cumulogous $[Cp_2Ti=CR_2]$ species [38]. There is a problem, however, namely that singly substituted titanacyclobutanes such as α,β-dimethyltitanacyclobutane react preferably under elimination of butene [141] instead of propene. Hence, 'productive' cleavage reactions are required (Scheme 4-9).

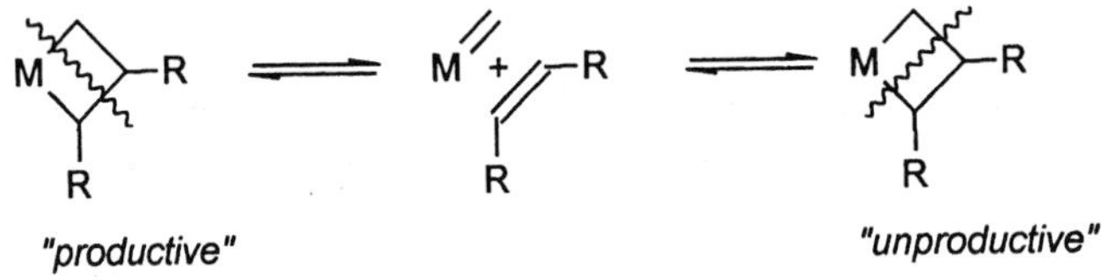

"productive"   "unproductive"

Scheme 4-9. *'Productive'* and *'unproductive'* cleavage of metallacyclobutanes.

A 'productive' cleavage occurs when highly strained cyclic olefins like 3,3-dimethylcyclopropenes are used in conjunction with Tebbe's reagent **3**. The resulting titanacycle **4** is capable of transferring α-substituted carbenes in carbonyl olefinations [131]. The reactive titanium carbene can be stabilized with PMe_3 or PMe_2Ph (**5**).

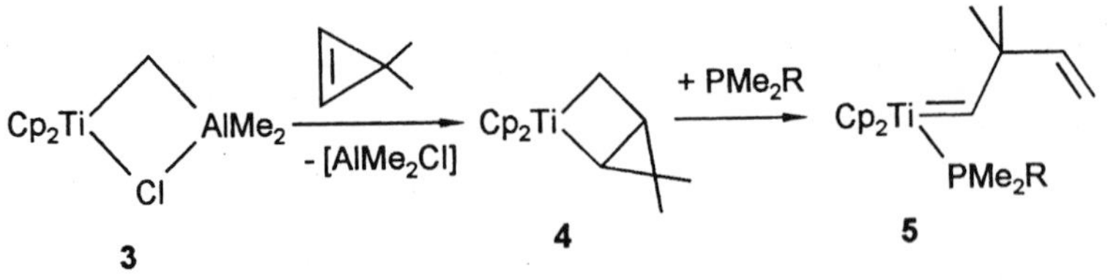

Analogous to many applications of salt metathesis reactions for generation of simple organometallic compounds, geminal dimetallic species can be used to prepare carbene complexes (path **d**). Butyllithium reacts under mild conditions with diphenylcyclopropene to yield 1,1'-dilithio-3,3-diphenylallene derivative, which can react with $[Cp_2TiCl_2]$ in the presence of PMe_3 to yield the first titanabutatriene derivative **6** [52].

The reaction of [Cp*$_2$Ti(η^2-C$_2$H$_4$)] with a diazoalkane leads to a titanium carbene complex that can be subsequently trapped (path **e**) [142]. The reaction of cyclopropenes with low-valent metal complex fragments leads to vinylcarbene complexes (path **f**). Fundamental studies were done by Binger *et al.*, who synthesized **7** in 72% yield [143].

If methylenecyclopropane is used instead of cyclopropenes, the bis-titanacumulene [Cp$_2$Ti(PMe$_3$)=C=C=Ti(PMe$_3$)Cp$_2$] **8** is formed. This cumulene is characterized by a linear [Ti=C=C=Ti] chain [144].

Table 4-6.   Selected analytical data of complexes with Ti–C multiple bonds (compare Scheme 4-10)

Compound		Distances [Å] Angle [°]	NMR data [ppm]		Ref.
	Crystal habit		^{1}H NMR, M=C(H)R	^{13}C NMR, M=C	
10	green crystals	Ti=C: 1.911(3) Ti=C-C: 158.7(2) Ti-C-H: 85(3)	12.0	280 ($^1J_{CH}$ = 95 Hz)	[119]
9		-	12.4	287.3	[119]
6	dark violett T_{dec} > 70 °C	-		264.9 (J_{PC} = 31.4 Hz)	[52]
8	dark green crystals $T_{dec.}$> 232 °C	Ti=C: 2.051(2) C=C: 1.253(2)		258.1	[144]
7	red powder, $T_{dec.}$ > 50 °C	-	12.5 -12.9	284 - 286 (J_{PC} = 29 Hz, J_{CH} = 121 Hz)	[143]
11	only in solution	-	11.8	285.2 (J_{PC} = 30.5 Hz, J_{CH} = 125.1 Hz)	[117]
12	-	-	12.3	312.9 (J_{PC} = 27 Hz, J_{CH} = 110 Hz)	[103]
5	only in solution	-	12.1	306.90 (J_{PC} = 26.6 Hz, J_{CH} = 111 Hz)	[131]

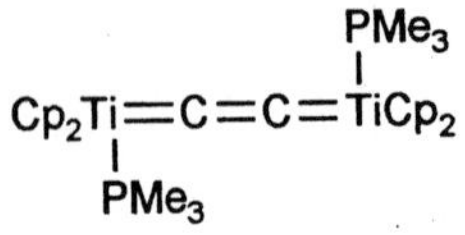

$$Cp_2Ti=C=C=TiCp_2$$

with PMe$_3$ above and PMe$_3$ below the central carbons

8

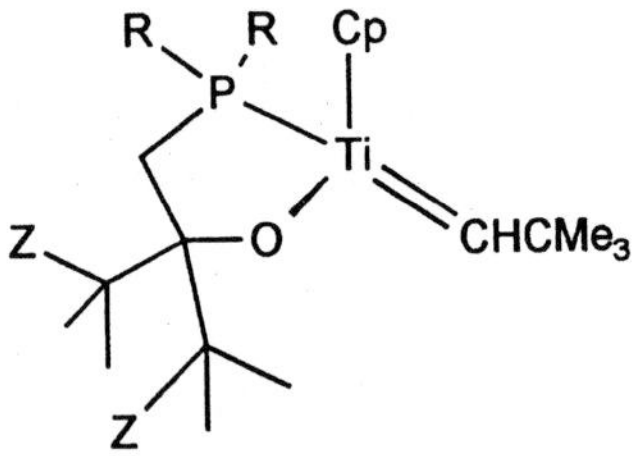

R = Ph; Z = CH$_3$ **9**
R = CH$_3$; Z + Z = 1,2-Phenylen **10**

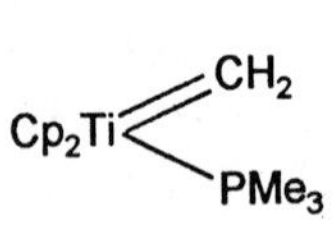

11

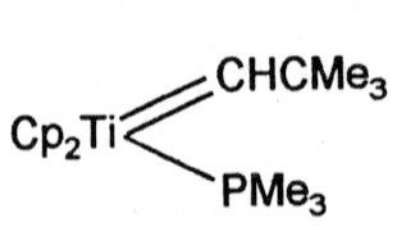

12

Scheme 4-10. Examples of isolated titanium carbene complexes.

The M–C bonds of a few structurally characterized $[L_nTi=CR_2]$ complexes are shortened to different extents depending on the ligand system. The Ti=C bond length (1.911(3) Å) in **10** is significantly shorter than that in the bisneopentyl starting material (2.120(4) Å) [119]. The Ti=C–C angle of 158.7(2)° and the Ti=C–H angle of 85(3)°, in combination with a Ti–H distance of 2.05(5) Å, indicate an α-agostic interaction, which is typical for d^0 alkylidenes. The ^1H resonance signals at low fields ($\delta = 11–13$) and the ^{13}C resonance signals of the carbene C atoms ($\delta = 265–313$) can be considered as characteristic signals of an alkylidene group.

4.3.6.2 Complexes Containing Ti–X Double Bonds (X: N, P, Chalcogenides)

α-H transfers can be used to generate multiple bonds between not only transition metals and carbon, but also between metals and other heteroatoms. In this way, synthetic building blocks incorporating Ti=Si [145], and Ti=N [146] can be obtained.

Permethyltitanocene hydride $[Cp*_2TiH]$ **13** reacts with elemental selenium or tellurium to give the products $[(Cp*_2Ti)_2(\mu\text{-}E)]$ (E = Se, Te) (**14**) [147]. These dinuclear compounds are paramagnetic and have D_{2d} (idealized) structures, as shown by X-ray structural analysis. They may be converted to diamagnetic dichalcogenides through further reaction with the appropriate chalcogen. Derivatives **15** are monomeric in the solid state and in solution.

E: Se, Te

In the context of these investigations it becomes obvious that in the series $[Cp*_2Ti=E]$, the singlet triplet gap is substantial. The Ti–O bond is very strong, and because oxygen is much smaller than tellurium, pyridine can get in to trap the terminal oxo derivative effectively [148].

The oxidation of $[Cp*_2Ti(N_2)]$ with N_2O in toluene gives a oxygen-bridged dimer (**16**), exhibiting a fulvene substructure [149]. On the other hand, the oxidation of $[Cp*_2Ti(\eta^2\text{-}C_2H_4)]$ with N_2O in the presence of pyridine as a Lewis base give the diamagnetic orange complex $[Cp*_2TiO(py)]$ (**17**). The Ti–O stretching vibration is observed at 852 cm^{-1} which shifts to 818 cm^{-1} on labeling with ^{18}O. The crystal structure confirms the presence of a terminal, very short Ti–O bond of 1.665(3) Å with substantial multiple-bond character [148]. This oxo compound **17** undergoes a [2+2] cycloaddition reaction with allene to give [Cp*₂TiOC(=CH₂)CH₂] (**18**), which contains a puckered TiOC₂ ring in the solid state. This titanacycle does not exchange with either free allene or pyridine on the NMR time scale and slowly converts to the ring-metallated product [Cp*(C₅Me₄CH₂)TiOC(=CH₂)CH₃] (**19**) in solution and to the enolate complex [Cp*₂Ti(H)(OC(=CH₂)CH₃] (**20**) in the presence of 1 atm H₂ [150].

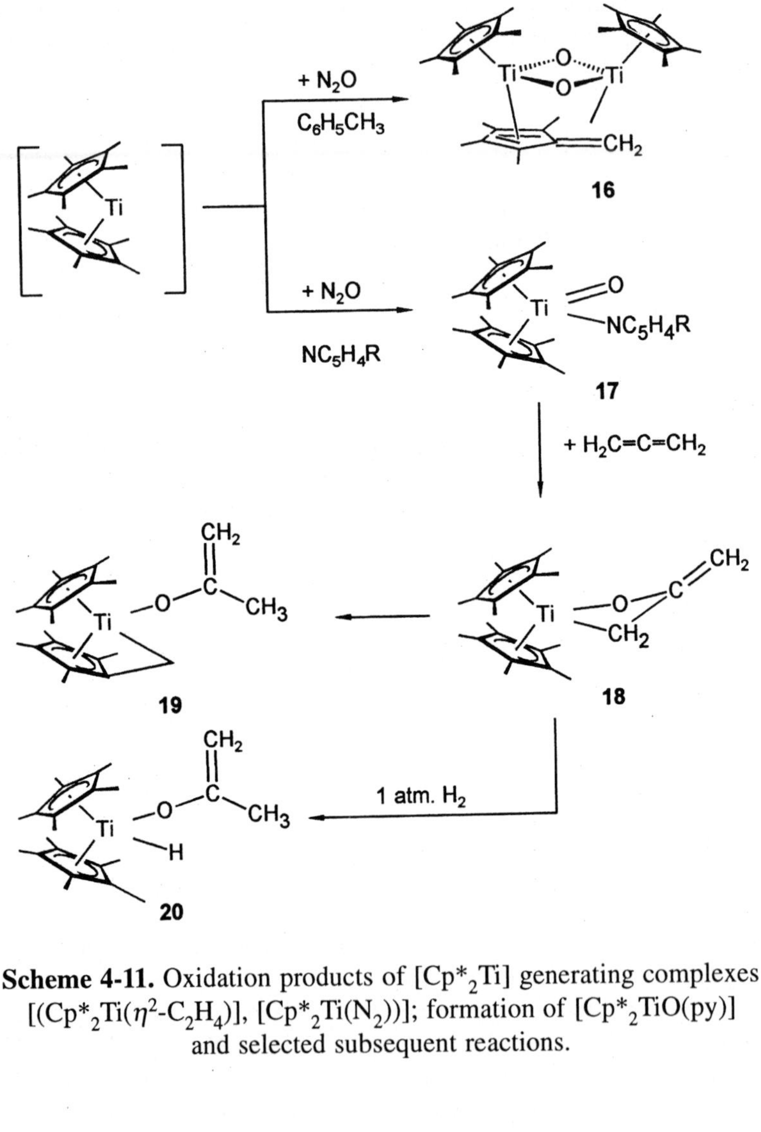

Scheme 4-11. Oxidation products of [Cp*₂Ti] generating complexes [(Cp*₂Ti(η^2-C₂H₄)], [Cp*₂Ti(N₂))]; formation of [Cp*₂TiO(py)] and selected subsequent reactions.

4.3.7 Metallocene Complexes of the Type [Cp$_2$TiL$_2$]$^{2+}$

4.3.7.1 Aqua–Titanocene Complexes

Cationic titanocene–aqua complexes can be readily prepared from the corresponding bishalides by reaction with silver complexes [151], or as shown earlier, by substitution reactions of weakly coordinated ligands in [Cp$_2$TiX$_2$] complexes (X = NO$_3^-$ [152, 153], ClO$_4^-$ [154]). By the latter method, the complexes [Cp$_2$Ti(H$_2$O)$_2$](NO$_3$)$_2$ and [Cp$_2$Ti(H$_2$O)$_2$](ClO$_4$)$_2$·3C$_4$H$_8$O are prepared. [Cp*$_2$Ti(H$_2$O)$_2$](CF$_3$SO$_3$)$_2$ becomes available from Cp*$_2$TiCl$_2$ and AgCF$_3$SO$_3$ in the presence of small amounts of water as dark–violet crystals. This complex can be handled in air and is soluble in moderately polar solvents such as CH$_2$Cl$_2$ and CH$_3$NO$_2$, but is not soluble in nonpolar solvents such as benzene. From solutions of [Cp*$_2$Ti(H$_2$O)$_2$](CF$_3$SO$_3$)$_2$ the two salts [Cp*$_2$Ti(OH)(H$_2$O)](CF$_3$SO$_3$)$_2$ · H$_2$O and [Cp*$_2$Ti(OH)(H$_2$O)](CF$_3$SO$_3$)$_2$ ·2 H$_2$O can be isolated. The [Cp*$_2$Ti(H$_2$O)$_2$]$^{2+}$ and the [Cp*$_2$Ti(OH)(H$_2$O)]$^+$ cations show the expected, nearly tetrahedral geometry. The Ti–OH$_2$ distances in the di-aqua cation are 2.06(1) and 2.09(1) Å. The Ti–OH$_2$ and Ti–OH distances in the aqua–hydroxo cations are 2.11 and 1.87 Å respectively [151].

The diaqua complexes of bent metallocenes tend to be moderately strong acids, and in this respect they are excellent catalysts in Diels–Alder reactions. For example, the reaction of 3-pentenone with isoprene in the presence of [Cp*$_2$Ti(H$_2$O)$_2$](CF$_3$SO$_3$)$_2$ as catalyst (CDCl$_2$, 25°C) proceeds 6 × 10^3 times faster than the thermal reaction [155].

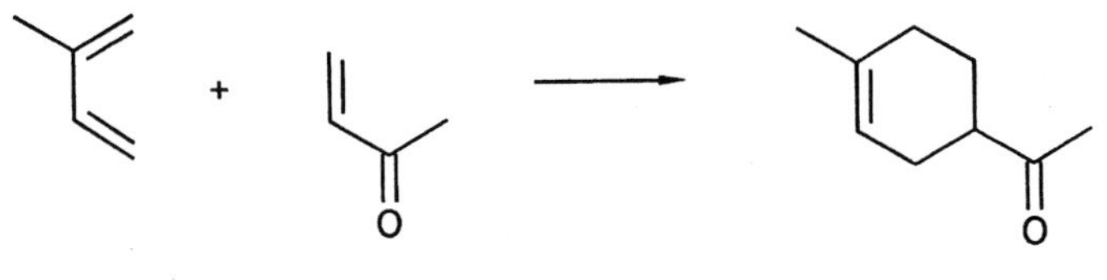

Scheme 4-12. Diels–Alder catalysis by [Cp*$_2$Ti(H$_2$O)$_2$](CF$_3$SO$_3$)$_2$.

The reaction of [Cp$_2$TiCl$_2$] with Na[BPh$_4$] in water leads to the aqua–metallocene derivative [Cp$_2$Ti(H$_2$O)$_6$][BPh$_4$], which undergo a facile reaction with methanol under loss of water. According to the single crystal X-ray diffraction study, one product turns out to be the salt [Ti$_3$(μ_3-O)(μ_2-OCH$_3$)$_3$(OCH$_3$)$_3$(η^5-C$_5$H$_5$)$_3$][BPh$_4$], in which the μ_3-O atom is located at the apex of a trigonal Ti$_3$O pyramid [156]. The aqueous chemistry of titanocenes was of interest for a long time, particularly in the field of cancerostatic activities [157–160]. However, complexes of titanocene are hydrolyzed in aqueous solution, leading to insoluble polymeric [(CpTiO)$_4$O$_2$]$_n$ at pH 5.5 [161, 162].

4.3.7.2 Titanocene Complexes with Highly Fluorinated Ligand Systems

The reaction of $[Cp_2TiCl_2]$ with two equivalents of $[AgEF_6]$ (E = As, Sb) in liquid sulfur dioxide led to the quantitative preparation of the hexafluoropnicogenate complexes $[Cp_2Ti(AsF_6)_2]$ and $[Cp_2Ti(AsF_6)_2]$ [163]. By reaction with the neutral Lewis base CH_3CN the cationic metallocene derivatives $[Cp_2Ti(CH_3CN)_n]^{2+}[AsF_6]^{2-}$ could be obtained [164]. If $[Cp_2TiF_2]$ is reacted with two equivalents of SbF_5 in liquid sulfur dioxide, $[Cp_2Ti(SbF_6)_2]$ is formed quantitatively. The equilibrium in SO_2 solution of $[Cp_2Ti(SbF_6)_2]$ with SbF_5 is completely shifted to $[Cp_2Ti(Sb_2F_{11})_2]$ [165].

$$Cp_2TiCl_2 \quad + \quad 2 \ AgEF_6 \quad \xrightarrow{SO_2} \quad Cp_2Ti(EF_6)_2 \quad + \quad 2 \ AgCl$$

$$Cp_2TiF_2 \quad + \quad 2 \ EF_5 \quad \xrightarrow{SO_2} \quad Cp_2Ti(EF_6)_2$$

Scheme 4-13. Synthesis of highly fluorinated titanocene complexes (E = As, Sb, Bi).

The preparation of the fluorine-coordinated metallocene hexafluorobismuthate complex $[Cp_2Ti(BiF_6)_2]$ proceeds from the reaction of $[Cp_2TiF_2]$ with BiF_5 in SO_2ClF in the presence of anhydrous HF, followed by recrystallization from liquid SO_2. IR investigations clearly indicate that the BiF_6 unit is not octahedral but is strongly distorted, and that both bridging (Ti⋯F⋯Bi) and nonbridging (Bi–F) fluorines are present [166].

4.3.7.3 Further Complexes with [Titanocene]$^{2+}$ (d^0) and [Titanocene]$^+$ (d^1) Units

The reaction of $[Cp_2Ti(CF_3SO_3)_2]$ with 2,2-bipyridine or 1,10-phenanthroline yields the salts $[Cp_2Ti(bipy)]^{2+}(CF_3SO_3)^-_2$ and $[Cp_2Ti(phen)]^{2+}(CF_3SO_3)^-_2$. X-ray structure determinations show that the Ti atoms, which are to a first approximation tetrahedrally coordinated, and the atoms of the bipy or phen chelate ligands, are nearly coplanar [167]. However, from reaction of $[Cp_2Ti(\eta^2\text{-}Me_3SiC{\equiv}CSiMe_3)]$ with trimethylammoniumtetraphenylborate cationic d^1-titanocene complexes $[\{Cp_2Ti(L)_2^+\}\{BPh_4^-\}]$ with L = THF (**21**) and pyridine (**22**) become available *via* 1e′-oxidation of the 14e′-Cp$_2$Ti-unit to paramagnetic Ti(III) complexes under evolution of hydrogen [168]. Complex **22** is one of the rare examples of such cationic-only neutral-ligand containing complexes without anionic ligands that have been characterized by X-ray structure analysis.

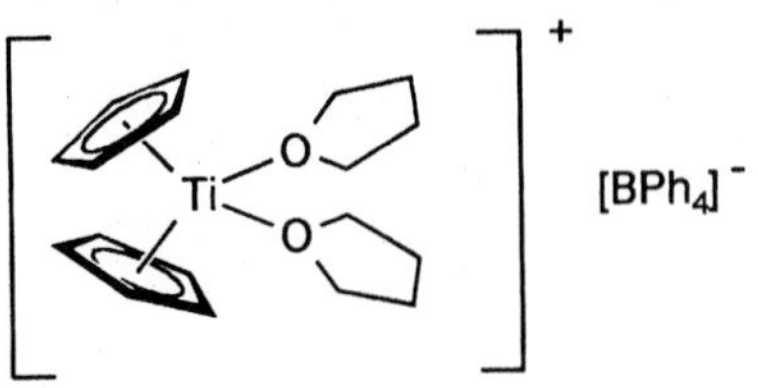

21 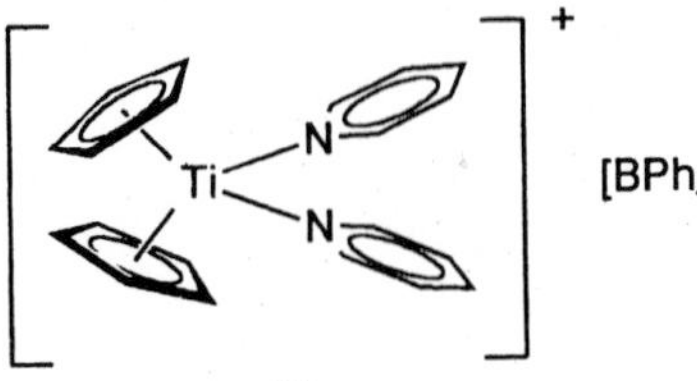   **22**

4.4 Reactions in the Coordination Sphere of Titanocenes

4.4.1 The free Titanocenes

For the rearrangement of $[(\eta^5\text{-}C_5H_5)_2Ti]$ (**23**) to the dimer **26**, Brintzinger and Bercaw suggested formation of an $\eta^1\text{-}C_5H_5$ titanium complex (**24**), followed by α-H abstraction to give the titanafulvalene **26** [169]. Neither **24** nor **25** was observed.

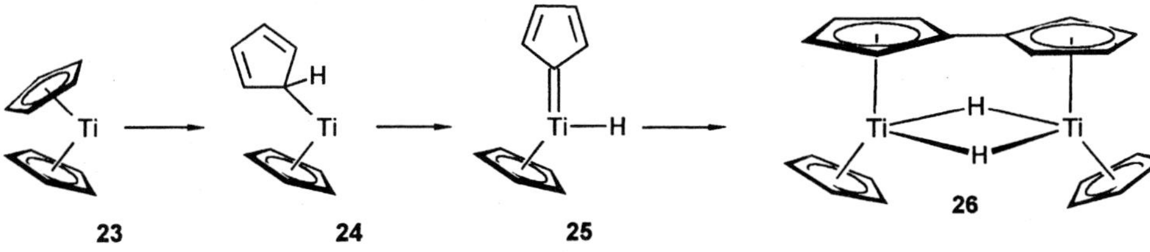

23   24   25

Scheme 4-14. Formation of
μ-(η^5:η^5-fulvalene)-di(μ-hydrido)-bis(η^5-cyclopentadienyl)titanium **26**.

Bright green '*titanocene*' (**26**) was first reported in 1956 to be obtained from the reaction of $TiCl_2$ with CpNa [1]. The same bright green material was later obtained by reduction of $[Cp_2TiCl_2]$ with sodium naphthalene in tetrahydrofuran [170] or with sodium metal in hydrocarbons [171, 172], or by hydrogenolysis of $[Cp_2TiMe_2]$ [173]. Its chemical properties were, however, different to the expected chemical behavior of monomeric titanocene and the structure of the compound was subject to extensive discussion. Strong evidence for the dimeric structure was given by Brintzinger and Bercaw [169] and the presence of the fulvalene ligand was established by an X-ray crystallographic study of the compound arising from interaction of **26** with Et_3Al [174]. The fulvalene structure of the compound was established on the basis of the ^{13}C NMR spectrum [175] and from the X-ray study of its di-(μ-hydroxyl) derivative [176]. Later, the X-ray study of a purple product arising from the reaction **26** with HCl revealed that it is its di(μ-chloro) derivative [177]. The mixed (μ-hydrido)–(μ-chloro) complex was also prepared [178] and its X-ray structure was determined [179]. Attempts to determine the structure of **26** directly by X-ray crystallographic studies have been unsuccessful for some time [174, 176]. In 1992, Mach, Antropiusova and Troyanov were able to solve the X-ray structure of **26** [90]. Selected structural data of fulvalene titanium complexes ('titanocene') are given in Table 4-7.

Table 4-7. Structural data in dititanium fulvalene $(C_{10}H_8)[(C_5H_5)TiX]_2$ complexes

Compound	Interatomic distances [Å]				Angles [°]			Ref.
	Ti-Ti	Ti-C(fulv)	Ti-C(Cp)	Ti-X	X-Ti-X	Ti-X-Ti	
26 $Cp_2Ti_2(C_{10}H_8)(H)_2 \cdot 1.5C_6H_6$	2.989	2.30-2.39	2.36	1.71	57	120	[90]
$Cp_2Ti_2(C_{10}H_8)(H)(Cl)$	3.125	2.32-2.39	2.36	2.17	92	93	[179]
				2.49		77.6	
$Cp_2Ti_2(C_{10}H_8)(Cl)_2$	3.638	2.37-2.40	2.38	2.52	81.3	92.5	[177]
$Cp_2Ti_2(C_{10}H_8)(OH)_2 \cdot THF$	3.195	2.34-2.47	2.41	2.07	75.4	100.7	[176]
$Cp_2Ti_2(C_{10}H_8)(H)(H_2AlEt_2)$	3.374	2.33-2.39	2.35	1.80	-	138	[174]

If the dimeric *'titanocene'* (**26**) is reacted with HCl the dichloro compound **27** is formed, from which organyl derivatives (**28**) are available in reaction with LiR (R = Me [180], R = Ph [181]) . The X-ray structure confirms the presence of the fulvalene bridge, in which the CpTiMe$_2$ fragments are *trans* with respect to the fulvalene group [180].

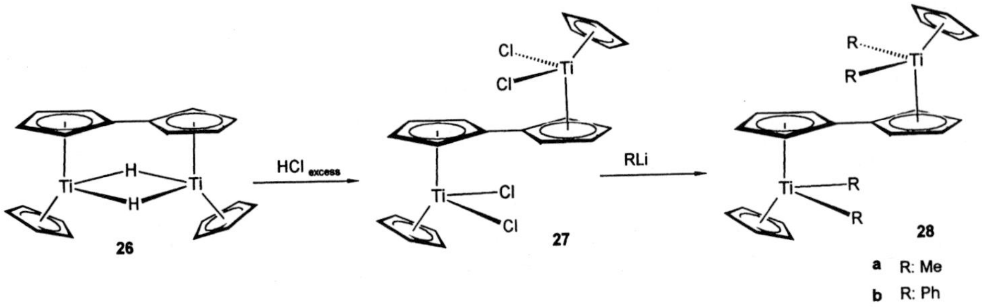

Scheme 4-15. Reactions of the green 'titanocene'.

From the dimeric *'titanocene'* **26** the air-stable compound $[CpTi(H_2O)](C_{10}H_8,O)]^{2+}(SO_3CF_3)_2^-$ as the first example of an ionic titanafulvene complex can be prepared in reaction with trifluoromethanesulfonic acid and small amounts of water [182].

In the chemistry of permethylated titanocenes, intramolecular C–H activation reactions dominate. When permethylated titanocene complexes are thermolyzed, the formation of carbene [39, 183] or aryne [184] intermediates occurs by H-transformation between the R groups, which react intramolecularly to form fulvene complexes *e.g.* **30**, the so called *'tuck in'* derivatives. Further heating leads to liberation of a further hydrocarbon molecule producing the bis *'tuck in'* complex **31** [41].

However, depending on the number of methyl groups (n) on the Cp rings, the formation of both fulvalene and fulvene complexes is observed. Those with $n = 0-3$ ($\mu\text{-}\eta^5\text{:}\eta^5$-fulvalene)(di-$\mu$-hydrido)bis($\eta^5$-cyclopentadienyltitanium) complexes and its methylated analogs are formed. For C_5Me_4H (Cp'''') complexes mixtures of fulvalene and fulvene complexes are observed [71].

Table 4-8. NMR data of fulvene titanium complexes

Complex	NMR data of the CH_2-fulvene group		
	^1H NMR (δ)	^{13}C NMR (δ)	Ref.
30 Cp*(η^6-$C_5Me_4CH_2$)TiCH$_3$	1.14 (1H, d, 4 Hz) 1.92 (1H, d, 4 Hz)	73.9, $J_{CH} = 150$ Hz	[183]
Cp''''(η^6-$C_5Me_3HCH_2$)TiCH$_3$	1.06 (1H, d, 4 Hz) 2.02 (1H, d, 4 Hz)	74.4, $J_{CH} = 150$ Hz	[185, 186]
(η^5-$C_2B_9H_{11}$)(η^6-$C_5Me_4CH_2$)TiCH$_3$	2.43 (s)	74.4, $J_{CH} = 155$ Hz	[187]
Cp*(η^6-$C_5Me_4CH_2$)TiCH=CH$_2$	1.24 (1H, d, 4 Hz) 1.78 (1H, d, 4 Hz)	76.2, $J_{CH} = 149$ Hz	[41]
31 Cp*(η^7-$C_5Me_3(CH_2)_2$)Ti	0.97, 0.90 (4H, dd, 4, 2 Hz)	67.4, $J_{CH} = 160$ Hz	[188, 41]
19 Cp*(η^6-$C_5Me_4CH_2$)TiOC(CH$_2$)CH$_3$	2.86 (d, 1H, 2.7 Hz) 2.08 (br, 1H)	72.6	[150]
Cp'(η^6-$C_5H_4CH_2$)TiC$_6$H$_5$	1.80, 2,18	80.9, $J_{CH} = 153$ Hz	[189]

The proton NMR spectra of the fulvene complexes shows two doublets with coupling constants $^2J_{HH} = 4$ Hz assigned to the diastereotopic hydrogens of the CH_2 group. This coupling constant is more consistent with the geminal sp^2 hydrogens of structure **B**, normally 0–3 Hz, than with the geminal sp^3 hydrogens of structure **A**, normally 12–15 Hz [41, 183–188]. The carbon of the =CH$_2$-fulvene group of **30** resonates at δ = 73.9 with $^1J_{CH}$ = 149–160 Hz, values consistent with sp^2-hybridized carbon. However, the crystal structure of the paramagnetic [Cp*(η^6-C$_5$Me$_4$CH$_2$)Ti] confirm the π-η^5: σ-η^1 bonding mode for the tetramethylfulvene ligand (**A**) [91b]. A diene like bonding mode is found by X-ray structure determination in the case of bis-fulvene complexes [91a].

4.4.2 Reactions of Ti–C σ Bonds

Our understanding of the chemistry of the transition metal–carbon σ bond is improved by investigations of the chemistry of 1-alkenyl complexes of electron poor transition metals. There is no other system known in which we can easily switch between the possible reaction pathways, depending on the nature of the metal, the ligands L and the 1-alkenylgroup. Only reductive elimination, α- and β-H elimination reactions give high selectivity. α-H Elimination from [Cp*$_2$Ti(CH=CH$_2$)R] derivatives leads to the versatile titana–allene intermediate [Cp*$_2$Ti=C=CH$_2$], from which a wide range of cycloaddition products of high thermal stability can be prepared. By comparison of the thermal stability of the vinyl, phenyl and alkyl complexes of titanium group metals, it becomes obvious that the vinyl derivative is the most reactive [190]. As well as α-elimination processes, β- and reductive elimination reactions are also observed. In each case, only one reaction pathway results in high selectivity (Scheme 4-16).

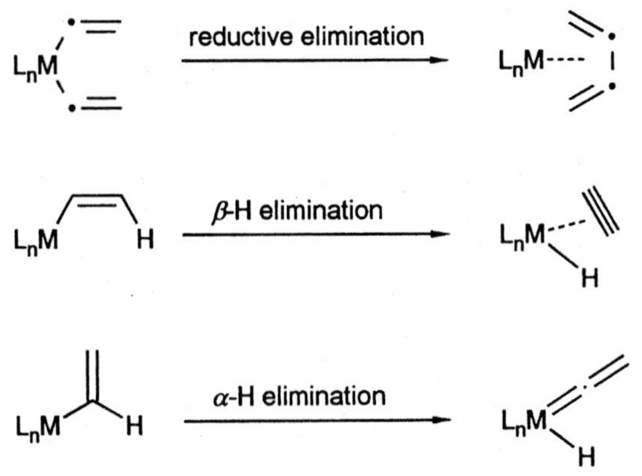

Scheme 4-16. Reaction pathways of 1-alkenyl transition metal compounds.

Thermodynamic considerations suggest that the reductive elimination of vinyl groups should be preferred, as is shown, for example, by *ab initio* calculations on Cl$_2$Ti(CH=CH$_2$)$_2$ model complexes (Scheme 4-17) [191].

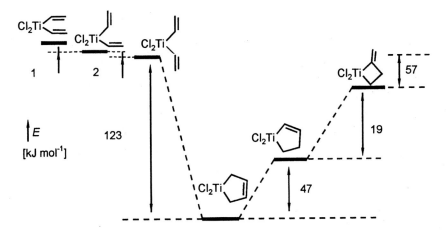

Scheme 4-17. Results from *ab initio* calculations for $[Cl_2Ti(CH=CH_2)_2]$
transformation products (pseudo potential basis set according
to Stevens, Basch, and Krauss).

Kinetic factors are able, however, to force the reaction to go in different directions, especially when the metal center is not readily reduced, when hydrocarbon ligands with acidic hydrogen atoms are employed, or when only limited rotation of 1-alkenyl groups is possible [39].

4.4.2.1 Reductive Elimination

Generally, reductive elimination reactions are the most preferred reactions of organometallic complexes from the thermodynamic point of view [191]. However, for alkyl derivatives, H-transformation reactions are kinetically preferred . In the case of 1-alkenyl compounds of the titanium group metals we find that there is a correlation between the rotational barriers of the 1-alkenyl ligand about the M–C σ bond and the observed reaction pathways (Scheme 4-18).

If free rotation is possible, reductive elimination products dominate. Due to the orientation of the acceptor orbitals of the bent metallocene fragment in the equatorial plane between the Cp-ligands, electron transfer from the C=C double bond to the transition metal center in this 16 electron complex becomes possible if the vinyl group is orientated perpendicular to it (Scheme 4-18, **A**). This rotameric orientation leads to a suitable transition state for reductive elimination, due to the differences in partial charge of the α-carbon atoms, as calculated by *ab initio* methods [191]. Generally, for d^0 systems, concerted reductive elimination reactions are symmetry forbidden, although reductive elimination might actually be possible, especially if charge transfer processes are involved [192, 193]. On the other hand, if

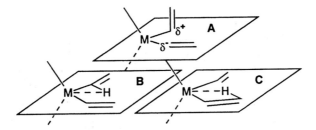

Scheme 4-18. Proposed transition states of reductive elimination **A**, α-H elimination **B** and β-H elimination **C** from Cp$_2$M alkenyl derivatives (M = Ti, Zr, Hf).

this rotation is hindered by using bulky ligands (Cp* instead of Cp [194], or substituted 1-alkenyl groups instead of the simple CH=CH$_2$-ligand [195]) C–H bond activation reactions become dominating (Scheme 4-18, **B**, **C**). Whereas the α-CH bond is activated in titanium complexes, β-CH activation occurs in zirconium complexes [37].

4.4.2.2 Oxidative Addition

A small number of well characterized oxidative addition reactions (by splitting the R–X bond) involving the [Cp$_2$Ti] unit are known. For example the [Cp*$_2$Ti] fragment, generated from [Cp*$_2$Ti(η^2-C$_2$H$_4$)], reacts by oxidative addition with CH$_3$I to form [Cp*$_2$TiI(Me)] [61]. Also, in reaction of [Cp*$_2$Ti], generated from [Cp*$_2$Ti(Me$_3$SiC≡CSiMe$_3$)], with methanol the hydride complex [Cp*$_2$Ti(H)OMe] can be isolated (δ Ti–H = 3.33, OMe = 4.01, Cp* = 1.89) [42]. This type of reaction proceeds only in the case of permethylated titanocenes. Using the nonmethylated derivative [Cp$_2$Ti(Me$_3$SiC≡CSiMe$_3$)], the liberation of hydrogen and a binuclear Ti(III) complex [Cp$_2$TiOR]$_2$ is observed [196]. By using alkynes other than the silylalkyne the protonation of the acetylene ligand to 1-alkenylcomplexes is observed [40].

$$>Ti \quad + \quad \overset{R}{\underset{X}{|}} \quad \longrightarrow \quad >Ti\overset{R}{\underset{X}{<}}$$

However, not only polar R–X can be added oxidatively to titanocene fragments. In the reaction of 'Cp$_2$Ti' with RC≡C–C≡CR' bimetallic complexes can be formed with or without splitting of the internal C–C single bond [31]. The obtained reaction products contain intact 1,4-disubstituted μ-η(1,3),η(2–4)-*trans,trans*-butadiene units between the two metal centers ('zig-zag-butadiynes') (**32**) [197], or alternatively, doubly σ,π-acetylide bridged complexes, Cp$_2$Ti(μ-η^1:η^2-C≡CR)(μ-η^1:η^2-C≡CR')MCp$_2$ (**33**) are obtained [198].

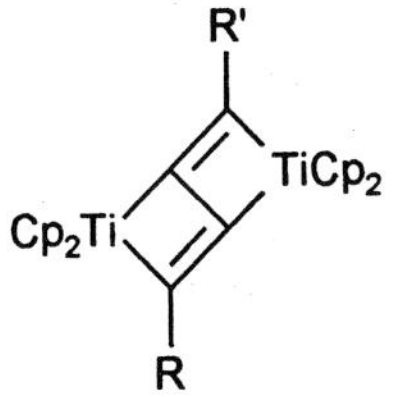

32 a: R = R' = ^{t}Bu
  b: R = SiMe$_3$, R' = ^{t}Bu
  c: R = SiMe$_3$, R' = Ph

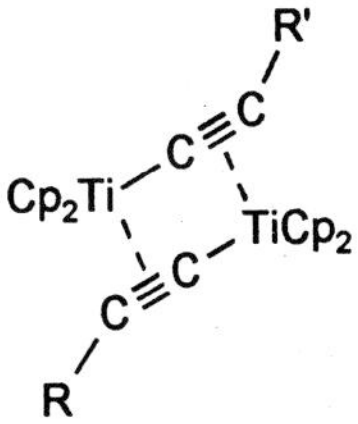

33: R = R' = SiMe$_3$

4.4.2.3 α-H Elimination

α-H eliminations proceed especially readily when there is significant steric crowding and where it is possible to transfer an α hydrogen atom to a carbanionic acceptor ligand. The reaction is intramolecular, and first-order reaction kinetics are observed for organometallic complexes of the titanium group metals. The breaking of the C–H bond is therefore shown to be the rate-determining step with either a cyclic transition state (path **A**) or a hydride intermediate (path **B**), where a titanium carbene complex, is formed as an intermediate (Scheme 4-19) [39].

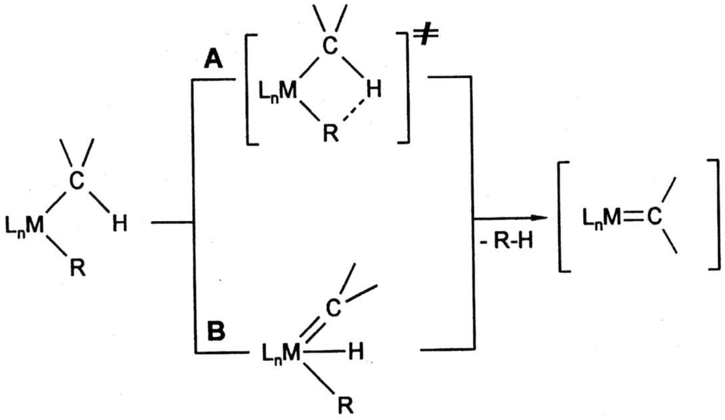

Scheme 4-19. The reaction course of α-H eliminations.

The activation enthalpy and entropy are characteristic of σ bond metathesis, which is identified by negative values for $\Delta S^{\neq}$, that is, by a high degree of order in the rate-determining step. A comparison of activation enthalpies (Table 4-9) shows that, as expected, α-H abstractions occur more readily for neopentyl species [103] than for methyl derivatives [183]. The values of the activation entropies indicate a higher degree of order, corresponding to a lower mobility, for neopentyl, aryl, and vinyl derivatives than for methyl or benzyl compounds. The strong isotope effects $k_{H/D}$, in addition to the negative $\Delta S^{\neq}$ values are in agreement with a transition state (path **A**) [118, 183, 184, 199–201]. In contrast to this, hydride intermediates (path **B**) show $k_{H/D}$ values close to 1 [202]. The half-life for the thermolysis of $[Cp_2Ti(CH_2CMe_3)_2]$ at 20°C is between 20 and 56 min. The $[Cp_2Ti=CHMe_3]$ formed can be trapped with phosphanes or used in further reactions [103]. The steric overloading favorable for α-eliminations can be achieved either by organic moieties (*e.g.* neopentyl groups) or by other bulky ligands [39].

Table 4-9. Selected kinetic parameters of H eliminations

Starting material	Leaving group	Intermediate or product	$\Delta H^{\neq}$ [kJ mol^{-1}]	$\Delta S^{\neq}$ [J mol^{-1}K^{-1}]	$k_{H/D}$	Ref.
$[Cp_2Ti(CH_3)_2]$	HCH$_3$	$[Cp_2Ti=CH_2]$	-	-	9-10	[118]
$[Cp_2Ti(CH_2CMe_3)_2]$	HCH$_2$CMe$_3$	$[Cp_2Ti=CHCMe_3]$	76.2(5)	-49.8(8)	-	[103]
$[Cp^*_2Ti(CH=CH_2)(CH_3)]$	HCH$_3$	$[Cp^*_2Ti=C=CH_2]$	87.9(5)	-21(4)	5.1	[184]
$[Cp^*_2Ti(C_6H_5)(CH_3)]$	HCH$_3$	$[Cp^*_2Ti(C_6H_4)]$	96.4(7)	-41(9)	5.7	[184]
$[Cp^*_2Ti(CH_3)_2]$ **(29)**	HCH$_3$	$[Cp^*_2Ti=CH_2]$	115.5(3)	-11.7(7)	2.9	[183]

The high selectivity of the methane elimination from the vinyltitanium compound $[Cp^*_2Ti(CH=CH_2)CH_3]$ is of particular value in the generation of the vinylidene intermediate $[Cp^*_2Ti=C=CH_2]$ **(35)** [184]. The spontaneous intramolecular H transfer between the two vinyl groups in $[Cp^*_2Ti(CH=CH_2)_2]$ **(34)** had been discovered earlier [37], but proceeds under conditions that prevent the isolation of the bisvinyl species. The resulting vinylidene–ethylene complex **36** reacts quantitatively to yield the methylenetitanacyclobutane **37** [190, 203].

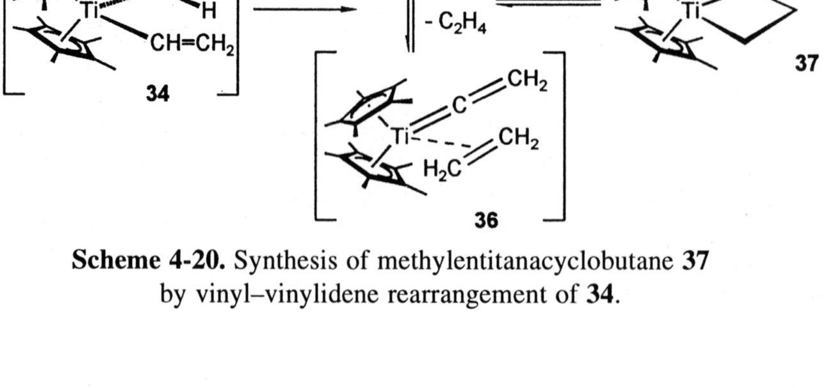

Scheme 4-20. Synthesis of methylentitanacyclobutane **37** by vinyl–vinylidene rearrangement of **34**.

The cyclic transition state necessary for the generation of **36** from **34** (Scheme 4-19, path **A**) can only be formed if the vinyl group is present as the so called *C–H inside* rotamer (Scheme 4-21, **B**). Nuclear Overhauser effects (NOE) measurements, however, show the orientation of the vinyl group in [Cp*$_2$Ti(CH=CH$_2$)CH$_3$] to be exclusively that of the *C–H outside* rotamer (Scheme 4-21, **A**).

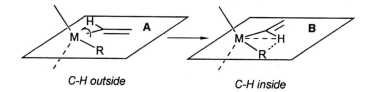

Scheme 4-21. Orientation of 1-alkenyl ligands.

The energy necessary for the rotation process in [Cp*$_2$Ti(CH=CH$_2$)$_2$], is found from MMX-force field calculation to be about 52.1 kJ mol^{-1}, compared to [Cp$_2$Ti(CH=CH$_2$)$_2$] for which a value of 18.3 kJ mol^{-1} is calculated [191]. Therefore it can be concluded that the rotation barrier is the main contribution to the activation energy of C–H elimination as determined by kinetic measurements for the process **38**→**35**→**39** (87.9(5) kJ mol^{-1}) [184]. The alternative elimination of ethylene and formation of a [Cp*$_2$Ti=CH$_2$] intermediate is not observed.

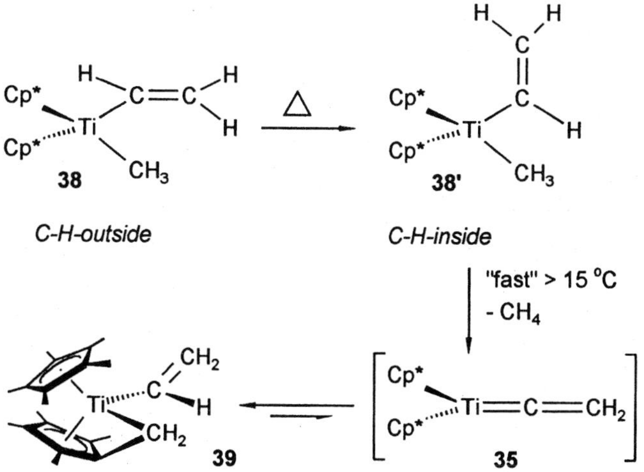

Selective α-H eliminations can also be used to generate a wide range of substituted alkenylating reagents [L$_n$Ti=CHR] (R = C$_6$H$_5$ [112], *t*Bu [119], SiMe$_3$ [111, 107]). For instance, during the thermolysis of [Cp$_2$Ti(CH$_2$C$_6$H$_5$)$_2$] [204], more toluene (α-H elimination product) is formed than dibenzyl (reductive elimination product; toluene:dibenzyl = 9.4:1) [112]. As the reaction of bicyclopropyltitanocene shows [110], even β-H containing compounds with strained ring systems can be used for 'alkylidencyclopropanation' in carbonyl olefination reactions.

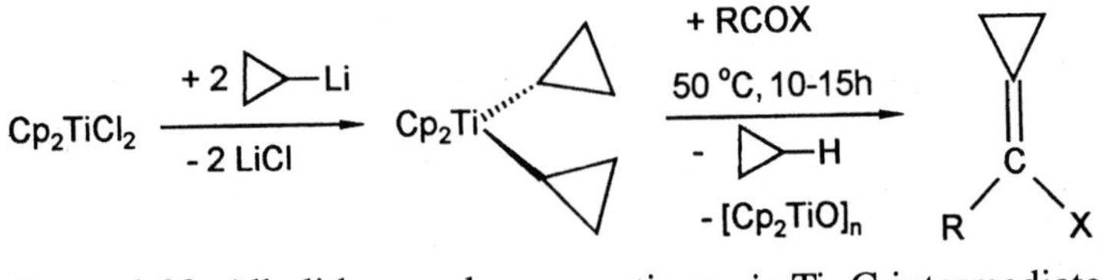

Scheme 4-22. Alkylidenecyclopropanations *via* Ti=C intermediates.

4.4.2.4 β-H Elimination

The major decomposition pathway for transition metal alkyls is the β-H elimination, which converts a metal alkyl into a hydrido metal alkene complex (Scheme 4-23).

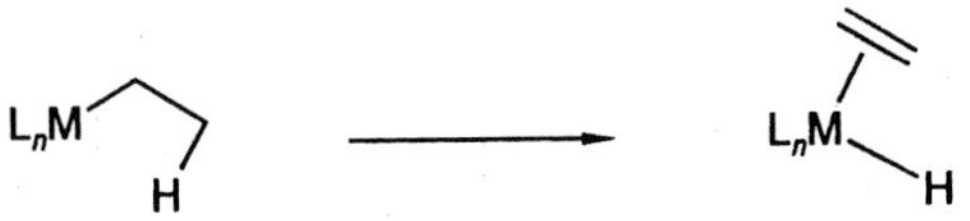

Scheme 4-23. β-H elimination.

Such examples are also known in the case of titanocene alkyl complexes. Attempts to prepare [Cp*$_2$Ti(Et)R] derivatives (**41**) by substitution of chlorine in [Cp*$_2$Ti(Et)Cl] (**40**) by salt metathesis with alkali-metal reagents RM were not successful. In general, there was no reaction below –10°C, but the products are too thermally unstable to allow isolation at higher temperatures. Reaction of **40** with MeLi, KCH$_2$Ph or LiCH=CH$_2$ resulted in the formation of the permethylated ethene adduct [Cp*$_2$Ti(η^2-C$_2$H$_4$)] (**42**) and RH in essentially quantitative yields [184].

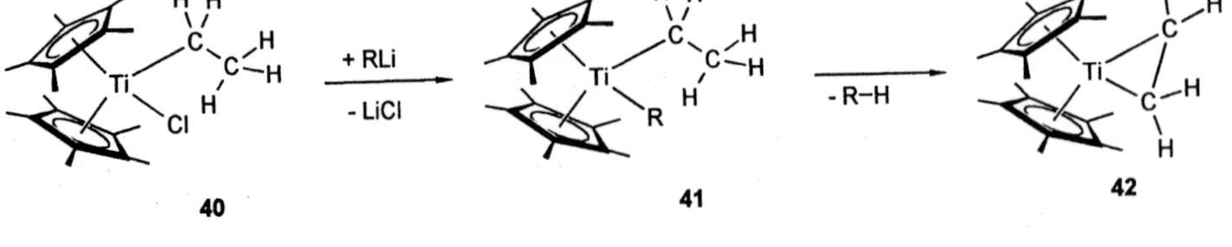

In order to determine whether the α-hydrogen atoms are involved, [(Cp*-d$_{15}$)$_2$Ti(CD$_2$CH$_3$)Cl] was prepared and treated with MeLi. Only CH$_4$ (1mol/Ti) was formed, so that hydrogen from the α position of the ethyl ligand or from the pentamethylcyclopentadienyl ligands can be ruled out as the source of the hydrogen atom used in the formation of methane. Apparently, β-hydrogen transfer from the ethyl ligand to the leaving group R in Cp*$_2$Ti(Et)R has a very low activation energy.

The thermolysis of [Cp*$_2$Ti(Ph)Me] followed first-order kinetics for at least 3 half-lives, and the reaction rate appeared to be independent of the concentration. The kinetic isotope effect observed ($k_{H/D}$ = 5.1 at 80°C) indicates that a phenylic C–H bond is broken in the rate-determining step. The aryne intermediate [Cp*$_2$Ti(η^2-C$_6$H$_4$)] formed can by trapped with carbon dioxide [184]. The intramolecular decomposition *via* an ordered transition state, *i.e.*, σ-bond metathesis, is in sharp contrast with the radical decomposition of other mixed alkyl/aryl compounds. The formation of the aryne intermediate [Cp*$_2$Ti(η^2-C$_6$H$_4$)] is analogous with the reported reactivity of [Cp$_2$Ti(C$_6$H$_5$)$_2$], which leads to the *o*-phenylene intermediate [Cp$_2$Ti(η^2-C$_6$H$_4$)] [205, 206].

4.4.3 Reactions of Ti=C Double Bonds

Kinetic studies and numerous trapping experiments have shown that thermally generated [L$_2$Ti=CH$_2$] and [L$_2$Ti=C=CH$_2$] species represent real intermediates [41, 51, 118]. In particular, the existence of the titana–allene **35** as a real intermediate, generated from **38** by methane elimination (5–20°C) or from **37** by ethylene liberation (70–100°C) [39, 190, 203, 207] can be proved by simple trapping experiments. Substrates such as ketones, alcohols, cumulenes and heterocumulenes do not react directly with the starting materials. Substitution products **45** or ring-opened derivatives **43** could not be detected in reactions of **37** and **38** with acidic substrates. This is an indication that cycloreversion or α-H transfer from **38** take place before electrophilic addition reactions can occur. However, strong acidic substrates like thiophenol lead to formation of products of type **43** [41].

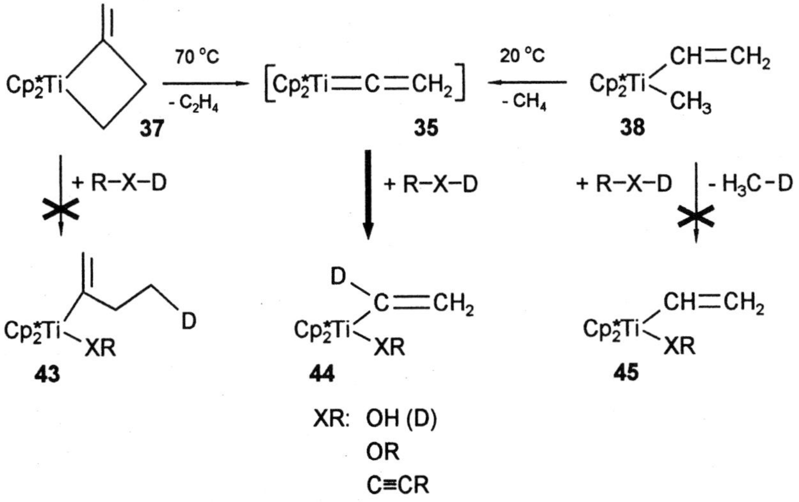

Scheme 4-24. Selective trapping experiments of [Cp*$_2$Ti=C=CH$_2$].

4.4.3.1 Reactions with Electrophiles

In view of the nucleophilic properties of the α-atoms in [L$_n$Ti=C] derivatives, rapid reactions with electrophiles are expected. If **35** is allowed to react with a stoichiometric amount of water, the monomeric vinyltitaniumhydroxide [Cp*$_2$Ti(CH=CH$_2$)OH] (δ_{OH} = 7.10; IR v_{OH} = 3657 cm^{-1}) is obtained in 86% yield as a yellow, microcrystalline product [41]. The selective protonation of the vinylidene species **35** can also be transferred to C–H [208] and N–H [209] acidic compounds. In such a way the reaction of **38** with benzophenonimine in *n*-hexane as solvent yields the azavinylidene [Cp*$_2$Ti(CH=CH$_2$)(N=CPh)$_2$] **46** in the form of light yellow crystals (55%, mp 190°C (dec.)) [209].

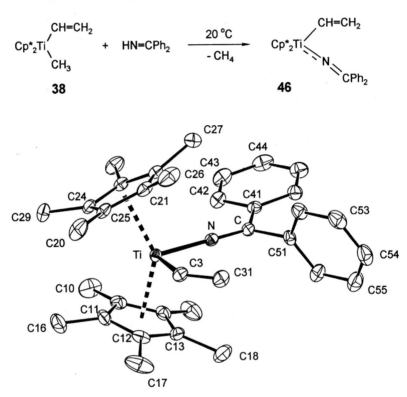

Fig. 4-2. Molecular structure of **46** in the solid state. Selected bond distances (Å) and angles (°): Ti–N 1.901(2), Ti–C3 2.159(2), C2–C31 1.319(3), N–C 1.259(3), Ti–N–C 179.1(2), Ti–C3–C31 131.8(2).

4.4.3.2 C–H Activation Reactions on [L$_n$Ti=C] Fragments

In order to coordinatively saturate the [L$_n$Ti=C] derivatives, intramolecular C–H activation reactions occur in many cases. For example, the titana–allene **35** spontaneously reacts to yield the dark green vinylfulvene derivative [Cp*(η^6-C$_5$Me$_4$CH$_2$)TiCH=CH$_2$] **39** [184]. This behavior is similar with related ligands [105, 119, 183]. The signal of the methyne hydrogen atom of the vinyl group in **39** is shifted to higher field (δ = 5.11) than in other vinyl compounds [210].

Starting from **47**, *via* the titanium carbene [Cp$_2$Ti=CH(CMe$_3$)] (**48**), C–H bonds of benzene or *p*-xylene can undergo intermolecular activation and [CpTi(CH$_2$CMe$_3$)R′] complexes are formed [103].

Complex **47** reacts in neat C$_6$D$_6$ to produce quantitatively the [1-^{2}H] neopentyl [^{2}H$_5$] phenyl complex **49** with the selective incorporation of one deuterium on the neopentyl α-carbon. Intermolecular activation of benzylic sp^3-CH bonds was also observed. This behavior is comparable to d^0 titanium imides [146, 211]. Overall, this reaction type corresponds to the reverse of the original formation reaction by α-H elimination.

4.4.3.3 [2+1] Additions

Due to its electronic structure, the vinylidene intermediate **35** is very useful in cycloaddition reactions [38, 191]. By using isonitriles an azabutatriene complex **50** is formed in the first reaction step in a [2+1] cycloaddition [212, 213]. Due to the high reactivity of the intermediate **50** further reaction occurs, forming the five membered metallacycle **51**, which exhibits a heteroradialene substructure [213], or by trapping with W(CO)$_6$ the metallacycle **52** [212].

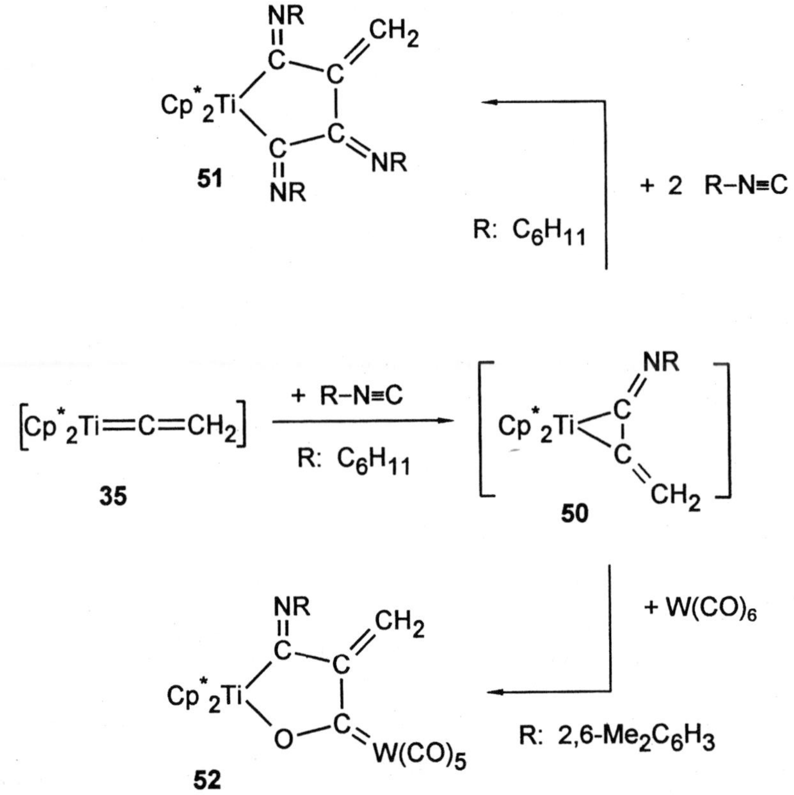

Scheme 4-25. [2+1] Addition reactions of isonitriles with [Cp*$_2$Ti=C=CH$_2$].

Reactions of metal–carbon multiple bond systems with isocyanides present a potential synthetic method for the generation of keteneimines [214]. The coupling of a carbene ligand with carbon monoxide can be demonstrated for isolable carbene complexes with bulky, chelating phosphanealkoxides and a Cp ligand. Ketene complexes are formed in this reaction [119].

4.4.3.4 Cycloaddition Reactions with Acetylenes

Titanacyclobutenes are found to be the products of [2+2] cycloadditions of metal carbenes with acetylenes [215, 216], as well as the ring-opening of cyclopropenes [217]. The reactivity of the titana–allene **35** with alkynes is particularly important with respect to the mechanisms of polymerizations of acetylenes [208]. Studies of the behavior of thermally generated **35** towards a number of alkynes have shown that the regiochemistry of the titanacyclobutene ring formation **53→54** is kinetically controlled and proceeds according to the polarity of the alkyne employed [208, 218].

If acidic (terminal) alkynes are used, a competing reaction occurs, namely the formation of vinyl titanium acetylides **55**.

The use of unsubstituted acetylene (HC≡CH), however, leads exclusively to the metallacyclobutene **54** (R = H), a thermally stable compound. Competing reactions of acetylenes and olefins in cycloadditions with **35** yield exclusively the metallacyclobutene derivatives. *Ab initio* calculations for the system [$Cl_2Ti=C=CH_2$ + $H_2C=CH_2$] and [$Cl_2Ti=C=CH_2$ + HC≡CH] indicate a preference of 70.9 kJ mol^{-1} [191]. Kinetic studies have shown that the primary alkyne insertion into a Ti–CH_3 bond with subsequent γ-elimination from **57** to give **58** could be an alternative to the direct [2+2] cycloaddition of the [Ti=C] species to the alkyne [109].

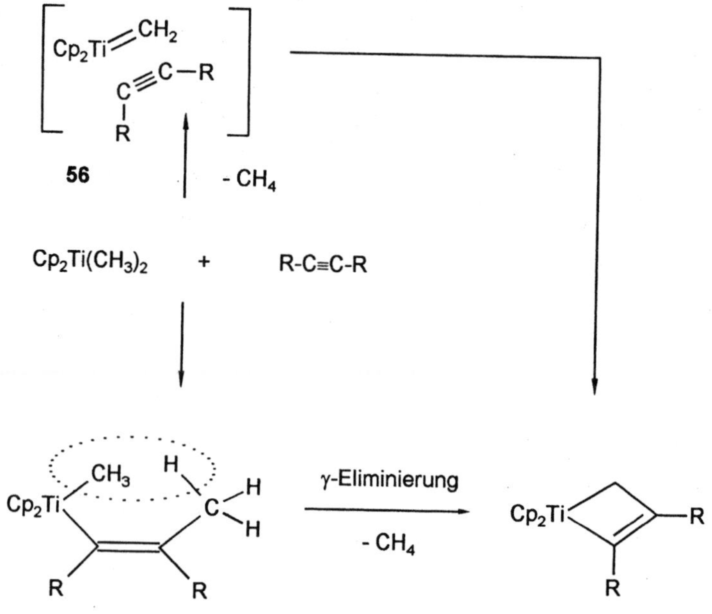

Scheme 4-26. Formation of titanacyclobutenes, by [2+2] cycloaddition reaction as well as by γ-elimination reactions from 1-alkenyltitanocene complexes.

The reactivity of titanacyclobutenes is characterized by their ability in principle to undergo cycloreversions **A** or electrocyclic ring-opening reactions **B**. Depending on the substitution patterns, the reaction goes by path **A** or **B** (Scheme 4-27).

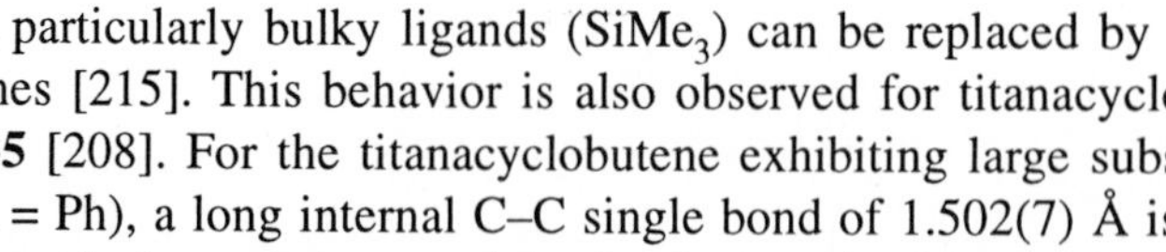

Scheme 4-27. Reactions of titanacyclobutenes.

Alkynes with particularly bulky ligands ($SiMe_3$) can be replaced by less sterically demanding ones [215]. This behavior is also observed for titanacyclobutenes generated from **35** [208]. For the titanacyclobutene exhibiting large substituents (α-R = $SiMe_3$, β-R = Ph), a long internal C–C single bond of 1.502(7) Å is found by X-ray structural analysis, and the reactivity is characterized by cycloreversion processes at higher temperatures. On the other hand, with smaller substituents (R = R = Me) the internal C–C single bond is found to be shorter, (1.434(4) Å) indicating a transition in the reactivity of the titanacyclobutene. Indeed, electrocyclic ring opening reactions are only observed for the nonsubstituted titanacyclobutene, leading to the formation of polyacetylene if an excess of acetylene is used [208]. This behavior corresponds to that of tantalacyclobutenes, which can be isolated in the form of vinylcarbene complexes by reaction with pyridine [219]. Reactions of the titana–allene intermediate with 1,3-diynes, yielding titanacyclobutenes in a regioselective manner. Only one regioisomer (59) is formed, containing the acetylide group in the α position of the titanacyclic ring. The regioselectivity is in accordance with the polarities of the diynes and stereochemical conditions in the cyclobutene ring [220].

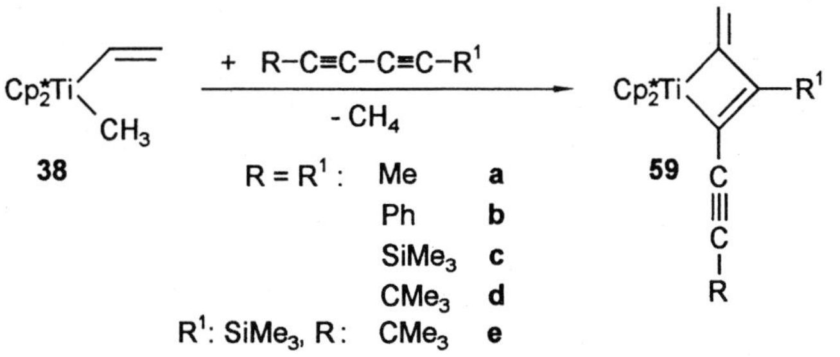

4.4.3.5 Synthesis and Reactivity of Metallaoxetanes

Metallaoxetanes **60** are discussed as essential intermediates in the reactions of carbenoid metal compounds with carbonyl derivatives (**60→61**). In addition, metallaoxetanes can be considered as intermediates in the oxidation of olefins as well as in the formation and deoxygenation of epoxides [221]. Generally, metallaoxetanes have been prepared by two principal synthetic routes, by reaction of a carbene with a carbonyl functional group [222, 223] or by oxygen insertion into a coordinated olefin [224]. Additionally, the reverse reaction of a terminal metal oxide fragment (Cp*$_2$Ti=O) and allene also leads to the formation of titanaoxetanes [148]. The formation of titanaoxetanes is also possible by the addition of two equivalents of dimethylsulfoxonium methylide to titanocene chloroacyl complexes *i.e.* the addition of a methylene fragment to a ketene complex [225].

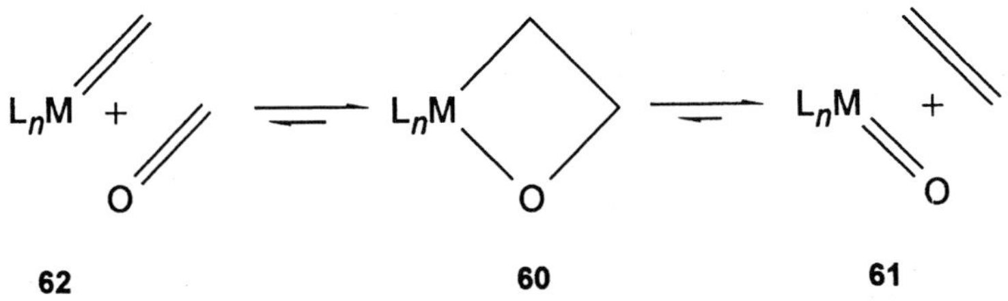

Scheme 4-28. Reactivity pattern of metalloxetanes.

The reactions of the vinylidene intermediate **35** with ketenes, isocyanates or carbon dioxide indicate the possibility of the synthesis of novel titanaoxetanes **63** with considerable thermal stability. The X-ray crystal structure analyses of **63b** and **63d** confirm the presence of planar, monomeric metallaoxetane rings.

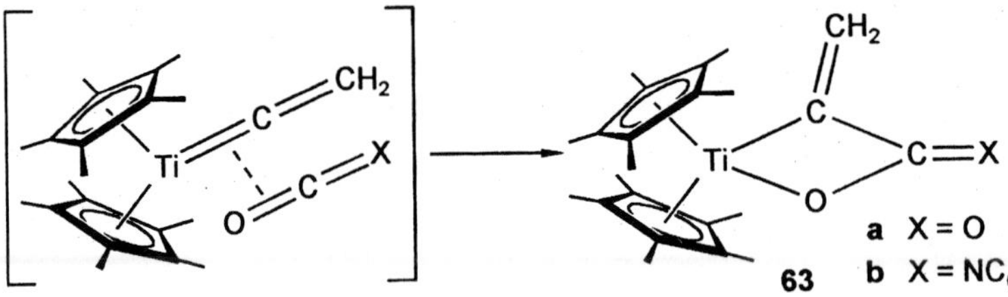

a X = O
b X = NC$_6$H$_{11}$
c X = NC$_6$H$_5$
d X = C(C$_6$H$_5$)$_2$
e X = C[C(CH$_3$)$_3$]$_2$

63

Table 4-10. Selected properties of isolated titanaoxetanes **63**

X		yield [%]	mp [°C]	Ref.
63a O	red crystals	86	110 (dec.)	[222]
63b NC$_6$H$_{11}$	red crystals	98	128-131	[222]
63c NC$_6$H$_5$	dark red crystals	94	130-132	[226]
63d C(C$_6$H$_5$)$_2$	black crystals	49	142 (dec 129)	[223]
63e C[C(CH$_3$)$_3$]$_2$	golden crystals	65	143	[222]

These cycloaddition products represent the first structurally characterized titanaoxetanes with a planar ring geometry. Considering that the driving force of carbonyl olefinations stems from the formation of a titanium oxide fragment [191, 227], the fact that planar oxetanes **63** can be isolated must be attributed to the reduced electrophilicity of the transition metal center due to the strongly basic Cp* ligands [53, 56]. This assumption is supported by the observation that titanaoxetanes similar to **63**, but with the unsubstituted Cp ligands **65**, have never been detected, even by spectroscopic means [228]. This reaction is very useful in the formation of substituted allenes **66**.

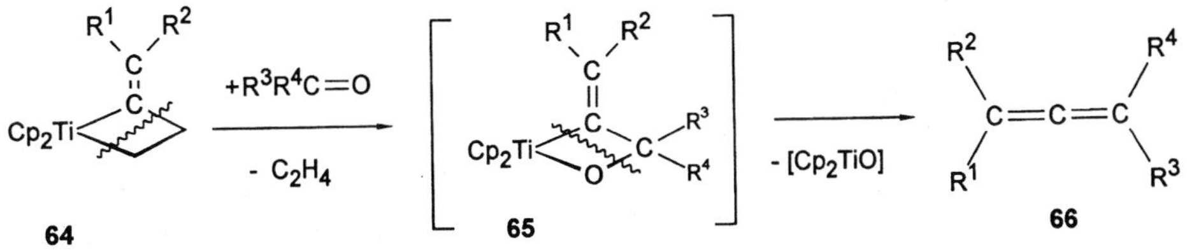

Scheme 4-29. Synthesis of substituted allenes from vinylidene intermediates and carbonyl compounds.

Depending on the resulting reaction pattern, these metallaoxetanes can be classified as *'classical'* (**60**→**61**) or *'nonclassical'* (**60**→**62**). Mass spectrometry shows the formation of [Cp*Ti=O] fragments for the metallaoxetanes **63**, so that their classification as *'classical'* oxetanes is justified. As well as the effect of the Cp* ligands, the position of the exocyclic double bond in the metallacyclic ring also has implications for the stability of metallaoxetanes **63**. If the double bond is in the β position, such as in cycloaddition products of [Cp*₂Ti=O] with allene, rapid ring opening occurs in solution, with the subsequent formation of fulvene enolates [150].

4.4.3.6 Synthesis of Heterodinuclear Carbene Complexes with a Titanaoxetane Substructure

The *'nonclassical'* metallaoxetanes can be obtained by the use of metal carbonyls as heterocumulene units in reactions with **35** [229]. In contrast to metal-mediated cycloadditions of metal carbonyls with arynes, dienes [230], olefins [63] and heteroolefins [231] to metallocene fragments of group 4 metals the corresponding reactions of Schrock carbene fragments with metal carbonyls are little known thus far. In a few cases, corresponding intermediates have either been observed spectroscopically [232] or their existence has been postulated based on the products obtained [233].

However, the four membered heterodimetallic carbene complexes **67** are easily prepared by trapping the vinylidene titanocene fragment **35** with the corresponding metal carbonyl [229]. The spectroscopic data for **67** clearly indicate the titanaoxetane substructure. By using the binuclear transition metal carbonyls Mn₂(CO)₁₀ and Re₂(CO)₁₀, a selective axial functionalization occurs. The ¹³C NMR chemical shifts of the β-carbon atoms in the titanacycle are in the expected range of Fischer–Carbene complexes [234].

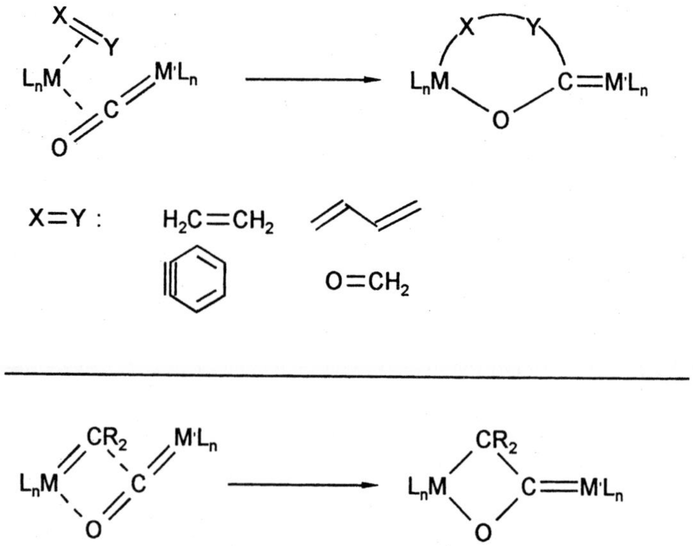

$X=Y$: $H_2C=CH_2$

$O=CH_2$

Scheme 4-30. Top: Metal mediated cycloadditions to metal carbonyls; bottom: [2+2] cycloadditions of $[L_nM=CR_2]$ to metal carbonyls. M = Ti, Zr; M' = Cr, Mo, W, Mn, Re, Fe, Rh.

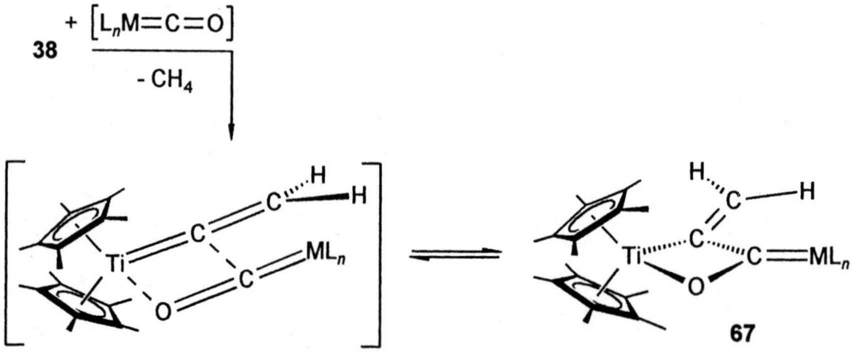

38 $+ [L_nM=C=O]$

$- CH_4$

67

	ML_n		ML_n
a	$Cr(CO)_5$	e	$Mn_2(CO)_9$
b	$Mo(CO)_5$	f	$Re_2(CO)_9$
c	$W(CO)_5$	g	$Fe(CO)_4$
d	$CpRe(CO)_2$	h	$Rh(CO)(acac)$

Table 4-11. Selected properties of nonclassical titanaoxetanes [Cp*$_2$TiOC(=ML$_n$)C=CH$_2$], obtained from [Cp*$_2$Ti=C=CH$_2$] and transition metal carbonyls [229]

ML$_n$		yield [%]	^{13}C NMR Ti-O-C= (δ)
67a Cr(CO)$_5$	dark green crystals	56	311.9
67b Mo(CO)$_5$	dark green crystals	32	295.3
67c W(CO)$_5$	dark green crystals	38	284.9
67d Mn(CO)$_4$Mn(CO)$_5$	brown green crystals	56	302.3
67e Re(CO)$_4$Re(CO)$_5$	violet crystals	38	259.3
67f ReCp(CO)$_2$	dark green crystals	45	257.5
67g Fe(CO)$_4$	red powder	37	291.8
67h Rh(CO)(acac)	black crystals	37	257.6 (d, $^1J_{\text{Rh-C}}$ = 52 Hz)

In the solid state, these *'nonclassical'* oxetanes are stable up to 130°C. Solutions, however, show a much higher reactivity, which is quite in contrast to the 'classical' oxetanes **63**. For example, the rapid release of the heterocumulene occurs even at low temperatures, especially when potential π-acceptor ligands (ethylene, isocyanates) or traces of moisture and oxygen are present. This means that the Fischer carbene complexes **67** with a titanaoxetane substructure represent another effective source of the Schrock carbene intermediate **35**.

A further nonclassical type of reaction behavior is observed for the titanaoxetanes **67**, which leads to an unusual destabilization of the titanium vinylidene moiety, causing it to undergo a rapid vinylidene–acetylene rearrangement, leading to the five membered titanacycles **68** [229].

This transformation illustrates the behavior of the vinylidene :C=CH$_2$ molecule in the gas phase. The five membered dinuclear complexes **68** are crystalline materials of high thermal stability (up to 220°C). In the ^{13}C NMR spectra, signals typical of carbene C atoms are observed. The structures of the titanaoxacyclic carbene complexes are confirmed by single crystal X-ray diffraction. A nearly planar five membered titanaoxacyclic ring can be observed. The Ti–O and the Cr=C or Mn=C bond lengths are in the normal range of cyclic Fischer–carbene complexes. This reaction type is only observed in the case of the *nonclassical* bimetallic oxetanes **67**. That means that a metal-mediated 1,2-H-shift must have taken place. In the first step, the vinylidene intermediate **69** is formed by cycloreversion of **67** followed by coordination of the metal–carbonyl to the C=C double bond of the vinylidene (**70**). This leads to a vinylidene–acetylene transformation (**71**). The coordinated acetylene then undergoes cycloaddition with the metal carbonyl [229].

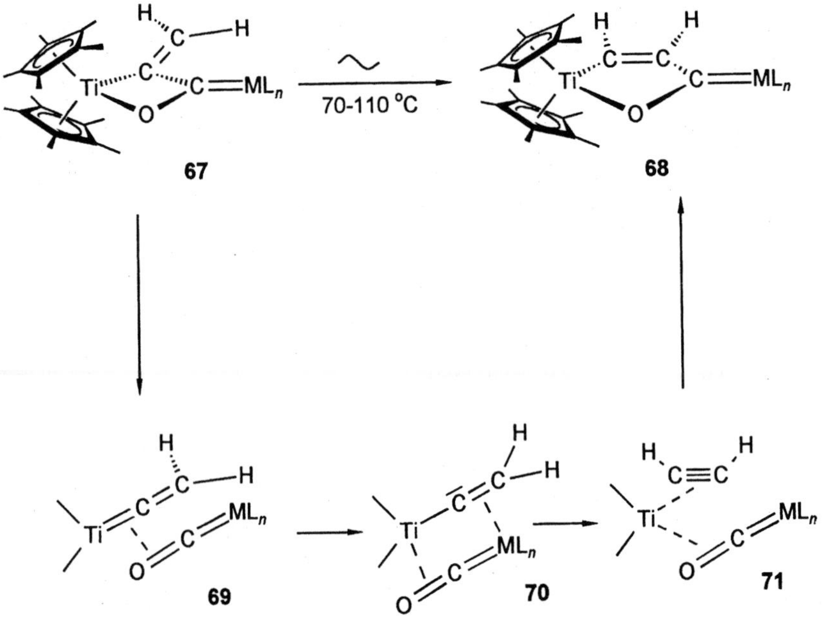

Scheme 4-31. Isomerization of nonclassical titanaoxetanes **67** to the five membered ring compounds **68**—an unexpected vinylidene–acetylene rearrangement.

4.4.3.7 Synthesis and Reactivity of Heterotitanacyclobutenes and -butanes

Cycloadditions of nitriles to Ti=C bond systems ought to lead to the formation of azatitanacyclobutenes (**73**). The electrophilicity of the titanium center, however, also influences the reaction behavior of the primary reaction product in this case [93], leading to products stemming from intermediary vinylimido derivatives **74** [113, 235].

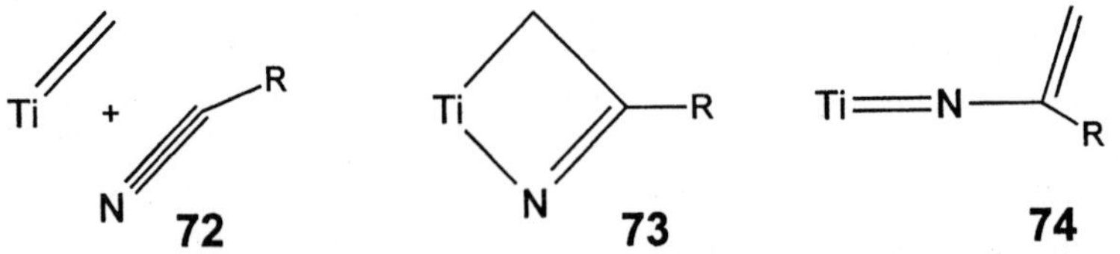

Using [Cp*$_2$Ti=C=CH$_2$] (**35**) instead of the [Cp$_2$Ti=CH$_2$] intermediate, the preparation of azatitanacyclobutenes **75** is possible [236]. In *n*-hexane [Cp*$_2$Ti(CH=CH$_2$)CH$_3$] (**38**) reacts with acetonitrile, pivalonitrile, cinnamonitrile or 2-*tert*-butyl-1-phosphaacetylene at room temperature yielding the products **75** as well as **76** and its isomer **77**.

R:		75
CH$_3$		a
C(CH$_3$)$_3$		b
trans-CHCHPh		c

The complexes **75b** and **75c** were isolated as red ashlars or needles in 53% and 38% yield, respectively. The isomer **76** was obtained as brown ashlars in 35% yield. The metal bonded λ^3 phosphorous atom in compound **76** shows a very low field shift in the ^{31}P{^{1}H} NMR spectrum ($\delta = 549$). The structures of **75b** and **76**, which were determined in solution using NMR spectroscopy, have been confirmed by X-ray crystal structure analyses. The structures show planar heterotitanacyclobutene rings. Alternating bond lengths of the metallacyclic ring-sequences C=C–C=X (X: N, P) are comparable to a conjugated π bond system like free butadiene. Furthermore the atomic distances Ti–X and C–X (X = N, P) and the small angles Ti–X–C prove the existence of an sp^2-hybridized X atom with a noncoordinated lone electron pair. This is in contrast to analogous but unstable compounds based on [Cp$_2$Ti=CH$_2$]. Their high reactivity can be regarded as a consequence of the participation of the nitrogen lone pair in bonding to the titanium center, leading to the widening of the Ti–N–C angle and subsequent opening of the azatitanacyclobutene ring, observed for example by Doxsee in the rearrangement *via* vinylimido intermediates [113]. Because of the higher basicity of the permethylated cyclopentadienyl ligands, there is reduced electrophilicity at the titanium center in **75** and **77**. An excess of nitrile and higher temperatures favor the insertion of another nitrile molecule into **75** to give **78**.

75

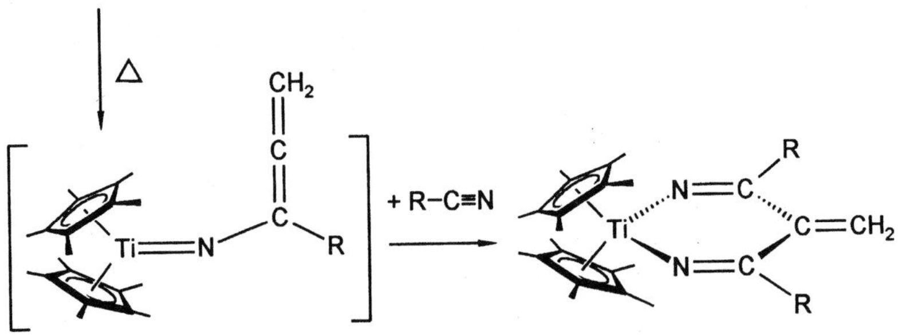

78

The different regioselectivities in reactions with nitriles and phosphaalkynes arises from the frontier orbital arrangements of the reactants. Whereas *end-on* coordination dominates in nitrile complexes, phosphaalkyne ligands prefer a *side-on* geometry (**B**, **C**) due to the π character of their HOMO (*i.e.* the π–n-separation is greater for phosphaalkynes than for nitriles). Starting from the primary *end on* coordination of the nitrile (**A**), the cycloaddition takes place forming only one regioisomer. From the experimental data it can be concluded that a primary η^2-coordination of the phosphaalkyne showing its alkyne-like behavior is very likely.

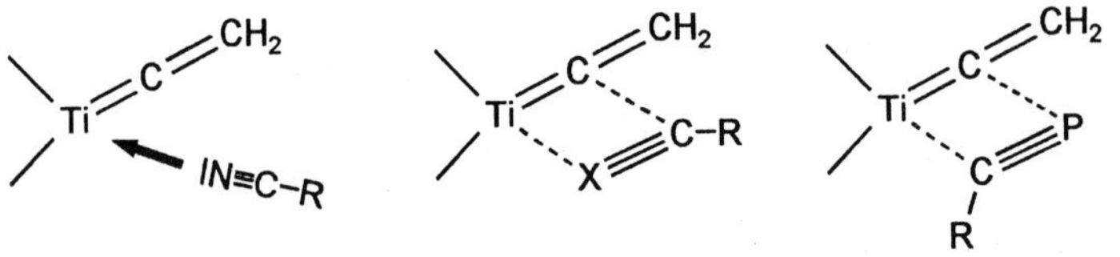

Scheme 4-32. *End-on* (**A**) and *side-on* coordination modes (**B**, **C**).

The behavior of the azatitanacyclobutenes **75** is quite different compared to the reactivity of azatitanacyclobutanes **79**, obtained from **38** and carbodiimides [218]. The azatitanacyclobutanes **79** are thermally stable up to 150°C. They are characterized by the ability to undergo cycloreversion reactions instead of the above-mentioned electrocyclic ring-openings.

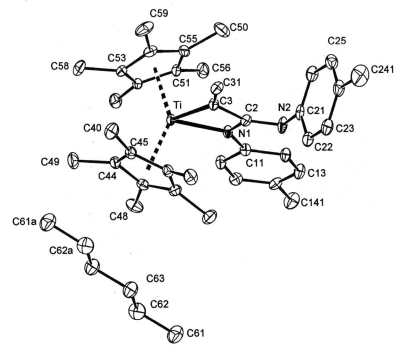

The structure of **79b** is confirmed by X-ray structure determination. The azatitana-cyclobutane ring is shown to be planar. The Ti–N distance is longer than in the azatitanacyclobutene **75b** (2.084(2) compared to 2.017(2) [236]).

Fig. 4-3. Molecular structure of **79b** . n-C$_6$H$_{14}$ in the solid state. Selected bond distance [Å] and angles [°]: Ti–N1 2.084(2), Ti–C3 2.129(3), C3–C31 1.329(4), C3–C2 1.505(4), C2–N2 1.287(4), C2–N1 1.374(4), N1–Ti–C3 65.8(1), Ti–C3–C2 91.5(2), C3–C2–N1 105.1(2), C2–N1–Ti 97.4(2), C2–N1–C11 124.6(3), C11–N1–Ti 135.5(2) [218].

The only primary regioisomer formed by cycloaddition of **35** with isothiocyanates is **80**, which contains the sulfur atom in the α position of the metallacyclic ring. When heated in the presence of pyridine, **80** can be isomerized into the other possible regioisomer **81** [237]. C=N-cycloaddition products are not observed. The quantitative isomerization **80→81** shows that the lower polarity of the C=S unit in the RNCS molecule allows the formation of a second isomer. In this respect, the behavior of titanathietanes is quite different from the behavior of titanaoxetanes, which normally react to give Ti=O fragments (*classical* behavior). In cycloaddition reactions of **35** with CS_2, only the regioisomer with the sulfur atom in the β-ring position is observed [238]. The inverse regiochemistry results from the difference in reactivity of carbonyl and thiocarbonyl groups with strong carbanionic molecules and shows the thiophilic character of the nucleophilic carbene intermediate **35** [239].

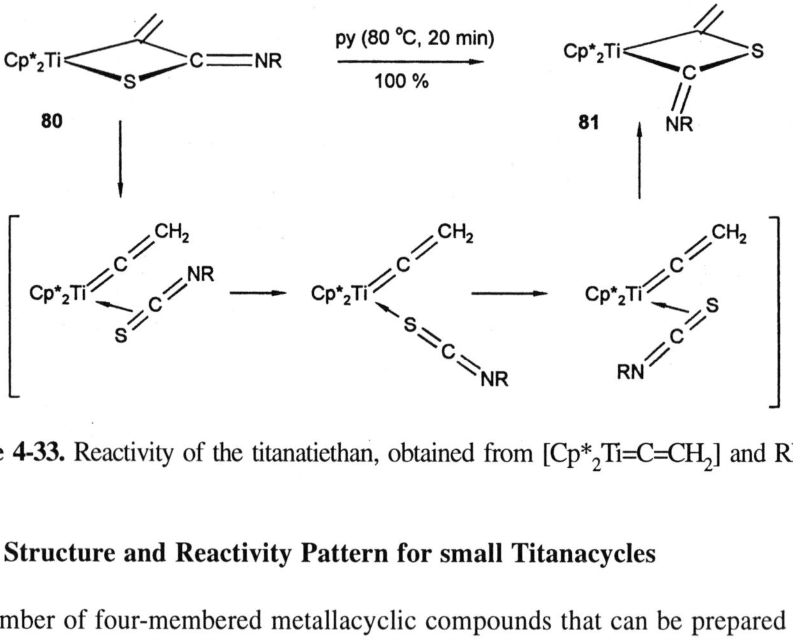

Scheme 4-33. Reactivity of the titanatiethan, obtained from $[Cp*_2Ti=C=CH_2]$ and RNCS.

4.4.3.8 Structure and Reactivity Pattern for small Titanacycles

The number of four-membered metallacyclic compounds that can be prepared from **35** allow a comparative discussion of their molecular structures (Table 4-12). All α-methylenetitanacycles are characterized by a planar four-membered ring. The bond angles indicate a stretching of the ring, which is reflected in the relatively short distances c (2.47 (**37**)–2.80 Å (**76**)). Particular in **37**, this leads to an extreme high-field shift of the corresponding signals in the NMR spectrum [37].

The planarity of the four-membered rings is lost if the exocyclic methylene group is located in the β instead of the α position (**18, 82**). The automerization equilibria corresponding to the ring folding (22.5° **18**; 33.0(9) **82**) could not be frozen out in NMR spectra by either **18** [150] or **82** [240]. Neither compound undergoes cycloreversion.

18   **82**

The Ti–C distances *a* and *b* in **37** are different (2.068(6) and 2.137(7) Å, respectively), in agreement with the different hybridizations of the carbon atoms. The equal C–C bond lengths (*f, e*) (1.520(10) and 1.521(10) Å, respectively) rule out a partially ring-opened structure (→ **35**). In the metallacyclobutene **54b** (R = Me), inner C–C bond (*e*), which can be conjugated, shortens significantly. Bulky substituents facilitate cycloreversions, which is reflected in longer distances *e* and *b* in **54c** (R$_\alpha$ = SiMe$_3$, R$_\beta$ = Ph). The exocyclic α-methylene group exhibit characteristic bond lengths of 1.318(8) Å for **63d** and 1.377(4) Å for **54b**. The increase of the angle *γ* in **37** to 152.4° is noteworthy. The distances b to the heteroatoms (O, S. P) are in the expected range, while for **75b** a slightly shortened Ti–N bond is found (*b* = 2.017(2) Å), which is in accordance with the observed reactivity in electrocyclic ring-opening, as well as the lengthening of the M–C bond (*a* = 2.134(2) Å). The azatitanacyclobutan **79b** shows a longer Ti–N bond (*b* = 2.129(3)) [218].

Table 4-12. Selected structural parameters of four-membered metallacyclic compounds

	Distances [Å]						Angles [°]						Ref.
	a	b	c	d	e	f	α	β	γ	δ	ε	φ	
54b	2.104(3)	2.109(3)	2.5	1.377(4)	1.434(4)	1.365(4)	139.8	68.0(1)	146.6(3)	87.8(2)	114.8(2)	89.4(2)	[208]
54c	2.102(6)	2.173(6)	2.5	1.322(8)	1.502(7)	1.352(8)	137.9	69.3(2)	149.0(5)	86.5(3)	116.7(5)	87.4(4)	[208]
79b	2.129(3)	2.084(2)	2.64	1.329(4)	1.505(4)	1.374(4)	139.5	65.8(1)	140.6(2)	91.5(2)	105.1(2)	97.4(2)	[218]
75b	2.134(2)	2.017(2)	2.48	1.337(3)	1.485(3)	1.290(3)	140.0	67.85(8)	145.4(2)	84.4(1)	113.2(2)	94.5(1)	[236]
76	2.161(5)	2.504(2)	2.8	1.326(8)	1.473(7)	1.701(5)	139.0	68.2(1)	133.1(4)	99.1(3)	111.6(3)	81.1(2)	[236]
63b	2.121(3)	1.983(2)	2.52	1.325(4)	1.477(4)	1.348(3)	140.8	67.6(1)	147.3(3)	87.2(2)	107.9(2)	96.7(2)	[222]
63d	2.119(6)	1.966(3)	2.53	1.318(8)	1.466(8)	1.362(7)	140.3	67.5(2)	139.1(5)	87.9(4)	106.9(5)	97.5(3)	[223]
18	2.074(4)	1.992(4)	2.47	1.318(9) β-exo-CH$_2$	1.428(7)	1.408(6)	139.0	68.7(2)	-	87.8(3)	-	91.7(3)	[150]
81	2.156(3)	2.466(1)	2.81	1.320(4)	1.484(5)	1.795(3)	138.9	70.45(9)	140.2(3)	99.5(2)	109.2(2)	80.9(1)	[237]
37	2.068(6)	2.137(7)	2.47	1.321(10)	1.521(10)	1.520(10)	138.5	83.1(4)	152.4	85.5	115.1	75.2	[38]
82	2.277(6)	2.258(7)	2.64	1.376(9) β-exo-CH$_2$	1.49(1)	1.51(1)	-	66.7(3)	-	86.1(4)	112.2(6)	86.4(4)	[240]

	X	Y	M
37	CH$_2$	CH$_2$	Ti
54b	C-Me	C-Me	Ti
54c	C-SiMe$_3$	C-Ph	Ti
79b	NC$_6$H$_4$CH$_3$	C=NC$_6$H$_4$CH$_3$	Ti
75b	N	C-tBu	Ti
76	P	C-tBu	Ti
63b	O	C=NC$_6$H$_{11}$	Ti
63d	O	C=CPh$_2$	Ti
81	S	C=NC$_6$H$_{11}$	Ti
18	O	C=CH$_2$	Ti
82	CH$_2$	C=CH$_2$	Zr

Remarkably, depending on the nature of the heteroatom, different reactions of the titanacycles are observed. In the case of the titanacyclobutane **37**, cycloreversion reactions dominate, forming the vinylidene intermediate **35**. In the case of the oxetanes **63**, metathesis reactions are observed in the mass spectrometer, whereas the bimetallic oxetanes **67** fragment to give the starting materials. Fast ring opening reactions occur in the case of the aza-titanacyclobutenes **75**, leading to products of insertion into the Ti–C bond, which exhibit the *exo* methylene group [236]. This behavior is explained by the orientation of the lone pair at the heteroatom towards the acceptor orbitals in the equatorial plane of the metallocene $Cp*_2Ti$ fragment. The best orbital overlap can be expected in the case of the azatitanacyclobutenes **75**, because the lone pair at the nitrogen atom and the lateral acceptor orbital of the titanium center are orientated in the same plane, leading to fast ring opening [191].

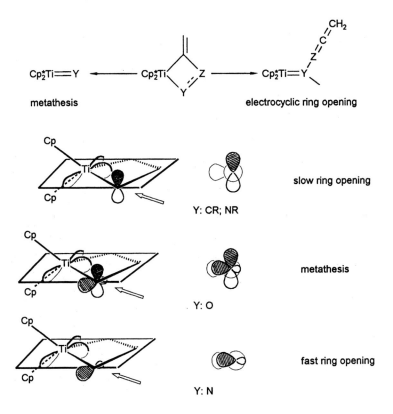

Scheme 4-34. Reaction behavior of different four membered titanacycles.

On the other hand, if the hybridisation of the nitrogen atom is different, as in the azatitanacyclobutanes **79**, no electrocyclic ring opening reaction is observed. In the molecule **79** the nitrogen lone pair is orientated perpendicular to the acceptor orbital on titanium [218]. A similar orientation of donor and acceptor orbitals is also expected in the case of titanacyclobutenes, leading to substituent controlled reactions. These reactivity patterns can also be observed in the insertion behavior of the titan-acycles (Scheme 4-35). In the case of the titanacyclobutane **37** [241] and cy-clobutenes **54** [208], insertion of small molecules like isonitriles are observed only into the Ti–C bond opposite the *exo* methylene group. For the azatitanacyclobutene **75**, a spontaneous ring enlargement by insertion of a further nitrile molecule is ob-served **75→78**, due to the activation of the Ti–C sp²-bond by the lone pair on the nitrogen center. In a similar manner the oxetanes **63** and **67** show insertion reac-tions into this Ti–C bond, forming mono or double insertion products [212, 242]. On the other hand, azatitanacyclobutanes **79** are inert to ring enlargement reactions with isonitriles even at higher temperatures [218].

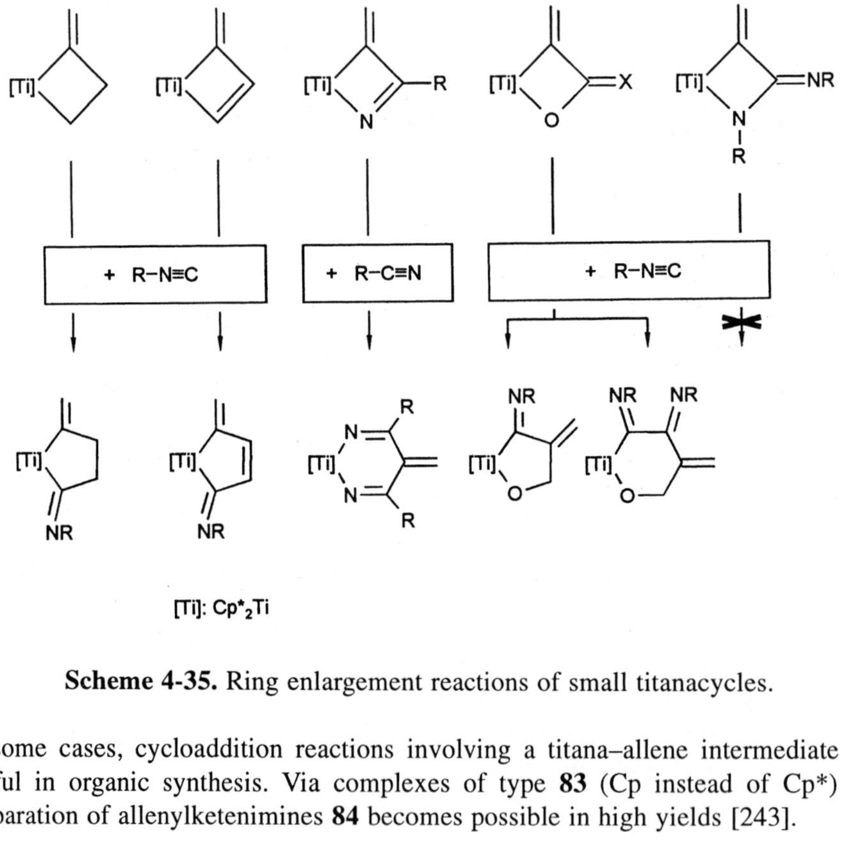

Scheme 4-35. Ring enlargement reactions of small titanacycles.

In some cases, cycloaddition reactions involving a titana–allene intermediate are useful in organic synthesis. Via complexes of type **83** (Cp instead of Cp*) the preparation of allenylketenimines **84** becomes possible in high yields [243].

4.4.3.9 Dimetallic Compounds Derived from Ti=C Species

Let's have a first look at the Tebbe reagent, the most important source of a titanium methylene compound. *Ab initio* calculations (STO-3G) on [H₂Ti=CH₂]/[ClAlH₂] show that for the Tebbe reagent, strong bonding of the Lewis acid [ClAlMe₂] must be assumed, in contrast to the much-discussed weak association [244]. The presence of a strong bond is also shown in the course of the reactions of the Tebbe reagent as compared to the Petasis or Grubbs variants [99]. Kinetic studies of the formation of the Tebbe reagent **87** by the reaction of [Me₃Al] with [Cp₂TiCl₂] (**85**) and its derivatives [97, 245] show an isotope effect $k_{H/D}$ of 2.9 and a large negative entropy value, which is in agreement with a cyclic transition state **86**. The polarization of the Ti–Cl bond by the aluminum center increases the acidity of the α-H atoms of the Ti-CH₃ group. This rules out the primary formation of a [Cp₂Ti=CH₂] species with subsequent complexation by [ClAlMe₂] during the course of Tebbe reagent formation.

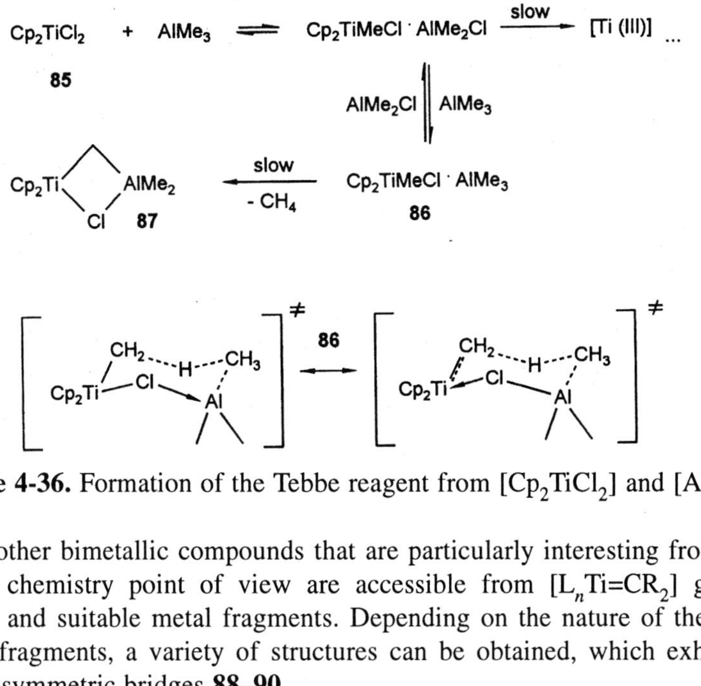

Scheme 4-36. Formation of the Tebbe reagent from [Cp₂TiCl₂] and [AlMe₃].

However, other bimetallic compounds that are particularly interesting from the coordination chemistry point of view are accessible from [LₙTi=CR₂] generating complexes and suitable metal fragments. Depending on the nature of the interacting metal fragments, a variety of structures can be obtained, which exhibit symmetric or asymmetric bridges **88–90**.

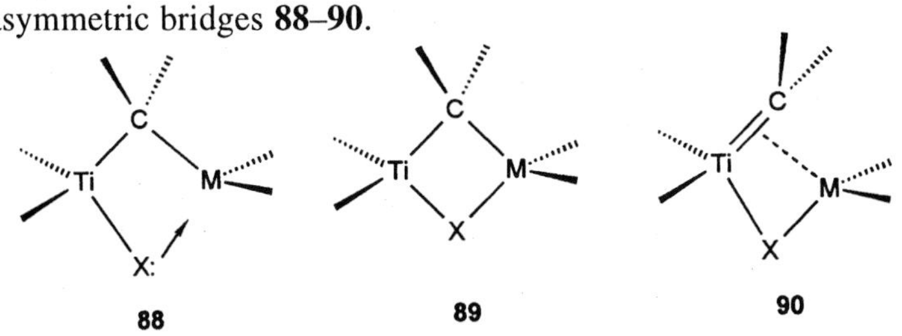

Scheme 4-37. Schematic drawings of methylene bridged dinuclear complexes.

Distinct carbenoid character is revealed by the reduced Ti–C distances and a notably low field shift of the μ-CH$_2$ signal [97, 116, 126, 246–250]. Table 4-13 summarizes selected examples. A participation of the mesomeric structure **90** is discussed in particular for the Ti–Pt compound [Cp$_2$Ti(μ-CH$_2$)(μ-Cl)Pt(PPhMe$_2$)(Me)] and [Cp$_2$Ti(μ-CH$_2$)(μ-CH$_3$)Pt(PPhMe$_2$)(Me)] [247]. For example, in contrast to dititanacyclobutane [Cp$_2$Ti(μ-CH$_2$)]$_2$ [126, 250], the silatitanacyclobutane [Cp$_2$Ti(μ-CH$_2$)(μ-CH$_2$)SiMe$_2$] [124, 125] shows no carbenoid activity.

Table 4-13. Selected structural and NMR data of dinuclear μ-methylene and related complexes [Cp$_2$Ti(μ-CH$_2$)(μ-X)ML$_n$]

	X	ML$_n$	Crystal habit Distances [Å], Angles [°]	NMR data δ (μ-CH$_2$)	δ(μ-CH$_2$)	Ref.
92	p-Me$_2$N Ph	Rh(cod)	dark green crystals Ti-Rh 2.827(1); Ti-CH$_2$ 2.076(4); Ti-C$_i$ 2.403(4); Rh-CH$_2$ 2.131(4) Ti-CH$_2$-Rh 84.4(1); CH$_2$-Ti-C$_i$ 89.9(1)	6.92 (s)	189.4 (dt)	[246]
	Cl	Pt(PPh Me$_2$)Me	red crystals Ti-Pt 2.962(2); Ti-CH$_2$ 2.066(18); Ti-Cl 2.427(5); Pt-CH$_2$ 2.112(17) Ti-CH$_2$-Pt 93.3(7)	8.09 (d)	179.2 (d)	[247]
	Me	Pt(PPh Me$_2$)Me	red-orange crystals Ti-Pt 2.776(1); Ti-CH$_2$ 2.115(7); Ti-CH$_3$ 2.395(8); Pt-CH$_2$ 2.078(7) Ti-CH$_2$-Pt 82.9(3)	7.42 (d)	180.0 (d)	[247]
91	Cl	Rh(cod)	red crystals Ti-CH$_2$ 2.018; Ti-Cl 2.493; Rh-CH$_2$ 2.133 Ti-CH$_2$-Rh 92.0	7.48 (s)	186.5 (td)	[248]
93	Me	Rh(cod)	orange-yellow crystals Ti-Rh 2.835(1); Ti-CH$_2$ 2.147(5); Ti-CH$_3$ 2.294(6)	7.24 (s)	185.4 (dt)	[249]
	CH$_2$	TiCp$_2$	crimson platelets	8.72 (s)	235.8 (t)	[126, 250]
	CH$_2$	SiMe$_2$	red-orange crystals Ti-CH$_2$ 2.146(3) [2.169(4)] [a]; CH$_2$-Ti-CH$_2$ 84.1(2) [83.7(2)]	2.50 (s)	70.55	[125]
87	Cl	AlMe$_2$	red crystals, CH$_2$- and Cl-group disordered	8.28 (s)	188.0	[97, 245]

[a] Two symmetry-independent molecules.

The μ-methylene titanium–platinum and –rhodium complexes do not only act as activated [Ti=CH$_2$] equivalents, but can undergo a variety of different subsequent reactions. For example, by varying the bridging ligand novel structural types can be obtained such as those with μ-phenyl or μ-methyl groups and thus providing information about the nature of these ligands. Compound **92**, the first example of a dinuclear phenyl species that is asymmetrically bridged by the *ipso*-C atom of the phenyl group [246], can be synthesized by reaction of **91** with phenyllithium derivatives.

cod = 1,5-cycloocatdiene

A unique feature of the Ti–Rh complex **93** is the bridging methyl group which forms a three-center 2-electron agostic bond with the titanium center [249]. Crystallographic and spectroscopic studies of this molecule have been carried out to show the static and dynamic behavior of this agostic interaction. The methyl group form an agostic interaction with the titanium atom. For the bridging methyl group, the Ti–H bond distance is 2.02(5) Å and the Ti–C bond distance is 2.294(6) Å. At room temperature the μ-CH$_3$ group shows a broad resonance at $\delta = -3.13$. When cooled to $-90°$C, this resonance is replaced by two resonances at $\delta = 1.28$ (d) and $\delta = -12.15$ (t, $J = 12.8$ Hz). The observed coalescence temperature is -40°C. In the ^{13}C NMR at room temperature the μ-CH$_3$ resonance is at $\delta = 49.7$ (dq, $J_{CH} = 114$, $J_{CRh} = 29$ Hz). At -92°C, the μ-CH$_3$ resonance was observed at $\delta = 51.1$. The coupling constant for the bridging proton is found to be 87.7 Hz; for the terminal protons 126.7 Hz. As expected, abnormally low coupling constants for the agostic proton are found [249].

The syntheses of vinylidene bridged complexes of the types **94–96** are of interest *e.g.* in the understanding of the ring expansion reaction **67→68** and the related possibilities of the formation of μ-vinylidene compounds as well as in the study of this bridging ligand. The majority of known μ-vinylidene complexes belong to the structural type **94** [251]. Some homo- and heterodinuclear μ-vinylidene complexes indicate the possibility of *side-on* bridges (**96**). A *semi-bridge* (**95**), which is characteristic for CO- or CS-bridged heterodinuclear complexes with early–late metal combinations [252] has only once been reported for comparable vinylidene complexes [253].

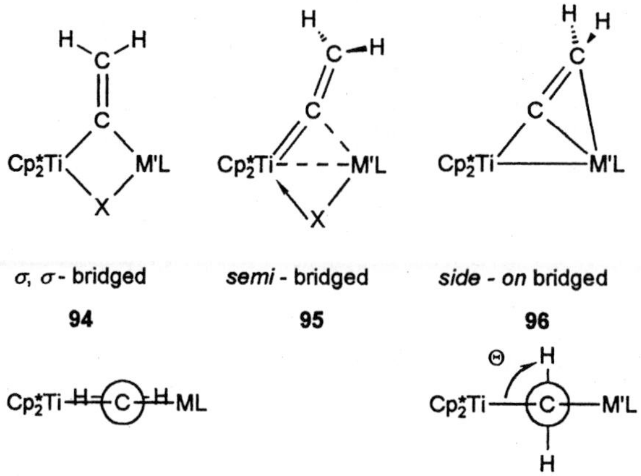

Scheme 4-38. Dimetallaethylene (94) as well as *semi-* and *side-on-*bridged structures (95 and 96, respectively) of dinuclear μ-vinylidene complexes.

When $[Cp*_2Ti(CH=CH_2)CH_3]$ (38) reacts with complexes of group 11 metals (97a–e), the μ-vinylidene complexes 98 can be isolated (40–50%) [254, 255]. In the complexes 98a–98e, the signal of the α-vinylidene carbon atom is shifted to very low field ($\delta = 300$–330; Table 4-14).

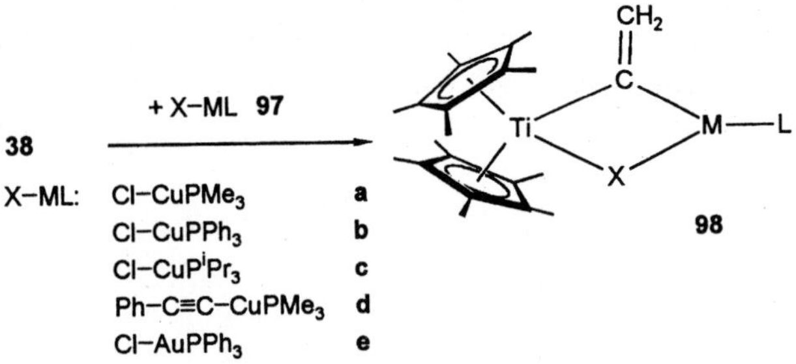

This is in accordance with a *semi-*bridging structure 95. The possibility of a *side-on* bridge (96) can be ruled out due to the orientation of the π-bonding planes in 98. In the case of *side-on* bridged vinylidene complexes (96) equal protons are expected in the ^1H NMR spectra.

Table 4-14. Selected NMR data of heterodinuclear μ-vinylidene complexes **98**

	Habit	mp	^{13}C NMR	^{1}H NMR			
		[°C]	$\delta(\mu\text{-}C=CH_2)$	$^{2}J_{HH}$ [Hz]	J_{PH} [Hz]	$\delta(\mu\text{-}C=CH_2)$	
98a	red crystals	93-95	330.1	9.4	5.2	6.16	5.70
98b	red crystals	137-140	328.7	9.3	5.9	6.16	5.72
98c	red crystals	145-147	329.8	9.1	4.9	6.31	5.79
98d	yellow crystals	180-182	329.2	8.6	2.9	6.16	6.00
98e	red crystals	109-110	300.5	-	-	6.71	5.62

4.4.4 Reactions Involving the Cp Ligand

Generally, the chemistry of titanocene fragments is characterized by the great stability of the η^5-Cp binding mode. In most organometallic reactions of transition metal complexes, the η^5-Cp ligand plays the role of spectator, staying tightly bound η^5 to the metal center throughout the course of the reaction. However, cyclopentadienyl ligand-transfer reactions have been known since 1956 when Wilkinson, Cotton, and Birmingham reported that the reaction of chromocene and ferrous chloride gave ferrocene [256]. The most detailed work with early transition metal complexes is that of Brubaker and co-workers [257–259]. Photolysis of [(η^5-C$_5$H$_5$)$_2$MCl$_2$] and [(η^5-C$_5$D$_5$)$_2$MCl$_2$] mixtures (M = Ti, V, Hf) results in intermolecular exchange of the cyclopentadienyl ligands. The fact that [(η^5-C$_5$H$_5$)TiCl$_3$] is not found in the absence of a sensitizer led Brubaker to suggest that in some cases a photochemically induced ring slippage may open up the coordination sites necessary for a bimolecular C$_5$H$_5$-bridged mechanism.

Other new investigations show that the Cp ligand is not as inert as many people believe, as cyclopentadienyl activation has been observed by several groups. Cp-activation seems to become possible if the titanium atom is part of a heteroatom containing metallacycle. For example, the formation of complex **103** is observed during the formation of **102**. Whereas **102** can be isolated in a pure form [260] **103** rearranges by Cp-activation to form **104** [261].

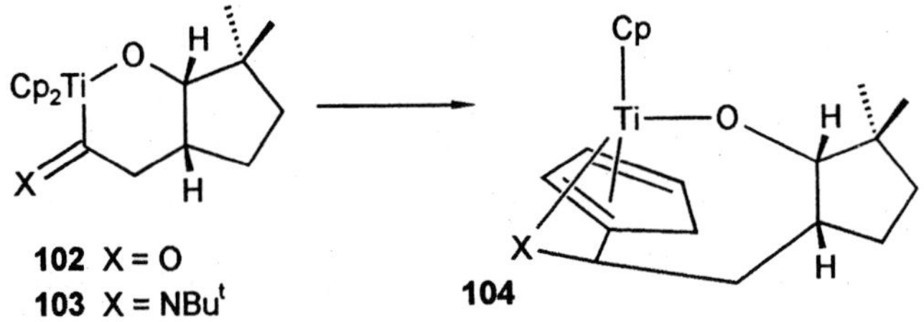

102 X = O
103 X = NBu^t

104

The formation of Cp-activation products from electron rich titanacyclopentadienes [262] like **105** is also observed [263], and further examples involving titanacyclopentadienes are known [262]. These reactions under formation of **106** are promoted by nonpolar solvents.

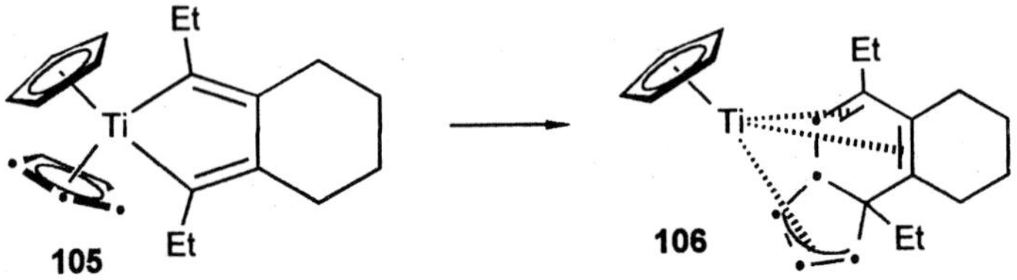

105

106

The formation of fulvenes as the product of reactions of aliphatic and aromatic ketones with [Cp$_2$Ti(PMe$_3$)$_2$] is discussed in terms of Cp-activation instead of a pinacol coupling reaction [264]. Also, examples of ring enlargements of the Cp* ligand are known. Reaction of [Cp*$_2$TiR] (R = CH$_3$, CH$_2$CMe$_3$) with CO gives labile acyl compounds [Cp*$_2$TiC(O)R]. They react with [(CpMo(CO)$_3$)$_2$] to yield complexes [Cp*$_2$Ti(η^2-COR)(μ-OC)MoCp(CO)$_2$], which undergo acyl C–O bond breakage to form benzene derivatives at room temperature. The benzene derivatives are formed through ring expansion of the Cp* ligand by incorporation of the R–C≡ fragment of the acyl group. The acyl oxygen is incorporated into a dimeric titanium oxo complex [265].

4.5 Analytical Data

4.5.1 Structural Data

The number of Cp$_2$TiL$_2$ compounds known in each oxidation state decreases in the order: four > three > two [266]. Although the comparison is obviously not strictly relevant, the Ti–Cp bond distance appear to show a slight increase with increasing formal oxidation state of the titanium atom. For example, the mean Ti–Cp distances for Ti(II) (2.041 Å), are shorter than those for Ti(III) (2.048 Å), and these are again somewhat smaller than those for Ti(IV), (2.058 Å) [266]. Comparison of the Ti–Cp distances for the mono- and bis-cyclopentadienyl compounds reveals that the mean values are similar for both. In general, the Ti–L distances increases with increasing van der Waals radius of the ligated atom: 2.003 Å (range 1.860–2.155 Å) for O (1.52 Å) < 2.007 Å (range 1.922–2.100 Å) for N (1.55 Å) < 2.359 (range 2.310–2.405 Å) for Cl (1.75 Å) < 2.429 Å (range 2.395–2.455 Å) for S (1.80 Å) < 2.497 Å (2.340–2.585 Å) for P (1.80 Å) [266]. Representative examples of titanocene complexes are summarized in Table 4-15. It becomes obvious that introduction of methyl groups on the Cp ligand leads to increasing Z–Ti–Z angle. Additionally, the Z–Ti–Z angle is smaller for Ti(IV) than for Ti(II) complexes. An extremely large Z–Ti–Z angle of 150.84° is found for [(PhC$_5$Me$_4$)$_2$TiH] [86].

Table 4-15. Selected structural data of representative titanocene complexes L$_2$TiX$_2$

Compound	Ti-Z [Å]	Ti-X [Å]	Z-Ti-Z [°]	X-Ti-X [°]	Ref.
Titanium (IV) complexes					
85 [Cp$_2$TiCl$_2$]	2.058	2.364(3)	130.89(5)	94.43(6)	[267]
[(C$_5$Me$_4$H)$_2$TiCl$_2$]	2.11	2.344	133.4	94.2	[268]
[Cp*CpTiCl$_2$]	2.07 (Cp) 2.10 (Cp*)	2.3518(9)	132.0	94.44(5)	[269]
[Cp*$_2$TiCl$_2$]	2.127(4) 2.128(4)	2.352(1) 2.346(1)	137.4(1)	92.94(4)	[270]
[Cp$_2$TiMe$_2$]	2.075 2.082	2.170(2) 2.181(2)	134.5	91.3(1)	[271]
Titanium (III) complexes					
[Cp*$_2$TiCl]	2.06	2.363	143.6	-	[272]
[(PhC$_5$Me$_4$)$_2$TiH]	-	1.768(15)	150.84	-	[86]
[Cp*$_2$TiCH$_2$CMe$_3$]	2.104(2) 2.094(2)	2.231(5)	139.4(3)	-	[273]
Titanium (II) complexes					
[Cp*$_2$Ti(η^2-C$_2$H$_4$)]	2.092	2.160(4)	143.6(4)	38.9(1)	[61]
[Cp*$_2$Ti(CO)$_2$]	2.06 2.07	2.01	147.9	83.3	[274]
[Cp$_2$Ti(CO)$_2$]	2.025	2.030(11)	138.6	87.9(6)	[275]

Z centroids of the cyclopentadienyl ligands.

4.5.2 NMR Data

While the rotation barrier about the M–Cp bond is usually small, restricted rotation in complexes containing substituted cyclopentadienyl ligands has been noted [18, 276]. As part of the characterization of substituted biscyclopentadienyl Ti(IV) compounds, ^{13}C NMR and ^{1}H NMR data indicate that the effects of both metal shielding and ring substitution can be taken into account by a simple additive relationship so that it is possible to accurately predict chemical shifts for these systems [277].

4.5.2.1 ^{1}H NMR Data

^{1}H NMR shifts of some titanocene complexes are given in Table 4-16 [277]. The assignment was based on the following considerations. The alkyl protons and cyclopentadienyl protons resonate in quite different regions. The ^{1}H NMR spectrum of the monoalkylated cyclopentadienyl ring shows two triplets. The ^{1}H NMR spectrum of [1,3-(Me$_2$C$_5$H$_3$)$_2$TiCl$_2$] consists of an AB$_2$ pattern while that of 1,3-(MeEtC$_5$H$_3$)$_2$TiCl$_2$ is an ABX pattern. The most surprising result is the apparent lack of cyclopentadienyl proton splitting in [1,2-(Me$_2$C$_5$H$_3$)$_2$TiCl$_2$]. A similar observation has been reported for titanium indenyl complexes. For the series [(Ind)$_2$TiX$_2$], an AB$_2$ pattern is observed where X = CH$_3$, but a singlet is observed for X = C$_6$H$_5$ [278]. In contrast, an AB$_2$ pattern was reported for [(Ind)$_2$Zr(C$_6$H$_5$)$_2$].

Table 4-16. ^{1}H NMR Data of selected Titanium (IV) compounds (measured in CDCl$_3$)

Compound	$\delta \beta$-H	δ ring	$\delta \alpha$-H	R in CpR	Ref.
85 [(C$_5$H$_5$)$_2$TiCl$_2$]		6.55			[277]
[(MeC$_5$H$_4$)$_2$TiCl$_2$]	6.38 (J_{HH} = 2.6 Hz)		6.30	2.30	[277]
[(EtC$_5$H$_4$)$_2$TiCl$_2$]	6.41 (J_{HH} = 2.5 Hz)		6.33	2.76 (J_{HH} = 7.6 Hz), 1.19	[277]
[(n-PrC$_5$H$_4$)$_2$TiCl$_2$]	6.41 (J_{HH} = 2.5 Hz)		6.31	2.68 (J_{HH} = 7.9 Hz), 1.60 (J_{HH} = 7.5 Hz), 0.94 (J_{HH} = 7.3 Hz)	[277]
[(1,2-Me$_2$C$_5$H$_3$)$_2$TiCl$_2$]	6.13		6.13	2.10	[277]
[(1,3-Me$_2$C$_5$H$_3$)$_2$TiCl$_2$]	6.21 (J_{HH} = 2.2 Hz)		5.98	2.14	[277]
[(C$_5$Me$_4$H)$_2$TiCl$_2$]		6.00		1.98, 2.05	[194]
[Cp*$_2$TiCl$_2$]				2.00	
[Cp*CpTiCl$_2$]*		5.95		1.72	

* C$_6$D$_6$

4.5.2.2 ^{13}C NMR Data

Selected ^{13}C NMR data are given in Table 4-17. The alkyl and the cyclopentadi-enyl carbon atoms again resonate in two different regions of the spectrum. The chemical shifts of the chain carbon atoms were assigned by comparison with chemical shifts in free alkanes. The ring carbon atoms were identified by compari-son to specifically deuterated species [277].

Table 4-17. ^{13}C NMR data of selected titanium(IV) Compounds ($CDCl_3$)

		ring			
Compound	1	2	3	R in CpR	Ref.
85 [(C_5H_5)$_2$TiCl$_2$]	120.4				[277]
[(MeC$_5$H$_4$)$_2$TiCl$_2$]$^{a)}$	134.2	123.4	115.8	16.5	[277]
[(EtC$_5$H$_4$)$_2$TiCl$_2$]$^{a)}$	140.8	121.5	116.0	24.0, 13.9	[277]
[(n-PrC$_5$H$_4$)$_2$TiCl$_2$]$^{a)}$	138.9	122.5	115.9	33.2, 23.7, 13.9	[277]
[(1,2-Me$_2$C$_5$H$_3$)$_2$TiCl$_2$]$^{b)}$	133.6	119.8	109.5	14.5	[277]
[(1,3-Me$_2$C$_5$H$_3$)$_2$TiCl$_2$]$^{c)}$	134.4	126.2	116.4	17.1	[277]
[(C$_5$Me$_4$H)$_2$TiCl$_2$]$^{d)}$	113.1	125.2	135.9	13.1, 14.8	

assignment:

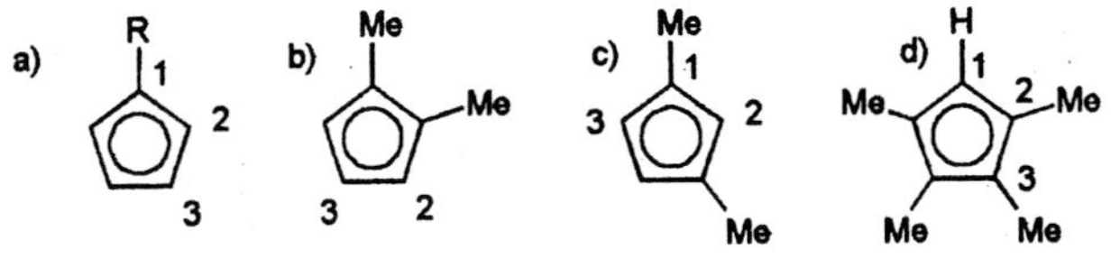

The C–H coupling constants of methyl groups depend on the amount of σ charac-ter at the carbon in the C–H bonds. To take a simple approximation the amount of σ character depends on the effective electronegativity of the group to which the methyl group is bonded; the higher the electronegativity of this group, the lower the   character of its bond with the methyl carbon. This leaves more s character in the C–H bonds, and hence, increasing the electronegativity of the group attached to the carbons should increase the C–H coupling constant observed for a methyl group. Conversely, decreasing the electronegativity of the group attached to carbon should lower the C–H coupling constant. Increasing the electron density at the car-bon to which the methyl group is bonded effectively lowers the electronegativity and should lower the C–H coupling constant. Indeed, increasing the electron densi-ty at the titanium center by adding methyl groups to the Cp ligand causes a lowering of the J_{CH} of the methyl group bonded to the Ti. Conversely, substitution of electron-withdrawing chlorides or a CF_3 group on the ring causes an increase in J_{CH} [129].

Table 4-18. C–H Coupling constants of methyl groups attached to a titanocene center [129]

Compound	J_{CH} [Hz]
$(F_3CC_5H_4)CpTiMeCl$	129.9
$(Cl_2C_5H_3)_2TiMeCl$	129.9
$(ClC_5H_4)CpTiMeCl$	129.1
$Cp_2TiMeCl$	128.9
$Cp'CpTiMeCl$	128.4
$Cp'_2TiMeCl$	128.2
$(Me_3C_5H_2)CpTiMeCl$	127.7
$Cp*CpTiMeCl$	127.2
$Cp*_2TiMeCl$	126.5

As well as the solution NMR techniques also solid state NMR measurements were used. Variable-temperature ^{13}C CP/MAS NMR spectroscopy of $(\eta^5\text{-}Cp)_2Ti(\eta^1\text{-}Cp)_2$ has shown that sigmatropic rearrangement of the σ-cyclopentadienyl rings occurs in the molecule in the crystalline state. Magnetization-transfer experiments at 165 K are consistent with [1,2] shifts being the main pathway for the rearrangement, but there is evidence for the possibility of [1,3] shifts as a minor pathway. These measurements, in conjunction with spectral line shape analysis in the exchange broadening regime, have been used to estimate Arrhenius activation parameters for [1,2] shifts of $E_a = 33.2 \pm 1.0$ kJ mol^{-1} and $A = 2.9 \times 10^{10}$ s^{-1}. The activation barrier is similar to that suggested by solution NMR studies, implying that control of the rearrangement is principally by electronic factors [279]. Also the structures and dynamics of a series of crystalline *ansa*-metallocenes, $[(C_5H_4)_2\text{-}C_2Me_4]TiX_2$ (X = F, Cl, Br, I) have been elucidated by joint application of solid state NMR and single crystal X-ray diffraction techniques [280].

4.5.2.3 Ti NMR Data

Both ^{49}Ti and ^{47}Ti resonances appear in the same spectrum, with the ^{49}Ti resonance 268.1 ppm downfield of the ^{47}Ti resonance. Due to its higher quadrupole moment the ^{47}Ti resonance is always broader than the ^{49}Ti resonance. Increasing methyl substitution of the rings causes an increasing downfield shift of both the ^{47}Ti and ^{49}Ti resonances [129]. The ^{49}Ti NMR chemical shifts of a series of titanocene derivatives have been shown to have an inverse relationship to the $Ti(2p_{3/2})$ binding energies of these same compounds as measured by X-ray photoelectron spectroscopy (ESCA). Replacement of one cyclopentadienyl (Cp) group in the titanocene difluorides, dichlorides, and dibromides by a pentamethylcyclopentadienyl (Cp*) moiety resulted in an average downfield shift in the ^{49}Ti NMR absorption of 148 ppm. Substitution of both Cp groups by Cp* groups resulted in an average downfield shift of 312 ppm [28].

Table 4-19. ^{49}Ti NMR chemical shifts (vs $TiCl_4$, δ 0.00) and $Ti(2p_{3/2})$ binding energies [28]

compound	^{49}Ti, $\delta^{a)}$	$Ti(2p_{3/2})^{b)}$ ± 0.1 eV
Cp_2TiF_2	-1051	457.5
$CpCp*TiF_2$	-964	457.0
$Cp*_2TiF_2$	-823	456.7
Cp_2TiCl_2	-773	456.9
Cp_2TiBr_2	-671	456.8
$CpCp*TiCl_2$	-601	456.6
$CpCp*TiBr_2$	-486	456.5
$Cp*_2TiCl_2$	-443	456.1
$Cp*_2TiBr_2$	-293	455.9

$^{)}$ $TiCl_4$ was used as external standard, NMR spectra were measured in $CDCl_3$.

$^{)}$ All binding energies are calibrated vs. C(1s) of polyethylene.

Titanium NMR spectroscopic data for a series of trichlorotitanium complexes bearing a cyclopentadienyl ligand substituted with tert-butyl and/or trimethylsilyl groups have also been collected. Chemical shift values imply that trimethylsilyl groups are weakly electron-donating, whereas the line-widths reflect the symmetry of the substitution pattern of the five-membered rings [281].

4.5.3 UV vis Spectra

The UV vis spectra of ring substituted titanocene dichlorides exhibit a weak absorbance between 500 and 560 nm, which in titanocene dichloride has been assigned to a symmetry-forbidden $A_1 \rightarrow A_2$ transition. This band moves to lower energy as electron donating substituents are added to the rings and to higher energy as electron-withdrawing substituents are added. The absorption spectra of the compounds with one ring substituent are qualitatively similar to the spectrum of Cp_2TiCl_2; in addition to the band between 500 and 560 nm, they exhibit a broad region of absorption between 260 and 410 nm. The compounds with more highly substituted rings, exhibit not only this broad region of absorption, but also a band between 470 and 490 nm [129].

Table 4-20. Absorption maxima of lowest energy transitions in the UV–vis spectra of ring-substituted titanocene dichlorides (3.3×10^{-3} M solutions in $CHCl_3$)

Compound	λ_{max} [nm]	Ref.
$Cp^*_2TiCl_2$	560	[129]
$(1,2,4\text{-}Me_3C_5H_2)_2TiCl_2$	536	[129]
$(Me_3SiC_5H_4)CpTiCl_2$	532	[129]
Cp'_2TiCl_2	528	[129]
$Cp'CpTiCl_2$	525	[129]
Cp_2TiCl_2	523	[129]
$(C_5H_4Cl)CpTiCl_2$	517	[129]
$(C_5H_4Cl)_2TiCl_2$	514	[129]
$(F_3CC_5H_4)CpTiCl_2$	513	[129]
$(1,2,4\text{-}Me_3C_5H_2)CpTiCl_2$	510	[129]

Using self-consistent-field-Xα-scattered-wave (SCF-Xα-SW) molecular orbital methods to investigate the electronic structures of Cp_2TiX_2 (X = F, Cl, Br) the lowest energy electronic transitions are predicted to be due to Cp→Ti charge-transfer. The halide→Ti charge-transfer transitions occur at higher energy. For X = I, Cp→Ti and I→Ti charge-transfer transitions are predicted to be very close in energy. That the fluoride, chloride, and bromide complexes have lowest energy exited states different from those of the iodide and methyl complexes is reflected in the photochemistry of the two type of complexes. In the former complexes, Cp–Ti bond cleavage results from low-energy irradiation; in the latter complexes, Ti–X bond cleavage occurs when the complexes are irradiated [282].

4.6 Selected Applications of Titanocene Complexes

Titanocene complexes have a large number of applications. These range from stoichiometric organic reactions to catalytic reactions in the field of polymerization and fine chemical synthesis [283, 284], which are explained in other chapters of this book. Selected examples, reduction of carbonyl compounds [285, 286], of epoxides [287], of olefins [288], and epoxidations [289–291], isomerizations [292], Diels–Alder reactions [293, 294], C–F activation reactions [295–297] as well as carbomagnezations [298–301], are summarized in Scheme 4-39.

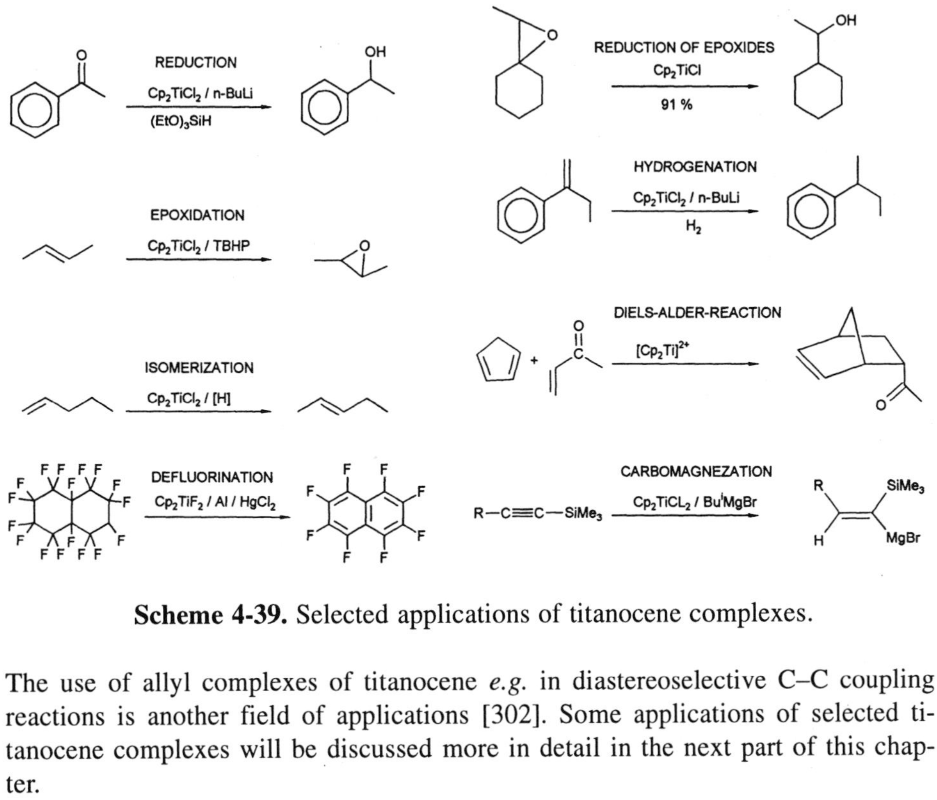

Scheme 4-39. Selected applications of titanocene complexes.

The use of allyl complexes of titanocene *e.g.* in diastereoselective C–C coupling reactions is another field of applications [302]. Some applications of selected titanocene complexes will be discussed more in detail in the next part of this chapter.

4.6.1 C–C Coupling Reactions

4.6.1.1 Dimerization of Ethylene

The selective dimerization of Ethylene, first described by Ziegler, uses a mixture of titanium tetrabutyl ester and trialkylaluminum [303].

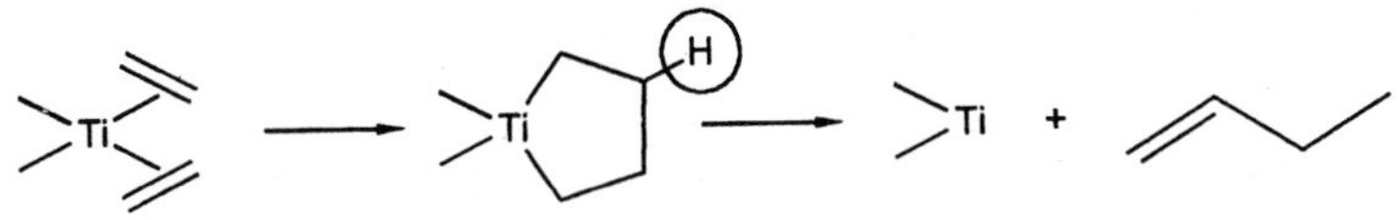

Interestingly, benzene and toluene solutions containing [Cp*$_2$Ti(η^2-C$_2$H$_4$)] (**42**) catalyze the specific conversion of ethylene to 1,3-butadiene and ethane under very mild conditions (25°C, ≤ 4 atm), although turnover numbers are exceedingly small [61]. As shown in Scheme 4-40, ethylene reacts reversibly with **42** in aromatic solvents to yield an equilibrium mixture containing the titanacyclopentan derivative **107**. In the combined dimerization / hydrogen-transfer reaction, butadiene and ethane are formed. The formation of butenes is not observed.

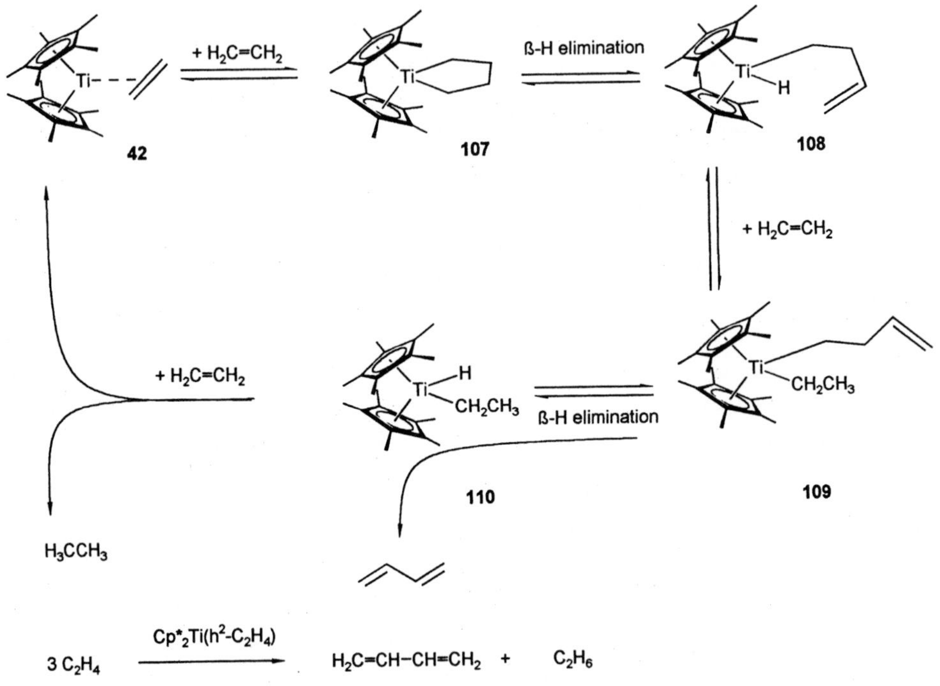

Scheme 4-40. Dimerization of ethylene to butadiene-1,3 and ethane.

4.6.1.2 Head to Tail Coupling Reactions of Acetylenes

The acetylene complex [Cp*$_2$Ti(η^2-Me$_3$SiC≡CSiMe$_3$)] catalyzed the linear head-to-tail dimerization of 1-alkynes. A selectivity of better than 98% and a turnover number in the range 1200–1500 mmol of 1-alkyne per mmol of titanium was obtained after 14 days at 30°C for 1-pentyne, 1-hexyne, cyclohexylethyne, phenylethyne and trimethylsilylethyne [304].

4.6.1.3 Cyclization Reactions

The intramolecular cycloaddition of the titanocene vinylidene complex, formed by dechloroalumination of **111** [305], to an alkene (or alkyne) affords bicyclic titana-cyclobutanes and **112** (or butenes) [306]. This cyclization has proven to be possible in all such complexes. The remaining carbon–metal bond in the metallacycles **112** offers great potential for further elaboration to give organic products. This is illustrated by complex **112,** which reacts with *N*-bromosuccinimide (NBS), to give the dibromide **115**. The insertion reaction of isonitrile affords the imino complex **113** which gives the aldehyde **114** on acidic work-up; alternative reaction with carbon monoxide followed by acidic work-up gave the ketone **117**, presumably *via* the ene–diolate complex **116** [306].

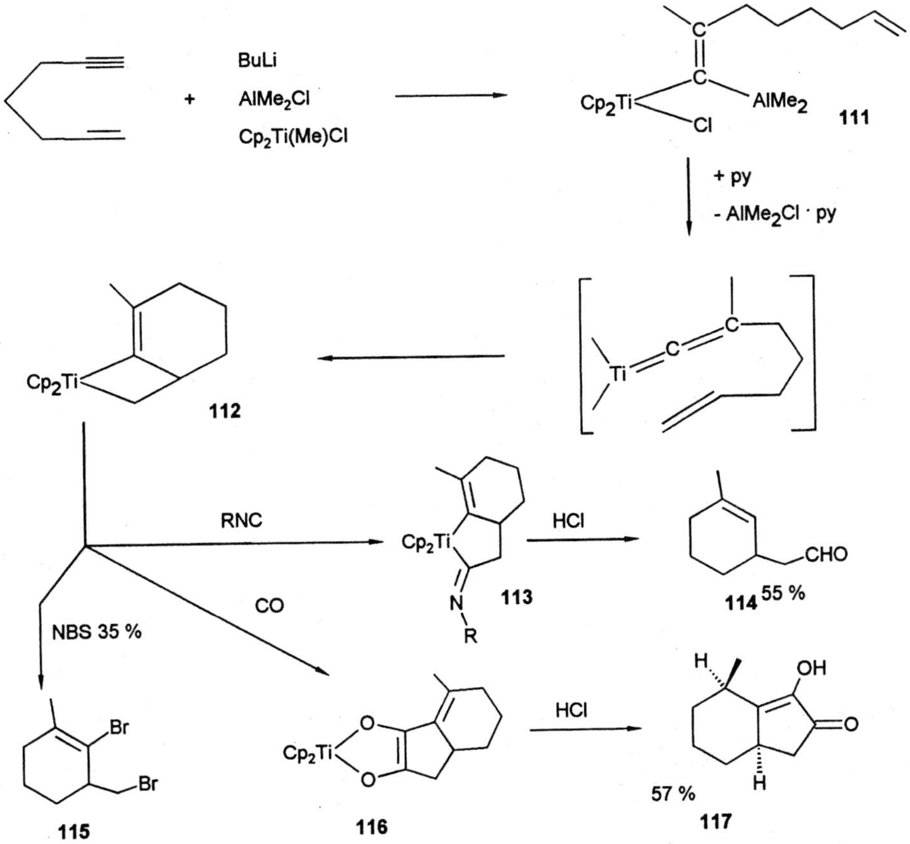

Scheme 4-41. Selected organic syntheses with a titanocene vinylidene intermediate.

4.6.2 Stoichiometric and Catalytic Reactions of Carbenoid Metal Complexes of Electron-Deficient Transition Metals

The most important applications of carbenoid complexes of electron-deficient transition metals lie in the areas of organic synthesis and catalytic reactions, such as carbonyl olefination (**c**) [98] the generation of cyclopropane derivatives by the oxidation of titanacyclobutanes (**a**) [307, 308] the preparation of heterocyclic four-membered ring compounds (**b**) [309, 310], and ring-opening polymerization [99, 311–316] to obtain special polymers, such as 'living' [132, 317] (**d**) or 'conducting' [318] (**e**) polymers (Scheme 4-42).

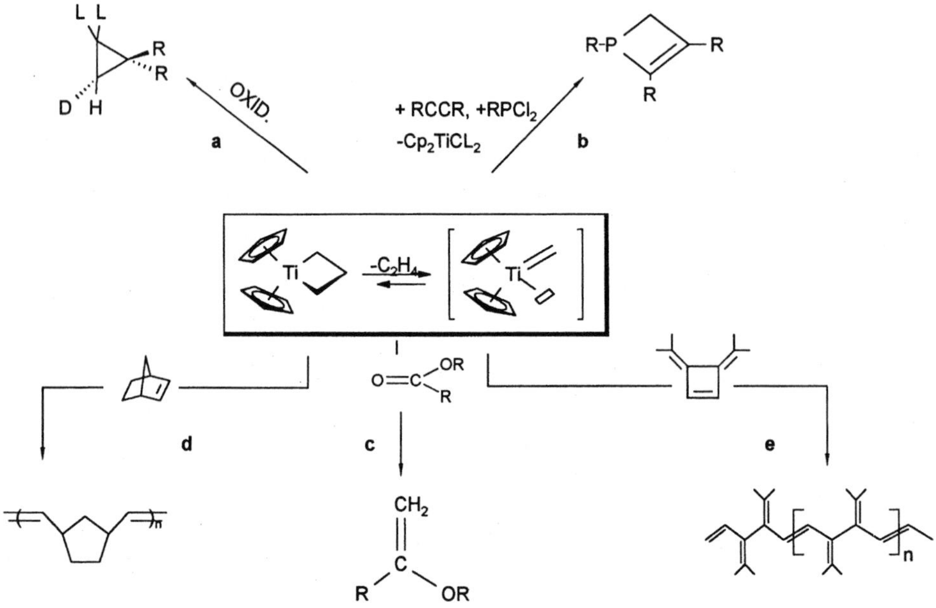

Scheme 4-42. Applications of titanacyclobutane derivatives.

4.6.2.1 Carbonyl Olefinations with Tebbe, Grubbs, and Petasis Reagents

Many carbonyl compounds, such as esters, ketones, and amides, can be methylated with various titanium-based reagents. The most important of these are the Tebbe reagent, titanacyclobutanes (Grubbs reagent) and titanocene alkyl complexes (Petasis reagent). The application of the Tebbe reagent in particular has found broad interest in organic chemistry, owing to its early discovery [98, 99]. The use of $[Cp_2TiMe_2]$ is particularly advantageous, as this compound is not pyrophoric, is relatively stable toward air and water, and no Lewis acidic aluminum by-products have to be separated off later, which enables its use in olefination of silyl esters, anhydrides, carbonates, and acylsilanes [101]. A number of different, sometimes contradicting, mechanisms [99] have been proposed for the reactions that occur, which will be discussed using the carbonyl olefination of esters as an example (Scheme 4-43). In the absence of an additional base, the reaction between the Tebbe reagent and the ester is first-order with respect to the titanium reagent employed.

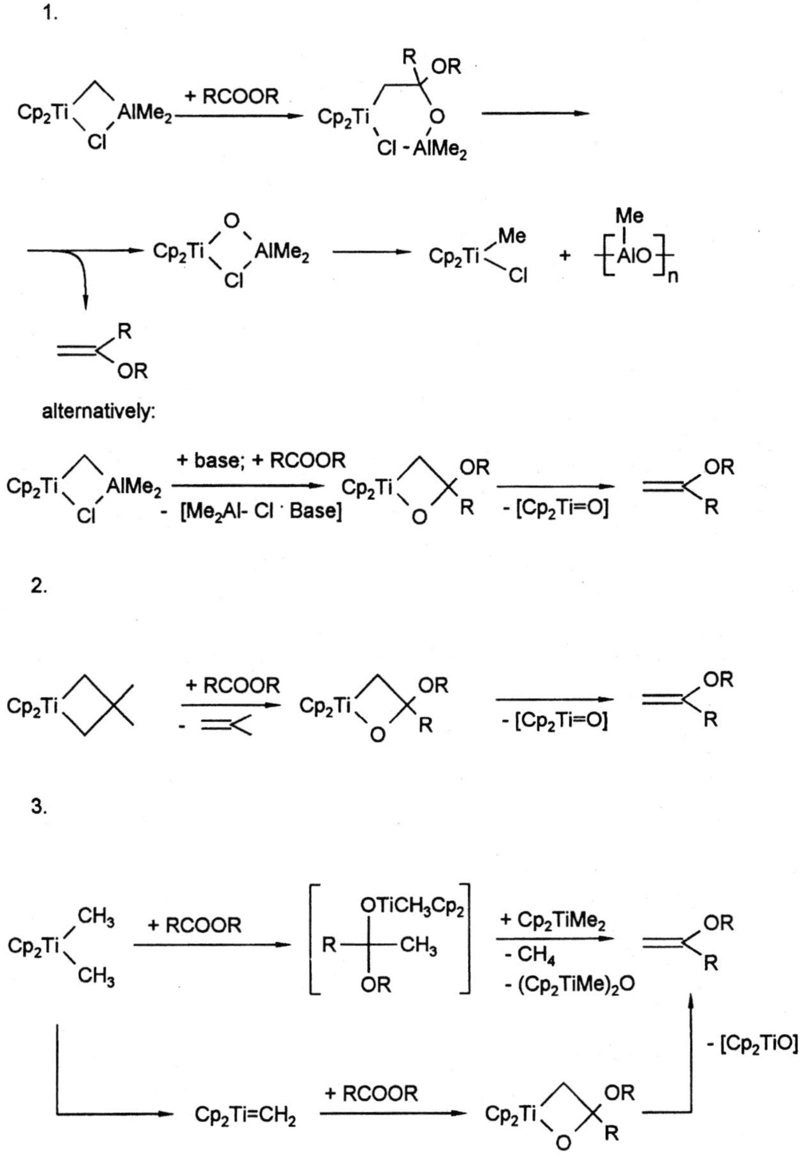

1.

alternatively:

2.

3.

Scheme 4-43. Mechanisms of the carbonyl olefinations 1. with Tebbe reagent, 2. titanacyclobutanes (Grubbs reagent), 3. with titanocene alkyls (Petasis reagent).

The strongly negative activation entropy observed is in agreement with a six-membered intermediate. In the presence of bases (pyridine), the reaction proceeds more rapidly and shows zero-order kinetics with respect to the ester and first-order kinetics with respect to the titanium reagent. The use of titanacyclobutanes in carbonyl olefination is characterized by the primary formation of a metallaoxetane, which spontaneously decomposes to give the desired olefin and a Ti=O compound [99]. Based on the observed H/D exchange, Petasis and Bzowej favored a methyl addition mechanism [115]; however, they reported deuterium scrambling in the reactions of esters with cyclopropyltitanocenes [110].

Detailed kinetic studies show that the olefination of esters with [Cp_2TiMe_2] proceeds *via* the titanium carbene [$Cp_2Ti=CH_2$] [118]. This is proved by: 1) the absence of H/D exchange or ^{13}C scrambling when labeled esters or titanium compounds are employed; 2) the finding of zero-order kinetics with respect to the ester, and first-order kinetics with respect to the titanium compound; 3) the reaction of ethyl acetate or dodecyl acetate with [$Cp_2Ti(CD_3)_2$] yielding a $k_{H/D}$ effect of 9–10; 4) esters with different electronic and steric environments exhibiting comparable reactivities. The stoichiometry required for the Petasis reagent (substrate : Ti = 1:2) can be explained by the necessity of reactions which trap the Ti=O fragment formed. The products were obtained in better yields with the Tebbe reagent than with the Wittig reagent. This is especially significant when sterically hindered ketones are to be olefinated. For the generation of special vinylsilanes, there are effective reagents in the form of [$Cp_2Ti(CH_2SiMe_3)_2$] and [$CpTi(CH_2SiMe_3)_3$] [111, 102]. The introduction of a benzylidene group can be successfully performed by using [$Cp_2Ti(CH_2C_6H_5)_2$] [112]. Substituents in the μ-position of the phenyl ring (Cl or F) increase the yield. For example, in this way aryl substituted enol ethers **118**, which also find use as bioactive prostaglandin analogs, can be formed quantitatively [112].

Disubstituted titanacyclobutenes **119** ($R_\alpha = R_\beta = CH_3$, CH_2CH_3) react with two equivalents of nitrile under mild conditions to yield diazacyclooctatetraenes, a result of nitrile insertion into the titanium–alkyl bond as well as into the alkenyl bond. Subsequent hydrochlorination of these products yields tetrasubstituted pyridines **124** [114]. Steric bulk leads to the exclusive formation of monoinsertion products. Insertions into the titanium–vinyl bond are favored over those into the metal–alkyl bond, when titanacyclobutenes with small substituents (methyl) are used. Diphenyltitanacyclobutenes, however, react with nitriles or ketones to yield solely products of monoinsertion into the titanium–alkyl bond, which can be worked up to yield unsaturated products **120** and **121**, respectively [114, 319]. Reactions of **119** with ketones can be used to generate dienes **122** [320], or after the reaction with RPCl$_2$, the phosphacyclobutenes **123** [309, 321].

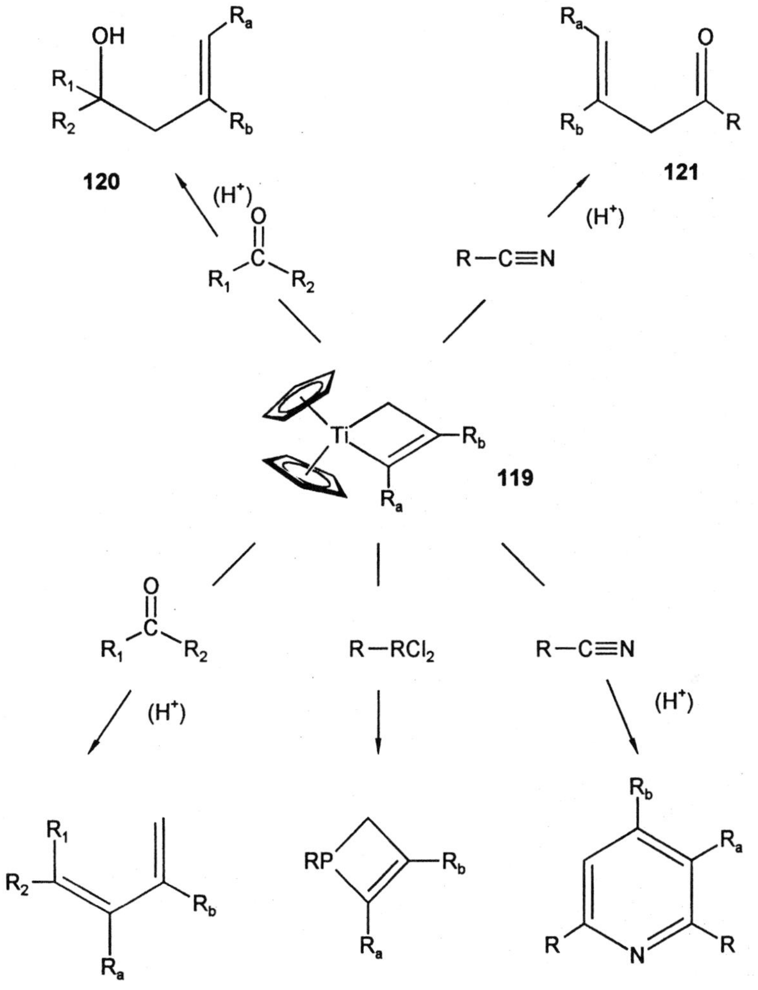

Scheme 4-44. Organic Synthesis employing titanacyclobutenes.

4.6.2.2 Ene Reactions

Reactions of Ti=C intermediates with carbonyl compounds can lead to carbonyl olefinations, as discussed. Enolizable ketones, on the other hand, can react in a

1,5-sigmatropic intramolecular H shift to yield titanium enolates [322–324], a reaction that is particularly noteworthy with respect to stereoselective C–C linkages. The majority of studies carried out to date have used NMR spectroscopy to characterize such substrates, and only recently have studies of the solid-state structures have been performed [207, 210, 325, 326]. When [Cp*$_2$TiMe$_2$] (**29**), [Cp*$_2$Ti(Me)CH=CH$_2$] (**38**), or [Cp*$_2$TiCH$_2$CH$_2$C=CH$_2$] (**37**) reacts with enolizable ketones, the titanocene enolates **125** [323] and **126** [210] are exclusively obtained.

The formation of products exhibiting Δ^1- and *E*-configurations is always preferred for both **125** and **126**. The observed high regio- and stereoselectivity results from the geometric conditions at the [Cp*$_2$Ti=C] intermediate **127**, which apparently only permit selected deprotonations of the acidic ketones. Theoretical studies show that for [Cp$_2$Ti=CH$_2$] fragments, the *end-on* coordination **128** is more favorable than side-on mode **129** by 0.96 eV [227]. This in turn apparently leads to the observed preference for formation of the enolate over the oxetane.

The structural data for enolates of type **126** show a distinct alkoxide character (Ti–O: 1.859(2) Å; Ti–O–C 165.9°) [210]. This reduces the nucleophilic properties of the methylene group and the oxygen in the enolate group, so that typical reactions with electrophiles [327, 328], such as methyl iodide or benzaldehyde, do not occur.

4.6.2.3 Catalytic Reactions

Catalytic applications of precursors for [Cp$_2$Ti=CR$_2$] intermediates in the context of ring-opening metathesis polymerization (ROMP) have already been summarized and discussed in various articles [314–316, 329, 330]. New studies show the effective use of [Cp$_2$TiMe$_2$], [CpTiMe$_3$], [CpTiClMe$_2$], and [Cp$_2$Ti(CH$_2$SiMe$_3$)$_2$] as carbenoid reagents [107]. The important advantages in these cases are similar to those already discussed for stoichiometric reactions, namely the easier preparation and handling of the corresponding titanium alkyl compounds as compared to Tebbe or Grubbs reagents. For example, the ring-opening polymerization of norbornene can be carried out at high temperatures in the presence of these reagents (Scheme 4-45).

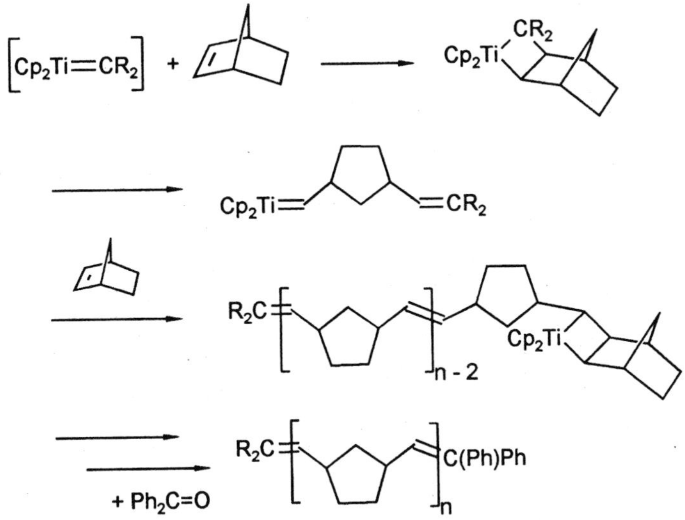

Scheme 4-45. The mechanism of the ring-opening polymerization of norbornene.

When THF is used as solvent, the reaction does not take place, even though carbonyl olefination proceeds more rapidly in THF than in toluene or hexane [115]. This effect is attributed to the nucleophilicity of the carbene C atom. Other studies show a slower rate of chain growth with THF or pyridine as solvents. In general, THF has only a small effect when [CpTiMe$_3$] and [CpTiMe$_2$Cl] are employed. In the case of [CpTi(CH$_2$SiMe$_3$)$_2$], though, THF nearly completely inhibits the ring-opening polymerization. The different ROMP activities correspond with the electronic structure of the [Ti=CR$_2$] derivatives according to studies of Cundari and Gordon [331–333]. These indicate, for example, that [Cp$_2$Ti=C(H)SiMe$_3$] is a particularly effective metathesis initiator. Furthermore, when electron-withdrawing substituents are present on the carbene C atom, increased activity can be expected, which is demonstrated by results obtained with titanocene benzyl complexes [112]. While unsubstituted [Cp$_2$Ti(CH$_2$Ph)$_2$] is suitable for carbonyl olefination, it is unable to initiate the ROMP reactions.

Titanocene vinylidene derivatives will be discussed as intermediates in catalytic reactions. Alt *et al.*, for example, have suggested a vinylidene intermediate **131** in the formation of *trans*-polyacetylene **134** from the acetylene complex **130** and excess acetylene [334].

This proposed mechanism is supported by the observation that it is possible, start-ing from methylenetitanancyclobutene [Cp*$_2$TiCH=CHC=CH$_2$] (**54a**), to achieve the polymerization of acetylene to form pure *trans*-polyacetylene [208]. The trans-formation of a metal-coordinated acetylene to a vinylidene (**130**→**131**) has not so far been experimentally verified in an electron-deficient transition metal complex.

4.7   Cp-Equivalents

In the following chapter some examples of ligands with similar properties to Cp as well as Cp* ligands will be given.

Unlike the titanocene fragment, which cannot exist as a stable molecule without further stabilizing ligands, the so called 'open' metallocenes can be isolated [335]. For example, titanium dichloride, generated *in situ* by the reduction of TiCl$_4$ with magnesium, reacts with potassium 2,4-dimethylpentadienide in THF at −78°C to yield the deep–green, diamagnetic 14-electron complex [Ti(η^5-C$_5$H$_5$Me$_2$-2,4)$_2$] (**135a**). The compound is pyrophoric and can be purified by sublimation [336]. The analogous reaction with K[C$_5$H$_5$Bu$^t_2$-2,4] gives the more stable complex [Ti(η^5-C$_5$H$_5$Bu$^t_2$-2,4)$_2$] (**135b**). The NMR spectrum suggests that the two pentadienyl lig-ands are oriented at an angle of *ca* 90° with respect to each other [337]. The 14-electron bis(dienyl) complexes readily form adducts **136** with carbon monoxide or phosphines (16-electron complexes) [338].

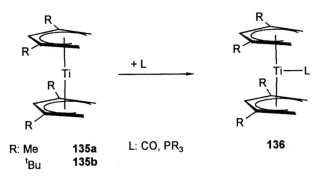

R: Me   **135a**   L: CO, PR$_3$
     ^{t}Bu   **135b**

136

In mixed CpTi(pentadienyl) complexes [CpTi(η^5-C$_5$H$_5$Me$_2$-2,4)PEt$_3$], molecular orbital calculations show that the titanium–pentadienyl bond is stronger than the titanium cyclopentadienyl bond. This is attributed to the fact that the pentadienyl ligand serves as a much better electron acceptor than cyclopentadienyl [339]. To a large extent this may be attributed to significant d bonding interactions between titanium d_{xy} and pentadienyl $\pi_4{}^*$ orbitals [340].

The alkane elimination reaction of $C_2B_9H_{13}$ with [Cp*TiMe$_3$] provides a simple route to the bent-metallocene complex [Cp*Ti(η^5-$C_2B_9H_{11}$)TiMe] (**137**) in base-free form [187]. Compound **137** is also related to [Cp*$_2$TiMe]$^+$ by replacement of a C_5Me_5 ligand with a (η^5-$C_2B_9H_{11}$)^2 ligand. EHMO calculations show that d^0 [(C$_5$R$_5$)(η^5-$C_2B_9H_{11}$)M(H)] species contain two empty low-lying metal-centered orbitals similar to those of the d^0 [(C$_5$R$_5$)$_2$M(H)]$^{n+}$ species, and structural comparisons indicate that the steric properties of the Cp* and [(η^5-$C_2B_9H_{11}$)]$^{2-}$ ligands are similar [341]. Thus, complexes of the type **137** are best described as 14-electron Ti(IV) species, analogous to [Cp*$_2$Ti(Me)]$^+$ and [Cp*$_2$Sc(Me)]. Consistent with this, **137** species display structural features and reactivity patterns characteristic of electrophilic metal complexes, including agostic interactions, ligand coordination (→**139**), insertion (→**140**) and intramolecular C–H activation (→**138**).

In addition to the examples mentioned above, bulky alkoxide ligands support organometallic chemistry at early d-block metal centers. In the case of group 4 metals, both complementary and novel stoichiometric reactivity have been developed at bis(aryloxide) metal complexes. The fragment [(ArO)$_2$M] **141** can be considered to be formally isoelectronic with [Cp$_2$M] if the aryl oxide oxygen atom donates a full six electrons to the metal center [342–345]. The spatial resemblance of cyclopentadienyl ligand and tri-*tert*-butylmethoxide (tritox) **142** [346] and the Si-analog complexes (silox) **143** have also been discussed [347, 348].

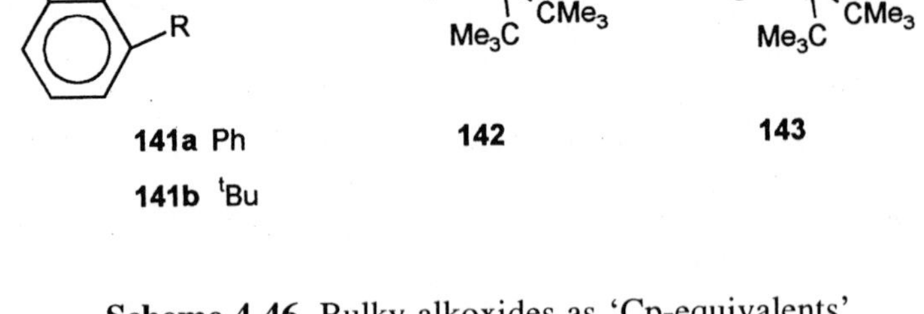

141a Ph
141b ^{t}Bu

142

143

Scheme 4-46. Bulky alkoxides as 'Cp-equivalents'.

In addition to the bulky alkoxides of titanium group metals, bulky amides are being investigated also in the hope that they may provide the next generation of rea-

gents and catalysts [349]. One class of amido groups is the sterically-demanding
N-tert-hydrocarbylanilide ligand, which stabilizes low coordinate transition metal
complexes [350]. Chelating amides like η^2-coordinated aminopyridinato ligands
with a 'maximal steric angle' of 144° [351] are similar to tripod ligands [352]. The
chelating bis(borylamido) ligand [MesBNCH$_2$CH$_2$NBMes$_2$]$^{2-}$, or (Ben)$^{2-}$ would be
a structurally and electronically unusual bidentate diamido ligand, since the two
mesityl groups should sterically protect two or three coordination sites in a plane
roughly perpendicular to the NCCN ligand backbone as a consequence of N–B π-
bonding [353]. Furthermore, the strongly π-accepting boryl groups should attenuate
the ability of the nitrogen atoms to donate electron density to the metal center and
therefore should yield complexes in which the metal is more electrophilic than in
traditional alkyl and arylamide complexes [353]. First examples of (Ben)Ti complexes **144** have been synthesized.

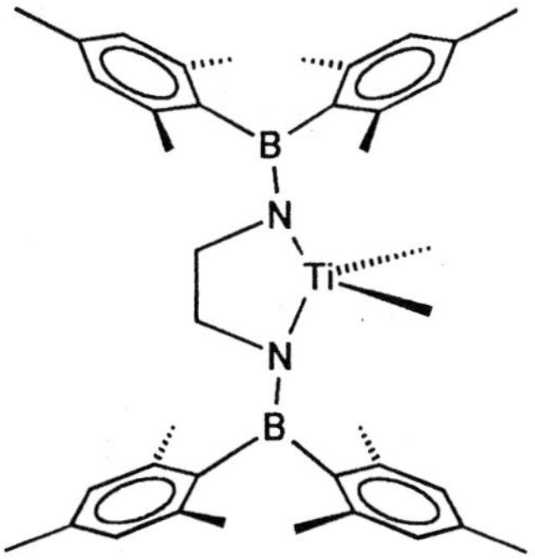

144 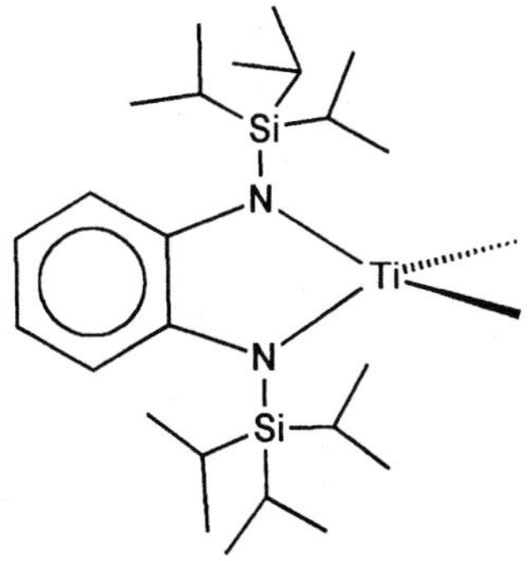   **145**

Scheme 4-47. Schematic drawing of titanium complexes containing the
[Mes$_2$BNCH$_2$CH$_2$NBMes$_2$]$^{2-}$ and the [o-C$_6$H$_4$(NSi^{i}Pr$_3$)$_2$]$^{2-}$ ligand.

From the dilithio reagent o-C$_6$H$_4$(NLiSi^{i}Pr$_3$)$_2$ new titanium (**145**) and zirconium
complexes have been prepared. X-ray crystallographic and variable-temperature
NMR studies show that these complexes possess a η^4-C$_6$H$_4$(NSi^{i}Pr$_3$)$_2$ ligand
(bonded to the metal *via* two carbon atoms in the phenylene group), which undergoes a η^4–η^2 fluxional process in solution [354]. A wide range of further examples
of nitrogen-based donors can be derived from the *N,N'*-bis(alkyl)benzamidinato
ligand [η-PhC(NR)$_2$] [355–357]. The benzamidinato ligands [η-PhC(NR)$_2$] have
been shown to bond to a transition metal through coordination of the central NCN
system, which is formally a three-electron donating group. It was noted that the [η-
PhC(NSiMe$_3$)$_2$] ligand has some steric similarities to the η-cyclopentadienyl ligand
[358].

Acknowledgments

I wish to express my particular thanks to many very active co-workers, Dr. Jürgen Oster, Dr. Javier Sang, Dr. Isabelle Strauß, Dipl.-Chem. Jürgen Heinrichs, and Dipl.-Chem. Martin Wagner, for their work, which has contributed significantly to this chapter. I am indebted to all my colleagues for constructive discussions and suggestions, in particular to Prof. Dr. Uwe Rosenthal (Rostock) and Dr. Uwe Böhme (Freiberg). I wish to express my gratitude to Mr. Edward Pritchard (York), for checking the English version of the manuscript. I especially wish to thank the Institute für Anorganische Chemie der RWTH Aachen, where I worked as a guest Professor and have carried out research for the last five years. I want to gratefully acknowledge that my research was supported by the Deutsche Forschungsgemeinschaft and the Fonds der Chemischen Industrie. The Bayer AG Leverkusen and the Degussa AG are also acknowledged for their generous financial support.

References

[1]   A. K. Fischer, G. Wilkinson, *J. Inorg. Nucl. Chem.* **1956**, *2*, 149-152.
[2]   M. Bochmann, *Comprehensive Organometallic Chemistry II; Titanium Complexes in Oxidation States +2 and +3* in *Comprehensive Organometallic Chemistry II* (Ed.: E. W. Abel, F. G. A. Stone, G. Wilkinson), Pergamon Elsevier Science Ltd, Oxford, **1995**, Vol. 4, p. 221.
[3]   M. Bochmann, *Comprehensive Organometallic Chemistry II; Titanium Complexes in Oxidation States +4* in *Comprehensive Organometallic Chemistry II* (Ed.: E. W. Abel, F. G. A. Stone, G. Wilkinson), Pergamon Elsevier Science Ltd, Oxford, **1995**, Vol. 4, p. 273.
[4]   G. P. Pez, J. N. Armor, *Adv. Organomet. Chem.* **1981**, *19*, 1-50.
[5]   R. F. Jordan, *Adv. Organomet. Chem.* **1991**, *32*, 325-387.
[6]   S. Collins, D. G. Ward, *J. Am. Chem. Soc.* **1992**, *114*, 5460-5462.
[7]   P. C. Wailes, R. S. P. Coutts, H. Weigold, *Organometallic Chemistry of Titanium, Zirconium, and Hafnium*, Academic Press, New York, **1974**.
[8]   J. M. O'Connor, C. P. Casey, *Chem. Rev.* **1987**, *87*, 307-318.
[9]   J. L. Calderon, F. A. Cotton, J. Takats, *J. Am. Chem. Soc.* **1971**, *93*, 3587-3591.
[10] J. L. Calderon, F. A. Cotton, B. G. DeBoer, J. Takats, *J. Am. Chem. Soc.* **1971**, *93*, 3592-3597.
[11] G. Wilkinson, J. M. Birmingham, *J. Am. Chem. Soc.* **1954**, *76*, 4281-4284.
[12] R. B. King, M. B. Bisnette, *J. Organomet. Chem.* **1967**, *8*, 287-297.
[13] J. E. Bercaw, R. H. Marvich, L. G. Bell, H. H. Brintzinger, *J. Am. Chem. Soc.* **1972**, *94*, 1219-1238.
[14] J. E. Bercaw, *J. Am. Chem. Soc.* **1974**, *96*, 5087-5095.
[15] E. W. Abel, S. Moorhouse, *J. Chem. Soc., Dalton Trans.* **1973**, 1706-1711.
[16] P. Jutzi, M. Kuhn, *J. Organomet. Chem.* **1979**, *173*, 221-229.
[17] S. Ciruelos, T. Cuenca, P. Gomez-Sal, A. Manzanero, P. Royo, *Organometallics* **1995**, *14*, 177-185.
[18] S. L. Hart, D. J. Dunclaf, J. J. Hastings, A. McCamley, P. C. Taylor, *J. Chem. Soc., Dalton Trans.* **1996**, 2843-2849.
[19] G. H. Llinas, M. Mena, F. Palacios, P. Royo, R. Serrano, *J. Organomet. Chem.* **1988**, *340*, 37-40.
[20] R. Beckhaus, J. Oster, B. Ganter, U. Englert, *Organometallics* **1997**, *16*, 3902-3909.

[21] G. A. Luinstra, J. H. Teuben, *J. Chem. Soc., Chem. Commun.* **1990**, 1470-1471.

[22] G. Chandra, M. F. Lappert, *J. Chem. Soc. A* **1968**, 1940-1945.

[23] G. M. Diamond, R. F. Jordan, J. L. Petersen, *Organometallics* **1996**, *15*, 4030-4037.

[24] J. N. Christopher, G. M. Diamond, R. F. Jordan, J. L. Petersen, *Organometallics* **1996**, *15*, 4038-4044.

[25] G. M. Diamond, R. F. Jordan, J. L. Petersen, *Organometallics* **1996**, *15*, 4045-4053.

[26] W. A. Herrmann, M. J. A. Morawietz, *J. Organomet. Chem.* **1994**, *482*, 169-181.

[27] J. M. Birmingham, *Adv. Organomet. Chem.* **1964**, *2*, 365-413.

[28] P. G. Gassman, W. H. Campbell, D. W. Macomber, *Organometallics* **1984**, *3*, 385-387.

[29] K. Chandra, R. K. Sharma, B. S. Garg, R. P. Singh, *Inorg. Chim. Acta* **1979**, *37*, 125-127.

[30] R. B. Grossman, S. L. Buchwald, *J. Org. Chem.* **1992**, *57*, 5803-5805.

[31] A. Ohff, S. Pulst, C. Lefeber, N. Peulecke, P. Arndt, V. V. Burlakov, U. Rosenthal, *Synlett* **1996**, 111-118.

[32] D. H. Berry, L. J. Procpio, P. J. Carroll, *Organometallics* **1988**, *7*, 570-572.

[33] J. M. Manriquez, D. R. McAlister, E. Rosenberg, A. M. Shiller, K. L. Williamson, S. I. Chan, J. E. Bercaw, *J. Am. Chem. Soc.* **1978**, *100*, 3078-3083.

[34] H. G. Alt, K.-H. Schwind, M. D. Rausch, U. Thewalt, *J. Organomet. Chem.* **1988**, *349*, C7-C10.

[35] S. A. Cohen, J. E. Bercaw, *Organometallics* **1985**, *4*, 1006-1014.

[36] R. Beckhaus, K.-H. Thiele, *J. Organomet. Chem.* **1986**, *317*, 23-31.

[37] R. Beckhaus, K.-H. Thiele, D. Ströhl, *J. Organomet. Chem.* **1989**, *369*, 43-54.

[38] R. Beckhaus, S. Flatau, S. I. Troyanov, P. Hofmann, *Chem. Ber.* **1992**, *125*, 291-299.

[39] R. Beckhaus, *Angew. Chem.* **1997**, *109*, 694-722; *Angew. Chem. Int. Ed. Engl.* **1997**, *36*, 686-713.

[40] H. G. Alt, G. S. Herrmann, M. D. Rausch, *J. Organomet. Chem.* **1988**, *356*, C50-C52.

[41] R. Beckhaus, J. Sang, J. Oster, T. Wagner, *J. Organomet. Chem.* **1994**, *484*, 179-190.

[42] R. Beckhaus, M. Wagner, V.V. Burlakov, W. Baumann, N. Peulecke, A. Spannenberg, R. Kempe, U. Rosenthal, *Z. anorg. allg. Chem.* **1998**, *624*, 129-134.

[43] J. W. Lauher, R. Hoffmann, *J. Am. Chem. Soc.* **1976**, *98*, 1729-1742.

[44] L. Zhu, N. M. Kostic, *J. Organomet. Chem.* **1987**, *335*, 395-411.

[45] V. C. Gibson, *J. Chem. Soc., Dalton Trans.* **1994**, 1607-1618.

[46] V. C. Gibson, *Angew. Chem.* **1994**, *106*, 1640-1648; *Angew. Chem. Int. Ed. Engl.* **1994**, *33*, 1565-1572.

[47] Lukens Jr., W. W., M. R. Smith III, R. A. Andersen, *J. Am. Chem. Soc.* **1996**, *118*, 1719-1728.

[48] E. Brady, W. Lukens, J. Telford, G. Mitchell, *Acta Cryst.* **1994**, *C51*, 558-560.

[49] G. L. Hillhouse, A. R. Bulls, B. D. Santarsiero, J. E. Bercaw, *Organometallics* **1988**, *7*, 1309-1312.

[50] J. Feldman, J. C. Calabrese, *J. Chem. Soc., Chem. Commun.* **1991**, 1042-1044.

[51] J. M. Hawkins, R. H. Grubbs, *J. Am. Chem. Soc.* **1988**, *110*, 2821-2823.

[52] P. Binger, P. Müller, R. Wenz, R. Mynott, *Angew. Chem.* **1990**, *102*, 1070-1071; *Angew. Chem. Int. Ed. Engl.* **1990**, *29*, 1037-1038.

[53] P. G. Gassman, D. J. Macomber, J. W. Hershberger, *Organometallics* **1983**, *2*, 1470-1472.

[54] J. L. Robbins, N. Edelstein, B. Spencer, J. C. Smart, *J. Am. Chem. Soc.* **1982**, *104*, 1882-1893.

[55] E. J. Miller, S. J. Landon, T. B. Brill, *Organometallics* **1985**, *4*, 534-538.

[56] P. G. Gassman, C. H. Winter, *Organometallics* **1991**, *10*, 1592-1598.

[57] A. Terpstra, J. N. Louwen, A. Oskam, J. H. Teuben, *J. Organomet. Chem.* **1984**, *260*, 207-217.

[58] B. E. Bursten, M. R. Callstrom, A. J. Cynthia, L. A. Paquette, M. R. Sivik, R. S. Tucker, C. A. Wartchow, *Organometallics* **1994**, *13*, 127-133.

[59] P. G. Gassman, P. A. Deck, C. H. Winter, D. A. Dobbs, D. H. Cao, *Organometallics* **1992**, *11*, 959-960.

[60] S. A. Rao, M. Periasamy, *J. Organomet. Chem.* **1988**, *352*, 125-131.

[61] S. A. Cohen, P. R. Auburn, J. E. Bercaw, *J. Am. Chem. Soc.* **1983**, *105*, 1136-1143.

[62] M. L. Steigerwald, W. A. Goddard III, *J. Am. Chem. Soc.* **1985**, *107*, 5027-5035.

[63]   K. Mashima, K. Jyodoi, A. Ohyshi, H. Takaya, *J. Chem. Soc., Chem. Commun.* **1986**, 1145-1146.

[64]   K. Mashima, H. Takaya, *Organometallics* **1985**, *4*, 1464-1466.

[65]   G. Fachinetti, C. Floriani, *J. Chem. Soc., Chem. Commun.* **1974**, 66-67.

[66]   G. Fachinetti, C. Floriani, F. Marchetti, M. Mellini, *J. Chem. Soc., Dalton Trans.* **1978**, 1398-1403.

[67]   B. Demerseman, R. Mahe, P. H. Dixneuf, *J. Chem. Soc., Chem. Commun.* **1984**, 1394-1395.

[68]   B. Demerseman, P. H. Dixneuf, *J. Chem. Soc., Chem. Commun.* **1981**, 665-666.

[69]   B. Demerseman, P. Le Coupanec, P. H. Dixneuf, *J. Organomet. Chem.* **1985**, *287*, C35-C38.

[70]   B. H. Edwards, R. D. Rogers, D. J. Sikora, J. L. Atwood, M. D. Rausch, *J. Am. Chem. Soc.* **1983**, *105*, 416-426.

[71]   V. Varga, K. Mach, M. Polasek, P. Sedmera, J. Hiller, U. Thewalt, S. I. Troyanov, *J. Organomet. Chem.* **1996**, *506*, 241-251.

[72]   V. B. Shur, V. V. Burlakov, M. E. Volpin, *J. Organomet. Chem.* **1988**, *347*, 77-83.

[73]   V. V. Burlakov, U. Rosenthal, P. V. Petrovsky, V. B. Shur, M. E. Volpin, *Organomet. Chem. USSR* **1988**, *1*, 953-954.

[74]   V. V. Burlakov, U. Rosenthal, R. Beckhaus, A. V. Polyakov, Y. T. Struchkov, G. Oeme, V. B. Shur, M. E. Volpin, *Organomet. Chem. USSR* **1990**, *3*, 476-477.

[75]   V. V. Burlakov, A. V. Polyakov, A. I. Yanovsky, Y. T. Struchkov, V. B. Shur, M. E. Vol'pin, U. Rosenthal, H. Görls, *J. Organomet. Chem.* **1994**, *476*, 197-206.

[76]   [a] F. H. Allen, O. Kennrad, D. G. Watson, L. Brammer, A. G. Orpen, R. Taylor, *J. Chem. Soc., Perkin Trans. II* **1987**, S1-S19. – [b] K. Krüger, Personal Communication. – [c] A.A. Espiritu, J.G. White, *Z. Kristall.* **1978**, *147*, 177-186.

[77]   R. Threlkel, *PhD Thesis,* California Institute of Technology **1980**.

[78]   U. Rosenthal, G. Oehme, V. V. Burlakov, P. V. Petrovskii, V. B. Shur, M. E. Volpin, *J. Organomet. Chem.* **1990**, *391*, 119-122.

[79]   D. J. Sikora, D. W. Macomber, M. D. Rausch, *Adv. Organomet. Chem.* **1986**, *25*, 317-379.

[80]   D. J. Sikora, K. J. Moriarty, M. D. Rausch, A. R. Bulls, J. E. Bercaw, V. D. Patel, A. J. Carty, *Inorg. Synth.* **1986**, *24*, 147-157.

[81]   L. B. Kool, M. D. Rausch, H. G. Alt, M. Herberhold, B. Wolf, U. Thewalt, *J. Organomet. Chem.* **1985**, *297*, 159-169.

[82]   G. T. Palmer, F. Basolo, L. B. Kool, M. D. Rausch, *J. Am. Chem. Soc.* **1986**, *108*, 4417-4422.

[83]   L. B. Kool, M. D. Rausch, H. G. Alt, M. Herberhold, U. Thewalt, B. Wolf, *Angew. Chem.* **1985**, *97*, 425-426; *Angew. Chem. Int. Ed. Engl.* **1985**, *24*, 394-395.

[84]   L. B. Kool, M. D. Rausch, H. G. Alt, M. Herberhold, B. Honold, U. Thewalt, *J. Organomet. Chem.* **1987**, *320*, 37-45.

[85]   L. B. Kool, M. D. Rausch, H. G. Alt, M. Herberhold in *Organometallic Syntheses* (Ed.: R. B. King, J. J. Eisch), Elsevier, Amsterdam, **1986**, Vol. 3, p. 13.

[86]   J. M. de Wolf, A. Meetsma, J. H. Teuben, *Organometallics* **1995**, *14*, 5466-5468.

[87]   J. W. Pattiasina, F. van Bolhuis, J. H. Teuben, *Angew. Chem.* **1987**, *99*, 342-343; *Angew. Chem. Int. Ed. Engl.* **1987**, *26*, 330-331.

[88]   S. R. Frerichs, B. K. Stein, J. E. Ellis, *J. Am. Chem. Soc.* **1987**, *109*, 5558-5560.

[89]   Y. You, S. R. Wilson, G. S. Girolami, *Organometallics* **1994**, *13*, 4655-4657.

[90]   a) S. I. Troyanov, H. Antropiusova, K. Mach, *J. Organomet. Chem.* **1992**, *427*, 49-55. - b) J. M. Fischer, W. E. Piers, V. G. Jr. Young, *Organometallics* **1996**, *15*, 2410-2412.

[91]   S. I. Troyanov, K. Mach, V. Varga, *Organometallics* **1993**, *12*, 3387-3389.

[92]   S. Xin, J. F. Harrod, E. Samuel, *J. Am. Chem. Soc.* **1994**, *116*, 11562-11563.

[93]   K. M. Doxsee, J. K. M. Mouser, J. B. Farahi, *Synlett* **1992**, 13-21.

[94]   C. Lamberth, *J. Prakt. Chem.* **1994**, *336*, 632-633.

[95]   S. H. Pine, G. S. Shen, H. Hoang, *Synthesis* **1991**, 165-167.

[96]   S. H. Pine, R. J. Pettit, G. D. Geib, S. G. Cruz, C. H. Gallego, T. Tijerina, R. D. Pine, *J. Org. Chem.* **1985**, *50*, 1212-1216.

[97]   K. C. Ott, E. J. M. deBoer, R. H. Grubbs, *Organometallics* **1984**, *3*, 223-230.

[98] S. H. Pine, *Carbonyl Methylenation and Alkylidation using Titanium-Based Reagents* in *Organic Reactions* (Ed.: L. A. Paquette), John Wiley & Sons, **1993**, Vol. 43, p. 1.

[99] R. H. Grubbs, R. H. Pine, *Comprehensive Organic Synthesis; Alkene Metathesis and Related Reactions* in *Comprehensive Organic Synthesis* (Ed.: B. M. Trost), Pergamon, New York, **1991**, Vol. 5, p. 1115.

[100] K. M. Doxsee, J. B. Farahi, *J. Am. Chem. Soc.* **1988**, *110*, 7239-7240.

[101] N. A. Petasis, S.-P. Lu, *Tetrahedron Lett.* **1995**, *36*, 2393-2396.

[102] J.-P. Begue, M. H. Rock, *J. Organomet. Chem.* **1995**, *489*, C7-C8.

[103] H. van der Heijden, B. Hessen, *J. Chem. Soc., Chem. Commun.* **1995**, 145-146.

[104] L. Scoles, R. Minhas, R. Duchateau, J. Jubb, S. Gambarotta, *Organometallics* **1994**, *13*, 4978-4983.

[105] J. A. van Doorn, H. van der Heijden, A. G. Orpen, *Organometallics* **1994**, *13*, 4271-4277.

[106] K. M. Doxsee, J. J. J. Juliette, K. Zientara, G. Nieckarz, *J. Am. Chem. Soc.* **1994**, *116*, 2147-2148.

[107] N. A. Petasis, D.-K. Fu, *J. Am. Chem. Soc.* **1993**, *115*, 7208-7214.

[108] K. M. Doxsee, J. J. J. Juliette, J. K. M. Mouser, K. Zientara, *Organometallics* **1993**, *12*, 4682-4686.

[109] N. A. Petasis, D.-K. Fu, *Organometallics* **1993**, *12*, 3776-3780.

[110] N. A. Petasis, E. I. Bzowej, *Tetrahedron Lett.* **1993**, *34*, 943-946.

[111] N. A. Petasis, I. Akritopoulou, *Synlett* **1992**, 665-667.

[112] N. A. Petasis, E. I. Bzowej, *J. Org. Chem.* **1992**, *57*, 1327-1330.

[113] K. M. Doxsee, J. B. Farahi, H. Hope, *J. Am. Chem. Soc.* **1991**, *113*, 8889-8898.

[114] K. M. Doxsee, J. K. M. Mouser, *Organometallics* **1990**, *9*, 3012-3014.

[115] N. A. Petasis, E. I. Bzowej, *J. Am. Chem. Soc.* **1990**, *112*, 6392-6394.

[116] B. J. J. van de Heisteeg, G. Schat, O. S. Akkerman, F. Bickelhaupt, *Tetrahedron Lett.* **1987**, *28*, 6493-6496.

[117] B. J. J. van de Heisteeg, G. Schat, O. S. Akkerman, F. Bickelhaupt, *J. Organomet. Chem.* **1986**, *310*, C25-C28.

[118] D. L. Hughes, J. F. Payack, D. Cai, T. R. Verhoeven, P. J. Reider, *Organometallics* **1996**, *15*, 663-667.

[119] J. A. van Doorn, H. van der Heijden, A. G. Orpen, *Organometallics* **1995**, *14*, 1278-1283.

[120] S. De Angelis, E. Solari, C. Floriani, A. Chiesi-Villa, C. Rizzoli, *Angew. Chem.* **1995**, *107*, 1200-1202; *Angew. Chem. Int. Ed. Engl.* **1995**, *34*, 1092-1094.

[121] M. D. Fryzuk, S. S. H. Mao, M. J. Zaworotko, L. R. MacGillivray, *J. Am. Chem. Soc.* **1993**, *115*, 5336-5337.

[122] A. Kabi-Satpathy, C. S. Bajgur, K. P. Reddy, J. L. Petersen, *J. Organomet. Chem.* **1989**, *364*, 105-117.

[123] F. J. Berg, J. L. Petersen, *Organometallics* **1989**, *8*, 2461-2470.

[124] B. J. J. van de Heisteeg, G. Schat, O. S. Akkerman, F. Bickelhaupt, *Organometallics* **1986**, *5*, 1749-1750.

[125] W. R. Tikkanen, J. Z. Liu, J. W. Egan Jr., J. L. Petersen, *Organometallics* **1984**, *3*, 825-830.

[126] K. C. Ott, R. H. Grubbs, *J. Am. Chem. Soc.* **1981**, *103*, 5922-5923.

[127] S. H. Pine, R. Zahler, D. A. Evans, R. H. Grubbs, *J. Am. Chem. Soc.* **1980**, *102*, 3270-3272.

[128] J. D. Meinhart, E. V. Anslyn, R. H. Grubbs, *Organometallics* **1989**, *8*, 583-589.

[129] W. C. Finch, E. V. Anslyn, R. H. Grubbs, *J. Am. Chem. Soc.* **1988**, *110*, 2406-2413.

[130] E. V. Anslyn, R. H. Grubbs, *J. Am. Chem. Soc.* **1987**, *109*, 4880-4890.

[131] L. R. Gilliom, R. H. Grubbs, *Organometallics* **1986**, *5*, 721-724.

[132] L. R. Gilliom, R. H. Grubbs, *J. Am. Chem. Soc.* **1986**, *108*, 733-742.

[133] T. Ikariya, S. C. H. Ho, R. H. Grubbs, *Organometallics* **1985**, *4*, 199-200.

[134] J. R. Stille, R. H. Grubbs, *J. Am. Chem. Soc.* **1983**, *105*, 1664-1665.

[135] J. B. Lee, K. C. Ott, R. H. Grubbs, *J. Am. Chem. Soc.* **1982**, *104*, 7491-7496.

[136] D. A. Straus, R. H. Grubbs, *Organometallics* **1982**, *1*, 1658-1661.

[137] J. B. Lee, G. J. Gajda, W. P. Schaefer, T. R. Howard, T. Ikariya, D. A. Straus, R. H. Grubbs, *J. Am. Chem. Soc.* **1981**, *103*, 7358-7361.

[138] T. R. Howard, J. B. Lee, R. H. Grubbs, *J. Am. Chem. Soc.* **1980**, *102*, 6876-6878.

[139] F. Bickelhaupt, *Angew. Chem.* **1987**, *99*, 1020-1035; *Angew. Chem. Int. Ed. Engl.* **1987**, *26*, 990-1005.

[140] G. L. Casty, J. M. Stryker, *J. Am. Chem. Soc.* **1995**, *117*, 7814-7815.

[141] D. A. Straus, R. H. Grubbs, *J. Mol. Catal.* **1985**, *28*, 9-25.

[142] J. L. Polse, R. A. Andersen, R. G. Bergman, *J. Am. Chem. Soc.* **1996**, *118*, 8737-8738.

[143] P. Binger, P. Müller, R. Benn, R. Mynott, *Angew. Chem.* **1989**, *101*, 647-648; *Angew. Chem. Int. Ed. Engl.* **1989**, *28*, 610-611.

[144] P. Binger, P. Müller, P. Phillipps, B. Gabor, R. Mynott, A. T. Herrmann, F. Langhauser, C. Krüger, *Chem. Ber.* **1992**, *125*, 2209-2212.

[145] E. Hengge, M. Weinberger, *J. Organomet. Chem.* **1993**, *443*, 167-173.

[146] J. L. Bennett, P. T. Wolczanski, *J. Am. Chem. Soc.* **1994**, *116*, 2179-2180.

[147] J. M. Fischer, W. E. Piers, T. Ziegler, L. R. MacGillivray, M. J. Zaworotko, *Chem. Eur. J.* **1996**, *2*, 1221-1229.

[148] M. R. Smith, P. T. Matsunaga, R. A. Andersen, *J. Am. Chem. Soc.* **1993**, *115*, 7049-7050.

[149] F. Bottomley, G. O. Egharevba, I. J. B. Lin, P. S. White, *Organometallics* **1985**, *4*, 550-553.

[150] D. J. Schwartz, M. R. Smith III, R. A. Andersen, *Organometallics* **1996**, *15*, 1446-1450.

[151] U. Thewalt, B. Honold, *J. Organomet. Chem.* **1988**, *348*, 291-303.

[152] H.-P. Klein, U. Thewalt, *Z. Anorg. Allg. Chem.* **1981**, *476*, 62-68.

[153] H.-P. Klein, U. Thewalt, *J. Organomet. Chem.* **1981**, *206*, 69-75.

[154] U. Thewalt, H.-P. Klein, *J. Organomet. Chem.* **1980**, *194*, 297-307.

[155] T. K. Hollis, N. P. Robinson, B. Bosnich, *J. Am. Chem. Soc.* **1992**, *114*, 5464-5466.

[156] H. Aslan, T. Sielisch, R. D. Fischer, *J. Organomet. Chem.* **1986**, *315*, C69-C72.

[157] M. L. McLaughlin, J. M. Cronan Jr., T. R. Schaller, R. D. Snelling, *J. Am. Chem. Soc.* **1990**, *112*, 8949-8952.

[158] H. Köpf, P. Köpf-Maier, *Nachr. Chem. Tech. Lab.* **1981**, *29*, 154-156.

[159] P. Köpf-Maier, H. Köpf, *J. Organomet. Chem.* **1988**, *342*, 167-176.

[160] J. H. Toney, T. J. Marks, *J. Am. Chem. Soc.* **1985**, *107*, 947-953.

[161] K. Döppert, *J. Organomet. Chem.* **1987**, *319*, 351-354.

[162] K. Döppert, U. Thewalt, *J. Organomet. Chem.* **1986**, *301*, 41-48.

[163] P. Gowik, T. Klapötke, *J. Organomet. Chem.* **1989**, *372*, 33-40.

[164] T. Klapötke, *Polyhedron* **1989**, *8*, 311-315.

[165] P. Gowik, T. Klapötke, U. Thewalt, *J. Organomet. Chem.* **1990**, *385*, 345-350.

[166] P. Gowik, T. Klapötke, *J. Organomet. Chem.* **1990**, *387*, C27-C30.

[167] U. Thewalt, K. Berhalter, *J. Organomet. Chem.* **1986**, *302*, 193-200.

[168] A. Ohff, R. Kempe, W. Baumann, U. Rosenthal, *J. Organomet. Chem.* **1996**, *520*, 241-244.

[169] H. H. Brintzinger, J. E. Bercaw, *J. Am. Chem. Soc.* **1970**, *92*, 6182-6185.

[170] G. W. Watt, L. J. Baye, F. O. Jr. Drummond, *J. Am. Chem. Soc.* **1966**, *88*, 1138-1140.

[171] J.-J. Salzmann, P. Mosimann, *Helv. Chim. Acta* **1967**, *7*, 1830-1836.

[172] E. E. van Tamelen, W. Cretney, N. Klaentsch, J. S. Miller, *J. Chem. Soc., Chem. Commun.* **1972**, 481-482.

[173] K. Clauss, H. Bestian, *Liebigs Ann. Chem.* **1962**, *654*, 8-19.

[174] L. J. Guggenberger, F. N. Tebbe, *J. Am. Chem. Soc.* **1973**, *95*, 7870-7872.

[175] A. Davison, S. S. Wreford, *J. Am. Chem. Soc.* **1974**, *96*, 3017-3018.

[176] L. J. Guggenberger, F. N. Tebbe, *J. Am. Chem. Soc.* **1976**, *98*, 4137-4243.

[177] G. J. Olthof, *J. Organomet. Chem.* **1977**, *128*, 367-373.

[178] H. Antropiusova, V. Hanus, K. Mach, *Transition Met. Chem.* **1978**, *3*, 121-122.

[179] E. G. Perevalova, I. F. Uranzowski, D. A. Lemenovskii, Y. L. Slovokhotov, Y. T. Struchkov, *J. Organomet. Chem.* **1985**, *289*, 319-329.

[180] T. Wöhrle, U. Thewalt, *J. Organomet. Chem.* **1993**, *456*, C21-C23.

[181] U. Thewalt, T. Wöhrle, *Z. Naturforsch.* **1993**, *48b*, 603-607.

[182] U. Thewalt, T. Wöhrle, *J. Organomet. Chem.* **1996**, *506*, 331-335.

[183] C. McDade, J. C. Green, J. E. Bercaw, *Organometallics* **1982**, *1*, 1629-1634.

[184] G. A. Luinstra, J. H. Teuben, *Organometallics* **1992**, *11*, 1793-1801.
[185] K. Mach, V. Varga, V. Hanus, P. Sedmera, *J. Organomet. Chem.* **1991**, *415*, 87-95.
[186] T. Vondrák, K. Mach, V. Varga, A. Terpstra, *J. Organomet. Chem.* **1992**, *425*, 27-39.
[187] C. Kreuder, R. F. Jordan, H. Zhang, *Organometallics* **1995**, *14*, 2993-3001.
[188] J. W. Pattiasina, C. E. Hissink, J. L. de Boer, A. Meetsma, J. H. Teuben, A. L. Spek, *J. Am. Chem. Soc.* **1985**, *107*, 7758-7759.
[189] G. Erker, U. Korek, Z. *Naturforsch.* **1989**, *44b*, 1593-1598.
[190] R. Beckhaus, J. Oster, J. Sang, I. Strauß, M. Wagner, *Synlett* **1997**, 241-249.
[191] R. Beckhaus, U. Böhme, unpublished results.
[192] B. Akermark, A. Ljungqvist, *J. Organomet. Chem.* **1979**, *182*, 59-75.
[193] J. M. Brown, N. A. Cooley, *Chem. Rev.* **1988**, *88*, 1031-1046.
[194] P. Courtot, V. Labed, R. Pichon, J. Y. Salaün, *J. Organomet. Chem.* **1989**, *359*, C9-C13.
[195] S. L. Buchwald, R. B. Nielsen, *Chem. Rev.* **1988**, *88*, 1047-1058.
[196] C. Lefeber, A. Ohff, A. Tillack, W. Baumann, R. Kempe, V. V. Burlakov, U. Rosenthal, H. Görls, *J. Organomet. Chem.* **1995**, *501*, 179-188.
[197] U. Rosenthal, A. Ohff, A. Tillack, W. Baumann, H. Goerls, *J. Organomet. Chem.* **1994**, *468*, C4-C8.
[198] U. Rosenthal, H. Görls, *J. Organomet. Chem.* **1992**, *439*, C36-C41.
[199] A. R. Bulls, W. P. Schaefer, M. Serfas, J. E. Bercaw, *Organometallics* **1987**, *6*, 1219-1226.
[200] L. H. Toporcer, R. E. Dessy, S. I. E. Green, *J. Am. Chem. Soc.* **1965**, *87*, 1236-1240.
[201] E. L. Motell, A. W. Boone, W. H. Fink, *Tetrahedron* **1978**, *34*, 1619-1626.
[202] N. M. Doherty, J. E. Bercaw, *J. Am. Chem. Soc.* **1985**, *107*, 2670-2682.
[203] R. Beckhaus, *J. Chem. Soc., Dalton Trans.* **1997**, 1991-2001.
[204] C. P. Boekel, J. H. Teuben, H. J. De Liefde Meijer, *J. Organomet. Chem.* **1974**, *81*, 371-377.
[205] I. S. Kolomnikov, T. S. Lobeeva, V. V. Gorbachevskaya, G. G. Aleksandrov, Y. T. Struchkov, M. E. Volpin, *J. Chem. Soc., Chem. Commun.* **1971**, 972-973.
[206] V. B. Shur, E. G. Berkovitch, L. B. Vasiljeva, R. V. Kudryavtsev, M. E. Volpin, *J. Organomet. Chem.* **1974**, *78*, 127-132.
[207] R. Beckhaus, *Organic Synthesis via Organometallics (OSM 4) Methylidentitanacyclobutane vs. Titanocene-Vinylidene - Versatile Building Blocks* in *Organic Synthesis via Organometallics (OSM 4), Proceedings of the Fourth Symposium in Aachen, July 15 to 18, 1992* (Ed.: D. Enders, H.-J. Gais, W. Keim), Vieweg Verlag, Braunschweig, **1993**, p. 131.
[208] R. Beckhaus, J. Sang, T. Wagner, B. Ganter, *Organometallics* **1996**, *15*, 1176-1187.
[209] R. Beckhaus, M. Wagner, R. Wang, Z. *Anorg. Allg. Chem.* **1998**, *624*, 277-280.
[210] R. Beckhaus, I. Strauß, T. Wagner, *J. Organomet. Chem.* **1994**, *464*, 155-161.
[211] C. C. Cummins, C. P. Schaller, G. D. Van Duyne, P. T. Wolczanski, A. W. E. Chan, R. Hoffmann, *J. Am. Chem. Soc.* **1991**, *113*, 2985-2994.
[212] R. Beckhaus, J. Oster, Z. *Anorg. Allg. Chem.* **1995**, *621*, 359-364.
[213] R. Beckhaus, I. Strauß, unpublished results.
[214] R. Aumann, *Angew. Chem.* **1988**, *100*, 1512-1524; *Angew. Chem. Int. Ed. Engl.* **1988**, *27*, 1456-1467.
[215] F. N. Tebbe, R. L. Harlow, *J. Am. Chem. Soc.* **1980**, *102*, 6149-6151.
[216] R. J. McKinney, T. H. Tulip, D. L. Thorn, T. S. Coolbaugh, F. N. Tebbe, *J. Am. Chem. Soc.* **1981**, *103*, 5584-5586.
[217] P. Binger, P. Müller, A. T. Herrmann, P. Philipps, B. Gabor, F. Langhauser, C. Krüger, *Chem. Ber.* **1991**, *124*, 2165-2170.
[218] R. Beckhaus, M. Wagner, R. Wang, *Eur. J. Inorg. Chem.*, **1998**, 253-256.
[219] K. C. Wallace, A. H. Liu, W. M. Davis, R. R. Schrock, *Organometallics* **1989**, *8*, 644-654.
[220] R. Beckhaus, J. Sang, U. Englert, U. Böhme, *Organometallics* **1996**, *15*, 4731-4736.
[221] K. A. Jørgensen, B. Schiøtt, *Chem. Rev.* **1990**, *90*, 1483-1506.
[222] R. Beckhaus, I. Strauß, T. Wagner, P. Kiprof, *Angew. Chem.* **1993**, *105*, 281-283; *Angew. Chem. Int. Ed. Engl.* **1993**, *32*, 264-266.
[223] R. Beckhaus, I. Strauß, T. Wagner, Z. *Anorg. Allg. Chem.* **1997**, *623*, 654-658.

[224] V. W. Day, W. G. Klemperer, S. P. Lockledge, D. J. Main, *J. Am. Chem. Soc.* **1990**, *112*, 2031-2033.

[225] S. C. Ho, S. Hentges, R. H. Grubbs, *Organometallics* **1988**, *7*, 780-782.

[226] I. Strauß, *Diplomarbeit,* RWTH Aachen **1992**.

[227] B. Schiøtt, K. A. Jørgensen, *J. Chem. Soc., Dalton Trans.* **1993**, 337-344.

[228] S. L. Buchwald, R. H. Grubbs, *J. Am. Chem. Soc.* **1983**, *105*, 5490-5491.

[229] R. Beckhaus, J. Oster, T. Wagner, *Chem. Ber.* **1994**, *127*, 1003-1013.

[230] G. Erker, *Angew. Chem.* **1989**, *101*, 411-426; *Angew. Chem. Int. Ed. Engl.* **1989**, *28*, 397-412.

[231] G. Erker, M. Mena, U. Hoffmann, B. Minjón, J. L. Petersen, *Organometallics* **1991**, *10*, 291-298.

[232] E. V. Anslyn, B. D. Santarsiero, R. H. Grubbs, *Organometallics* **1988**, *7*, 2137-2145.

[233] G. Proulx, R. G. Bergman, *J. Am. Chem. Soc.* **1993**, *115*, 9802-9803.

[234] K. H. Dötz, H. Fischer, P. Hofmann, F. R. Kreissl, U. Schubert, K. Weiss, *Transition Metal Carbene Complexes*, Verlag Chemie, Weinheim **1983**.

[235] K. M. Doxsee, J. B. Farahi, *J. Chem. Soc., Chem. Commun.* **1990**, 1452-1454.

[236] R. Beckhaus, I. Strauß, T. Wagner, *Angew. Chem.* **1995**, *107*, 738-740; *Angew. Chem. Int. Ed. Engl.* **1995**, *34*, 688-690.

[237] R. Beckhaus, J. Sang, T. Wagner, U. Böhme, *J. Chem. Soc., Dalton Trans.* **1997**, 2249-2255.

[238] R. Beckhaus, J. Sang, unpublished results.

[239] J. March, *Advanced Organic Chemistry*, John Wiley & Sons, New York, **1985**.

[240] G. E. Herberich, C. Kreuder, U. Englert, *Angew. Chem.* **1994**, *106*, 2589-2590; *Angew. Chem. Int. Ed. Engl.* **1994**, *33*, 2465-2466.

[241] R. Beckhaus, C. Zimmermann, T. Wagner, E. Herdtweck, *J. Organomet. Chem.* **1993**, *460*, 181-189.

[242] R. Beckhaus, I. Strauß, unpublished.

[243] R. D. Dennehy, R. J. Whitby, *J. Chem. Soc., Chem. Commun.* **1992**, 35-36.

[244] M. M. Francl, W. J. Hehre, *Organometallics* **1983**, *2*, 457-459.

[245] U. Klabunde, F. N. Tebbe, G. W. Parshall, R. L. Harlow, *J. Mol. Catal.* **1980**, *8*, 37-51.

[246] J. W. Park, L. M. Henling, W. P. Schaefer, R. H. Grubbs, *Organometallics* **1991**, *10*, 171-175.

[247] F. Ozawa, J. W. Park, P. B. Mackenzie, W. P. Schaefer, L. M. Henling, R. H. Grubbs, *J. Am. Chem. Soc.* **1989**, *111*, 1319-1327.

[248] P. B. Mackenzie, R. J. Coots, R. H. Grubbs, *Organometallics* **1989**, *8*, 8-14.

[249] J. W. Park, P. B. Mackenzie, W. P. Schaefer, R. H. Grubbs, *J. Am. Chem. Soc.* **1986**, *108*, 6402-6404.

[250] B. J. J. van de Heisteeg, G. Schat, O. S. Akkerman, F. Bickelhaupt, *Organometallics* **1985**, *4*, 1141-1142.

[251] M. I. Bruce, *Chem. Rev.* **1991**, *91*, 197-257.

[252] H. P. Kim, S. Kim, R. A. Jacobson, R. J. Angelici, *J. Am. Chem. Soc.* **1986**, *108*, 5154-5158.

[253] R. Boese, M. A. Huffmann, K. P. C. Vollhardt, *Angew. Chem.* **1991**, *103*, 1542-1543; *Angew. Chem. Int. Ed. Engl.* **1991**, *30*, 1463-1464.

[254] J. Oster, *Dissertation,* RWTH Aachen **1996**.

[255] R. Beckhaus, J. Oster, R. Wang, *Organometallics,* **1998**, *17*, in press.

[256] G. Wilkinson, F. A. Cotton, J. M. Birmingham, *J. Inorg. Nucl. Chem.* **1956**, *2*, 95-113.

[257] E. Vitz, C. H. Jr. Brubaker, *J. Organomet. Chem.* **1974**, *82*, C16-C18.

[258] M. H. Peng, C. H. Jr. Brubaker, *J. Organomet. Chem.* **1977**, *135*, 333-337.

[259] J. G.-S. Lee, C. H. Jr. Brubaker, *Inorg. Chim. Acta* **1977**, *25*, 181-184.

[260] W. E. Crowe, A. T. Vu, *J. Am. Chem. Soc.* **1996**, *118*, 1557-1558.

[261] W. E. Crowe, A. T. Vu, *J. Am. Chem. Soc.* **1996**, *118*, 5508-5509.

[262] U. Rosenthal, C. Lefeber, P. Arndt, A. Tillack, W. Baumann, R. Kempe, V. V. Burlakov, *J. Organomet. Chem.* **1995**, *503*, 221-223.

[263] A. Tillack, W. Baumann, A. Ohff, C. Lefeber, A. Spannenberg, R. Kempe, U. Rosenthal, *J. Organomet. Chem.* **1996**, *520*, 187-193.

[264] R. Gleiter, W. Wittwer, *Chem. Ber.* **1994**, *127*, 1797-1798.

[265] E. J. M. De Boer, J. De With, *J. Organomet. Chem.* **1987**, *320*, 289-293.

[266] D. Cozak, M. Melnik, *Coord. Chem. Rev.* **1986**, *14*, 53-99.

[267] A. Clearfield, D. K. Warner, C. H. Saldarriaga-Molina, R. Ropal, I. Bernal, *Can. J. Chem.* **1975**, *53*, 1622-1629.

[268] S. I. Troyanov, V. B. Rybakov, U. Thewalt, V. Varga, K. Mach, *J. Organomet. Chem.* **1993**, *447*, 221-225.

[269] R. D. Rogers, M. M. Benning, L. K. Kurihara, K. J. Moriarty, M. D. Rausch, *J. Organomet. Chem.* **1985**, *293*, 51-60.

[270] T. C. McKenzie, R. D. Sanner, J. E. Bercaw, *J. Organomet. Chem.* **1975**, *102*, 457-466.

[271] U. Thewalt, T. Wöhrle, *J. Organomet. Chem.* **1994**, *464*, C17-C19.

[272] J. W. Pattiasina, H. J. Heeres, F. Van Bolhuis, A. Meetsma, J. H. Teuben, A. L. Spek, *Organometallics* **1987**, *6*, 1004-1010.

[273] G. A. Luinstra, L. C. ten Cate, H. J. Heeres, J. W. Pattiasina, A. Meetsma, J. H. Teuben, *Organometallics* **1991**, *10*, 3227-3237.

[274] D. J. Sikora, M. D. Rausch, R. D. Rogers, J. L. Atwood, *J. Am. Chem. Soc.* **1981**, *103*, 1265-1267.

[275] J. L. Atwood, K. E. Stone, H. G. Alt, D. C. Hrncir, M. D. Rausch, *J. Organomet. Chem.* **1977**, *132*, 367-375.

[276] T. E. Bitterwolf, A. C. Ling, *J. Organomet. Chem.* **1977**, *141*, 355.

[277] J. H. Davis, H.-N. Sun, D. Redfield, G. D. Stucky, *J. Magn. Res.* **1980**, *37*, 441-448.

[278] E. Samuel, M. D. Rausch, *J. Am. Chem. Soc.* **1973**, *95*, 6263-6266.

[279] S. J. Heyes, C. M. Dobson, *J. Am. Chem. Soc.* **1991**, *113*, 463-469.

[280] A. J. Edwards, N. J. Burke, C. M. Dobson, K. Prout, S. J. Heyes, *J. Am. Chem. Soc.* **1995**, *117*, 4637-4653.

[281] A. Hafner, J. Okuda, *Organometallics* **1993**, *12*, 949-950.

[282] M. R. M. Bruce, A. Kenter, D. R. Tayler, *J. Am. Chem. Soc.* **1984**, *106*, 639-644.

[283] M. T. Reetz, *Titanium in Organic Synthesis-A Manual* in *Organometallics in Synthesis* (Ed.: M. Schlosser), John Wiley & Sons Ltd, Chichester, **1994**, p. 195.

[284] F. Küber, *Metallocenes as a Source of Fine Chemicals* in *Applied Homogenous Catalysis with Organometallic Compounds* (Ed.: B. Cornils, W. A. Herrmann), VCH, Weinheim, **1996**, Vol. 2, p. 893.

[285] C. A. Willoughby, S. L. Buchwald, *J. Am. Chem. Soc.* **1992**, *114*, 7562-7564.

[286] A. Viso, N. E. Lee, S. L. Buchwald, *J. Am. Chem. Soc.* **1994**, *116*, 9373-9374.

[287] T. V. RajanBau, W. A. Nugent, M. S. Beattie, *J. Am. Chem. Soc.* **1990**, *112*, 6408-6409.

[288] J. F. Harrod, S. S. Yun, *Organometallics* **1987**, *6*, 1381-1387.

[289] R. A. Sheldon, *Synthesis of Oxiranes* in *Applied Homogenous Catalysis with Organometallic Compounds* (Ed.: B. Cornils, W. A. Herrmann), VCH, Weinheim, **1996**, Vol. 1, p. 411.

[290] R. L. Halterman, T. M. Ramsey, *Organometallics* **1993**, *12*, 2879-2880.

[291] S. L. Colletti, R. L. Halterman, *J. Organomet. Chem.* **1993**, *455*, 99-106.

[292] W. A. Herrmann, *Double-Bond Isomerization of Olefins* in *Applied Homogenous Catalysis with Organometallic Compounds* (Ed.: B. Cornils, W. A. Herrmann), VCH, Weinheim, **1996**, Vol. 2, p. 980.

[293] T. K. Hollis, N. P. Robinson, B. Bosnich, *Organometallics* **1992**, *11*, 2747-2748.

[294] J. B. Jaquith, J. Guan, S. Collins, S. Wang, *Organometallics* **1995**, *14*, 1079-1081.

[295] J. L. Kiplinger, T. G. Richmond, C. E. Osterberg, *Chem. Rev.* **1994**, *94*, 373-431.

[296] J. L. Kiplinger, T. G. Richmond, *J. Am. Chem. Soc.* **1996**, *118*, 1805-1806.

[297] J. L. Kiplinger, T. G. Richmond, *Chem. Commun.* **1996**, 1115-1116.

[298] M. T. Didiuk, C. W. Johannes, J. P. Morken, A. H. Hoveyda, *J. Am. Chem. Soc.* **1995**, *117*, 7097-7104.

[299] F. Sato, *J. Organomet. Chem.* **1985**, *285*, 53-64.

[300] Y. Gao, F. Sato, *J. Chem. Soc., Chem. Commun.* **1995**, 659-690.

[301] Y. Gao, F. Sato, *J. Chem. Soc., Chem. Commun.* **1995**(6), 659-660.

[302] S. Collins, W. P. Dean, D. G. Ward, *Organometallics* **1988**, *7*, 2289-2293.

[303] Y. Cauvin, H. Oliver, *Reactions of unsaturated compounds: Polymerization, Oligomerization, and Coplymerization of Olefins - Dimerization and Codimerization* in *Applied Homogenous*

Catalysis with Organometallic Compounds (Ed.: B. Cornils, W. A. Herrmann), VCH, Weinheim, **1996**, Vol. 1, p. 258.

[304]   V. Varga, L. Petrusova, J. Cejka, V. Hanus, K. Mach, *J. Organomet. Chem.* **1996**, *509*, 235-240.

[305]   T. Yoshida, E. Negishi, *J. Am. Chem. Soc.* **1981**, *103*, 1276-1277.

[306]   R. D. Dennehy, R. J. Whitby, *J. Chem. Soc., Chem. Commun.* **1990**, 1060-1062.

[307]   M. J. Burk, W. Tumas, M. D. Ward, D. R. Wheeler, *J. Am. Chem. Soc.* **1990**, *112*, 6133-6135.

[308]   M. J. Burk, D. L. Staley, W. Tumas, *J. Chem. Soc., Chem. Commun.* **1990**, 809-810.

[309]   W. Tumas, J. A. Suriano, R. L. Harlow, *Angew. Chem.* **1990**, *102*, 89-90; *Angew. Chem. Int. Ed. Engl.* **1990**, *29*, 75-76.

[310]   W. Tumas, J. C. Huang, P. E. Fanwick, C. P. Kubiak, *Organometallics* **1992**, *11*, 2944-2947.

[311]   R. R. Schrock, *Pure Appl. Chem.* **1994**, *66*, 1447-1454.

[312]   R. H. Grubbs, W. Tumas, *Science* **1989**, *243*, 907-915.

[313]   J. Feldman, R. R. Schrock, *Prog. Inorg. Chem.* **1991**, *39*, 1-74.

[314]   A. J. Amass, *Compr. Polym. Sci.* **1989**, *4*, 109-134.

[315]   W. J. Fast, *Compr. Polym. Sci.* **1989**, *4*, 135-142.

[316]   R. H. Grubbs, L. Gilliom, *NATO ASI Ser., Ser. C* **1987**, *215* (Recent Adv. Mech. Synth. Aspects Polym.), 343-352.

[317]   W. Risse, R. H. Grubbs, *J. Mol. Catal.* **1991**, *65*, 211-217.

[318]   T. M. Swager, R. H. Grubbs, *J. Am. Chem. Soc.* **1987**, *109*, 894-896.

[319]   J. D. Meinhart, R. H. Grubbs, *Bull. Chem. Soc. Jpn.* **1988**, *61*, 171-180.

[320]   K. M. Doxsee, J. K. M. Mouser, *Tetrahedron Lett.* **1991**, *32*, 1687-1690.

[321]   K. M. Doxsee, G. S. Shen, C. B. Knobler, *J. Am. Chem. Soc.* **1989**, *111*, 9129-9130.

[322]   L. Clawson, S. L. Buchwald, R. H. Grubbs, *Tetrahedron Lett.* **1984**, *25*, 5733-5736.

[323]   C. P. Gibson, D. S. Bem, *J. Organomet. Chem.* **1991**, *414*, 23-32.

[324]   S. H. Bertz, G. Dabbagh, C. P. Gibson, *Organometallics* **1988**, *7*, 563-565.

[325]   P. Veya, C. Floriani, A. Chiesi-Villa, C. Rizzoli, *Organometallics* **1993**, *12*, 4892-4898.

[326]   M. D. Curtis, S. Thanedar, W. M. Butler, *Organometallics* **1984**, *3*, 1855-1859.

[327]   P. Berno, C. Floriani, A. Chiesi-Villa, C. Guastini, *Organometallics* **1990**, *9*, 1995-1997.

[328]   D. A. Evans, F. Urpi, T. C. Somers, J. S. Clark, M. T. Bilodeau, *J. Am. Chem. Soc.* **1990**, *112*, 8215-8216.

[329]   K. J. Ivin, *Olefin Metathesis*, Academic, New York, **1983**.

[330]   B. M. Novak, W. Risse, R. H. Grubbs, *Adv. Polym. Sci.* **1992**, *102*, 47-72.

[331]   T. R. Cundari, M. S. Gordon, *Organometallics* **1992**, *11*, 55-63.

[332]   T. R. Cundari, M. S. Gordon, *J. Am. Chem. Soc.* **1992**, *114*, 539-548.

[333]   T. R. Cundari, M. S. Gordon, *J. Am. Chem. Soc.* **1991**, *113*, 5231-5243.

[334]   H. G. Alt, H. E. Engelhardt, M. D. Rausch, L. B. Kool, *J. Organomet. Chem.* **1987**, *329*, 61-67.

[335]   R. D. Ernst, *Chem. Rev.* **1988**, *88*, 1255-1291.

[336]   J.-Z. Liu, R. D. Ernst, *J. Am. Chem. Soc.* **1982**, *104*, 3737-3739.

[337]   R. D. Ernst, J. W. Freeman, P. N. Swepston, D. R. Wilson, *J. Organomet. Chem.* **1991**, *402*, 17-25.

[338]   R. D. Ernst, J. W. Freeman, L. Stahl, D. R. Wilson, A. M. Arif, B. Nuber, M. L. Ziegler, *J. Am. Chem. Soc.* **1995**, *117*, 5075-5081.

[339]   E. Melendez, A. M. Arif, M. L. Ziegler, R. D. Ernst, *Angew. Chem.* **1988**, *100*, 1132-1134; *Angew. Chem. Int. Ed. Engl.* **1988**, *27*, 1099-1111.

[340]   I. Hyla-Kryspin, T. E. Waldman, E. Melendez, W. Trakarnpruk, A. M. Arif, M. L. Ziegler, R. D. Ernst, R. Gleiter, *Organometallics* **1995**, *14*, 5030-5040.

[341]   T. P. Hanusa, *Polyhedron* **1982**, *1*, 663-665.

[342]   J. E. Hill, G. Balaich, P. E. Fanwick, I. P. Rothwell, *Organometallics* **1993**, *12*, 2911-2924.

[343]   G. J. Balaich, J. E. Hill, S. A. Waratuke, P. E. Fanwick, I. P. Rothwell, *Organometallics* **1995**, *14*, 656-665.

[344]   G. J. Balaich, P. E. Fanwick, I. P. Rothwell, *Organometallics* **1994**, *13*, 4117-4118.

[345] I. P. Rothwell, *Acc. Chem. Res.* **1988**, *21*, 153-159.

[346] T. V. Lubben, P. T. Wolczanski, G. D. van Duyne, *Organometallics* **1984**, *3*, 977-983.

[347] R. E. Lapointe, P. T. Wolczanski, G. D. van Duyne, *Organometallics* **1985**, *4*, 1810-1818.

[348] K. J. Covert, P. T. Wolczanski, *Inorg. Chem.* **1989**, *28*, 4565-4567.

[349] R. K. Minhas, L. Scoles, S. Wong, S. Gambarotta, *Organometallics* **1996**, *15*, 1113-1121.

[350] A. R. Johnson, W. M. Davis, C. C. Cummins, *Organometallics* **1996**, *15*, 3825-3835.

[351] T. L. Brown, K. J. Lee, *Coord. Chem. Rev.* **1993**, *128*, 89-116.

[352] R. Kempe, S. Brenner, P. Arndt, *Organometallics* **1996**, *15*, 1071-1074.

[353] T. H. Warren, R. R. Schrock, W. M. Davis, *Organometallics* **1996**, *15*, 562-569.

[354] K. Aoyagi, P. K. Gantzel, K. Kalai, T. D. Tilley, *Organometallics* **1996**, *15*, 923-927.

[355] H. C. S. Clark, G. N. Cloke, P. B. Hitchcock, J. B. Love, *J. Organomet. Chem.* **1995**, *501*, 333-340.

[356] R. Gomez, R. Duchateau, A. N. Chernega, J. H. Teuben, F. T. Edelmann, M. L. H. Green, *J. Organomet. Chem.* **1995**, *491*, 153-158.

[357] J. R. Hagedorn, J. Arnold, *J. Am. Chem. Soc.* **1996**, *118*, 893-894.

[358] M. Wedler, F. Knösel, F. T. Edelmann, U. Behrens, *Chem. Ber.* **1992**, *125*, 1313-1318.

5 Zirconocenes

Ei-ichi Negishi and Jean-Luc Montchamp

5.1 Introduction

This chapter primarily deals with the formation and reactions of zirconocene derivatives with emphasis on fundamental reaction patterns and attendant mechanistic aspects of synthetically interesting reactions. This area of chemistry has rapidly and enormously grown over the past two decades, and it is no longer practical to even touch on all important topics. Fortunately, there are a few monographs [1, 2] and a large number of pertinent chapters including Chapters 7–10 of this book and reviews in journals [3–40]. So, the authors' attention is mainly focused on providing a coherent overview of the subject and pertinent references. For those topics that are not properly represented in this chapter, the readers are referred to pertinent chapters and reviews which are classified and listed at the end of this chapter for this very purpose.

Zirconium discovered in 1824 [41] occurs to the extent of 0.022% in the lithosphere and is roughly as abundant as carbon. It is less abundant than titanium (0.63%) but roughly 40 times as abundant as hafnium (0.0053%) [1]. A hundred and thirty years after its discovery, Wilkinson and his coworkers reported zirconocene dihalides, Cp_2ZrX_2, where $Cp = \eta^5\text{-}C_5H_5$ and $X = Cl$ or Br, as some of the earliest examples of organozirconium compounds [42]. Despite the fact that Zr is one of the several least expensive transition metals ($11 mol^{-1} of 98% pure $ZrCl_4$, Strem Chemicals) along with Fe, Cu, Mn, and Ti, the use of Zr and organozirconium compounds in organic synthesis had been virtually unknown until the 1970s, with the exception of occasional uses of Zr salts as somewhat exotic Lewis acid catalysts in the Friedel–Crafts reactions [43] and generally inferior substitutes for Ti salts in the Ziegler–Natta polymerization [44]. Thus, Zr had mainly remained as a basic scientific curiosity. Triggered by some epoch-making discoveries and developments, ten or a dozen of which are listed in Table 5-1, Zr has become over the last two decades one of the most widely used transition metals in organic synthesis. While it still lags behind Cu and Pd in terms of versatility and frequency of use, it now appears to be as widely used as several others, such as Ti, Cr, Fe, Co, Rh, and Ni. It is striking that all of the reactions listed in Table 5-1 involve Cp_2Zr derivatives including those containing modified Cp derivatives.

Table 5-1. Some Notable Discoveries and Developments in the Use of Organozirconium Compounds in Organic Synthesis

Year	Major Author	Discoveries and Developments
1954	G. Wilkinson[42]	Preparation of Cp_2ZrX_2 (X = Cl, Br)
1970 & 1971	P. C. Wailes[45, 46]	Preparation of $Cp_2Zr(H)Cl$ and its hydrozirconation
1974	J. Schwartz[47]	Systematic study of hydrozirconation for organic synthesis
1974	W. Kaminsky[48]	Alkene polymerization catalyzed by Cp_2ZrCl_2 and methyl-aluminoxanes
1978	E. Negishi[49,50]	Controlled carboalumination of alkynes catalyzed by Cp_2ZrCl_2
1979	G. Erker[51]	Benzyne-zirconocenes
1980 & 1981	G. Erker[52] & A. Nakamura[53]	Conjugated diene-zirconocenes
1983	U. M. Dzhemilev[54]	Ethylmagnesiation of alkenes catalyzed by Cp_2ZrCl_2
1985	E. Negishi[55]	Cp_2Zr-promoted enyne bicyclization
1986	E. Negishi[56,57]	Generation of alkene-zirconocenes from dialkylzirconocenes and their systematic investigation
1986	S. L. Buchwald[58,59]	Systematic investigation of benzyne-zirconocenes and alkyne-zirconocenes

Since the chemistry of chiral metallocenes and the Ziegler–Natta polymerization are discussed in Chapters 8–10, their discussion in this chapter is kept at a minimum. Limitation of space does not permit discussion of all of the important topics. As a consequence, discussion of many pioneering and basic studies not emphasizing synthetic applications is also kept at a minimum. For extensive discussion of the basic aspects of zirconocene chemistry, the readers are referred to a few monographs and reviews [1–40].

The great majority of the currently known $ZrCp_2$ derivatives are Zr(IV) and Zr(II) compounds. Some Zr(III) species, such as those shown in Scheme 5-1, are known [60, 61], and the number of such compounds is slowly but steadily increasing. Their formation and structure are of obvious scientific interest. At present, however, there is virtually no synthetically useful chemistry of these compounds known. In fact, they often represent unwanted byproducts to be avoided in the reactions of

Cp$_2$Zr(IV) and/or Cp$_2$Zr(II) compounds, and they are significant at present primarily within this context, although some useful new chemistry dealing with Zr(III) compounds, such as those shown in Scheme 5-1, is conceivable.

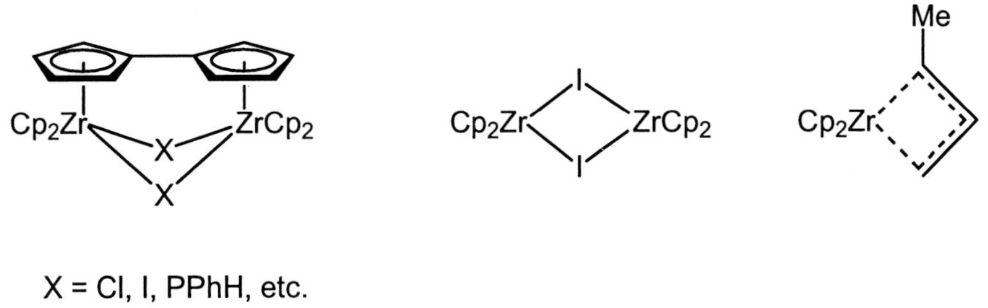

X = Cl, I, PPhH, etc.

Scheme 5-1

Various structural types are conceivable for Cp$_2$Zr(IV) and Cp$_2$Zr(II) compounds, and examples of most of them are known. Some representative structural types are shown in Table 5-2.

The following general comments are in order with regards to the structures shown in Table 5-2. First, formal charges are shown in some cases, but they are frequently omitted, in part, for the sake of simplicity. For example, one may argue that Cp$_2$Zr(PMe$_3$)$_2$ might be better represented by Cp$_2$Zr^{2-}(P$^+$Me$_3$)$_2$ without considering backdonation. However, it appears reasonable to omit all formal charges for simplicity in cases where the species shown are overall neutral. Second, some of the structures shown in Table 5-2, such as the structures IV-1 and XV-1 (Scheme 5-2), can represent two resonance structures of the same species. In fact, there have been no cases where two isomeric compounds should be separately represented by IV-1 and XV-1. On the other hand, many experimental results and observations have supported the notion that IV-1 and XV-1 may be considered as two resonance structures of the same species, which may be viewed as either alkene–Zr^{II}Cp$_2$ complexes or zirconacyclopropanes containing a Zr^{IV} atom and that either may be interchangeably used. It appears useful and reasonable to compare IV-1 with the classical bent-bond structure of cyclopropane and XV-1 with its Walsh structure [62]. Admittedly, some such species may be better represented by IV-1, while others by XV-1, but no attempts are made to discuss such a subtle issue here. Third, it is potentially very important to realize that any coordinatively unsaturated species, *i.e.*, 16- and 14-electron species in Table 5-2, can exist as dimeric, oligomeric, or polymeric aggregates. Fourth, such compounds can also form complexes with Lewis bases. For the sake of simplicity, however, such structures are not shown in Table 5-2. As stated earlier, the formation and interconversion of these Cp$_2$Zr(IV) and Cp$_2$Zr(II) species as well as the attendant organic transformations are the main topics of this chapter.

Table 5-2. Structural Types of Cp$_2$Zr(IV) and Cp$_2$Zr(II) Compounds

Cp$_2$Zr(IV) Compounds

Type	Structure	Examples (Other Structural Types or Comments)
I	Cp$_2$Zr$\overset{IV}{\underset{16e}{\diagdown}}\begin{smallmatrix}a\\b\end{smallmatrix}$	Cp$_2$ZrXY, Cp$_2$ZrH$_2$, Cp$_2$Zr(H)X, Cp$_2$ZrRX, Cp$_2$ZrR^{1}R^2, etc.
II	Cp$_2$Zr$\overset{IV}{\underset{16e}{=}}$a	Cp$_2$Zr=CR^{1}R^2, Cp$_2$Zr=NR, Cp$_2$Zr=O, etc.
III	Cp$_2$Z̄r$\overset{\cdot IV}{\underset{18e}{\equiv}}$a	Cp$_2$Z̄r≡CR, Cp$_2$Z̄r≡N Cp$_2$Z̄r≡O$^+$ etc.
IV	Cp$_2$Zr$\overset{IV}{\underset{16e}{\diagdown}}\begin{smallmatrix}a\\\vert\\b\end{smallmatrix}$	Cp$_2$Zr$\diagdown\!\!\diagup\begin{smallmatrix}C\\C\end{smallmatrix}$, Cp$_2Zr\diagdown\!\!\diagup\begin{smallmatrix}C\\N\end{smallmatrix}$, Cp$_2Zr\diagdown\!\!\diagup\begin{smallmatrix}C\\O\end{smallmatrix}$, etc.
V	Cp$_2$Zr$\overset{IV}{\underset{16e}{\diagdown}}\begin{smallmatrix}a\\\Vert\\b\end{smallmatrix}$	Cp$_2$Zr$\diagdown\!\!\!\Vert\begin{smallmatrix}C\\C\end{smallmatrix}$, Cp$_2Zr\diagdown\!\!\!\Vert\begin{smallmatrix}C\\N\end{smallmatrix}$, Cp$_2Zr\diagdown\!\!\!\Vert\begin{smallmatrix}C\\O\end{smallmatrix}$+, etc.
VI	Cp$_2$Zr$\overset{IV}{\underset{16e}{\diamond}}\begin{smallmatrix}a\\b\\c\end{smallmatrix}$	$\left(\text{Cp}_2\text{Zr}\diamond\begin{smallmatrix}a\\c\end{smallmatrix}b , \text{Cp}_2\text{Z̄r}\diamond\begin{smallmatrix}a\\c\end{smallmatrix}b , \text{Cp}_2\text{Z̄r}\diamond\begin{smallmatrix}a\\c\end{smallmatrix}b , \text{etc.} \right)$
VII	Cp$_2$Zr$\overset{IV}{\underset{16e}{\pentagon}}\begin{smallmatrix}a-b\\d-c\end{smallmatrix}$	$\left(\text{Cp}_2\text{Zr}\pentagon\begin{smallmatrix}a-b\\d-c\end{smallmatrix} , \text{Cp}_2\text{Zr}\pentagon\begin{smallmatrix}a=b\\d-c\end{smallmatrix} , \text{Cp}_2\text{Zr}\pentagon\begin{smallmatrix}a=b\\d=c\end{smallmatrix} , \text{etc.} \right)$
VIII	Cp$_2$Zr$\overset{IV}{\underset{16e}{\bigcirc}}$	(six-membered and larger zirconacycles)
IX	Cp$_2$Zr$\overset{IV+}{\underset{14e}{-}}$a	Cp$_2$Zr$^+$–C$\diagdown$–, Cp$_2$Zr$^+$–N$\diagup$, Cp$_2$Zr$^+$–O– , etc.
X	Cp$_2$Z̄r$\overset{\cdot IV}{\underset{18e}{\diagdown}}\begin{smallmatrix}a\\b\\c\end{smallmatrix}$	(acyclic or cyclic zirconates)

Cp$_2$Zr(II) Compounds

Type	Structure	Examples (Other Structural Types or Comments)
XI	$\overset{\|}{\text{Cp}_2\text{Zr}}$ 14e	(hypothetical parent compound)
XII	$\overset{\|}{\text{Cp}_2\text{Zr}}$—a 16e	(σ complexes and zirconates)
XIII	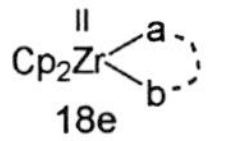 $\overset{\|}{\text{Cp}_2\text{Zr}}\overset{a}{\underset{b}{\diagdown}}$ 18e	(acyclic and cyclic σ complexes and zirconates)
XIV	$\overset{\|}{\text{Cp}_2\text{Zr}}$=a 18e	(complexes and zirconates)
XV	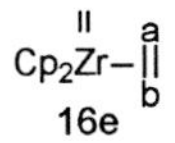 $\overset{\|}{\text{Cp}_2\text{Zr}}-\overset{a}{\underset{b}{\|}}$ 16e	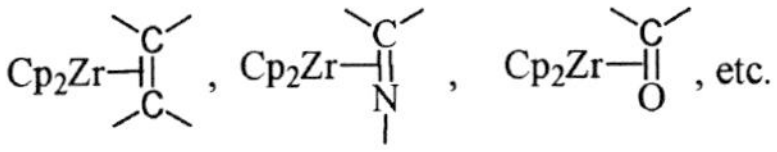 $\text{Cp}_2\text{Zr}-\overset{C}{\underset{C}{\|}}$, $\text{Cp}_2\text{Zr}-\overset{C}{\underset{N}{\|}}$, $\text{Cp}_2\text{Zr}-\overset{C}{\underset{O}{\|}}$, etc.
XVI	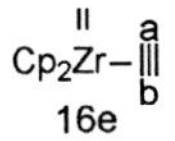 $\overset{\|}{\text{Cp}_2\text{Zr}}-\overset{a}{\underset{b}{\|\|}}$ 16e	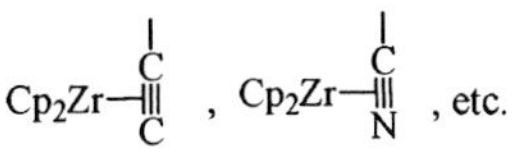 $\text{Cp}_2\text{Zr}-\overset{C}{\underset{C}{\|\|}}$, $\text{Cp}_2\text{Zr}-\overset{C}{\underset{N}{\|\|}}$, etc.

$$\text{Cp}_2\text{Zr}\overset{\text{IV}}{\underset{}{\diagdown}}\overset{C}{\underset{C}{|}} \quad \longleftrightarrow \quad \text{Cp}_2\text{Zr}-\overset{\text{II}}{\underset{}{\|}}\overset{C}{\underset{C}{}}$$

IV-I   XV-I

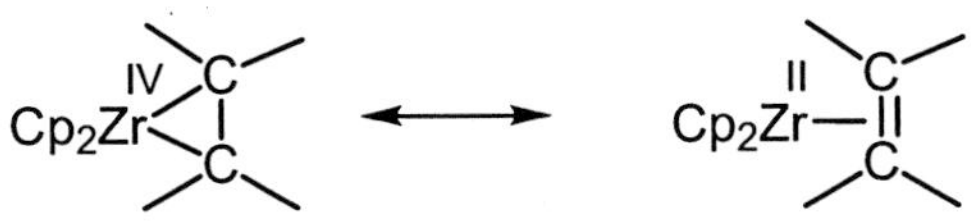

Scheme 5-2

It is striking that, in virtually all transformations discussed in this chapter, the Cp$_2$Zr unit remains intact. It is, however, important to note that the Cp$_2$Zr unit is, in fact, more fragile and chemically labile than might be thought. For example, treatment of Cp$_2$ZrCl$_2$ with 2 equivalents of *n*-BuLi is known to give Cp$_2$Zr(Bu-*n*)$_2$ which serves as a convenient precursor to a 'ZrCp$_2$' equivalent [56, 57]. On the other hand, the use of 3 or more equivalents of *n*-BuLi produces a mixture of CpZr(Bu-*n*)$_3$ and LiCp which is in equilibrium with a minor amount of a mixture of Cp$_2$Zr(Bu-*n*)$_2$ and LiBu-*n* [63] (Scheme 5-3). As might be expected, the facile displacement of a Cp group does not occur with *n*-butylmagnesium halides under comparable conditions. Conversion of Cp$_2$Zr derivatives into CpZr derivatives can provide useful and convenient routes to the latter, since their direct formation is often problematical due to competitive formation of Cp$_2$Zr derivatives. A representative example of such reactions is treatment of Cp$_2$ZrCl$_2$ with Cl$_2$ to give CpZrCl$_3$ [64] (Scheme 5-4).

$$Cp_2ZrCl_2 \quad \xrightarrow{\text{2 LiBu-}n} \quad Cp_2Zr(Bu\text{-}n)_2 \quad \underset{\xleftarrow{\hspace{1cm}}}{\overset{\text{LiBu-}n}{\rightleftharpoons}} \quad CpZr(Bu\text{-}n)_3 + LiCp$$

Scheme 5-3

$$Cp_2ZrCl_2 \quad \xrightarrow{\text{Cl}_2, \text{hv}} \quad CpZrCl_3$$

Scheme 5-4

In cases where the desirable transformations in high yields are observed with con-
servation of the Cp_2Zr unit, they obviously take place faster than any undesirable
side reactions. On the other hand, it has become abundantly clear that, in cases
where the desired reactions are comparatively slow, various unwanted side reactions, reversible or irreversible, can divert the reaction courses, and our knowledge
of such side reactions becomes critically important.

5.2 Preparation of Cp₂Zr Compounds

As in the synthesis of any other organometals, there are a dozen or so methods
[65] to be considered for the synthesis of Cp_2Zr compounds [6]. Several of them
listed in Table 5-3 are particularly important. Transmetallation and hydrozirconation convert $Cp_2Zr(IV)$ compounds into other $Cp_2Zr(IV)$ compounds. On the other
hand, oxidative addition and oxidative coupling convert $Cp_2Zr(II)$ compounds into
$Cp_2Zr(IV)$ compounds. If the latter reaction is considered as complexation, the
products may be viewed as $Cp_2Zr(II)$ compounds. Similar $Cp_2Zr(II)$-to-$Cp_2Zr(II)$
conversions may be achieved by complexation with *n*-donors, *e.g.*, PMe_3 and CO.
((Table 5-3))

5.2.1 Transmetallation

Transmetallation and hydrozirconation are the two most general methods for the
synthesis of $Cp_2Zr(IV)$ compounds, and they are often complementary with each
other. For the preparation of $Cp_2Zr(IV)$ compounds by transmetallation, the most
frequently used starting materials are Cp_2ZrX_2 and $Cp_2Zr(X)R$, where X = Cl or
Br, and the latter compounds are most frequently prepared by hydrozirconation
with $Cp_2Zr(H)Cl$ and by transmetallation with Cp_2ZrX_2. The starting Cp_2ZrX_2,
where X = Cl, Br, and I, can be prepared by the treatment of the respective ZrX_4
with 2 equivalents of CpM, where M = MgCl, MgBr, Li, Na, *etc.* [1, 42]. Currently, Cp_2ZrCl_2 is commercially available at a reasonable price. Since $ZrBr_4$ and ZrI_4
are very much more expensive than $ZrCl_4$, it may be more economical to prepare

Table 5-3. Methods of Preparation of Cp_2Zr Compounds

1. **Transmetallation:** $Cp_2Zr(IV) \longrightarrow Cp_2Zr(IV)$

$$Cp_2ZrX_2 \xrightarrow{\ MR\ } Cp_2Zr(X)R \xrightarrow{\ MR\ } Cp_2ZrR_2$$

2. **Zr-H Exchange:** $Cp_2Zr(IV) \longrightarrow Cp_2Zr(IV)$

$$Cp_2ZrX_2 \xrightarrow{\ HR\ } Cp_2Zr(X)R \xrightarrow{\ HR\ } Cp_2ZrR_2$$

3. **Hydrozirconation:** $Cp_2Zr(IV) \longrightarrow Cp_2Zr(IV)$

4. **Oxidative Addition:** $Cp_2Zr(II) \longrightarrow Cp_2Zr(IV)$

5. **Oxidative Coupling:** $Cp_2Zr(II) \longrightarrow Cp_2Zr(IV)$
   (or Complexation: $Cp_2Zr(II) \longrightarrow Cp_2Zr(II)$)

6. **Complexation with *n*-Donors:** $Cp_2Zr(II) \longrightarrow Cp_2Zr(II)$

$$Cp_2ZrL_n \xrightarrow{\ mL\ } Cp_2ZrL_{n+m}$$

Cp_2ZrBr_2 or Cp_2ZrI_2 from Cp_2ZrCl_2 by transhalogenation. Treatment of Cp_2ZrCl_2 with KI in acetone [66] gives only an 80/20 mixture of the iodide and chloride. The reaction of Cp_2ZrCl_2 with BI_3 gives a purer Cp_2ZrI_2 [67]. The use of Me_3SiI in place of BI_3 [68] is a convenient alternative. However, one of the most convenient methods involves treatment of Cp_2ZrCl_2 with n-BuLi (2 equiv.) at −78°C followed by iodinolysis with I_2 [56] (Scheme 5-5).

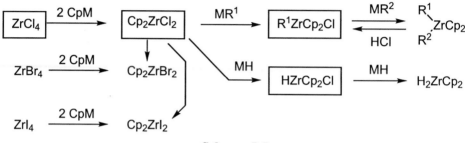

Scheme 5-5

Substitution of one or two Cl atoms of Cp_2ZrCl_2 with carbon groups by transmetallation (Scheme 5-5) is, in principle, favorable in cases where M is more electropositive than Zr, or more specifically $ZrCp_2Cl$ or $ZrCp_2R$. Thus, alkali metals, e.g., Li, Na, and K, as well as Group 2, 12, and 13 metals, e.g., Mg, Zn, and Al, participate in this reaction. However, some subtle but critical differences should be noted. For example, organolithiums and organomagnesiums readily dialkylate Cp_2ZrCl_2 to give Cp_2ZrR_2 in most cases. In fact, it is often not practical to monoalkylate Cp_2ZrCl_2 with RLi or RMgX (X = Cl, Br, or I), although some Grignard reagents, e.g., t-BuMgCl, cleanly reacts with Cp_2ZrCl_2 in a 1:1 manner to give i-$BuZrCp_2Cl$ in high yield [69, 70] (Scheme 5-6). One unwanted side reaction with RLi is the formation of $CpZrR_3$ as discussed earlier (Scheme 5-3). This does not appear to be a serious problem with RMgX. Dialkylzirconocenes, or more generally diorganylzirconocenes, are, in most cases, thermally labile, and they undergo synthetically interesting reactions, as discussed later.

Selective monoalkylation of Cp_2ZrCl_2 is still an unsolved problem in many cases. For example, a somewhat tedious and indirect method shown in Scheme 5-6 has remained as one of the standard methods for the preparation of $MeZrCp_2Cl$ [71]. Although possible, there does not appear to be a very convenient method for the preparation of $ArZrCp_2Cl$. In such cases, selective C–Cl exchange reaction of Cp_2ZrR_2 and/or thermal comproportionation between Cp_2ZrR_2 and Cp_2ZrCl_2 might prove to be useful. In most of those cases where hydrozirconation is applicable, $RZrCp_2Cl$ in which R is alkyl or alkenyl are best prepared by hydrozirconation. The reaction of organoalanes with Cp_2ZrCl_2 is a delicate equilibrium reaction. For example, treatment of Cp_2ZrCl_2 with one equivalent of Me_3Al proceeds partially to give $MeZrCp_2Cl$ and Me_2AlCl, and there has been no indication for the formation of Me_2ZrCp_2 under a variety of conditions [50, 72] (Scheme 5-6). On the other hand, it is feasible to transfer an alkenyl group from Zr to Al [50, 73, 74] (Scheme 5-6). These results indicate that transmetallation between Zr and Al is very much ligand-dependent in a more or less predictable manner. Thus, the alkylating ability of organoaluminums increases in the order: $RAlCl_2 < R_2AlCl < R_3Al < R_4Al^-$. For example, transfer of alkenyl groups from Zr to Al discussed above can be reversed by the use of alkenylaluminates [74] (Scheme 5-6). Good understanding of the Zr–Al equilibria is particularly important in dealing with addition and polymerization reactions of alkenes and alkynes catalyzed by Zr–Al reagents.

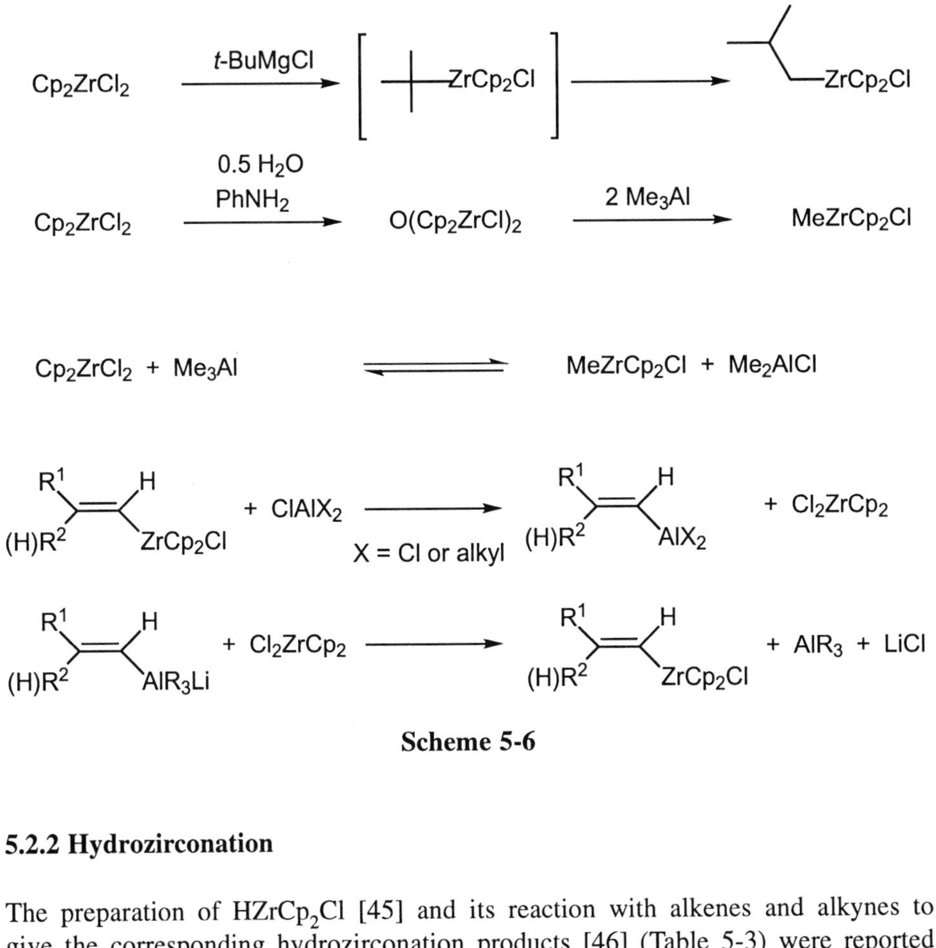

$$Cp_2ZrCl_2 \xrightarrow{t\text{-BuMgCl}} \left[\quad \text{+ZrCp}_2\text{Cl} \right] \longrightarrow \quad \text{—ZrCp}_2\text{Cl}$$

$$Cp_2ZrCl_2 \xrightarrow[\text{PhNH}_2]{0.5\ H_2O} O(Cp_2ZrCl)_2 \xrightarrow{2\ Me_3Al} MeZrCp_2Cl$$

$$Cp_2ZrCl_2 + Me_3Al \rightleftharpoons MeZrCp_2Cl + Me_2AlCl$$

$$\underset{(H)R^2}{\overset{R^1}{\diagdown}}\!=\!\underset{ZrCp_2Cl}{\overset{H}{\diagup}} + ClAlX_2 \xrightarrow[X = Cl\ or\ alkyl]{} \underset{(H)R^2}{\overset{R^1}{\diagdown}}\!=\!\underset{AlX_2}{\overset{H}{\diagup}} + Cl_2ZrCp_2$$

$$\underset{(H)R^2}{\overset{R^1}{\diagdown}}\!=\!\underset{AlR_3Li}{\overset{H}{\diagup}} + Cl_2ZrCp_2 \longrightarrow \underset{(H)R^2}{\overset{R^1}{\diagdown}}\!=\!\underset{ZrCp_2Cl}{\overset{H}{\diagup}} + AlR_3 + LiCl$$

Scheme 5-6

5.2.2 Hydrozirconation

The preparation of HZrCp₂Cl [45] and its reaction with alkenes and alkynes to give the corresponding hydrozirconation products [46] (Table 5-3) were reported first by Wailes *et al.* However, it was Schwartz [47] who demonstrated its usefulness in organic synthesis by converting organozirconium derivatives into organic products. The reaction shares some synthetically important characteristics with other facile and clean hydrometallation reactions, such as hydroboration [75] and hydroalumination [76, 77]. Thus, the reaction is generally clean and high-yielding, involving essentially 100% *cis* addition. However, a few noteworthy differences among them should be noted. The scope of hydrozirconation with respect to both substrate structure and chemoselectivity lies somewhere between those of hydroboration and hydroalumination (Table 5-4 and Scheme 5-7). The facile isomerization of secondary and tertiary alkylzirconium derivatives into primary alkyl derivatives can occur even at or below room temperature [47], and it is far more facile than that of B or Al. Alkenylzirconiums can also isomerize in the presence of a catalytic amount of $HZrCp_2Cl$, and a regioselectivity of $\geq 90\%$ has been attained without loss of stereoisomeric purity in the reaction of $MeC \equiv CR$, where R is primary, secondary, or tertiary alkyl [78]. It is also noteworthy that conjugated dienes can be converted to homoallylzirconium derivatives in which Zr is bonded to the less hindered of the two terminal carbon atoms of the diene unit [79]. Halogens, ethers, and metal-substituted alcohols are tolerated. Carbonyl compounds, such as aldehydes, ketones, carboxylic acids, and esters are reduced to alcohols, while nitriles are reduced to aldehydes [8]. However, the relative reactivity of various functional groups can be significantly modified, as shown in Scheme 5-8.

Table 5-4. Scope of Hydrometallation Reactions with Respect to Substrate Structures

Substrate structure	Metal		
	B	Zr	Al
RC/CH	+	+	+
RC/CR	+	+	+
$RCH=CH_2$	+	+	$+^a$
$R_2C=CH_2$	+	+	sluggish
$RCH=CHR$	+	$+^{a,b}$	sluggish
$R_2C=CHR$	+	$+^b$	?
$R_2C=CR_2$	+	-	?

^{a} Sluggish unless catalyzed. ^{b} Accompanied by Zr migration.

Despite the ease with which alkylzirconium derivatives regiochemically isomerize, alkyl groups, such as Et and *n*-alkyl, cannot be readily displaced by an external alkene, suggesting that the regioisomerization process must be nondissociative. It has, in fact, been difficult to free terminal alkenes from alkylzirconiums. The use of the stoichiometric amount of Ph_3CBF_4 was necessary in one successful case [8].

A wide variety of solvents may be used for hydrozirconation. However, ethers, such as THF, morpholine, and especially oxetane, appear to be some of the most

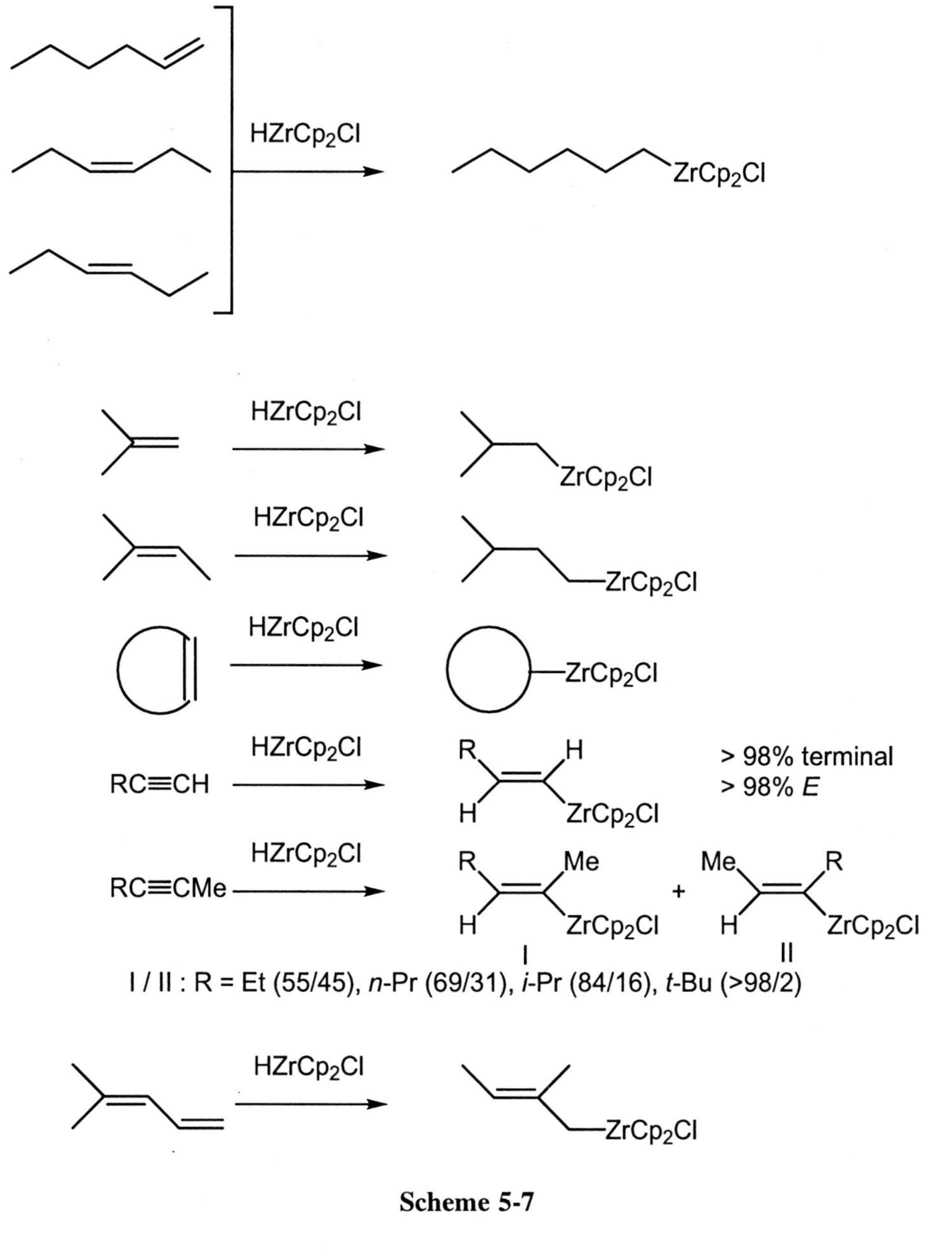

Scheme 5-7

favorable solvents [10], although less polar solvents, such as benzene and toluene, have also been widely used.

The preparation of HZrCp$_2$Cl free of any other species has been accomplished by the treatment of Cp$_2$ZrCl$_2$ with various metal hydrides, such as LiAlH$_4$ [45], LiAlH(OBu-t)$_3$ [45, 82], and Red-Al, *i.e.,* NaAlH$_2$(OCH$_2$CH$_2$OCH$_3$)$_2$ [83], fol-

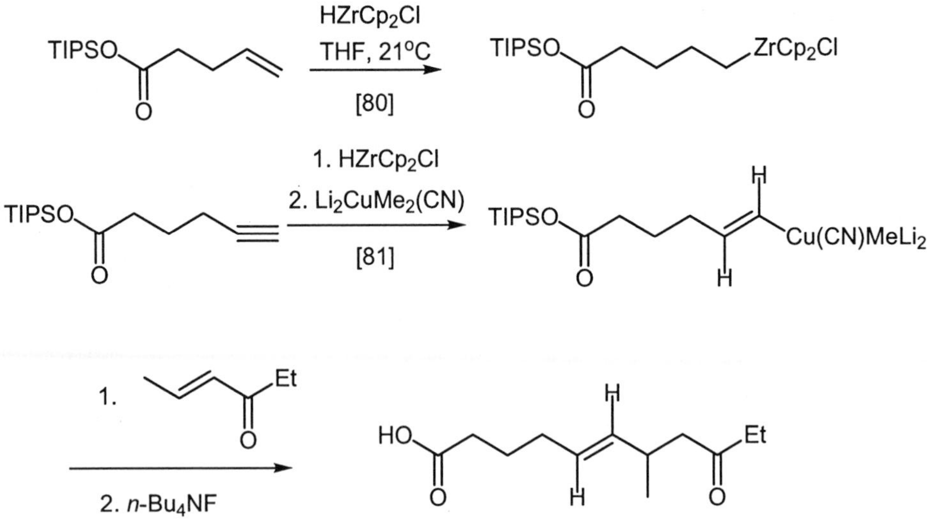

Scheme 5-8

lowed by purification. The low solubility of $HZrCp_2Cl$ in common organic solvents permits a simple, if rather tedious, method of its purification by filtration under an inert atmosphere. One common problem has been competitive formation of Cp_2ZrH_2, which can cause various unwanted side reactions, *e.g.*, those shown in Scheme 5-9. This problem can be largely avoided by treating Cp_2ZrH_2 with CH_2Cl_2 to regenerate $HZrCp_2Cl$ [84, 85].

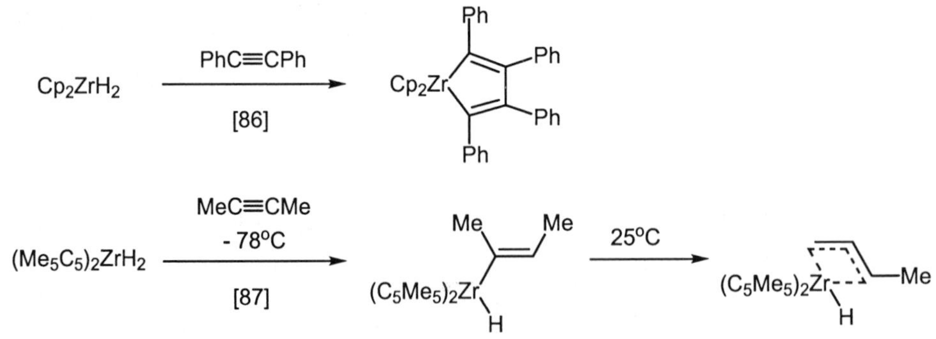

Scheme 5-9

In cases where there are no complications, $HZrCp_2Cl$ equivalents can be *in situ* generated by the treatment of Cp_2ZrCl_2 with *t*-BuMgCl, $LiAlH_4$, Red-Al, and $LiBEt_3H$ [69, 70, 88]. Another useful variation of hydrozirconation is a Zr-catalyzed hydroalumination with *i*-Bu$_3$Al and a catalytic amount of Cp$_2$ZrCl$_2$ [89] (Scheme 5-10). While these variations can provide some convenient alternatives to the use of preformed and purified HZrCp$_2$Cl, they must not be considered to be the same as pure HZrCp$_2$Cl. For example, whereas hydrozirconation of 5-decyne with pure HZrCp$_2$Cl gives essentially one isomer, that with *i*-Bu$_3$Al and a catalytic amount of Cp$_2$ZrCl$_2$ produces two regioisomers in an essentially 1:1 ratio [90] (Scheme 5-10). These and other recent results indicate that the Zr-catalyzed hydroalumination reaction clearly involves hydrozirconation but that it must be promoted by Al, which clearly is also responsible for the regioisomerization only to an adjacent position. This interesting observation can be readily understood by assuming the intermediacy of an allene derivative and an exclusive placement of Zr to its central carbon atom. These aspects of hydrozirconation are becoming clear only within the last few years, and further clarifications are undoubtedly forthcoming.

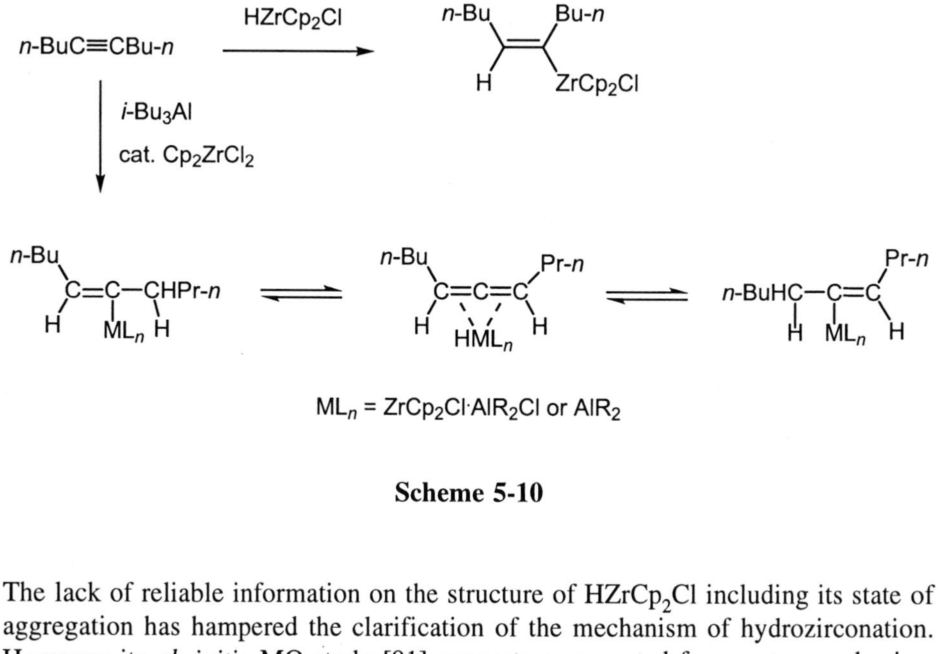

Scheme 5-10

The lack of reliable information on the structure of HZrCp$_2$Cl including its state of aggregation has hampered the clarification of the mechanism of hydrozirconation. However, its *ab initio* MO study [91] suggests a concerted four-center mechanism involving a frontal coplanar attack between H and Cl of HZrCp$_2$Cl by alkenes and alkynes, which is very analogous to those suggested for other facile *cis* hydrometallation reactions, such as hydroboration [75] (Scheme 5-11). An activation energy of 0–15 kcal mol^{-1} has been suggested for the reaction of ethylene. In addition to an earlier review by Schwartz [8] several other reviews have discussed hydrozirconation [5, 6, 9, 10].

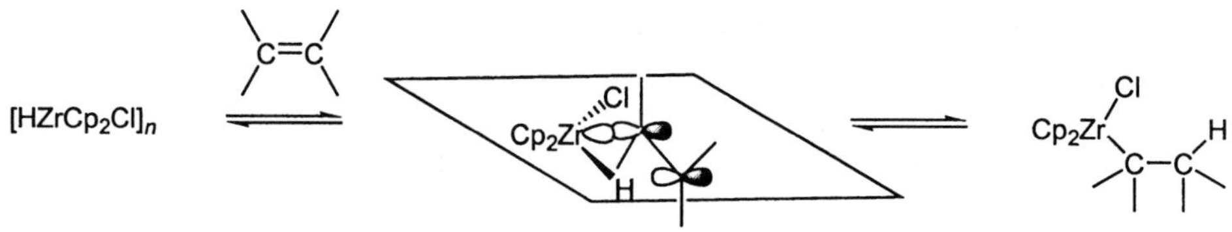

[HZrCp₂Cl]ₙ ⇌   ⇌

Scheme 5-11

5.2.3 Oxidative Addition

Oxidative addition is a potentially important process for the preparation of organozirconium derivatives. One prototypical example of this reaction was accidentally discovered in an unsuccessful attempt to bicyclize diallyl ether with *n*-
Bu₂ZrCp₂. The reaction instead produced an oxidative addition product which must
have been formed *via* oxidative coupling–elimination [92] (Scheme 5-12). This reaction has since been developed into a convenient method for converting allyl and
propargyl ethers into allylzirconiums [93, 94] and allenylzirconiums [95], respectively (Scheme 5-12). A review of this reaction with pertinent references has recently been published [11].

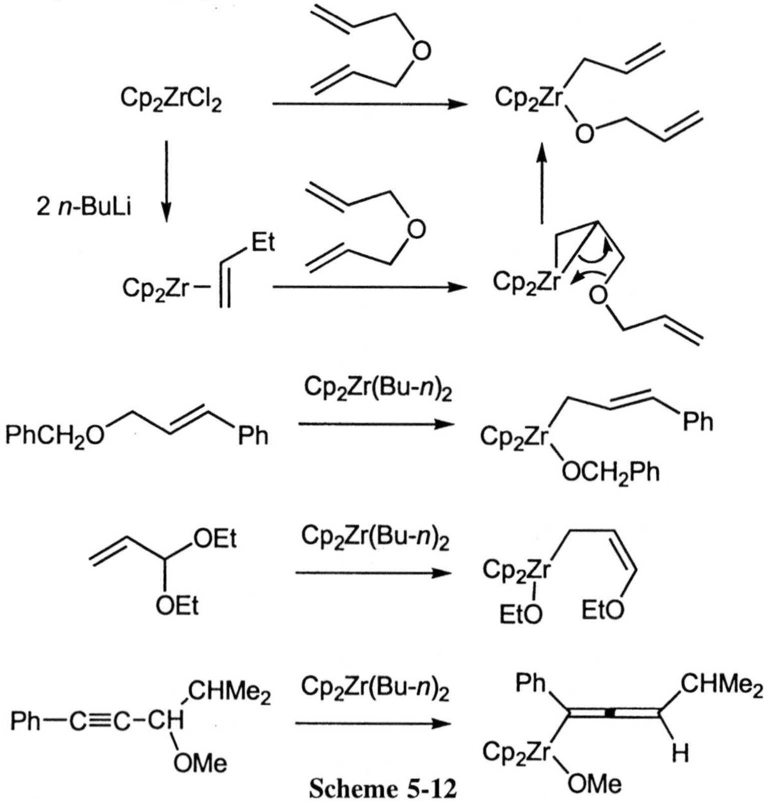

Scheme 5-12

The scope of the oxidative addition reaction of 'Cp$_2$Zr' derivatives has been significantly expanded by the recent development of oxidative addition of alkenyl chlorides to give alkenylzirconocene chlorides, which also appears to proceed *via* oxidative coupling–elimination [96]. Its ability to afford alkenylzirconocene derivatives corresponding to Markovnikov addition to alkynes is particularly noteworthy (Scheme 5-13).

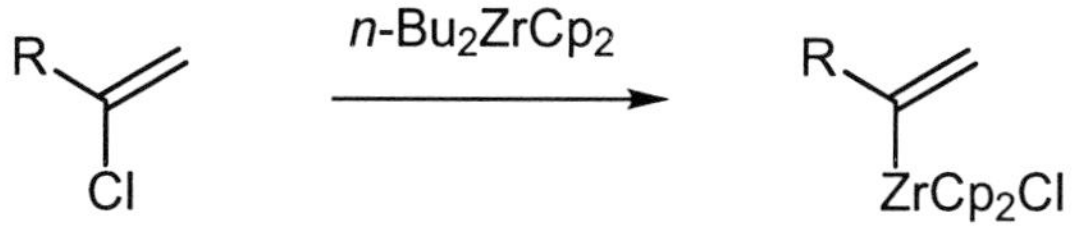

Scheme 5-13

It appears promising to explore the oxidative addition reactions of aryl and benzyl derivatives. As yet, no satisfactory procedures involving these derivatives appear to have been developed.

5.2.4 Oxidative Coupling and Complexation

Bonding in π-complexes of transition metal species is best described in terms of the Dewar–Chatt–Duncanson model [97, 98] involving π-donation by a π-ligand and backdonation by a metal. According to the theory, a metal-containing species must have at least one filled nonbonding orbital, *i.e.*, d^2, d^4,... but not d^0, to afford stable π-complexes. However, even d^0 complexes may serve as Lewis acids to accept an electron pair from a π-ligand to form unstable transient species. If a pair of σ-electrons are available for interaction (or backdonation), stable species may be formed. This indeed amounts to the well-accepted mechanism for concerted addition reactions, such as hydrozirconation. Since Cp$_2$Zr^{IV} derivatives are d^0 complexes, they may interact with π-compounds to produce only transient π-complexes that can be further converted to stable addition products (Scheme 5-14). On the other hand, Cp$_2$Zr^{II} derivatives are d^2 complexes which may contain one or two *n*-ligands or one π-ligand (Scheme 5-15). The parent Cp$_2$Zr is a 14-electron species which can, in principle, exist as either a singlet (Ia) or a triplet (Ib), but it does not appear to have been detected and characterized as such. All four types of 14- and 16-electron Cp$_2$Zr^{II} species shown in Scheme 5-15 are, in principle, capable of reacting with π-donors to produce π-complexes, but the following comments are in order. Complexes represented by Cp$_2$Zr^{II}L$_2$, where L is an *n*-donor, are 18-electron species with no low-lying empty orbital. So, their π-complexation must proceed with loss of one or both *n*-donors (L). More intricate is the reaction of π-complexes containing a Cp$_2$Zr^{II} moiety. Since they are 16-electron species with no *n*-electron pair available for interaction, their reaction with a π-donor must involve either π-ligand displacement or addition to produce five-membered zirconacycles (Scheme 5-16). Furthermore, the latter process can serve as a mechanistic path for π-displacement, as detailed later. For the sake of simplicity, alkene–ZrCp$_2$ complexes are shown in Scheme 5-16, but similar processes involving other π-ligands have also been observed.

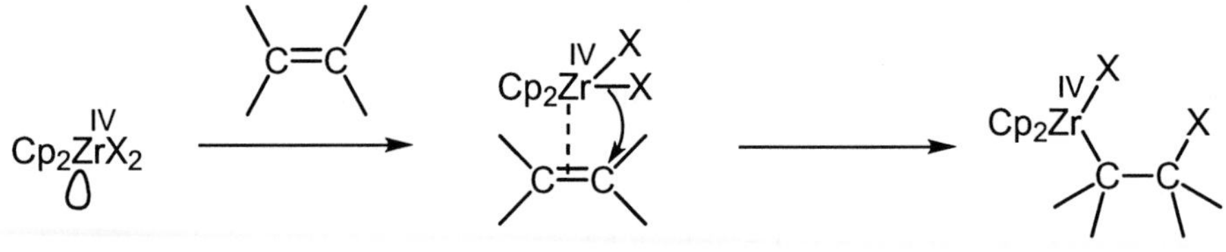

Scheme 5-14

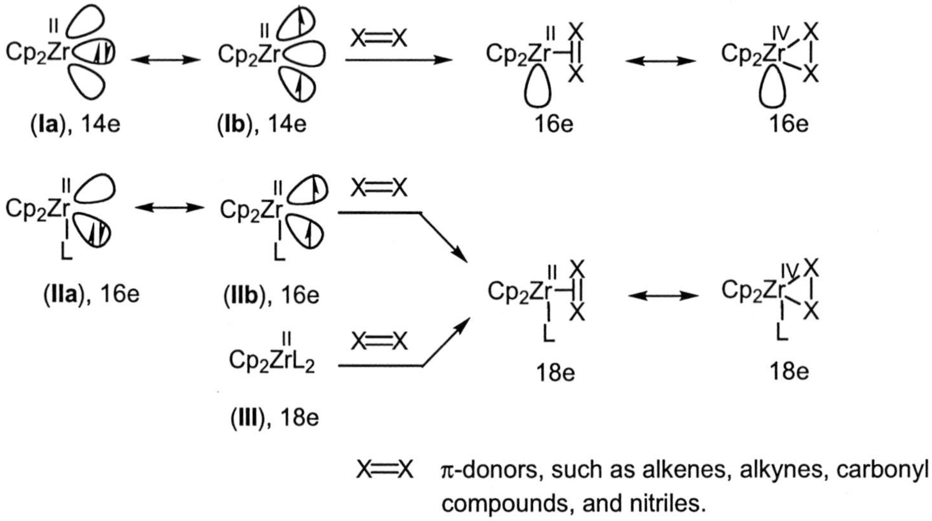

X═X π-donors, such as alkenes, alkynes, carbonyl compounds, and nitriles.

Scheme 5-15

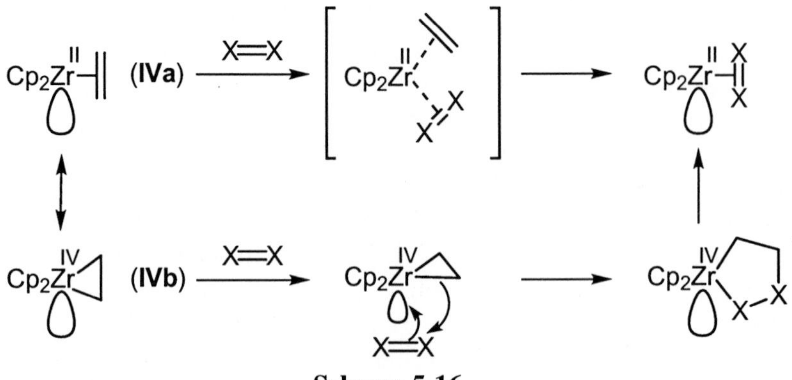

Scheme 5-16

The preparation of the required Cp_2Zr^{II} reagents has been achieved by either reduction of Cp_2ZrX_2, where X is Cl, Br, I, *etc.*, with metals [99] or β-H abstraction of organozirconocene derivatives. An early example of the reduction of Cp_2ZrCl_2 and Cp_2HfCl_2 with Na and naphthalene produced some reduced products represented by $(C_{10}H_{10}M)_n$, where M is Zr or Hf. Formation of Cp_2Zr as a transient species may occur but has never been experimentally demonstrated. In the presence of PhC≡CPh, the reaction produced the corresponding zirconacyclopentadiene [100] (Scheme 5-17). In the presence of *n*-donors such as CO and PMe_3, Cp_2ZrCl_2 has been converted to $Cp_2Zr(CO)_2$ [101] and $Cp_2Zr(PMe_3)_2$ [102] respectively (Scheme 5-18), and their structures have been fully characterized by spectroscopic methods including X-ray analysis [103, 104].

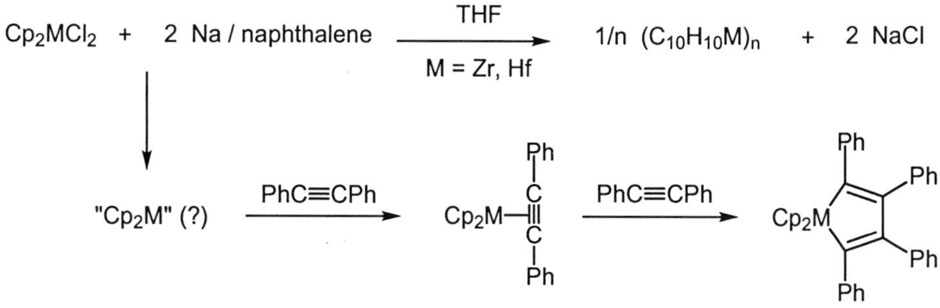

Scheme 5-17

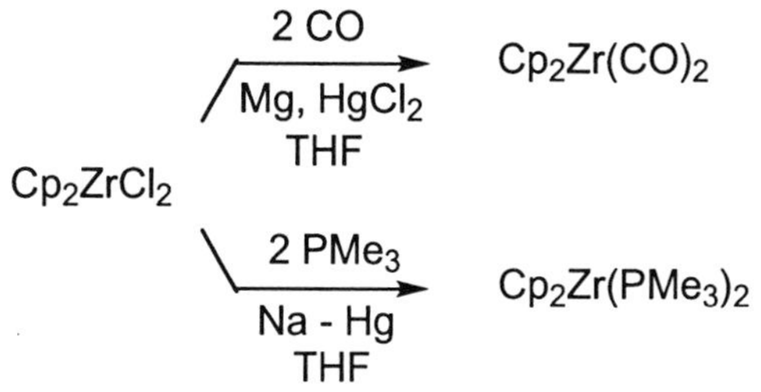

Scheme 5-18

These Cp_2Zr^{II} complexes tend to be thermally and chemical labile. For example, $Cp_2Zr(PMe_3)_2$ can be converted to a dimeric Zr^{III} species [104] (Scheme 5-19). For further information of these complexes, pertinent reviews [3, 27, 36, 105] should be consulted.

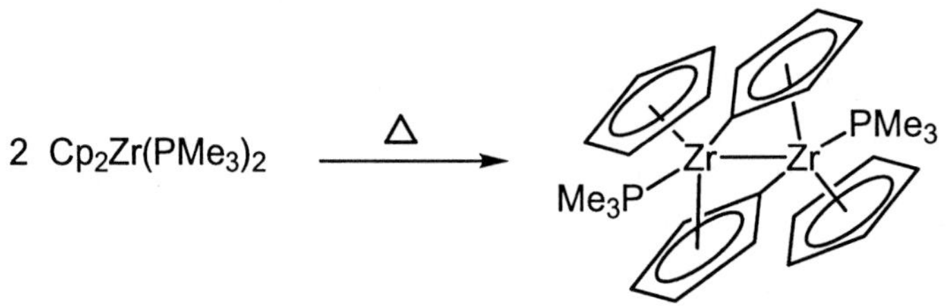

$$2 \text{ Cp}_2\text{Zr(PMe}_3)_2 \xrightarrow{\Delta}$$

Scheme 5-19

From the viewpoint of organic synthesis, the β-H abstraction reaction of mono- and diorganylzirconocenes represents the most convenient route to $\text{Cp}_2\text{Zr}^{II}$ derivatives at present. The formation of $\text{Cp}_2\text{Zr}^{II}$ complexes containing benzynes [51, 58, 106], alkynes [107–109], alkenes [107, 110, 111], as well as various heteroalkynes [112] and heteroalkenes [113, 114] have been generated by this method. There are two convenient protocols. In one, dialkylzirconocenes, such as $n\text{-Bu}_2\text{ZrCp}_2$ [56, 57, 107] and $t\text{-Bu}(i\text{-Bu})\text{ZrCp}_2$ [115], are *in situ* generated and decomposed in the presence of π-compounds, such as alkenes and alkynes (Negishi protocol). In this process, alkyl groups serve as precursors to temporary π-ligands which are to be displaced by externally introduced π-compounds [35] (Scheme 5-20). In the other, the

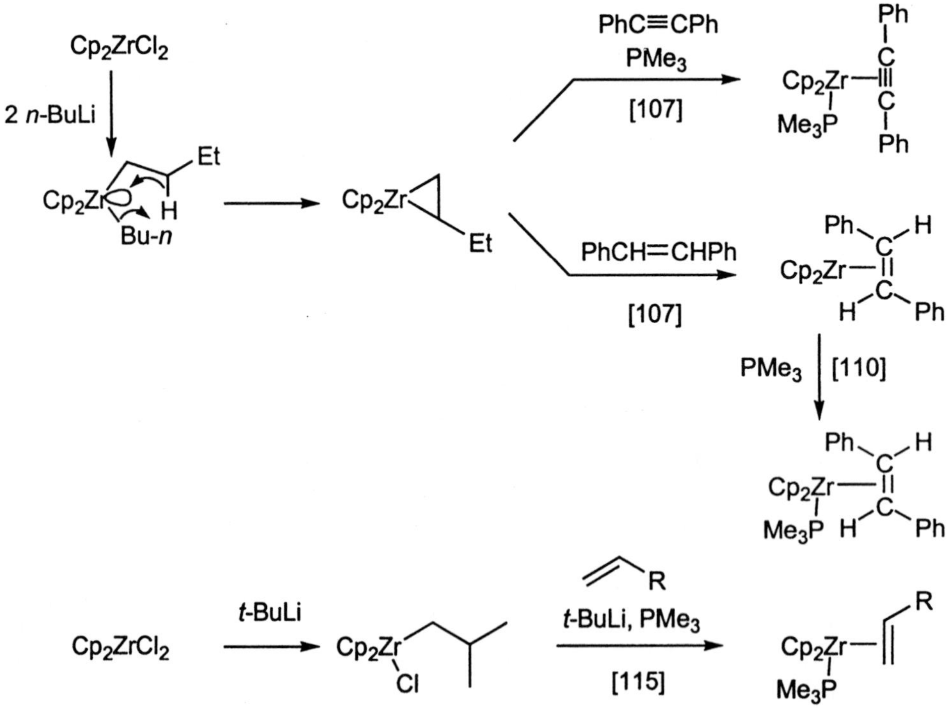

Scheme 5-20   R = H, Et, *n*-Hex, Cy, Ph

final π-ligands are *in situ* generated by the decomposition of diorganylzirconocenes (Erker–Buchwald protocol). No external π-compound is used for the preparation of Cp_2Zr^{II} π-complexes. In the original Erker procedure [51], one of the two organic groups was sacrificed. This practical difficulty can be avoided by the Buchwald procedure in which an inexpensive Me group serves as the β-H abstracter [29, 58, 108] (Scheme 5-21).

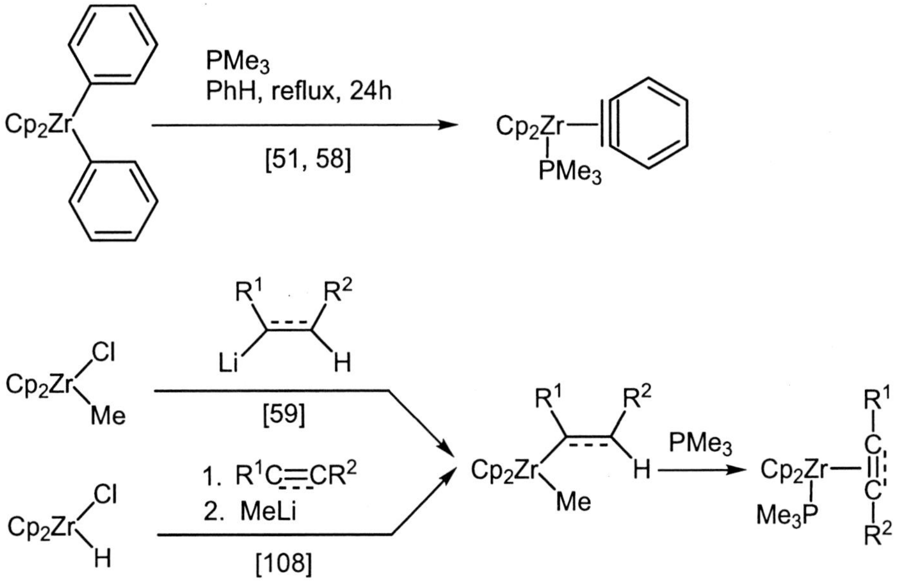

Scheme 5-21

In general, unstabilized 16-electron Cp_2Zr^{II} complexes with π-donors are thermally and chemically labile, although some, such as Cp₂Zr–stilbene [107], have been spectroscopically characterized. However, they can be stabilized by *n*-donors, such as PMe₃, CO, and amines. Alternatively, Cp_2Zr^{II} π-complexes containing heteroalkynes and heteroalkenes can stabilize themselves through dimerization, oligomerization, and polymerization. Many of the stabilized complexes have been fully characterized by spectroscopic methods including X-ray crystallography. Some representative examples of Cp_2Zr^{II} π-complexes are shown in Table 5-5.

Table 5-5. Representative Examples of Cp_2Zr^{II} π-Complexes Characterized by X-Ray Analysis

Complex	Reference	Complex	Reference
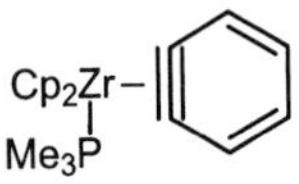 Cp₂Zr—, Me₃P	[58]	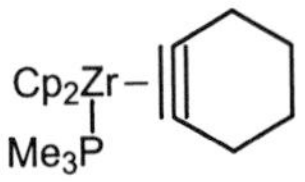 Cp₂Zr—, Me₃P	[59]
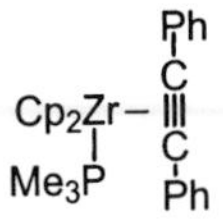 Cp₂Zr—, Me₃P	[107]	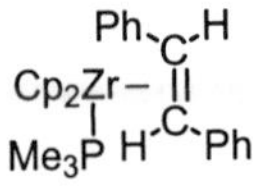 Cp₂Zr—, Me₃P	[110]
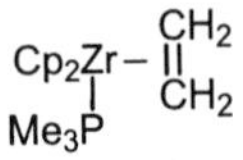 Cp₂Zr—, Me₃P	[111, 116, 117]	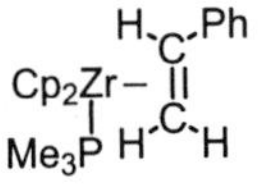 Cp₂Zr—, Me₃P	[111]
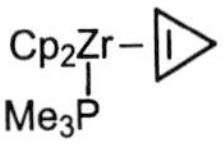 Cp₂Zr—, Me₃P	[29]	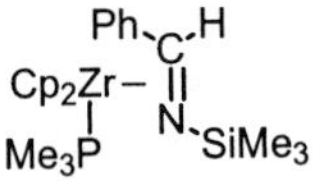 Cp₂Zr—, Me₃P	[114]
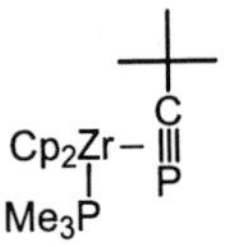 Cp₂Zr—, Me₃P	[112]	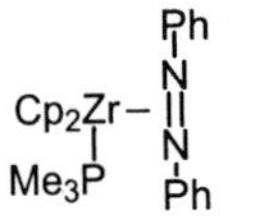 Cp₂Zr—, Me₃P	[118]

It should be pointed out here that, in some cases, transmetallation reactions of Cp_2Zr^{IV} derivatives provide convenient alternatives to the preparation of Cp_2Zr^{II} π-complexes, as indicated by the examples shown in Scheme 5-22. Yet another alternative for the preparation of Cp_2Zr π-complexes or carbonyl compounds is migratory insertion reactions of acylzirconium derivatives [118–122] (Scheme 5-23).

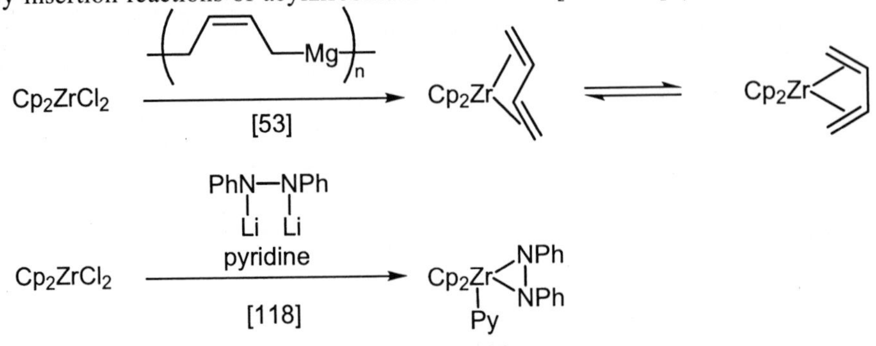

Scheme 5-22

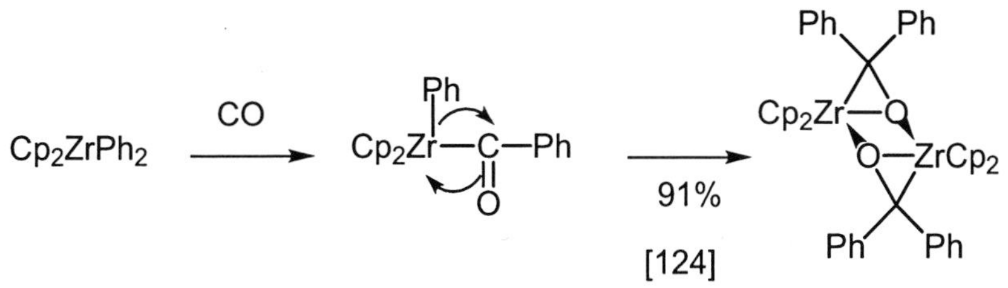

Scheme 5-23

5.3 Reactions and Synthetic Applications of Monoorganylzirconocene Derivatives

Monoorganylzirconocene derivatives represented by RZrCp₂Cl, where R is alkyl, allyl, allenyl (or propargyl), and alkenyl, are now widely available *via* transmetallation, hydrozirconation, or oxidative addition. Those containing a benzyl or aryl group are currently less readily available, and their chemistry as well as that of RC≡CZrCp₂Cl remains largely unexplored. The majority of known monoorganylzirconocene chlorides are thermally stable at ambient temperatures. However, they are sensitive to both air and moisture, and they must be handled under an inert atmosphere of Ar or nitrogen. The structure of RZrCp₂Cl is distorted tetrahedral. For example, the Cp–Zr–Cp and Me–Zr–Cl angles of MeZrCp₂Cl are 134 and 95°, respectively, and the Me–Zr and Cl–Zr bond lengths are 2.36 and 2.49°, respectively [125] (Scheme 5-24). It is important to note that an empty valence shell orbital which can act as an acidic or electrophilic center may be available in any of the three regions indicated by the orbital lobes lying in the plane defined by C, Zr, and Cl (Scheme 5-24). As will be clear from the discussions that follow, this empty orbital plays a pivotal role in the reactions of Cp₂Zr derivatives including RZrCp₂Cl. Since the chemistry of RZrCp₂Cl has been extensively and repeatedly reviewed [5, 6, 8, 10], only a brief overview is presented below.

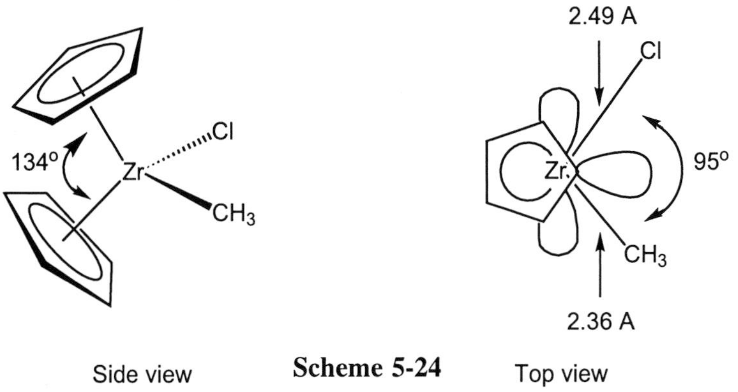

Side view   **Scheme 5-24**   Top view

5.3.1 Protonolysis, Halogenolysis, Oxidation, and Related Carbon–Heteroatom Bond Formation Reactions

5.3.1.1 Protonolysis and Deuterolysis

Cleavage of σ Zr–C bonds occurs readily upon treatment with H_2O or dilute acids, while the Cp–Zr bond usually survives mild protonolysis conditions. The use of D_2O or more reliably $DCl–D_2O$ permits replacement of Zr with D. Together with iodinolysis discussed below, deuterolysis provides a generally reliable method for establishing the presence of the C–Zr bond. Protonolysis or deuterolysis of Zr–$C_{sp}2$ bonds proceeds with retention of configuration [126], while the stereochemistry of the corresponding reaction of Zr–$C_{sp}3$ bonds does not appear to have been well established. In contrast with protonolysis, catalytic hydrogenation of RZrCp$_2$Cl is sluggish and not very clean [127].

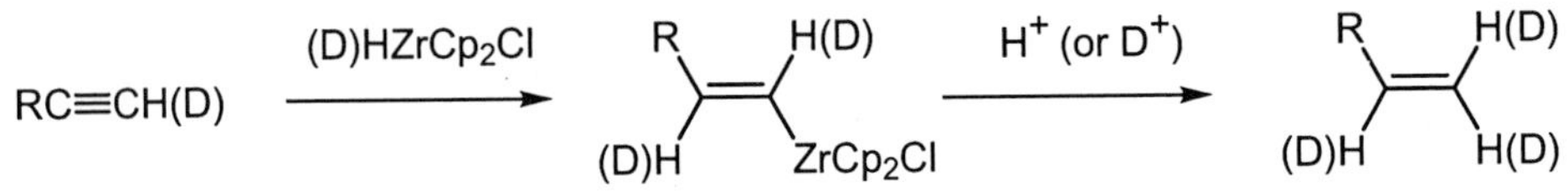

$$RC\equiv CH(D) \quad \xrightarrow{\text{(D)HZrCp}_2\text{Cl}} \quad \underset{(D)H \quad ZrCp_2Cl}{\overset{R \quad\quad H(D)}{>\!=\!<}} \quad \xrightarrow{H^+ \text{ (or } D^+)} \quad \underset{(D)H \quad H(D)}{\overset{R \quad\quad H(D)}{>\!=\!<}}$$

Scheme 5-25

5.3.1.2 Halogenolysis

Iodinolysis of RZrCp$_2$Cl with I_2 in THF occurs readily and widely at or below room temperature. And yet, it generally preserves the Cp–Zr bond [47, 78]. Brominolysis with Br_2 is also facile, but it is often advantageous to use NBS in place of Br_2. Chlorinolysis is more sluggish, but favorable results may be obtained with NCS [78] or PhICl$_2$ [47]. Iodinolysis and brominolysis of both Zr–$C_{sp}2$ and Zr–$C_{sp}3$ bonds have proceeded with retention of configuration (Scheme 5-26). Brominolysis of some homoallyl derivatives suffer from skeletal rearrangement [79].

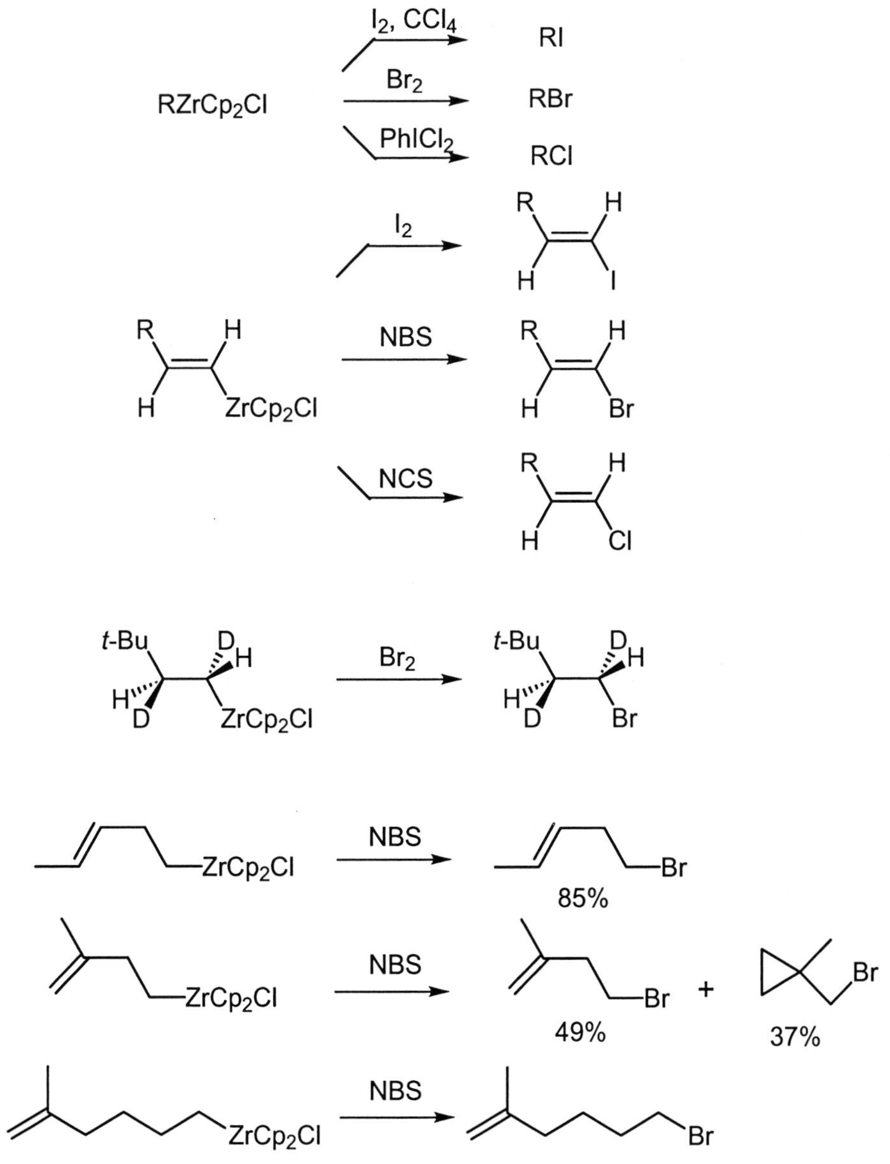
I_2, CCl_4
RI
Br_2
RZrCp2Cl
RBr
$PhICl_2$
RCl
I_2
NBS
NCS
R
H
ZrCp2Cl
I
Br
Cl
t-Bu
D
ZrCp2Cl
Br
85%
49%
37%
+

Scheme 5-26

5.3.1.3 Oxidation

Alkylzirconocene derivatives can be converted to alcohols with H_2O_2–NaOH, *t*-BuOOH, *m*-chloroperbenzoic acid, or O_2 [128]. On the other hand, conversion of alkenylzirconocenes to aldehydes and ketones cannot be readily achieved [47]. While oxidation reactions in some cases may proceed stereospecifically, those involving radicals are nonspecific. Clearly, additional investigations are necessary for oxidation of $RZrCp_2Cl$ to be synthetically useful. Other heteroatom reagents that react with $RZrCp_2Cl$ include SO_2 and NO [129, 130].

5.3.2 Carbon–Carbon Bond Formation *via* Polar Reactions of $RZrCp_2Cl$ with Carbon Electrophiles

5.3.2.1 Reactions of Allylzirconocenes with Aldehydes

Alkyl- and alkenylzirconiums are generally of low nucleophilicity towards common organic electrophiles, such as those shown in Scheme 5-27. Of the six types of electrophiles shown, only acyl halides are sufficiently reactive towards alkyl- and alkenylzirconocenes, but the reaction is not general and reliable. An improved procedure involving Zr-to-Al transmetallation discussed later is a superior alternative. The only known exception is the reaction of allylzirconocenes with aldehydes [131, 132] (Scheme 5-28). Dienezirconocenes, which may also be viewed as zirconacyclopent-3-enes, readily react with aldehydes, ketones, esters, and even nitriles as well as nonpolar π-compounds, such as alkenes and alkynes [133, 134] (Scheme 5-29 and 30). Some of these reactions hold considerable promise as practically useful synthetic tools, as exemplified by those shown in Scheme 5-31 [11].

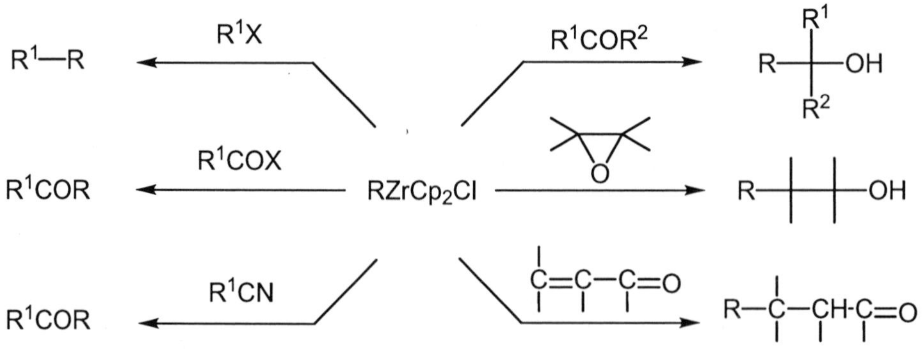

Scheme 5-27

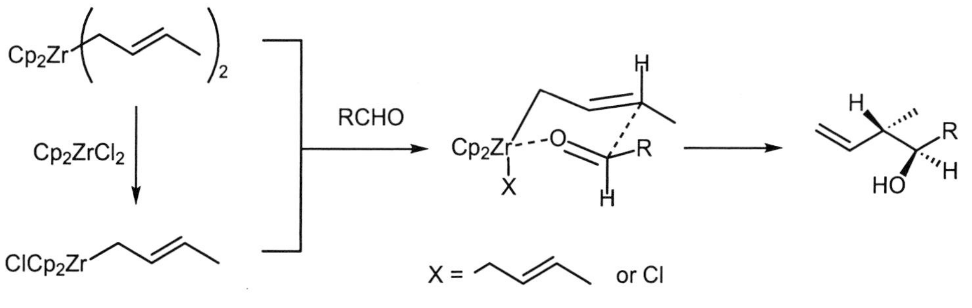

Scheme 5-28

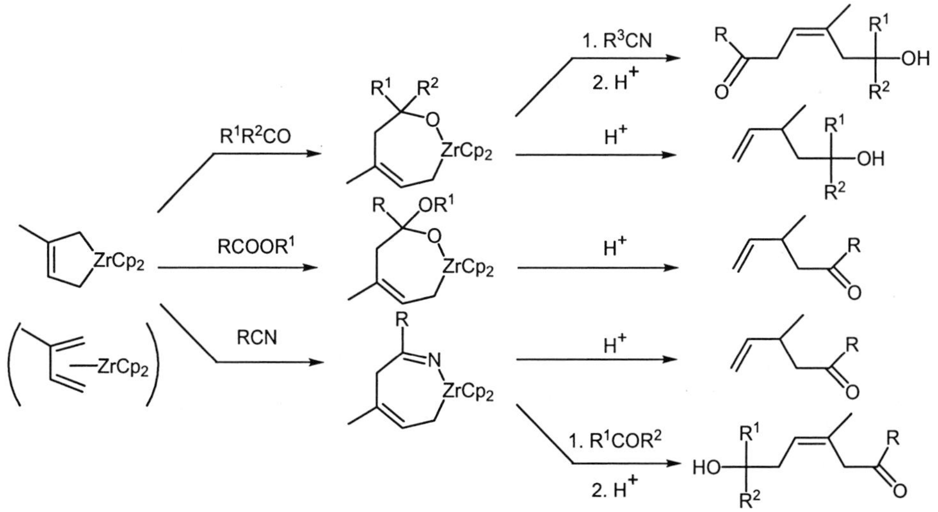

Scheme 5-29

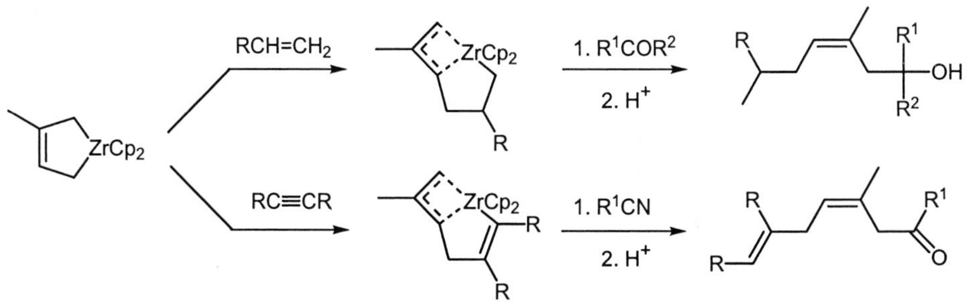

Scheme 5-30

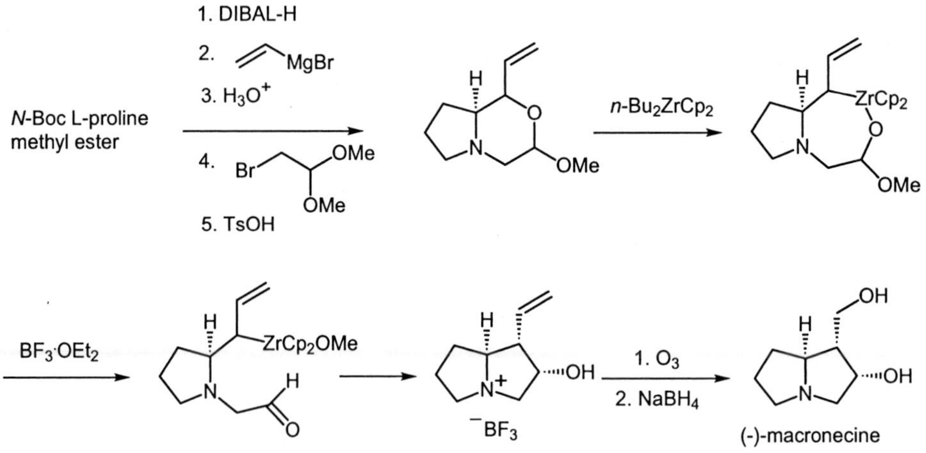
1. DIBAL-H
2. MgBr
3. H_3O^+
N-Boc L-proline methyl ester
4. Br OMe OMe
5. TsOH
H
O
N
OMe
n-Bu2ZrCp2
ZrCp2
BF3·OEt2
ZrCp2OMe
H
O
N+
OH
BF3
1. O3
2. NaBH4
OH
(-)-macronecine

Scheme 5-31

5.3.2.2 Reactions of Cationic Alkyl- and Alkenylzirconocenes with Aldehydes, Epoxides, Ketones, and Nitriles

Treatment of organylzirconocene chlorides with a halogen scavenger, such as $AgClO_4$, has been shown to significantly boost their reactivity towards aldehydes [135] (Scheme 5-32). The use of $AgAsF_6$ [136] in place of $AgClO_4$ is not only safer but also more effective in some cases.

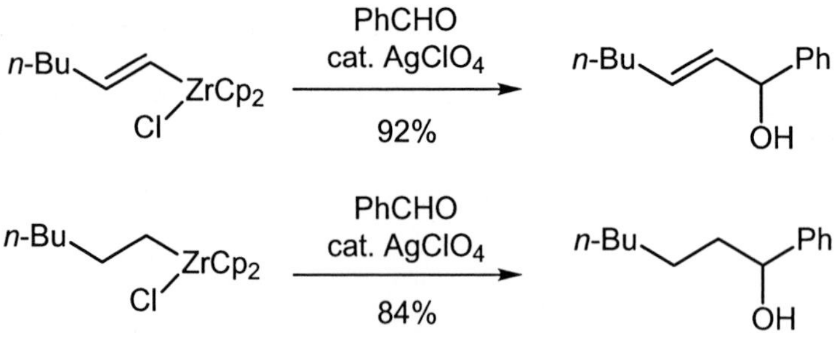
n-Bu
ZrCp2
Cl
PhCHO
cat. AgClO4
92%
Ph
OH
n-Bu
ZrCp2
Cl
PhCHO
cat. AgClO4
84%
Ph
OH

Scheme 5-32

The same strategy is also effective in the reaction with epoxides [137] (Scheme 5-33). This reaction has been shown to involve Zr-catalyzed isomerization of epoxides to aldehydes.

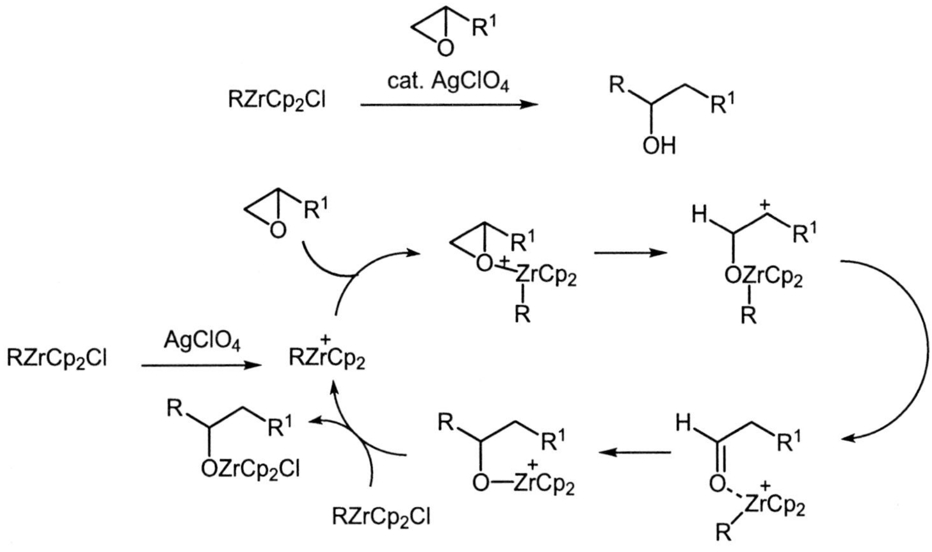

Scheme 5-33

Cationic monoalkylzirconocenes, such as MeZr$^+$Cp$_2$(THF)$^-$BPh$_4$, react even with ketones and nitriles in similar manners [16, 138] (Scheme 5-34).

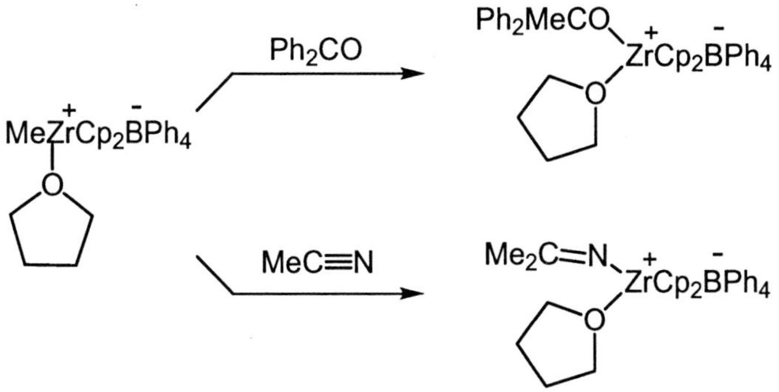

Scheme 5-34

5.3.3 Carbon–Carbon Bond Formation *via* Migratory Insertion Reactions of RZrCp$_2$Cl

Organometals of low intrinsic nucleophilicity, such as those containing B and transition metals, may resort to other patterns of processes for carbon–carbon bond formation than those shown in Scheme 5-27. Three such processes available to a wide variety of organotransition metals are shown in Scheme 5-35. It should be pointed out that reductive elimination, in fact, provides a means of achieving cross coupling, acylation, and conjugate addition shown in Scheme 5-27, as discussed later.

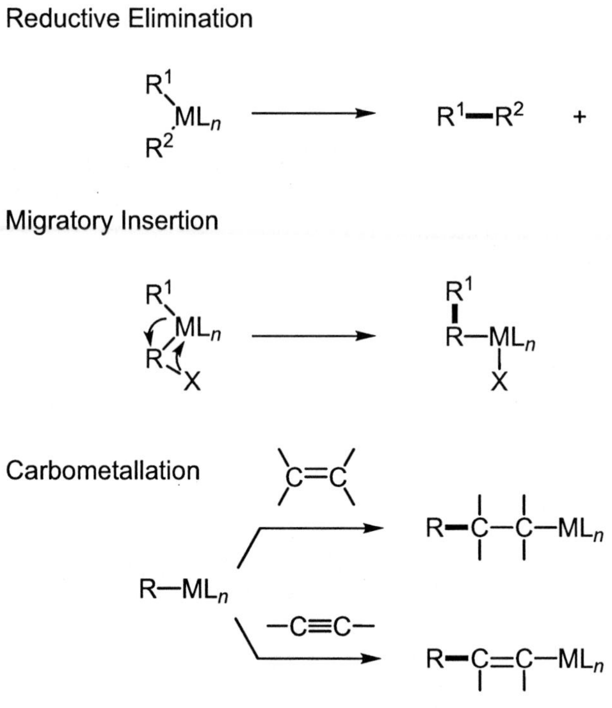

Reductive Elimination

$$ \begin{matrix} R^1 \\ \diagdown \\ \quad ML_n \\ \diagup \\ R^2 \end{matrix} \longrightarrow \quad R^1\!-\!R^2 \quad + \quad ML_n $$

Migratory Insertion

$$ \begin{matrix} R^1 \\ \diagdown \\ \quad ML_n \\ \diagup \\ R\diagup\diagdown X \end{matrix} \longrightarrow \begin{matrix} R^1 \\ | \\ R\!-\!ML_n \\ | \\ X \end{matrix} $$

Carbometallation

$$ R\!-\!ML_n \quad \xrightarrow{\;C=C\;} \quad R\!-\!\overset{|}{C}\!-\!\overset{|}{C}\!-\!ML_n $$

$$ \xrightarrow{\;-C\equiv C-\;} \quad R\!-\!\overset{|}{C}\!=\!\overset{|}{C}\!-\!ML_n $$

Scheme 5-35

5.3.3.1 Carbonylation

The reaction of RZrCp$_2$Cl with CO, typically 1–3 atm, readily produces the corresponding RCOZrCp$_2$Cl which can be converted to aldehydes and carboxylic acid derivatives [126, 139] (Scheme 5-36).

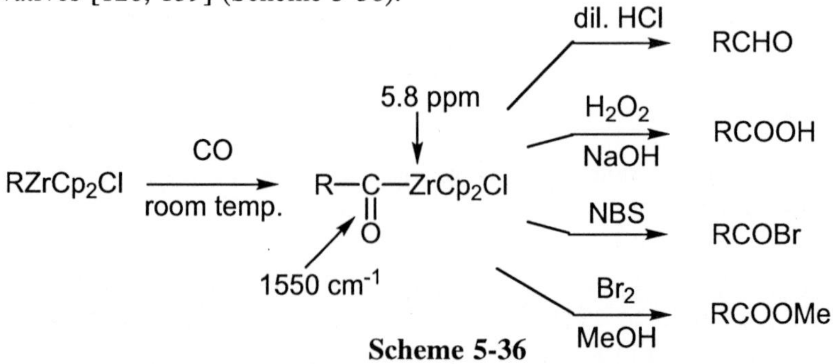

$$ RZrCp_2Cl \xrightarrow[\text{room temp.}]{CO} R\!-\!\underset{\underset{\displaystyle O}{\|}}{C}\!-\!ZrCp_2Cl $$

5.8 ppm

1550 cm^{-1}

dil. HCl → RCHO

$\dfrac{H_2O_2}{NaOH}$ → RCOOH

NBS → RCOBr

$\dfrac{Br_2}{MeOH}$ → RCOOMe

Scheme 5-36

Cyclic diorganylzirconocenes [51, 140] and certain dialkylzirconocenes, *e.g.*, Me_2ZrCp_2 [141], also readily undergo carbonylation to give ketones. In some cases, the organozirconium products before protonolysis have been shown to be zirconium enolates [17, 140].

5.3.3.2 Isonitrile Insertion

Isonitriles are useful alternatives to CO for some one-carbon homologation reactions where CO is not very satisfactory. The preparation of α,β-unsaturated aldehydes shown in Scheme 5-37 is one such example [142]. Another interesting reaction is the cyanation of alkenes shown in Scheme 5-38 [143].

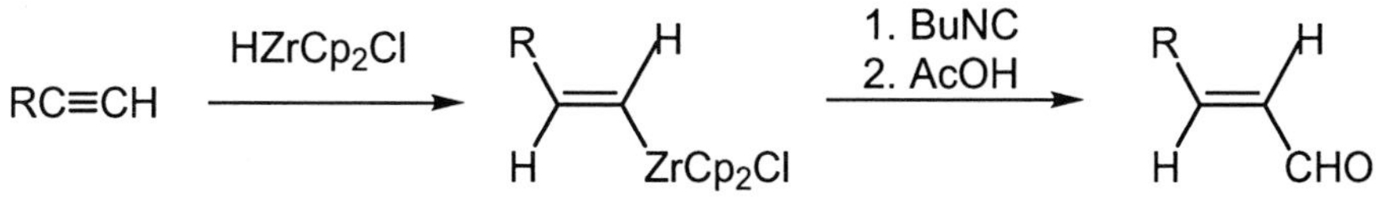

Scheme 5-37

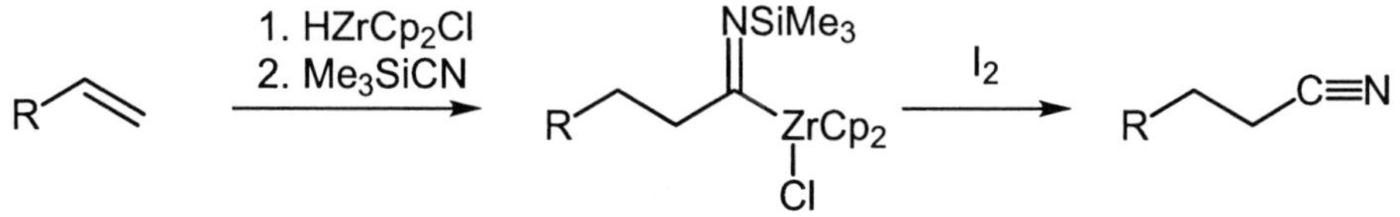

Scheme 5-38

5.3.3.3 Other Migratory Insertion Reactions

Although the potential scope of the migratory insertion reactions of organozirconium compounds appears to be very broad, its current scope is largely limited to carbonylation and related reactions with isonitriles. Nonetheless, there are some scattered reports on other migratory insertion reactions as shown in Scheme 5-39 [144] and Scheme 5-40 [145]. This topic clearly deserves to be further explored.

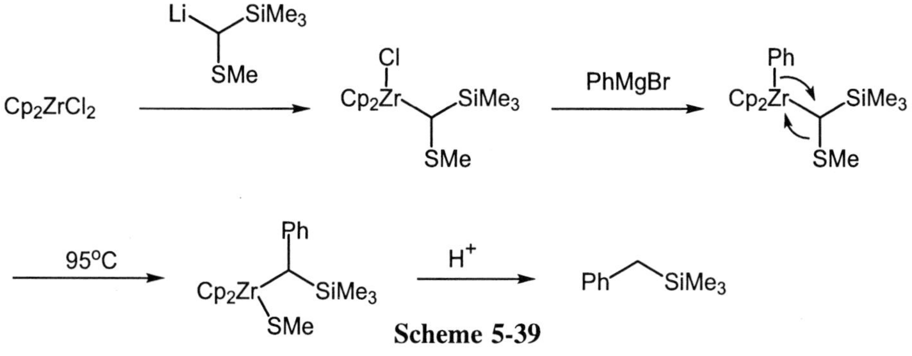

Scheme 5-39

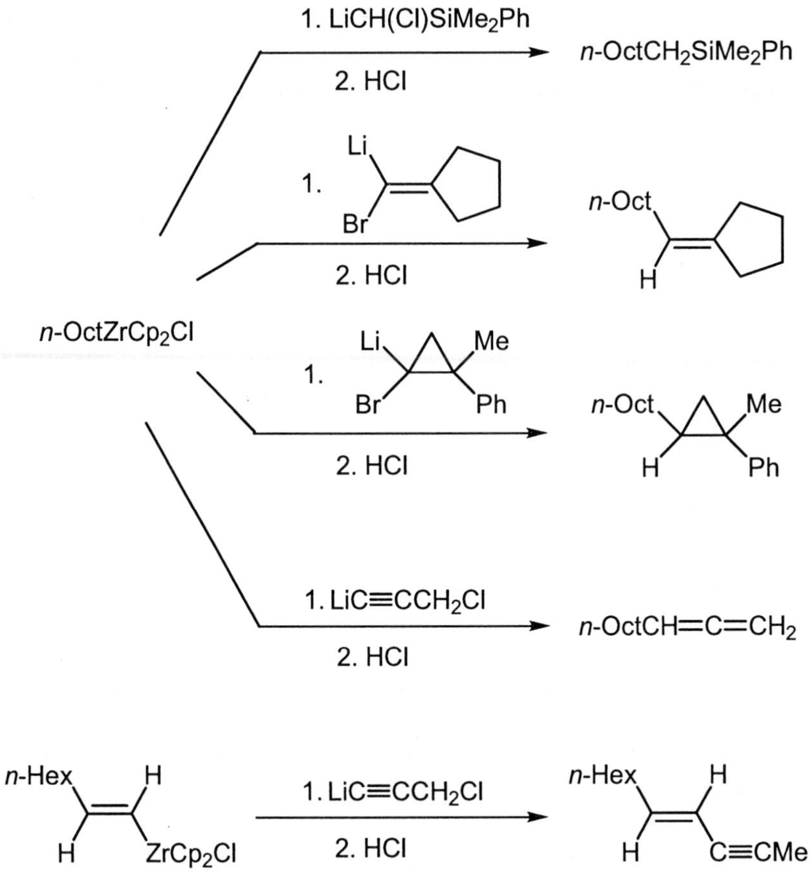

Scheme 5-40

5.3.4 Stoichiometric Carbozirconation of Monoorganylzirconocenes

Carbozirconation, *i.e.*, addition of the C–Zr bond to alkenes and alkynes, has been implicated in a number of Zr-catalyzed carbometallation reactions, most notably the Ziegler–Natta olefin polymerization [44] and the Zr-catalyzed carboalumination of alkynes [49, 50] as well as stoichiometric and catalytic carbometallation reactions involving cyclic diorganylzirconocenes, such as those shown in Scheme 5-30. And yet, clear-cut examples of the stoichiometric carbozirconation of monoorganylzirconocenes are rather rare. The reaction of *n*-PrC≡CAlMe$_2$ with MeZrCp$_2$Cl may represent one of the earliest examples [146] (Scheme 5-41). Another significant example is the methylzirconation with a cationic reagent, MeZr$^+$Cp$_2$(THF)$^-$ BPh$_4$ [138] (Scheme 5-42). In both of these examples, cationic methylzirconocene species may be implicated as reactive carbometallating agents. Without such activation, monoorganylzirconocenes other than allyl and related derivatives may not be capable of undergoing facile carbometallation. Either catalysis with metal compounds and other Lewis acids or stoichiometric transmetallation would be required, as discussed later.

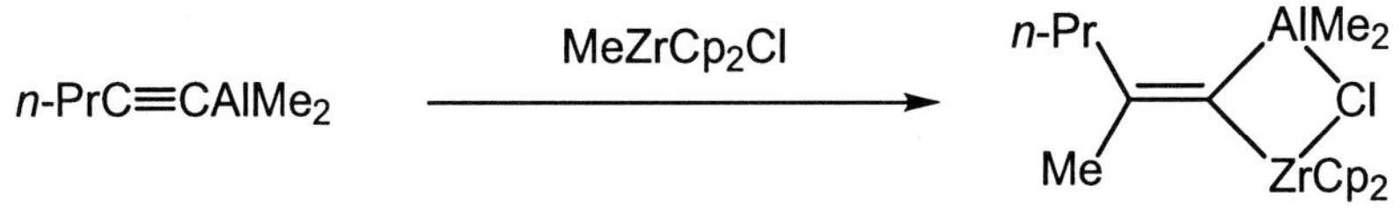

Scheme 5-41

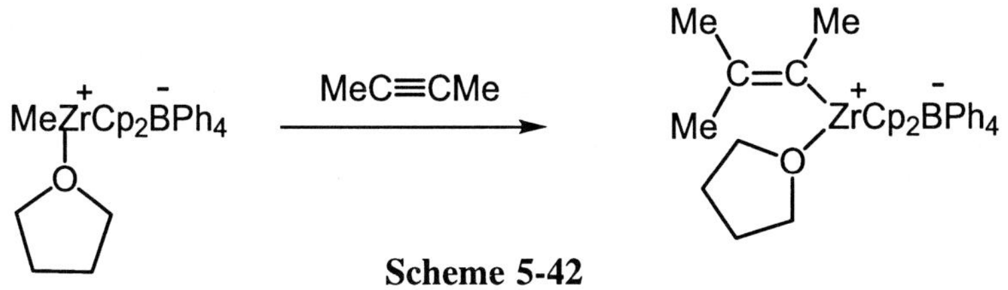

Scheme 5-42

5.3.5 Reactions of Monoorganylzirconocenes *via* Transmetallation

One generally applicable and potentially powerful notion for expanding the scope of any organometallic chemistry is to consider various transmetallation processes. In Sect. 5.2.1, conversion of various organometals into organozirconiums is discussed. In this section, the reverse transmetallation processes, *i.e.*, Zr-to-other metals, are discussed. As discussed earlier (Sect. 5.2.1) in the opposite context, conversion of RZrCp$_2$Cl into RML$_n$ *via* transmetallation with XML$_n$, where X is Cl, Br, or an O group, should be favored in cases where Zr, or more specifically ZrCp$_2$Cl, is more electropositive than ML$_n$. In view of the electronegativity index of 1.2–1.4 for Zr [65], transfer of a carbon group from Zr to a variety of metals including B (2.0), Si (1.8), Sn (1.7–1.8), Hg (1.4–1.9), Cu (1.75–1.9), Ni (1.75–1.8), and Pd (1.35–2.2), whose electronegativity indexes are shown in parenthesis, is expected to be favorable. In dealing with transmetallation, however, the following points should be kept in mind. First, the direction of transmetallation may depend on both the number and nature of ligands in cases where the electronegativity indexes of the two participating metals are similar. Second, the products of transmetallation may in some cases be better represented as 'ate' complexes containing both metals. Third, in cases where the organometallic reagents are generated and consumed *in situ*, the required transmetallation needs not to be thermodynamically favorable so long as the overall transformation is favorable. There are the following three scenarios for the reactions of monoorganylzirconocenes *via* transmetallation: (1) stoichiometric conversion into other organometals, (2) reactions of monoorganylzirconocenes catalyzed by other metal compounds, and (3) reactions of other organometals catalyzed by zirconocene derivatives which involve transient formation of monoorganylzirconocenes.

5.3.5.1 Stoichiometric Conversion of Monoorganylzirconocenes into Other Organometals

Treatment of $RZrCp_2Cl$ with metal halides and alkoxides containing Al [73, 89], B [147–151], Hg [152], and Sn [153, 154] has led to the formation of organometals containing these metals as discrete and Zr-free species (Scheme 5-43). Once these compounds are formed, their subsequent reactions have little or nothing to do with Zr. It nonetheless provides a useful way of expanding the synthetic value of $RZrCp_2Cl$. Conversion of alkenylzirconocene chlorides with $AlCl_3$ produces the corresponding alkenyldichloroalanes which react readily with acyl halides [73]. The parent alkenylzirconocene chlorides do not readily react with acyl chlorides to give α,β-unsaturated ketones. Even dialkylaluminum chlorides react readily with alkenylzirconocene chlorides to produce alkenylalanes in high yields [89]. On the other hand, the use of trialkylalanes lead to an equilibrium reaction [72]. With alkenyltrialkylaluminates, clean transfer of the alkenyl group from Al to Zr has been observed [74].

X = Cl (24%), OEt (85%)

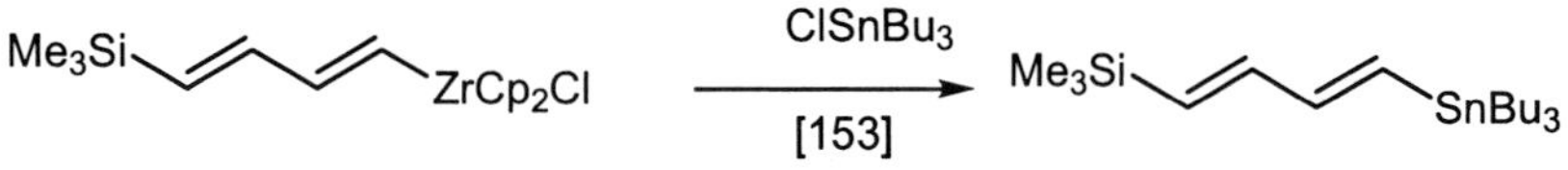

Me₃Si $\diagdown$/$\diagup$/$\diagdown$ ZrCp₂Cl $\xrightarrow[\text{[153]}]{\text{ClSnBu}_3}$ Me₃Si $\diagdown$/$\diagup$/$\diagdown$ SnBu₃

Scheme 5-43

In many other cases, the formation of new organometals and their structure have not been fully established. It is even likely that many of them are bimetallic 'ate' complexes. Despite such ambiguities, the synthetic usefulness of the stoichiometric transmetallation involving Cu in conjugate addition [155] (Scheme 5-44) and Zn in aldehyde addition [156–158] (Scheme 5-45) is unquestionable. More recently, both of these reactions have been shown to be catalytic in Cu [159, 160] (Scheme 5-46) and Zn [161], respectively (Scheme 5-47).

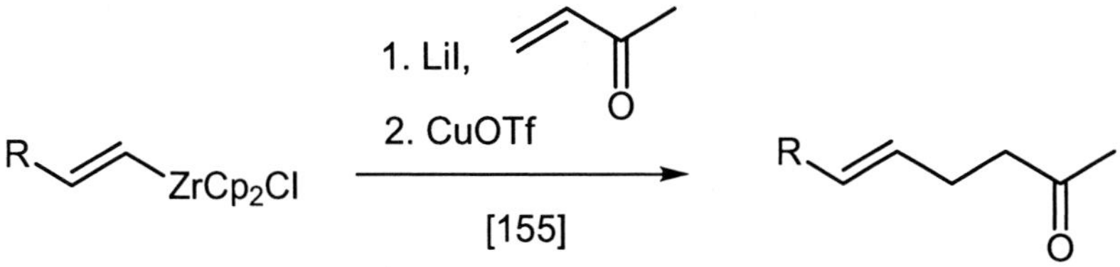

R $\diagdown$/$\diagup$ ZrCp₂Cl $\xrightarrow[\text{[155]}]{\text{1. LiI,} \quad \text{2. CuOTf}}$ R $\diagdown$/$\diagup$/$\diagdown$/$\diagup$ O

Scheme 5-44

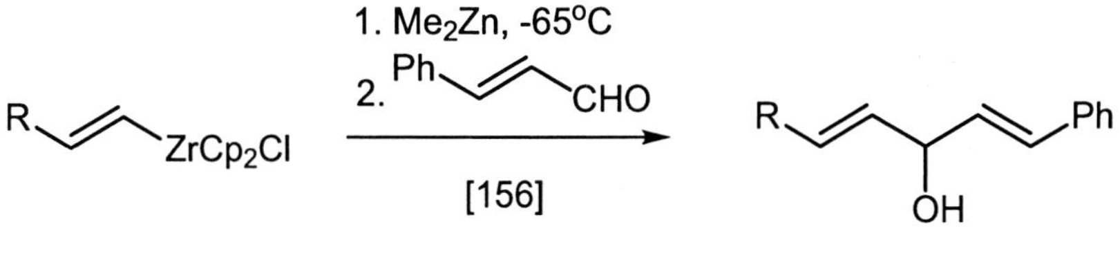

R $\diagdown$/$\diagup$ ZrCp₂Cl $\xrightarrow[\text{[156]}]{\text{1. Me}_2\text{Zn, -65°C} \quad \text{2. Ph} \diagdown\diagup \text{CHO}}$ R $\diagdown$/$\diagup$/$\diagdown$ Ph, OH

Scheme 5-45

R¹ $\diagdown$/$\diagup$ ZrCp₂Cl $\xrightarrow[\text{[159, 160]}]{\text{cat. Li}_2\text{CuMe}_2\text{(CN), Et}_3\text{SiO}}$

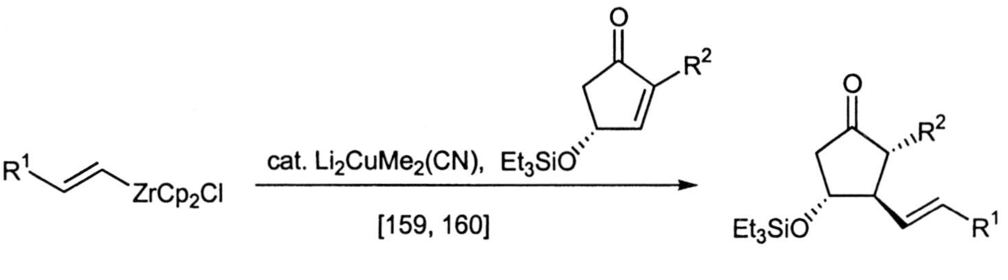

Scheme 5-46

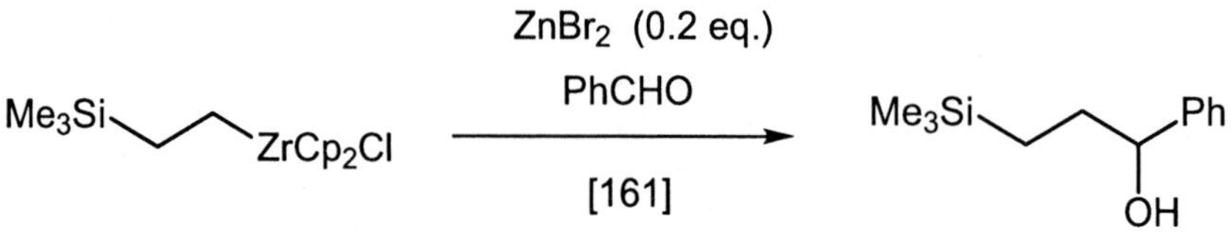

Scheme 5-47

5.3.5.2 Reactions of Monoorganylzirconocenes Catalyzed by Metal Complexes

In 1977 alkenylzirconocene chlorides were found to react with aryl iodides and bromides in the presence of a catalytic amount of a Ni–phosphine complex [162], which appears to represent the first example of the reactions of organylzirconocenes catalyzed by other metals. Catalysis by Pd–phosphine complexes was shown to be more general and applicable to alkenyl–alkenyl, alkenyl–alkynyl, and alkenyl–allyl coupling as well [163, 164] (Scheme 5-48). It was also found that, in cases where alkenylzirconocene chlorides are sterically demanding, addition of ZnX_2 (X = Cl or Br) can significantly accelerate the desired cross coupling. Dou-

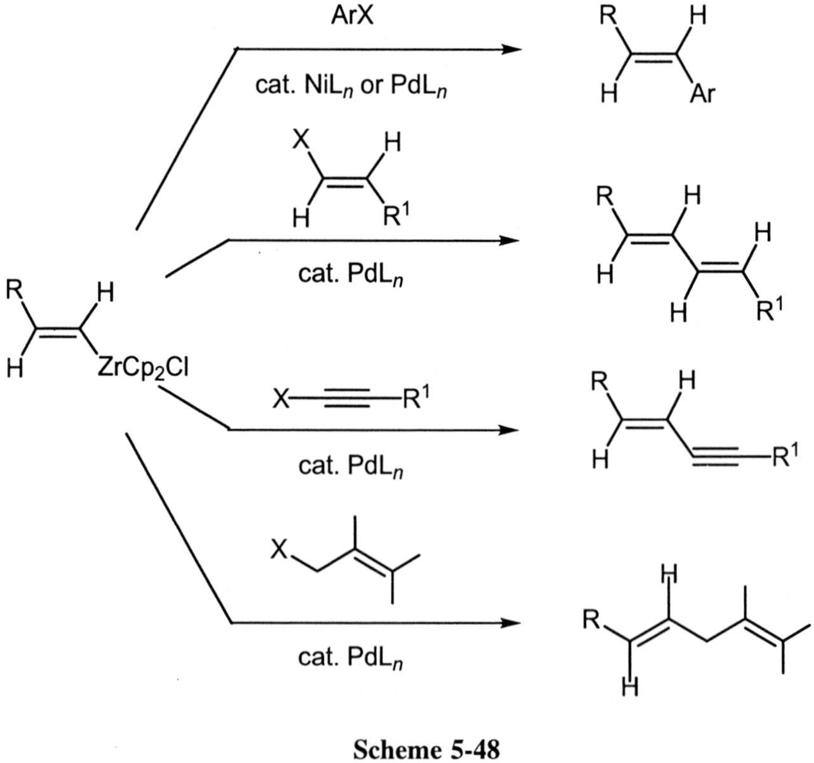

Scheme 5-48

ble transmetallation, *i.e.,* Zr→Zn→Pd or Ni, was suggested as a plausible path (Scheme 5-49). The applicability of the Ni catalysis was immediately extended to conjugate additions, for which a one-electron transfer mechanism was proposed [165–167] (Scheme 5-50). Recent examples of catalysis by Cu [159, 160, 168] and Zn [161] salts (Schemes5- 46 and 47) have further expanded the scope of the metal-catalyzed reactions of organylzirconocenes.

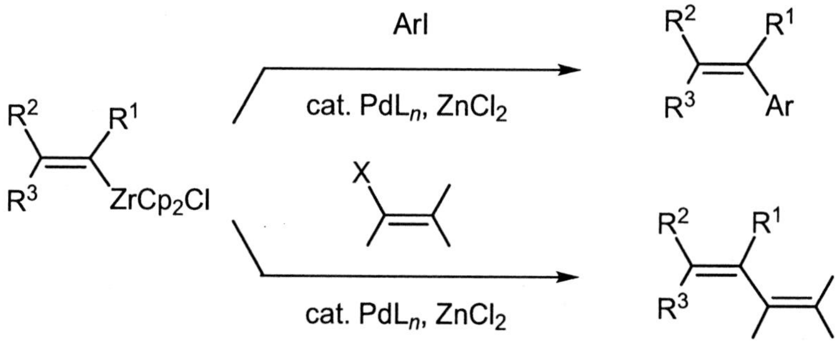

Scheme 5-49

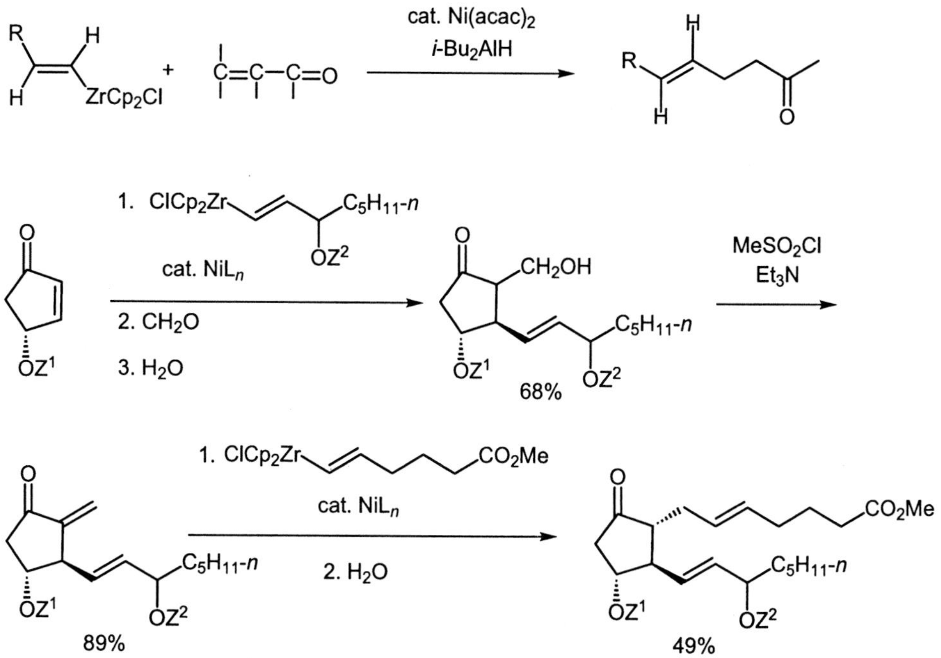

$Z^1 = CH_2Ph$ $Z^2 = SiMe_2Bu\text{-}t$

Scheme 5-50

It is noteworthy that, despite the intrinsically low nucleophilicity of monoorganylzirconocenes towards carbon nucleophiles, all of the polar carbon–carbon bond formation processes shown in Scheme 5-27 are now practically useful, and stoichiometric or catalytic transmetallation as well as activation through ionization are largely responsible for the versatility of organylzirconocenes as 'nucleophilic' reagents.

5.3.6 Organometallic Reactions Catalyzed by Zirconocene Derivatives

In the preceding section, the reactions of the stoichiometric quantity of organylzirconocenes in the presence of metal catalysts are discussed. Conversely, various reactions involving the stoichiometric use of metal compounds have been shown to be catalyzed by zirconocenes. In most cases, Cp_2ZrCl_2 and its derivatives serve as catalysts or catalyst precursors. Most of the widely used main group metals including Al, B, Mg, Si, and Zn have participated in such reactions, and most of these reactions appear to involve the intermediacy of organylzirconocenes. Some of these reactions that are thought to proceed *via* monoorganylzirconocene halides are discussed in this section.

5.3.6.1 Zirconium-Catalyzed Carboalumination and Carbozincation of Alkynes

Aside from the Ziegler–Natta-type alkene polymerization catalyzed by zirconocene derivatives [44], most notably the Kaminsky protocol [48], the methylalumination of alkynes by Me_3Al and a catalytic amount of Cp_2ZrCl_2 [49, 50] (Scheme 5-51) appears to represent the first example of the organic synthetic reactions catalyzed by a zirconocene derivative. The reaction is generally high yielding, essentially 100% *syn*-stereoselective, and about 95% regioselective for methylalumination of terminal alkynes. Although some related reactions using Me_3Al-Cp_2TiCl_2 [169] and R_2Zn-Cp_2ZrI_2 [170] were also discovered, they are of much limited synthetic utility. Various heterofunctionalities, such as halogens, alcohols, and amines, can be accommodated [171], but in some cases, they lead to side reactions including synthetically useful cyclobutenation [172–176] and *anti*-carbometallation [177]. Once alkenylalanes are generated, they undergo essentially all of the known reactions of organoalanes and organoalanates, the latter of which can be readily generated by treating the former with alkyllithiums [77, 178]. It suffices to point out here that various carbon–heteroatom bond formation reactions (Scheme 5-52), conventional polar reactions for carbon–carbon bond formation (Scheme 5-53), and the Ni- or Pd-catalyzed cross-coupling reactions (Scheme 5-54) proceed satisfactorily in many cases. Collectively, this new methodology has provided the most selective, predictable, and hence reliable route to trisubstituted alkenes which represent (*E*)-

trisubstituted isoprene units of a wide variety of terpenoids, and many complex natural products of terpenoid origin, such as those shown in Scheme 5-55, have been synthesized using this reaction in some critical steps. Allylalanes and benzylalanes [179] undergo similar carbometallation reactions.

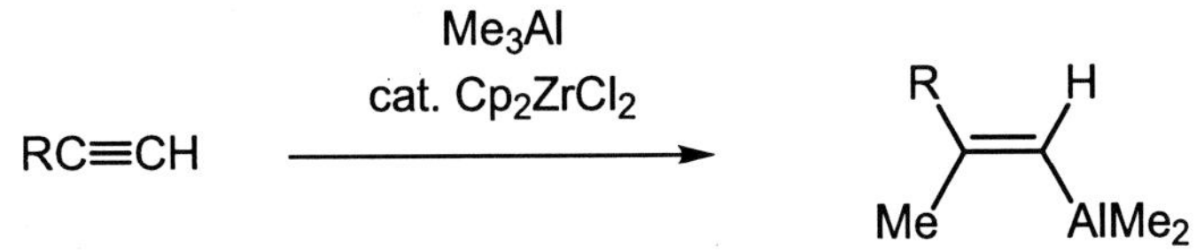

Scheme 5-51

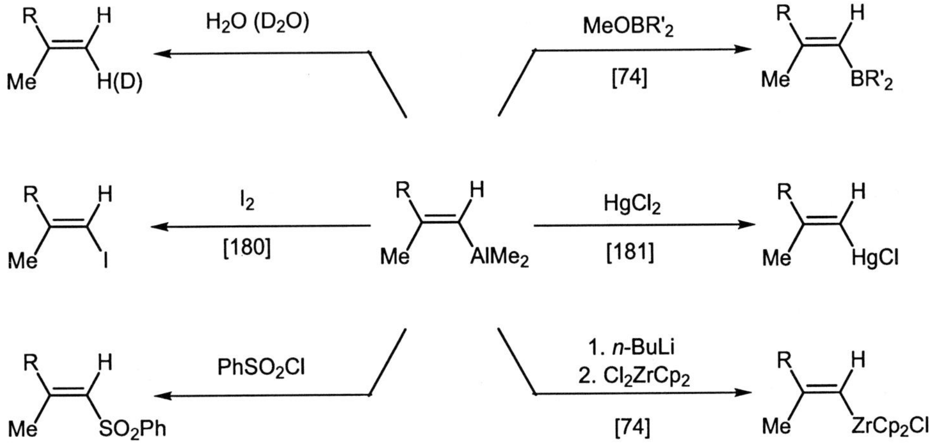

Scheme 5-52

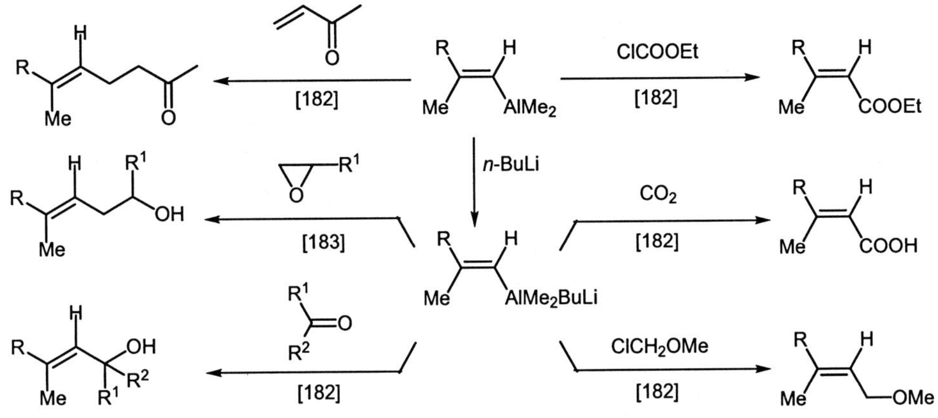

Scheme 5-53

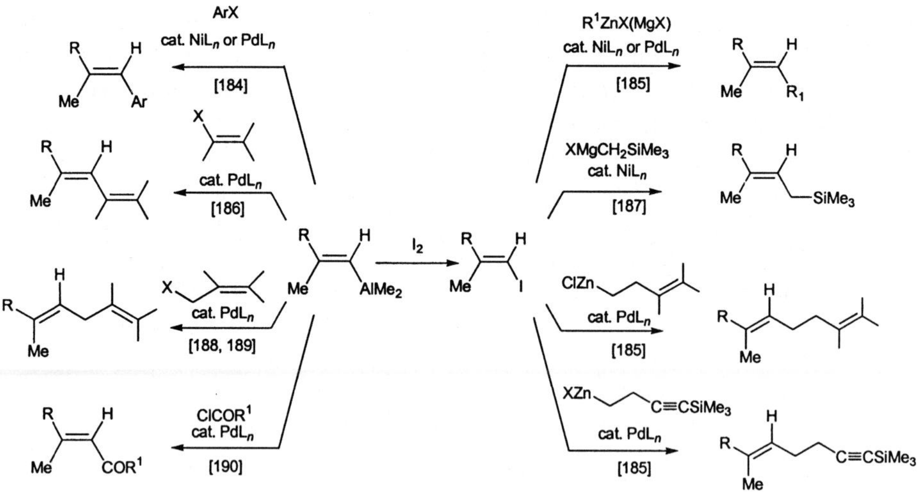

Scheme 5-54

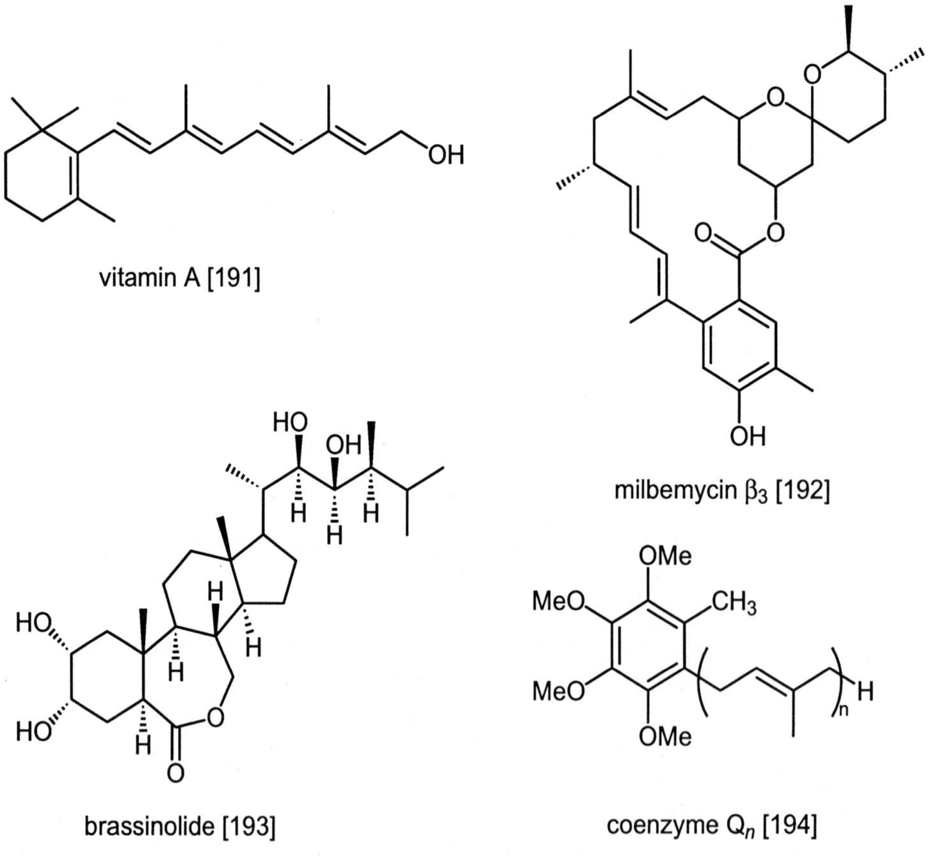

vitamin A [191]

milbemycin β₃ [192]

brassinolide [193]

coenzyme Qₙ [194]

Scheme 5-55

The reaction of Me$_3$Al and Cp$_2$ZrCl$_2$ unquestionably generates an equilibrium quantity of MeZrCp$_2$Cl but not Me$_2$ZrCp$_2$ under the methylalumination conditions [50, 72]. Taken together with the stoichiometric methylzirconation shown in Scheme 5-41, it appeared plausible to propose a methylzirconation mechanism in which a cationic MeZr$^+$Cp$_2$ derivative is the active reagent (Scheme 5-56). And yet, a mixture of Me$_2$AlCl and Cp$_2$ZrCl$_2$, which does not undergo Me–Cl exchange detectable by NMR spectroscopy, is nearly as effective as Me$_3$Al-Cp$_2$ZrCl$_2$. In fact, the use of dialkylchloroalanes (R$_2$AlCl) in place of R$_3$Al is the key to achieving straightforward noncyclic alkylalumination with ethyl- and higher alkyl-containing alanes containing β-hydrogens. For these cases, direct C–Al bond addition promoted by Zr, as opposed to C–Zr bond addition promoted by Al, appears to be more plausible [50, 72] (Scheme 5-56). Despite these ambiguities, however, it appears certain that these reactions proceed *via straightforward, concerted carbon-metal bond addition* to alkynes involving some activated 'cationic' Zr or Al species. These straightforward addition processes should be distinguished from and contrasted with mechanistically more convoluted and complex cyclic carbozirconation processes discussed later.

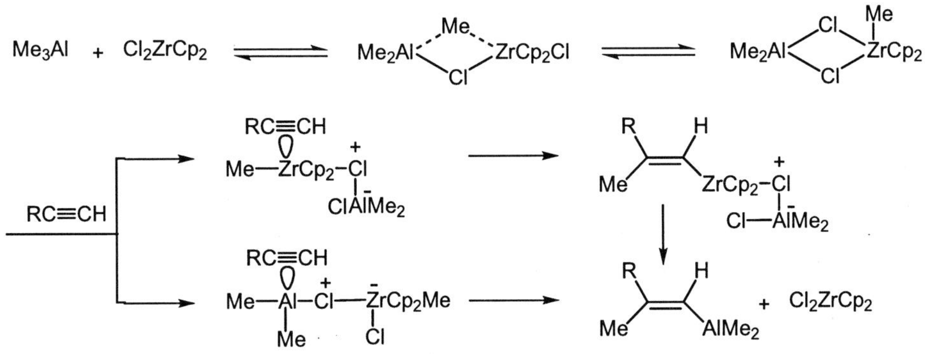

Scheme 5-56

5.3.6.2 Zirconium-Catalyzed Hydroalumination and Hydroboration of Alkenes and Alkynes

Although a mixture of *i*-Bu$_2$AlH and Cp$_2$ZrCl$_2$ does undergo hydrozirconation, the reaction is only stoichiometric in Zr [89]. On the other hand, the *i*-Bu$_3$Al-Cp$_2$ZrCl$_2$ reagent system undergoes a Zr-catalyzed hydroalumination of alkenes [89] and alkynes [90], as discussed in Sect. 5.2.2. These reactions clearly involve a version of hydrozirconation and Zr→Al transmetallation to produce net hydroalumination products, and the hydrozirconation step is clearly promoted by Al, which appears to be also responsible for undesirable side reactions, such as regioisomerization (Scheme 5-10). Although the mechanism of the critical Al-promoted hydrozirconation step remains to be further clarified, the overall transformation proceeds as shown in Scheme 5-57.

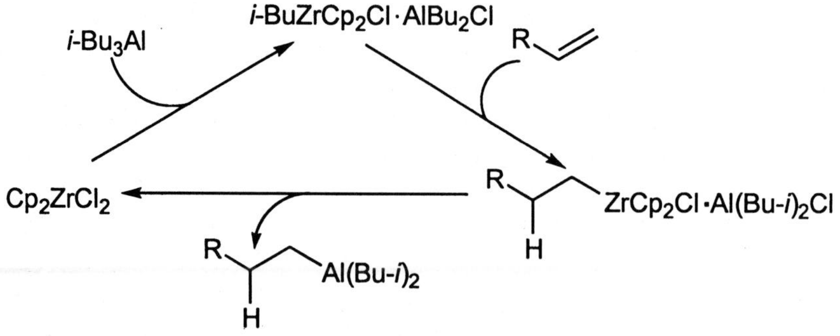

Scheme 5-57

A Zr-catalyzed hydroboration of alkynes reported recently [161] further extends the scope of the Zr-catalyzed hydrometallation (Scheme 5-58).

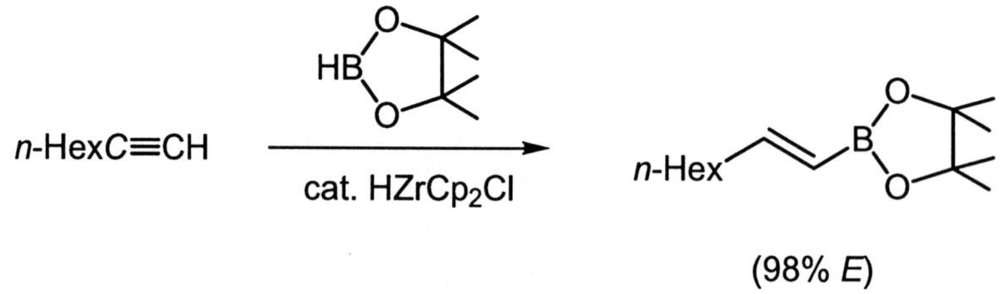

(98% *E*)

Scheme 5-58

5.3.6.3 Zirconium-Catalyzed Enantioselective Carboalumination of Alkenes

Early attempts at methylalumination of 1-alkenes with Me$_3$Al and Cp$_2$ZrCl$_2$ were largely unsuccessful. Reinvestigation of the reaction of 1-octene with Me$_3$Al (1 equiv.) and 8 mol % of Cp$_2$ZrCl$_2$ has revealed that the two major products obtained after protonolysis are 2-(*n*-hexyl)-1-decene (59%) and 2-methyl-1-octene (18%), the yield of the desired 2-methyloctane being <2% [195]. Evidently, the desired methylmetallation takes place, but the isoalkylzirconocene product in conjunction with an alane readily hydroaluminate alkenes, as exemplified by the reaction shown in Scheme 5-57. In view of these results, it is rather surprising that the reaction of 1-alkenes with 1 equiv. of Me$_3$Al and 8 mol % of dichlorobis(1-neomenthylindenyl)zirconium [196] in (CH$_2$Cl)$_2$ has produced in 70–90% yields the desired methyl-aluminated products readily oxidizable with O$_2$ to give the corresponding alcohols. Moreover, the reaction has proceeded in 65–85% ee (Scheme 5-59). Evidently, the use of certain bulky and/or chiral Cp derivatives as ligands must slow down β-dehydrometallation relative to the desired carbometallation. The selective formation of the *R* isomer observed with a wide range of 1-alkenes and other experimental results are consistent with a mechanism represented by Scheme 5-60, in which a C_2-symmetric, cationic methylzirconium derivative containing two 1-neomenthylindenyl groups interacts with the *re* face of 1-alkenes [195]. This mechanism clearly favors direct addition of the C–Zr bond to alkenes, which is promoted by Al. However, it should be emphasized that seemingly minor variations in certain reaction parameters can lead to drastic mechanistic changes and that any mechanistic extrapolation can lead to false conclusions.

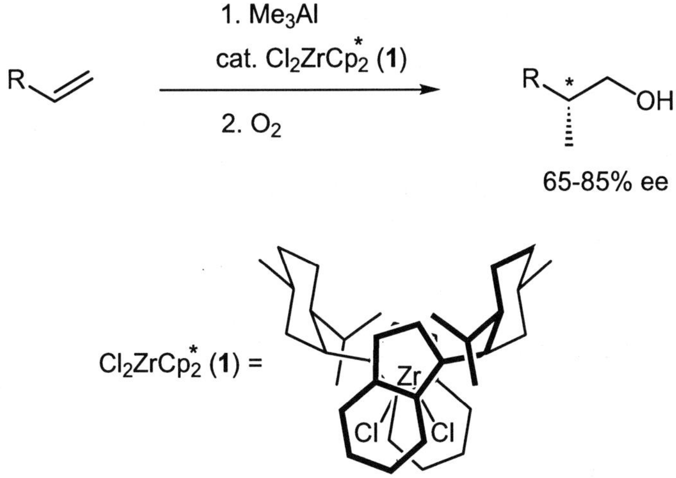

Scheme 5-59

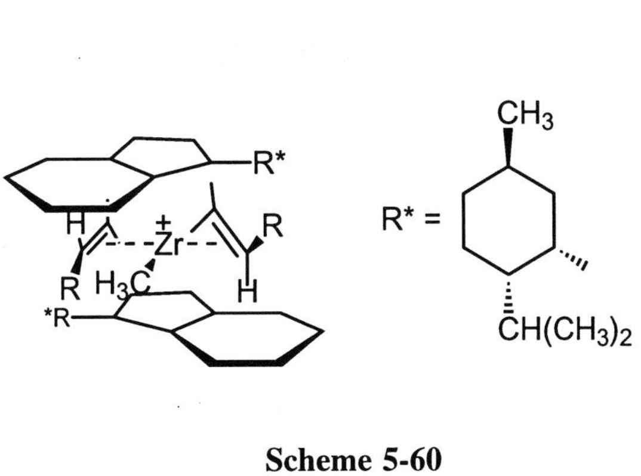

Scheme 5-60

The reaction of trialkylalanes containing Et and high *n*-alkyl groups with 1-alkenes in the presence of Cp$_2$ZrCl$_2$ in nonpolar solvents is known to produce aluminacyclopentanes rather than the straightforward alkylalumination products [197] (Scheme 5-61). In principle, this cyclic carboalumination can proceed enantioselectively. However, the ee figures have been limited to 30–35% [198]. As suspected, a complete mechanistic change was observed, when nonpolar solvents, *e.g.*, hexanes, were replaced by polar solvents, *e.g.*, CH$_2$Cl$_2$, (CH$_2$Cl)$_2$, and CH$_3$CHCl$_2$. Thus, the reaction of 1-decene with Et$_3$Al in the presence of Cp$_2$ZrCl$_2$ in (CH$_2$Cl)$_2$ was a totally acyclic process with no indication of the formation of aluminacyclopentanes, even though competitive dehydrometallation and hydrometallation were significant side reactions. Here again, the use of bulky and chiral 1-neomenthylindene in place of cyclopentadiene not only suppressed dehydrometallation but also led to the formation of the desired alkylmetallation products in 60–90% yields in 90–95% ee [198] (Scheme 5-61). Some representative results with Me$_3$Al and Et$_3$Al are summarized in Table 5-6.

Some other Zr-catalyzed enantioselective alkene carbometallation reactions have also been developed [199, 200]. These reactions must involve cyclic carbometallation process with Cp$_2$Zr^{II} species, as discussed later.

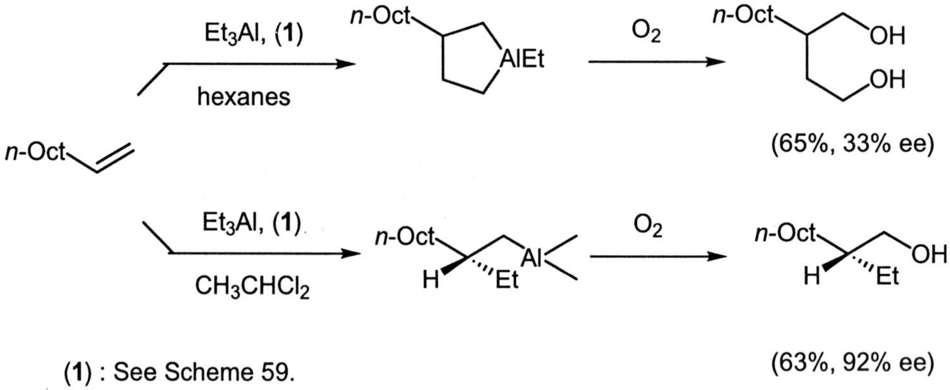

Scheme 5-61

Table 5-6. Zirconium-Catalyzed Enantioselective Alkylalumination of Monosubstituted Alkenes[a]

Alkene	Methylalumination		Ethylalumination	
	Yield (%)	ee (%)	Yield (%)	ee (%)
nHexCH=CH$_2$	88	72	&	&
n-BuCH=CH$_2$	&	&	74	93
i-BuCH=CH$_2$	92	74	77	90
c-HexCH=CH$_2$	80	65	low	&
t-BuCH=CH$_2$	low	&	&	&
PhCH$_2$CH=CH$_2$	77	70	69	93
HO(CH$_2$)$_4$CH=CH$_2$	79	75	88	90
Et$_2$N(CH$_2$)$_3$CH=CH$_2$	68	71	56	95
Me$_2$Si(CH$_2$CH=CH$_2$)$_2$[b]	81	74	66	96

[a] The reactions were run with 8 mol % of (**1**) in Scheme 59 in (CH$_2$Cl)$_2$, CH$_3$CHCl$_2$, or CH$_2$Cl$_2$, and the yield figures are for the oxidized alcohols which were uniformly (*R*)-isomers.

[b] The structure of the oxidation product was: Me$_2$Si〈〉''''''OH

R(Me or Et)

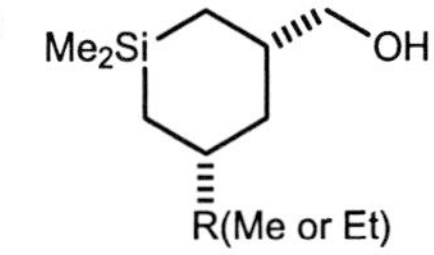

5.4   Reactions Involving π-Complexes of Zirconocenes or Three-Membered Zirconacycles

The chemistry of Cp$_2$Zr^{II} compounds had not been extensively developed until recently. Thus, for example, a comprehensive review of the organometallic chemistry of Zr published in 1974 [1] allocated less than a page on this subject. Pioneering investigations in the 1970s and early 1980s including noteworthy contributions by Bercaw [87, 201, 202] and Erker [17, 28, 51, 52, 64, 124, 141, 203–209] led to some fundamentally significant findings. However, it was not until the 1985–87 period that the significance of the Cp$_2$Zr^{II} chemistry as a versatile new tool for the synthesis of both simple and complex organic compounds was recognized and exploited. Development of the bicyclization of enynes promoted by 'Cp$_2$Zr(II)' to give zirconabicycles convertible to bicyclic enones *via* carbonylation in 1985 [55] was a breakthrough inducing numerous other related studies discussed in this section. Recognition that dialkylzirconocenes, in particular *n*-Bu$_2$ZrCp$_2$, can serve as convenient precursors to 'Cp$_2$Zr(II)' was a practical, experimental breakthrough [56], since the presence of π-compounds, such as alkenes, alkynes, carbonyl compounds, and nitriles, provides a simple route to their π-complexes with Cp$_2$Zr(II). Alternatively, treatment of alkyl- and alkenylzirconocene halides with an appropriate organometal, *e.g.*, MeLi, and treatment of MeZrCp$_2$Cl with various organolithiums and Grignard reagents provide convenient routes to alkyne–, alkene– and benzyne–ZrCp$_2$ complexes without the use of alkynes, alkenes, and benzynes as starting compounds [58, 59]. These modifications of earlier related processes reported since 1986 also represented an experimental breakthrough. The synthetic significance of the Cp$_2$Zr(II) chemistry mainly developed since 1985 easily rivals that of RZrCp$_2$Cl discussed in Sect. 5.3.

5.4.1 Formation of π-Complexes of Zirconocenes

Alkenes including dienes, alkynes including benzynes, carbonyl compounds, and imines are the four most commonly used and synthetically important classes of π-compounds. Their π-complexes with ZrCp$_2$ are accessible by various methods, of which those shown in Table 5-7 are significant.

Table 7. Representative Methods for the Formation of π-Complexes of Zirconocene

Method	Pattern[a]	Applicability[b]			
		Alkene	Alkyne	Carbonyl	Imine
β-H Abstraction	$Cp_2Zr \cdots \rightarrow Cp_2Zr$	+	+	c	+
Complexation	"Cp$_2$Zr" $\rightarrow$ Cp$_2$Zr (Cp$_2$Zr)	+	+	+	+
Ring Contraction	$Cp_2Zr \rightarrow Cp_2Zr$	+	+		
Transmetallation	Cp$_2$ZrCl$_2$ $\rightarrow$ Cp$_2$Zr	d		c	
Migratory Insertion	$Cp_2Zr \rightarrow Cp_2Zr$	c	e	+	+

[a] X, Y = C, N, and/or O. [b] + indicates that the method is applicable. [c] Not clarified. [d] Known for conjugated dienes. [e] See Scheme

β-Hydrogen abstraction of diorganylzirconocenes and related monoorganylzirconocenes is currently by far the most convenient route to π-complexes of zirconocene. Some representative examples of the Negishi protocol [107, 110]
(Scheme 5-62) [115] (Scheme 5-63) and the Buchwald protocol [108] (Scheme 5-
64) [59] (Scheme 5-65) are shown below.

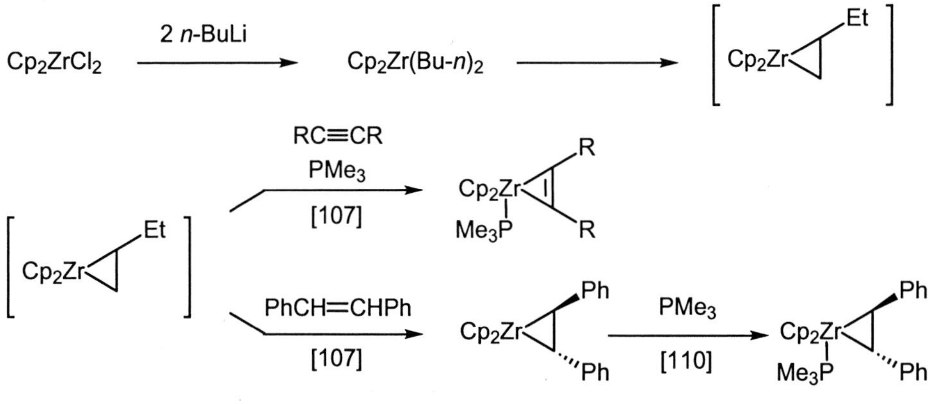

Scheme 5-62

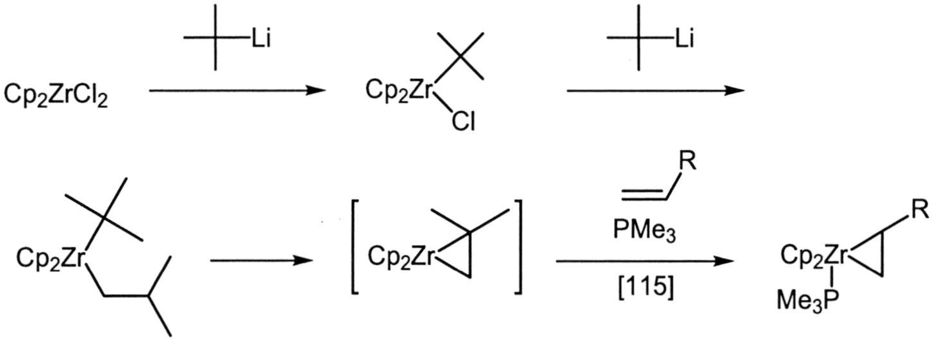

Scheme 5-63

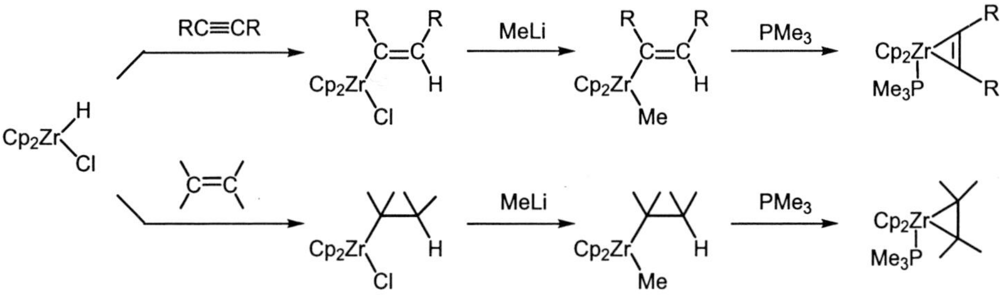

Scheme 5-64

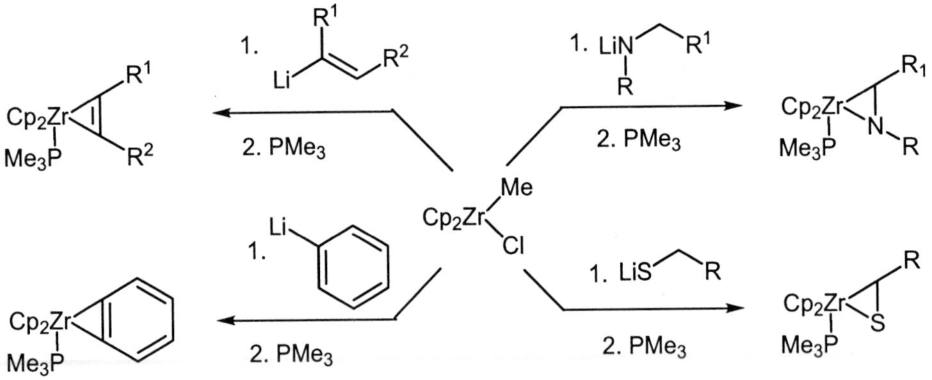

Scheme 5-65

In cases where the starting π-compounds are readily available, the Negishi protocol provides the most convenient route. On the other hand, the Buchwald protocol is applicable even to those cases where π-compounds are exceedingly unstable, *e.g.*, cyclohexyne [59], benzynes [58], and cyclic allenes and cumulenes [210–212]. The two protocols are complementary with each other in many other respects as well. The mechanistic details of these reactions have been reasonably well clarified. For example, $Cp_2Zr(Bu-n)_2$, generated at low temperature by transmetallation, decomposes smoothly at or above room temperature by a nondissociative, unimolecular process. The presence of a base, such as PMe_3, generally has little or no effect on the kinetics. It merely traps an unstable 16e π-complex. A concerted intramolecular mechanism shown in Scheme 5-66 [57, 213, 214] appears to be plausible, although its more elaborate version involving the transient formation of 18e alkene(alkyl)(hydrido)zirconocene and other minor variations may not be rigorously ruled out at this point. As in the β-dehydrozirconation of $RZrCp_2Cl$, β-agostic interaction must be of crucial importance. Interestingly, those dialkylzirconocenes containing a β-CH_3 group, such as Et, *s*-Bu, and *t*-Bu, decompose most readily, followed by those containing a β-CH_2 group, *e.g.*, *n*-Bu, which, in turn, are much more reactive than those containing only β-CH groups, *e.g.*, *i*-Bu_2ZrCp_2 [214]. The observed order of reactivity, *i.e.*, *s*-Bu ≈ *t*-Bu ≥ Et > *n*-Bu > *i*-Bu, has little to do with the overall steric requirements. A very high reactivity of $PhCH_2CH_2$ indicates that some electronic factors are also important. The order presented above does not appear to agree with the relative order of effectiveness of $RZrCp_2Cl$ as a hydrozirconating agent. And yet, the same β-agostic interaction must play a crucial role in both reactions. A plausible explanation of the puzzling discrepancy is that, whereas the hydride-transfer hydrozirconation is an equilibrium reaction, the β-H abstraction reaction of dialkylzirconocenes is essentially irreversible, even though the latter reaction formally involves cleavage and formation of one each of C–H and C–Zr bonds and hence should be essentially thermoneutral. This reaction must be largely entropically driven, and the relative reactivity must be kinetically governed. On the other hand, the reversible hydride-transfer hydrozirconation must be thermodynamically controlled, which makes isoalkyl groups highly effective. The unexpectedly sluggish alkene displacement shown in parentheses must be attributable to the coordinative saturation of chloro(hydrido)alkenezirconocenes.

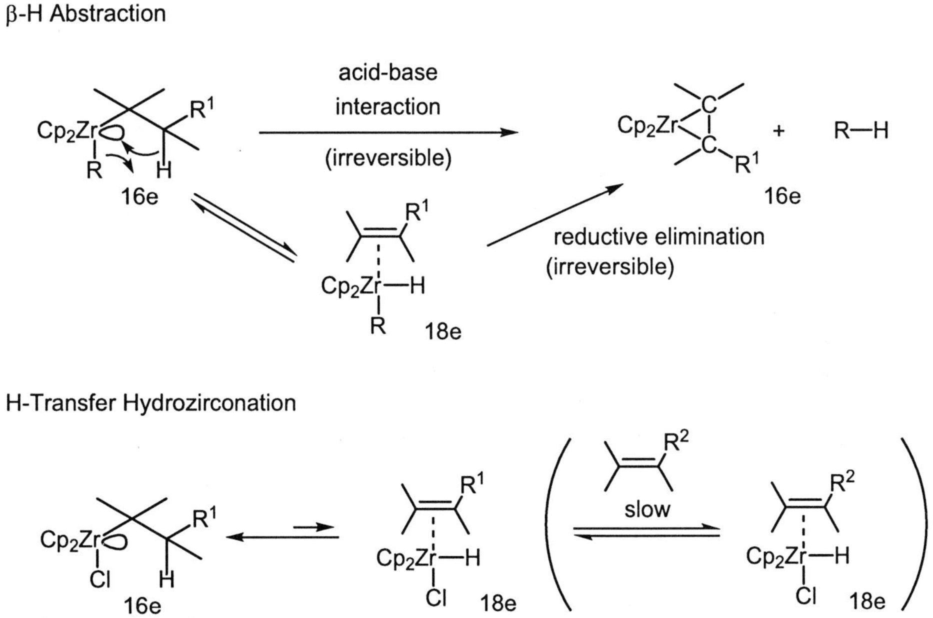

β-H Abstraction

H-Transfer Hydrozirconation

Scheme 5-66

The ring contraction route to π-complexes of zirconocene must involve interaction between the C_β–$C_{\beta'}$ bond and an empty orbital of Zr [35], as discussed later in detail. In this connection, it now is reasonably clear that π-complexation *via* alkene displacement in the Negishi protocol must proceed in most cases by a ring expansion-ring contraction mechanism (Scheme 5-67).

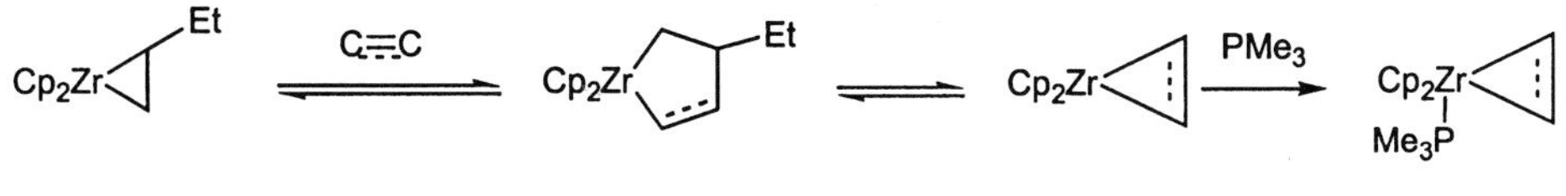

Scheme 5-67

The π-complexation method with 'Cp$_2$Zr(II)' other than alkene–ZrCp$_2$ discussed above tends to be more complex and generally inferior. Zirconocene reagents may be isolated and purified as Cp$_2$Zr(CO)$_2$, Cp$_2$Zr(PMe$_3$)$_2$, and so on. These complexes are 18e species, and their complexation reactions are generally sluggish. The current scope of the transmetallation and migratory insertion routes to π-complexes of zirconocene is essentially limited to those discussed in Sect. 5.2.1. However, they may prove to be much more useful and complementary with the other methods. For example, treatment of Cp_2ZrCl_2 with 3 equiv. of LiC≡CR first generates $LiCp_2Zr(C≡CR)_3$ which then undergo an unprecedented migratory insertion to give alkyne–ZrCp$_2$ complexes, displaying a novel transformation of Cp_2Zr derivatives which also promises to be synthetically useful [215] (Scheme 5-68).

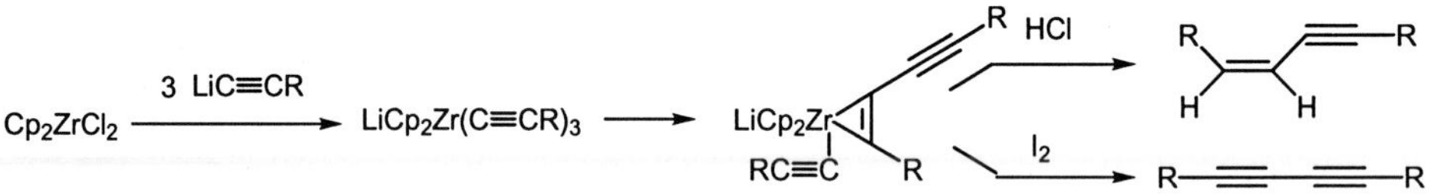

Scheme 5-68

5.4.2 Stoichiometric Reactions of π-Complexes of Zirconocenes (or Three-Membered Zirconacycles)

5.4.2.1 Protonolysis and Halogenolysis

Treatment of π-complexes of zirconocenes with proton acids can be complicated by simultaneous formation of diprotonated compounds and dissociated π-compounds [216]. To achieve dissociation of π-ligands in a cleaner manner, the use of I_2 should be considered [92, 216] (Scheme 5-69). The I_2 treatment method is also applicable to release carbonyl compounds from zirconaoxiranes [216] (Scheme 5-70).

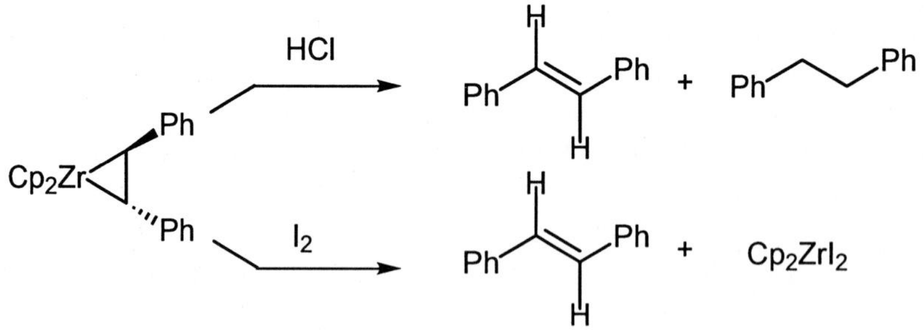

Scheme 5-69

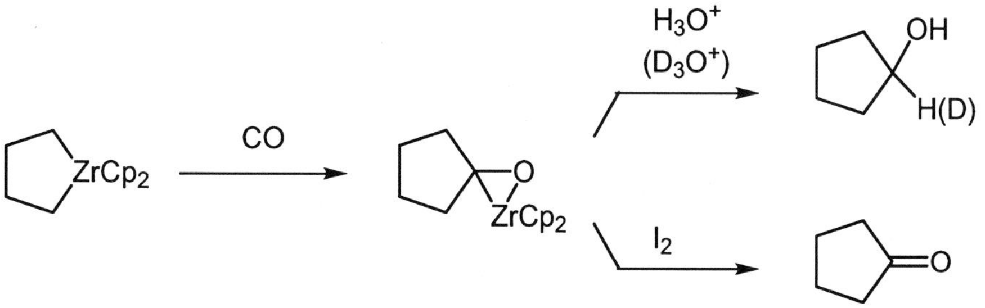

Scheme 5-70

5.4.2.2 Ring Expansion Reactions

The synthetically most attractive transformations of three-membered zirconacycles are their ring expansion reactions producing five-membered zirconacycles. Those involving alkenes and alkynes are formally carbozirconation reactions, and those involving carbonyl compounds, imines, and nitriles are formally polar addition reactions. These two classes of reactions may share some common mechanistic features, but they may also display mutually distinguishing features. For the carbometallation processes of 16e species a concerted four-center addition mechanism may be tentatively proposed (Scheme 5-71). Regardless of precise mechanisms, 16e π-complexes with one empty orbital may be expected to be more reactive in most, if not all, of these reactions than their 18e counterparts. Even when we consider only alkenes, alkynes, carbonyl compounds, and imines, there are four homo-coupling combinations and six cross-coupling combinations. Other structural variations, such as conjugated dienes in place of alkenes, benzynes in place of alkynes, thiocarbonyl compounds in place of carbonyl compounds and nitriles in place of imines, lead to an even larger number of binary combinations for the synthesis of a wide range of organic compounds. Another attractive synthetic strategy is to link two π-functional groups with tethers of various lengths and structures. In the ZrCp$_2$-promoted π-π coupling reactions, 'pair'-selectivity and regioselectivity as well as stereoselectivity in some cases provide challenging synthetic problems. In cases where the two π-compounds can be linked with an appropriate tether, the 'pair'-selectivity and regioselectivity problems may be alleviated. Furthermore, the intramolecular coupling process can be kinetically very favorable. Here again, a large number of binary combinations of π-functional groups are conceivable, but the current scope of the ZrCp$_2$-promoted bicyclization appears to be still largely limited to the three binary combinations of alkenes and alkynes (Scheme 5-72). A notable example of other binary combinations is the ZrCp$_2$-promoted bicyclization of alkenyl and alkynyl hydrazones [217] (Scheme 5-73).

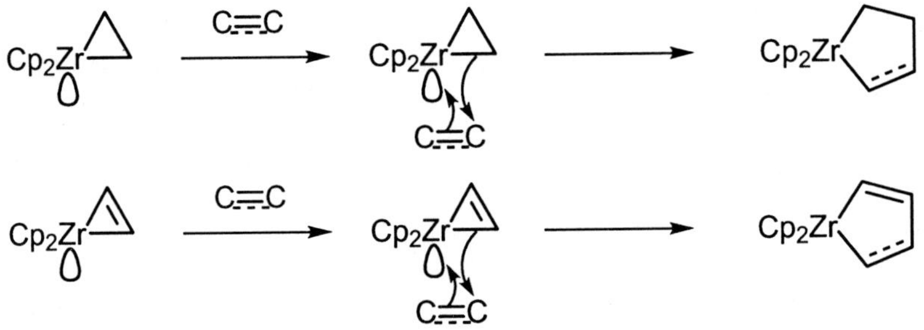

Scheme 5-71

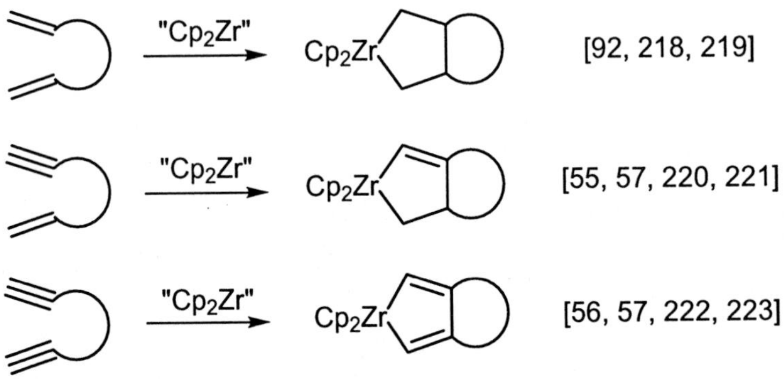

Scheme 5-72

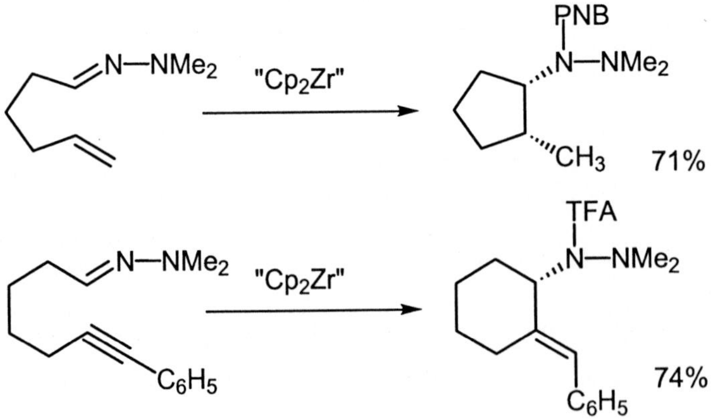

Scheme 5-73

Discussion of all of the known Cp$_2$Zr-promoted coupling reactions is not practical. Only some seminal and representative references of intermolecular coupling reactions are presented in Table 5-8.

Table 5-8. Some Representative References for Intermolecular Coupling Reactions of π-Compounds Promoted by Cp$_2$Zr

Second π-Compound \ First π-Compound	Alkene	Alkyne	Carbonyl compound	Imines
Alkene	[216, 224, 225]	[59, 107, 108, 226]	[227]	[114]
Alkyne	[228, 229]	[58, 59, 107, 108, 230-234]	[122, 235]	[114]
Carbonyl Compounds	[236]	[58, 59, 107, 108]	[202]	[114]
Nitrile	----	[59]	----	[114, 237]

In each of the intermolecular coupling reactions, the 'pair'-selectivity and regioselectivity can be significant issues, as detailed below for the alkenyl–alkenyl, alkenyl–alkynyl, and alkynyl–alkynyl coupling reactions. At least five different types of 'pair'-selectivity patterns have been observed, as summarized in Table 5-9 [35].

Table 5-9. Some Distinctive Features of Various Types of Ring Expansion Reactions of Alkene-Zirconocene Complexes

Type	Displacement of initial alkene	"Pair"-selectivity	Equilibrating processor or not
I	no	very high (~100%)	no
II	no	high	yes
III	yes	homo-coupling	*a*
IV	yes	high	*b*
V	*b*	low	yes

a Not pertinent. *b* Depends on the substrates.

Some representative examples of the five different types of reactions involving alkenes and alkynes are shown in Scheme 5-74. The following common characteristics should also be noted. Firstly, alkyl groups prefer to be β to Zr, presumably to

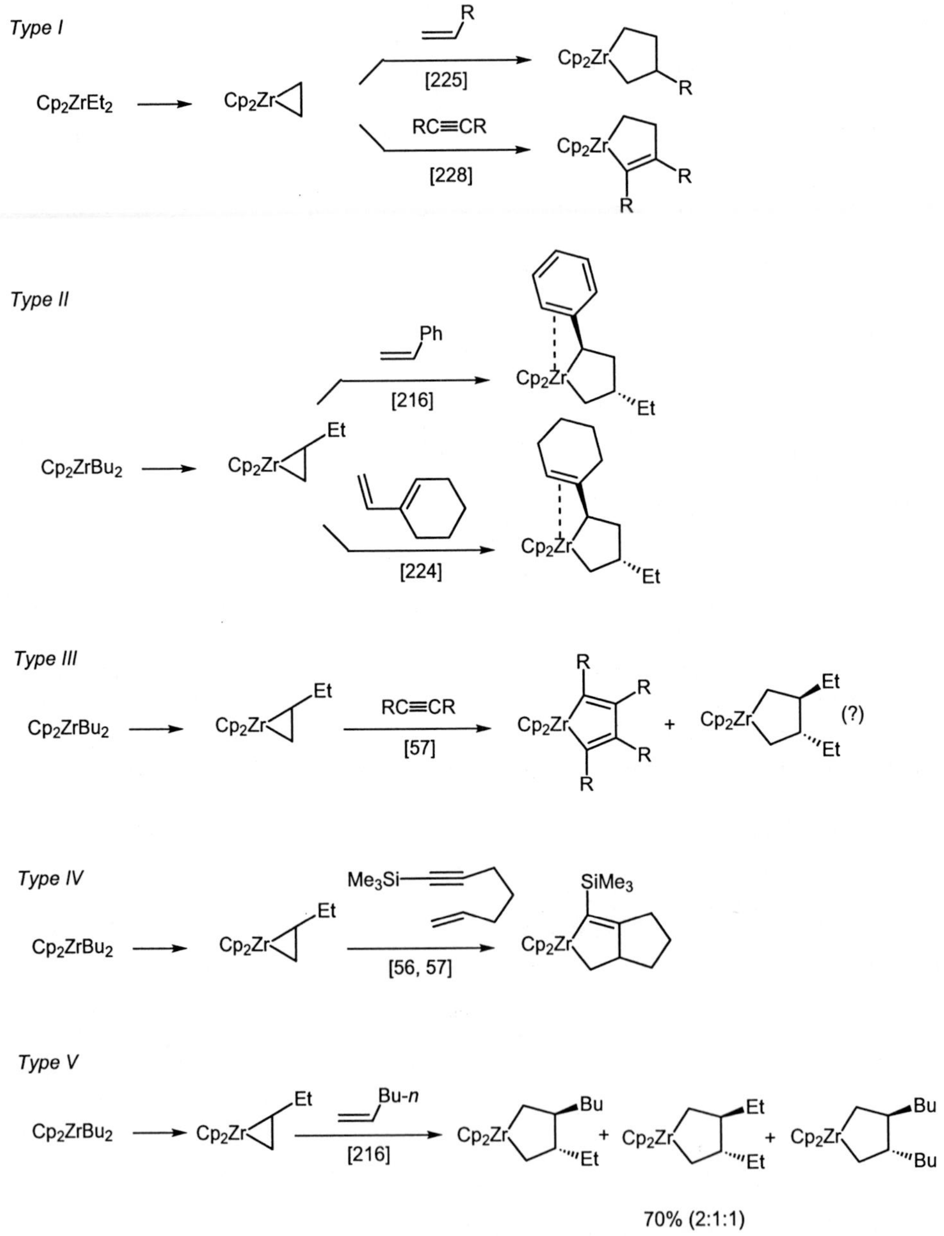

Type I

Type II

Type III

Type IV

Type V

70% (2:1:1)

Scheme 5-74

minimize steric hindrance, while aryl, alkenyl, and alkynyl groups favor the α position, presumably due to favorable benzylic, allylic, and propargylic interaction with Zr. The α/β ratio for aryl is typically 90/10, while that for alkenyl is ≥95%. Secondly, two substituents of the two π-compounds prefer to be *trans* to each other in the zirconacycles, presumably for steric reasons.

The ZrCp$_2$-promoted bicyclization (Scheme 5-72) has proved to be highly useful for the construction of complex bicyclic and polycyclic compounds. It shares some common features with the Co-based Pauson–Khand reaction [238, 239]. However, significant differences presented below should be clearly noted. Firstly, the Pauson–Khand reaction appears to be limited to enyne bicyclization to produce cyclic enones, whereas the Zr-based methodology is applicable to the bicyclization of dienes and diynes as well. Furthermore, the organometallic products can be generated as discrete products which can be further converted to various types of compounds including cyclic enones. So, the Zr-based methodology is more versatile. Secondly, the Pauson–Khand reaction is normally run at high temperatures, whereas the Zr-promoted reactions proceed readily at or below room temperature. Thirdly, while the Pauson–Khand reaction has recently been made catalytic [240], the catalytic version of the Zr-based bicyclization largely remains to be developed. Finally, the enantioselective versions of both of these reactions are also to be developed. Fortunately, both can proceed diastereoselectively, placing most of the substituents on the *exo* side of the bicyclic skeleton [241, 243]. The initial kinetic argument [241] appears to be untenable, and it must be a largely thermodynamic phenomenon. Another striking feature is the high *trans* selectivity in the bicyclization of 1,6-heptadienes [92, 218, 243], despite the fact that 1,7-octadienes under the same conditions give predominantly the *cis*-fused products. The product from 1,7-octadiene was shown to be a kinetic product which can be thermally isomerized to the *trans*-fused isomer [219]. The *trans*-selectivity is readily rationalized on the basis of the stereochemistry observed in the intermolecular alkene–alkene coupling, and it offers a novel and attractive route to *trans*-fused bicyclo[3.3.0]octane derivatives which are otherwise difficult to prepare [243] (Scheme 5-75).

The Zr-promoted enyne bicyclization has been applied to efficient and selective syntheses of natural products, notably phorbol (**1**) [244], pentalenic acid (**2**) [242], iridomyrmecin (**3**) [245], dendrobine (**4**) [246], and 7-*epi*-β-bulnesene (**5**) [247] (Scheme 5-76).

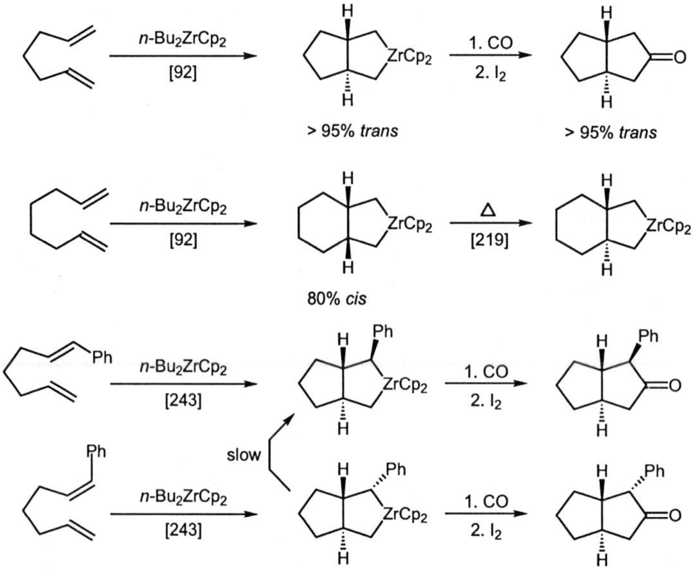

Scheme 5-75

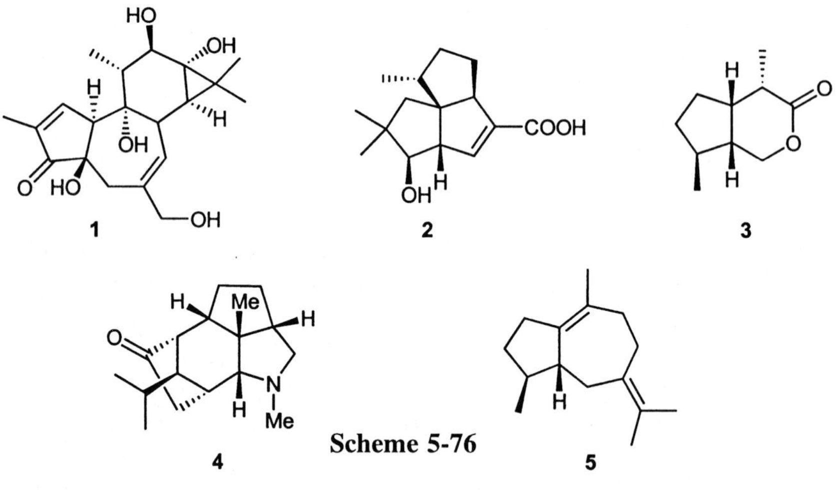

Scheme 5-76

5.4.2.3 Regioisomerization, Stereoisomerization, σ-Bond Metathesis, and Other Stoichiometric Reactions of Three-Membered Zirconacycles

In addition to synthetically useful reactions discussed above, many other reactions of three-membered zirconacycles are conceivable, and some have been observed. In some cases, they represent unwanted side reactions, but some of these reactions may prove to be useful. In any event, good knowledge of them is essential to successful use of the Cp_2Zr^{II} chemistry.

In some diene bicyclization reactions, a multipositional double bond migration to give conjugated diene–ZrCp$_2$ complexes has been observed. A set of representative and contrasting results are shown in Scheme 5-77 [248, 249]. A number of factors including steric requirements of the two alkene groups and tether structure appear to be involved, and the reaction courses are more or less predictable and interpreted in terms of a dynamic scheme shown in Scheme 5-78. Migration of Zr occurs *via* 1,3-shift rather than 1,2-shift.

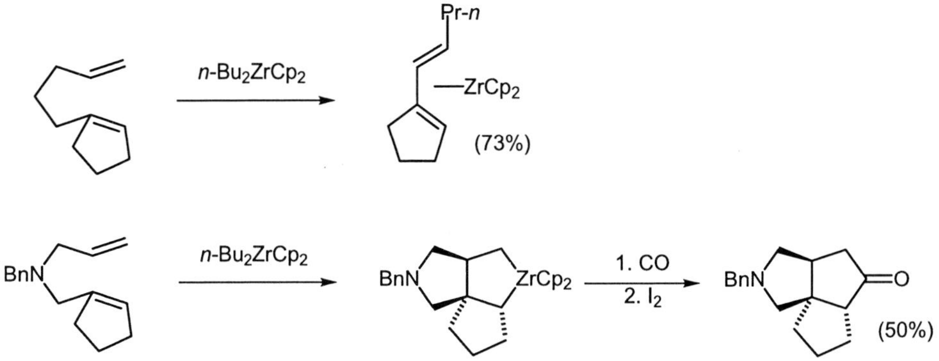

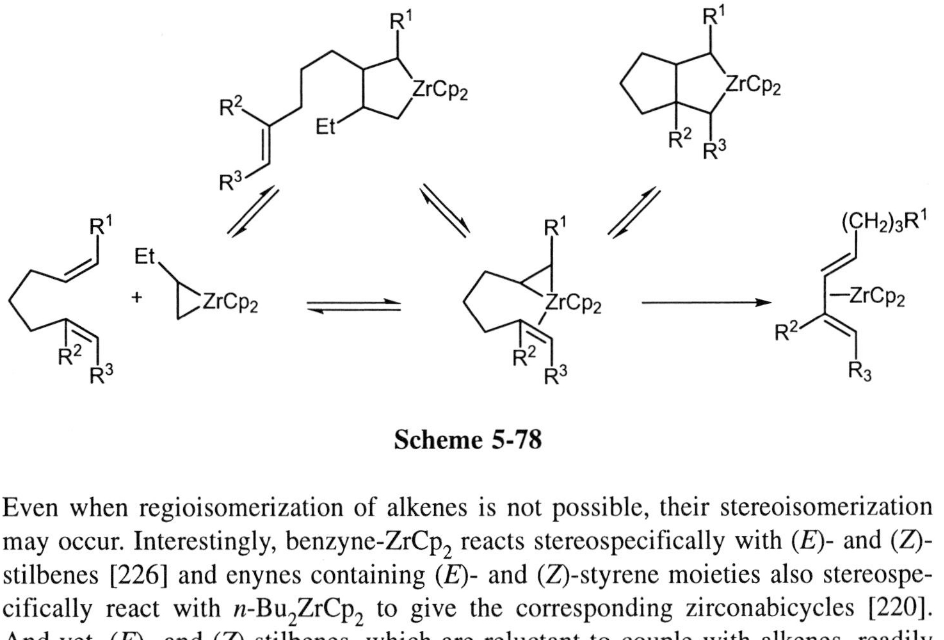

Scheme 5-78

Even when regioisomerization of alkenes is not possible, their stereoisomerization may occur. Interestingly, benzyne-ZrCp$_2$ reacts stereospecifically with (*E*)- and (*Z*)-stilbenes [226] and enynes containing (*E*)- and (*Z*)-styrene moieties also stereospecifically react with *n*-Bu$_2$ZrCp$_2$ to give the corresponding zirconabicycles [220]. And yet, (*E*)- and (*Z*)-stilbenes, which are reluctant to couple with alkenes, readily undergo ZrCp$_2$-catalyzed Z-to-E isomerization [107], and a slower but similar isomerization was also shown in Scheme 5-75. The isomerization of (*Z*)-stilbene was shown to proceed by a very intriguing polar mechanism [243], suggesting that alkene–ZrCp$_2$ complexes are more prone to heterolysis than alkyne–ZrCp$_2$ complexes (Scheme 5-79).

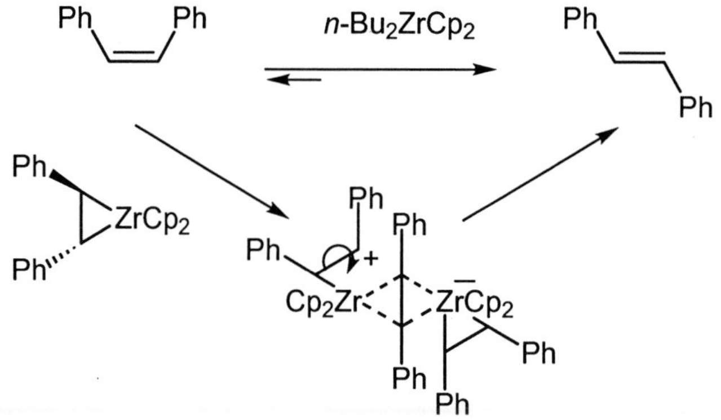

Scheme 5-79

It has recently become increasingly clear that zirconacyclopropanes readily react with various σ bonds to undergo σ-bond metathesis. For example, the reaction of $Cp_2ZrEt(CH_2=CH_2)MgBr$ with Cp_2ZrMe_2 produces in 43% yield $(Cp_2ZrMe)_2(CH_2CH_2)$, which is thought to proceed as shown in Scheme 5-80 [250]. It appears certain that many more σ-bond metathesis reactions of three-membered zirconacycles will be found in the future.

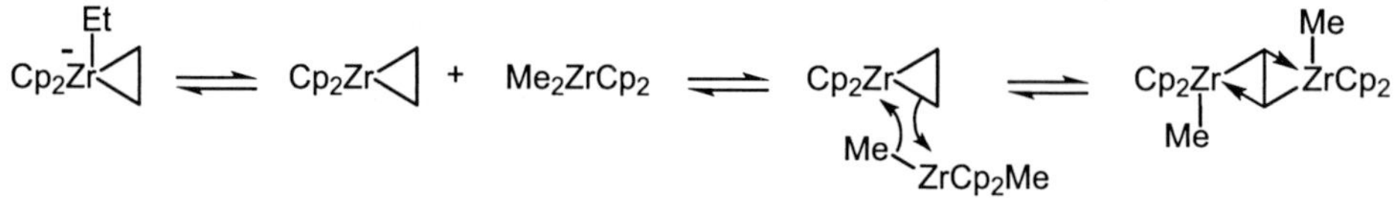

Scheme 5-80

5.4.3 Stoichiometric Reactions of Five-Membered Zirconacycles

In some respects, five-membered zirconacycles behave as diorganylzirconocenes. Thus, protonolysis and iodinolysis tend to proceed normally. In many other respects, however, five-membered zirconacycles exhibit their own chemical properties. The five-membered cyclic structure effectively blocks β-dehydrozirconation which makes them thermally more stable than many of the acyclic diorganylzirconocenes. On the other hand, the very cyclic structure must be responsible for their strong tendency to undergo ring contraction to generate three-membered zirconacycles, which can then undergo a variety of reactions discussed in the previous section. One of the earliest examples is an unexpected ring-opening reaction of a zirconabicycle with MeCN [57] (Scheme 5-81).

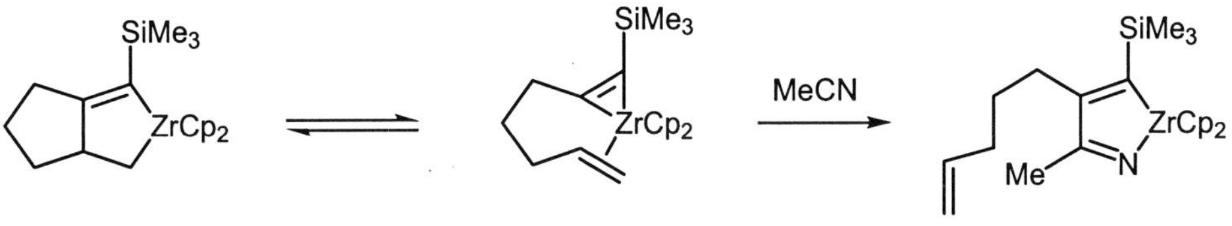

Scheme 5-81

The ring-contraction reaction most certainly is the microscopic reversal of the car-bometallative ring-expansion reaction, and the agostic interaction involving the C_β–$C_{\beta'}$ bond must be critically responsible. It is also reasonable to envision a ring expansion–ring contraction tandem as a plausible mechanism for π-substitution on Zr (Scheme 5-67). More important from the synthetic viewpoint is that the ring contraction–ring expansion tandem promises to offer a number of synthetically useful routes to three- and five-membered zirconacycles that are otherwise incon-venient, expensive, and/or difficult to prepare. Some representative examples are shown in Scheme 5-82 [228, 251, 252].

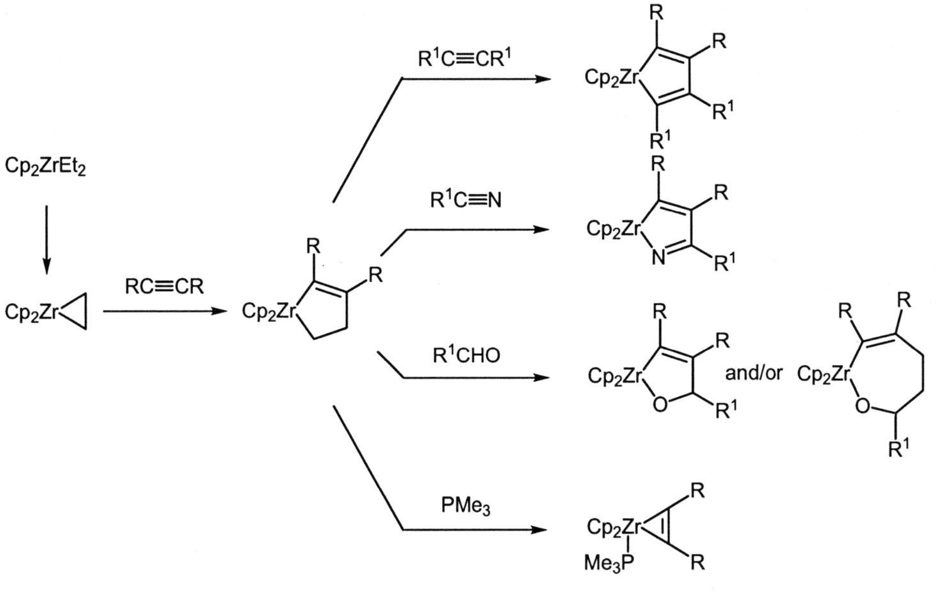

Scheme 5-82

A skeletal rearrangement reaction shown in Scheme 5-83 [253] must be mechanis-tically closely related to the displacement reactions shown in Scheme 5-82, and it must proceed *via* a series of ring contraction and ring-expansion steps.

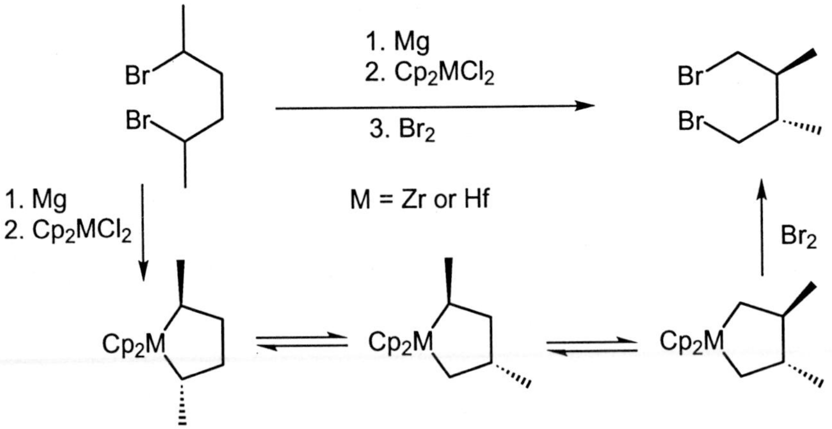

Scheme 5-83

σ-Bond metathesis is another potentially general pattern for various reactions of five-membered zirconacycles, even though its current scope is still very narrow. The reaction of PhC≡CPh with H_2ZrCp_2 to give a dimeric product shown in Scheme 5-84 represents one of the earliest examples [127] and has been interpreted in terms of σ-bond metathesis [5].

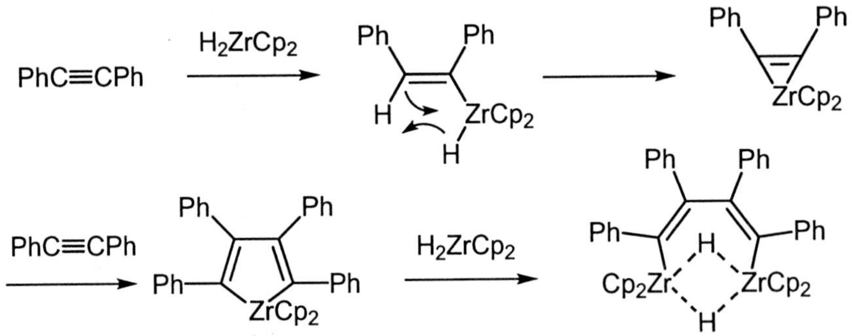

Scheme 5-84

Opening of five-membered zirconacycles *via* transmetallation is another potentially general and significant σ-bond metathesis process (Scheme 5-85). The reaction of zirconacyclopentanes with EtMgBr in the presence of PMe_3 gives regioselectively 2-ethylalkylmagnesium bromides and $Cp_2Zr(CH_2=CH_2)(PMe_3)$ most probably *via* transmetallation followed by β-H abstraction [225] (Scheme 5-86). As discussed earlier, abstraction of a β-H atom from the Et group should be strongly favored over the Mg-containing alkyl group which is either a β-CH_2 or β-CH group. However, it is not clear what factors favor the observed regioselectivity. Despite this ambiguity, the σ-bond metathesis tandem consisting of transmetallation and β-H abstraction undoubtedly plays a crucial role in the Dzhemilev ethylmagnesiation of alkenes catalyzed by Cp$_2$ZrCl$_2$ [54], as detailed later. A related reaction with alkyllithiums proceeds with displacement of one Cp group [63] (Scheme 5-87).

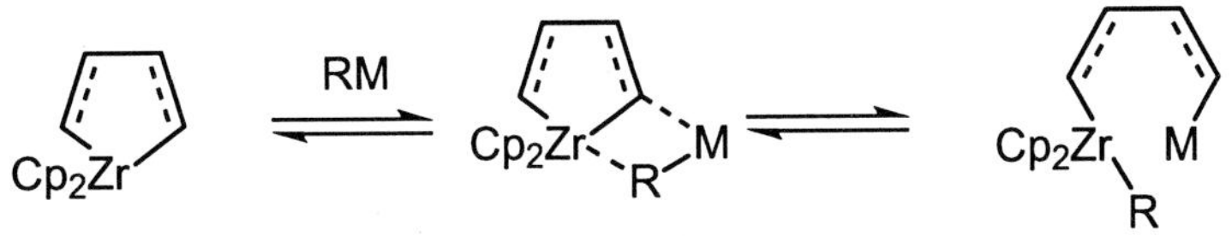

Scheme 5-85

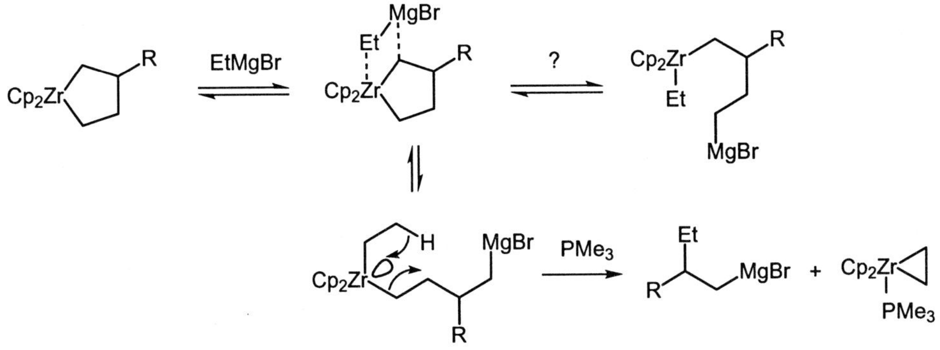

Scheme 5-86

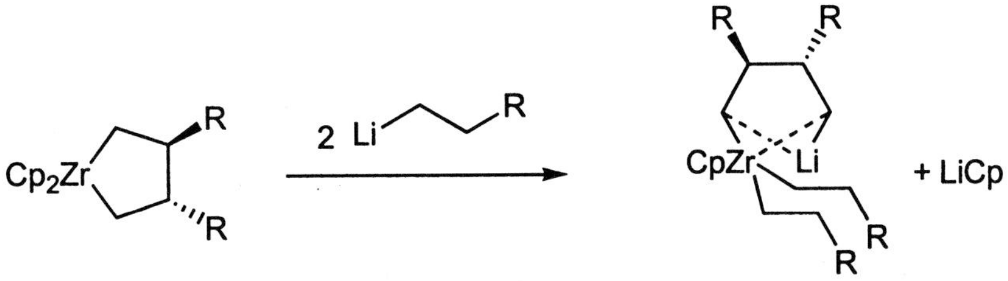

Scheme 5-87

5.4.4 Catalytic Reactions Involving Zirconacycles and Related Derivatives

Several catalytic reactions involving monoorganylzirconocene halides are discussed in Sect. 5.3.6. In this section, those catalytic reactions that involve π-complexes of ZrCp$_2$ or three-membered zirconacycles as well as their derivatives are presented. Stereoisomerization of stilbene catalyzed by 'ZrCp$_2$' discussed in Sect. 5.4.2.3

(Scheme 5-79) is an example of this class of reactions. However, it is a series of discoveries and recognition of cyclic mechanisms involving C–H activation *via* β-agostic interaction that have opened up an exciting new chapter in the organozirconium chemistry, which not only is of basic scientific interest but also of considerable synthetic potential.

5.4.4.1 Zirconium-Catalyzed Carbomagnesiation Reactions Involving C–H Activation

As discussed in Sect. 5.3.6, the Zr-catalyzed methylalumination of alkynes discovered in 1978 [49] is thought to involve a straightforward concerted addition of a Me–metal bond to alkynes, irrespective of its precise mechanism. Several years later, Dzhemilev [54] reported a seemingly related reaction involving Zr-catalyzed ethylmagnesiation of alkenes (Scheme 5-88), which was also thought to involve a straightforward carbometallation, even though the failure in attempts to achieve the corresponding methylmagnesiation was puzzling. Through systematic investigations of three stoichiometric reactions, *i.e.*, (1) β-H abstraction reaction of Et$_2$ZrCp$_2$ to form ethylene-ZrCp$_2$ [110], (2) its reaction with 1-alkenes to produce 'pair'-selectively and regioselectively 3-alkylzirconacyclopentanes [225], and (3) their reaction with EtMgBr to afford 2-ethylalkylmagnesium bromides and ethylene-ZrCp$_2$ *via* regioselective transmetallation and -H abstraction discussed in the previous section [225], the mechanism of the Zr-catalyzed ethylmagnesiation was accidentally clarified [225, see also 254] (Scheme 5-89). A few related suggestions were made [255–257], but they did not lead to full mechanistic clarification. Whereas titanocene derivatives readily undergo α C-H activation [258–260], zirconocenes are reluctant to participate in α-H abstraction. This readily explains the failure in attempts to observe methylmagnesiation. Higher alkylmagnesium halides do undergo a similar reaction, but it is less clean [261]. This can also be readily understood if one reviews the contrasting behavior of ethylene-ZrCp$_2$ and 1-alkene-ZrCp$_2$ and the relative order of reactivity of various alkyl groups in β-H abstraction (Sect. 5.4)

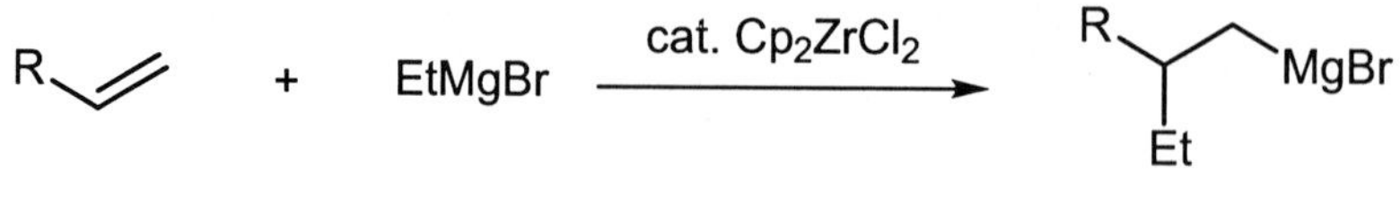

Scheme 5-88

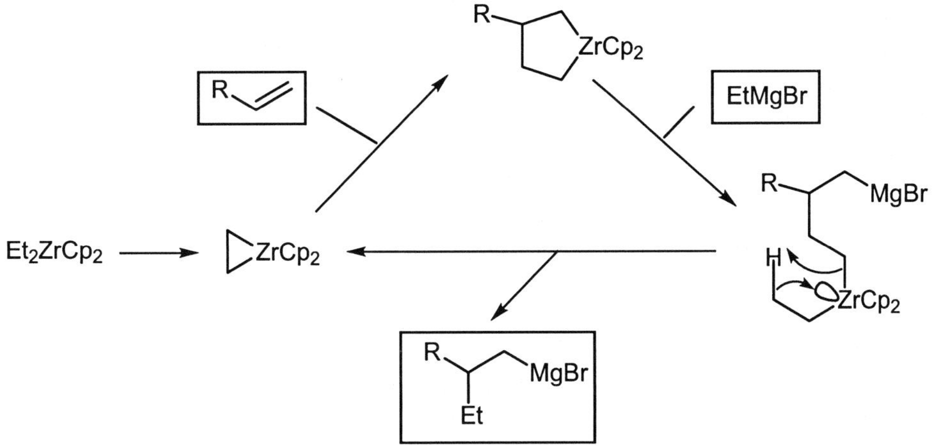

Scheme 5-89

The corresponding Zr-catalyzed reactions of nonconjugated dienes with alkylmagnesium reagents proceed similarly, but the courses of these reactions depend on various reaction parameters. With EtMgBr, dienes act as two monoenes in one, and the Dzhemilev ethylmagnesiation takes place at each vinyl group [262] (Scheme 5-90). The reaction of 1,6-heptadiene with *n*-BuMgCl in THF catalyzed by Cp$_2$ZrCl$_2$ predominantly gives 1-magnesiomethyl-2-methylcyclopentane, while the use of *n*-Bu$_2$Mg in place of *n*-BuMgCl and/or Et$_2$O in place of THF leads to 1,2-bis(magnesiomethyl)cyclopentanes [256, 262–265, See also 266] (Scheme 5-91), and some other dienes also reacted similarly. The mechanism of the monomagnesium derivative must closely resemble that of the ethylmagnesiation of 1-alkenes (Scheme 5-92), whereas double transmetallation of zirconacyclopentanes with alkylmagnesiums must be involved in the formation of 1,2-bis(magnesiomethyl)cyclopentanes.

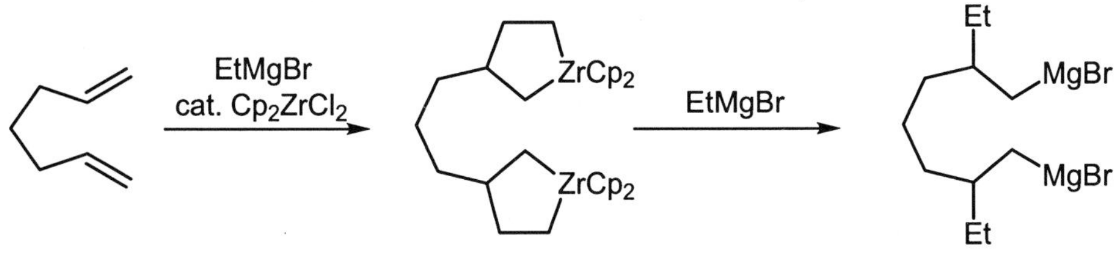

Scheme 5-90

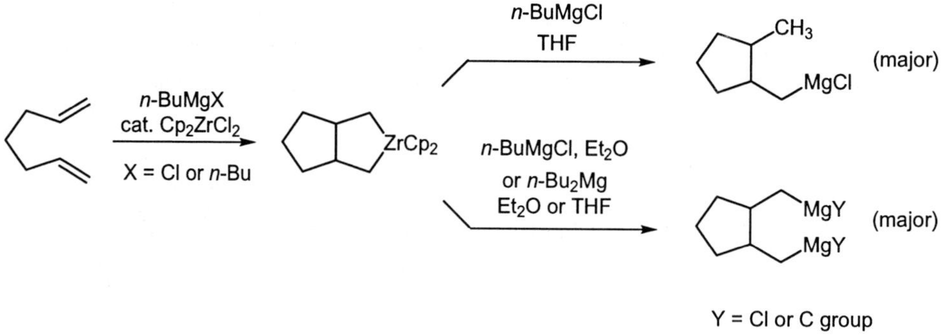

Scheme 5-91

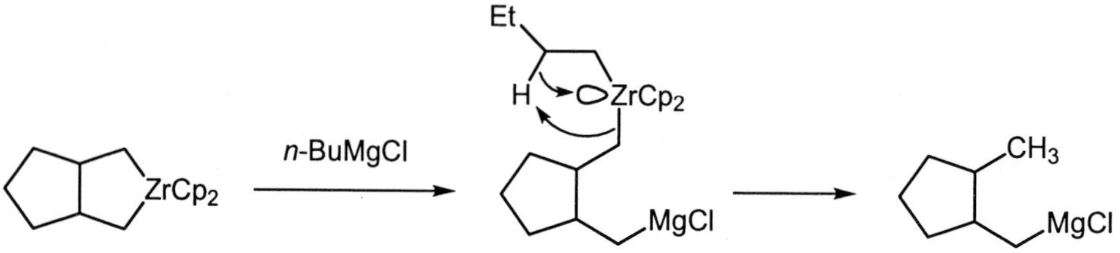

Scheme 5-92

A potentially very useful variation of the ethylmagnesiation reaction is an ethyl-magnesiation–elimination tandem process to convert allyl ethers into 3-ethyl-1-alk-enes [267, 268] (Scheme 5-93).

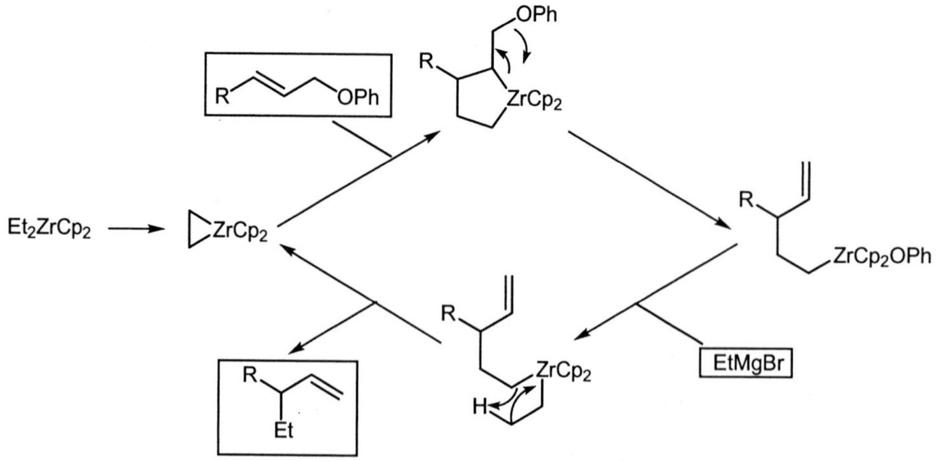

Scheme 5-93

Attempts to observe Zr-catalyzed carbomagnesiation of alkynes have been mostly disappointing. A notable exception is the Zr-catalyzed reaction of conjugated diynes which gives the expected ethylmagnesiation products [269] (Scheme 5-94).

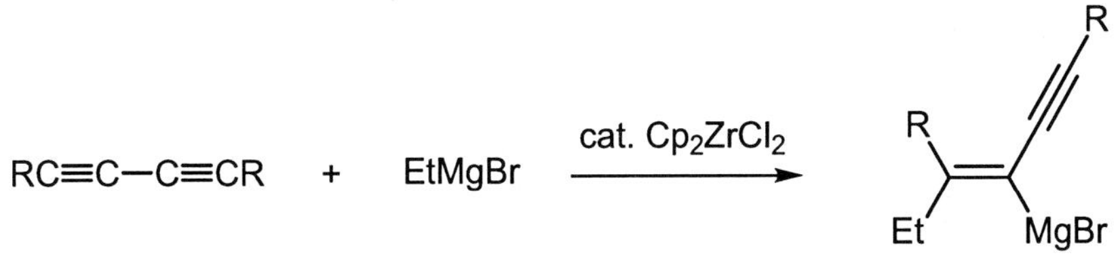

Scheme 5-94

One very significant variation of the Zr-catalyzed carbomagnesiation of alkenes is its enantioselective version, which may be related but distinct from the Zr-catalyzed enantioselective carboalumination of alkenes involving straightforward, acyclic C–Zr bond addition (Sect. 5.3.6.3). Two notable examples are the Zr-catalyzed ethylmagnesiation followed by β-elimination of allylic ethers and amines [199] (Scheme 5-95) and its more recent modification producing alkylmagnesium derivatives [200] (Scheme 5-96). For further details, the readers are referred to Chapter 10.

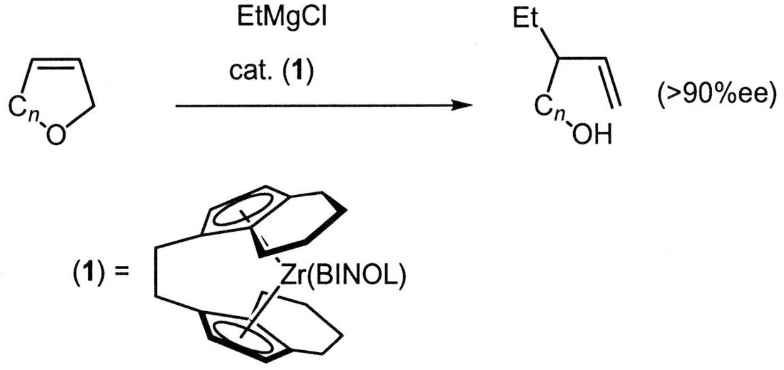

Scheme 5-95

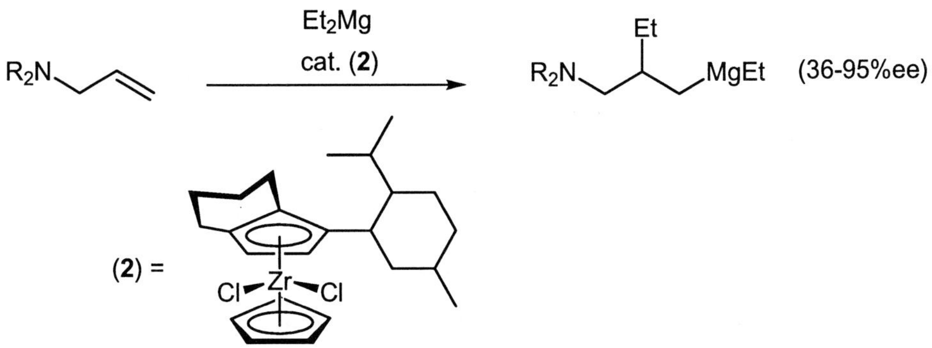

Scheme 5-96

5.4.4.2 Zirconium-Catalyzed Carboalumination of Alkynes *via* C–H Activation

Clarification of the mechanism of the Zr-catalyzed ethylmagnesiation as a cyclic process involving C–H activation was a significant turning point, which necessitated reinvestigation of all Zr-catalyzed carbometallation reactions, as the overall equation showing the addition of alkyl–metal bond to alkenes and alkynes should no longer be taken as an indication of straightforward four-centered addition reactions. Indeed, reinvestigation of the Zr-catalyzed carboalumination of alkynes with ethyl- and higher alkylalanes [49] has revealed a number of unexpected and highly intriguing results, as represented by the following results with ethylalanes. Firstly, the reaction of alkynes with Et_2AlCl and Cp_2ZrCl_2 shows no sign of cyclic processes. In polar solvents, *e.g.*, $(CH_2Cl)_2$, which are essential to observing favorable results, it is a satisfactory reaction of synthetic utility, producing *cis* addition products in high yields. This reaction can also accommodate higher *n*-alkyl groups [90]. On the other hand, the corresponding reaction of Et_3Al and Cp_2ZrCl_2 is mostly or nearly exclusively cyclic [270] and proceeds better in nonpolar solvents, *e.g.*, hexanes, than in $(CH_2Cl)_2$ [90] (Scheme 5-97). Secondly, a diethylation mechanism proceeding through Et_2ZrCp_2, ethylene-$ZrCp_2$, and a zirconacyclopentene intermediate shown in Scheme 5-98 is not operative, as the reaction of 5-decyne and Et_3Al (3 equiv.) is not catalyzed by the zirconacyclopentene derived from 5-decyne and Et_2ZrCp_2, generated by treating Cl_2ZrCp_2 with 2 equiv. of EtMgBr. However, addition of $ClAlEt_2$ (1 equiv. relative to Cl_2ZrCp_2) induces the desired catalytic process. Clearly, an active form of Cl or Br not sequestered as part of Mg salts is necessary. Thirdly, the three stoichiometric reactions summarized in Scheme 5-99 give three discrete major products. In the absence of 5-decyne, these three reactions were reported to give discrete bimetallic complexes [271–275] (Scheme 5-100). Assuming a 1:1 correlation between the results shown in Scheme 5-99 and 100, a carbozirconation mechanism accommodating all three reactions may be proposed (Scheme 5-101). Direct carbozirconation is tentatively favored to accommodate the D distribution pattern for the formation of the trideuterated product. Fourthly, treatment of $EtZrCp_2Cl$, generated by hydrozirconation of ethylene with $HZrCp_2Cl$, with 1 equiv. of Et_3Al provides a clean and high-yielding route to the five-membered bimetallic reagent, which was thought to be the carbozirconating agent in the catalytic reaction (Scheme 5-102). Indeed, its reaction with 5-decyne smoothly gives the same organoalane product and $EtZrCp_2Cl$ under the stoichiometric conditions (Scheme 5-103), and it also induces the same catalytic reaction. All these results rather strongly support a novel bimetallic and monoalkylative mechanism shown in Scheme 5-104. In the critical step producing the five-membered bimetallic complex, three components consisting of (1) $EtZrCp_2$ for crucial agostic interaction, (2) Et_3Al but not Et_2AlCl to provide an Et base, and (3) one Cl atom not sequestered by Mg and hence available for linking Zr and Al are thought to be necessary.

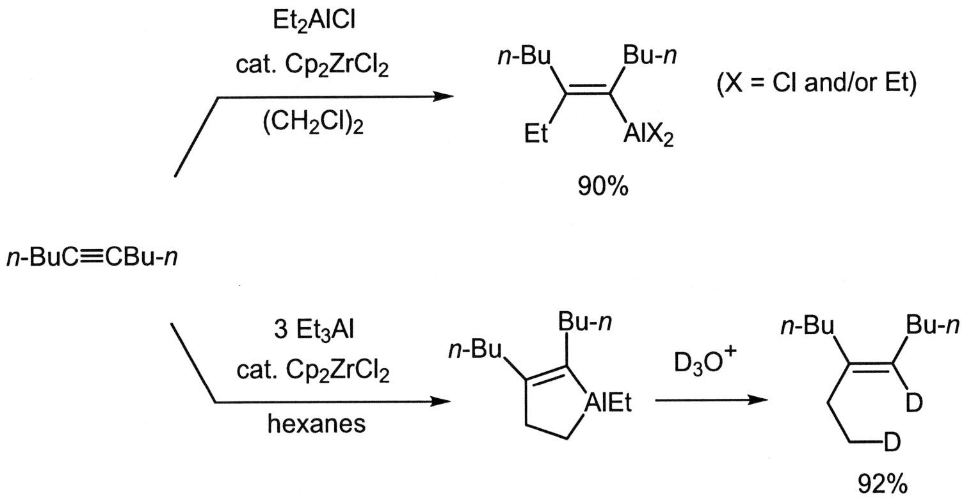

Et$_2$AlCl
cat. Cp$_2$ZrCl$_2$

(CH$_2$Cl)$_2$

n-Bu, Bu-n

Et, AlX$_2$

(X = Cl and/or Et)

90%

n-BuC≡CBu-n

3 Et$_3$Al
cat. Cp$_2$ZrCl$_2$

hexanes

n-Bu, Bu-n

AlEt

D$_3$O$^+$

n-Bu, Bu-n

D

D

92%

Scheme 5-97

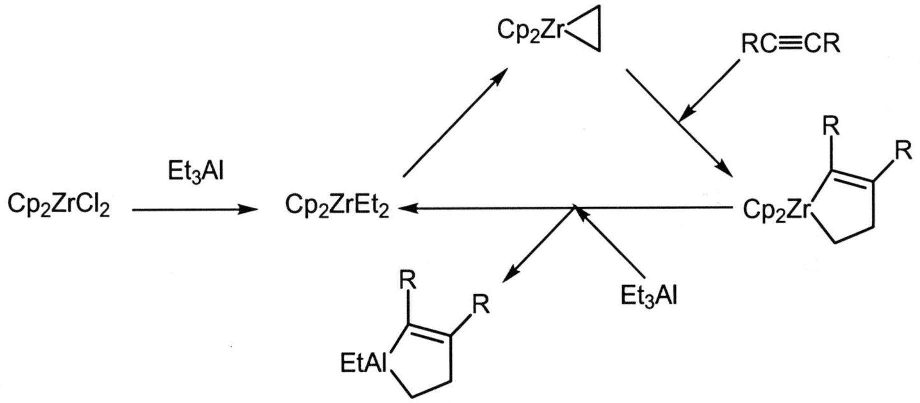

Cp$_2$Zr◁

RC≡CR

Cp$_2$ZrCl$_2$ →(Et$_3$Al)→ Cp$_2$ZrEt$_2$

R, R

Cp$_2$Zr

R, R

EtAl

Et$_3$Al

Scheme 5-98

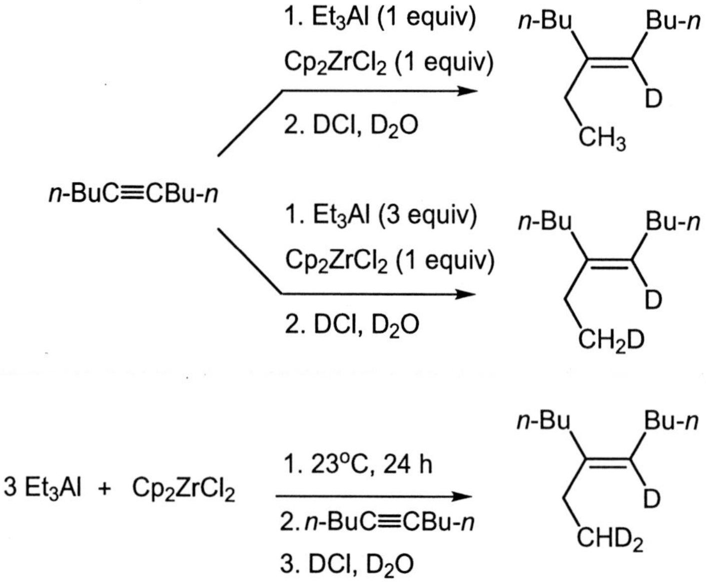
1. Et3Al (1 equiv)
Cp2ZrCl2 (1 equiv)
2. DCl, D2O
n-BuC≡CBu-n
1. Et3Al (3 equiv)
Cp2ZrCl2 (1 equiv)
2. DCl, D2O
n-Bu
Bu-n
D
CH3
CH2D
3 Et3Al + Cp2ZrCl2
1. 23°C, 24 h
2. n-BuC≡CBu-n
3. DCl, D2O
CHD2

Scheme 5-99

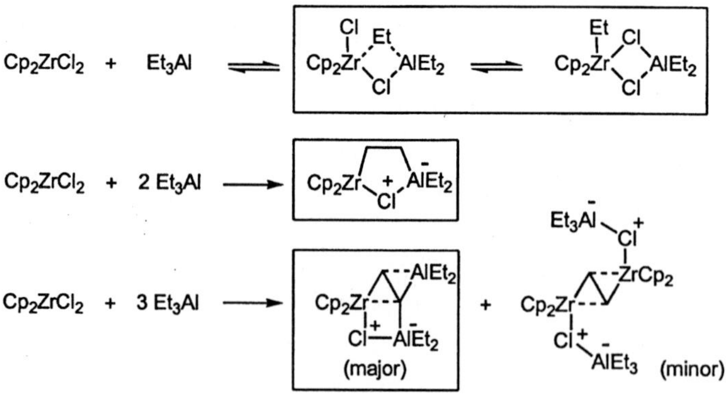
Cp2ZrCl2 + Et3Al
Cp2ZrCl2 + 2 Et3Al
Cp2ZrCl2 + 3 Et3Al
AlEt2
Cp2Zr
(major)
AlEt3
(minor)
ZrCp2

Scheme 5-100

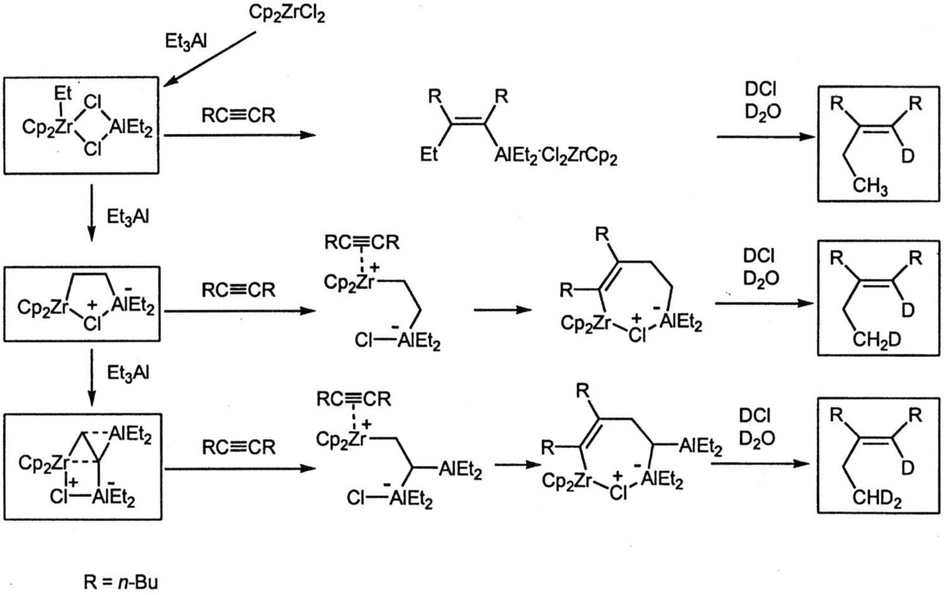

Scheme 5-101

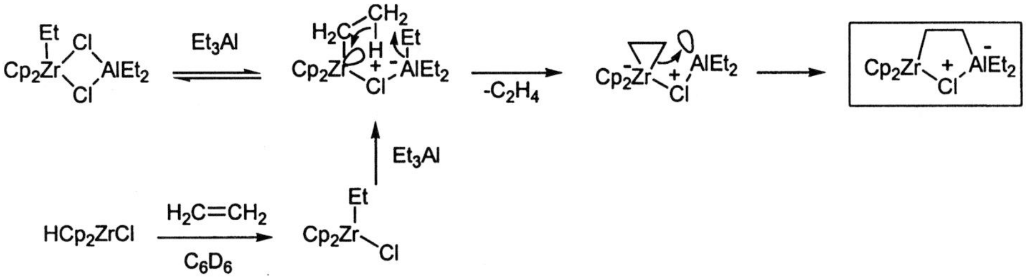

Scheme 5-102

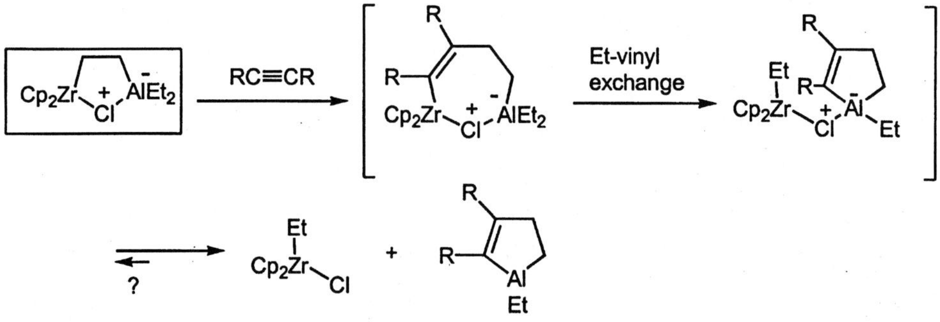

Scheme 5-103

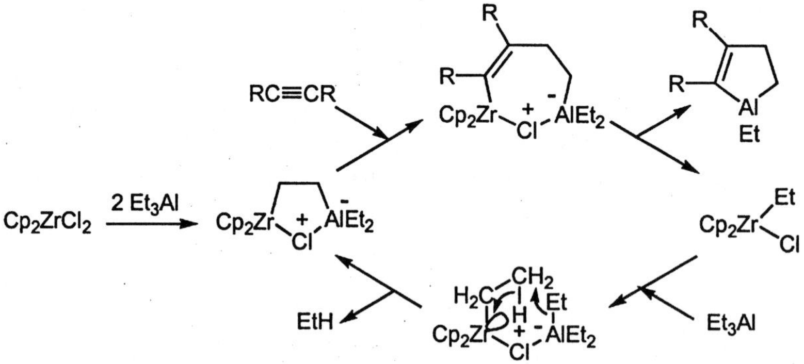

Scheme 5-104

5.4.4.3 Other Zirconium-Catalyzed Reactions Involving Cp₂Zr(II) Complexes

Three-membered zirconacycles can react with various σ-bonded species to undergo cleavage of C–Zr bonds. Their protonolysis and iodinolysis (Sect. 5.4.2.1), and transmetallation with organometals containing Mg (Scheme 5-80) and Al (Scheme 5-102) are but a few examples of such reactions. It is therefore not surprising to find that three-membered zirconacycles can serve as catalysts or catalyst precursors in the reactions of σ-bonded compounds. Hydrogenation [127, 276–279] and hydrosilation [280–282] of alkenes (Scheme 5-105) have indeed been shown to be catalyzed by 1-butenezirconocene and related compounds.

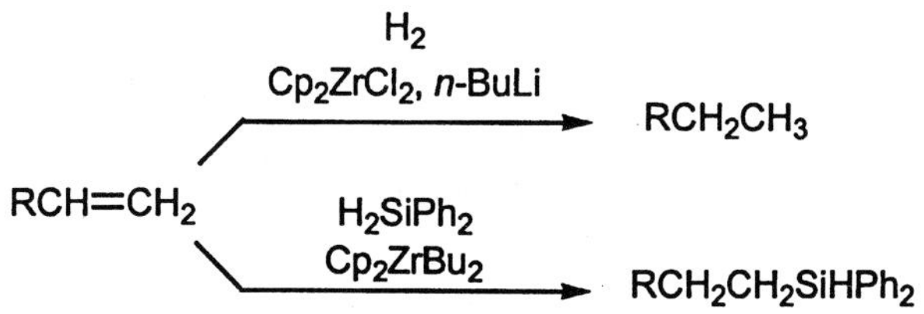

Scheme 5-105

Besides being potentially useful, their mechanisms present interesting puzzles of fundamental significance. In one study, for example, the hydrosilation reaction was thought to proceed *via* (1) oxidative addition of H_2SiPh_2 to alkene-ZrCp$_2$, (2) hydrozirconation or silylzirconation, and (3) reductive elimination [280] (Scheme 5-106). The feasibility of oxidative addition and the ability of monomeric oxidative addition product to undergo hydrosilation with alkenes have been experimentally demonstrated. And yet, some of the other proposed steps, such as silylzirconation and formation of C–Si bonds *via* reductive elimination, are not experimentally supported. Mechanistic schemes involving σ-bond metathesis have also been suggested [35, 282, See also 283, 284]. Despite firm experimental indications for oxidative addition of Cp$_2$Zr to H–Si bonds, it appears worthwhile to examine the feasibility of σ-bond metathesis mechanisms such as Scheme 5-107.

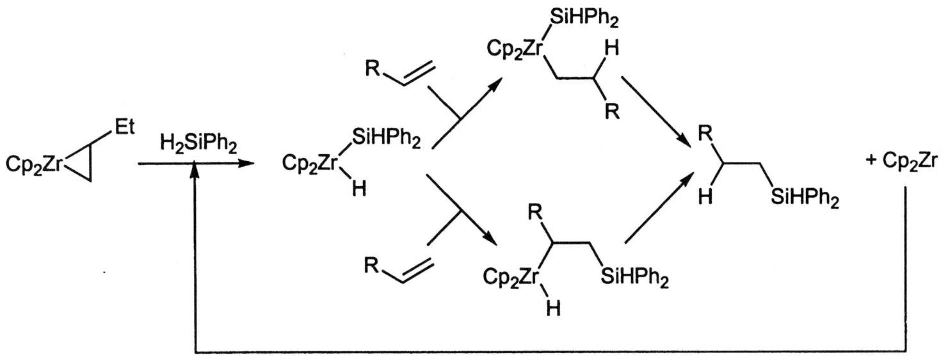

Scheme 5-106

Clearly, further explorations of fundamental transformations of Cp$_2$Zr complexes, especially their reactions with σ-bonded compounds are highly desirable, and many additional stoichiometric and catalytic reactions involving such transformations will be found.

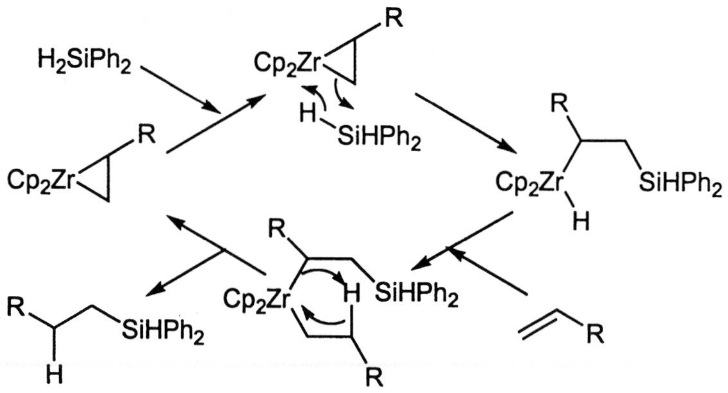

Scheme 5-107

5.5 Conclusion

The chemistry of Cp$_2$Zr derivatives is now almost half a century old. Developments mainly in the areas of hydrozirconation (from 1971), Zr-mediated carbometallation reactions represented by the Kaminsky polymerization (from 1974), and Zr-catalyzed controlled carboalumination (from 1978) have led to explosive investigations of RZr^{IV}Cp$_2$ derivatives discussed in Sect. 5.3. The scientific and practical significance of Cp$_2$Zr chemistry has been substantially elevated by similarly explosive investigations and applications of Cp$_2$Zr^{II} chemistry discussed in Sect. 5.4 mainly in the 1980s and 1990s. Undoubtedly, these areas will be further explored and developed. For example, such fundamentally important processes as oxidative addition, reductive elimination, and σ-bond metathesis of Cp$_2$Zr derivatives remain largely unexplored and poorly understood. Their migratory insertion chemistry with the notable exception of carbonylation is still poorly developed. Many of the topics covered in this chapter are also being rewritten, as their enantioselective versions are developed. Most of the structural types shown in Table 5-2 are covered in this chapter. However, several of them, notably Cp$_2$Zr=a (Types II and XIV), Cp$_2$Zr≡a (Type III), four-membered zirconacycles (Type VI), are hardly covered despite some potentially significant investigations including carbonyl–olefination reactions of 1,1-dimetalloalkanes and -alkenes and Cp$_2$Zr–carbene complexes [146, 285–287] and chemistry of imidozirconocene derivatives [288–296]. Admittedly, it is not possible to properly represent all pertinent topics in a single chapter. The readers are referred to a few monographs [1,2] and various reviews listed in Table 5-10 according to their topics.

Table 5-10. Classified List of Reviews

Section number	Topic	Reference number (Senior author)
All sections	General	[3] (Cardin), [4] (Cardin), [5] (Negishi-Takahashi), [6] (Negishi-Takahashi)
5.2	Preparation of Zirconocene Derivatives	
5.2.1	Transmetallation	[7] (Farina)
5.2.2	Hydrozirconation	[8] (Schwartz), [9] (Labinger), [10] (Wipf)
5.2.3	Oxidative addition	[11] (Hanzawa-Taguchi)
5.2.4	Oxidative coupling (Complexation)	[12] (Buchwald)
5.3	Reactions of $RZrCp_2X$	
5.3.2.1	Allyl $ZrCp_2X$	[11] (Hanzawa-Taguchi), [13] (Yasuda-Nakamura), [14] (Yasuda-Nakamura)
5.3.2.2	R^+ZrCp_2	[15] (Jordan), [16] (Jordan)
5.3.3	Migratory insertion	[17] (Erker), [18] (Gambarotta)
5.3.4	Stoichiometric carbozirconation	[16] (Jordan)
5.3.5	Transmetallation	[7] (Farina)
5.3.5.2	Catalysis by other metals	[19] (Negishi)
5.3.6	Catalysis by zirconocenes	[20] (Dzhemilev), [21] (Dzhemilev)
5.3.6.1	Carboalumination of alkynes	[22] (Negishi), [23] (Dzhemilev), [24] (Negishi), [25] (Negishi)
5.3.6.3	Enantioselective carboalumination	[25] (Negishi), [26] (Erker)
5.4	$Cp_2Zr(II)$ chemistry	[27] (Pez-Armor), [28] (Erker), [29] (Buchwald), [30] (Buchwald), [31] (Nugent), [32] (Negishi), [33] (Negishi), [34] (Buchwald), [35] (Negishi-Takahashi), [36] (Binger), [37] (Broene)
5.4.1	Formation of π-complexes	[12] (Buchwald), [38] (Rosenthal)
5.4.2	Stoichiometric reactions of π-complexes	[12] (Buchwald), [24] (Negishi), [39] (Dzhemilev)
5.4.3	Stoichiometric reactions of zirconacycles	[21] (Dzhemilev), [25] (Negishi)
5.4.4.1	Carbomagnesiation	[40] (Hoveyda)

References

Monographs on Cp₂Zr Chemistry
[1]   P. C. Wailes, R. S. P. Courtts and H. Weigold, *Organometallic Chemistry of Titanium, Zirconium, and Hafnium*, Academic Press, New York, **1974**, 302 pp.
[2]   D. J. Cardin, M. F. Lappert, and C. L. Raston *Chemistry of Organozirconium and Hafnium Compounds*, John Wiley & Sons, New York, **1986**, 451 pp.

Reviews
[3]   D. J. Cardin, M. F. Lappert, C. L. Raston, and R. I. Riley in *Comprehensive Organometallic Chemistry*, Vol. 1, G. Wilkinson, F. G. A. Stone, and E. W. Abel, Eds., Pergamon Press, Oxford, **1982**, 549.
[4]   D. J. Cardin, M. F. Lappert, C. L. Raston, and R. I. Riley, in *Comprehensive Organometallic Chemistry*, Vol. 1, G. Wilkinson, F. G. A. Stone, and E. W. Abel, Eds., Pergamon Press, Oxford, **1982**, 559.
[5]   E. Negishi and T. Takahashi, *Aldrichim. Acta*, **1985**, *18*, 31.
[6]   E. Negishi and T. Takahashi, *Synthesis*, **1988**, 1.
[7]   V. Farina in *Comprehensive Organometallic Chemistry II*, E. W. Abel, F. G. A. Stone, G. Wilkinson, Eds., Pergamon Press, Oxford, **1995**, Vol. 12, p. 161.
[8]   J. Schwartz and J. A. Labinger, *Angew. Chem., Int. Ed. Engl.*, **1976**, *15*, 333.
[9]   J. A. Labinger in *Comprehensive Organic Synthesis*, Vol. 8, B. M. Trost, I. Fleming, Eds., Pergamon Press, Oxford, **1991**, 667.
[10] P. Wipf and H. Jahn, *Tetrahedron*, **1996**, *52*, 12853.
[11] Y. Hanzawa, H. Ito, and T. Taguchi, *Synlett.*, **1995**, 299.
[12] S. L. Buchwald and R. D. Broene, in *Comprehensive Organometallic Chemistry II*, E. W. Abel, F. G. A. Stone, G. Wilkinson, Eds., Pergamon Press, Oxford, **1995**, Vol. 12, p. 771.
[13] H. Yasuda, K. Tatsumi, and A. Nakamura, *Acc. Chem. Res.*, **1985**, *18*, 120.
[14] H. Yasuda and A. Nakamura, *Rev. Chem. Intermediates*, **1986**, *6*, 365.
[15] R. F. Jordan, *Adv. Organomet. Chem.*, **1991**, *32*, 325.
[16] A. S. Guram and R. F. Jordan in *Comprehensive Organometallic Chemistry II*, E. W. Abel, F. G. A. Stone, G. Wilkinson, Eds., Pergamon Press, Oxford, **1995**, Vol. 4, p. 589.
[17] G. Erker, *Acc. Chem. Res.*, **1984**, *17*, 103.
[18] J. Jubb, J. Fong, D. Richeson, and S. Gambarotta in *Comprehensive Organometallic Chemistry II*, E. W. Abel, F. G. A. Stone, G. Wilkinson, Eds., Pergamon Press, Oxford, **1995**, Vol. 4, p. 543.
[19] E. Negishi, *Acc. Chem. Res.*, **1982**, *15*, 340.
[20] U. M. Dzhemilev, O. S. Vostrikova, and G. A. Tolstikov, *Russ. Chem. Rev.*, **1990**, *59*, 1157.
[21] U. M. Dzhemilev, *Tetrahedron*, **1995**, *51*, 4333.
[22] E. Negishi, *Pure Appl. Chem.*, **1981**, *53*, 2333.
[23] U. M. Dzhemilev, O. S. Vostrikova, and G. A. Tolstikov, *J. Organomet. Chem.*, **1986**, *304*, 17.
[24] E. Negishi, *Acc. Chem. Res.*, **1987**, *20*, 65.
[25] E. Negishi and D. Y. Kondakov, *Chem. Soc. Rev.*, **1996**, *26*, 417.
[26] G. Erker, *Pure Appl. Chem.*, **1992**, *64*, 393.
[27] G. P. Pez and J. N. Armor, *Adv. Organomet. Chem.*, **1981**, *19*, 1.
[28] G. Erker, C. Krüger, and G. Müller, *Adv. Organomet. Chem.*, **1985**, *24*, 1.
[29] S. L. Buchwald and R. B. Nielsen, *Chem. Rev.*, **1988**, *88*, 1047.
[30] S. L. Buchwald and R. A. Fisher, *Chim. Scripta*, **1989**, *29*, 417.
[31] W. A. Nugent, T. V. RajanBabu, and D. F. Taber, *Chim. Scripta*, **1989**, *29*, 439.
[32] E. Negishi, *Chim. Scripta*, **1989**, *29*, 457.
[33] E. Negishi in *Comprehensive Organic Synthesis*, Vol. 5, L. A. Paquette, Ed., Pergamon Press, Oxford, **1991**, 1163.
[34] R. D. Broene and S. L. Buchwald, *Science*, **1993**, *261*, 1696.

[35]   E. Negishi and T. Takahashi, *Acc. Chem. Res.*, **1994**, *27*, 124.

[36]   P. Binger and S. Podubrin in *Comprehensive Organometallic Chemistry II*, E. W. Abel, F. G. A. Stone, G. Wilkinson, Eds., Pergamon Press, Oxford, **1995**, Vol. 4, p. 439.

[37]   R. D. Broene in *Comprehensive Organometallic Chemistry II*, E. W. Abel, F. G. A. Stone, G. Wilkinson, Eds., Pergamon Press, Oxford, **1995**, Vol. 12, p. 313.

[38]   A. Ohff, S. Pulst, C. Lefeber, N. Peulecke, P. Arndt, V. V. Burkalov, and U. Rosenthal, *Synlett.*, **1996**, 111.

[39]   U. M. Dzhemilev, R. M. Sultanov, and R. G. Gaimaldinov, *J. Organomet. Chem.*, **1995**, *491*, 1.

[40]   A. H. Hoveyda and J. P. Morken, *Angew. Chem., Int. Ed. Engl.*, **1996**, *35*, 1262.

Original papers and others

[41]   J.J. Berzelius, *Ann. Chim. Phys.* **1824**, *26*, 43.

[42]   G. Wilkinson and J. M. Birmingham, *J. Am. Chem. Soc.*, **1954**, *76*, 4281.

[43]   G. A. Olah, *Friedel-Crafts and Related Reactions*, Interscience, New York, **1963**, Vol. 1, 1031 pp.

[44]   J. Boor, *Ziegler-Natta Catalysis and Polymerization*, Academic Press, New York, **1978**, 670 pp.

[45]   P. C. Wailes and H. Weigold, *J. Organomet. Chem.*, **1970**, *24*, 405.

[46]   P. C. Wailes, H. Weigold, and A. P. Bell, *J. Organomet. Chem.*, **1971**, *27*, 373.

[47]   D. W. Hart and J. Schwartz, *J. Am. Chem. Soc.*, **1974**, *96*, 8115.

[48]   W. Kaminsky, H.-J. Vollmer, E. Heins, and H. Sinn, *Makromol. Chem.*, **1974**, *175*, 443.

[49]   D. E. Van Horn and E. Negishi, *J. Am. Chem. Soc.*, **1978**, *100*, 2252.

[50]   E. Negishi, D. E. Van Horn, and T. Yoshida, *J. Am. Chem. Soc.*, **1985**, *107*, 6639.

[51]   G. Erker and K. Kropp, *J. Am. Chem. Soc.*, **1979**, *101*, 3659.

[52]   G. Erker, J. Wicher, K. Engel, F. Rosenfeldt, W. Dietrich, and C. Kruger, *J. Am. Chem. Soc.*, **1980**, *102*, 6346.

[53]   H. Yasuda, Y. Kajihara, K. Mashima, K. Lee, and A. Nakamura, *Chem. Lett.*, **1981**, 519.

[54]   U. M. Dzhemilev, O. S. Vostrikova, and R. M. Sultanov, *Izv. Akad. Nauk SSSR, Ser. Khim.*, **1983**, 218.

[55]   E. Negishi, S. J. Holmes, J. M. Tour, and J. A. Miller, *J. Am. Chem. Soc.*, **1985**, *107*, 2568.

[56]   E. Negishi, F. E. Cederbaum, and T. Takahashi, *Tetrahedron Lett.*, **1986**, *27*, 2829.

[57]   E. Negishi, S. J. Holmes, J. M. Tour, J. A. Miller, F. E. Cederbaum, D. R. Swanson, and T. Takahashi, *J. Am. Chem. Soc.*, **1989**, *111*, 3336.

[58]   S. L. Buchwald, B. T. Watson, and J. C. Huffman, *J. Am. Chem. Soc.*, **1986**, *108*, 7411.

[59]   S. L. Buchwald, R. T. Lum, and J. C. Dewan, *J. Am. Chem. Soc.*, **1986**, *108*, 7441.

[60]   E. J. Ryan in *Comprehensive Organometallic Chemistry II*, E. W. Abel, F. G. A. Stone, G. Wilkinson, Eds., Pergamon Press, Oxford, **1995**, Vol. 4, 465-480.

[61]   F. Soleil and R. Choukroun, *J. Am. Chem. Soc.*, **1997**, *119*, 2938.

[62]   A. D. Walsh, *Trans. Faraday, Soc.*, **1949**, *45*, 179

[63]   D. Y. Kondakov and E. Negishi, *J. Chem. Soc., Chem. Comm.*, **1996**, 963.

[64]   G. Erker, K. Berg, L. Treschanke, and K. Engel, *Inorg. Chem.*, **1982**, *21*, 1277.

[65]   E. Negishi, *Organometallics in Organic Synthesis*, Vol. 1, J. Wiley & Sons, New York, **1980**, 532 pp.

[66]   A. F. Reid and P. C. Wailes, *J. Organomet. Chem.*, **1964**, *2*, 329.

[67]   P. M. Druce, B. M. Kingston, M. F. Lappert, T. R. Spalding, and R. C. Sriniastava, *J. Chem. Soc., A*, **1969**, 2106.

[68]   Unpublished results by E. Negishi and S. J. Holmes.

[69]   E. Negishi, J. A. Miller, and T. Yoshida, *Tetrahedron Lett.*, **1984**, *25*, 3407.

[70]   D. R. Swanson, T. Nguyen, Y. Noda, and E. Negishi, *J. Org. Chem.*, **1991**, *56*, 2590.

[71]   J. R. Surtees, *J. Chem. Soc., Chem. Comm.*, **1965**, 567.

[72]   E. Negishi and T. Yoshida, *J. Am. Chem. Soc.*, **1981**, *103*, 4985.

[73]   D. B. Carr and J. Schwartz, *J. Am. Chem. Soc.*, **1977**, *99*, 638.

[74]   E. Negishi and L. D. Boardman, *Tetrahedron Lett.*, **1982**, *23*, 3327.

[75]   A. Pelter, K. Smith, and H. C. Brown, *Borane Reagents,* Academic Press, New York, **1988,** 503 pp.

[76]   G. Wilke and H. Müller, *Ann. Chem.,* **1960,** *629,* 222.

[77]   G. Zweifel and J. A. Miller, *Org. React.,* **1984,** *32,* 1.

[78]   D. W. Hart, T. F. Blackburn, and J. Schwartz, *J. Am. Chem. Soc.,* **1975,** *97,* 679.

[79]   C. A. Bertelo and J. Schwartz, *J. Am. Chem. Soc.,* **1976,** *98,* 262.

[80]   P. Wipf, W. J. Xu, J. H. Smitrovich, R. Lehmann, and L. M. Venanzi, *Tetrahedron,* **1994,** *50,* 1935.

[81]   B. H. Lipshutz, C. Lindsley, and A. Bhandari, *Tetrahedron Lett.,* **1994,** *35,* 4669.

[82]   P. C. Wailes and H. Weigold, *Inorg. Synth.,* **1979,** *19,* 223.

[83]   D. B. Carr and J. Schwartz, *J. Am. Chem. Soc.,* **1979,** *101,* 3521.

[84]   S. L. Buchwald, S. J. La Maire, R. B. Nielsen, B. T. Watson, and S. M. King, *Tetrahedron Lett.,* **1987,** *28,* 3895.

[85]   S. L. Buchwald, S. J. La Maire, R. B. Nielsen, B. T. Watson, and S. M. King, *Org. Synth.,* **1992,** *71,* 77.

[86]   D. G. Bickley, N. Hao, P. Bougeard, B. G. Sayer, R. C. Burns, and M. J. McGlinchey, *J. Organomet. Chem.,* **1983,** *246,* 257.

[87]   C. McDade and J. E. Bercaw, *J. Organomet. Chem.,* **1985,** *279,* 281.

[88]   B. H. Lipshutz, R. Keil, and E. L. Ellsworth, *Tetrahedron Lett.,* **1990,** *31,* 7257.

[89]   E. Negishi and T. Yoshida, *Tetrahedron Lett.,* **1980,** *21,* 1501.

[90]   E. Negishi, D. Y. Kondakov, D. Choueiry, K. Kasai, and T. Takahashi, *J. Am. Chem. Soc.,* **1996,** *118,* 9577.

[91]   J. Endo, N. Koga, and K. Morokuma, *Organometallics,* **1993,** *12,* 2777.

[92]   C. J. Rousset, D. R. Swanson, F. Lamaty, and E. Negishi, *Tetrahedron Lett.,* **1989,** *30,* 5105.

[93]   H. Ito, T. Taguchi, and Y. Hanzawa, *Tetrahedron Lett.,* **1992,** *33,* 7873.

[94]   M. Chino, T. Matsumoto, and K. Suzuki, *Synlett.,* **1994,** 359.

[95]   H. Ito, T. Nakamura, T. Taguchi, and Y. Hanzawa, *Tetrahedron Lett.,* **1992,** *33,* 3769. See also (a) H. Ito, T. Taguchi, and Y. Hanzawa, *Tetrahedron Lett.,* **1992,** *33,* 1295. (b) H. Ito, T. Taguchi, and Y. Hanzawa, *J. Org. Chem.,* **1993,** *58,* 774. (c) H. Ito, Y. Ikeuchi, T. Taguchi, Y. Hanzawa, and M. Shiro, *J. Am. Chem. Soc.,* **1994,** *116,* 5469. (d) H. Ito, T. Nakamura, T. Taguchi, and Y. Hanzawa, *Tetrahedron,* **1995,** *51,* 4507.

[96]   T. Takahashi, M. Kotora, R. Fischer, Y. Nishihara, and K. Nakajima, *J. Am. Chem. Soc.,* **1995,** *117,* 11039.

[97]   M. J. S. Dewar, *Bull. Soc. Chim. Fr.* **1951,** 18, C71.

[98]   J. Chatt and L. A. Duncanson, *J. Chem. Soc.,* **1953,** 2939.

[99]   G. W. Watt and F. O. Drummond, *J. Am. Chem. Soc.,* **1966,** *88,* 5926.

[100]   G. W. Watt and F. O. Drummond, *J. Am. Chem. Soc.,* **1970,** *92,* 826.

[101]   D. J. Sikora, K. J. Moriarty, and M. D. Rausch, *Inorg. Synth.,* **1990,** *28,* 249.

[102]   M. D. Fryzuk, T. S. Hadad, and D. J. Berg, *Coord. Chem. Rev.,* **1990,** *99,* 137.

[103]   J. L. Atwood, R. D. Rogers, W. E. Hunter, C. Floriani, G. Fachinetti, and A. Chiesi-Villa, *Inorg. Chem.,* **1980,** *19,* 3812.

[104]   J. B. Kool, M. D. Rausch, H. G. Alt, M. Herberhold, B. Honold, and U. Thewalt, *J. Organomet. Chem.,* **1987,** *320,* 37.

[105]   D. J. Sikora, D. W. Macomber, and M. D. Rausch, *Adv. Organomet. Chem.,* **1986,** *25,* 318.

[106]   S. L. Buchwald, B. T. Watson, R. T. Lum, and W. A. Nugent, *J. Am. Chem. Soc.,* **1987,** *109,* 7137.

[107]   T. Takahashi, D. R. Swanson, and E. Negishi, *Chem. Lett.,* **1987,** 623.

[108]   S. L. Buchwald, B. T. Watson, and J. C. Huffman, *J. Am. Chem. Soc.,* **1987,** *109,* 2544.

[109]   S. L. Buchwald, R. T. Lum, R. A. Fisher, W. M. Davis, *J. Am. Chem. Soc.,* **1989,** *111,* 9113.

[110]   T. Takahashi, M. Murakami, M. Kunishige, M. Saburi, Y. Uchida, K. Kozawa, T. Uchida, D. R. Swanson, and E. Negishi, *Chem. Lett.,* **1989,** 761.

[111]   P. Binger, P. Müller, R. Benn, A. Rufinska, B. Gabor, C. Krüger, and P. Betz, *Chem. Ber.,* **1989,** *122,* 1035.

[112] P. Binger, B. Biedenbach, A. T. Herrmann, F. Langhauser, P. Betz, R. Goddard, and C. Krüger, *Chem. Ber.*, **1990**, *123*, 1617.

[113] S. L. Buchwald and R. B. Nielsen, *J. Am. Chem. Soc.*, **1988**, *110*, 3171.

[114] S. L. Buchwald, B. T. Watson, M. W. Wannamaker, and J. C. Dewar, *J. Am. Chem. Soc.*, **1989**, *111*, 4486.

[115] D. R. Swanson and E. Negishi, *Organometallics*, **1991**, *10*, 825.

[116] H. G. Alt, C. E. Denner, U. Thewalt, and M. D. Rausch, *J. Organomet. Chem.*, **1988**, *356*, C83.

[117] T. Takahashi, M. Tamura, M. Saburi, Y. Uchida, and E. Negishi, *J. Chem. Soc., Chem. Comm.*, **1989**, 852.

[118] P. J. Walsh, F. J. Hollander, and R. G. Bergman, *J. Organomet. Chem.*, **1992**, *428*, 13.

[119] D. A. Straus and R. H. Grubbs, *J. Am. Chem. Soc.*, **1982**, *104*, 5499.

[120] E. J. Moore, D. A. Straus, J. Armantrout, B. D. Santarsiero, R. H. Grubbs, and J. E. Bercaw, *J. Am. Chem. Soc.*, **1983**, *105*, 2068.

[121] S. C. H. Ito, D. A. Straus, J Armantrout, W. P. Schaefer, and R. H. Grubbs, *J. Am. Chem. Soc.*, **1984**, *106*, 2210.

[122] R. M. Waymouth, B. D. Samtarsiero, R. J. Coots, M. J. Bronikowski, and R. H. Grubbs, *J. Am. Chem. Soc.*, **1986**, *108*, 1427.

[123] P. J. Walsh, F. J. Hollander, and R. G. Bergman, *J. Organomet. Chem.*, **1992**, *428*, 13.

[124] G. Erker and F. Rosenfeldt, *J. Organomet. Chem.*, **1982**, *224*, 29.

[125] W. E. Hunter, D. C. Hrncir, R. V. Bynum, R. A. Penttila, and J. L. Atwood, *Organometallics*, **1983**, *2*, 750.

[126] J. A. Labinger, D. W. Hart, W. E. Seibert, III, and J. Schwartz, *J. Am. Chem. Soc.*, **1975**, *97*, 3851.

[127] P. C. Wailes, H. Weigold, and A. P. Bell, *J. Organomet. Chem.*, **1972**, *43*, C32.

[128] T. F. Blackburn, J. A. Labinger, and J. Schwartz, *Tetrahedron Lett.*, **1975**, 3041.

[129] P. C. Wailes, H. Weigold, and A. P. Bell, *J. Organomet. Chem.*, **1971**, *33*, 181.

[130] P. C. Wailes, H. Weigold, and A. P. Bell, *J. Organomet. Chem.*, **1971**, *34*, 155.

[131] Y. Yamamoto and K. Muruyama, *Tetrahedron Lett.*, **1981**, *22*, 2895.

[132] K. Mashima, H. Yasuda, K. Asami, and A. Nakamura, *Chem. Lett.*, **1983**, 219.

[133] H. Yasuda, Y. Kajihara, K. Nagasuna, K. Mashima, and A. Nakamura, *Chem. Lett.*, **1981**, 719.

[134] M. Akita, H. Yasuda, and A. Nakamura, *Chem. Lett.*, **1983**, 217.

[135] H. Maeta, T. Hashimoto, T. Hasegawa, and K. Suzuki, *Tetrahedron Lett.*, **1992**, *33*, 5965.

[136] K. Suzuki, T. Hasegawa, T. Imai, H. Maeta, and S. Ohba, *Tetrahedron*, **1995**, *51*, 4483.

[137] P. Wipf and W. Xu, *J. Org. Chem.*, **1993**, *58*, 825.

[138] R. F. Jordan, C. S. Bajgur, R. Willett, and B. Scott, *J. Am. Chem. Soc.*, **1986**, *108*, 7410.

[139] C. A. Bertelo and J. Schwartz, *J. Am. Chem. Soc.*, **1975**, *97*, 228.

[140] J. M. Manriquez, D. R. McAlister, R. D. Sanner, and J. E. Bercaw, *J. Am. Chem. Soc.*, **1978**, *100*, 2716.

[141] G. Erker and F. Rosenfeldt, *J. Organomet. Chem.*, **1980**, *188*, C1.

[142] E. Negishi, D. R. Swanson, and S. R. Miller, *Tetrahedron Lett.*, **1988**, *29*, 1631.

[143] S. L. Buchwald and S. J. La Maire, *Tetrahedron Lett.*, **1987**, *28*, 295.

[144] E. A. Mintz, A. S. Ward, and D. S. Tice, *Organometallics*, **1985**, *4*, 1308.

[145] E. Negishi, K. Akiyoshi, B. O'Connor, K. Takagi, and G. Wu, *J. Am. Chem. Soc.*, **1989**, *111*, 3089.

[146] T. Yoshida and E. Negishi, *J. Am. Chem. Soc.*, **1981**, *103*, 1276.

[147] T. E. Cole, R. Quintanilla, and S. Rodewald, *Organometallics*, **1991**, *10*, 3777.

[148] T. E. Cole, S. Rodewald, and C. L. Watson, *Tetrahedron Lett.*, **1992**, *33*, 5295.

[149] T. E. Cole, R. Quintanilla, B. M. Smith, and D. Hurst, *Tetrahedron Lett.*, **1992**, *33*, 2761. See also R. Quintanilla and T. E. Cole, *Tetrahedron*, **1995**, *51*, 4297.

[150] T. E. Cole and R. Quintanilla, *J. Org. Chem.*, **1992**, *57*, 7366.

[151] S. Y. Lee and R. G. Bergman, *Tetrahedron*, **1995**, *51*, 4255.

[152] R. A. Budnik and J. K. Kochi, *J. Organomet. Chem.*, **1976**, *116*, C3.

[153] M. D. Fryzuk, G. S. Bates, and C. Stone, *Tetrahedron Lett.*, **1986**, *27*, 1537.
[154] S. Kim and K. H. Kim, *Tetrahedron Lett.*, **1995**, *36*, 3725.
[155] M. Yoshifuji, M. J. Loots, and J. Schwartz, *Tetrahedron Lett.*, **1977**, 1303.
[156] P. Wipf and W. Xu, *Tetrahedron Lett.*, **1994**, *35*, 5197.
[157] P. Wipf and W. Xu, *Org. Synth.*, **1997**, *74*, 205.
[158] P. Knochel and R. D. Singer, *Chem. Rev.*, **1993**, *93*, 2117.
[159] B. H. Lipshutz and E. L. Ellsworth, *J. Am. Chem. Soc.*, **1990**, *112*, 7440.
[160] B. H. Lipshutz and R. Keil, *J. Am. Chem. Soc.*, **1992**, *114*, 7919.
[161] B. Zheng and M. Srebnik, *J. Org. Chem.*, **1995**, *60*, 3278. See also S. Pereira and M. Srebnik, *Organometallics*, **1995**, *14*, 3127.
[162] E. Negishi and D. E. Van Horn, *J. Am. Chem. Soc.*, **1977**, *99*, 3168.
[163] N. Okukado, D. E. Van Horn, W. L. Klima, and E. Negishi, *Tetrahedron Lett.*, **1978**, 1027.
[164] E. Negishi, N. Okukado, A. O. King, D. E. Van Horn, and B. I. Spiegel, *J. Am. Chem. Soc.*, **1978**, *100*, 2254.
[165] M. J. Loots and J. Schwartz, *J. Am. Chem. Soc.*, **1977**, *99*, 8045.
[166] M. J. Loots and J. Schwartz, *Tetrahedron Lett.*, **1978**, 4381.
[167] J. Schwartz, M. J. Loots, and H. Kosugi, *J. Am. Chem. Soc.*, **1980**, *102*, 1333.
[168] B. H. Lipshutz and M. Segi, *Tetrahedron*, **1995**, *51*, 4407.
[169] D. E. Van Horn, L. F. Valente, M. J. Idacavage, and E. Negishi, *J. Organomet. Chem.*, **1978**, *156*, C20.
[170] E. Negishi, D. E. Van Horn, T. Yoshida, and C. L. Rand, *Organometallics*, **1983**, *2*, 563.
[171] C. L. Rand, D. E. Van Horn, M. W. Moore, and E. Negishi, *J. Org. Chem.*, **1981**, *46*, 4093.
[172] E. Negishi, L. D. Boardman, J. M. Tour, H. Sawada, and C. L. Rand, *J. Am. Chem. Soc.*, **1983**, *105*, 5344.
[173] L. D. Boardman, V. Bagheri, H. Sawada, and E. Negishi, *J. Am. Chem. Soc.*, **1984**, *106*, 6105.
[174] E. Negishi, L. D. Boardman, H. Sawada, V. Bagheri, A. T. Stoll, J. M. Tour, and C. L. Rand, *J. Am. Chem. Soc.*, **1988**, *110*, 5383.
[175] E. Negishi, F. Liu, D. Choueiry, M. M. Mohamud, A. Silveira, Jr., and M. Reeves, *J. Org. Chem.*, **1996**, *61*, 8325.
[176] F. Liu and E. Negishi, *Tetrahedron Lett.*, **1997**, *38*, 1149.
[177] S. Ma and E. Negishi, *J. Org. Chem.*, **1997**, *62*, 784.
[178] T. Mole and E. A. Jeffery, *Organoaluminum Compounds,* Elsevier, Amsterdam, **1972**, 465 pp.
[179] J. A. Miller and E. Negishi, *Tetrahedron Lett.*, **1984**, *25*, 5863.
[180] E. Negishi, D. E. Van Horn, A. O. King, and N. Okukado, *Synthesis*, **1979,** 501.
[181] E. Negishi, K. P. Jadhav, and N. Daotien, *Tetrahedron Lett.*, **1982**, *23*, 2085.
[182] N. Okukado and E. Negishi, *Tetrahedron Lett.*, **1978**, 2357.
[183] M. Kobayashi, L. F. Valente, E. Negishi, W. Patterson, and A. Silveira, Jr., *Synthesis*, **1980**, 1034.
[184] E. Negishi and S. Baba, *J. Chem. Soc., Chem. Comm.*, **1976**, 596.
[185] E. Negishi, L. F. Valente, and M. Kobayashi, *J. Am. Chem. Soc.*, **1980**, *102*, 3298.
[186] S. Baba and E. Negishi, *J. Am. Chem. Soc.*, **1976**, *98*, 6729.
[187] E. Negishi, F. T. Luo, and C. L. Rand, *Tetrahedron Lett.*, **1982**, *23*, 27.
[188] H. Matsushita and E. Negishi, *J. Am. Chem. Soc.*, **1981**, *103*, 2882.
[189] E. Negishi, H. Matsushita, and N. Okukado, *Tetrahedron Lett.*, **1981**, *32*, 2715.
[190] E. Negishi, V. Bagheri, S. Chatterjee, F. T. Luo, J. A. Miller, and A. T. Stoll, *Tetrahedron Lett.*, **1983**, *24*, 5181.
[191] E. Negishi and Z. Owczarczyk, *Tetrahedron Lett.*, **1991**, *32*, 6683.
[192] D. R. Williams, B. A. Barner, K. Nishitani, and J. G. Phillips, *J. Am. Chem. Soc.*, **1982**, *104*, 4708.
[193] S. Fung and J. B. Siddall, *J. Am. Chem. Soc.*, **1980**, *102*, 6580.
[194] B. H. Lipshutz, G. Bulow, R. F. Lowe, and K. L. Stevens, *J. Am. Chem. Soc.*, **1996**, *118*, 5512.
[195] D. Y. Kondakov and E. Negishi, *J. Am. Chem. Soc.*, **1995**, *117*, 10771.

[196] G. Erker, M. Aulbach, M. Knickmeier, D. Wingbermuhle, C. Kruger, M. Nolte, and S. Werner, *J. Am. Chem. Soc.*, **1993**, *115*, 4590.

[197] V. M. Dzhemilev, A. G. Ibragimov, A. P. Zoltarev, R. R. Muslukhov, and G. A. Tolstikov, *Izv. Akad. Nauk SSSR, Ser. Khim.*, (Engl. Transl.), **1989**, 194; **1991**, 2570.

[198] D. Y. Kondakov and E. Negishi, *J. Am. Chem. Soc.*, **1996**, *118*, 1577.

[199] J. P. Morken, M. T. Didiuk, and A. H. Hoveyda, *J. Am. Chem. Soc.*, **1993**, *115*, 6997.

[200] L. Bell, R. J. Whitby, R. V. H. Jones, and M. C. H. Standen, *Tetrahedron Lett.*, **1996**, *37*, 7139.

[201] D. R. McAlister, D. K. Erwin, and J. E. Bercaw, *J. Am. Chem. Soc.*, **1978**, *100*, 5966.

[202] J. E. Bercaw, D. H. Berry, A. J. Jircitano, and K. B. Mertes, *J. Am. Chem. Soc.*, **1982**, *104*, 4712.

[203] G. Erker, *J. Organomet. Chem.*, **1977**, *134*, 189.

[204] G. Erker and K. Kropp, *J. Organomet. Chem.*, **1980**, *194*, 45.

[205] F. Rosenfeldt and G. Erker, *Tetrahedron Lett.*, **1980**, *21*, 1637.

[206] G. Erker, K. Kropp, J. L. Atwood, and W. E. Hunter, *Organometallics*, **1983**, *2*, 1555.

[207] G. Erker, K. Engel, U. Dorf, J. L. Atwood, and W. E. Hunter, *Angew. Chem., Int. Ed. Engl.*, **1982**, *21*, 914.

[208] G. Erker, K. Engel, J. L. Atwood, and W. E. Hunter, *Angew. Chem., Int. Ed. Engl.*, **1983**, *22*, 494.

[209] G. Erker and U. Dorf, *Angew. Chem., Int. Ed. Engl.*, **1983**, *22*, 777.

[210] J. Yin, K. Abboud, and W. M. Jones, *J. Am. Chem. Soc.*, **1993**, *115*, 3810.

[211] J. Yin, K. Abboud, and W. M. Jones, *J. Am. Chem. Soc.*, **1993**, *115*, 8859.

[212] J. Yin and W. M. Jones, *Tetrahedron*, **1995**, *51*, 4395.

[213] E. Negishi, D. R. Swanson, and T. Takahashi, *J. Chem. Soc., Chem. Comm.*, **1990**, *18*, 1254.

[214] E. Negishi, T. Nguyen, J. P. Maye, D. Choueiry, N. Suzuki, and T. Takahashi, *Chem. Lett.*, **1992**, 2367.

[215] K. Takagi, C. J. Rousset, and E. Negishi, *J. Am. Chem. Soc.*, **1991**, *113*, 1440.

[216] D. R. Swanson, C. J. Rousset, E. Negishi, T. Takahashi, T. Seki, M. Saburi, and Y. Uchida, *J. Org. Chem.*, **1989**, *54*, 3521.

[217] M. Jensen and T. Livinghouse, *J. Am. Chem. Soc.*, **1989**, *111*, 4495.

[218] W. A. Nugent and D. F. Taber, *J. Am. Chem. Soc.*, **1989**, *111*, 6435.

[219] M. Akita, H. Yasuda, H. Yamamoto, and A. Nakamura, *Polyhedron*, **1991**, *10*, 1.

[220] E. Negishi, D. R. Swanson, F. E. Cederbaum, and T. Takahashi, *Tetrahedron Lett.*, **1987**, *28*, 917.

[221] T. V. RajanBabu, W. A. Nugent, D. F. Taber, and P. J. Fagan, *J. Am. Chem. Soc.*, **1988**, *110*, 7128.

[222] W. A. Nugent, D. L. Thorn, and R. L. Harlow, *J. Am. Chem. Soc.*, **1987**, *109*, 2788.

[223] D. F. Taber and J. P. Louey, *Tetrahedron*, **1995**, *51*, 4495.

[224] E. Negishi and S. R. Miller, *J. Org. Chem.*, **1989**, *54*, 6014.

[225] T. Takahashi, T. Seki, Y. Nitto, M. Saburi, C. J. Rousset, and E. Negishi, *J. Am. Chem. Soc.*, **1991**, *113*, 6266.

[226] K. Kropp and G. Erker, *Organometallics*, **1982**, *1*, 1246.

[227] G. Erker, U. Dorf, P. Czisch, and J. L. Petersen, *Organometallics*, **1986**, *5*, 668.

[228] T. Takahashi, M. Kageyama, V. Denisov, R. Hara, and E. Negishi, *Tetrahedron Lett.*, **1993**, *34*, 687.

[229] T. Takahashi, Z. Xi, C. J. Rousset, and N. Suzuki, *Chem. Lett.*, **1993**, 1001.

[230] H. G. Alt and M. D. Rausch, *J. Am. Chem. Soc.*, **1974**, *96*, 5936.

[231] S. Thaneder and M. F. Farona, *J. Organomet. Chem.*, **1982**, *235*, 65.

[232] S. L. Buchwald and R. B. Nielsen, *J. Am. Chem. Soc.*, **1989**, *111*, 2870.

[233] R. Hara, Z. Xi, M. Kotora, C. Xi, and T. Takahashi, *Chem. Lett.*, **1996**, 1003.

[234] U. Rosenthal, A. Ohff, W. Baumann, A. Tillack, H. Görls, V. B. Burlakov, and V. B. Shur, *J. Organomet. Chem.*, **1994**, *484*, 203.

[235] V. Skibbe and G. Erker, *J. Organomet. Chem.*, **1983**, *241*, 15.

[236] T. Takahashi, N. Suzuki, M. Hasegawa, Y. Nito, K. Aoyagi, and M. Saburi, *Chem. Lett.*, **1992**, 331. See also N. Suzuki, D. Y. Kondakov, M. Kageyama, M. Kotora, R. Hara, and T. Takahashi, *Tetrahedron*, **1995**, *51*, 4519.

[237] C. Lefeber, A. A. Tillack, W. Baumann, R. Kempe, V. V. Burlakov, and U. Rosenthal, *Organometallics*, **1995**, *14*, 3090.

[238] I. U. Khand, G. R. Knox, P. L. Pauson, W. E. Watts, and M. I. Foreman, *J. Chem. Soc., Perkin Trans. 1*, **1973**, 977.

[239] N. E. Schore in *Comprehensive Organic Synthesis*, Vol. 5, L. A. Paquette, Ed., Pergamon Press, Oxford, **1991**, 1037.

[240] B. Y. Lee, Y. K. Chung, N. Jeong, Y. Lee, and S. H. Hwang, *J. Am. Chem. Soc.*, **1994**, *116*, 8793.

[241] E. C. Lund and T. Livinghouse, *J. Org. Chem.*, **1989**, *54*, 4487. See also B. L. Pagenkopf, E. C. Lund, and T. Livinghouse, *Tetrahedron*, **1995**, *51*, 4421.

[242] G. Agnel and E. Negishi, *J. Am. Chem. Soc.*, **1991**, *113*, 7424.

[243] E. Negishi, D. Choueiry, T. B. Nguyen, D. R. Swanson, N. Suzuki, and T. Takahashi, *J. Am. Chem. Soc.*, **1994**, *116*, 9751.

[244] P. A. Wender and F. E. McDonald, *J. Am. Chem. Soc.*, **1990**, *112*, 4956.

[245] G. Agnel, Z. Owczarczyk, and E. Negishi, *Tetrahedron Lett.*, **1992**, *33*, 1543.

[246] M. Mori, N. Uesaka, and M. Shibasaki, *J. Org. Chem.*, **1992**, *57*, 3519. See also (a) M. Mori, F. Saitoh, N. Uesake, K. Okamura, and T. Date, *J. Org. Chem.*, **1994**, *59*, 4993. (b) N. Uesaka, F. Saitoh, M. Mori, M. Shibasaki, K. Okamura, and T. Date, *J. Org. Chem.*, **1994**, *59*, 5633. (c) M. Mori, N. Uesaka, F. Saitoh, and M. Shibasaki, *J. Org. Chem.*, **1994**, *59*, 5643. (d) T. Honda, S. Satoh, and M. Mori, *Organometallics*, **1995**, *14*, 1548. (e) F. Saitoh, M. Mori, K. Okamura, and T. Date, *Tetrahedron*, **1995**, *51*, 4439.

[247] E. Negishi, S. Ma, T. Sugihara, and Y. Noda, *J. Org. Chem.*, **1997**, *62*, 1922.

[248] J. P. Maye and E. Negishi, *Tetrahedron Lett.*, **1993**, *34*, 3359.

[249] E. Negishi, J. P. Maye, and D. Choueiry, *Tetrahedron*, **1995**, *51*, 4447.

[250] T. Takahashi, K. Kasai, N. Suzuki, K. Nakajima, and E. Negishi, *Organometallics*, **1994**, *13*, 3413.

[251] C. Copˌret, E. Negishi, Z. Xi, and T. Takahashi, *Tetrahedron Lett.*, **1994**, *35*, 695.

[252] T. Takahashi, D. Y. Kondakov, Z. Xi, and N. Suzuki, *J. Am. Chem. Soc.*, **1995**, *117*, 5871.

[253] T. Takahashi, T. Fujimori, T. Seki, M. Saburi, Y. Uchida, C. J. Rousset, and E. Negishi, *J. Chem. Soc., Chem. Comm.*, **1990**, *18*, 182.

[254] D. P. Lewis, R. J. Whitby, and R. V. H. Jones, *Tetrahedron*, **1995**, *51*, 4541.

[255] A. H. Hoveyda and Z. Xu, *J. Am. Chem. Soc.*, **1991**, *113*, 5079.

[256] K. S. Knight and R. M. Waymouth, *J. Am. Chem. Soc.*, **1991**, *113*, 6268.

[257] D. P. Lewis, D. M. Muller, R. J. Whitby, and R. V. H. Jones, *Tetrahedron Lett.*, **1991**, *32*, 6797.

[258] F. N. Tebbe, G. W. Parshall, and G. S. Reddy, *J. Am. Chem. Soc.*, **1978**, *100*, 3611.

[259] K. A. Brown-Wensley, S. L. Buchwald, L. Cannizzo, L. Clawson, S. Ho, D. Meinhardt, J. R. Stille, D. Straus, and R. H. Grubbs, *Pure Appl. Chem.*, **1983**, *55*, 1733.

[260] K. C. Ott, E. J. M. deBoer, and R. H. Grubbs, *Organometallics*, **1984**, *3*, 223.

[261] C. J. Rousset, E. Negishi, N. Suzuki, and T. Takahashi, *Tetrahedron Lett.*, **1992**, *33*, 1965.

[262] E. Negishi, C. J. Rousset, D. Choueiry, J. P. Maye, N. Suzuki, and T. Takahashi, *Inorg. Chem. Acta*, submitted.

[263] U. Wischmeyer, K. S. Knight, and R. M. Waymouth, *Tetrahedron Lett.*, **1992**, *33*, 7735.

[264] K. S. Knight, D. Wang, R. M. Waymouth, and J. Ziller, *J. Am. Chem. Soc.*, **1994**, *116*, 1845.

[265] N. Uesaka, M. Mori, K. Okamura, and T. Date, *J. Org. Chem.*, **1994**, *59*, 4542.

[266] K. H. Shaugnessy and R. M. Waymouth, *J. Am. Chem. Soc.*, **1995**, *117*, 5873.

[267] N. Suzuki, D. Y. Kondakov, and T. Takahashi, *J. Am. Chem. Soc.*, **1993**, *115*, 8485.

[268] K. S. Knight and R. M. Waymouth, *Organometallics*, **1994**, *13*, 2575.

[269] T. Takahashi, K. Aoyagi, V. Denisov, N. Suzuki, D. Choueiry, and E. Negishi, *Tetrahedron Lett.*, **1993**, *34*, 8301.

[270] U. M. Dzhemilev, A. G. Ibragimov, A. P. Zoltarev, R. R. Muslukov, and G. A. Tolstikov, *Izv. Akad. Nauk SSSR, Ser. Khim.*, **1991**, 2570.

[271] H. Sinn and G. Oppermann, *Angew. Chem., Int. Ed. Engl.*, **1966**, *5*, 962.

[272] H. Sinn and E. Kolk, *J. Organomet. Chem.*, **1966**, *6*, 373.

[273] W. Kaminsky and H. Sinn, *Liebigs Ann. Chem.*, **1975**, 424.

[274] W. Kaminsky and H. Sinn, *Liebigs Ann. Chem.*, **1975**, 438.

[275] W. Kaminsky, J. Kopf, H. Sinn, and H. Vollmer, *Angew. Chem., Int. Ed. Engl.*, **1976**, *15*, 629.

[276] T. Takahashi, N. Suzuki, M. Kageyama, Y. Nitto, M. Saburi, and E. Negishi, *Chem. Lett.*, **1991**, 1579.

[277] S. Couturier and B. Gautheron, *J. Organomet. Chem.*, **1978**, *157*, C61.

[278] R. Choukroun, M. B. Bert, and D. Gervais, *J. Chem. Soc., Chem. Comm.*, **1986**, 1317.

[279] R. Waymouth and P. Pino, *J. Am. Chem. Soc.*, **1990**, *112*, 4911.

[280] T. Takahashi, M. Hasegawa, N. Suzuki, M. Saburi, C. J. Rousset, P. E. Fanwick, and E. Negishi, *J. Am. Chem. Soc.*, **1991**, *113*, 8564.

[281] M. R. Kesti, M. Abdulrahman, and R. M. Waymouth, *J. Organomet. Chem.*, **1991**, *417*, C12.

[282] M. R. Kesti and R. M. Waymouth, *Organometallics*, **1992**, *11*, 1095.

[283] H. G. Woo, R. H. Heyn, and T. D. Tilley, *J. Am. Chem. Soc.*, **1992**, *114*, 5698.

[284] T. D. Tilley, *Acc. Chem. Res.*, **1993**, *26*, 22.

[285] F. W. Hartner, Jr., J. Schwartz, and S. M. Clift, *J. Am. Chem. Soc.*, **1983**, *105*, 640.

[286] F. W. Hartner, Jr., S. M. Clift, J. Schwartz, *Organometallics*, **1987**, *6*, 1346.

[287] R. Beckhaus, *Angew. Chem., Int. Ed. Engl.*,**1997**, *36*, 686.

[288] P. J. Walsch, F. J. Hollander, and R. G. Bergman, *J. Am. Chem. Soc.*, **1988**, *110*, 8729.

[289] P. J. Walsh, F. J. Hollander, and R. G. Bergman, *J. Am. Chem. Soc.*, **1990**, *112*, 894.

[290] P. J. Walsh, A. M. Baranger, and R. G. Bergman, *J. Am. Chem. Soc.*, **1992**, *114*, 1708.

[291] A. M. Baranger, P. J. Walsh, and R. G. Bergman, *J. Am. Chem. Soc.*, **1993**, *115*, 2753.

[292] P. J. Walsh, F. J. Hollander, and R. G. Bergman, *Organometallics*, **1993**, *12*, 3705.

[293] K. E. Meyer, P. J. Walsh, and R. G. Bergman, *J. Am. Chem. Soc.*, **1994**, *116*, 2669.

[294] K. E. Meyer, M. J. Walsh, and R. G. Bergman, *J. Am. Chem. Soc.*, **1995**, *117*, 974.

[295] S. Y. Lee and R. G. Bergman, *J. Am. Chem. Soc.*, **1995**, *117*, 5877.

[296] S. Y. Lee and R. G. Bergman, *J. Am. Chem. Soc.*, **1996**, *118*, 6396.

6 Group 5 and Group 6 Metallocenes

Pascual Royo and Evelyn Ryan

6.1 Introduction

In this Chapter the chemistry of group 5 and 6 bis(cyclopentadienyl) complexes is presented, summarizing those synthetic, structural and reactivity aspects which illustrate the chemical behavior of these types of compounds. This study does not pretend to be an exhaustive collection of literature references on this subject, although the references given cover the period up to the end of 1996. The reader is advised to consult the excellent reviews published in the first and second editions of *Comprehensive Organometallic Chemistry* for more detailed lists of references and previous contributions in the field [1, 2]. A more recent review on 'open shell' organometallics, also including sandwich and half-sandwich cyclopentadienyl metal derivatives, has been published by Poli [3], and other publications [4] reviewing particular topics include compounds of the types described in this Chapter.

According to the latest IUPAC recommendations [5] we use the name *'metallocene'* only to describe real 'sandwich' metal complexes containing two *parallel* cyclopentadienyl rings coordinated to the central atom, and no other component, although even here the two cyclopentadienyl rings are not exactly parallel but slightly bent due to attractive van der Waals forces between the rings [6, 7]. All those compounds formed by coordination of one or more additional substituents, which show structures with bent cyclopentadienyl rings are called *'bent-bis(cyclopentadienyl)'* or simply *'bis(cyclopentadienyl)'* metal complexes (the Cp-M-Cp bend angle, θ, being that formed at the metal centre by the two perpendiculars from the centroids of the cyclopentadienyl rings).

The theoretical explanation of the behavior of the bis(cyclopentadienyl) compounds with bend angles between 180° and 120° is based on the Lahuer and Hoffman calculations [8], lately extended to compounds with narrower Cp–M–Cp angles down to 110° [9, 10].

Most of the compounds described in this Chapter have metal centers with 15 to 18 electrons. As shown in Figure 6-1, twelve of these electrons are accommodated in the six low-lying, essentially cyclopentadienide ligand bonding orbitals, therefore the crucial orbitals which determine the structure of these complexes are the three e_2 and a_1 orbitals of the metallocene compounds, which provide the $1a_1$, b_2 and $2a_1$ orbitals for the bent-bis(cyclopentadienyl) complexes. Differences may be interpreted in terms of the different occupation of these three mainly metal-based orbitals by the remaining 1 to 6 electrons. We also consider where appropriate some metal systems with a higher total number of electrons, particularly those with 19.

Energy differences between these important three orbitals depend not only on the bend angle but also on factors associated with the nature of the metal, its oxidation state or valence number, and the electron-donating or -withdrawing character of the substituents bonded directly to the metal and indirectly to the cyclopentadienyl rings. The spin-pairing energy required to obtain a lower spin-state system, when possible, is therefore also related to these factors.

A useful new approach to the formal classification of covalent compounds has been published by M. L. H. Green [11]. In this Chapter [MCp$_2$] is a core fragment common to all complexes; the ligand notations defined by Green are generally used and similar methodology is followed to derive anionic and cationic complexes from their neutral precursors.

Fundamental differences between compounds are related to the number of substituents bonded to the [MCp$_2$] fragment and to the bonding capacity of the ligands. Compounds discussed here are therefore classified accordingly in four types: **Type A** compounds are metallocenes with no additional ligand, which are structurally and chemically different from the bent-bis(cyclopentadienyl) complexes; compounds containing mono-functional single-atom donor ligands are classed as **Type B**, including therefore complexes with: Lewis acids, mono-functional mono-, bi- and tri-dentate X ligands, and single-atom donor neutral mono-, bi- and tri-dentate L ligands; complexes with bi-functional single-atom donor ligands such as $=CR_2$, $=NR$ (linear and bent), $=O$, $=S$, $=C=CR_2$ and related systems are identified as **Type C**; complexes formed by side-bonded η^2 and η^3 ligands appear as **Type D**. Compounds of Types C and D are presented separately although they belong to some of the three groups studied for compounds of Type B and could be treated on the same basis.

Formal d^n configurations, oxidation states or valence numbers are frequently given, although sometimes these do not provide a complete, unequivocal description, and further discussion is included where necessary to aid good understanding of the real chemical behavior.

6.2   Type A Metallocene Compounds

Compounds of this stoichiometry, free from additional substituents bonded to the metal, have structure A (Figure 6-1) and are unlike any other type of complex described in this Chapter. They are structurally classified as true metallocenes containing parallel or almost parallel cyclopentadienyl rings in a typical sandwich disposition. The e_2 orbitals are degenerate and lower in energy than the a_1 orbital for all the group 5 and 6 metal complexes (Figure 6-1).

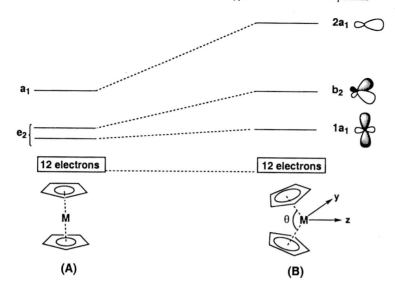

Figure 6-1

Qualitative variation of available molecular orbitals on going from metallocenes to
bent-dicyclopentadienyl metal complexes

The neutral group 5 and 6 metallocenes [MCp$_2$] with 15 and 16 electrons respectively are discussed in Sect. 6.2.1. The related anionic and cationic metallocenes, with a total number of electrons between 15 and 17, obtained from redox reactions of the neutral compounds, are discussed in Sect. 6.2.2.

6.2.1 Neutral Metallocenes

6.2.1.1 Group 5 Metallocenes [MCp$_2$]

Synthesis

Most of the synthetic methods used to obtain the 15-electron vanadocenes [VCp$_2$] (**1–3**) involve metathesis of vanadium(II) halides using alkali cyclopentadienides to transfer the cyclopentadienyl rings (Scheme 6-1).

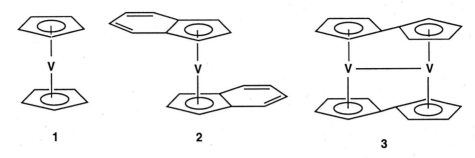

1   **2**   **3**

Vanadium(II) species used in situ were obtained by reducing [VCl$_3$(THF)$_3$] with agents such as Li[AlH$_4$] [12], Zn [13, 14] or more frequently the alkali metal cyclopentadienide when used in excess under reflux [15, 16]. Other vanadium(II) compounds such as [VCl$_2$(THF)$_6$] or better [V$_2$Cl$_3$(THF)$_6$]$_2$ [Zn$_2$Cl$_6$] [17] have also been used. Reduction of [VCp$_2$X$_2$] and [VCp$_2$X] (X = Cl, Me) with Li[AlH$_4$] [18] or aluminum slurries [19] prepared by metal vapor condensation with organic solvent has been used to prepare vanadocene. Complexes containing mixed cyclopentadienyl rings were isolated by the metathetical reaction of [CpVCl(THF)] with alkali metal cyclopentadienides [20].

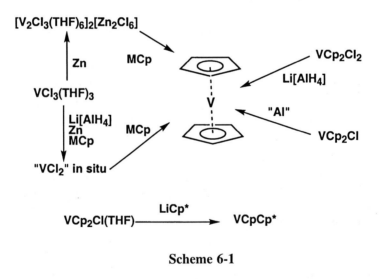

Scheme 6-1

The reported vanadocenes include compounds containing variously substituted cyclopentadienyl rings such as C$_5$H$_5$, C$_5$D$_5$, C$_5$H$_4$R (R = Me, Pr^i, Bu^n, Bu^t, Et), 1-Me-3-Ph-C$_5$H$_3$, 1,3-Ph$_2$-C$_5$H$_3$, C$_5$Me$_5$, C$_5$Me$_4$Et, C$_5$H$_4$EMe$_3$ (E = C, Si, Ge, Sn) and η^5-indenyl (**2**) [1,2,16]. The dinuclear bis(fulvalene) derivative

[V$_2$(fulvalene)$_2$] (**3**) [21], a dark purple pyrophoric solid, and complexes containing mixed cyclopentadienyl rings [VCp(C$_5$H$_4$R)] (R = CMe$_2$Ph, CHMePh) have been isolated.

All of these compounds were thermally stable, air sensitive, red to purple oils or crystals; those containing substituted rings had lower melting points and were more soluble and more air sensitive.

The niobium and tantalum derivatives were less stable. Niobocene was formed by H abstraction and spontaneous H$_2$ reductive elimination from [NbCp$_2$H$_3$] [22] and by reduction of [NbCp$_2$Cl$_2$] with sodium naphthalenide [23]. Substituted niobocenes (**4A**), formed in solution [24] by reduction of their dihalides with 2 equiv. of Na/Hg, are thought to exist in equilibrium with their fulvene isomers (**4B**) (Scheme 6-2). Their structures were assigned on the basis of their EPR spectra, and reaction of the permethylated niobocene with sulfur, trapping both forms in the products, has provided chemical evidence for these structures [25].

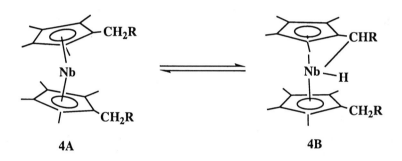

4A   **4B**

Scheme 6-2

Structure

Spin-restricted and spin-polarized calculations[26] with S=3/2 made for [VCp$_2$] by SCF-DV-Xα methods resulted in an e$_{2g}$-a$_{1g}$ energy separation (Δ = 0.25 eV) in agreement with experimental values determined by optical spectroscopy [27] and the observed magnetic moment of 3.78 μ_B, consistent with the presence of three unpaired electrons. Similarly, three unpaired electrons in [V(C$_5$Me$_5$)$_2$] consistent with a ^{4}A ground state were suggested by magnetic susceptibility measurements and EPR spectroscopy. [15].

Paramagnetic ^{1}H and ^{13}C NMR spectra [28] and UV–Vis spectra [14] have been recorded for the EMe$_3$-ring-substituted vanadocenes. However, the dinuclear difulvalene complex **3** was reported as diamagnetic.

The nature of the ring disorder causing the disagreement between previous X-ray [29] and electron diffraction [30] studies of [VCp$_2$] was explained by an X-ray structural determination in the temperature interval 108–357 K [31]. A crystallographic determination of [V(C$_5$Me$_5$)$_2$] has shown a sandwich structure with parallel planar cyclopentadienyl rings at an average V–C$_{ring}$ distance of 1.91(5)Å [15], and a similar sandwich disposition was found for [V(C$_9$H$_7$)$_2$] [17], although the presence of two different orientations in the crystal lattice prevented its resolution. All these vanadocenes are very reactive as expected for 15-electron metallocene compounds [2]. They behave as carbene-like fragments, which readily undergo oxidative addition to many organic functional groups. They form unstable complexes with N$_2$, O$_2$, H$_2$ and methane which can only be studied at low temperature and which are easily and rapidly oxidized to higher oxidation states [32]. In all of these reactions, they lose their sandwich structure through the introduction of additional substituents in the coordination sphere, giving bent bis(cyclopentadienyl) complexes.

Niobocene has a simple ten-line EPR spectrum in THF with A$_{Nb}$ = 103 G, which excludes the possibility of solvent coordination [22]. The permethylated niobocenes [Nb(C$_5$Me$_4$R)$_2$] (R = Me, Et) have more complex EPR spectra which indicate the presence of a metallocene species (**4A**) with A$_{Nb}$ = 100 G, together with a second isomer (**4B**) which results from the C–H activation and oxidative addition of one of the methyl cyclopentadienyl groups [24]. Both isomers shown in Scheme 6-2 were characterized by their chemical reactivity, described in later sections [33].

Applications

Vanadocene and related compounds are catalysts for acetylene [34] and 1,3-diene [35] polymerization and for deuteration [36] of CH$_4$ by D$_2$ or C$_2$D$_4$.

6.2.1.2 Group 6 Metallocenes [MCp$_2$]

Synthesis

The 16-electron neutral chromocene derivatives, [CrCp$_2$] (**5**) are usually obtained [1,2] by reaction of a previously reduced chromium(II) species such as CrCl$_2$ or [Cr$_2$(μ-OAc)$_4$] with alkali metal cyclopentadienides (Scheme 6-3). Used in excess under reflux, these cyclopentadienides can also act as reducing agents for [CrCl$_3$(THF)$_3$]. Similar reactions with appropriate alkali cyclopentadienide salts have been used to prepare ring-substituted [37, 38], indenyl (**6**) [39], and fulvalene dinuclear (**7**) [40] derivatives, and related complexes containing bridged bis(cyclopentadienyl) ligands (**8**) [41].

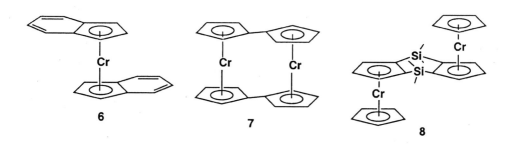

Scheme 6-3

Reported complexes include those with two equal or mixed ring-substituted ligands such as C_5H_5, C_5H_4R, $C_5H_4EMe_3$ (E = Si, Ge, Sn), C_5HPh_4, C_5Me_5, C_5Me_4Et, η^5-indenyl, $C_{10}H_8$ (fulvalene) and μ-[(SiMe$_2$)$_2$(C$_5$H$_3$)$_2$].[2]

The chromium derivatives, obtained as red solids, are more accessible, thermally more stable and less reactive than those of the heavier metals. Photochemically induced elimination of neutral ligands from [MCp$_2$(CO)] (M = Mo, W) and [WCp$_2$(C$_2$H$_4$)], or similar reductive elimination from [MCp$_2$H$_2$] (M = Mo, W), [WCp$_2$(CH$_3$)H] or [WCp$_2$(Bcat)H] (Bcat = C$_6$H$_4$O$_2$), has been used to obtain the heavier metallocenes [MCp$_2$] (M = Mo, W) which have only been studied in low temperature matrices or as transient species. [42, 43, 44, 45].

Structure

Chromocenes are stable enough to allow their investigation by different structural techniques and chemical reactions. The orbitally degenerate ground state of 16-electron chromocenes leads to temperature-dependent magnetic moments greater than the spin-only value, although the orbital angular momentum is partly quenched by Jahn–Teller effect. Unsubstituted chromocene has a magnetic moment [46] of 3.27 μ_B and values of between 2.90 and 3.20 μ_B have been found [38] for substituted chromocenes. Various studies conclude [46, 47] that the triplet ground state for chromocenes is $^3E_{2g}$ and corresponds to a $e_{2g}^3a_{1g}^1$ configuration, leading to a significant orbital contribution. Their structural behavior has been studied by IR [48], gas phase absorption [49] and electron transmission [50] spectroscopy. Coupling of the spins of both chromocene halves due to metal–metal interaction rather than superexchange has been demonstrated by variable temperature ^1H NMR for the difulvalene dinuclear complex (**7**) [40]. The ^1H and ^{13}C NMR spectra observed for variously substituted mononuclear complexes and the dinuclear complex **8** give broad and paramagnetically shifted resonances [51, 52] and the ^{29}Si paramagnetic shift has been measured for $[Cr(C_5H_4SiMe_3)_2]$ [53].

The molecular structure of chromocene has been determined by electron diffraction [30] and X-ray diffraction analysis [54] and variously substituted chromocenes have also been characterized by X-ray diffraction studies [38, 41]. All of these reported structures show a sandwich disposition with metal–ring-centroid distances of around 1.80 Å. A single crystal X-ray diffraction analysis of $[\{Cr(Indenyl)_2\}_2]$ shows it to be a dimer containing η^5- and μ-η^3-indenyl groups with metal-ring-centroid distances of 1.999 Å and 2.127 Å respectively [55].

The molybdenum and tungsten compounds have only been studied when trapped in inert argon matrices at low temperature, in order to prevent their spontaneous transformation into bent bis(cyclopentadienyl) compounds with η^1–η^5-bridging rings or fulvene structures. The expected two unpaired electrons for the 16-electron molybdenum and tungsten complexes $[MCp_2]$ (M = Mo, W), corresponding to a $^3E_{2g}$ ground state, have been inferred from their UV and magnetic circular dichroism spectra [42–44].

Chromocene is a very reactive compound which loses its cyclopentadienyl rings easily when treated with diols, carboxylic acids, acetylacetone or ethylenediamine [56, 57] and has a low capacity to coordinate additional ligands [58].

Applications

The heterogeneous catalyst generated by Union Carbide [59] by reacting $[CrCp_2]$ with silica requires neither a cocatalyst nor an activator, and displays high activity in ethylene polymerization at low pressure and low temperature. The system is very sensitive to hydrogen transfer to the polymer chain giving polyethylene of specific melt index. Although many analytical and spectroscopic studies have been carried out, the structure of the active species and its oxidation state are not known. It has been demonstrated however that only some mono(cyclopentadienyl) chromium(III) species are catalytically active in solution [60]. Recent solid-state NMR studies conclude that the catalyst contains at least one mononuclear and two dinuclear surface-attached chromium complexes with one or more cyclopentadienyl rings [61]; molecular model compounds and NMR spectra have been used to analyze their nature.

6.2.2 Ionic Metallocenes

A summary of the redox behavior of vanadocenes and chromocenes is shown in Scheme 6-4.

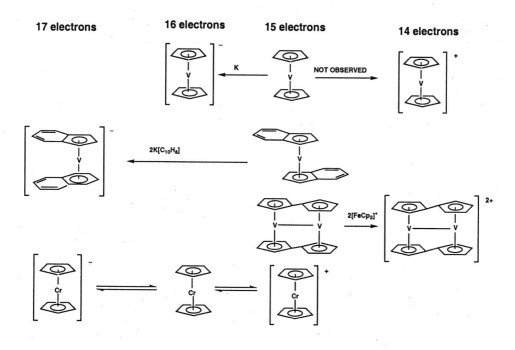

Scheme 6-4

6.2.2.1 Group 5 Cationic Metallocenes [MCp₂]⁺

14-electron [MCp$_2$]⁺ group 5 metallocene cations do not exist and all attempts made to oxidize the neutral metallocenes failed [14, 62]. Similarly, no base-free [MCp$_2$]$^{2+}$ dicationic 14-electron group 6 metallocenes nor any base-free cationic 13-electron complex of either group 5 or group 6 metals have ever been observed. The formation of the unstable 13-electron dication [Cp$_2$V]$^{2+}$ was proposed by rapid disproportionation of the cation [VCp$_2$Cl]⁺ resulting from the electrochemical oxidation (–0.08 V) of [Cp$_2$VCl] [63].

Exceptionally, the diamagnetic difulvalene complex [V$_2$(fulvalene)$_2$] is oxidized by one equivalent of [FeCp$_2$][PF$_6$] to yield purple black crystals of the very air sensitive salt [V$_2$(fulvalene)$_2$][PF$_6$], which is paramagnetic with one unpaired electron [64]; its EPR spectrum shows that the free electron is coupled to only one metal atom. Oxidation of this product by a second equivalent gives the diamagnetic, air sensitive purple salt $[V_2(fulvalene)_2][PF_6]_2$.

6.2.2.2 Group 5 Anionic Metallocenes [MCp$_2$]$^-$

Reduction of vanadocene by potassium gives $K[VCp_2]$ [21], a highly reactive 16-electron metallocene which easily loses one cyclopentadienyl ligand when it reacts with CO to give the more stable 18-electron monocyclopentadienyl derivative $[VCp(CO)_4]$ [65]. $K[VCp_2]$ is a paramagnetic compound probably having the same spin triplet ground state $^3E_{2g}$ as the isoelectronic chromocene derivative. The related anionic vanadium(I) mixed ring sandwich compound $[V(\eta^5\text{-Ind})(\eta^6\text{-naphtha-}$ lene)]$^-$ was obtained via the reduction of $[V(Ind)_2]$ with potassium naphthalenide [66]. Similar mixed-ring neutral $[VCp(\eta^6\text{-Arene})]$ complexes were isolated by reaction of $[(VCp)_2(\mu\text{-benzene})]$ with LiCp followed by addition of alcohol [67].

6.2.2.3 Group 6 Cationic Metallocenes [MCp$_2$]$^+$

A quasi-reversible oxidation to the 15-electron chromocenium cation $[CrCp_2]^+$ with a potential of $E_p = -0.67$ V was observed in an electrochemical study of chromocene $[CrCp_2]$ [68]. Stable solutions can be prepared by electrochemical oxidation in the presence of PF_6^-. Salts of solid paramagnetic $[CrCp_2]^+$ with three unpaired electrons were also isolated with various anions; stability was dependent on the anion. The iodide obtained by chemical oxidation with allyl iodide is pyrophoric whereas the salts of $[CrCpCl_3]^-$ and $[CrCp(CO_3)]^-$ are stable and the molecular structure of the latter was determined by X-ray diffraction [69]. Greater stability is conferred by bulky cyclopentadienyl ligands such as C_5Me_5 and C_5Me_4Et and the salts can be isolated by oxidation of the appropriate chromocenes by $AgPF_6$ [14]. The chromocenium cation has a spin quartet ground state and its NMR ^1H and ^{13}C paramagnetic shifts have been determined [62].

Related molybdenum and tungsten derivatives are less known, a remarkable exception being $[Mo(C_5Ph_5)_2]^+$, isolated as its Br_3^- partially spin-paired paramagnetic salt ($\mu_{eff} = 3.5 \ \mu_B$) [70].

6.2.2.4 Group 6 Anionic Metallocenes [MCp$_2$]$^-$

Electrochemical reduction of chromocene shows quasi-reversible formation of the anionic compound $[CrCp_2]^-$ which is persistent on the electrochemical time scale [68]. Reduction and reoxidation potentials of $E_p = -2.44$ V and $E_p = -2.26$ V were measured, however none of these 17-electron, anionic group 6 metallocenes have ever been isolated.

6.3   Type B-Bent bis(Cyclopentadienyl) Metal Complexes with Mono-Functional Single-Atom Donor Monoanionic and Neutral Ligands

Bonding of one or more additional substituents to the metal center leads to structures with bent bis(cyclopentadienyl) rings (Figure 6-1B). The bonding capacity of the bent $[MCp_2]$ fragment, and therefore the structural and chemical behavior of these compounds, was rationalized by Lahuer and Hoffmann [8] and this theoretical study is used to classify the different types of complexes described in the following sections.

6.3.1   Complexes with One Coordinated Ligand

When only one additional monoanionic (X) or neutral (L) ligand is coordinated to the metal center the resulting species for group 5 and 6 metals belong to the following types (Table 6-1):

Table 6-1. Different types of group 5 and 6 $[MCp_2X]$ and $[MCp_2L]$ metal complexes

Total Number of electrons	16	17	18
Configuration	d^2	d^3	d^4
	$[(M5)Cp_2X]$	$[(M5)Cp_2L]$ $[(M6)Cp_2X]$	$[(M6)Cp_2L]$
	$[(M5)Cp_2L]^+$ $[(M6)Cp_2X]^+$	$[(M5)Cp_2X]^-$ $[(M6)Cp_2L]^+$	$[(M6)Cp_2X]^-$ $[(M5)Cp_2L]^-$

18-Electron Compounds

When only one additional σ bonding ligand is coordinated to a metal with a C_{2v} symmetry, the $2a_1$ metal based orbital is used for bonding interaction with the σ orbital of the ligand, whereas the other two orbitals $1a_1$ and b_2 remain essentially non-bonding, occupied by the d^4 metal electrons. This has been predicted as the most stable structure for d^4 and high-spin d^2 metal complexes [8]. However other less symmetrical structures with the ligand located in the equatorial plane but forming an angle α with the twofold axis (**9**) are favoured for d^0, d^1, low-spin d^2 and d^3 metal compounds,8 although it has been calculated for d^0 complexes that the distorsion α angle also depends on the $s^n d^{3-n}$ ground-state configuration of the metal [71].

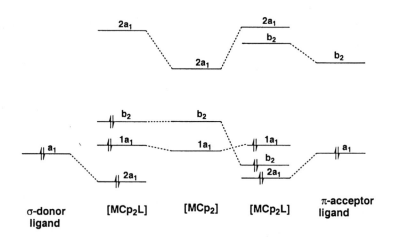

Figure 6-2

Qualitative diagram for [MCp$_2$L] d^4 metal complexes
with σ-donor and π-acceptor ligands

Electron-rich d^4 metal complexes are stabilized when the coordinated substituent is a π-acceptor ligand, whose empty π* orbital of b$_2$ symmetry can participate in a π-bonding interaction with a metal orbital of the same symmetry, releasing part of the electron density of the metal (Figure 6-2). It should be expected therefore that formation of d^4 metal compounds of this type will be favored by π-acid ligands which can alleviate the excess electron density on the metal by adopting a symmetrical disposition of the ligand in the equatorial plane of the cyclopentadienyl rings.

6.3.1.1 Neutral Group 6 [MCp$_2$L] Metal Complexes

As mentioned in Section 6.2.1, chromocene has a low capacity to coordinate one additional ligand,[58] which would lead to an 18 electron bent *bis*(cyclopentadienyl) system such as [Cp$_2$Cr(CO)] (**10**). The estimated formation constant of this adduct is K ≈ 50 atm^{-1} at 0°C and 2% of uncomplexed chromocene is always present. This resistance is mainly due to the energy required to bend the rings (21 kcal · mol^{-1}) [72], therefore factors which reduce ring-ring repulsive forces by distorsion of the ground-state geometry will favour coordination and for this reason the incorporation of a heteroannular ethylene bridge between the rings (**11**) provides a more stable CO adduct [35]. Complex (**11**) and the related tertiary-butyl isocyanide adduct were conveniently prepared via the reaction of the calcium *ansa*-metallocene with chromium dichloride in the presence of the ligand [73].

Related molybdenum and tungsten complexes are not accessible by this direct preparative method due to the thermal instability of their metallocene precursors. The carbonyl adducts [MCp$_2$(CO)] (M = Mo, W) were isolated by trapping the transient metallocene (obtained by the reduction of [MCp$_2$Cl$_2$] with Na/Hg) [74] with CO, or by thermal or photochemical decomposition of the dihydrides [MCp$_2$H$_2$] or alkyl derivatives in the presence of CO. They are resistant to further carbonylation except in the presence of oxidants to give [WCp(η^3-Cp)(CO)$_2$], where formation of an otherwise unstable 20 electron system is avoided by slippage of one of the cyclopentadienyl rings [75 ,76]. The same behaviour was also found for [Mo(η^5-

Indenyl)$_2$(dppe)],[37] for which the exchange between the two η^5-η^3 coordination modes is rapid even at low temperature. Reaction of [WCp$_2$(CO)] with carbon tetrachloride results in electrophilic addition to one of the rings to give [WCp(η^4-C$_5$H$_5$-exo-CCl$_3$)Cl(CO)] [77].

Related phosphine adducts were obtained by similar reactions and [MoCp$_2$(PR$_3$)] was isolated by deprotonation of the cationic hydride [MoCp$_2$H(PR$_3$)]$^+$ [78]. They are labile compounds in which the ligand is easily displaced when treated with ketenes [79]. The isocyanide complexes [MoCp$_2$(CNR)] (R = Me, Et, Bu^t) were isolated by reduction of the cationic metal(IV) isocyanide derivatives [MoCp$_2$X(CNR)]$^+$ (X= halide, H, Me) with Na/Hg, or by substitution in the side-bonded acetonitrile compound [MoCp$_2$(η^2-NCR)] [80]. The x-ray molecular structure of the t-butyl derivative reveals that the ligand is symmetrically located in the equatorial plane and the θ bend angle of the cyclopentadienyl rings is very open (147.7°), typical for this type of complex and even larger than values observed for related compounds (142.8° for the corresponding Mo-PMe$_3$ complex). The extensive bending of the isocyanide ligand with a C-N-C angle of 139.5(4)°, the short Mo-C and the long C-N bond distances of 1.997(4) Å and 1.193(4) Å respectively are consistent with a strong π-back bonding interaction.

6.3.1.2 Anionic Group 5 [MCp$_2$L]$^-$ and Group 6 [MCp$_2$X]$^-$ Metal Complexes

Although the coordination of one π-acceptor ligand to an anionic vanadocene [VCp$_2$]$^-$ species would presumably lead to a stable 18-electron vanadium(I) [VCp$_2$L]$^-$ system, reaction of the electron-rich metallocene with neutral ligands always takes place with loss of one of the cyclopentadienyl ligands. The neutral adducts [VCp$_2$L] (L = CO, CNCy) undergo reversible one-electron reduction to the anionic species [VCp$_2$L]$^-$ as evidenced by cyclic voltametry, but they cannot be obtained on bulk electrolytic reduction, which only leads to their decomposition to give vanadium metal [1].

The isoelectronic group 6 anionic species [MCp$_2$X]$^-$ have only seldom been observed since they easily lose the X$^-$ ligand to give the more stable neutral metallocene. A rare example of this type of species is the molybdenum hydride derivative [K(18-crown-6)(μ-H) MoCp$_2$], which results from reaction of [MoCp$_2$H$_2$] with KH in the presence of 18-crown-6, and whose x-ray structure shows a bend angle for the cyclopentadienyl rings of 153.26° with a hydride Mo-H-K bridge described as a two-electron three-center covalent bond [81].

Related species with π-acid ligands [MCp$_2$L]$^-$ and more than 18-electrons have however been observed. An electrochemical study of chromocene under CO reveals that reduction to the 19-electron anionic species [Cp$_2$Cr(CO)]$^-$ is slightly more demanding than that of the sandwich chromocene, with a reduction potential E$_p$ = -2.54 V. The resulting anion is short lived and substitution of one ring by CO

[68] occurs rapidly to give [CpCr(CO)$_3$]$^-$. Similar behaviour was also observed for the *ansa-* complex containing a Me$_2$C-CMe$_2$ bridge between the two cyclopentadienyl rings, which decays giving an irreversible reduction, although none of these compounds were isolated.

17-Electron Compounds

17-electron complexes with only one additional mono-functional substituent coordinated to the metal centre are essentially represented by the group 5 d^3 metal derivatives [MCp$_2$L] with neutral ligands L, whereas the related d^3 group 6 metal compounds [MCp$_2$X] with one monodentate anionic ligand X are rather rare and scarcely studied. In addition we mention some aspects of the chemistry of related anionic group 5 [MCp$_2$X]$^-$ and cationic group 6 [MCp$_2$L]$^+$ derivatives.

For d^3 metal compounds with one σ-bonding ligand which uses the a$_1$ orbital to accommodate the σ-bonding electron-pair, the three electrons have to occupy the 1a$_1$ and b$_2$ orbitals (Figure 6-2) to give paramagnetic compounds with one unpaired electron and a fundamental spin-doublet state; these are coordinatively unsaturated and electron-rich systems. When a π-acceptor ligand is used the empty orbital of the ligand can interact with the occupied b$_2$ metal orbital providing stronger metal-ligand bonds and equally leaving one unpaired electron (Figure 6-3).

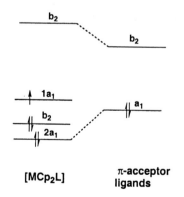

Figure 6-3

Electron configuration for [MCp$_2$L] d^3 metal complexes with π-acceptor ligands

6.3.1.3 Neutral Group 5 [MCp$_2$L] and Group 6 [MCp$_2$X] Metal Complexes

We have seen that vanadocene has basic character and behaves as a carbene-like unit. It therefore resists coordination of nucleophilic σ-donor ligands, but reacts easily with π-acceptor ligands leading to paramagnetic 17-electron *bis*(cyclopentadienyl) vanadium complexes [VCp$_2$L]. As a carbene unit it reacts easily with many organic unsaturated and functionalized groups to give the corresponding oxidative addition d^1 metal products which are discussed in Section 6.5.2. In this section we are concerned with essentially d^3 complexes containing monodentate, usually π-acid, ligands.

[VCp$_2$] reacts rapidly with CO at room temperature to give the adduct [VCp$_2$(CO)] (**12**), a dark brown paramagnetic solid with one unpaired electron (μ_{eff} = 1.76 μ_B) [82]; (**12**) has a very low ν(C-O) stretching frequency of 1881 cm^{-1} (Nujol), which is indicative of a strong V-C bond with a high level of π-back donation. The more electron donating pentamethylcyclopentadienyl ligands increase the basicity of the related [VCp$^*_2$] complex [15], which undergoes the same reaction with CO leading to a similar paramagnetic red-maroon compound (μ_{eff} = 1.71 μ_B), with an even lower ν(C-O) stretching frequency of 1845 cm^{-1} (Nujol). At low CO pressure, both compounds dissociate CO in an equilibrium with the sandwich metallocene [83]. The EPR spectra of their ^{13}C-CO derivatives show hyperfine couplings which demonstrate that the unpaired electron is delocalized onto the ^{13}CO ligand. This is consistent with the SCF-Xα-DV calculations which show a ^{2}A$_1$ ground state with an a$_1$ type HOMO [83], the odd electron occupying a hybrid d orbital localized between the cyclopentadienyl rings.

Similar carbonylation takes place with the dinuclear *bis*(fulvalene) complex leading to green crystals of [V$_2$(CO)$_2$(fulvalene)$_2$] (**13**) whose IR spectrum suggests a *cis*- dicarbonyl structure [1]. A similar reaction between CO and [V(Indenyl)$_2$] affords a structurally characterized dicarbonyl compound in which ring slippage of one of the indenyl ligands occurs to give η^3-coordination, in order to avoid a formally 19-electron configuration [84]. The x-ray molecular structure of [VCp$^*_2$(CO)] shows an average V-Cp distance of 1.928 Å with a θ bend angle of 153.6° and a very short V-CO bond distance of 1.879 Å, with the linear carbonyl ligand symmetrically placed in the equatorial plane. Similar reactions with isocyanides have also been studied but do not result in simple coordination of the ligand [15].

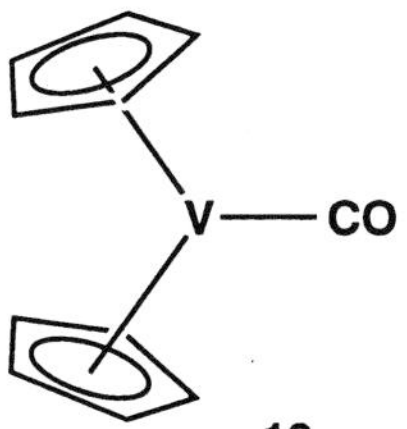

12

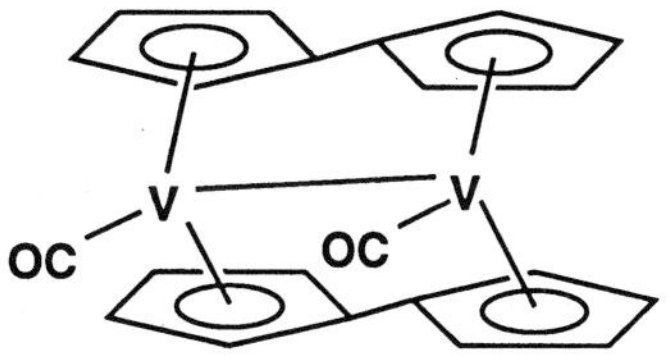

13

The 17-electron cyclohexylisocyanide complex (**14**) is formed as a yellow-brown paramagnetic (μ_{eff} = 1.78 μ_B) solid by coordination of the ligand as the first step in a reaction which proceeds with coordination of a second molecule of the ligand and further dealkylation of the isocyanide moiety to finally give the stable 18-electron [VCp*$_2$(CN)(CNR)] complexes (Scheme 6-5). This transformation is more rapid with CNBu^t, for which the intermediate adduct could not be isolated. Evidence for the formation of a 19-electron intermediate species [VCp*$_2$L$_2$] is supported by the related reaction of [VCp*$_2$(CO)] with CNBu^t in which the CO adduct [VCp*$_2$(CN)(CO)] is formed. The greater lability of the carbonyl ligand facilitates its dissociation to give the final 16-electron compound [VCp*$_2$(CN)].

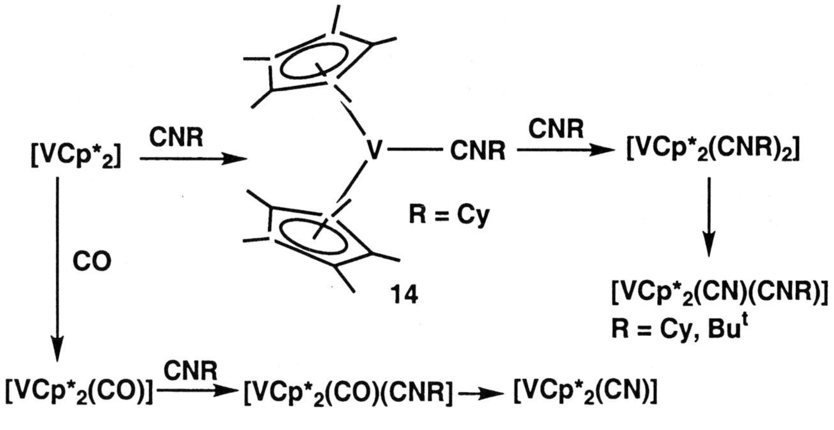

Scheme 6-5

Similar complexes [VCp(η^5-C$_5$H$_4$R)(CO)] with mixed cyclopentadienyl ligands have been reported as the products of the thermal decomposition of 18-electron monocyclopentadienyl complexes [VCp(η^4-C$_5$H$_5$R) (CO)$_2$] (R = allyl, phenyl) [85]. The reactions proceed with dissociation of CO, elimination of dihydrogen and transformation of the η^4- into the η^5-coordinated ligand.

Related niobium and tantalum complexes are not known because the corresponding niobocene and tantalocene are not accessible. Recently, a niobium derivative of this type [NbCp$_2$(CO)] has been reported as an intermediate species observed in the reduction of [NbCp$_2$Cl(CO)] with excess sodium amalgam [86]. Voltametric studies of the 18-electron niobium(III) complexes [NbCp$_2$ClL] (L = CO, PMe$_3$, CNBu^t) showed irreversible reduction waves with peak potentials of -2.36, -3.02 and -2.69 V respectively. The ESR spectrum of the reduced carbonyl species showed a 10-line signal centred at g = 2.097 with splitting due to ^{93}Nb and <a>$_{Nb}$ = 1.72 G, consistent with the presence of the radical [NbCp$_2$(CO)] which is persistent for 5-10 min. at ambient temperature. Similar reactions are observed for complexes with L = PMe$_3$ and CNBu^t; they react finally with Hg to give Nb-Hg complexes.

An unusual, crystallographically characterized, molydenum complex (**17**) obtained from the reaction of (**15**) with ethylene at 1 atm was reported (Scheme 6-6) [87]. An unexpected feature of the reaction was the formation of the second fulvalene unit which required the removal of one equivalent of dihydrogen, thought to be via hydrogenation of ethylene, and (**16**) was a proposed intermediate. The Mo-Mo distance was slightly longer than expected for a metal-metal single bond. The same reaction using 10 atm of ethylene gave (**18**) and no (**17**) was detected.

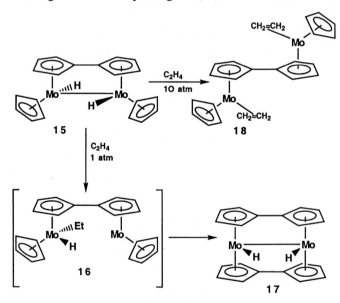

Scheme 6-6

6.3.1.4 Anionic Group 5 [MCp$_2$X]$^-$ and Cationic Group 6 [MCp$_2$L]$^+$ Metal Complexes

Anionic group 5 d^3 metal compounds [MCp$_2$X]$^-$ do not exist . It has been observed that the electrochemical reduction of [VCp$_2$Cl] (-1.63 V) is followed by rapid loss of chloride and conversion into vanadocene.

Although the neutral group 6 d^3 metal compounds [MCp$_2$X] could not be isolated, formation of their related cationic species [MCp$_2$L]$^+$ has been observed in electrochemical studies. Oxidation of neutral chromocene under CO leads to the formation of the 17-electron cationic adduct which instantaneously decays to the sandwich cation, losing CO (see section 6.2.2.3) [68]. The measured potential difference implies that the equilibrium constant for the formation of [CrCp$_2$(CO)]$^+$ is about 1000 times smaller than that for the sandwich complex, and therefore only a small fraction of the carbonyl cation remains in solution under 1 atm of CO. For reasons already discussed the *ansa- bis*(cyclopentadienyl) carbonyl cation [Cr{(C$_5$H$_4$)CMe$_2$CMe$_2$ (C$_5$H$_4$)}(CO)]$^+$ does not lose CO on the electrochemical time scale even in the absence of an excess of CO, although attempts to isolate salts of this cation were unsuccessful [88].

Similar molybdenum and tungsten derivatives have not been isolated.

16-Electron Compounds

An important group of compounds of this type are the neutral group 5 d^2 metal complexes [MCp$_2$X] with one anionic (X) ligand. Related ionic compounds are the cationic group 5 [MCp$_2$L]$^+$ and group 6 [MCp$_2$X]$^+$ complexes with neutral (L) or anionic (X) ligands respectively.

As shown in Figure 6-2, one of the three equatorial metal-based orbitals (2a$_1$) of the [MCp$_2$] fragment becomes a molecular bonding orbital used to coordinate the anionic or neutral substituent. The two (d^2) electrons for [MCp$_2$X] and [MCp$_2$L]$^+$ group 5 metal compounds can occupy the other two (1a$_1$ and b$_2$) orbitals in three different ways, leading to two possible spin states: singlet and triplet (Figure 6-4).

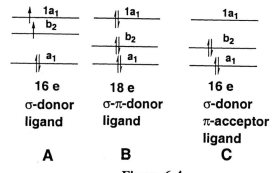

Figure 6-4

Electron configurations for [MCp2X] and [MCp2L]$^+$ d^2 metal complexes with σ- and π-donor/acceptor ligands

Occupation of these two orbitals depends on their relative energy, which is drastically changed if one of them (b_2) is used for π-bonding with π-donor or π-acceptor ligands. The use of a single σ-donor ligand would lead to paramagnetic 16-electron compounds (**A**) with two unpaired electrons in the $1a_1$ and b_2 orbitals. The use of one σ-donor and strong π-donor (3 electron donor) ligand would give an additional stabilized occupied b_2 orbital, leading to 18-electron compounds (**B**). The use of one σ- donor and strong π-acceptor ligand would stabilize the metal-based b_2 orbital which would then be occupied by the two (d^2) electrons to give diamagnetic 16-electron compounds (**C**), which could be easily converted into 18-electron systems similar to (**B**), occupying the vacant $1a_1$ orbital by coordination of a second preferably π-acceptor ligand.

This is the basis of the very different chemical behaviour observed for *bis*(cyclopentadienyl) vanadium(III) derivatives compared with the related niobium and tantalum(III) compounds. With σ-donors, vanadium shows a marked tendency to form paramagnetic compounds with two unpaired electrons, and strong π-donor ligands are required to form diamagnetic 18-electron systems. However, when a strong π-acceptor ligand L is used, the spin-singlet state is accessible, but then a second ligand coordinates to achieve the more favourable 18 electron system. Similar behaviour is also observed for niobium and tantalum(III) compounds but, in contrast, they show a higher tendency to give diamagnetic spin-singlet compounds even when σ-donors are present.

From extended Hückel calculations [8] the most stable minimum energy structure for paramagnetic compounds locates the substituent in the symmetrically central position of the equatorial plane, with an angle $\alpha = 0$ with the C_2 axis of the [MCp$_2$] fragment. The presence of a free pair of electrons in compounds exhibiting a singlet-spin state would tend to distort this geometry and to locate the substituent at an angle $\alpha > 0$.

From a structural point of view, all the vanadium(III) compounds reported exhibit the same bent *bis*(cyclopentadienyl) disposition with the substituent occupying the central position in the equatorial plane, and this is a very important, abundantly represented, group of vanadium(III) 16-electron compounds. The estimated energy gap between the two orbitals is narrow (0.3 to 0.6 eV), making spin-pairing very unfavourable in these complexes [89]. Furthermore, their tendency to coordinate one additional ligand to give the 18-electron derivatives is less pronounced than for the heavier niobium and tantalum metals, because the required spin-pairing energy is too unfavourable.

6.3.1.5 Neutral d^2 Vanadium Compounds [VCp$_2$X]

The simplest of this group are the neutral complexes which result from the coordination of only one monodentate anionic ligand.

Vanadium(III) halides are useful starting materials for a wide variety of other derivatives. Most synthetic methods which use these halides are based on the one-electron oxidation of the neutral sandwich vanadocenes with various oxidizing agents. Reactions of [VCp$_2$] with alkyl halides, methylene dihalides, hydrogen chloride, halogen (Br$_2$, I$_2$), lead dichloride and [VCp$_2$X$_2$] have been extensively used to prepare monohalides or pseudohalides [VCp$_2$X] (**19**) [1,2,90].   A particularly interesting method discussed in Section 6.3.1.3 is the oxidative dealkylation of the 17-electron isocyanide complexes [VCp*$_2$(CNR)] on addition of CO [15]. More recently, PCl$_3$ has been reported as a convenient chlorinating agent leading to mixtures of alkyl-substituted *bis*(cyclopentadienyl) vanadium(III) and (IV) halides in ratios which depend on the bulkiness of the alkyl substituent [16]. An alternative method (Scheme 6-7) uses TlCp to transfer two cyclopentadienyl rings to VCl$_3$ [91]. Some particular complexes containing substituted cyclopentadienyl ligands have also been synthesized by reacting VBr$_3$(THF)$_3$ with an appropriate lithium cyclopentadienide [37].

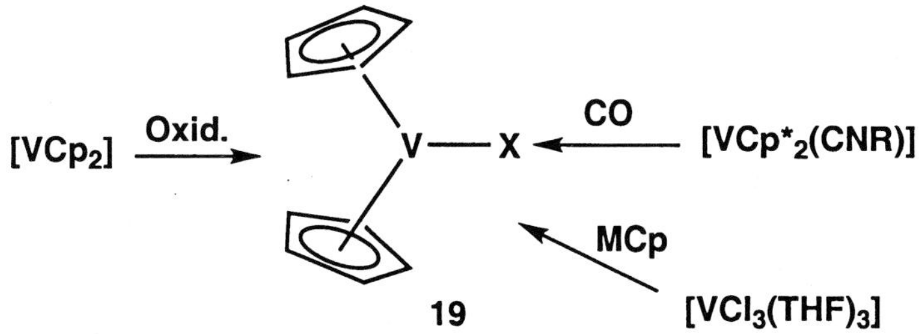

Scheme 6-7

All of these halides are very air sensitive and reactive paramagnetic compounds with two unpaired electrons and little tendency to dimerize; their unfavourable electron-pairing energy hinders the coordination of one additional ligand to give diamagnetic 18-electron compounds.

The alkyl complexes [VCp$_2$R] (**20**) are one of the products of the reaction of vanadocene with alkyl halides, together with the corresponding vanadium(III) halide, but a better preparative method for (**20**) is alkylation of the monohalides [VCp$_2$Cl], or the dihalides [VCp$_2$Cl$_2$] which are simultaneously reduced (Scheme 6-8).

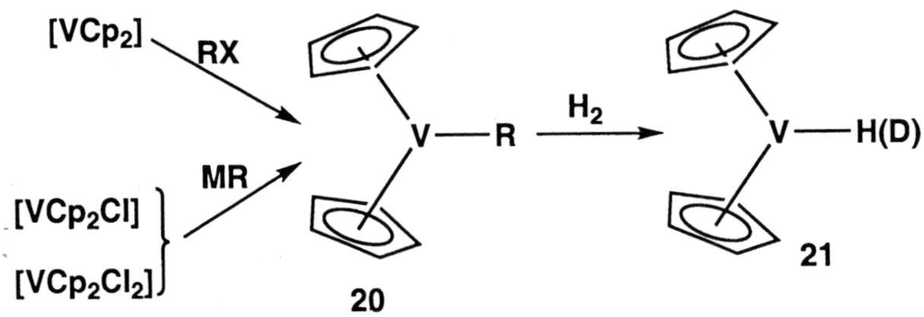

[VCp$_2$] ⟶ RX

[VCp$_2$Cl]
[VCp$_2$Cl$_2$] ⟶ MR

20   V—R

H$_2$ ⟶

21   V—H(D)

Scheme 6-8

A great variety of alkyl, aryl, alkenyl and alkynyl complexes with unsubstituted, substituted and permethylated cyclopentadienyl rings have been isolated [1,2,3].

The hydride and deuteride complexes (**21**) of the permethylated rings have also been obtained by hydrogenolysis of the methyl complex [92], whereas its reaction with Li[BEt$_3$H] proceeds with reduction to the vanadocene.

All of these compounds are highly coloured, air sensitive paramagnetic solids with magnetic moments consistent with the presence of two unpaired electrons, obeying the Curie-Weiss law. They have the predicted bent cyclopentadienyl symmetrical geometry which has been found for the cyanide derivative [VCp*$_2$(CN)] by X-ray diffraction [15], showing an average V-Cp* distance of 1.97 Å with a Cp*-V-Cp* angle of 151.5 ° and the linear cyanide ligand symmetrically disposed in the equatorial plane; the V-CN bond distance of 2.07 Å indicates a weak double bond and the triple C-N bond distance is 1.17 Å.

The chemical behaviour of compounds with typical σ-donor ligands such as alkyls and hydrides is the best illustration of the reluctance of vanadium to give spin-singlet systems by coordinating an additional ligand to afford diamagnetic 18-electron compounds, and is supported by the following observations:

a) Alkyl radicals with potential π-systems are always coordinated as η^1-ligands occupying only one coordination site; this is the behaviour observed for allyl and cyclopentadienyl ligands [93]. An excellent demonstration of this behaviour was reported by Nieman and Teuben in the study of the reactivity of the allyl [VCp$_2$(η^1-allyl)] and alkyl [VCp$_2$R] (R = alkyl, phenyl) derivatives [85]. They all coordinate ligands with good acceptor properties, such as CO, CNR and CO$_2$, but they do not react with nitriles due to their limited acceptor properties. All of these compounds react with their coordinated isocyanides or CO$_2$ with migration of the appropriate allyl, alkyl and phenyl groups to give the η^1-coordinated acyl, iminoacyl (**22**) and carboxylate (**23**) complexes; the iminoacyl-nitrogen and the carboxylate-oxygen donor atoms lack acceptor properties to induce the spin-pairing which would lead to their η^2-coordination. The same reaction also occurs with CO to give the η^1-acyl complex, but a second molecule of the strong π-acceptor ligand CO is coordinated to give the 18-electron species [VCp$_2$\{η^1-C(O)R\}(CO)].

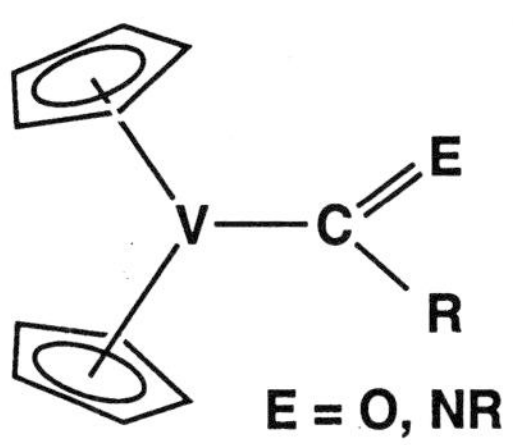

22 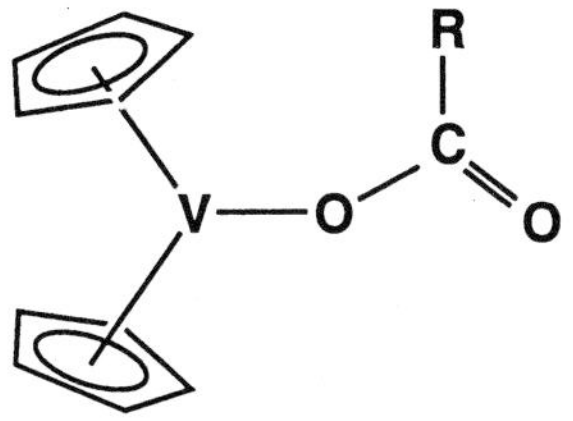   **23**

E = O, NR

b) Alkyl radicals with substituted donor groups able to act as chelate or bridging bidentate ligands, such as o-aminephenyl and o-aminebenzyl, give only mononuclear paramagnetic species ($\mu_{eff.} = 2.6\ \mu_B$), and their proposed stability cannot be due to any interaction with the amino substituent [94].

c) Alkyls with β-hydrogen containing radicals are thermally stable and do not transform into the olefin hydride by β-elimination [95], and the related permethylated complex [V(C$_5$Me$_5$)$_2$H] does not react with ethylene at pressures up to 17 atm [92].

d) The unique examples of transformation into diamagnetic 18-electron compounds are the reactions with CO discussed in Section 6.3.2.1.

e) Without exception, the hydride and all of the alkyl vanadium(III) compounds reported are paramagnetic with magnetic moments of 2.77 μ_B for [V(C$_5$Me$_5$)$_2$Me] and 2.81 μ_B for [V(C$_5$Me$_5$)$_2$H] [92].

Vanadium(III) complexes with potential π-donor anionic ligands such as [VCp$_2$(SR)] (**24**) have also been isolated by one-electron oxidation of vanadocenes with RSSR; they too are paramagnetic compounds [96].

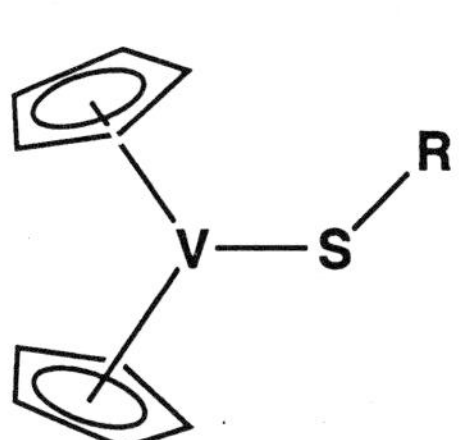

24 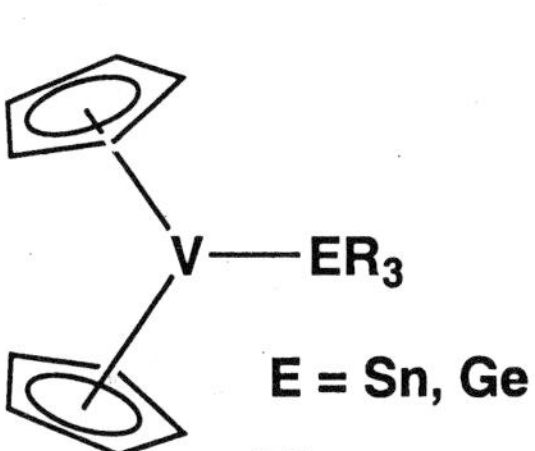   **25**

E = Sn, Ge

Exceptional behaviour is observed for stannyl and germyl derivatives (**25**) which are obtained by reaction of vanadocene with an excess of SnHEt$_3$ or [Cd(GeR$_3$)$_2$] (R = Et, Ph) respectively [1]. They can also be prepared by reaction of [VCp$_2$(CH$_2$SiMe$_3$)$_2$] or [VCp$_2$Me$_2$] with HER$_3$ (E = Si, Ge), and by reacting [VCp$_2$Cl] with Li[GePr$^i_3$]. Unusually, these compounds are monomeric and diamagnetic.

26   **27**

Stronger π-donor ligands may force spin-pairing, as observed [1] for the bridging imido and terminal amido complexes. The dinuclear imido complex was isolated by one-electron oxidation of vanadocene with $Si(N_3)Me_3$, which takes place with evolution of N_2 leading to the dinuclear complex $[Cp_2V\{\mu\text{-}N(SiMe_3)\}VCp_2]$ (**26**), whereas the mononuclear amido derivative (**27**) is prepared by reaction of $[VCp_2(CH_2SiMe_3)]$ with $NH(SiMe_3)_2$ or $[VCp_2Cl]$ with $Li[N(SiMe_3)_2]$. These seem to be the only π-donor ligands able to give diamagnetic compounds [97].

6.3.1.6 Neutral d^2 Niobium and Tantalum Compounds [(Nb,Ta)Cp$_2$X]

In contrast to the vanadium(III) complexes described above, the *bis*(cyclopentadienyl)niobium and tantalum(III) complexes show a greater tendency to adopt an 18-electron singlet-spin state, not requiring the cooperation of strong acceptor ligands because simple σ-donor ligands can produce the spin-pairing re-quired to give stable diamagnetic complexes. Similar behaviour is observed with π-acceptor ligands, which stabilize the b$_2$ orbital, favouring the 18-electron spin-paired state. Formation of a true 16-electron [NbCp$_2$Me] (**28**) paramagnetic species occurs by photolysis of a toluene solution of [NbCp$_2$Me(CO)] [98]; the dark blue solution is fairly stable at room temperature but decomposes on heating, possibly via a rearrangement to a methylidene hydride, behaviour typically observed for [MCp$_2$R] (M = Nb, Ta) complexes.

28   **29**

Preparation of metal(III) halides by reduction of THF suspensions of [Cp$_2$MCl$_2$] with Na/Hg has been reported [99 ,100 ,101 ,102]. In solution they must exist as adducts with one coordinated solvent molecule which is easily lost by evaporation under vacuum, whereas in the solid they are probably dimeric compounds with μ-halide bridges [103]. The increased lipophilicity conferred by cyclopentadienyl rings with bulky TMS substituents gave compounds which were soluble in non-donor solvents, where they probably also exist as dimers, although molecular weights were not measured [104 ,105 ,106]. These metal(III) halides have a great tendency to coordinate one additional ligand, and several metal halides can even accomplish this task through the formation of halide bridges [102]. They cannot therefore be considered as 16-electron d^2 metal complexes and are better studied in the following sections 6.3.2.1.

The niobium and tantalum hydrides [MCp$_2$H] (M = Nb, Ta) have only been generated photochemically and studied by IR and UV spectroscopy in matrices at low temperature, although their multiplicity state could not be determined [107]. A related niobium(III) hydride has been proposed as consistent with the electrochemical behaviour observed for [Nb(C$_5$H$_4$SiMe$_3$)$_2$H$_3$] [108].

Singular examples are the niobium(III/II) complexes [NbCp]$_2$(μ-fulvalene)(μ-X)] (**29**) (X = Cl, Br, SCN) for which the x-ray crystal structure of the chloro derivative has been determined [109].

6.3.1.7 Cationic Group 5 d^2 Metal Compounds [MCp$_2$L]$^+$

The monocationic species [Cp$_2$ML]$^+$ (M = V, Nb, Ta) are further examples of 16-electron group 5 metal complexes containing only one additional neutral (L) ligand. These compounds are very rare due to the even higher tendency of the metal to coordinate one additional ligand to give the more stable 18-electron compounds [Cp$_2$ML$_2$]$^+$ discussed in Section 6.3.2.1. The use of σ-donor ligands such as acetone, ether, pyridine and acetonitrile gives the green, very air sensitive and paramagnetic cationic 16-electron complexes [VCp$_2$L]$^+$ (**30**) with two unpaired electrons. These were obtained from the reaction of neutral *bis*(cyclopentadienyl) vanadium(III) chloride [VCp$_2$Cl] with NaBPh$_4$ in the presence of the ligands [110], or from the oxidation of either neutral sandwich vanadocene or decamethylvanadocene by [FeCp$_2$]PF$_6$ [14].

30   **31**

Oxidation of the dinuclear *bis*(fulvalene) complex with $[FeCp_2][PF_6]$ in acetonitrile gives air sensitive brown crystals of $[V_2(NCMe)_2(fulvalene)_2][PF_6]_2$ (**31**) which is a paramagnetic compound with $\mu_{eff} = 3.0\ \mu_B$; its x-ray crystal structure shows a V-V distance of 3.329 Å, slightly longer than that expected for a single bond.[1] Two related carbonyl cations are the brown paramagnetic $[V_2(CO)_2(fulvalene)_2]^+$ ($\mu_{eff} = 1.7\ \mu_B$) and the orange dication $[V_2(CO)_2(fulvalene)_2]^{2+}$ which has a low measured magnetic moment ($\mu_{eff} = 0.9\text{-}2.1\ \mu_B$), probably due to decomposition [1].

However, in all other cases the use of π-acceptor ligands such as CO, isocyanides and phosphines always leads to diamagnetic 18-electron derivatives. Similar species are also present in solution when vanadocene is electrochemically oxidized in acetonitrile, as evidenced by a reversible one-electron wave, whereas the same oxidation of decamethylvanadocene gives a cyclic voltammogram with no reversible waves [14].

6.3.1.8 Cationic Group 6 d² Metal Compounds [MCp₂X]⁺

These cationic species cannot be isolated for the reasons given for the group 5 metal derivatives. In a recent report on 18-electron cationic tungsten complexes, the 16-electron cations were proposed as transient species [111]. The intermediate reversible formation of $[WCp_2H]^+$ was demonstrated by the complete exchange of coordinated acetonitrile in $[WCp_2H(NCMe)]^+$ when treated with deuterated acetonitrile. Similarly, formation of an intermediate species $[WCp_2Me]^+$ was proposed to explain the quantitative transformation of $[WCp_2Me(NCMe)]^+$ via C-H bond activation of one of the cyclopentadienyl rings, and $[WCp_2Et]^+$ was proposed to be the intermediate insertion product of the reaction between $[WCp_2H(\eta^2\text{-}C_2H_4)]$ and I_2 to give $[WCp_2(=CHCH_3)I]^+$ [112]   Formation of a similar molybdenum salt $[MoCp_2H]^+$ salt has been suggested to explain the deprotonation of $[MoCp_2H_2]$ by $CH_2=PMe_3$, which finally gives $[MoCp_2(PMe_3)]$ [113].   However, these types of cationic compound have never been isolated for any of the group 6 metals.

6.3.2 Complexes with Two Coordinated Ligands

All the possible formulations and electron configurations for group 5 and 6 metal compounds containing two additional anionic (X) or neutral (L) ligands with between 16- and 18-electrons are summarized in Table 6-2. All of the d⁰ metal compounds are cationic derivatives of the higher valent metals. Neutral complexes fall into three important groups: i) d² group 5, ii) d² group 6 and iii) d¹ group 5 metal complexes. In this section we are mainly concerned with these complexes and their ionic derivatives, although other 16-electron compounds are also discussed.

Table 6-2. Different types of [MCp₂L₂], [MCp₂XL] and [MCp₂X₂] groups 5 and 6 metal complexes

Total Number of electrons	16	17	18
Configuration	d⁰	d¹	d²
	[(M5)Cp₂X₂]⁺	[(M5)Cp₂X₂]	[(M5)Cp₂XL]
	[(M5)Cp₂XL]²⁺		[(M6)Cp₂X₂]
	[(M6)Cp₂X₂]²⁺	[(M5)Cp₂XL]⁺	
		[(M5)Cp₂L₂]²⁺	[(M5)Cp₂L₂]⁺
		[(M6)Cp₂X₂]⁺	[(M5)Cp₂X₂]⁻
			[(M6)Cp₂XL]⁺
			[(M6)Cp₂L₂]²⁺

When two additional σ-donor ligands are coordinated to the metal, contributing their $a_1 + b_2$ orbitals, two of the metal based orbitals $2a_1$ and b_2 engage in the bonding interaction (Figure 6-5), whereas the nature and energy of the nonbonding a_1 metal orbital is very sensitive to the angle φ between the two ligands (**32**) and their π-acceptor capacity [8]. φ values in the ranges 94° - 97° for d⁰, 85° - 88° for d¹ and 76° - 82° for d² metal complexes have been predicted, and are in reasonably good agreement with the observed values.

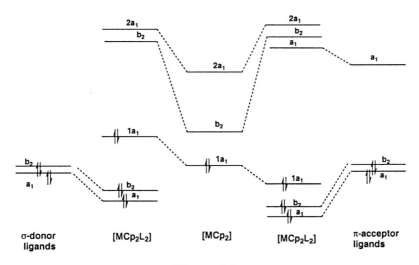

Figure 6-5

Qualitative diagram for group 5 [MCp₂XL] and group 6 [MCp₂X₂] d² metal complexes with σ-donor and π-acceptor ligands

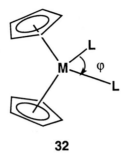

32

18-Electron Compounds

The simplest neutral complexes with 18 electrons are the d^2 vanadium, niobium and tantalum(III) derivatives [Cp$_2$MXL] containing either one anionic (X) and one neutral (L) monodentate ligand or one monoanionic bidentate ligand, and the d^2 chromium, molybdenum and tungsten(IV) complexes [Cp$_2$MX$_2$] with either two monodentate anions or one bidentate dianion. In all of these diamagnetic compounds the nonbonding orbital a_1 is occupied by one pair of electrons, which is still available for π-bonding when one of the substituents is a π-acceptor ligand, but no interaction can take place with π-donor ligands, which would orientate their occupied π-orbital perpendicular to the equatorial plane. On the other hand the free pair of a_1 electrons can also be donated to an appropriate Lewis acid. Consequently these compounds can have a basic character and from an structural point of view the bend angle θ between the two cyclopentadienyl ligands varies between 120° and 150°.

6.3.2.1 Neutral Group 5 d^2 Metal Compounds [MCp$_2$XL]

As discussed in Section 6.3.1, the central atom of 16-electron [Cp$_2$MX] complexes possesses two available orbitals which can accommodate the remaining two electrons, leading to either a spin-singlet or -triplet system. An examination of the known chemistry of group 5 metals shows that the spin-pairing energy of the two metal-based orbitals is lower for the heavier niobium and tantalum atoms than for vanadium, although estimates of this energy have not been reported. In fact, most of the vanadium complexes show a very low tendency to coordinate one additional ligand, indicating that the difference in energy between the two metal orbitals is clearly smaller than the spin-pairing energy, whereas the reverse is true for the heavier metals for which the formation of [Cp$_2$MXL] species is favoured, with the exception of the niobium and tantalum(III) derivatives described in Section 6.3.1.

Only when strong π-acceptor ligands are used it is possible to isolate 18-electron vanadium [Cp$_2$VXL] complexes, which easily revert to their paramagnetic precursors. Examples are the reactions with isocyanides (Scheme 6-5), which finally give the cyanide complexes [Cp$_2$V(CN)(CNR)] (**33**). However, isocyanides react with vanadium (III) halides [Cp$_2$VX] to give cationic complexes, but neutral complexes cannot be isolated in this way. This is only possible when CO is used: [Cp*$_2$VI] readily reacts with CO in nonpolar solvents to give [Cp*$_2$VI(CO)] (**34**), which regenerates the iodide on gentle heating of the solution under vacuum [15]. Similarly, reaction of the hydride (**21**) or methyl (**20**) derivatives with CO leads respectively to the diamagnetic green hydrido [Cp*$_2$VH(CO)] (**34**) or acyl carbonyl (**35**), which loses CO when stored at -40°C over several weeks [92].

The huge number of known niobium and tantalum [Cp$_2$MXL] compounds demonstrates the great variety of substituents (Table 6-3) which include unsubstituted and substituted cyclopentadienyl rings and an extensive range of anionic X and neutral L ligands. In this section we are concerned only with complexes containing monofunctional single atom donor ligands whereas complexes with η^2-coordinated ligands will be discussed in Section 6.5.

Table 6-3. Different types of substituents for [MCp₂XL] group 5 metal complexes

Cyclopentadienyl ligand	Anionic ligand	Neutral ligand
C_5H_5	H	CO
C_5H_4Me	Cl	CNR
C_5Me_5	Br	PR_3
$C_5H_4SiMe_3$	NMe_2	olefines
$C_5H_3(SiMe_3)_2$	$N(Tol)(CH=NTol)$	alkynes
$C_5H_4{}^tBu$	OR	CO_2
$C_5H_4CHMePh$	SR	CS_2
$C_5H_4{}^iPr$	PR_2	isocyanate
Indenyl	$P(O)PhH$	ketene
	$P(X)R_2$ (X=O,S)	ketenimine
	μ-$(PR_2)M(CO)_n$	ketone
	μ-$[P(S)R_2]ML_n$	aldehyde
	μ-$(OC)M(CO)_n$	SR
	CN	$P(OR)_3$
	R	μ-$ClML_n$
	$SiMe_2H$	
	$SiMe_3$	
	SnR_3	
	μ-$SnCl_2$	

In all of these compounds the metal centre possesses one pair of free electrons which can be used for π-donation and therefore π-acid ligands are particularly effective in providing strong metal-ligand bonds. Inspection of Table 6-3 shows that most of the ligands used to stabilize this type of compound have empty acceptor orbitals and formation of π-bonds plays an important role.

Halides

When *bis*(cyclopentadienyl) metal(IV) halides were reduced in donor solvents (THF) in the absence of additional ligands, the resulting species [MCp₂X(solvent)] (M = Nb, Ta) formed in solution could not be isolated as solvent removal under vacuum also eliminates the labile σ-coordinated ligand, giving compounds which probably reach the more stable 18-electron system by dimerization, although molecular weight measurements were not reported [100,102]. Evidence for the dimeric nature of such compounds was provided by an X-ray diffraction study of

[Nb(C$_5$H$_4$SiMe$_3$)$_2$Cl]$_2$ (**36**), which shows the angle between the chlorine atoms $\varphi \approx$ 75° [114]. The role played by the second molecule coordinated through a μ-chloro bridge in the dimer can also be taken by other metal halides, indicating the high tendency of these compounds to achieve an 18 electron system. For example, reduction of [MCp$_2$Cl$_2$] (M = Nb, Ta) with one half equivalent of Na/Hg affords purple compounds [MCp$_2$Cl(μ-ClMClCp$_2$)] (**37**) which contain a μ-chloro bridging metal(IV) halide as ligand [99,102], and which can be transformed by reaction with AgClO$_4$ into the cationic complexes proposed to have two μ-chloro bridges [MCp$_2$(μ-Cl)$_2$MCp$_2$)]$^+$ (**38**). The μ-chloro bridging ligand can be replaced by a heterometal halide which leads to different heterodinuclear complexes [MCp$_2$(μ-Cl)(μ-X)M'Y$_2$)] (**39**) (X = Cl, Br; M'Y$_2$ = AlCl$_2$, FeCl$_2$, TlCl$_2$, Tl(C$_6$F$_5$)$_2$).

Similarly, reduction of the *bis*(cyclopentadienyl) metal(IV) and (V) halides with various reducing agents in the presence of π-acceptor ligands is the method of synthesis generally used to obtain stable carbonyl, isocyanide, phosphine and phosphite halocomplexes [MCp$_2$XL] (**40**) [L = CO, CNR, PR$_3$, P(OR)$_3$)] which are the most extensively studied compounds in this category [100, 106, 115, 116 , 117, 118, 119, 120, 121]; they can also be prepared by metathesis of the hydride via reaction with HCl, CHCl$_3$, RCOCl and MeI (Scheme 6-9) [122 ,123].

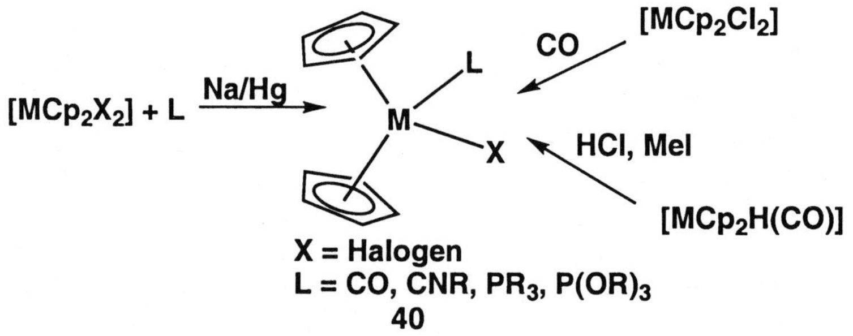

$$[MCp_2X_2] + L \xrightarrow{Na/Hg}$$

[MCp$_2$Cl$_2$]

CO

L

M

X

HCl, MeI

[MCp$_2$H(CO)]

X = Halogen
L = CO, CNR, PR$_3$, P(OR)$_3$
40

Scheme 6-9

Hydrides

Hydride complexes with η^1-coordinated ligands [MCp$_2$HL] (**41**) are generally obtained on treatment of the reactive trihydrides with the ligand. The mononuclear derivatives [MCp$_2$H{R$_2$P(CH$_2$)$_n$PR$_2$}] (M = Nb or Ta, R = Me or Ph, n = 1 or 2) were obtained in this way whereas reaction of [MCp$_2$H$_3$] with dppp (dppp = Ph$_2$P(CH$_2$)$_3$PPh$_2$) gave the bimetallic dihydride [(NbCp$_2$H)$_2$(μ-dppp)] or [{TaCp$_2$H}$_2$] [124]. The ability of the mononuclear complexes to act as metalloligands which can coordinate to a second metal centre to give heterodinuclear complexes has been explored. They react with [M'(CO)$_4$L$_2$] (M = Cr, Mo, W; L= norbornadiene or piperidine) to give the unsymetrically bridged complexes (**42**). An alternative method extracts BH$_3$ by reacting the boranate derivatives with NMe$_3$ in the presence of the π-acid ligand (Scheme 6-10) [102,122]. This is also the method used to isolate the *bis*-indenyl niobium (**43**) derivatives containing different ligands such as PMe$_2$Ph, pyridine [125], whereas the reaction with CO gives the mono-indenyl tetracarbonyl complex, and with allene gives the corresponding allyl derivative. [Nb{(η^5-C$_5$H$_4$SiMe$_3$)$_2$}$_2$H(CO)] forms adducts with coinage metal cations giving heteronuclear cationic complexes with μ-(hydrido-M'-hydrido) bridging the two Nb centres [126].

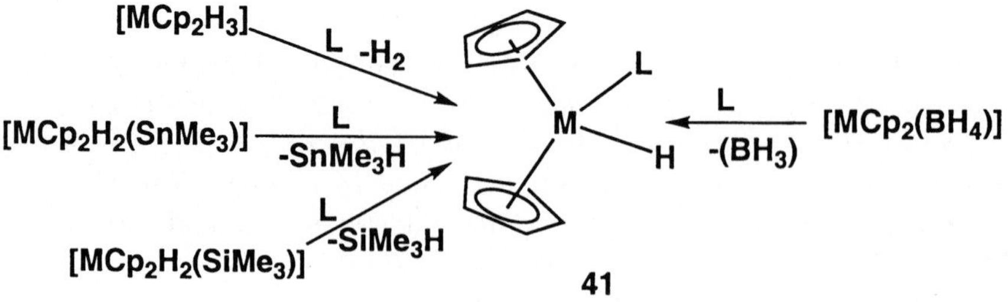

[MCp$_2$H$_3$]

L   -H$_2$

[MCp$_2$H$_2$(SnMe$_3$)] $\xrightarrow[-SnMe_3H]{L}$

[MCp$_2$H$_2$(SiMe$_3$)]

L   -SiMe$_3$H

M

L

L   -(BH$_3$)

H

[MCp$_2$(BH$_4$)]

41

Scheme 6-10

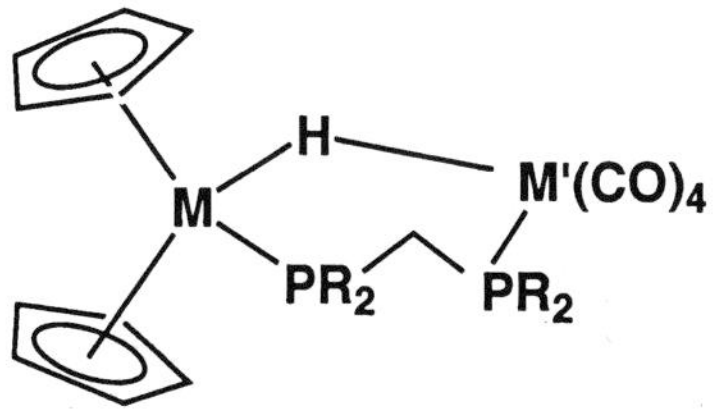

42

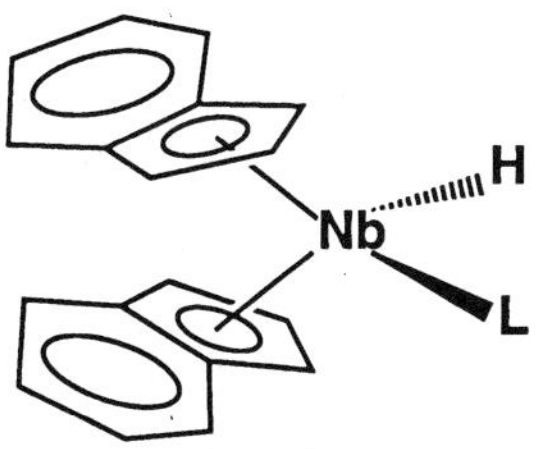

43

Alkyls

Two alternative methods may be used to prepare alkyl complexes [MCp$_2$RL] (**44**) (Scheme 6-11). The reduction of haloalkylmetal(IV) complexes has been used to prepare some of these compounds [98,127], but has very few applications due to the limited accessibility of the monoalkylated derivatives.

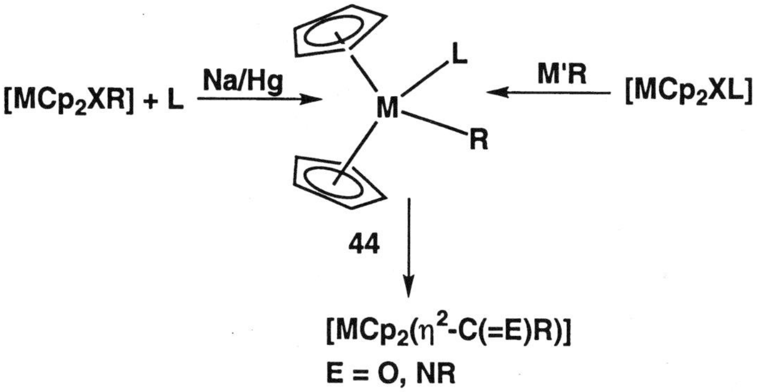

Scheme 6-11

A more useful method is the alkylation of the halometal(III) complexes with various alkylating agents, usually Li and Mg reagents [117]. When CO and CNR are the ligands present in the coordination sphere, further migration of the alkyl group may take place to finally give the corresponding η^2-acyl or η^2-iminoacyl derivatives. This reaction is unfavourable because it involves substitution of a π-acceptor ligand CO or CNR by a σ-oxygen or nitrogen donor ligand, and has not been observed for carbonyl derivatives [118], however it readily occurs in isocyanide complexes [119].

Complexes with Other Anionic Ligands

Related compounds with different anionic substituents such as -OR, -SR, -NR$_2$, -NC, -SiR$_3$, -SnR$_3$, -PR$_2$, -P(O)R$_2$ and -P(S)R$_2$ are also prepared by metathesis of the halides with appropriate reagents [2]. The reactivity of the trihydrides and hydrido-silyl and -stannyl derivatives with various ligands has been extensively studied. As shown in Scheme 6-12, hydrido-silyl and -stannyl complexes eliminate SiR$_3$H or SnR$_3$H when treated with ligands to give the corresponding adducts and this was the method used to prepare the silyl complex with PMe$_3$ (**45**) [125]. The silyl complexes [TaCp$_2$(L)(SiR$_3$)] (L = PMe$_3$, CO) activate the C-H bonds of unhindered arenes, reacting in neat arenes (benzene, toluene or *m*-xylene) to give equilibrium mixtures of aryl and silyl complexes; however, the sterically hindered complex [TaCp$_2$(PMe$_3$)(SiBu$^t_2$H)] gave [TaCp$_2$(PMe$_3$)Ar] as the sole product of the reaction [128].

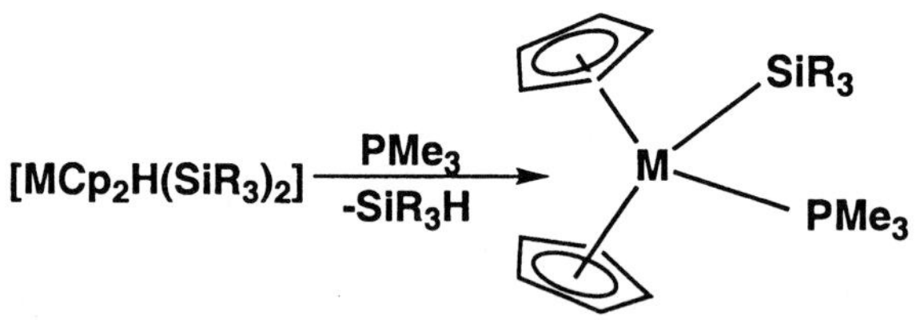

$$[MCp_2H(SiR_3)_2] \xrightarrow[\text{-SiR}_3H]{\text{PMe}_3} $$

45

Scheme 6-12

New thiolate derivatives were synthesized in high yield by reaction of the iodo derivatives with alkali thiolates [123], and various phosphido derivatives [NbCp$_2$(PR$_2$)(CO)] were obtained by insertion of PR$_2$Cl into Nb-H bonds of [NbCp$_2$H(CO)] followed by reaction of the resulting cationic species (**46**) with KOH to give the phosphido complexes (**47**) (Scheme 6-13) [129, 130, 131, 132, 133, 134]. The phosphido group in these complexes is a σ-donor ligand and the orientation of the phosphorus lone pair is at ca. 74° with respect to the equatorial plane in order to minimise the electron repulsion, defined as the 'gauche effect' [135], of the occupied 1a$_1$ metal orbital, which is partially used for π-back bonding to the carbonyl ligand. Related compounds with mixed cyclopentadienyl rings and chiral metal centres [136], and complexes derived from diphosphine precursors have also been reported [137]. Reactions of the phosphido complexes with H$_2$O$_2$ and S$_8$ lead to similar compounds with -P(O)R$_2$ and -P(S)R$_2$ ligands (**48**) [134].

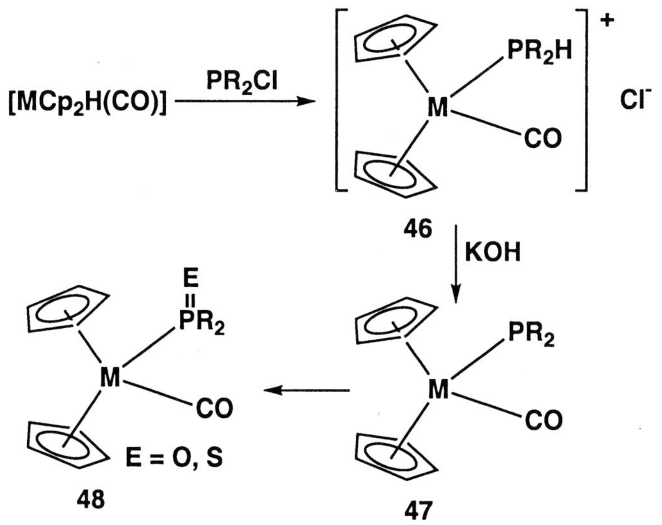

Scheme 6-13

This type of complex, with anionic ligands such as thiolate and phosphide which have one additional lone pair of electrons still available for bonding, has been extensively explored as a metallo-ligand able to coordinate to a second metal centre to give heterodinuclear complexes with $Fe(CO)_4$, $Fe(CO)_3(PMe_2Ph)$, $M(CO)_nL_{5-n}$ (M = Cr, Mo, W), and $MnCp(CO)_2$ with μ-phosphido and in some cases also with μ-carbonyl bridges [123, 131, 132, 134, 138, 138, 139, 140, 141, 142, 143].

The heterodinuclear complexes $[MCp_2(CO)\{M'(CO)_n\}]$, in which the role of the anion (X) is played by a metal-metal bonded system, are related compounds which were isolated by reaction of the trihydrides or carbonyl hydrides with the corresponding hetero-metal carbonyls [144, 145, 86]. They were also generated from photochemical reactions of the dihydride in the presence of the hetero-metal species [146].

All the above complexes contain one π-acceptor ligand which is required to withdraw the electron density of the metal lone pair and few examples of complexes without such π-acceptor ligands have been reported. The trimethylacetate complex $[NbCp_2(OOCCMe_3)]$ was isolated by reaction of the boranate niobium(III) derivative with the acid, and was characterized by IR spectroscopy as a monomeric species containing a terminal η^1-coordinated carboxylate group [147]. Its tendency to coordinate π-acid ligands explains its reactivity with diphenylacetylene to give an adduct characterized by X-ray diffraction [148], although it does not react with pyridine or triphenylphosphine. More recently the formate complex (**49**) was reported as an air sensitive oily product resulting from the reaction of the trihydride with CO_2 [149]; similar acetate and acetylacetonate derivatives were prepared by metathesis of the chloride with the corresponding thallium salts and were characterized as species containing chelate ligands, however transformation of the chelate into a monodentate ligand (50) readily takes place on addition of π-acceptor ligands such as CO, isocyanides and other η^2-ligands (Scheme 6-14).

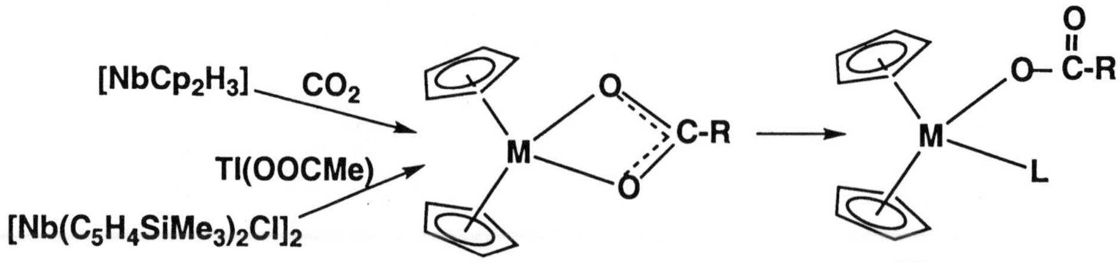

Scheme 6-14

6.3.2.2 Cationic Group 5 d^2 Metal Compounds [MCp$_2$L$_2$]$^+$

Replacement of the anionic ligand in the neutral group 5 metal complexes [MCp$_2$XL] gives access to the related cationic species [MCp$_2$L$_2$]$^+$. For vanadium, coordination of two neutral ligands to give cationic 18 electron systems is only possible when labile σ-donor anions and strong π-acceptor ligands are used [15,17]. [VCp$_2$X] (Cp = C$_5$H$_5$, C$_5$Me$_5$; X = I, CN, SR) partially reacts with CO leading to an equilibrium between the adduct and the starting material. 18-electron compounds are only formed under appropriate conditions, and [V(Indenyl)$_2$X] does not react at all with CO unless NaBPh$_4$ is present.

Niobium and tantalum have a higher tendency to coordinate neutral ligands in order to complete their 18-electron system; displacement of halide by CO or phosphine is not easy and cationic compounds are not directly accessible. The first of these compounds was the dicarbonyl complex (51) reported by Brintzinger [150] and several methods have been used to prepare complexes of this type, most of which contain η^2-coordinated groups as π-acceptor ligands (Section 6.5). A convenient method involves protonation of the metal(III) phosphides [MCp$_2$(PR$_2$)L] (L = CO, PR$_3$) with HCl, or methylation with MeI, to give cationic complexes [MCp$_2$(CO)(PHR$_2$)]$^+$ or [MCp$_2$(CO)(PMeR$_2$)]$^+$ respectively [131]. Alternatively, reduction of [MCp$_2$Cl$_2$] in the presence of dmpe with various reducing agents leads to the [MCp$_2$(dmpe)]$^+$ complex [151].

An unusual example of this type of compound is the carbonyl vinylidene complex (52), which results from the oxidation of the acetylene niobium(III) derivative [NbCp$_2$\{η^2-MeC≡C(COOMe)\}] by [FeCp$_2$][BPh$_4$] via a mechanism proposed to involve intermediate oxidation to a niobium(V) species [152]. Oxidative loss of a methoxide group from the acetylene methylcarboxylate substituent generates the carbonyl ligand required to stabilize the coordinated vinylidene, which behaves as a π-acceptor ligand in the resulting cationic species.

6.3.2.3 Anionic Group 5 d^2 Metal Compounds [MCp$_2$X$_2$]$^-$

We have observed that strong π-acceptor ligands are generally required to alleviate the high electron density in these niobium and tantalum(III) complexes. Therefore, as expected, derivatives formed by two essentially σ-donor anionic ligands coordinated to the metal centre are unable to afford stable compounds. In fact most of the reported dihalide derivatives [NbCp$_2$Cl$_2$]$^-$ have only been observed as transient species resulting from electrochemical reduction of [NbCp$_2$Cl$_2$] or [Nb(C$_5$H$_4$SiMe$_3$)$_2$Cl$_2$] [63,153,154,155], which finally leads to the anionic dimer or [Nb(C$_5$H$_4$SiMe$_3$)$_2$Cl]$_2$ respectively. Similar electrochemical reduction of several dialkyl complexes leads to [NbCp$_2$R$_2$]$^-$ in a reversible process with E = -2.08 V for R = CH$_2$Ph [156], and E = -1.63 V for R = [CH(SiMe$_3$)]$_2$o-C$_6$H$_4$ [157]. Stable anionic dialkyl complexes [MCp$_2$R$_2$]$^-$ (R$_2$ =2 CH$_2$PPh$_2$, (2-CH$_2$C$_6$H$_4$)$_2$ (53) have however been isolated as salts with the bulky cation [Na(18-crown-6)(THF)$_2$]$^+$ from the chemical reduction of metal(IV) dihalides with Na[C$_{10}$H$_8$] [158,159]; their X-ray molecular structures show the expected conformation with bend angles θ of 136° for R$_2$ = 2CH$_2$PPh$_2$ and φ angles between the alkyl substituents of 74.2(2)° and 80.1(11)° respectively.

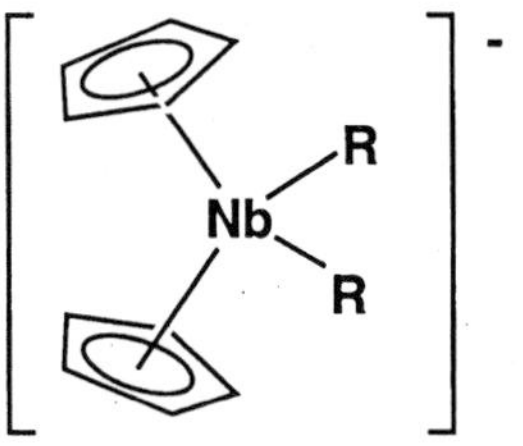

R = CH₂PPh₂
(2-CH₂-C₆H₄)₂

53

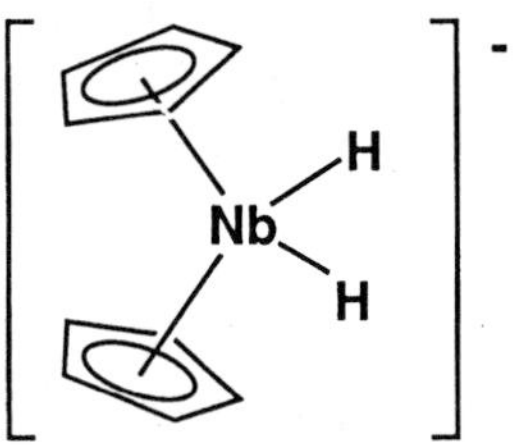

54

More extensive studies [14,15,16] have been made with anionic dihydride complexes $[NbCp_2H_2]^-$ (**54**) obtained by reaction of $[NbCp_2Cl_2]$, the trihydrides $[NbCp_2H_3]$ or the niobium(III) boranate $[NbCp_2(BH_4)]$ with excess of the alkali hydrides in the presence of crown-ethers [160], or by further addition of lanthanide salts [161]. Similar compounds with stannyl substituents can also be obtained by reaction of the stannyl-hydrides with butyllithium, always in the presence of crown-ethers to provide bulky cations [162]. Salts of the $[MCp_2H_2]^-$ anion containing different cations, such as Li^+, Na^+, K^+, $Li(12\text{-crown-4})^+$, $Na(15\text{-crown-5})^+$, $Na(benzo\text{-}15\text{-crown-5})^+$, $K(18\text{-crown-6})^+$, $Yb(diglyme)^{2+}$, $Ce(THF)_6^{3+}$ have been isolated and the molecular structure of some characterized by X-ray diffraction. All of them contain hydrogen bridges between the group 5 transition metal and the alkali or lanthanide atom, with φ angles between the substituents of 86(1)° (H-Nb-H) to 87.13(3)° (Sn-Nb-Sn) and θ bend angles between 139.4° and 144.0°.

The dinuclear anionic hydride complexes containing one bridging η^5-η^1-cyclopentadienyl ring between the two metal atoms (**55**) or substituted by a siloxide group (**56**) are particularly interesting. They were obtained [14,18,19,20,22] by reduction of $[NbCp_2Cl_2]$ with an excess of Na, or by sodium naphtalenide reduction of the Nb-Nb bonded (η^5-η^1-cyclopentadienyl) hydride intermediate (obtained from the reduction of $[NbCp_2Cl_2]$ by NaH in an attempt to prepare the corresponding niobocene, which decomposes spontaneously by activation of the ring C-H bond and hydrogen transfer to the metal).

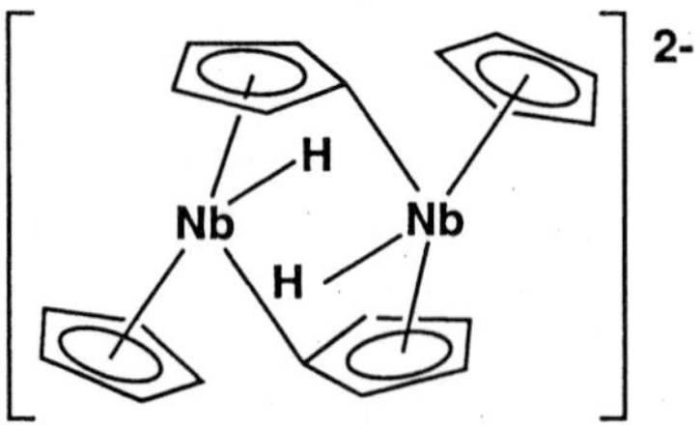

55

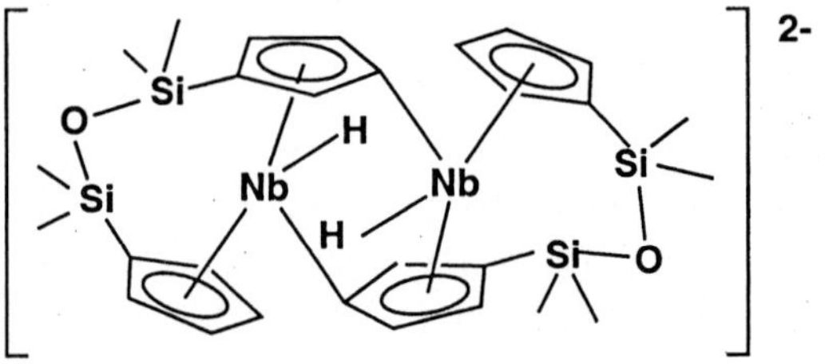

56

Addition of the carbonate dianion to the electro-generated species {Nb(η^5-C$_5$H$_3$RR')$_2$Cl} (R = H, R' = SiMe$_3$; R = R' = SiMe$_3$) gave the anionic carbonato niobium(III) derivative [Nb(η^5-C$_5$H$_3$RR')$_2$\{η^2-OC(O)O\}]$^-$, which underwent electrochemical oxidation to give the corresponding neutral paramagnetic niobium(IV) complex [163].

6.3.2.4 Neutral Group 6 d^2 Metal Compounds [MCp$_2$X$_2$]

These are the most important type of molybdenum and tungsten derivatives, represented by an enormous number of complexes which are the result of the considerable efforts made by different groups of researchers. They are 18-electron compounds, isoelectronic with the *bis*(cyclopentadienyl) group 5 metal(III) complexes [MCp$_2$XL], with a free pair of electrons which has a significant effect in their chemical behaviour.

Halides and hydrides

Several synthetic methods have been used to prepare particular compounds [1,2], but the first entrance to the dihalide compounds [MCp$_2$Cl$_2$] (**57**) was the reaction of MoCl$_5$ with excess of NaCp [1]; a more convenient method giving higher yields uses reactions of halogenating reagents such as chloroform with the dihydrides [MCp$_2$H$_2$] (**58**), which can be isolated in yields of around 40% by reaction of MoCl$_5$ or WCl$_6$ with NaCp.DME and NaBH$_4$ [164,165]. Hydrogenation of (**15**) gave [Mo$_2$(μ-η^5:η^5-C$_{10}$H$_8$)Cp$_2$H$_4$] which was structurally characterized [87].

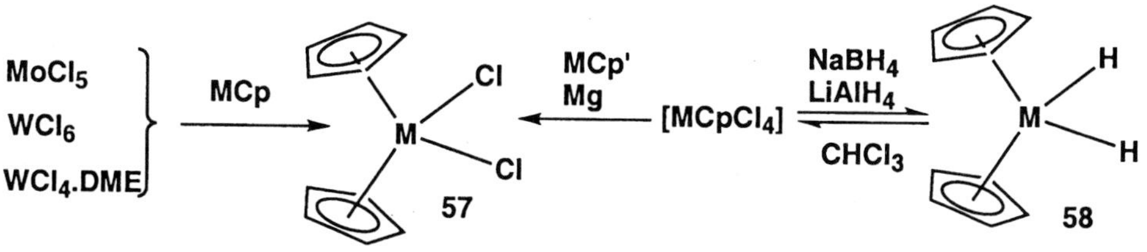

Scheme 6-15

An improved synthesis of [WCp$_2$Cl$_2$] used the reaction of [WCl$_4$(DME)] with LiCp, and [WCp$_2$H$_2$] was obtained by reaction of Li[AlH$_4$] with the dichloride (Scheme 6-12) [166]. The dihydrides were also prepared by co-condensation of the metal atoms and CpH at low temperature, and by reaction of the dihalides with

Na[BH$_4$]. The reverse transformation of the dihydrides into the dihalides was effected by reaction with CHCl$_3$ or other appropriate halogenating agents, but not with halogens X$_2$ since they cause oxidation to higher valent compounds. The ansa-metallocene dichlorides (59) [167], (60) and (61) [168] were obtained in a similar way from reactions of the meta(IV) chlorides with alkali metal salts of the appropriate ansa-cyclopentadienyl ligand. Complexes (59), (60) and (61) were used to prepare hydrido, sulfido, alkyl and aryl derivatives. X-Ray molecular structures of (60) and (61) show the expected reduction in the bend angle θ associated with sterically rigid ansa-bridged complexes.

M = Mo or W, R = (SiMe$_2$)$_2$O 59
M = Mo or W, R = CMe$_2$ 60
M = Mo, R = C(CH$_2$)$_4$ 61

62

63

An alternative synthetic route requiring more steps but leading to higher yields involves the preparation of [MCp(CO)$_3$H] (directly accessible in high yield from [M(CO)$_6$] and CpH) followed by oxidation with PCl$_5$ giving [MCpCl$_4$] [169]. Reduction of these metal(V) complexes with magnesium or other reducing agents and simultaneous reaction with alkali cyclopentadienides gives access to the dihalides; this was the method used to prepare [WCp*$_2$Cl$_2$] [170] and is a particularly useful method for the preparation of compounds with mixed cyclopentadienyl rings. The reaction of [MCp$_2$Cl$_2$] with [M'PPh$_2$(CO)$_5$] (M' =Cr, or W) leads to ring substitution [171] and the formation of several bimetallic chlorohydride derivatives (62) (M = W); the substitution is followed by orthometallation to (63) when M = Mo [172 , 173].

These metal(IV) compounds have one pair of free electrons which are responsible for the Lewis basic properties of the dihydride complexes, although not chemically significant for the dihalides and other dianionic derivatives due to their higher electronegativities. The location of this electron pair and its chemical role have been the subject of many theoretical and structural studies [1,2]. The narrow H-M-H φ angle of 75.5° in [MoCp$_2$H$_2$], found by neutron diffraction, reveals that it cannot be located between the two hydride ligands but occupies the 1a$_1$ orbital extended along the y axis; this is in good agreement with the results of theoretical calculations.

Dihydride compounds can be protonated to give the cationic trihydride deriva-tives [MCp$_2$H$_3$]$^+$ (Section 6.3.3) and many of their reactions can be interpreted in terms of interactions due to their Lewis basicity. Several stable adducts [MCp$_2$H$_2$. A] with different Lewis acids can be isolated, such as those with A = BF$_3$, AlR$_3$ and many transition metal fragments in which the donor-acceptor metal-metal bond is frequently acompanied by μ-H and/or μ-CO bridges. In some cases simple hy-dride transfer occurs to give the corresponding products or the ionic salts of the cationic hydride [MCp$_2$HL]$^+$.

Alkyls

Dihydride and dihalide complexes are convenient starting materials for the prepa-ration of many related derivatives. Monoalkylated hydrides [MCp$_2$HR] can be iso-lated by insertion of alkenes (CHR=CH$_2$) or alkynes (CR≡CH) with activating substituents (R = CN, COOMe) into one M-H bond of the dihydrides; insertion into the second M-H bond is possible only in a few limited cases, such as the reac-tion which gives [MoCp$_2${C(CN)=CHCN}$_2$]. A particularly interesting group of alkyl hydrides are those which result from the activation of ring or alkyl-ring C-H bonds when methods designed to prepare metallocenes are applied. All of these re-actions, which have been extensively studied, proceed with transfer of hydrogen to the metal and formation of η^1-η^5-cyclopentadienyl bridged compounds (**64**) or η^5-η^5-fulvalene hydrides (**17**). Interesting C-H activations occur in reactions of [MCp$_2$H$_2$] with BH$_3$.THF leading to various boranate derivatives [174]. Protona-tion of the alkyl hydrides with HX affords the corresponding alkyl derivatives [MCp$_2$RX] (X = Cl, OTf, SPh, SH) [175]. Dialkyl complexes can be isolated by alkylation of the dihalides with various alkali, magnesium and aluminium alkylat-ing agents, and the same reactions can be used to obtain metallacyclobutanes, met-allacyclobutabenzenes, metallasilacyclobutanes and other related metallacycles. Different behaviour was observed for certain magnesium alkyls such as allyl, 2-methylallyl and benzyl [176], which in the presence of MgCl$_2$ attack the cyclopen-tadienyl ring and, after elimination of chloride and hydrogen transfer to the ring, give ring-substituted chloro hydrides [MCp(C$_5$H$_4$R)ClH]. After protonation they are transformed into the dichlorides which can be treated in the same way to give [M(C$_5$H$_4$R)$_2$ClH] and [M(C$_5$H$_4$R)$_2$Cl$_2$] derivatives. Similar cyclopentadienyl ring-substituted compounds have been isolated by silylation of the allylic double bond with silylalkyl magnesium reagents [177]. Several differences between the ansa-bridged compounds and their non-bridged analogues were noted [169]. Photolysis of [M{CMe$_2$(η^5-C$_5$H$_4$)$_2$}H$_2$] in benzene gave [M{CMe$_2$(η^5-C$_5$H$_4$)$_2$}PhH] for M=Mo but not for M = W, whereas the reverse situation applies in the unbridged complexes. In addition the dihydrides and the alkyl or aryl hydrides were reported to be much more stable towards reductive elimination than their non-bridged ana-logues; this behavior was thought to reflect the inability of the 16 electron interme-diate [M{CMe$_2$(η^5-C$_5$H$_4$)$_2$}] to adopt a relatively stable parallel ring structure [178].

64   65   66

Other derivatives

The dihydride complexes are convenient starting materials for a wide range of derivatives. For example, the diboryl complex [WCp$_2$(Bcat')$_2$] (65) (Bcat' = O$_2$C$_6$H$_3$-
4-Bu^t, or O$_2$C$_6$H$_3$-3,5-Bu$^t_2$) was obtained from the photochemical reaction of the
dihydride with cat'BBcat'. The hydrido boryl complex [WCp$_2$H(Bcat)] (66) (Bcat
= O$_2$C$_6$H$_4$) was obtained from the reaction of chlorocatecholborane with
Li[WCp$_2$H], catecholborane being eliminated from (66) under photochemical conditions. X-Ray molecular structures obtained for (65) and (66) have geometries
which are consistent with the presence of a weak dπ-pπ metal-boryl interaction
[179,180,181].

Alkoxide and carboxylate hydrides [MCp$_2$H(OR)] and [MCp$_2$H(OOCR)] are
best prepared via reactions of the alkyl hydrides [MCp$_2$HR] with the corresponding
alcohol or carboxylic acid [182]. When the same reactions are carried out using the
dihydrides, the resulting protonated complex [MCp$_2$H$_3$]$^+$ can only be transformed
by exposure to light in the presence of O$_2$ as a hydrogen acceptor, to initially give
monosubstituted [MCp$_2$H(OR)] or [MCp$_2$H(OOCR)] and finally disubstituted compounds [MCp$_2$(OR)$_2$] and [MCp$_2$(OOCR)$_2$] under appropriate conditions [183].

Thiolate complexes [MCp$_2$(SR)$_2$] have been isolated by reaction of the dichlorides with thiols [2,184], and the influence of the alkyl chain length and steric effects on the Mo-S bond enthalpies has been studied [185].

Silanes react photochemically with group 6 metal dihydrides to give hydride
complexes [MCp$_2$H(SiR$_3$)], and with the silene tungsten complex (67) to give *bis*-
(silyl) compounds [WCp$_2$(SiMe$_3$)(SiR$_3$)] [186]. Similarly the germyl silyl complexes [WCp$_2$(SiMe$_3$)(GeR$_3$)] (68) were obtained from the reaction of (67) with ger-
manes [187]. Crystallographic studies show that complexes with a chloride substituent on the YR group (Y = Si or Ge; R = Me$_2$Cl or Pr$^i_2$Cl) experienced considerable shortening of the M-Y bond accompanied by lengthening of the Y-Cl bond, resulting from π-back-bonding from the metal to the Y-Cl σ-bond. Thermolysis of
(68) gave an unusual dinuclear germylidene complex (69) with a three coordinate
germanium moiety bridging the two tungsten centres [188]. Reactions of
[MCp$_2$H$_2$] with tin alkyl halides afforded disubstituted [MCp$_2$(SnR$_3$)$_2$] and monosubstituted [MCp$_2$H(SnR$_3$)] stannyl complexes [189]. The molecular structure of the

di(chlorodimethylstannyl) complex (**70**) shows a Sn-Mo-Sn bond angle of 83.2(1)°, in the range observed for Y-W-Y in the silyl (86.0(1)° - 97.3(1)°) and silyl germyl (85.9(1)°) complexes.

Molybdenum-magnesium bonded compounds (**71**) have been isolated by reaction of the molybdenum dihydride with magnesium dimethyl in the presence of TME-DA [190].

Bidentate dianions

An important group of compounds with dianionic bidentate ligands are accessible in several ways. A particularly interesting synthetic method uses the *cis*- stilbene derivative [MCp$_2$(η^2-PhCH=CHPh)], which acts as a precursor of the metallocene fragment which is then transformed into a series of metal(IV) derivatives by oxidative coupling of various unsaturated substrates [191, 192]; it also reacts with non-enolizable ketones giving kelated diolate compounds [MCp$_2$(-OCR$_2$CR$_2$O-)].

Related complexes can also be obtained directly from hydride derivatives with evolution of H$_2$. For example, reaction of [MoCp$_2$H$_2$] with CO$_2$ gives the side-bound η^2-CO$_2$ intermediate which gradually disproportionates to give the carbonyl and the η^2-coordinated carbonate complexes $[MoCp_2(\eta^2\text{-}CO_3)]$ **(72)** [193]; the corresponding tungsten complex is obtained by reaction of $[WCp_2HPh]$ with CO_2 in wet acetone [183]. An unusual reaction takes place when the tungsten dihydride reacts with diphenylketene giving a product which rearranges to complex **(73)** [194].   Similar reactions of the dihydrides with CS_2 and SO_2 give the cyclic compounds with bidentate dianions $[MCp_2(\text{-SCS-})]$ and $[MCp_2(\text{-O(SO}_2\text{)S-})]$ respectively [195].   Several chelate structures with oxygen donor ligands such as $[MoCp_2(\text{-O\{PO(OH)\}O-}]$ were proposed to explain the species observed in aqueous solution during $[MoCp_2Cl_2]$ promoted phosphoester bond cleavage reactions [196].

72   **73**

An alternative synthetic strategy giving access to compounds of this type uses [2 + 2] cycloadditions to the M=O bond of the oxo metal complexes $[MCp_2O]$ discussed in section 6.4. Their reactions with ketenes, ketenediimines and isocyanates give complexes of the general type $[WCp_2\{\text{-O(C=E)E'-}\}]$ (E = CPh_2, NR, O; E' = O, NR) [197]. The related molybdenum dimer $[MoCp_2\{\text{-O(MoO}_2\text{)O-}\}]$ results from irradiation of $[MCp_2(O)]$ in the presence of oxygen or by reaction of $[MCp_2Cl_2]$ with $[MoO_4]^{2-}$ [198].

Cyclic compounds with sulphur and selenium donors such as $[MCp_2\{\text{-S(E)E'-}\}]$ (E = S_2, C=NTol) and $[MCp_2\{\text{-Se (o-C}_6\text{H}_4\text{)Se-}\}]$ have also been reported [199, 200]. Photolysis of $[WCp_2H_2]$ with thiophene gave mixtures of the CS_2 insertion product **(74)** and the thienyl derieative $[WCp_2(C_4H_3S)]$ in ratios dependent on the irradiation time, **(74)** being the initial product [201]. The x-ray molecular structure of **(74)** showed a nonplanar metallathiacyclic ring in which the bonding was thought to be localized. A similar reaction of the molybdenum dihydride gave only the thienyl complex.

74   **75**

A series of metallaheterocycles was obtained from reactions of the tungsten carbonyl derivative [WCp$_2$(CO)] with heteroallenes [202]. Reactions of elemental chalcogens (S, Se or Te), Me$_3$NO or Me$_3$PE (E = S or Se) with the disilenes [MCp$_2$(η^2-Me$_2$Si=SiMe$_2$)] afforded symmetrical metallacycles (**75**) (M = Mo; E = S, NSiMe$_3$, SiMe$_2$: M = W; E = O, S, Se, Te, SiMe$_2$) [203, 204]. Varying amounts of the unsymmetrical insertion products were also obtained from reactions with phosphine chalcogenides.

6.3.2.5 Cationic Group 6 d^2 Metal Compounds [MCp$_2$XL]$^+$

This important group of cationic complexes is represented by a large number of molybdenum and tungsten derivatives which are easily accessible because the presence of one pair of free electrons in the neutral [MCp$_2$X$_2$] compounds favours substitution of at least one of the anionic substituents, mainly by π-acceptor ligands.

Substitution in metal(IV) complexes [MCp$_2$X$_2$] (X = Cl, Br, I, SR) may be achieved by addition of ligands, halide abstraction with thallium salts and electrochemical or chemical oxidation with FeCp$_2^+$ and NO$^+$ salts. It has been demonstrated that electrochemical oxidation of [MCp$_2$X$_2$] (X = Cl, Br, I, SR) in nitrile solvents leads to the metal(V) cationic species [MCp$_2$X$_2$]$^+$ [205], which undergoes oxidatively induced reductive elimination of X$_2$ to give the nitrile-coordinated cationic metal(IV) complexes. On this basis, chemical oxidation in nitrile solvents was the method used to isolate nitrile coordinated cationic species [MCp$_2$X(NCMe)]$^+$ (**76**) [206], which are appropriate precursors for the substitution of different π-acceptor ligands such as CO or phosphines. Direct substitution of iodide by simple addition of isocyanides affords the cationic complexes [MCp$_2$H(CNR)]$^+$ [80, 207], and related halide compounds can also be obtained by abstraction of halide in the presence of the ligand. Photochemical reactions of [MInd(η^4- C$_5$H$_6$)(CO)$_2$]$^+$ gave the hydride [MIndCpH(CO)]$^+$ which was easily transformed into the corresponding halides on addition of chloroform or methyl iodide [208]. The high tendency to coordinate π-acceptor ligands is shown by the easy electrophilic attack on the coordinated CO$_2$ in [MoCp$_2$(η^2-CO$_2$)] to form the carbonyl derivative [209]; reaction with HCl or SiClMe$_3$ gives [MoCp$_2$Cl(CO)]$^+$; reaction with RCOCl gives [MoCp$_2$X(OOCR)(CO)]$^+$; reaction with [CoH(CO)$_4$] gives the heterodinuclear compound [MoCp$_2$H(CO)]$^+$ [Co(CO)$_4$]$^-$ [210], which is further transformed into [MCp$_2$(μ-H)(μ-CO)Co(CO)$_3$] [Co(CO)$_4$].

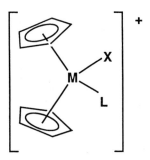

X = H, R, Cl, Br, I, SR
L = NCR, PR$_3$, CO

76

Dissociation of iodide takes place more easily than chloride or bromide, and tungsten complexes are more resistent to substitution. A solution of [MoCp$_2$HI] in acetonitrile rapidly establishes an equilibrium with its solvolysis product [MoCp$_2$H(NCMe)]$^+$ which is completely shifted by addition of TlBF$_4$, as evidenced by NMR spectroscopy; similar displacement of iodide takes place by simple addition of phosphines, whereas this reaction is slower for chloride and bromide and does not take place for related tungsten derivatives [211].

Halide species such as [WCp$_2$Br(SMe$_2$)]$^+$ and the dimer [MoCp$_2$(μ-SR)]$_2$$^{2+}$ are easily accessible from [WCp$_2$Cl$_2$] [212], and have been used to prepare *bis*(cyclopentadienyl) tungsten complexes with chiral metal centers [WCpCp'BrH] via nucleophilic attack on one of the cyclopentadienyl rings, with migration of hydrogen to the metal and elimination of the sulphide ligand [213, 214]. The dithiocarbamate complex [WCp$_2$(η^2-S$_2$CNHMe)]$^+$ was also obtained by a rare hydrogentransfer reaction from [WCp$_2$(SH)$_2$] [215]. Corresponding alkyl complexes [MCp$_2$R(NCMe)]$^+$ (R= Me, CH$_2$Ph, SR) were obtained by elimination of alkane on addition of NH$_4$PF$_6$ to the dialkyl complexes [215], or by migration of phenyl to the methylidene group of the intermediate [WCp$_2$(CH$_2$)Ph]$^+$ obtained by deprotonation of the methyl derivative with the trityl radical dimer [216].

Isocyanide complexes can be obtained by alternative methods (Scheme 6-16). The molybdenum isocyanide complexes [MoCp$_2$X(CNMe)]$^+$ (X = halide, H, Me) (**77**) were isolated by reaction of the corresponding cyanide complex wih LiMe [217], and the methyl complexes [MoCp$_2$MeL]$^+$ (L = PR$_3$, CNR) were isolated by reaction of the neutral adducts [MoCp$_2$L] with MeI [80]. Irradiation of [WCp$_2$H(C$_2$H$_4$)]$^+$ in acetonitrile yielded the tungsten hydride [WCp$_2$H(CNMe)]$^+$ (**78**), whereas [WCp$_2$Et(CNMe)]$^+$ was obtained after stirring the reaction mixture in the dark for one week [218]. Reaction of the dihydride with trifluoromethanesulfonate in the presence of ligand gave [WCp$_2$HL]$^+$ (L = NCR or THF) [219]. A relatively short W-O bond length (2.194 Å) was observed in the THF compound, which was thought to result from strong σ bonding between the cationic metal centre and the THF without any W-O π–bonding.

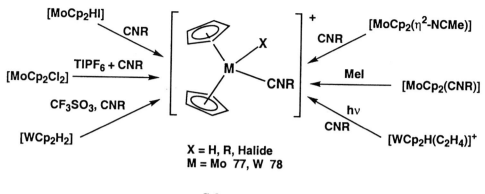

X = H, R, Halide
M = Mo 77, W 78

Scheme 6-16

6.3.2.6 Dicationic Group 6 d² Metal Compounds [MCp₂L₂]²⁺

When the solutions of $[MCp_2X(NCMe)]^+$ described above were treated with an appropriate excess of CO and phosphines the dicationic complexes $[MCp_2L_2]^{2+}$ (L = CO, PR_3, CNR) were obtained [80, 207, 208]. Reaction of $[MCp(\eta^4-C_5H_6)(CO)_2]^+$ with Ph_3C^+ also gave similar carbonyl adducts [220, 221]. Crystallographic studies show η^5-coordination of the Cp moieties in $[MoCpInd(CO)(NCMe)]^{2+}$ whereas η^3-coordination of the indenyl groups was observed in analogous neutral carbonyl species [209]. Complex (78), although stable in the solid state or in acetonitrile, reacts in diethyl ether or acetone solution with loss of H_2 to give the structurally characterized, dinuclear C-H activation product (79) [219].

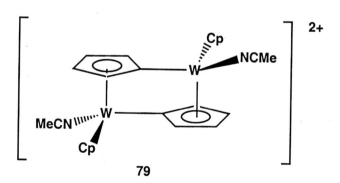

79

17-Electron Compounds

The most simple neutral complexes with 17 electrons are the d^1 vanadium(IV), niobium(IV) and tantalum(IV) derivatives. The HOMO accommodating the unpaired electron of all these paramagnetic compounds is the a_1, predominantly nonbonding, d_{z2} metal orbital which has a variable contribution from the d_{x2-y2} atomic orbital. The other two orbitals are used for bonding with the two anionic substituents (Figure 6-5)

All of these complexes are paramagnetic with effective magnetic moments corresponding to the presence of one unpaired electron. Delocalization of the unpaired electron on the ligands has been measured by the A_{iso} and g parameters in the EPR spectra and coupling with the six methyl protons has been demonstrated for the dimethyl-vanadium complex.

It is convenient, not only for structural but also for synthetic purposes, to distiguish between two types of compounds: a) those having independent monoanionic ligands and cyclopentadienyl ligands with typical bend angles θ between 130° and 135°, and the additional two ligands located in the equatorial plane at angles between 87° and 95°; b) those better considered as metallacycles where the two metal-bonded atoms are also bonded together, which have θ angles larger than 135° and the additional two ligands located in the equatorial plane at angles between 35° and 37°. For these metallacyclic compounds an increase in d_{z2} and a decrease in d_{x2-y2} character has been observed from a reduction of the vanadium hyperfine coupling in their EPR spectra. With this electronic distribution these compounds show neither basic nor acidic behaviour.

Compounds of type a) are included here whereas those of type b) are discussed in Section 6.5.

6.3.2.7 Neutral Group 5 d^1 Metal Compounds [MCp$_2$X$_2$]

Halides

Most of the synthetic methods for vanadium(IV) dihalides use metathetical reactions of the tetrahalides VX$_4$ with reagents which can transfer the cyclopentadienyl ligands, such as alkali cyclopentadienides or MgCpCl, or reactions of the dichlorides [VCp$_2$Cl$_2$] which exchange chloride for other anionic ligands.

Mixed dihalides have also been obtained by oxidation of [VCp$_2$Br] with ClCPh$_3$ and more recently the use of PCl$_3$ led to a mixture of [V(C$_5$H$_4$R)$_2$Cl] and [V(C$_5$H$_4$R)$_2$Cl$_2$] (R = Me, Pr^i, Bu^t) [16]. An extensive series of vanadium compounds containing variously substituted cyclopentadienyl and bridged *bis*(cyclopentadienyl) rings has been reported [1, 2, 222], and related pseudohalide complexes can be obtained by metathesis with the corresponding silver salts. Electrochemical studies of the dihalides show that reduction to [VCp$_2$Cl$_2$]$^-$ is accompanied by rapid loss of chloride to give the vanadium(III) chloride, whereas oxidation leads to the unstable cationic species [VCp$_2$Cl$_2$]$^+$.

The *bis*(cyclopentadienyl) niobium and tantalum(IV) dihalides are a large group of readily accessible compounds since the reducing properties of alkali cyclopentadienides and also of SnBu^{n}Cp (reagents frequently used in excess) lead directly to these d¹ metal compounds; mechanistic studies of this reduction have been reported [1]. The formation of difficult to separate mixtures may be avoided by addition of one equivalent of a reducing agent such as Na/Hg and then stoichiometric amounts of the Cp-transfer agent.

The following alternative methods are available for particular purposes:

a) Simple metathesis using the easy to isolate $NbCl_4(THF)_2$ with LiCp is a clean reaction used for many of these compounds.

b) Metathesis of monocyclopentadienyltantalum(IV) halides with different alkali cyclopentadienides has been used in particular to prepare compounds containing mixed Cp rings.

c) Metathesis of the anions in previously isolated *bis*(cyclopentadienyl)metal compounds has been used where other methods cannot be employed.

d) Reaction of the trihydrides with alkyl halides takes place with elimination of H_2 and alkane to give the dihalide.

e) Oxidation of metal(III) compounds such as $[M(C_5H_4SiMe_3)_2X]_2$ with alkyl halides or $[NbCp*_2(\eta^2\text{-}BH_4)]$ with HCl has been used to prepare less accessible complexes.

Method a) using the thallium salt of the ligand has more recently been used to prepare the *ansa*-niobium complex $[NbCl_2\{SiMe_2(C_5H_4)_2\}]$ **(80)** [223].

80

These dihalide complexes are frequent starting materials in a wide range of metathetical reactions.

Alkyls

Alkylation of vanadium dihalides with the usual alkylating agents in polar solvents (THF) generally causes reduction to vanadium(III) alkyls [VCp₂R], although vanadium(IV) dialkyls can be easily isolated by alkylation with lithium, magnesium, aluminium or cadmium reagents in toluene or dichloromethane. Metallacyclic dialkyls have also been isolated by similar methods. Haloalkyl complexes are usually prepared by reaction of the vanadium(III) alkyls [VCp₂R] with halogenating oxidants such as $CuCl_2$, CuCl and AgCl, and the chloromethyl complex

[VCp$_2$MeCl] can be isolated by reaction of the dimethyl with a stoichiometric amount of hydrogen chloride.

Alkylation with alkali metal, magnesium, aluminium, zinc and other reagents has also been extensively used to prepare dialkyl and alkyl halide derivatives of niobium and tantalum, although selective monoalkylation requires the use of particular alkylating agents or reverse reactions of the dialkyls with appropriate reagents. Reported alkyl complexes include Me, CH$_2$SiMe$_3$, CH$_2$Bu^t, CH$_2$Ph, Ph, η^1-C$_5$H$_5$, C(CF$_3$)=CH(CF$_3$) amongst others in compounds with both unsubstituted and substituted cyclopentadienyl rings [2]. Phosphinomethyl complexes have been also used as metallo-ligands in the preparation of heterodinuclear compounds [160].

An important group of related derivatives are the metallacyclo-pentane, -pentene and -pentadiene dialkyls [158, 224].

Hydrides

The metal dihydrides are not easily synthesized and very few examples of hydride derivatives have been reported. Niobium and tantalum trihydrides are typically easily transformed into metal(III) hydrides losing H$_2$, and the easy reduction of the assumed metal(IV) dihydrides by hydride-transfer reagents makes these compounds very difficult to isolate. [Nb(C$_5$H$_4$R)$_2$H(μ-H)]$_2$ (R = CH$_2$SiMe$_3$, CH$_2$Bu^t) has been electrochemically generated and its structure proposed on the basis of ESR studies [108]. The cyclopentadienyl-bridged hydride complex [{Nb(η^5-η^1-C$_5$H$_4$)(η^5-C$_5$H$_5$)H}$_2$] is one of the few hydrides which does not spontaneously rearrange into a fulvalene derivative.[225]

Other derivatives

A large number of vanadium-oxygen-, -sulphur- and -selenium-bonded compounds with independent terminal ligands and cyclic dianionic ligands can be isolated by metathetical reactions or oxidative addition of the appropriate reagents to vanadocene.

Related niobium and tantalum derivatives with different donor-atom ligands include silyl, alkoxo, thioalkoxo, thiocyanide, dilkylphosphide, amide, carbonate, bridging-imide etc. Most of them can be prepared by appropriate metathetical reactions of the dihalides or from the trihydrides. Metal(IV) *bis*(cyclopentadienyl) derivatives have been analysed in detail [1,2]. The crystallographically characterized amido niobium complex (**81**) with a planar metallacyclic ring was obtained from the reaction of Scheme 6-17 [226]. Several metallacyclic complexes with sulfur donor ligands have recently been reported and their electrochemical behaviour investigated [227].

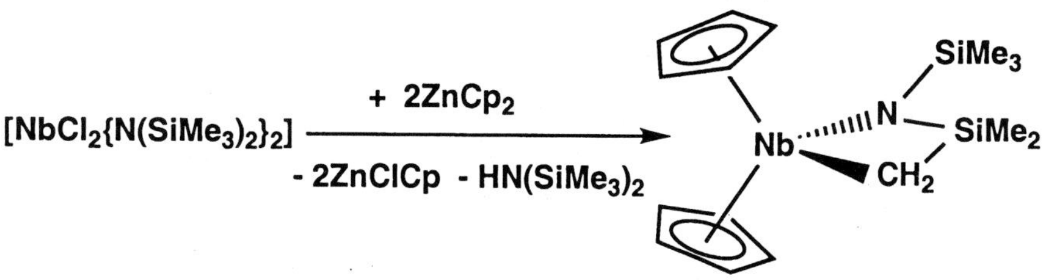

Scheme 6-17

6.3.2.8   Cationic Group 5 d^1 Metal Compounds [MCp$_2$XL]$^+$, [MCp$_2$L$_2$]$^{2+}$

Paramagnetic monocationic niobium halide complexes [MCp$_2$XL]$^+$ were isolated by oxidation of the metal(III) derivatives with Ag[BF$_4$] [116,120] or [FeCp$_2$][PF$_6$] [228] and by metathesis of the metal(IV) derivatives [116, 229]. They have also been generated by electrochemical oxidation of hydrido- [110] and halo- [116] niobium(III) compounds.

Related fulvalene dinuclear cationic compounds [{NbCp(NRR')}$_2$(μ-fulva-lene)]$^{2+}$ (R = H, R' = Ph, Tol; R = Et, R´= Tol) and the protonated μ-hydride complex [{NbCp(NHR)}$_2$(μ-H) (μ-fulvalene)]$^{3+}$ have been reported [109].

Dicationic complexes [MCp$_2$LL']$^{2+}$ have also been isolated by displacement of the second halide from [MCp$_2$XL]$^+$ with different monodentate ligands [116] and directly from [MCp$_2$Cl$_2$] by reaction with thallium salts (PF$_6$ or BF$_4$) in the presence of methylisocyanide [230] or bidentate nitrogen [231] donor ligands.

6.3.2.9   Cationic Group 6 d^1 Metal Compounds [MCp$_2$X$_2$]$^+$

Few examples of this type of paramagnetic metal(V) cationic species have been reported. Oxidation of [MCp$_2$Cl$_2$] with WF$_6$ in liquid SO$_2$ allows the WF$_6^-$ and W$_4$F$_{18}^-$ salts of the [MCp$_2$X$_2$]$^+$ cation to be isolated [232]. Electrochemical oxidation of [MCp$_2$Cl$_2$] in acetonitrile generates the cation [MoCp$_2$Cl$_2$]$^+$ which in the presence of ultrasound promotes the formation of two electron oxidation product [MoCp$_2$X(NCMe)]$^{2+}$ at a considerably lower voltage than in the absense of ultrasound [233].

Paramagnetic metal(V) dialkyls are an important group of compounds in which the α-elimination processes responsible for the transformation of metal(IV) dialkyls induced by trityl salts have been extensively studied. Electrochemical studies show that the metal (IV) dialkyls [MCp$_2$R$_2$] are easily oxidized to the monocations [MCp$_2$R$_2$]$^+$ with a more negative potential for R = Me (-0.27 V) than for higher alkyls (ca. -0.21 V) [234] and this oxidation can be effected at a preparative level by chemical oxidation. Mechanistic studies using various combinations of alkyls [MCp$_2$MeR']$^+$ (R = Me, Et, Ph) indicate that reactions of these species with Ph$_3$C-CPh$_3$ take place by preferential α-abstraction from the methyl group, leading to metal(IV) cationic species [MCp$_2$RL]$^+$; only when the reaction of [MCp$_2$MePh]$^+$ is carried out in CH$_2$Cl$_2$ is the chloro derivative [MCp$_2$Cl(CH$_2$Ph]$^+$ isolated [217].

Oxidation of [WCp$_2$(OMe)Me] with ferrocenium hexafluorophosphate gave the structurally characterized dicationic, d^1-d^1 dinuclear complex (82), in which shortening of the W-O bond (W-O = 1.904(1) Å) was observed, suggesting (together with its electronic spectrum and magnetic measurements) a bond order of about 1.5 [235]. Spin pairing of the d^1 metal centers through a π–interaction involving the oxo bridge was proposed to explain its diamagnetism in both solution and the solid state. Complex (82) is photosensitive and disproportionates in acetonitrile to give the terminal oxo complex [WCp$_2$(O)Me]$^+$ (83) and the d^2 solvent adduct [WCp$_2$Me(NCMe)]$^+$.

82   83

16-Electron Compounds

These cationic 16-electron complexes have two orbitals 2a$_1$ and b$_2$ involved in bonding with the two anionic substituents and one vacant orbital 1a$_1$ (refer to Figure 6.5)

6.3.2.10   Cationic Group 5 d^0 Metal Compounds [MCp$_2$X$_2$]$^+$, [MCp$_2$XL]$^{2+}$

Most of the reported d^0 cationic complexes are 16-electron compounds. The first cationic group 5 metal(V) derivatives were reported as products of the reaction of the disulphur compounds [MCp$_2$(η^2-S$_2$) Me] with MeI and [CPh$_3$][BF$_4$] which gives [MCp$_2$MeI]$^+$, and presumably in the presence of H$_2$O also gives the μ-oxo dinuclear $[(MCp_2Me)_2(\mu\text{-O})]^{2+}$ complexes [236]. Similar halo complexes $[MCp_2ClX]^+$ were isolated as the Br_3^- and I_3^- salts [237] by oxidation of $[MCp_2ClX]$ with an excess of halogen (X = Br, I), and by oxidation of $[MCp_2Cl(SiMe_3)]$ [238] or $[MCp_2X_2]$ with various oxidizing agents such as AsF_5 [239] or $[FeCp_2]^+$ [240,241,242]; the oxidations in DMF give the dicationic species $[MCp_2X(DMF)]^{2+}$. They were also generated by electrochemical oxidation of the metal(IV) dihalides [243]. Related alkyl [160,244,245,246] and phosphide [247] complexes have also been isolated. The alkoxide complexes (84) and (85) were obtained via the carbon-carbon or carbon-hydrogen bond formation reactions of Scheme 6-18; (84) was hydrolised to the Lewis acid adduct (86) [248]. Oxidation of $\{Nb(\eta\text{-}C_5Me_4Et)_2\}$ (prepared in situ by Na/Hg reduction of the dichloride) with excess HPF_6 yielded the structurally characterized difluoride $[Nb(\eta\text{-}C_5Me_4Et)_2F_2]^+$ under spontaneous evolution of H_2 [249].

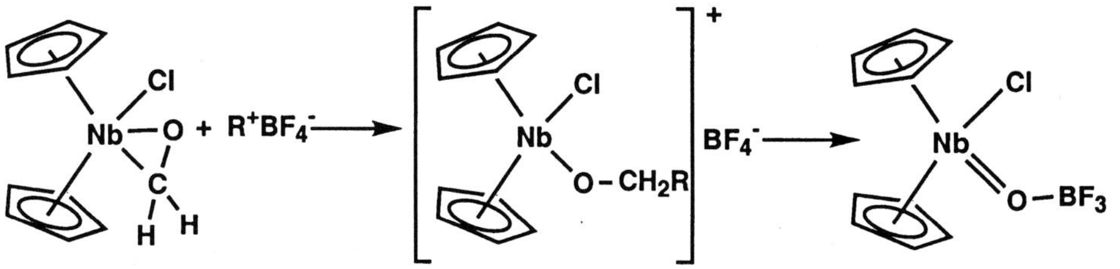

R = CPh₃ 84
R = H   85

Scheme 6-18

Analogous complexes with niobium-gold bonds $[NbCp_2\{Au(PPh_3)_2\}]^+$ have been isolated from the reaction of the trihydrides with gold(I) salts [250].

6.3.2.11 Dicationic Group 6 d⁰ Metal Compounds $[MCp_2X_2]^{2+}$

The oxidation of $[MCp_2X_2]$ with very strong oxidants such as AsF_5, SbF_5, or $[(FCN)_3F]^+[BF_4]^-$ gives the dicationic species $[MCp_2X_2]^{2+}$ (M = Mo, W; X = Cl, Me) [247, 251, 252].

6.3.3 Complexes with Three Coordinated Ligands

18-Electron Compounds

The only neutral complexes with three additional substituents bonded to the metal centre are the 18 electron species derived from d^0 vanadium, niobium and tantalum(V) metals (Table 6-4). All the other d^0 group 5 and 6 metal compounds are cationic species. For these compounds all three metal based valence orbitals are used to bond the three σ-donor ligands (Figure 6-6).

Table 6-4. Different types of $[MCp_2L_3]$, $[MCp_2XL_2]$ and $[MCp_2X_2L]$ groups 5 and 6 metal complexes

Total Number of electrons	18
Configuration	d^0

$$[(M5)Cp_2X_3]$$

$$[(M5)Cp_2X_2L]^+$$
$$[(M5)Cp_2XL_2]^{2+}$$
$$[(M6)Cp_2X_3]^+$$
$$[(M6)Cp_2X_2L]^{2+}$$

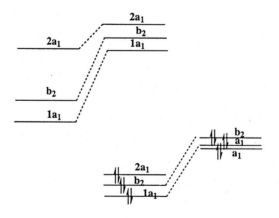

Figure 6-6

Qualitative diagram for $[MCp_2X_3]$ d^0 metal complexes with σ-donor ligands

6.3.3.1 Neutral Group 5 d^0 Metal Compounds [MCp$_2$X$_3$]

Halides

Almost a complete list of homo- and hetero-trihalide complexes with equal and mixed unsubstituted and substituted cyclopentadienyl rings have been isolated [1,2]. The presence of three rather bulky donor atoms in the equatorial plane between the two rings makes the coordination sphere too crowded and therefore these compounds have a high tendency to lose a halide ligand to give cationic species with fewer than 18 electrons. Isolation of neutral trihalides usually requires the use of non-coordinating solvents. Furthermore, the reducing properties of alkali cyclopentadienides, used to transfer the rings to the metal(V) halides, lead to the metal(IV) dihalides [MCp$_2$Cl$_2$], which are the most convenient complexes to prepare, store and use as starting materials for the preparation of trihalides [238].

These complexes and their related silyl, alkoxides and phosphides are easily hydrolized to the μ-oxo dimers [(MCp$_2$X$_2$)$_2$(μ-O)] (X = halide, SiPh$_3$, PR$_2$, OR). An alternative method of synthesis is the oxidation of the corresponding metal(IV) compounds by oxygen.

Hydrides

The trihydride is one of the most important and useful group 5 metal derivatives, however its preparation does not follow a simple direct path because the method involving the reaction of MCl$_5$ with NaBH$_4$ always leads to undesirably low yields. Better yields were obtained by using Na[AlH$_2$(OCH$_2$CH$_2$OMe)$_2$] (vitride or red-al) or reacting [MCp$_2$Cl$_2$] with Li[AlH$_4$]; the mechanism of the latter has been investigated [2].

Many structural studies have been carried out on trihydride compounds, including x-ray and neutron diffraction and photoelectron spectroscopy. The NMR behaviour of the niobium derivatives is particularly interesting; short T$_1$ values and large (up to 100 Hz at room temperature) temperature dependent ^{1}J$_{H-H}$ coupling constants have been reported, in contrast to the low temperature independent values (ca. 10 Hz) usually observed for the tantalum derivatives. Extensive NMR experiments, isotopic labeling studies and theoretical calculations have been undertaken to investigate this unusual NMR behaviour [253, 254, 255], which has been attributed to either a combination of magnetic and exchange couplings [254] or to quantum mechanical exchange couplings [255] as a simple manifestation of the quantum mechanical motion associated with the hydrogen nuclei in the trihydride structure. These observations indicate exchange with structures containing η^2-coordinated dihydrogen, consistent with the easy loss of hydrogen observed on heating or by addition of ligands [256, 257].

Similar elimination reactions take place with trimethylsilane affording the *mono* and *bis*(trimethylsilyl) hydrido derivatives [258, 259, 260, 125]. Likewise, *mono* and *bis*(trimethylstannyl) hydride complexes were obtained [261]. Phosphido [262] and arsenido [263] derivatives were obtained from reactions with halo-phosphines and -arsenes (Scheme 6-19). A dihydro phosphido niobium complex was however not isolated and the reaction yielded only the phosphine hydride [NbCp$_2$H(PHPH$_2$)] [264]. The structurally characterized boryl complexes (**87a**) and (**87b**) were obtained as an easily separated isomeric mixture from the reaction of Li[WCp$_2$H] with chlorocatecholborane [265]. Mixed hydride- halide, silyl, and amide complexes have also been reported [2].

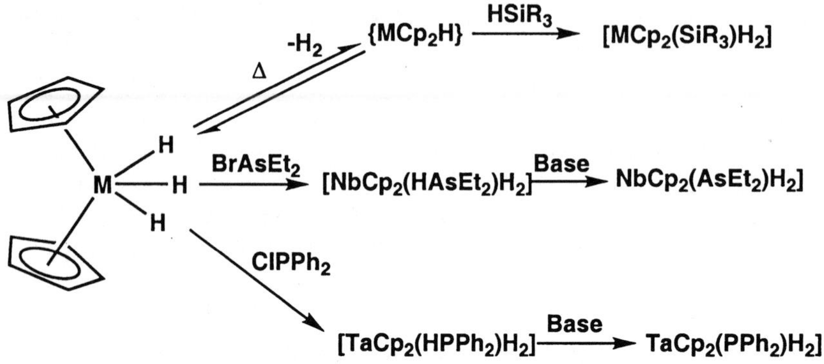

Scheme 6-19

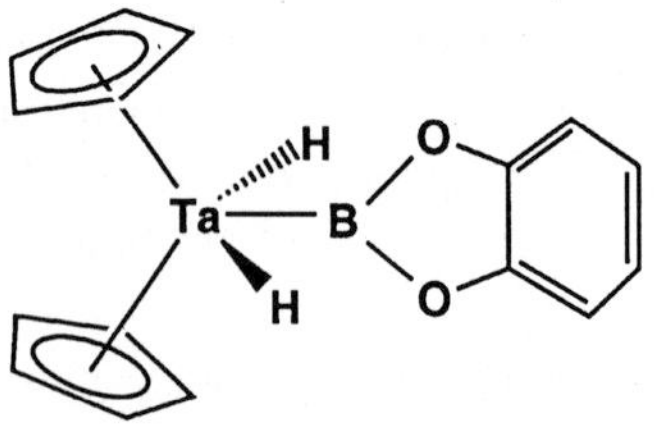

87a

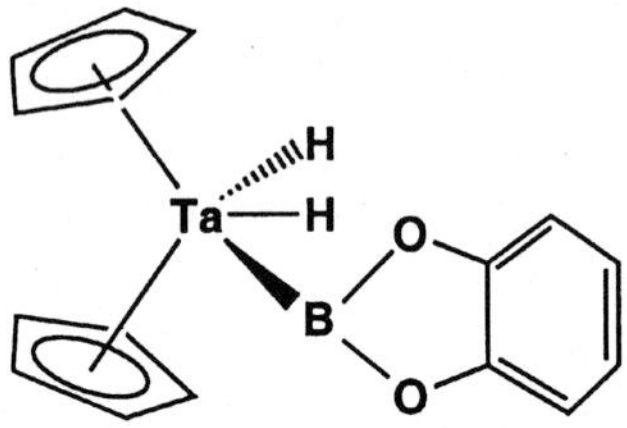

87b

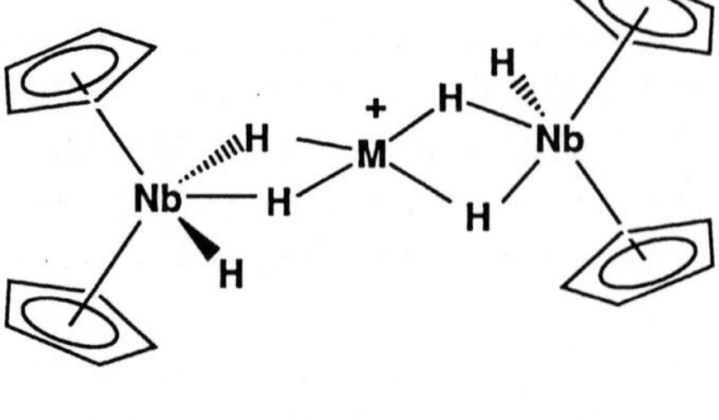

88

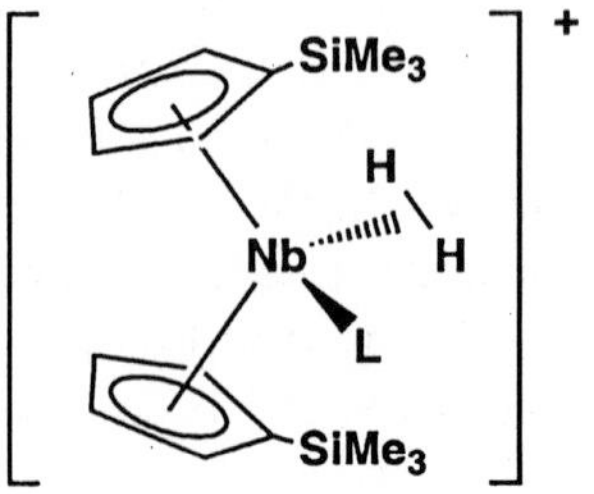

89

The trihydrides react with Lewis basic coinage metal cations [e.g. $Cu(MeCN)_4^+$, $Ag(THF)^+$, $Au(THF)^+$] to give hydrido bridged heterometallic species of type (88), several of which have been crystallographically characterized [266 ,267].

Alkyls

Bis(cyclopentadienyl) trialkyl complexes are not known and very few di- and mono-alkyl derivatives have been reported, most of which are methyl compounds. The dimethyl niobium complex $[NbCp_2Me_2(\eta^1-S_2CNEt_2)]$ was prepared by reaction of the niobium(IV) dimethyl derivative with $R_2NC(S)S-SC(S)NR_2$ [268], whereas the same reaction with $(RO)_2P(S)S-SP(S)(OR_2)$ gave the paramagnetic niobium(IV) compound with elimination of both methyl groups. Similarly the monoalkyl complexes $[NbCp_2Bu^n\{CH(SH)(SMe)\}(\eta^1-BH_4)]$ [269], $[NbCp_2Me(\eta^2-S_2CNMe_2)]$ $[S_2CNMe_2]$ [269], $[NbCp_2MeHX]$ (X = SiH_3, PH_2, OMe, SMe, NMe_2), [245] the metallaoxacyclic compounds $[MCp_2Me(\eta^2-O,C)-OCHR-CH_2-)]$ [101,270] and the vinyl derivative $[TaCp_2(CH=CH_2)(SBu)H]$ [271] have been isolated.

6.3.3.2 Cationic Group 5 d^0 Metal Compounds $[MCp_2X_2L]^+$, $[MCp_2XL_2]^{2+}$ and $[MCp_2L_3]^{3+}$

Few complexes of this type have been reported. Monocationic niobium complexes $[NbCp_2Cl_2(Me_2CO)]^+$ were obtained when acetone solutions of the trihalo complexes $[NbCp_2Cl_2X]$ (X = Cl, I) were treated with one equivalent of Ag^+ salts, and the related dicationic compounds $[NbCp_2ClL_2]^{2+}$ resulted from the reaction with two equivalents of Ag^+ salts in acetone or in the presence of bidentate nitrogen donor ligands such as bipy and phen. Even the tricationic complex $[NbCp_2Cl_2(dien)]^{3+}$ was isolated using the tridentate ligand and a stoichiometric amount of the silver salt [273].

Steric congestion in the coordination sphere, as found in tris-ligand *bis*(cyclopentadienyl) complexes, probably explains why cationic tantalum compounds have only been isolated as dihydrides with one neutral ligand $[TaCp_2H_2(PR_3)]^+$ $[PR_3$ = PMe_2Ph, $P(OMe)_3]$, via protonation of the appropriate Ta(III) hydride [274]. Protonation of $[TaCp_2H(CO)]^+$ (Cp = Cp or $(\eta-C_5H_4Bu^t)$) gave a mixture of isomers: the classical dihydride and an η^2-dihydrogen derivative $[TaCp_2(\eta^2-H_2)(CO)]^+$ [275]. The niobium derivatives were obtained by protonation of $[Nb(\eta-C_5H_4SiMe_2)_2HL]$ (L = CO) with $[NH_3R]PF_6$ (R = H or Ph). Appropriate reaction conditions and HCF_3CO_2 as protonating agent also gave the side bonded species identified as (89) by NMR spectroscopy [272]. Electrochemical reduction of the trihydrides in THF gave $[NbCp_2H_2(THF)]^+$ (Cp = $C_5H_4SiMe_3$, $C_5H_4CMe_3$) [108].

6.3.3.3 Cationic Group 6 d⁰ Metal Compounds [MCp₂X₃]⁺

The best example of this type of compound is the cationic trihydride obtained by protonation of the dihydride. Deuteration of [WCp*₂H₂] demonstrated that D⁺ is firstly bonded in the central position and subsequently slowly isomerizes to occupy an external coordination site [273]. The NMR spectra for [MCp₂H₃]⁺ (M = Mo, W) correspond to an AB₂ spin system with $^1J_{H-H}$ = 8.5 Hz for tungsten and $^1J_{H-H}$ = 450 Hz for molybdenum due to exchange coupling [254]. Exchange couplings in the cationic trihydrides have been theoretically analyzed and the calculated values were in good agreement with the experimental results and are consistent with an η^2-dihydrogen transition state [274].

The Lewis basicity of the trihydrides [MCp₂H₃]⁺ is comparable to that of Et₂O and adducts with different metal salts have been isolated.[2] Reactions with CuCl, Fe₂(CO)₉ or Co₂(CO)₈ give salts of the [CuCl₂]⁻, [HFe₃(CO)₁₀]⁻ or [Co(CO)₄]⁻ anions respectively, whereas μ-H bridges are formed with [RhH₂(acetone)₂(PPh₃)₂]; activation of the ring C-H bond of one of the cyclopentadienyl ligands takes place with [IrH₂(acetone)₂(PPh₃)₂]; hydrogen transfer occurs with [ZrCp₂X(COR)] to give [Cp₂M-CHR-O-ZrXCp₂]. The reducing properties of the cationic trihydride compounds have been investigated in reactions with ketones and alcohols.

6.4 Type C - Complexes with Single-Atom Donor Dianionic Ligands

6.4.1 18-Electron Compounds

The dianionic ligand in these compounds provides a double bond using the a₁ and b₂ metal orbitals, giving the same electron configurations discussed in Section 6.3.2 for related [MCp₂X₂] group 5 and 6 metal complexes (Table 6-2). Accordingly, one or two electrons are located in the third 1a₁ orbital of the neutral paramagnetic [MCp₂E] group 5 and diamagnetic group 6 metal complexes respectively. When this orbital is used to bond one additional substituent, the resulting 18-electron systems correspond to the neutral [MCp₂EX] and cationic [MCp₂EL]⁺ group 5 and [MCp₂EX]⁺ group 6 metal complexes. These are the most important types of compounds described in this section, where paramagnetic [MCp₂E] complexes are limited to the neutral d¹ vanadium compounds.

6.4.1.1 Neutral Group 5 d⁰ Metal Compounds [MCp₂EX]

Bis(cyclopentadienyl) complexes with one dianionic and one monoanionic ligand are more stable and more accessible than those with three monoanionic ligands discussed in Section 6.3. Many are formed spontaneously in order to avoid three substituents. These complexes constitute several families comprising a large number of compounds.

Oxo, Sulfido, Selenido and Tellurido Complexes

The high oxophilicity of the group 5 metals is apparent from the variety and number of complexes formed. Dinuclear polinuclear and cluster complexes with μ-oxo bridges are numerous. Although considerably less common, a significant number of compounds with terminal M=O bonds can be obtained, in particular via oxidation of 18-electron, electron rich metal(III) derivatives (40) (Cp variously substituted) [99, 118, 120, 275, 276]. Both oxo and peroxo products can result from O_2 oxidations; [NbCp₂X(CO)] gave either an oxo [NbCp₂(=O)Cl] (X = Cl) or a peroxo derivative (X = Me). Less electron-accepting ligands (L = PMe₃ or CNR) also favoured the peroxo complex [118,120,117]. The oxo complex was however obtained from [NbCp₂X(CNR)] using t-butyl hydroperoxide as oxidising agent or by oxygen abstraction from the peroxo complex using PPh₃ [120]. Careful control of the amount of oxygen used in the reaction shows it to be a multistep process involving the intermediate formation of both μ-oxo and peroxide species [280]. The μ-oxo species was in fact generated via a conproportionation reaction between [NbCp₂(=O)Cl] and [NbCp₂X)]. The oxo derivatives [Nb(η-C₅H₄Me)₂(=O)R] (90) (R = CH₂Buᵗ,CH₂Ph or Me) were obtained via thermolysis of the appropriate carbon dioxide precursor [277].

90

Likewise the tantalum (III) derivative **(91)** used in situ was a convenient precursor to **(92)** and **(93)** Scheme 6-20 [278]. **(92)** was also prepared by treatment of tantalum alkylydine complexes with water [279, 280].

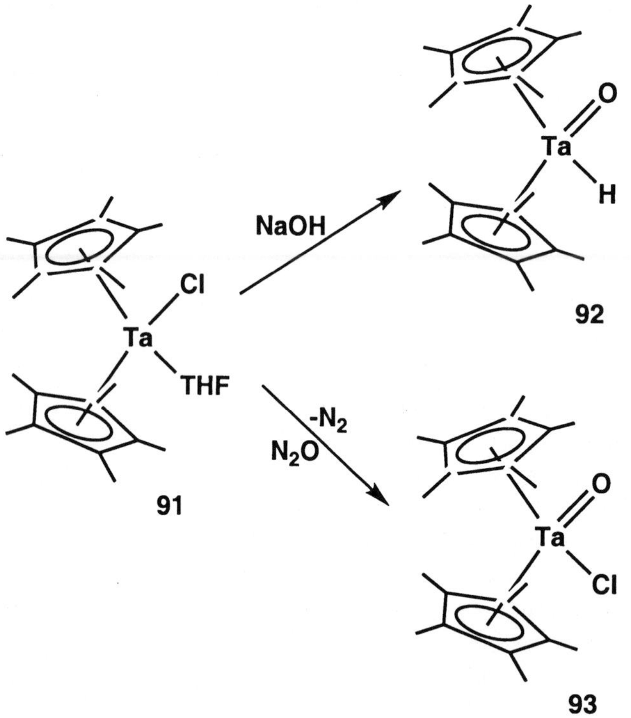

Scheme 6-20

The heterobimetallic species **(94-96)** were conveniently prepared by the reactions shown in Scheme 6-21 [281 ,282]. The alkyl derivatives [MCp$_2$(=O)Me] (M = Nb or Ta) were obtained by reaction of [NbCp$_2$(=O)Cl] with LiR or "TaCp$_2$Me", generated in situ, with an appropriate oxirane; both have been structurally characterised [283, 284].

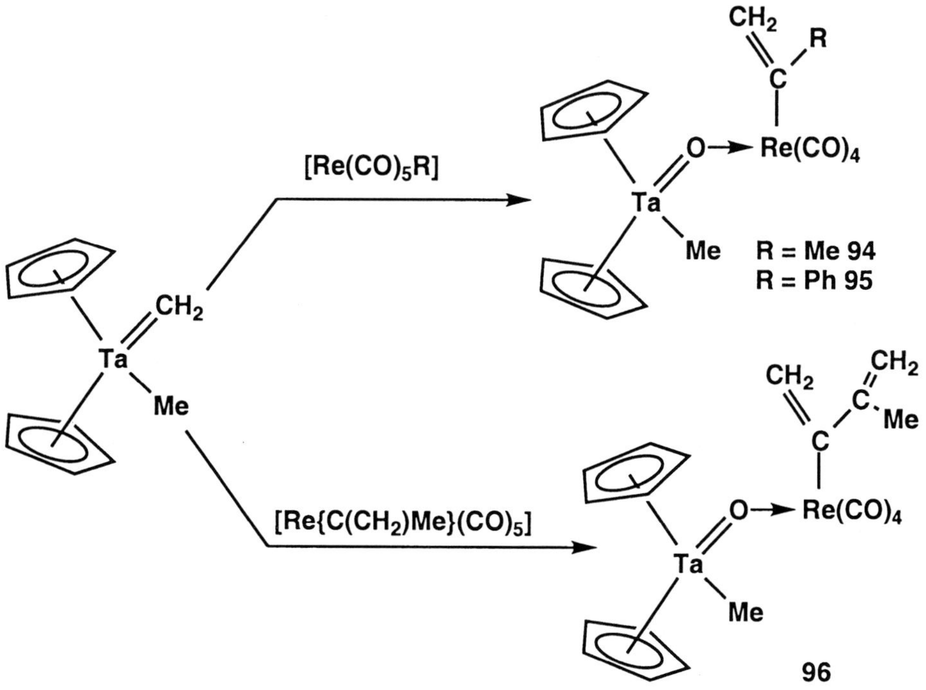

Scheme 6-21

Insight into the nature of the metal oxygen bond has been gained through ir stretching frequencies and x-ray structural characterisations. Relatively low v(M=O) values of ca 850 cm^{-1} are indicative of a bond order of less than 3; a bond order of between 2 and 3 is suggested by the metal-oxygen bond lengths (M=O: [Nb(η-C$_5$H$_4$SiMe$_3$)$_2$(=O)Me] [118] 1.720(2) Å; [NbCp$_2$(=O)Cl] [285] 1.737(6) Å; [TaCp*$_2$(=O)H] [284] 1.72(3) Å; **(90)** [281] 1.741 Å). The bond lengths in **(95)** (Ta=O 1.761(11) Å and Re-O 2.147(11) Å) were consistent with a Ta=O double bond with a Re-O donor-acceptor interaction.

In common with oxygen, sulfur can adopt a variety of bonding modes including η-S$_2$, μ-S sulfido bridges as well as terminal M=S. Reaction of the niobecene(III) boranate with S$_8$ gave [Nb(η-C$_5$Me$_4$Et)$_2$(=S)SH] (Nb-SH 2.381(1) Å and Nb=S 2.339(1) Å) together with the η-S$_2$ complex [286]. A series of sulfido permethylt-antalocene derivatives **(97)** were reported (Scheme 6-22) and their capacity to ef-fect catalytic hydrodesulfurization was investigated [272, 287]. Hydrogenation of **(97)** under forcing conditions gave **(98)**; further hydrogenation to **(99)** did not however take place even at 200°C. In fact reaction of **(99)** with H$_2$S at room tem-perature gave **(98)**, making their use in hydrodesulfurization unlikely. [Ta(C$_5$H$_4$Bu^t)$_2$(=S)I] or [Ta(C$_5$H$_4$Bu^t)$_2$(=S)H] were obtained from reactions of the persulfido hydride with MeI or PPh$_3$ [288].

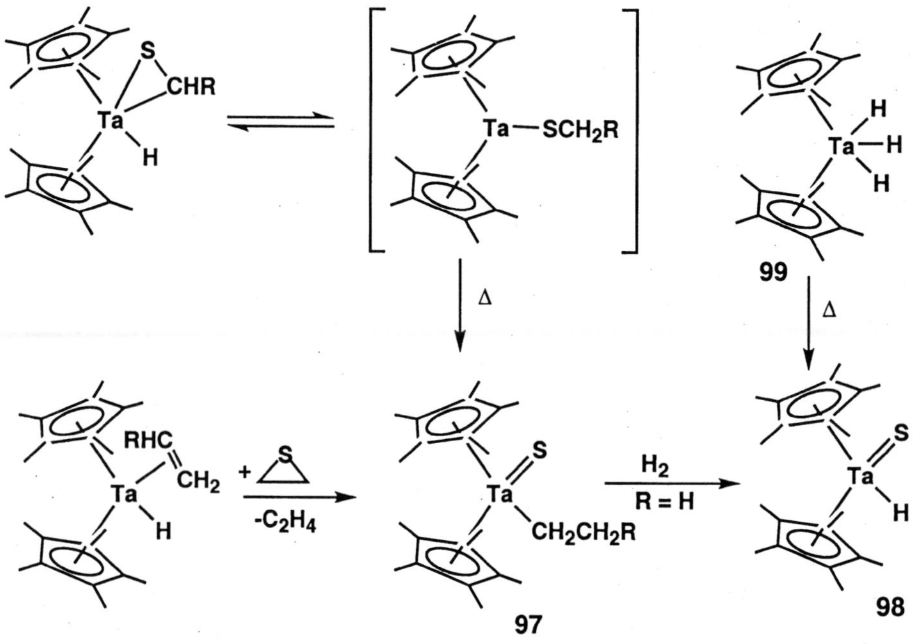

Scheme 6-22

In a reaction analogous to that discussed for oxygen, heteroatom abstraction from thiiranes has provided a convenient route to the sulfido complexes **(100)** and **(101)**, Scheme 6-23. The structure of **(102)** was proposed on the basis of spectroscopic and chemical evidence, although the presence of a species with two Ta(IV) centres was not ruled out [287,288]. Irradiation of a mixture of [TaCp*(η^6-C$_5$Me$_4$CH$_2$)(η^2-S$_2$)] **(103)** and [TaCp*$_2$(η^2-S$_2$)H] **(104)** by sunlight gave [TaCp*(η^6-C$_5$Me$_4$CH$_2$S)(=S)] and [TaCp*$_2$(=S)SH] which were separated by chromatography [289].

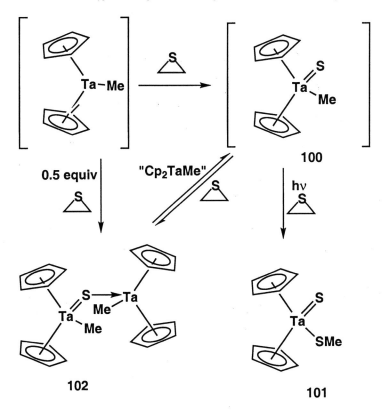

100

102

101

Scheme 6-23

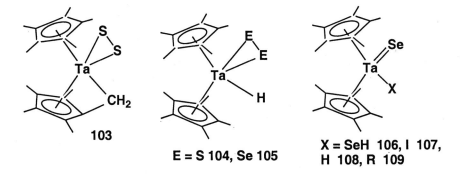

103

E = S **104**, Se **105**

X = SeH **106**, I **107**,
H **108**, R **109**

A series of selenido and tellurido complexes analogous to those of sulfur have been reported [290]: $[TaCp*_2(\eta^2\text{-}Se_2)H]$ **(105)** was converted into $[TaCp*_2(=Se)(SeH)]$ **(106)** in the presence of light; reaction of either **(105)** or **(106)** with MeI gave $[TaCp*_2(=Se)I]$ **(107)**; selenido- alkyl complexes $[TaCp*_2(=Se)R]$ (R = H **(108)**, Me or CH_2SiMe_3 **(109)**) were obtained via reactions of **(107)** with either alkyllithium or Grignard reagents; **(108)** was also obtained upon reaction of either **(105)** or **(106)** with PMe_3. The tellurido-hydride $[TaCp*_2(=Te)H]$ was obtained in a similar way from $[TaCp*_2(\eta^2\text{-}Te_2)H]$ and from heating $[TaCp*_2(\eta^2\text{-}TeCH_2)H]$ [291]. Ta=Se bond lengths of 3.29(2) Å and 3.372(1) Å were reported for **(108)** and **(109)**. Low field ^{77}Se NMR chemical shifts (δ 1990-2363) ppm were reported for the selenido complex (*cf* δ +54 to -595 ppm for singly bonded species) [294].

Imides

Relatively few complexes containing terminal imido groups have been reported, and most have been prepared using metal(III) derivatives as starting materials. In common with the oxide and sulfide complexes, $[TaCp_2(=NR)Me]$ (R = Me **(110)**, Bu^n or Ph) was obtained via nitrogen abstraction from aziranes [287, 288]. Imido complexes were also obtained via dinitrogen extrusion from azido complexes and **(110)** was isolated by this route from the phenylazido complexes $[TaCp_2(\eta^1\text{-}N_3Ar)Me]$ [292]. Reaction of $[TaCp*_2Cl(THF)]$ with N_3Ph gave upon thermolysis the imide $[TaCp*_2Cl(=NPh)]$ **(111)** and with LiNHR gave $[TaCp*_2H(=NR)]$ (R = Ph **(112)** or Bu^t); complex **(112)** was also obtained by treatment of $[TaCp*_2(=CH_2)H]$ with aniline (P = Ph) [282, 284]. Reaction of $[TaCp*_2Cl_2]$ with sodium amide gave $[TaCp*_2H(=NH)]$.

X = Me 110
X = Cl 111
X = H   112

Complexes (111) and (112) have been structurally characterised and in both com-pounds the Ta-N-C angle [176.4° (111) and 177.8° (112)] is virtually linear, indi-cating that the nitrogen atom is sp hybridized (often indicative of Ta≡N). The Ta-N distance [1.799(4) Å (111), 1.831(10) Å (112)] is however longer than expected for a triple bond and a bond order of between 2 and 3 is thus proposed. Molecular or-bital calculations on (112) show that a triple bond is not required to justify the lin-earity of the system [293].

Reaction of metallocene trihydrides with diazoalkanes gave the related diamag-netic η^1-diazoalkane complexes of scheme 6-24; the reaction was thought to pro-ceed through the intermediate (113) [294, 295, 296]. X-Ray structural determina-tions of (114), (115) and (116) reveal Nb-N bond lengths of *ca* 1.85 Å and a slight-ly bent Nb-N-N angle (*ca* 168 ° for (114) and (115) and an average 165° for (116)) with bond orders of slightly less than two suggested.

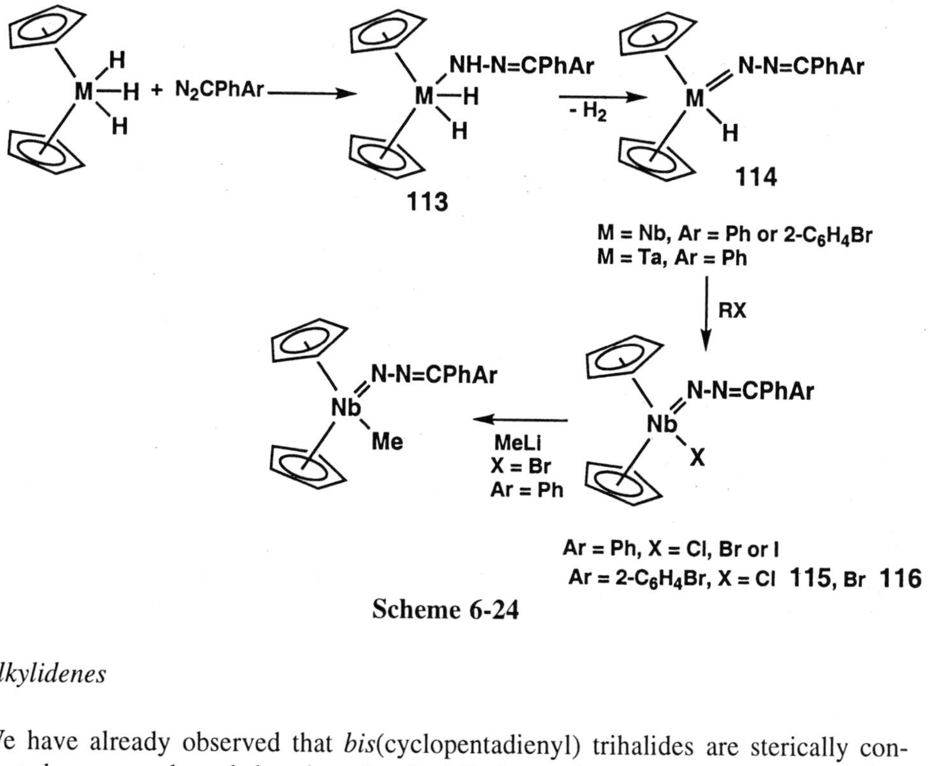

Scheme 6-24

Alkylidenes

We have already observed that *bis*(cyclopentadienyl) trihalides are sterically con-gested compounds and that the related trialkyl complexes do not exist. The tanta-lum alkylidene compounds were the first reported examples of "Schrock carbenes" and since then a huge number of alkylidene compounds have been synthesised. Most *bis*(cyclopentadienyl) derivatives [TaCp$_2$(=CH$_2$)R] were prepared by inter-molecular α-hydrogen abstraction using an external base or intramolecular α-hy-drogen transfer from an alkyl substituent. These compounds have been extensively reviewed [1, 2, 297].

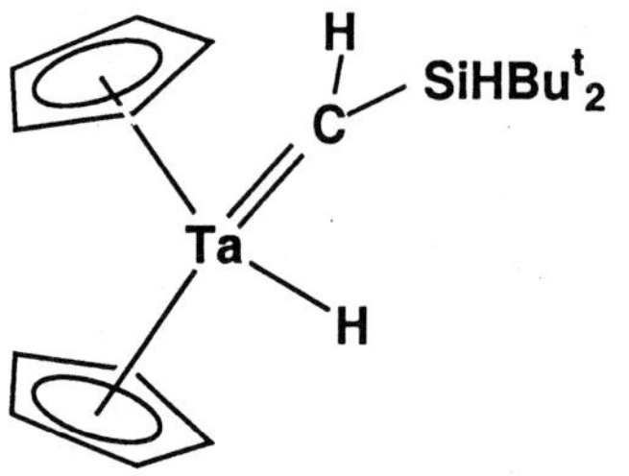

117

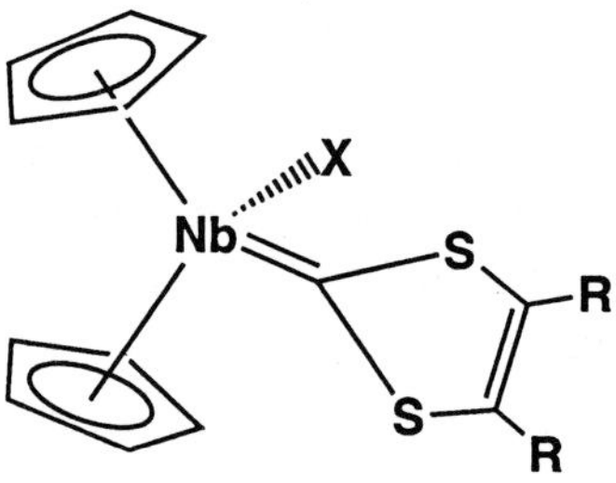

118

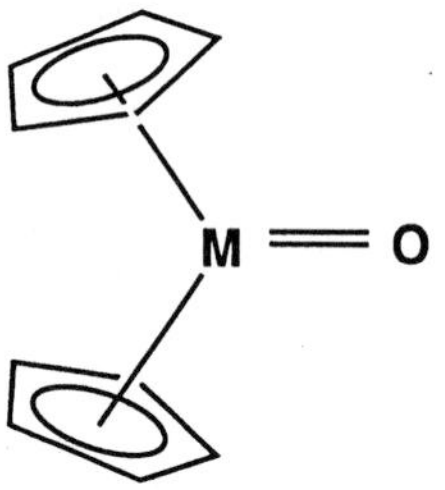

119

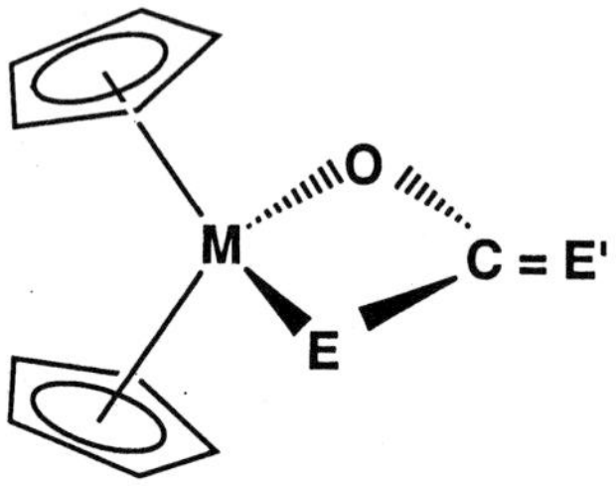

120

The silyl substituted alkylidene complex [TaCp$_2$\{=C(H)SiH(Bu^t)$_2$\}H] (**117**) was obtained from the reaction of [TaCp$_2$(=CH$_2$)Me] with the hindered silane H$_2$SiBu$^t_2$ whereas its reaction with unhindered silanes gave a mixture of the alkene derivative [TaCp$_2$(η^2-CH$_2$=CH$_2$)Me] and the hydrido silyl complex [TaCp$_2$(SiR$_3$)$_2$H]; the former reaction in benzene afforded the analogous phenyl methylene complex [298]. A number of other synthetic strategies have been employed including oxidation of the tantalum(III) complexes [TaCp$_2$XL] (X = Me, L = CO [283]; X = Cl, L = THF [282]) or thermal rearrangement of η^3-allyl derivatives to give vinylidine complexes [299]; several of those used to synthesise pentamethylcyclopentadienyl tantalum derivatives are summarized in Scheme 6-25. The η^2-CS$_2$ carbon disulfide complexes (Cp = η-C$_5$H$_4$SiMe$_3$) react with activated alkenes RC≡CR (COOMe or COOBu^t) to give the 1,3-dithiol-2-ylidene derivatives (**118**) (X = Cl, Et,CH$_2$CH$_2$Ph) [300]. An extensive series of heterobimetallic species with methylene bridges joining tantalum to various late transition metals has been obtained from reactions of [TaCp$_2$(=CH$_2$)Me] with an appropriate late transition metal complex [2, 301, 302].

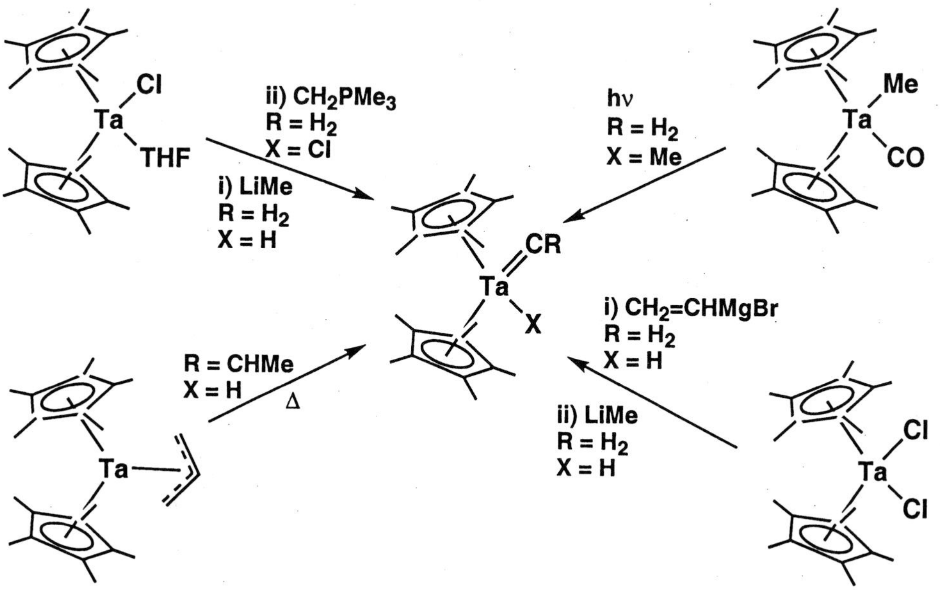

Scheme 6-25

6.4.1.2 Cationic Group 5 d⁰ Metal Compounds[MCp₂EL]⁺

Oxidation of $[Nb(\eta\text{-}C_5H_4Bu^t)_2(CH_2Ph)_2]$ with $AgBPh_4$ gave the thermally stable cationic niobium benzylidene complex $[Nb(\eta\text{-}C_5H_4Bu^t)_2(=CHPh)(THF)]^+$ [303].

6.4.1.3 Neutral Group 6 d² Metal Compounds [MCp₂E]

These group 6 d² metal compounds are loosely related to the $[MCp_2X_2]$ derivatives discussed in Section 6.3.2. They too have an 18 electron closed shell metal centre with two orbitals implied in M-E bonding, in this case to give a σ-π double bond with the dianionic ligand and the third a_1 orbital occupied by the pair of free electrons. Only strong π-donor ligands can stabilize this type of coordinatively unsaturated compound and not many of them have been reported. The simplest examples which illustrate their behaviour are the oxo derivatives $[MCp_2(=O)]$ (**119**) with various Cp ligands $(C_5H_5, C_5Me_5, C_5H_4R)$, obtained from the reaction of $[MCp_2Cl_2]$ with aqueous KOH [165, 304]. The molybdenum complex was also afforded by the photochemical reaction of $[MoCp_2H_2]$ with water [305]. The M=O bond length of 1.72 Å and IR stretching frequencies $v(M=O)$ of between 800 and 860 cm⁻¹ (cf $v(M\equiv O) \approx 930$ - 1000 cm⁻¹) measured in (**119**) indicate a bond order of about two [293, 306]. Molecular orbital calculations are broadly in agreement with this bonding picture [307]. The M=O bond is thus relatively weak and fairly reactive: the oxo ligand of [WCp*$_2$(=O)] exchanges very rapidly with H$_2{}^{18}$O; it reacts with H$_2$ to give [WCp*$_2$ H$_2$] and conversion of one of the rings into an alkoxo and η^1-coordinated ligand occurs in its reaction with O$_2$ or H$_2$O$_2$ to give respectively [WCp*(O-Cp*)(O)$_2$] or [WCp*(η^1-Cp*)(O)$_2$] [310]. The M=O undergoes [2+2] cycloaddition reactions with a range of electrophilic substrates [308, 198]. Reaction with a series of metal carbonyl complexes (CO)ML$_n{}^+$ (M' = Mn, Re, Fe, Ru, or Pt) gave the heterodinuclear compounds [MCp$_2$\{(η^2-O,O)C=M'L$_n$)\}] and cycloadditions of several heterocumulenes gave (**120**) (M = Mo, E = NR, E' = O; M = W, E = NR or O, E' = CPh$_2$, NR or O). Reactions with reagents such as MeOH, HX, Me$_3$SiCl or carboxylic acids gave [WCp$_2$X$_2$)] (X = OMe, Cl, RCOO).

Related sulfido derivatives have not been isolated and there is only indirect evidence for their formation, \{MoCp$_2$(S)\} was thought to be the reactive intermediate which undergoes cycloaddition, analogous to that observed with the oxo derivatives in the reaction of [MoCp$_2$S$_2$] with di-p-tolylcarbodiimide [309].

A series of imidomolybdenum complexes [Mo(η-C$_5$H$_4$R)$_2$(NR')] (**121**) and their mixed ring analogues [MoCp(η-C$_5$H$_4$R)(NR')] have been prepared from the reaction of a suitable mono(cyclopentadienyl) imido complex with an appropriate sodium cyclopentadienide [310, 311]. The molecular structure of (**121**), (R = Bu^t) implies sp hybridisation at the nitrogen atom (M=N 1.738(2) Å; Mo-N-C 177.7(2)°), however as with (**112**) a triple bond is not necessary to account for the near linearity of the Mo-N-C moiety, and photoelectron spectra and molecular orbital calculations support a bond order of about two with a ligand based lone pair. Reactions of (**121**) with HCl gas or LiAlH$_4$ gave the corresponding dichlorides and dihydrides (Scheme 6-26).

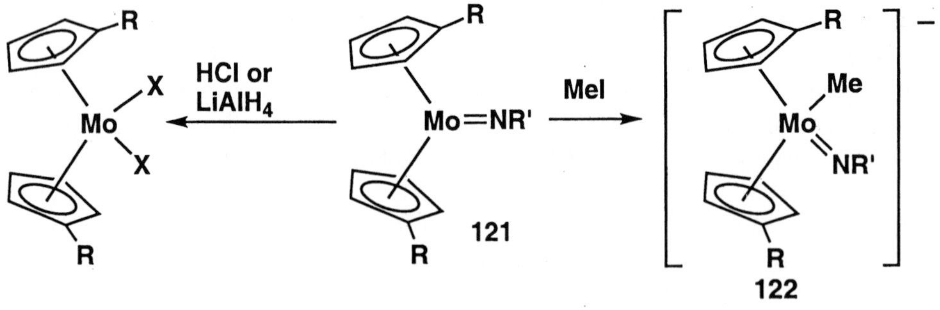

Scheme 6-26

Reaction of $[MCp_2HLi]_4$ with $PCl_2(PR)$ yielded the terminal phosphinidene complexes $[MCp_2(PR)]$ (M = Mo or W, R = 2,4,6-Bu^{t}C$_6$H$_2$ [312]; M = W, Cp = Cp*, R = NH(2,4,6-Bu^{t}C$_6$H$_2$) [313]); extremely low-field ^{31}P NMR shifts of between 666 and 799 ppm were reported.

The alkylidene tungsten derivative$[WCp_2(=CHOZrHCp*_2)]$ was isolated from the reaction of the zirconium dihydride derivative $[ZrCp*_2H_2]$ with $[WCp_2(CO)]$ [314].

6.4.1.4   Cationic Group 6 d^0 Metal Compounds $[MCp_2EX]^+$

Few examples of this type of complex have been reported. Photolysis of [$WCp_2(NCMe)Me]^+$ under O_2 yielded the structurally characterized oxo complex $[MCp_2(=O)Me]^+$ (Mo=O 1.700(11) Å) [236]; the ethyl analogue was prepared via photolysis of [$WCp_2(CH_2CH_2)H]^+$ exposed to air [315]. $[MCp_2(=S)X]^+$ was obtained by decomposition of the η^2-thiosulphate derivative $[MCp_2(\eta^2-(-S-SO_2-O-)]$ [196]. Methylation of (121) with MeI gave (122) (Scheme 6-26) [314, 315]. For R = Bu^t, the molecular structure of (122), in common with (121), shows an almost linear Mo-N-C angle (176.2(7)°) and a Mo=N distance of 1.704(8) Å. Oxidation of $[WCp_2H(\eta^2-CH_2=CH_2)]$ with I_2 yielded $[WCp_2(=CHMe)I]^+$ [112].

6.4.2   17-Electron Compounds

6.4.2.1   Neutral d^1 Vanadium Compounds $[VCp_2E]$

The imido derivatives $[VCp_2(NR)]$ (R = SiMe$_3$ or SiPh$_3$) and $[VCp*_2\{NR]$ (R = Ph, 2,6-Me$_2$(C$_6$H$_3$), 2-Ph(C$_6$H$_3$) or (C$_6$F$_5$)) obtained from reactions of organic azides with vanadocene are examples of this type of paramagnetic complex (Scheme 6-27) [316, 317, 318] when bulky azides (R = SiMe$_3$ or SiPh$_3$) were used with permethylated vanadocene, the azido complex was formed.

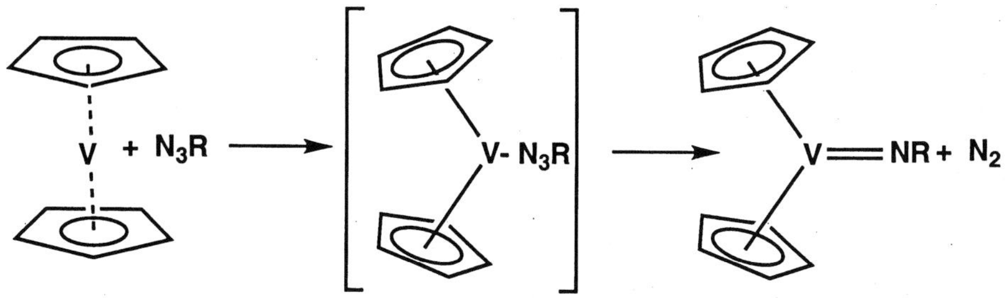

Scheme 6-27

Relatively low ν(MN) stretching frequencies, in the range 923-968cm^{-1}, were reported [319]. X-ray crystal structures show that the V-N-R angle is almost linear (*ca* 179 °) implying that the nitrogen is sp hybridized, with short V=N distances of between 1.665 Å and 1.730(5) Å indicative of multiple bonding. In addition to the σ-bond, molecular orbital calculations for [VCp$_2$(NPh)] show two V-N π-bonds with a nonbonding HOMO localized on vanadium. Measured μ_{eff} values of 1.56-1.79 μ_B are consistent with the presence of one unpaired electron. The complexes are thus formally 19e V(IV) paramagnetic species. Cyclic voltammograms show quasi-reversible oxidations to 18e vanadium (V) complexes. One cyclopentadienyl ligand was readily lost on heating.

6.5. Type D - Complexes with η^2-Side-Bonded Ligands

6.5.1 18-Electron Compounds

6.5.1.1 Neutral Group 5 d^2 Metal Compounds [MCp$_2$(η^2-X-L)]

Compounds formed by monoanionic bidentate ligands fall into this category, although relatively few of these metal(III) derivatives have been isolated.

Alkylation of the chloro isocyanide derivatives [Nb(C$_5$H$_4$SiMe$_3$)$_2$Cl(CNR')] with LiR gave the corresponding iminoacyl complexes (**123**) (R = Me, R' = Cy, Ph; R = CH$_2$SiMe$_3$, R'= Cy) [119]. Insertion of the isocyanide moiety into the Nb-R bond of the presumed alkyl isocyanide intermediate occurs spontaneously, except in the case of the *tert*-butyl isocyanide derivative when the alkyl isocyanide complex was isolated. An alternative synthetic route (R = R' = Ph) involved Na/Hg reduction of the ketenimine complex [Nb(η^5-C$_5$H$_4$SiMe$_3$)$_2$(η^2-(C,N)-PhN=C=CPh)] followed by protonation [320].

123

The η^2-dithioformate **(124)** was obtained via CS_2 insertion into the M-H bond of $[Nb(C_5H_4SiMe_3)_2HL)]$ (L = $P(OR)_3$) with substitution of L by sulfur (Scheme 6-28) [321]; when L = CO insertion of CS_2 occurs, but without the following ligand substitution, to give the η^1-dithioformate **(125)**.

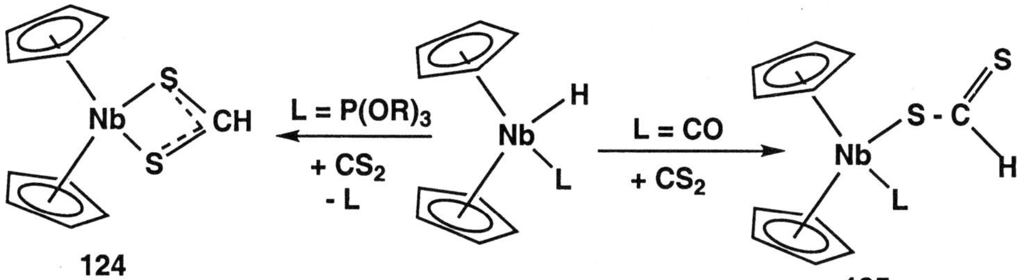

Scheme 6-28

One of the tautomeric forms of the heterodinuclear niobium derivative $[NbCp_2(\eta^2\text{-} C,O)(CH_2\text{-}O\text{-}ZrHCp^*_2]$ belongs to this group and was prepared by reacting $[NbCp_2R(CO)]$ with zirconocene dihydride [157].

6.5.1.2   Neutral Group 5 Metal Complexes $[MCp_2X(\eta^2\text{-}C\text{-}C)]$ or $[MCp_2X(\eta^2\text{-}C\text{-}Y)]$ (Y = N, O, or S)

Complexes with ligands which formally contain either a carbon-carbon multiple bond or a carbon- heteroatom double bond interacting with the metal are discussed here. The bonding in this category can be ambiguous and therefore we have not specified a general metal electron configuration. Some complexes are best described as M(III) d^2 whilst others sit more comfortably in the M(V) d^0 category. The neutral or dianionic nature of the ligand occupying one or two coordination sites respectively depends on the degree of π-back donation from the metal to the empty π^* orbitals of the ligand, and is influenced by the other substituents bonded to either the metal or the ligand. As an illustration, structure **(126a)** shows a formally M(III) alkene adduct whilst structure **(126b)** shows a complex with considerable M(V) character implying a significant π-bonding contribution from the metal to the ligand. The hydride alkenes and alkynes are important and widely studied groups in this general category.

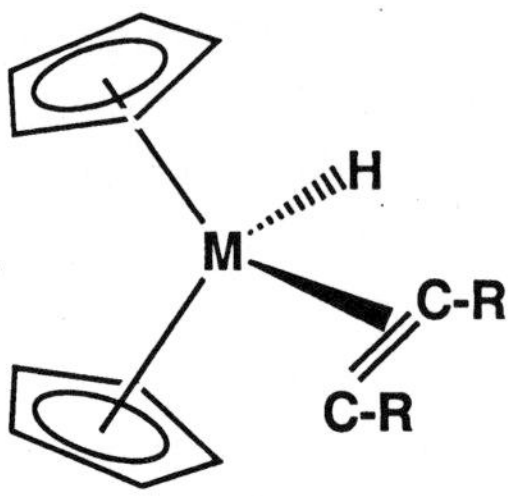

126a

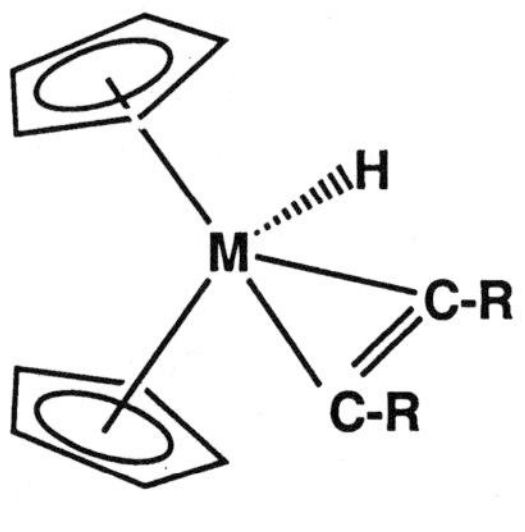

126b

Alkene complexes

A convenient preparative method (Scheme 6-29) for hydride alkene complexes is the Grignard reagent MgClR reduction of a metal(IV) dichloride, probably proceeding through the intermediate formation of the metal(III) alkyl compound [Cp$_2$MR] which finally gives the alkene hydride via β-hydrogen elimination [282, 322, 323]. The same compounds were also obtained from the reaction of Cp*$_2$NbH$_3$ with the corresponding alkene. Both *endo-* and *exo-* isomers were found in complexes where Cp = Cp, whereas the *endo-* isomer was favoured for steric reasons in complexes where Cp = Cp* [258]. Spin-saturation transfer NMR experiments have demonstrated that these insertion-elimination processes take place via species containing agostic M-H-C bonds [324, 325]. Ligand promoted alkene insertion into the M-H bond allows the isolation of the intermediate alkyl metal(III) derivatives, and the kinetics of this reaction have been extensively explored [258, 326].

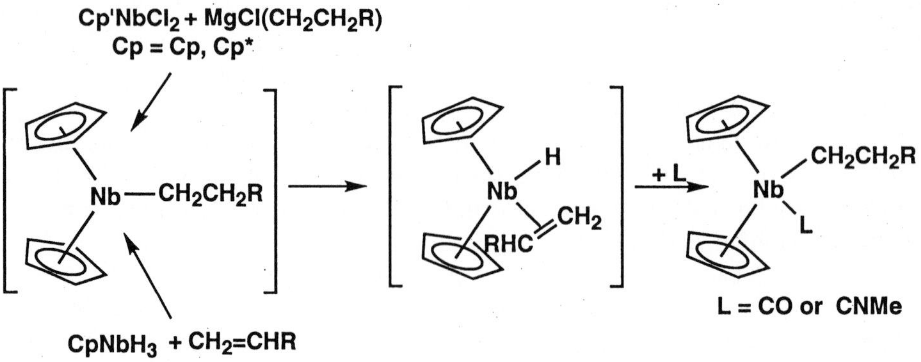

Scheme 6-29

Both methods were used to prepare [Nb(η⁵-C₅H₄SiMe₃)₂(η²-PhCH=CH₂)H], with the first producing exclusively the *endo-* isomer, while the second gave a mixture of *endo-* and *exo-* forms [327]. The analogous *endo-bis*(indenyl) derivative was obtained from the reaction of [Nb(Ind)₂(μ-H₂BH₂)] with styrene in the presence of NMe₃; the complex is fluxional with both dissociative and non-dissociative mechanisms proposed to account for the observed NMR exchange [126].

Reaction of a trihydride complex with the activated alkyne RC≡CR' (R = R' = CO₂Me or CO₂Buᵗ; R = CO₂Me, R' = H) gave not the alkyne hydride as might be expected, but the alkene hydride (**128**) via the insertion product (**127**) (Scheme 6-30) [328].

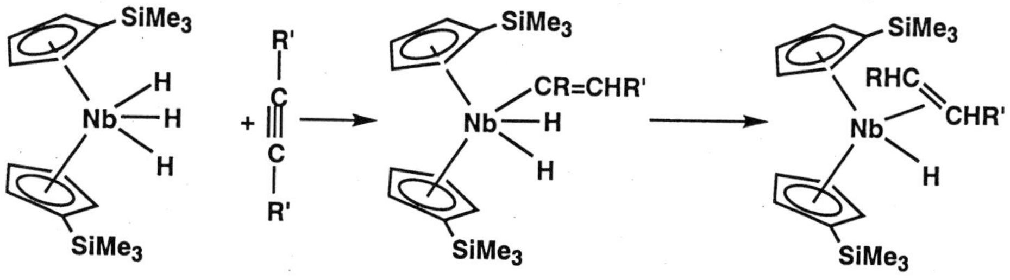

Scheme 6-30

A series of alkenyl complexes [Nb(η-C₅H₄SiMe₃)₂(η¹-C(R)=CH(R')L] (L = CNR or CO) were obtained from the reaction of appropriate metal(III) hydride isocyanide or carbonyl complexes with similar activated alkynes via alkyne insertion into the metal hydrogen bond. Reaction of the appropriate alkene with [NbCp₂(H₂)(SnMe₃)] afforded the alkene stannyl derivatives [NbCp₂(η²-H₂C=CHR)(SnMe₃)] (R = Ph or CMe=CH₂) [262]. Alkyne derivatives were obtained in a similar way.

X-Ray molecular structures of several of these alkene derivatives show lengthening of the ligand C=C bond in the range 1.429(7) Å to 1.4486(6) Å (cf uncoordinated ethene 1.337(2) Å) indicative of metal to alkene back donation introducing significant (**126b**) character.

Alkyne complexes

Several methods for the preparation of alkyne metal(III) complexes have been reported (Scheme 6-31):

a) A series of halide derivatives [NbCp₂X(RC≡CR')] (X = Cl, Cp = Cp or η-C₅H₄SiMe₃, R = Ph or H, R' = Ph or H; X = Br, I, Cp = η-C₅H₄SiMe₃, R = R' =

Ph; X = Cl, Cp = η-C$_5$H$_4$SiMe$_3$, R^1 = H, Me or COOMe, R^2 =COOMe) have been isolated by reduction of [MCp$_2$Cl$_2$] with various reducing agents, usually Na/Hg, in the presence of the alkyne [106, 117, 329]. Reduction of (80) gave the *ansa*-derivative [Nb{SiMe$_2$(C$_5$H$_4$)$_2$}Cl (RC≡CR)] (R = Me or Ph) [224].

b) Hydride and alkyl derivatives have usually been prepared by reaction of the hydride alkene complexes with an appropriate alkyne. When the alkyne has bulky substituents, a simple substitution takes place to give the hydride alkyne derivatives, whereas less bulky substituents induce hydride migration to give the alkyl insertion products. For example, reactions of [MCp$_2$H(CH$_2$=CHMe)] with RC≡CH (R = Bu^t, SiMe$_3$) lead to the simple substitution products [MCp$_2$H(RC≡CH)] [330]. Monosubstituted acetylenes RC≡CH (R = Me, Et, Ph) however react with [MCp$_2$H(CH$_2$=CH$_2$)] to give a mixture of *endo*- and *exo*- [MCp$_2$Et(RC≡CH)], or the ethyl complex [MCp$_2$Et(MeC≡CMe)] is obtained when using MeC≡CMe. The same method was used to prepare the benzyne complex [TaCp*$_2$H(η^2-C$_6$H$_4$)] [245].

c) The direct thermal reaction of the trihydride with an appropriate alkyne provides a route to [NbCp$_2$H(RC≡CR)] (R = Et, Ph, SiMe$_3$) [331]. This method was unsuitable for the methyl derivative, which was obtained by photolitic exchange of alkyne from the reaction of [NbCp$_2$H(Me$_3$SiC≡CSiMe$_3$)] with MeC≡CMe.

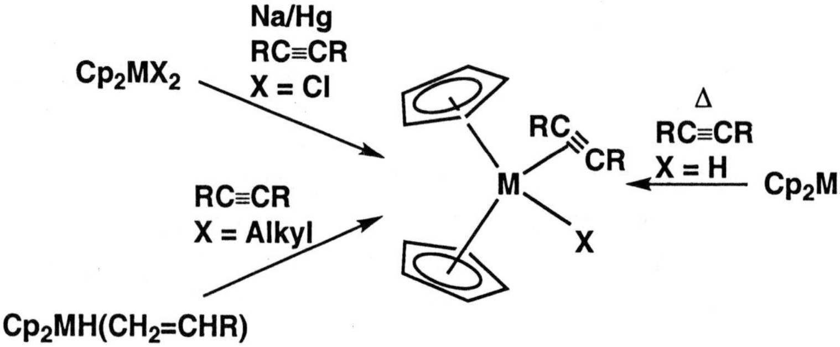

Scheme 6-31

In common with the alkene complexes discussed above, molecular structures show that lengthening of the C≡C bonds (1.230 - 1.288Å) occurs in the coordinated alkyne ligand, together with a significant deviation from linearity in the C≡C-R angles (137.7 - 146.7°), imparting a high degree of metallacyclopropene character [117, 224, 334]. The complexes [Nb(η-C$_5$H$_4$SiMe$_3$)Cl(η^2-R^{1}C≡CR^2)](R^1 = H, Me or COOMe, R^2 =COOMe) and [NbCpH(η^2-HC≡CR)]^{334} (R = SiMe$_3$ or Bu^t) were reported as Nb(V) species.

Complexes with (η^2-CY$_2$) ligands (Y = O or S)

Complexes with η^2-CO$_2$ were prepared via chemical or electrochemical reduction of a dihalide complex in the presence of CO$_2$ [281, 332, 333, 334]. An alternative approach was the oxidation with O$_2$ of the carbonyl ligand in the [NbCp$_2$R(CO)] (Cp variously substituted, R = CH$_2$Ph, CH$_2$Bu^t, CH$_2$SiMe$_3$, Me) derivatives [335]; labelling experiments show the additional oxygen atom in the resulting [NbCp'$_2$R(η^2-CO$_2$)] products is derived from O$_2$. Complexes are generally stable but lose CO when heated at 60°C to give the oxo compound **(90)**; all exhibited a strong v (C=O) band in their IR spectra at ca. 1700 cm^{-1}, indicative of η^2-coordinated CO$_2$. The chloride complexes [NbCp$_2$Cl(η^2-CO$_2$)] (Cp = Cp', Cp*), generated electrochemically and identified from their IR spectra (v (C=O) 1730 cm^{-1}), contain labile CO$_2$ moieties as evidenced by their FD-MS spectra [121].

A range of similar compounds [MCp$_2$R(η^2-CS$_2$)] were prepared by a) carbon disulfide promoted hydride-alkene insertion (Cp = Cp or Cp') [304, 336]; b) photochemical substitution [340]; c) reaction of a metal (III) derivative with carbon disulfide [106], Scheme 6-32. Complex **(129)** was obtained cleanly when reaction a) was carried out in nonpolar solvents, however the use of polar solvents gave **(129)** together with the dithioformato deriveative [NbCp'$_2$(CS$_2$H)(η^2-CH$_2$=CH$_2$R)] and free alkene. Complex **(129)** can act as a ligand through its uncoordinated sulfur atom and several heterobimetallic species have been reported [325, 340] it reacts with RI to give [MCp$_2$R(η^2-CS-SR)]$^+$ [270].

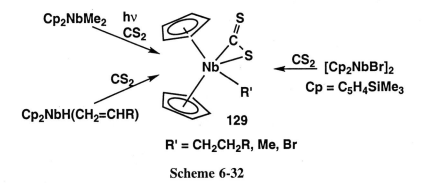

R' = CH$_2$CH$_2$R, Me, Br

Scheme 6-32

Complexes with (η^2-C,Y) ligands (Y = N, O or S)

A wide range of complexes with (η^2-C,Y) ligands **(130-132)** have been reported [2]. The most widely used synthetic route to aldehydes **(130)** Y = O [337], isocyanates **(131)** [106], ketenes **(132)** (Y = O) [101, 338, 339, 340, 341, 342] and ketimines(**132** Y = NR) [324, 343, 344] has been the reaction of a metal(III) chloride [MCp$_2$Cl]$_2$ (obtained via the reduction of [MCp$_2$Cl$_2$] [106] or via the reaction of NbCl$_3$(DME) with LiCp [101]) with an appropriate free ligand. The aldehyde (**130**) (Y = O, R^1 = R^2 = H) was also obtained from precursors such as [Nb(η-C$_5$H$_4$SiMe$_3$)Cl(L)] (L = CO or phosphine), and its hydrido analogue was obtained by the reduction of the chloro derivative in the presence of ethanol via the d^1 Nb(IV) radical intermediate {Nb(η-C$_5$H$_4$SiMe$_3$) (η^2-CH$_2$O)} [341]. Hydrido ketone, aldehyde and thioaldehyde complexes were also prepared from the reactions of alkylidenes [TaCp*$_2$H(=CR)] (R = H$_2$ or =CH$_2$) with MeOH or HSCH$_2$R; the reaction was thought to proceed via β−hydrogen elimination from a transient methoxide or sulphide {TaCp*$_2$H(CHR)(YCH$_2$R')} (Y = O or S) species to give the hydrido aldehyde or thioaldehyde complexes [TaCp*$_2$H(η^2-O=CH$_2$)] [283]   or [TaCp*$_2$H(η^2-S=CHR')] [272, 345]. The alkyl complexes [TaCp*$_2$Me(η^2-O=CHR)] were obtained in a similar way [271]. The telluroformaldehyde derivative [TaCp*$_2$H(η^2-Te=CHR)] was prepared by reaction of the tantalum alkenyl complex with Te in the presence of PMe$_3$ [295], and the selenoaldehyde [TaCp*$_2$H(η^2-Se=CHR)] was prepared by reaction of [TaCp*$_2$I(=Se)] with LiMe.[294]Reaction of the cationic enacyl complex [Nb(η-C$_5$H$_4$SiMe$_3$)Cl(η^2-O,C)O=CCR=CMeH)]$^+$ with KOBu^t gave the E-vinylketene complex (R = Ph or Et) [346].

Y = O, S•
130

131

Y = O, NR
132

When more than one coordination mode is possible, as in the heterocumulenes, the oxophilicity of the group 5 metals determines the preferential coordination in the order O > N > C, so that the resulting products always contain η^2-(O,C)-ketenes, η^2-(N,C)-ketenimines or η^2-(O,C)-isocyanates. X-Ray molecular structures of the aldehydes, ketenes and ketimines (Table 6-5) show that considerable lengthing of the C-C and C-Y bonds in the π complexed η^2-ligand occurs, compared to the free cumulenes (cf C=O 1.18(1) Å, C=C 1.29(1) Å measured for dimesityl ketene [347]), with lengths approaching those of C-Y single bonds (cf C-O 1.333 Å), thus the complexes are best characterized as Nb(V) d^0 species. Likewise the telluroformaldehyde complex was thought to have substantial Ta(V) tellurirane character [295].

Table 6-5. Selected X-Ray diffraction data on neutral aldehydes, ketenes and ketimines, and their cationic acyl and iminoacyl analogues.

Complex	$Y-C^1-C^2$ (°)	C^1-Y (Å)	C^1-C^2 (Å)	Nb-Y (Å)	Nb-C (Å)	Ref
(130) (Y = O, $R^1 = R^2$ = H)	N/A	1.322(5)	N/A	2.048(3)	2.257(4)	341
(132) (X = Cl, Y = O, R^1 = Et, R^2 = Ph)	135.3(4)	1.302(5)	1.346(6)	2.093(3)	2.135(4)	101
(132) (X = Br, Y = O, $R^1 = R^2$ = Ph)	133.2(3)	1.303(4)	1.351(5)	2.082(2)	2.170(5)	342
(132) (X = Cl, Y = NPh, $R^1 = R^2$ = Ph)	140.0(4)	1.333(6)	1.370(6)	2.140(4)	2.161(4)	324
(135) (X = Cl, Y = O, R^1 = Et, R^2 = Ph)	127.1(5)	1.242(5)	1.506(8)	2.233(4)	2.121(5)	343
(135) (X = Cl, Y = NPh, R^1 = Et, R^2 = Ph)	135.7(6)	1.23(1)	1.51(1)	2.157(4)	2.170(5)	347
(135) (X = F, Y = NPh, $R^1 = R^2$ Ph$_2$)	128.9(5)	1.253(7)	1.517(8)	2.148(5)	2.180(5)	347

Allyl Complexes [MCp$_2$(η^3-allyl)]

The η^3-allyl complexes of d^2 niobium and tantalum can be viewed as a particular type of [Cp$_2$MXL] complex if the allyl substituent is considered as an η^1-η^2-lig-and. We have already observed in Section 6.3.1.5 that paramagnetic 16-electon va-nadium complexes resist the coordination of an additional ligand because of the high spin pairing energy required, and for this reason vanadium only has η^1-allyl derivatives. In niobium and tantalum allyl complexes, all three metal based d orbit-als are involved in bonding, always giving 18 electron η^3-allyl complexes.

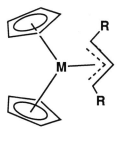

133

Several methods have been used to prepare group 5 metal allyl derivatives (**133**); several are based on a strategy which involves the insertion of allene or dienes into the Nb-H bond of metal(III) hydrides [MCp$_2$HL] or their precursors, after displace-ment of the ligand. The reaction of [TaCp$_2$H$_3$] (Cp = Cp or Cp*) with butadiene, for example, takes place with elimination of H$_2$ [303,348]; [NbCp$_2$H(CH$_2$=CH$_2$)] however reacts with dienes eliminating ethene (the tantalum derivatives were not obtained via this route) [349] and [Nb(Indenyl)$_2$(η^2-H$_2$BH$_2$)] eliminates BH$_3$ when treated with diene in the presence of NMe$_3$ [126]. Reactions of the dichlorides with allylmagnesium halides or 2-butenyl magnesium halides were used to obtain the derivatives [TaCp*$_2$(η-C$_3$H$_5$)] and [MCp$_2$(η-C$_3$H$_4$Me)] (M = Nb or Ta) [353, 303]. With the exception of allene all the reactions with dienes can lead to the formation of various *syn-anti* and *endo-exo* isomers and the regio- and stereo-selectivity of these processes was studied by labelling experiments and NMR spectroscopy [353]. The allyl complexes have been reported as fairly stable, scarcely reactive compounds, as expected for 18 electron metal systems.

6.5.1.3 Anionic Group 5 d^2 Metal Compounds [MCp$_2$(η^2-C,Y)]$^+$ and [MCp$_2$X(η^2-C,Y)]$^-$

Electrochemical reduction of the ketene complexes (**132**) afforded the radical anions [Nb(η-C$_5$H$_4$SiMe$_3$)$_2$Cl{η^2(C,O)-(OCCR^{1}R^2)}]$^-$ (R^1 = R^2 = Ph; R^1 = Et, R^2 = Ph) characterised by their typical 10 line ESR (g_{iso} 2.000 or 1.997, a(Nb^{93}) 16 G) spectra and IR spectroscopy. Low g_{iso} values (cf. [NbCp$_2$X] ca. 100 G) are often found in η^2-π-complexed ligands [345].

Electrochemical two electron reduction of the ketimine (**132** R^1 = R^2 = Ph) afforded the cationic species [Nb(η-C$_5$H$_4$SiMe$_3$)$_2${η^2(C,N)-(NPhCCPh$_2$)}]$^+$ [324].

6.5.1.4 Neutral Group 6 d^2 Metal Compounds [MCp$_2$(η^2-X-X)]

Metal complexes with η^2-coordinated ligands where the two donor atoms are bound together are considered in this section as d^2 metal compounds [MCp$_2$(η^2-X-X)], which result from oxidative addition of an unsaturated system to the metallocene fragment, although they could also be viewed as [MCp$_2$(η^2-L)] type d^4 metal compounds with unsaturated side-bonded neutral π-complexed ligands.

Metal complexes with η^2-coordinated acetonitrile [MCp$_2$(η^2-N≡CMe)] can be isolated either by reduction of [MoCp$_2$Cl$_2$] or by decomposition of [WCp$_2$HMe], in acetonitrile [350, 351, 352].

In addition to its reaction with ethene to give (**18**), the bimetallic fulvalene hydride complex (**15**) when treated with but-2-yne gave a mixture of [Mo$_2$(μ-η:η-C$_{10}$H$_8$)Cp$_2$(MeC≡CMe)$_2$] and [Mo$_2$(μ-η:η-C$_{10}$H$_8$)Cp$_2$(μ-MeC≡CMe)]; in the latter the alkyne group is thought to bridge the two metals. In contrast, reaction with hex-3-yne afforded only the unbridged analogue [87].

The synthetically useful Z-stilbene complex [MCp$_2$(-(Z)-(PhCH=CHPh)], obtained via insertion of PhC≡CPh into the Mo-H bonds of the dihydride [MCp$_2$H$_2$], has provided a route to a variety of η^2 derivatives [353, 354]. Isomerisation of the Z-isomer to the E-isomer is achieved in the presence of catalytic amounts of maleic anhydride. The E-isomer is significantly less labile and is not readily substituted [355]. The labile *cis*-stilbene adduct is readily replaced by alkynes to give $[MCp_2(\eta^2-RC{\equiv}CR)]$ and by other π-acceptor ligands, such as heteroalkenes $[MoCp_2(\eta^2-X-X)]$ (X-X = $PhCOCF_3$, Ph_2Cl, $PhCOCO_2Me$, PhCHO, PhCHNPH, Ph_2CS H_2CO_2 [359, 356 , 357], CO_2 [361]); it also provides access to a range of metal(IV) compounds by oxidative coupling reactions (Section 6.3). The structurally characterized complex $[MoCp_2(\eta^2-CO_2)]$ [361] was also observed as an intermediate in the reaction of the dihydride with CO_2 (Section 6.3.2.4) [194].

The related $[MoCp_2(\eta^2-C,S)CS_2)]$ complex was formed in the unusual reaction of its isomer $[MoCp_2(\eta^2-S,S)CS_2)]$ with alkynes which are simultaneously transformed into the metallacyclic dialkyls $[MoCp_2(\eta^2-C,C)C(=S)SCR=CR)]$ [358].

The silene $[WCp_2(\eta^2-CH_2{=}SiMe_2)]$ [359] and disilene $[MCp_2(\eta^2-SiMe_2{=}SiMe_2)]$ (M = Mo or W) derivatives [360] important compounds for their synthetic applications, were obtained from $[WCp_2Cl(RSiMe_2Cl)]$ (R = CH_2 or $SiMe_2$) when treated with magnesium. Their Si-Si [2.260(3) Å] or Si-C [1.800(8) Å] bond lengths lie midway between single and double bonds. These compounds are very reactive and useful in the preparation of an extensive range of silyl derivatives and metalla-sila- and disila-cyclo-butanes and -pentanes [187, 204, 205].

6.5.1.5 Cationic Group 6 d² Metal Compounds $[MCp_2(\eta^2-X-L)]^+$

This category comprises cationic metal(VI) complexes $[MCp_2(\eta^2-X-L)]^+$ with one anionic and one neutral ligand linked together, such as acyls.

The molybdenum acyl complex $[MoCp_2(\eta^2-CMe{=}NR)]^+$ was obtained on heating the cationic isocyanide $[MoCp_2Me(CNR)]I$ derivative [80]. Reactions of $[MoCp_2Br_2]$ with either a diarylazenine or diarylamidine yielded the triazenido $[MoCp_2(ArNNNAr)]^+$ and amidinato $[MoCp_2(ArNCRNAr)]^+$ (Ar = Ph, *p*-MeOPh, *p*-MePh, *p*-FPh; R = CH, CMe, CPh) respectively, and their redox chemistry was investigated [361, 362].

Reaction of either the cationic dimethyl complex $[WCp_2Me_2]^+$ with the Ph_3C radical or the neutral dimethyl $[WCp_2Me_2]$ with Ph_3C^+ resulted in their conversion to the alkene hydride derivative $[WCp_2H(\eta^2-H_2C{=}CH_2)]$; the reaction occurred via initial formation of an alkylidine species (which could be trapped on addition of phosphine as a phosphorus ylide) followed by methylene insertion into the M-Me bond to give the 16 electron alkyl intermediate, and finally β-hydrogen elimination to give the product [363, 364, 365, 366]. The mechanism of the final stage was further studied by deuterium labelling of isopropylmagnesium bromide in its reaction with $[WCp_2Cl_2]$ [111].

An example of a dicationic group 6 metal complex is the molybdenum derivative $[MoCp_2(NCMe)(\eta^2-(NH{=}CHMe)]^{2+}$ which resulted from protonation of the side-bound acetonitrile complex $[MoCp_2(\eta^2-N{\equiv}CMe)]$ with HBF_4. Et_2O [354]. Its x-ray molecular structure showed bending at the protonated N and C atoms with a very short N=C distance of 1.096(16) Å (*cf.* N≡C 1.200(10) Å in the acetonitrile complex [355]). Both ligands were replaced on addition of PMe$_3$ to give [MoCp$_2$(PMe$_3$)$_2$]$^{2+}$.

6.5.1.6   Neutral Group 5 d^0 Metal Compounds [MCp$_2$(η ^{2}-X-X)Z]

Complexes of the type [MCp$_2$(η^2-X-X)Z] with ligands considered to be η^2-dianionic species, such as η^2-O$_2$, η^2-ONMe, η^2-S$_2$, η^2-Se$_2$, or η^2-Te$_2$, are members of this class and can be clearly defined as 18 electron group 5 metal d^0 derivatives.

Oxidation of metal(III) complexes has been widely used to prepare oxo derivatives with teminal M=O ligands and carbonyl coomplexes are the most convenient substrates to make this transformation. However, oxidation of related metal(III) isocyanides frequently leads to the formation of η^2-peroxides and the complexes [NbCp$_2$X(η^2-O$_2$)] (X = Cl, Me; Cp = C$_5$H$_5$, η-C$_5$H$_4$SiMe$_3$ or ?-C$_5$H$_3$(SiMe$_3$)$_2$) were conveniently prepared by this method [120]. The methyl derivative was also obtained by oxidation of [NbCp$_2$Me(CO)] (Cp = Cp), whereas oxidation of the chloro carbonyl derivative gave the terminal oxo-species [118]; hydrogen peroxide oxidation of the dichloride complex [NbCp$_2$X$_2$] has also been used [367]. Reaction of [NbCp$_2$Cl]$_2$ (Cp = η-C$_5$H$_4$SiMe$_3$) with excess O$_2$ gave a mixture of the terminal oxide and the peroxide (section 6.4.1.1) [280]. The tantalum   derivatives [TaCp*$_2$R(η^2-O$_2$)] (R = Me, Et, Pr^n, Ph or CH$_2$Ph) were synthesized by O$_2$ oxidation of either the alkylidene complexes [TaCp*$_2$H(=CHR)] (R = H or Ph), the benzylidine hydride [TaCp*$_2$H(η^2-C$_6$H$_4$)]   or the alkene hydride complexes [TaCp*$_2$H(CH$_2$=CHR)] (R = H or Me) [368]. The peroxo complexes react with triphenyl phosphine to give the terminal oxo derivatives [MCp$_2$X(=O)] (M = Nb or Ta, X = Cl or Me) and triphenyl phosphine oxide. An analogous complex [TaCp*$_2$Me(η^2-ONMe)] was isolated by oxidation of [TaCp*$_2$Me$_2$] with NO [369]. Analogous h^2-disulfido, -selenido or -tellurido derivatives were obtained from reactions of the dialkyl, alkyl hydride or trihydride derivatives with S$_8$, Se or Te; sulfur extraction from thiiranes was also used (Scheme 6-33) [237,287,288,292,294,295]. Similar reactions with either [NbCp*$_2$(BH$_4$)] or [Nb(η^2-C$_5$H$_4$Bu^t)$_2$H$_3$)] afforded the polysulfide complexes [{NbCp$_2$(η^2-S$_2$)}(μ-S$_5$)] (Cp = Cp* or (η^2-C$_5$H$_4$Bu^t) [290, 370]. The fulvene derivatives [NbCp*(η^6-C$_5$Me$_4$R)(η^2-S$_2$)] [R = CH$_2$ or CH$_2$-S (134)] together with the hydrido and thio derivatives were observed in reactions of S$_8$ with {NbCp*}. Likewise, the tantalum fulvene derivative [TaCp*(η^6-C$_5$Me$_4$CH$_2$)H$_2$] reacted with sulfur to give a mixture of (103) and (104); the tantalum analogue of (134) was obtained upon irradiation of this mixture in the presence of sulfur [293, 25]. [M(η-C$_5$H$_4$Bu^t)H(η^2-S$_2$)] (M = Nb or Ta) reacts with sulfur to give the corresponding thio derivative and with MeI to give the iodo derivatives [371]. Heterobimetallic complexes formed by addition of Cr(CO)$_5$ fragments to one or both of the sulfur atoms have been reported [372].

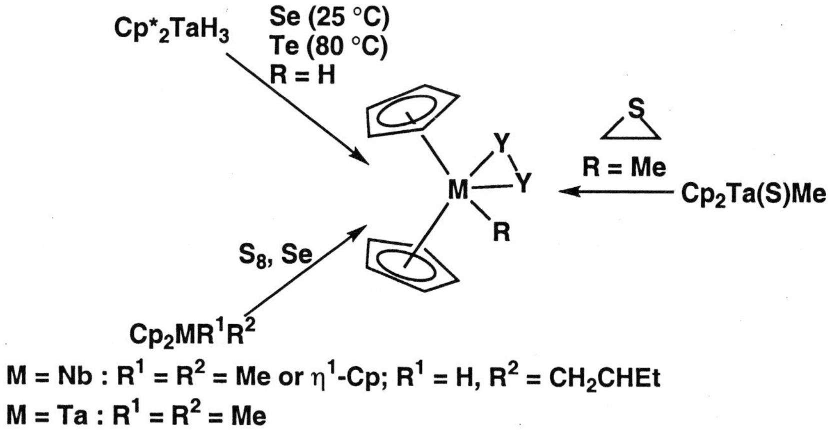

Scheme 6-33

6.5.1.7 Cationic Group 5 d⁰ Metal Compounds [MCp₂X₂L]⁺ and [MCp₂ (η²-X-L)Z]⁺

Most of the cationic derivatives which fall into this category were prepared from complexes with (η²-C,Y) ligands (Section 6.5.1.2). Cationic acyl and iminoacyl derivatives (**135**) were obtained by protonation of appropriate ketene complexes with $HBF_4.OEt$, or ketimine complexes with HBF_4. Protonation occurs at the β-carbon of (**131**) to give (**135**) as stable diamagnetic solids. An extensive series of complexes in each class have been reported: acyl derivatives [343,344,350] [Nb(η-$C_5H_4SiMe_3$)₂(η²-C(O)CHR¹R²)X]⁺ [(X = Cl: R¹ = Me, R² = Ph, Me or CH_2Bu^t; R¹ = Ph, R² = Ph, Et or CH_2Bu^t) (X = H: R¹ = Ph, R² = Et or Me; R¹ = Me, R² = CH_2Bu^t)]; iminoacyl derivatives [347, 348] [Nb(η-$C_5H_4SiMe_3$)₂(η²-C(NR)CHR¹R²)X]⁺ [(X = Cl or Br: R¹ = R² = Ph; R¹ = Ph, R² = Me or Ph) (X = F, R¹ = R² = Ph)]. Selected bond distances and angles for structurally characterized examples are given in Table 6-5.

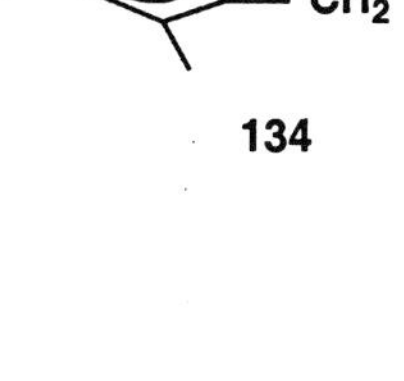

134

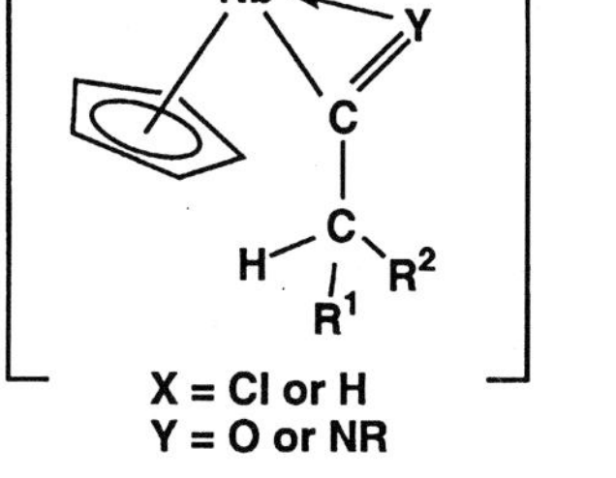

X = Cl or H
Y = O or NR

135

Electrochemical or chemical oxidation of the Nb(IV) ketimine complex (**136**) in the presence of NCR or CNR gave the air stable cationic ketimine complex (**137** R = Me, Et or Ph, L = MeCN; R = Ph, L = Bu^{t}CN or PhCN) (Scheme 6-34). Complex (**137**) reacted with water or MeOH to give the corresponding cationic hydrido or methoxy iminoacyl complexes [373]. The C-N (1.33(1) Å) and C-C (1.36(2) Å) bond distances are similar to those found in the neutral ketimines but shorter and longer respectively than those measured for the cationic iminoacyl complexes (Table 6-5).

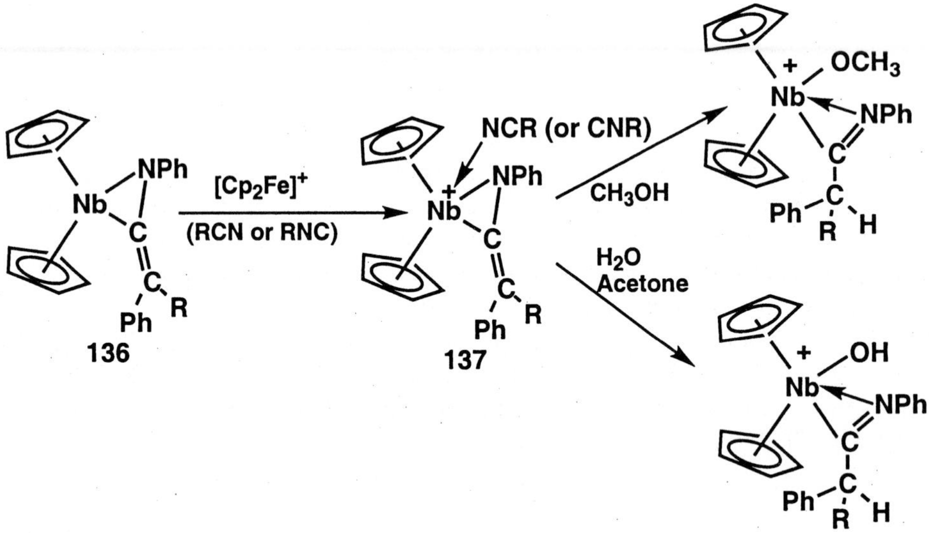

Scheme 6-34

Oxidation of the d^1 metal derivative [Nb(η-C$_5$H$_4$SiMe$_3$)$_2$(η^2(C,C)-(RC≡CR')]$^+$ with [FeCp$_2$][BPh$_4$] yielded the cationic alkyne complexes [Nb(η-C$_5$H$_4$SiMe$_3$)$_2$(NCMe)(η^2-(RC≡CR')]$^+$ (R = R' = Me, Ph, COOMe; R = COOMe, R' = Me (**138**)), in a reaction analogous to that of scheme 6-34 [153]. An X-ray molecular structural determination of (**138**) shows lengthening of the C-C bond (1.30(1) Å) which is only slightly shorter than a typical carbon-carbon double bond. The observed ν(C=C) stretching frequencies of between 1750 and 1850 cm^{-1} and the non-linearity in the C^2-C^1-R angle (*ca.* 141°) indicate a high degree of sp^2 character in the coordinated carbon atoms. The COOMe substituent is located in the expected *endo-* position and the actonitrile ligand is almost linear with Nb-N-C and N-C-Me angles (ca. 177°). Both the structural and IR evidence strongly support the view that the alkyne behaves as a strong π-acceptor ligand, giving rise to a high degree of metallacyclopropene character in the complex.

138

Although these complexes can be formally considered as either d^2 alkyne or d^0 niobacyclopropene derivatives, the latter description is more in accord with the structural and spectroscopic evidence. Its d^0 character is further supported by the unusual oxidation of the same starting complex which under appropriate conditions gives the cationic d^0 bimetallic derivative (**70**).

Alkylation of $[NbCp_2R^1(\eta^2\text{-}CS_2)]$ with R^2X (X = Br or I) yielded $[MCp_2R^1(\eta^2(C,S)\text{-}CSSR^2)]^+$ (R^1 = Bu or Me, R^2 = Me, Et, Pr^i, CH2COOEt, allyl) and with CH_2I_2 yielded the bimetallic complex $[\{NbCp_2R^1(\eta^2(C,S)\text{-}CSS)\}_2(\mu\text{-}CH_2)]^+$ [270, 340].

6.5.2 17-Electron Compounds

6.5.2.1 Neutral d^1 Vanadium Compounds $[VCp_2(\eta^2\text{-}X\text{-}X)]$

Addition of various unsaturated substrates to vanadocene (a carbene-like complex) yields a diverse group of complexes (see also section 6.2.1.1). Activated alkenes with electron-withdrawing groups such as propenal, fumarates and diethyl maleate react with vanadocene at room temperature affording vanadacyclopropane derivatives $[VCp_2(\eta^2\text{-}R_2C\text{-}CR_2)]$; similar reactions occur with alkynes to give vanadacyclopropene complexes $[VCp_2(\eta^2\text{-}RC=CR)]$ [374]. Their chemical and structural behaviour was as expected for typical vanadium (IV) derivatives [1,2]. They are paramagnetic compounds with magnetic moments of ca. 1.7-1.8 μ_B consistent with the presence of one unpaired electron and exhibited typical vanadium(IV) EPR spectral parameters. In common with the niobium and tantalum alkene complexes discussed in section 6.5.1.2, molecular structures show that lengthening of the C-C bonds [1.468 Å in the maleate derivative (only slightly shorter than expected for a single bond) and between 1.269 and 1.276 Å in the coordinated alkynes] occurs in the coordinated alkene or alkyne ligand. This, together with a significant deviation from linearity in the alkyne C-C-R angles (142 - 143°), is consistent with the presence of either cyclopropane or cyclopropene ligands; the ν(C=C) stretching vibration of between 1750 and 1800 cm^{-1} is also consistent with a metallacyclopropene derivative.

Similar reactions occured between vanadocene and a range of unsaturated substrates including aldehydes [375], ketenes, ketimines [376], thiobenzophenone [377] thioketenes [378, 379] and isothiocyanate [380] to give the corresponding cyclometallated product. In common with the niobium and tantalum derivatives, magnetic susceptibility measurements (μ_{eff}= 1.74 - 1.77 μ_B), IR, ESR, and crystallographic data were consistent with vanadium(IV) d^1 metal derivatives.

6.5.2.2 Niobium Compounds [NbCp$_2$(η^2-L-L)] and [NbCp$_2$(η^2-C,Y)]

Electrochemical or sodium amalgam reduction of the ketene or ketimine complexes (131 R^1 = R^2 = Ph) afforded the paramagnetic metal(IV) derivatives [Nb(η-C$_5$H$_4$SiMe$_3$)$_2${η^2(C,Y)-(YCCPh$_2$)}] (Y = NPh or O) [324,377]. Their ESR spectra each comprised a ten line signal (g$_{iso}$ 1.997 and 1.999, a(^{93}Nb) 18.1 G and 11.0 G respectively). Similar results were obtained for the aldehydes [Nb(η-C$_5$H$_4$SiMe$_3$)$_2${η^2(C,O)-(OCHR)}] (R = H or CHPh$_2$) (g$_{iso}$ 2.003, a(^{93}Nb) 10.3 G and 8.3 G respectively) [341, 381]. No hyperfine splitting was however observed for the CS$_2$ complex [Nb(η-C$_5$H$_4$SiMe$_3$)$_2${η^2(C,S)-(CS$_2$)}] (g$_{iso}$ 2.004). Extended Hückel calculations were consistent with the ESR data, and show that in the aldehyde, ketene, ketinimene and alkyne complexes the unpaired spin density is mainly localised on the metal, whereas in the carbon disulfide complex there is a high degree of delocalisation onto the ligand [385].

Likewise, reduction of [Nb(η-C$_5$H$_4$SiMe$_3$)Cl(η^2-R^{1}C≡CR^2)] afforded the neutral paramagnetic alkyne derivatives [M(η-C$_5$H$_4$SiMe$_3$)$_2${η^2-(CR^1≡CR^2)}] (R^1 = R^2 = Ph; R^1 = H, R^2 = Ph; R^1 = H, R^2 = COOMe; R^1 = Me, R^2 = COOMe; R^1 = R^2 = COOMe) [333,153]. Their ESR spectra showed a characteristic ten line signal with the small hyperfine coupling constants typically found in d^1 complexes with η^2 π-bonded ligands (g$_{iso}$ 2.0026 - 20140; a(^{93}Nb) 12.7 G - 16.5 G). Although these compounds could also be considered as d^3 metal complexes with one side-bonded neutral ligand [MCp$_2$(η^2-L-L)], their chemical and structural behaviour is better understood if they are classified as d^1 metal derivatives.

References

[1] Wilkinson G, Stone F G. A, Abel E W, *Comprehensive Organometallic Chemistry*, Pergamon, Oxford, **1982**.

[2] Wilkinson G, Stone F G A, Abel E W, *Comprehensive Organometallic Chemistry II*, Pergamon, Oxford, **1995**.

[3] Poli R, *Chem. Rev.* **1996**, *96*, 2135.

[4] Wigley D E, *Prog. Inorg. Chem.* **1994**, *42*, 239.

[5] IUPAC Commission (CNIC) meeting, Maryland, USA, **1996**.

[6] Hollis T K, Burdett J K, Bosnich B, *Organometallics*, **1993**, *12*, 3385.

[7] Timofeeva T V, Lii J-H, Allinger N L, *J. Am. Chem. Soc.* **1995**, *117*, 7452.

[8] Lauher J W, Hoffmann R, *J. Am. Chem. Soc.* **1976**, *98*, 1729.

[9] Bürgi T, Berke H, Wingbermühle D, Psiorz, C, Fox T, Knickmeier M, Berlekamp M, Fröhlich R, Erker G, *J. Organomet. Chem.* **1995**, *497*, 149.

[10] Fernández F J, Gómez-Sal P, Manzanero A, Royo P, Jacobsen H, Berke H, *Organometallics*, **1997**, *16*, 1553.

[11] Green M L H, *J. Organomet. Chem.* **1995**, *500*, 127.

[12] Handir K, Holecek J, Klikorka J, *Z. Chem.* **1979**, *19*, 265.

[13] Köhler F H, Prössdorf W, *Z. Naturforsch. Teil B*, **1977**, *32*, 1026.

[14] Robbins J L, Edelstein N, Spencer, B, Smart J C, *J. Am. Chem. Soc.* **1982**, *104*, 1882

[15] Gambarotta S, Floriani C, Chiesi-Villa A, Guatini C, *Inorg. Chem.* **1984**, *23*, 1739.

[16] Belot J A, McCullough R D, Rheingold A L, Yap G P A, *Organometallics* **1996**, *15*, 5062.

[17] Bocarsley J R, Floriani C, Chiesi-Villa A, Guastini C, *Inorg. Chem.* **1987**, *26*, 1871.

[18] Birmingham J M, Fischer A K, Wilkinson G, *Naturwissenschaften* **1955**, *42*, 96.

[19] Lindsell W E, Parr R A, *Polyhedron* **1986**,*6*, 197.

[20] Jonas K, Rüsseler W, Krüger C, Raabe E, *Angew. Chem.* **1986**, *98*, 905.

[21] Jonas K, Wiskamp V, *Z.Naturforsch.* **1983**, *38b*, 1113.

[22] Elson I H, Kochi J K, *J. Am. Chem. Soc.* **1975**,*97*, 1262.

[23] Nesmeyanov A N, Lemenovskii D A, Fedin V P, Perevalova E G, *Dokl. Akad. Nauk SSSR (english Transl.)* **1979**, *245*, 142.

[24] Brunner H, Gehart G, Meier W, Wachter J, Riedel A, Elkrami S, Mugnier Y, Nuber B, *Organometallics* **1994**, *13*, 135.

[25] Brunner H, Gehart G, Meier W, Wachter J, Burgemeister T, *J. Organomet. Chem.* **1995**, *493*, 163.

[26] Kowaleski R M, Basolo F, Osborne J H, Trogler W C, *Organometallics*, **1988**, *7*, 1425.

[27] Warren K D, *Struct. Bonding* **1976**, *127*, 45.

[28] Köhler F H, Geike W A, *J. Organomet. Chem.* **1987**, *328*, 35.

[29] Antipin M Y, Lobkovskii E B, Semenenko K N, Soloveichik G L, Struchkov Y T, *J. Struct. Chem. (Engl. Transl.)* **1979**, *20*, 810.

[30] Gard E, Haaland A, Novak D P, Seip R, *J. Organomet. Chem.* **1975**, *88*, 181.

[31] Antipin M Y, Boese R, *Acta Crystallogr. B Struct. Sci.* **1996**, *52*, 314.

[32] Lokshin B V, Greenwald I I, *J. Mol. Struct.* **1990**, *222*, 11.

[33] Gordetsov A S, Latyaeva V N, Zimina S V, Levakova E Y, Cherkasov V K, Moseeva E M, Skobeleva S E, *Russ. Chem. Bull.* **1996**, *45*, 1214.

[34] Tsumura R, Hagihara N, *Bull. Chem. Soc. Jpn.* **1964**, *37*, 1889.

[35] Ricci G, Panagia A, Porri, L, *Polymer.* **1996**, *37*, 363.

[35] Grigoryan E A, Dyachkovskii F S, Zhuk S Ya, Vyshinskaya L I, *Kinet. Catal. (Engl. Transl.)* **1978**, *19*, 1860.

[37] Köhler F H, Doll K H, Prössdorf W, *J. Organomet. Chem.* **1982**, *224*, 341.

[38] Castellani M P, Geib S J, Rheingold A L, Trogler W C, *Organometallics* **1987**,*6*, 1703.

[39] O'Hare D, Murphy V J, Kaltsoyannis N, *J. Chem. Soc. Dalton Trans.* **1993**, 383.

[40] Köhler F H, Doll K H, Prössdorf W, Müller J, *Angew. Chem. Int. Ed. Engl.* **1982**, *21*, 151.

[41] Atzkern H, Hiermeier J, Kanellakopulos B, Köhler F H, Müller G, Steigelmann O, *J. Chem. Soc. Chem. Commun.* **1991**, 997.

[42]   Perutz R N, Scaiano J C, *J. Chem. Soc. Chem. Commun.* **1984**, 457.
[43]   Chetwynd-Talbot J, Grebenik P, Perutz RN, *Inorg. Chem.* **1982**, *21*, 3647.
[44]   Cox P A, Grebenik P, Perutz R N, Robinson R D, Grinter R, Stern D R, *Inorg. Chem.* **1983**, *22*, 3614.
[45]   Hartwig J F, He X, *Angew. Chem. Int. Ed. Engl.* **1996**, *35*, 315
[46]   Gordon K R, Warren K D, J *Organomet. Chem.* **1976**, *117*, C27.
[47]   Gordon K R, Warren K D, *Inorg. Chem.* **1978**, *17*, 987.
[48]   Aleksanyan V T, Greenwald I I, *J. Mol. Struct.* **1982**, *90*, 35.
[49]   Ketkov S Y, Domrachev G A, *Inorg. Chim. Acta* **1990**, *178*, 233.
[50]   Giordan J C, Moore J H, Tossell J A, Weber J, *J. Am. Chem. Soc.* **1983**,*105*, 3431.
[51]   Köhler F H, Geike W, *J. Magn. Reson.* **1983**, *53*, 297.
[52]   Hebendanz N, Köhler F H, Scherbaum F, Schlesinger B, *Magn. Reson. Chem.* **1989**, *27*, 798.
[53]   Blümel J, Hofmann P, Köhler F H, *Magn. Reson. Chem.* **1993**, *31*, 2.
[54]   Flower K R, Hitchcock P B, *J. Organomet. Chem.* **1996**, *507*, 275.
[55]   Heinemann O, Jolly P W, Krüger C, Verhovnik G P J, *Organometallics*, **1996**, *15*, 5462.
[56]   Cotton F A, Feng X, Kibala P A, Matusz M, *J. Am. Chem. Soc.* **1988**, *110*, 2807.
[57]   Kalousova J, Benes L, Votinsky J, *Coll. Czech. Chem. Commun.* **1986**, *51*, 314.
[58]   Wong K L T, Brintzinger H H, *J. Am. Chem. Soc.* **1975**, *97*, 5143.
[59]   Karol F J, Karapinka G L, Wu C, Dow A. W, Johnson R N,Carrick W L, *J. Polym. Sci. A1*, **1972**, *10*, 2621.
[60]   Thomas B J, Noh S K, Schulte G K, Sendlinger S C, Theopold K H, *J. Am. Chem. Soc.* **1991**, *113*, 893.
[61]   Schnellbach M, Köhler F H, Blümel J, *J. Organomet. Chem.* **1996**, *520*, 227.
[62]   Blümel J, Hebendanz N, Hudeczek P, Köhler F H, Strauss W *J. Am. Chem. Soc.* **1992**, *114*, 4223.
[63]   Mugnier Y, Moise C, Laviron E, *Nouv. J. Chim.* **1982**, *6*, 197.
[64]   Smart J C, Pinsky B L, *J. Am. Chem. Soc.* **1980**, *102*, 3663.
[65]   Jonas K, Wiskamp V, Tsay Y-H, Krüger C, *J. Am. Chem. Soc.* **1983**, *105*, 5480.
[66]   Jonas, K, Rüsseler W, Krüger C, Raabe E, *Angew. Chem. Int. Ed. Engl.* **1986**, *25*, 928.
[67]   Jonas K, Rüsseler W, Angermund K, Krüger C, *Angew. Chem. Int. Ed. Engl.* **1986**, *25*, 927.
[68]   van Raaij E U, Mönkeberg S, Kiesele H, Brintzinger H H, *J. Organomet. Chem.* **1988**, *356*, 307.
[69]   Pasynskii A A, Eremenko I L, Abdullaev A S, Orazsakhatov B, Nefedov S E, Stomakhina E E, Ellert O G, Katser S B, Yanovskii A I, Struchkov Y T, *Zh. Neorg. Khim.* **1990**, *35*, 2257.
[70]   Hübel W, Merényi R, *J. Organomet. Chem.* **1964**, *2*, 213.
[71]   Bierwagen E P, Bercaw J E, Goddard III W A, *J. Am. Chem. Soc.* **1994**, *116*, 1481.
[72]   Simpson K M, Rettig M F, Wing R M, *Organometallics* **1992**,*11*, 4363.
[73]   Foo D M J, Shapiro P, *Organometallics*, **1995**, *14*, 4957.
[74]   Thomas J L, *J. Am. Chem. Soc.* **1973**, *95*, 1838.
[75]   Mönkeberg S, van Raaij E, Kiesele H, Brintzinger H H, *J. Organomet. Chem.* **1989**, *365*, 285.
[76]   Poli R, Mattamana . P, Falvello L R, *Gazz. Chim. Ital.* **1992**, *122*, 315.
[77]   Jernakoff P, Fox J R, Cooper N J, *J. Organomet. Chem.* **1996**, *512*, 175.
[78]   Ito T, Tokunaga T, Minato M, Nakamura T, *Chem. Lett.* **1991**, 1893.
[79]   Galante J M, Bruno J W, Hazin P N, Folting K, Huffman J C, *Organometallics*, **1988**, *7*, 1066.
[80]   Martins A M, Calhorda M J, Romão C C, Völkl C, Kiprof P, Filippou A C, *J. Organomet. Chem.* **1992**, *423*, 367.
[81]   Bandy J A, Berry A, Green M L H, Perutz R N, Prout K, Verpeaux J-N, *J. Chem. Soc. Chem. Commun.* **1984**, 729.
[82]   Calderazzo F, Fachinetti G, Floriani C, *J. Am. Chem. Soc.* **1974**, *96*, 3695.
[83]   Kowaleski R M, Basolo F, Trogler W C, Gedridge R W, Newbound T D, Ernst R D, *J. Am. Chem. Soc.* **1987**, *109*, 4860.
[84]   Kowaleski R M, Rheingold A L, Trogler W C, Basolo F, *J. Am. Chem. Soc.* **1986**, *108*, 2460.
[85]   Nieman J, Teuben J H, *J. Organomet. Chem.* **1985**, *287*, 207.
[86]   Thiyagarajan B, Michalczyk L, Bollinger J C, Bruno J W, *Organometallics*, **1996**, *15*, 2588.

]87]   Green M L H, Mtetwa S B, Sella A, Chernega A N, *J. Chem. Soc. Daton Trans.* **1994**, 201.

[88]   Schwemlein H, Zsolnai L, Huttner G, Brintzinger H H, *J. Organomet. Chem.* **1983**, *256*, 285.

[89]   Green J C, Payne H. P, Teuben J H, *Organometallics* **1983**, *2*, 203.

[90]   de Liefde Meijer H J, Jellinek F, *Inorg. Chim. Acta* **1970**, *4*, 651.

[91]   Manzer L E, *J. Organomet. Chem.* **1976**, *110*, 291.

[92]   Curtis C J, Smart J C, Robbins J L, *Organometallics* **1985**,*4*, 1283.

[93]   Siegert F W, de Liefde Meijer H J, *J. Organomet. Chem.* **1968**, *15*, 131.

[94]   Ytsma D, Hartsuiker J G, Teuben J H, *J. Organomet. Chem.* **1974**, *74*, 239.

[95]   Bouman H, Teuben J H, *J. Organomet. Chem.* **1976**, *110*, 327.

[96]   Fachinetti G, Floriani C, *J. Chem. Soc. Dalton Trans.* **1974**, 2433.

[97]   Razuvaev G A, Latyaeva V N, Gladyshev E N, Lineva A. N, Krasil'nikova E. V. *Dokl. Akad. Nauk. SSSR (Engl. Transl.)* **1975**, *223*, 481.

[98]   Otto E E H, Brintzinger H H, *J. Organomet. Chem.* **1979**, *170*, 209.

[99]   Antiñolo A, Fajardo M, Otero A, Royo P, *J. Organomet. Chem.* **1982**, *234*, 309.

[100]   Acedo L, Otero A, Royo P, *J. Organomet. Chem.* **1983**, *258*, 181.

[101]   Fermin M C, Hneihen A S, Maas J J, Bruno J W, *Organometallics* **1993**,*12*, 1845.

[102]   Antiñolo A, Fajardo M, Otero A, Royo P, *J. Organomet. Chem.* **1983**, *246*, 269.

[103]   Deutsch P P, Maguire J A, Jones W D, Eisenberg R, *Inorg. Chem.* **1990**, *29*, 686.

[104]   Antiñolo A, García-Lledó S, Martínez de Ilarduya J, Otero A, *J. Organomet. Chem.* **1987**, *335*, 85.

[105]   Nabaoui H, Mugnier Y, Fakhr A, Laviron E, Antiñolo A, Jalón F A, Fajardo M, Otero A, *J. Organomet. Chem.* **1989**, *375*, 67.

[106]   Antiñolo A, Fajardo M, Jalón F A, López Mardomingo C, Otero A, Sanz-Bernabé C, *J. Organomet. Chem.* **1989**, *369*, 187.

[107]   Baynham R F G, Chetwynd-Talbot J, Grebenik P, Perutz R N, Powell M H A, *J. Organomet. Chem.* **1985**, *284*, 229.

[108]   Roullier L, Lucas D, Mugnier Y, Antiñolo A, Fajardo M, Otero A, *J. Organomet. Chem.* **1991**, *412*, 353.

[109]   Lemenovskii D A, Tsikalova M V, Konde S A, Perevalova E G, *Koord. Khim.* **1983**, *9*, 1060.

[110]   Fachinetti G, Del Nero S, Floriani C, *J. Chem. Soc. Dalton Trans.* **1976**, 1046.

[111]   McNally J P, Cooper N J, *Organometallics,* **1988**, *7*, 1704.

[112]   Miller G A, Cooper N J, *J. Am. Chem. Soc.* **1985**, *107*, 709.

[113]   Fiederling K, Grob I, Maalisch W, *J. Organomet. Chem.* **1983**, *255*, 299.

[114]   Antiñolo A, Espinosa P, Fajardo M, Gómez-Sal P, López-Mardomingo C, Martín-Alonso A, Otero A, *J. Chem. Soc. Dalton Trans.* **1995**, 1007.

[115]   Serrano R , Royo P, *J. Organomet. Chem.* **1983**, *247*, 33.

[116]   Urbanos F A, Mena M, Royo P, Antiñolo A, *J. Organomet. Chem.* **1984**, *276*, 185.

[117]   Antiñolo A, Gómez-Sal P, Martínez de Ilrduya J, Otero A, Royo P, Martínez Carrera S, García Blanco S, *J. Chem. Soc. Dalton Trans.* **1987**, 975.

[118]   Antiñolo A, Martínez de Ilarduya J M, Otero A, Royo P, Manotti Lanfredi A M, Tiripicchio J, *Chem. Soc. Dalton Trans.* **1988**, 2685.

[119]   Martínez de Ilarduya J M, Otero A, Royo P, *J. Organomet. Chem.* **1988**, *340*, 187.

[120]   Gómez M, Martínez de Ilarduya J M, Royo P, *J. Organomet. Chem.* **1989**, *369*, 197.

[121]   Elkrami S, Mourad Y, Mugnier Y, Antiñolo A, del Hierro I, Garcia-Yuste S, Otero A, Fajardo M, Brunner H, Gehart G, Wachter J, Amaudrut J, *J. Organomet. Chem.* **1995**, *498*, 165.

[122]   Bell R A, Cohen S A, Doherty N M, Threlkel R S, Bercaw J E, *Organometallics* **1986**, *5*, 972.

[123]   Kubicki M M, Oudet P, Martin C, Barré C, *J. Chem. Soc. Dalton Trans.* **1995**, 3699.

[124]   Bonnet G, Oudet P, Moïse C, *J. Organomet. Chem.* **1995**, *487*, 105.

[125]   Green M L H, Hughes A K, *J. Organomet. Chem.* **1996**, *506*, 221.

[126]   Green M L H, Hughes A K, *J. Chem. Soc. Dalton Trans.* **1992**, 527.

[127]   Antiñolo A, Carrillo F, Chaudret B, Fajardo M, García-Yuste S, Lahoz F, Lanfranchi M, Lopez J A, Otero A, Pellinghelli M A, *Organometallics* **1995**, 14, 1297.

[128]   Hitchcock P B, Lappert M F, Milrre C R C, *J. Chem. Soc. Dalton Trans.* **1981**, 180.

[129] Jiang Q, Pestana D C, Carroll P J, Berry D H, , *Organometallics* **1994**, *13*, 3679.
[130] Bonnet G, Kubicki M M, Moïse C, Lazzaroni R, Salvadori P, Vitulli G, *Organometallics* **1992**, *11*, 964.
[131] Bonnet G, Lavastre O, Leblanc J C, Moise C, *New J. Chem.* **1988**, *12*, 551.
[132] Lavastre O, Bonnet G, Leblanc J C, Moïse C, *Polyedron* **1995**, *2*, 307.
[133] Oudet P, Kubicki M M, Moïsse C, *Organometallics* **1994**, *13*, 4278.
[134] Challet S, Kubicki M M, Leblanc J C, Moïse C, Nuber B, *J. Organomet. Chem.* **1994**, *483*, 47.
[135] Challet S, Lavastre O, Moïse C, Leblanc J C, Nuber B, *New J. Chem.* **1994**, *18*, 1155.
[136] Buhro W E, Zwick B. D, Georgiou S, Hutchinson J P, Gladysz J H, *J. Am. Chem. Soc.* **1988**, *110*, 2427.
[137] Bonnet G, Lavastre O, Leblanc J C, Moïse C, Vitulli G, *J. Organomet. Chem.* **1988**, *347*, C21.
[138] Oudet P, Moïse C, Kubicki M M, *Inorg. Chim. Acta* **1996**, *247*, 263.
[139] Oudet P, Perrey D, Bonnet G, Moïse C, Kubicki M M, *Inorg. Chim. Acta* **1995**, *237*, 79.
[140] Boni G, Sauvageot P, Moïse C, *J. Organomet. Chem.* **1995**, *489*, C32.
[141] Boni G, Sauvageot P, Marpeaux E, Moïse C, *Organometallics* **1995**, *14*, 5652.
[142] Challet S, Leblanc J C, Moïse C, *New J. Chem.* **1995**, *19*, 1139.
[143] Sauvageot P, Blacque O, Kubicki M M, Juge S, Moïse C, *Organometallics* **1996**, *15*, 2399.
[144] Barre C, Kubicki M M, Leblanc J C, Moïse C, *Inorg. Chem.* **1990**, *29*, 5244.
[145] Moise C, Reynoud J F, Leblanc J C, Broussier R, *J. Organomet. Chem.* **1982**, *240*, C15.
[146] Skripkin Y V, Pasynskii A A, Kalinnikov V T, Porai-KoshitsL K, Minacheva A, Antayshkina A S, Ostrikova V N, *J. Organomet. Chem.* **1982**, *231*, 205.
[147] Nakajima T, Takaya M, Shimizu I, Wakatsuki Y, *Organometallics,* **1995**, *14*, 5598.
[148] Pasynskii A A, Skripkin Y V, Kalinnikov V T, *J. Organomet. Chem.* **1978**, *150*, 51.
[149] Pasynskii A A, Skripkin Y V, Eremenko I L, Kalinnikov V T, Aleksandrov G G, Struchkov Y T, *J. Organomet. Chem.* **1979**, *165*, 39.
[150] Antiñolo A, Fajardo M, García-Yuste S, del Hierro I, Otero A, Elkrami S, Mourad Y, Mugnier Y, *J. Chem. Soc. Dalton Trans.* **1995**, 3409.
[151] Eckart E H O, Brintzinger H H, *J. Organomet. Chem.* **1978**, *148*, 29.
[152] Bond A M, Bixler J W, Mocellin E, Datta S, James E J, Wreford, S. *Inorg. Chem.* **1980,** *19*, 1760.
[153] Antiñolo A, Otero A, Fajardo M, García-Yebra G, Gil-Sanz R, López-Mardomingo C, Martín A, Gómez-Sal P, *Organometallics* **1994**, *13*, 4679.
[154] Kukharenko S V, Soloveichik G L, Strelets V V, *Metalloorg. Khim.* **1990**, *3*, 88.
[155] Fakhr A, Mugnier Y, Broussier R, Gautheron B, *J. Organomet. Chem.* **1984**, *269*, 53.
[156] Fakhr A, Mugnier Y, Broussier R, Gautheron B,Laviron E, *J. Organomet. Chem.* **1986**, *317*, 201.
[157] Threlkel R S, Bercaw J E, *J. Am. Chem. Soc.* **1981**, *103*, 2650.
[158] Lappert M F, Raston C L, Skelton B W, White A H, *J. Chem. Soc. Dalton Trans.* **1984**, 893.
[159] Bailey S I, Engelhardt L M, Leung W P, Raston C L, Ritchie I M, White A H, *J. Chem. Soc. Dalton. Trans.* **1985,** 1747.
[160] Tueting D. R, Olmstead M M, Schore N E, *Organometallics* **1992**, *11*, 2235.
[161] Lemenovskii D A, Nifant'ev I E, Urazowski I F, Perevalova E G, Timofeeva T V, Slovokhotov Yu L, Struchkov Yu T, *J. Organomet. Chem.* **1988**, *342*, 31.
[162] Green M L H, Hughes A K, Michaelidou D M, Mountford P, *J. Chem. Soc. Chem. Commun.* **1993**, 591
[163] Green M L H, Hughes A K, Mountford P, *J. Chem. Soc. Dalton Trans.* **1991**, 1699.
[164] El Krami S, Mourad Y, Lucas D, Mugnier Y, Antiñolo A, Fajardo M, Garcia-Yuste S, Otero A, *J. Organomet. Chem.* **1996**, *525*, 125.
[165] Silavwe N D, Castellani M P, Tyler D R, *Inorg. Synth.* **1992**, *29*, 204.
[166] Green M L H, McCleverty J A, Pratt L, Wilkinson G, *J. Chem. Soc.* **1961**, 4854.
[167] Persson C, Andersson C, *Organometallics* **1993**, *12*, 2370.
[168] Mise T, Maeda M, Nakajima T, Kobayashi K, Shimizu I, Yamamoto Y, Wakatsuki Y, *J. Organomet. Chem.* **1994**, *473*, 155.

[169] Labella L, Chernega A, Green M L H, *J. Organomet. Chem.* **1995**, *485*, C18.

[170] Murray R C, Blum L, Liu A H, Schrock R R, *Organometallics* **1985**, *4*, 953.

[171] Parking G, Bercaw J E, *Polyhedron* **1988**, *7*, 2053.

[172] Rigny S, Bakhmutov I, Nuber B, Leblanc J-C, Moïse C, *Inorg. Chem.* **1996**, *35*, 3202.

[173] Rigny S, Leblanc J-C, Moïse C, Nuber B, *J. Chem. Soc. Chem. Commun.* **1995**, 45.

[174] Rigny S, Leblanc J-C, Nuber B, Moïse C, *J. Chem. Soc. Dalton Trans.* **1997**, 1187.

[175] Grebenik P D, Green M L H, Kelland M A, Leach J B, Mountford P, *J. Chem. Soc. Chem. Commun.* **1989**, 1397.

[176] Cariou M, Kubicki M M, Kergoat R, Gomes de Lima L C, Scordia H, Guerchais J E, *Inorg. Chim. Acta* **1985**, *104*, 185.

[177] Forschner T C, Cooper N J, *J. Am. Chem. Soc.* **1989**, *111*, 7420.

[178] Forschner T C, Corella II J A , Cooper N J, *Organometallics* **1990**, *9*. 2478.

[179] Smith J A, Brintzinger H H, *J. Organomet. Chem.* **1981**, *218*, 159.

[180] Hartwig J F, He X, *J. Am. Chem. Soc.* **1994**, *116*, 3661.

[181] Hartwig J F, He X, *Organometallics*, **1996**, *15*, 5350.

[182] Hartwig J F, He X, *Angew. Chem. Int. Ed. Engl.* **1996**, *35*, 315.

[183] Ito T, Sugimoto S, Ohki T, Nakano T, Osakada K, *J. Organomet. Chem.* **1992**, *428*, 69.

[184] Ito T, Sugimoto S, Nakano T, *J. Chem. Soc. Dalton Trans.* **1987**, 1857.

[185] Fu E, Granell J, Green M L H, Lowe V J, Marder S R, Saunders G C, Tuddenham M, *J. Organomet. Chem.* **1988**, *355*, 205.

[186] Calhorda M J, Carrondo M A A F. de C. T, Dias A. R, Domingos A M T S, Martinho Simoes J A, Teixeira C, *Organometallics* **1986**, *5*, 660.

[187] Koloski T S, Pestana D C, Carroll P J, Berry D H, *Organometallics* **1994**, *13*, 489.

[188] Figg L K, Carroll P J, Berry D H, *Organometallics* **1994**, *13*, 209.

[189] Figg L K, Carroll P J, Berry D H, *Angew. Chem. Int. Ed. Engl.* **1996**, *35*, 435.

[190] Bel'skii V K, Protskii A N, Bulychev B M, Soloveichik G L, *J. Organomet. Chem.* **1985**, *280*, 45.

[191] Mackey O N D, Morley C P, *Polyhedron* **1992**, *11*, 389.

[192] Okuda J, Herberich G E, Raabe E, Bernal I, *J. Organomet. Chem.* **1988**, *353*, 65.

[193] Petersen J L, Egan Jr J W, *Inorg. Chem.* **1981**, *20*, 2883.

[194] Belmore K A, Vanderpool R A, Tsai J-C, Khan M A, Nicholas K M, *J. Am. Chem. Soc.* **1988**, *110*, 2004.

[195] Herberich G E, Linn K, *J. Organomet. Chem.* **1991**, *418*, 409.

[196] Kubas G J, Ryan R R, *Inorg. Chem.* **1984**, *23*, 3181.

[197] Kuo L Y, Kuhn S, Ly D *Inorg. Chem.* **1995**, *34*, 5341.

[198] Pilato R S, Housmekerides C E, Jernakoff P, Rubin D, Geoffroy G L, Rheingold A L, *Organometallics* **1990**, *9*, 2333.

[199] Adam G J S, Green M L H, *J. Organomet. Chem.* **1981**, *208*, 299.

[200] Bruce A E, Bruce M R M, Sclafani A, Tyler D R, *Organometallics* **1984**, *3*, 1610.

[201] Köpf H, Klapötke J, *Organomet. Chem.* **1986**, *310*, 303.

[202] Jones W D, Chin R M, Crane T W, Baruch D M, *Organometallics,* **1994**, *13*, 4448.

[203] Jernakoff P, Cooper N J, *J. Am. Chem. Soc.* **1989**, *111*, 7424.

[204] Berry D H, Chey J C, Zipin H S, Carroll P J, *Polyhedron*, **1991**, *10*, 1189.

[205] Hong P, Damrauer N H, Carroll P J, Berry D H, *Organometallics*, **1993**, *12*, 3698.

[206] Kotz J C, Vining W, Coco W, Rosen R, Dias A R, Garcia M H, *Organometallics* **1983**, *2*, 68.

[207] Calhorda M J, Carrondo M A A F de C T, Dias A R, Domingos A. M. T, Duarte M T L S, García M H, Romão C C, *J. Organomet. Chem.* **1987**, *320*, 63.

[208] Calhorda A M, Dias A R, Duarte M T, Martins A M, Matias P M, Romão C C, *J. Organomet. Chem.* **1992**, *440*, 119.

[209] Ascenso J R, de Azevedo C G, Gonçalves I S, Herdwick E, Moreno D S, Pessanha M, Romão C C, J *Organometallics* **1995**, *14*, 3109.

[210] Tsai J-C, Khan M, Nicholas K M, *Organometallics* **1989**, *8*, 2967.

[211] Tsai J-C, Wheeler R A, Khan M, Nicholas K M, *Organometallics* **1991**, *10*, 1344.

[212] De Azevedo C G, Dias A R, Martins A M, Romão C C, *J. Organomet. Chem.* **1989**, *368*, 57.
[213] Crabtree R H, Dias A R, Green M L H, Knowles P J, *J. Chem. Soc. A* **1971**, 1350.
[214] McNally J P, Glueck D, Cooper N J, *J. Am. Chem. Soc.* **1988**, *110*, 4838.
[215] McNally J P, Cooper N J, *J. Am. Chem. Soc.* **1989**, *111*, 4500.
[216] 183A Rauchfuss T B, Ruffing C J, *Organometallics* **1982**, *1*, 400.
[217] Jernakoff P, Cooper N J, *J. Am. Chem. Soc.* **1984**, *106*, 3026.
[218] Calhorda A M, Dias A R, Martins A M, Romão C C, *Polyhedron* **1989**, *8*, 1802.
[219] Carmichael A J, McCamley A *J. Chem. Soc. Dalton Trans.* **1995**, 3125.
[220] Carmichael A J, McCamley A *J. Chem. Soc. Dalton Trans.* **1997**, 93.
[221] Ascenso J R, de Azevedo C G, Gonçalves I S, Herdwick E, Moreno D S, Romão C C, Zühkle J, *Organometallics* **1994**, *13*, 429.
[222] Gonçalves I S, Romão C C, *J. Organomet. Chem.* **1995**, *486*, 155.
[223] Dorer B, Prosenc M-H, Reif U, Brintzinger H H, *Organometallics* **1994**, *13*, 3868.
[224] Antiñolo A, Martinez-Ripoll M, Mugnier Y, Otero A, Prashar S, Rodriguez A M, *Organometallics* **1996**, *15*, 3241.
[225] Bristow G S, Lappert M F, Martin T R, Atwood J L, Hunter W F, *J. Chem. Soc. Dalton Trans.* **1984**, 399.
[226] Lemenovskii D A, Urazovskii I. F, Nifant'ev I E, Perevalova E G, *J. Organomet. Chem.* **1985**, *292*, 217.
[227] Bott S G, Hoffmann D M, Rangarajan S P, *J. Chem. Soc. Dalton Trans.* **1996**, 1979.
[228] Gyon F, Amadrut J, Mercier M-F, Shimizu K, *J. Organomet. Chem.* **1994**, *465*, 187.
[229] Lucas D, Mugnier Y, Antiñolo A, Otero A, Fajardo M, *J. Organomet. Chem.* **1992**, *435*, C3.
[230] Sánchez C, Vivien D, Sala Pala J, Viard B, Guerchais J E, *J. Chem. Soc. Dalton Trans.* **1981**, 64.
[231] Carrondo M A A F de C T, Morais J, Romão C C, Romão M J, Veiros L F, *Polyhedron* **1993**, *12*, 765.
[232] Calhorda M J, Carrondo M A A F de C T, Bram A, Olsen P N, Dias A. R, Freitas A M, García M H, Piedade M F M, J *Organomet. Chem.* **1992**, *426*, 195.
[233] Klapötke T M, Schulz A, Cameron T S, Bakshi P K *J. Organomet. Chem.* **1993**, *463*, 115.
[234] Compton R G, Eklund J C, Page S D, Rebbitt T O *J. Chem. Soc. Dalton Trans.* **1995**, 389.
[235] Dias A. R, Garcia M H, Martins A M, Pinheiro C I, Romão C.C, Veiros L F, *J. Organomet. Chem.* **1987**, *327*, C59.
[236] Jernakoff P, Fox J R, Hayes J C, Lee S, Foxman B M, Cooper N J, *Organometallics*, **1995**, *14*, 4493.
[237] Migot J L, Sala-Pala J, Guerchais J E, *J. Organomet. Chem.* **1983**, *243*, 427.
[238] Antiñolo A, Fajardo M, Otero A, Royo P, *J. Organomet. Chem.* **1984**, *265*, 35.
[239] Arnold J, Don Tilley T, Rheingold A., L, Geib S J, *Organometallics* **1987**, *6*, 473.
[240] Gowik P, Klapötke T, *J. Organomet. Chem.* **1989**, *368*, 35.
[241] Antiñolo A, Fajardo M, Otero A, Puerta M C, Mugnier Y, *Polyhedron* **1989**,*8*, 1848.
[242] Antiñolo A, Fajardo M, Otero A, Mugnier Y, Nabaoui H, Mourad H P, *J. Organomet. Chem.* **1991**,*414*, 155.
[243] Hunter J A, Lindsell W E, McCullough K J, Parr R A, Scholes M L, *J. Chem. Soc. Dalton Trans.* **1990**, 2145.
[244] Fakhr A, Mugnier Y, Broussier R, Gautheron B, *J. Organomet. Chem.* **1985**, *279*, C15.
[245] Parkin G, Bunel E, Burger B. J, Trimmer M S, van Asselt A, Bercaw J E, *J. Mol. Catal.* **1987**, *41*, 21.
[246] Thiele K H, Kubak W, Sieler J, Borrmann H, Simon A, *Z. Anorg. Allg. Chem.* **1990**, *587*, 80.
[247] Gowik P, Klapötke T, *J. Organomet. Chem.* **1989**, *375*, C20.
[248] Ol'dekop, Yu A, Knizhnikov V A. *Zh. Obshch. Khim.* **1981**, *51*, 1723.
[249] Thiyagaragan B, Kerr M E, Bollinger J C, Young Jr V G, Bruno J W, *Organometallics* **1997**, 16, 1331.
[250] Brunner H, Gehart G, Meier W, Wachter J, *Organometallics*, **1994**, *13*, 134.
[251] Fajardo M, Gómez-Sal M P, Royo P, Martínez-Carrera S, García Blanco S, *J. Organomet. Chem.* **1986**, *312*, C44.

[252] Gowik P, Klapötke T, White P, *Chem. Ber.* **1989**, *122*, 1649.
[253] Schulz A, Klapötke T, *J. Organomet. Chem.* **1994**, *480*, 195.
[254] Heinekey D M, *J. Am. Chem. Soc.* **1991**, *113*, 6074.
[255] Zilm K W et al, *J. Am. Chem. Soc.* **1990**, *112*, 920.
[256] Antiñolo A, Carrillo F, Fernández-Baeza J, Otero A, Fajardo M, Chaudret B, *Inorg. Chem.* **1992**, *31*, 5156.
[257] Tebbe F N, Parshall G W, *J. Am. Chem. Soc.* **1971**, *93*, 3793.
[258] Doherty N M, Bercaw J E, *J. Am. Chem. Soc.* **1985**, *107*, 2670.
[259] Berry D H, Koloski T S, Carroll P J, *Organometallics* **1990**, 9, 2952.
[260] Nikonov G I, Kuzmina L, Lemenovskii D A, Kotov V V, *J. Am. Chem. Soc.* **1995**, *117*, 10133.
[261] Antiñolo A, Carrillo F, Fajardo M, Otero A, Lanfranchi M, Pellinghelli M A, *Organometallics* **1995**, 14, 1518.
[262] Green M L H, Hughes A K, Mountford P, *J. Chem. Soc. Dalton Trans.* **1991**, 1407.
[263] Nikonov G I, Kuzmina L, Mountford P, Lemenovskii D A, *Organometallics* **1995**, *14*, 3588.
[264] Nikonov G I, Lorberth J, Harms K, Lemenovskii D A, *Inorg. Chem.* **1995**, *34*, 2461.
[265] Nikonov G I, Lemenovskii D A, Lorbeth J, *Organometallics* **1994**, *13*, 3127.
[266] Lantero D R, Motry D H, Ward D L, Smith M R III, *J. Am. Chem. Soc.* **1994**, *116*, 10811.
[267] Antiñolo A, Carrillo-Hermosilla F, Chaudret B, Fajardo M, Fernández-Baeza J, Lanfranchi M, Limbach H-H, Maurer M, Otero A, Pellinghelli M A, *Inorg. Chem.* **1996**, 35, 7873.
[268] Antiñolo A, Carrillo, Chaudret B, Fajardo M, Fernández-Baeza J, Lanfranchi M, Limbach H-H, Maurer M, Otero A, Pellinghelli M A, *Inorg. Chem.* **1994**, *33*, 5163.
[269] Sala-Pala J, Migot J L, Guerchais J E, Le Gall L, Grosjean F, *J. Organomet. Chem.* **1983**, *248*, 299.
[270] Amaudrut J, Kadmiri A, Sala-Pala J, Guerchais J E, *J. Organomet. Chem.* **1984**, *266*, 53.
[271] Whinnery Jr. L L, Henling L M, Bercaw J E, *J. Am. Chem. Soc.* **1991**, *113*, 7575.
[272] Nelson J E, Parkin G, Bercaw J E, *Organometallics* **1992**, *11*, 2181.
[273] Royo P, Gòmez-Carrera J.A, *An. Quím.* **1984**, *80B*, 428.
[274] Leboeuf J-F, Lavastre O, Leblanc J-C, Moise C, *J. Organomet. Chem.* **1991**, *418*, 359.
[275] Sabo-Etienne S, Chaudret B, el Makarim H A, Barthelat J-C, Daudey J-P, Moïse C, Leblanc J-C, *J. Am. Chem. Soc.* **1994**, *116*, 9335.
[276] Jalón F A, Otero A, Manzano B R, Villaseñor, Chaudret B, *J. Am. Chem. Soc.* **1995**, *117*, 10123.
[277] Parking G, Bercaw J E, *J. Chem. Soc. Chem. Commun.* **1989**, 255.
[278] Camanyes S, Maseras F, Moreno M, Lledos J M, Lluch J M, Bertrán J, *J. Am. Chem. Soc.* **1996**, *118*, 4617.
[279] Wiberg N, Häring N W, Schubert U, *Z. Naturforsch. B* **1980**, 35, 599.
[280] Gambarotta S, Chiesi-Villa A, Guastini C, *J. Organomet. Chem.* **1984**, *270*, C49.
[281] Osborne J H, Rheingold A L, Trogler W C, *J. Am. Chem. Soc.* **1985**, *107*, 7945.
[282] Osborne J H, Rheingold A L, *Inorg. Chem.* **1985**, *24*, 3098.
[283] Broussier R, Olivier J D, Gautheron B, *J. Organomet. Chem.* **1983**, *251*, 307.
[284] Thiyagaragan B, Kerr M E, Bruno J W, *Inorg. Chem.* **1995**, *34*, 3444.
[285] Fu P-f, Khan M A, Nicholas K M, *J. Organomet. Chem.* **1996**, *506*, 49.
[286] Antonelli D M, Schaefer W P, Parkin G, Bercaw J E, *J. Organomet. Chem.* **1993**, *462*, 213.
[287] van Asselt A, Burger B J, Gibson V C, Bercaw J E, *J. Am. Chem. Soc.* **1986**, *108*, 5347
[288] Parkin G, van Asselt A, Leahy D J, Whinnery L, Hua N G, Quan W, Henling L M, Schaefer W P, Santarsiero B D, Bercaw J E, *Inorg. Chem.* **1992**, *31*, 82.
[289] Proulx G, Bergman RG, *J. Am. Chem. Soc.* **1996**, *118*, 1981.
[290] Proulx G, Bergman R G, *J. Am. Chem. Soc.*, **1993**, *115*, 9802.
[291] Proulx G, Bergman R G, *J. Am. Chem. Soc.*, **1994**, *116*, 7953.
[292] Prolux G, Bergman R G, *Organometallics*, **1996**, *15*, 133.
[293] Reingold A L, Strong J B, *Acta Crystallogr. Sect. C*, **1991**, 47, 1963
[294] Brunner H, Gehart G, Meier W, Wachter J, Nuber B, *J. Organomet. Chem.* **1993**, *454*, 117.
[295] Brunner H, Kubicki M M, Leblanc J C, Moise C, Volpato F, Wachter J, *J. Chem. Soc. Chem. Commun.* **1993**, 851.

[296] Bach H-J, Brunner H, Wachter J, Kubicki M M, Leblanc J-C, Moise C, Volpato F, Nuber B, Ziegler M L, *Organometallics*, **1992**, *11*, 1403-1407.
[297] Brunner H, Wachter J, Gehart G, Leblanc J-C, Moise C, *Organometallics*, **1996**, *15*, 1327.
[298] Shin J H, Parkin G, *Organometallics*, **1995**, *14*, 1104.
[299] Shin J H, Parkin G, *Organometallics*, **1994**, *13*, 2147.
[300] Prolux G, Bergman R G, *Organometallics*, **1996**, *15*, 684.
[301] Jorgensen K A, *Inorg. Chem.* **1993**, *32*, 1521.
[302] Lemenovskii D A, Putala M, Nikonov G I, Kazennova N B, Yufit D S , Struchov Y T, *J. Organomet. Chem.* **1993**, *454*, 123.
[303] Lemenovskii D A, Putala M, Nikonov G I, Zinin N B, Kazennova N B, Struchov Y T, *J. Organomet. Chem.* **1993**, *452*, 87.
[304] Lemenovskii D A, Nikonov G I, Brusova G P, Kuzmina L G, Stankovie E, Putala M, *J. Organomet. Chem.* **1995**, *496*, 227.
[305] Schrock R R, 'Reactions of coordinated ligands' ed Braterman, Plenum, New York, **1986**.
[306] Berry D H, Koloski T S, Carroll P J, *Organometallics*, **1990**, *9*, 2952.
[307] Gibson V C, Parkin G, Bercaw J E, *Organometallics*, **1991**, *10*, 220.
[308] Antiñolo A, del Hierro I, Fajardo M, García-Yuste S, Otero A, Blaque O, Kubicki M M, Amaudrut J, *Organometallics* **1996**, *15*, 1966.
[309] Butts M D, Bergman R G, *Organometallics*, **1994**, *13*, 2668.
[310] Butts M D, Bergman R G, *Organometallics*, **1994**, *13*, 1899.
[311] Duncalf D J, Harrison R J, McCamley A, Royan B W, *J. Chem. Soc. Chem. Commun.* **1995**, 2421.
[312] Silavwe N D, Chang M Y, Tyler D R, *Inorg Chem.* **1985**, *24*, 4219.
[313] Yoon M, Tyler D R, *J. Chem. Soc. Chem. Commun.* **1997**, 639.
[314] Parkin G, Bercaw J E, *J. Am. Chem. Soc.* **1989**, *111*, 391.
[315] Silavwe N D, Bruce M R M, Philbin C E, Tyler D R, *Inorg. Chem.* **1988**, *27*, 4669.
[316] a) Jernakoff P, Geoffroy G L, Rheingold A L, Geib S J, *J. Chem. Soc. Chem. Commun.* **1987**, 1610. b) Pilato R S, Geoffroy G L, *J. Chem. Soc. Chem. Commun.* **1987**, 1287.
[317] Pilato R S, Eriksen K A, Stiefel E I, Rheingold A L, *Inorg. Chem.* **1993**, *32*, 3799.
[318] Green J C, Green M L H, James J T, Konidarkis P C, Maunder G H, Mountford P, *J. Chem. Soc. Dalton Trans.* **1992**, 1361.
[319] Green M L H, Konidarkis P C, Michaelidou D M, Mountford P, *J. Chem. Soc. Dalton Trans.* **1995**, 155.
[320] Hitchcock P B, Lappert M F, Leung W-P, *J. Chem. Soc. Chem. Commun.* **1987**, 1282.
[321] Niecke E, Hein J, Nieger M, *Organometallics*, **1989**, *8*, 2290.
[322] Wolczanski P T, Threlkel R S, Santarsiero B D, *Acta Crystallogr., Sect. C,* **1983**, *39*, 1330.
[323] Miller G A, Cooper N J, *J. Organomet. Chem.* **1997**, *528*, 151.
[324] Antiñolo A, Fajardo M, Mardomingo C L, Otero A, Mourad Y, Mugnier Y, Sanz-Aparicio J, Fonseca I, Florencio F, *J. Organomet. Chem.* **1990**, *9*, 2919.
[325] Antiñolo A, Carillo F, Fajardo M, Garcia-Juste S, Otero A, *J. Organomet. Chem.* **1994**, *482*, 93.
[326] Klazinga H, Teuben, J H, *J. Organomet. Chem.* **1980**, *194*, 309.
[327] Gibson V C, Bercaw J E, Bruton W J, Sanner R D, *Organometallics*, **1986**, *5*, 976.
[328] Bercaw J E, Burger B J, Green M L H, Santarsiero B D, Sella A, Trimmer M S, Wong L-L, *J. Chem. Soc. Chem. Commun.* **1989**, 734.
[329] Burger B J, Santarsiero B D, Trimmer M S, Bercaw J E, *J. Am. Chem. Soc.* **1988**, *110*, 3134.
[330] Klazinga A H, Teuben J H, *J. Organomet. Chem.* **1980**, *192*, 75.
[331] Antiñolo A, Carrillo F, Garcia-Yuste S, Otero A, *Organometallics*, **1994**, *13*, 2761.
[332] Antiñolo A, Carrillo-Hermosilla F, Chaudret B, Fajardo M, García-Yuste S, Lanfranchi M, Lopez J A, Otero A, Pellinghelli M A, Prashar S, Villaseñor E, *Organometallics* **1996**, *15*, 5507.
[333] Antiñolo A, Fajardo M, Galakhov M, Gil-Sanz R, López-Mardomingo C, Otero A, Lucas D, Chollet H, Mugnier Y, *J. Organomet. Chem.* **1994**, *481*, 27.

[334] Yasuda H, Yamamoto H, Arai T, Nakamura A, Chen J, Kai Y, Kasai N, *Organometallics* **1991**, *10*, 4058.

[335] Herberich G E, Mayer H, *Organometallics*, **1990**, *9*, 2655.

[336] Fakhr A, Mugnier Y, Rouiller L, Broussier R, Gautheron B, Laviron E, *New J. Chem.* **1988**, *12*, 213.

[337] Fu P F, Khan M A, Nicholas K M, *Organometallics*, **1992**, *11*, 2607.

[338] Bristow G S, Hitchcock P B, Lappert M F, *J. Chem. Soc. Chem. Commun.* **1982**, 1145.

[339] Fu P F, Khan M A, Nicholas K M, *J. Am. Chem. Soc.* **1992**, *114*, 6579.

[340] Amaudrut J, Sala-Pala J, Guerchais J E, Mercier R, Douglade J, *J. Organomet. Chem.* **1982**, *235*, 301.

[341] Thiyagarajan B, Michalczyk L, Bollinger J C, Huffmann J C, Bruno J W, *Organometallics*, **1996**, *15*, 1989.

[342] Antiñolo A, Otero A, Fajardo M, López, Mardomingo C, Lucas D, Mugnier Y, Lanfranchi M, Pellingheilli MA, *J. Organomet. Chem.*, **1992**, *435*, 55.

[343] Bruno J W, Fermin M C, Halfon S E, Schulte G K, *J. Am. Chem. Soc.* **1989**, *111*, 8738.

[344] Fermin M C, Thiyagarajan B, Bruno J W, *J. Am. Chem. Soc.* **1993**, *115*, 974.

[345] Savaranamuthu A, Bruce A E, Bruce M R M, Fermin M C, Hneihen A S, Bruno J W, *Organometallics*, **1992**, *11*, 2190

[346] Kerr M, Fermin M C, Bruno J W, *J. Chem. Soc. Chem. Commun.* **1996**, 1221.

[347] Antiñolo A, Fajardo M, Gil-Sanz R, Mardomingo C M, Martín-Villa P, Otero A, Kubicki M M, Mugnier Y, El Krami S, Mourad Y, *Organometallics*, **1993**, *12*, 381.

[348] Antiñolo A, Fajardo M, Mardomingo C M, Martín-Villa P, Otero A, Kubicki M M, Mourad Y, Mugnier Y, *Organometallics*, **1991**, *10*, 3435.

[349] Nelson J E, Bercaw J E, Marsh R E, Henling L M. *Acta Crystallogr., Sect. C*, **1992**, *48*, 1023.

[350] Kerr M E, Bruno J W, *J. Am. Chem. Soc.* **1997**, *119*, 3183.

[351] Biali S E, Gozin M, Rappoport Z, *J. Phys. Org. Chem.* **1989**, *2*, 271.

[352] Bunker M J, De Cian A, Green M L H, Moreau J J E, Siganporia N, *J. Chem. Soc. Dalton Trans.* **1980**, 2155.

[353] Yasuda H, Arai T, Okamoto T, Nakamura A, *J. Organomet. Chem.* **1989**, *361*, 161.

[354] McGilligan B S, Wright T C, Wilkinson G, Motevalli M, Hursthouse M B, *J. Chem. Soc. Dalton Trans.* **1988**, 1737.

[355] Wright T C, Wilkinson G, Motevalli M, Hursthouse M B, *J. Chem. Soc. Dalton Trans.* **1986**, 2017.

[356] Chetcuti P A, Knobler C B, Hawthorne M F, *Organometallics* **1988**, *7*, 650.

[357] Nakamura A, Otsuka S, *J. Am. Chem. Soc.* **1972**, *94*, 1886.

[358] Herberich G E, Okuda J, *Chem. Ber.* **1984**, *117*, 3112.

[359] Okuda J, Herberich G E, *Organometallics* **1987**, *6*, 2331.

[360] Herberich G E, Okuda J, *Angew. Chem. Int. Ed. Engl.* **1985**, *24*, 402.

[361] Gambarotta S, Floriani C, Chiesi-Villa A, Guastini C, *J. Am. Chem. Soc.* **1985**, *107*, 2985.

[362] Conan F, Sala-Pala J, Guerchais J E, Li J, Hoffmann R, Mealli C, Mercier R, Toupet L, *Organometallics*, **1989**, *8*, 1929.

[363] Koloski T S, Carroll P J, Berry D H, *J. Am. Chem. Soc.* **1990**, *112*, 6405.

[364] Berry D H, Chey J C, Zipin H S, Carroll P J, *J. Am. Chem. Soc.* **1990**, *112*, 452.

[365] Dias A R, Queirós M A, *J. Organomet. Chem.* **1990**, *390*, 193.

[366] Queirós M A, Simão J E J, Dias A R, *J. Organomet. Chem.* **1987**, *329*, 85.

[367] Hayes J C, Jernakoff P, Miller G A, Cooper N J, *Pure Appl. Chem.* **1984**, *56*, 25.

[368] Hayes J C, Pearson G D N, Cooper N J, *J. Am. Chem. Soc.* **1981**, *103*, 4648.

[369] Hayes J C, Cooper N J, *J. Am. Chem. Soc.* **1982**, *104*, 5570

[370] Jernakoff P, Cooper N J, *Organometallics*, **1986**, *5*, 747.

[371] Bkouche-Waksman I, Bois C, Sala-Pala J, Guerchais J. E, *J. Organomet. Chem.* **1980**, *195*, 307.

[372] van Asselt A, Trimmer M S, Henling L M, Bercaw J E, *J. Am. Chem. Soc.* **1988**, *110*, 8254.

[373] Middleton A R, Wilkinson G, *J. Chem. Soc. Dalton Trans.* **1980**, 1888.

[374]   Brunner H, Klement U, Wachter J, Tsunoda M, Leblanc J C, Moise C, *Inorg. Chem.,* **1990**, *29*, 584.

[375]   Leblanc J-C, Moïse C, Volpato F, Brunner H, Gehart G, Wachter J, Nuber B, *J. Organomet. Chem.* **1995**, *485*, 237.

[376]   Brunner H, Gehart G, Leblanc J-C, Moïse C, Nuber B, Stubenhofer B, Volpato F, Wachter J, *J. Organomet. Chem.* **1996**, *517*, 47.

[377]   Antiñolo A, Fajardo M, Gil-Sanz R, López-Mardomingo C, Otero A, *Organometallics,* **1994**, *13*, 1200.

[378]   Morán M, Santos-Garcia J J, Masaguer J R, Fernández V, *J. Organomet. Chem.* **1985**, *295*, 327.

[379]   Gambarotta S, Floriani C, Chiesi-Villa, Guastini C, *Organometallics.* **1986**, *5*, 2425.

[380]   Sielsisch T, Behrens U, *J. Organomet. Chem.* **1984**, *272*, C40

[381]   Pasquali M, Leoni P, Floriani C, Chiesi-Villa, Guastini C, *Inor. Chem.* **1983**, *22*, 841.

[382]   Drews R, Wormsbächer D, Behrens U, *J. Organomet. Chem.* **1984**, *272*, C40.

[383]   Benecke J, Drews R, Behrens U, Edelmann F, Keller K, Roesky H W, *J. Organomet. Chem.* **1987**, *320*, C31.

[384]   Gambarotta S, Fiallo M L, Floriani C, Chiesi-Villa, Guastini C, *Inorg. Chem.* **1984**, *23*, 3532.

[385]   Antiñolo A, Fajardo M, de Jesús E, Mugnier Y, *J. Organomet. Chem.* **1994**, *470*, 127.

7 Half-Sandwich Complexes as Metallocene Analogs

Jun Okuda and Thomas Eberle

7.1 Introduction

The introduction of group 3 and 4 metallocenes has revolutionized the area of Ziegler–Natta catalysis [1]. For the first time in the history of this industrially important process, critical polymerization parameters such as activity, molecular weight, polydispersity and microstructure of the resulting polyolefins can be controlled by structurally well-defined and on the molecular level modifiable metal complexes. Moreover, the use of metallocenes as homogeneous polymerization catalysts has dramatically improved the understanding of mechanistic features such as the nature of the active sites and the influence of ligand structure on the regio- and stereoselectivity [2]. One of the many advantages generally associated with the bis(cyclopentadienyl) ligand systems, however, occasionally turns into a disadvantage: the characteristic and highly consistent electronic and steric situation within the bent metallocene unit [3] has long been recognized to cause substantial steric blocking of the metal-centered reaction site. Enhancement of reactivity is observed when the two ring ligands are 'tied back' by a dimethylsilanediyl link as in many Brintzinger-type *ansa*-metallocene complexes. But even in such cases the 'wedge' of the metallocene moiety still turns out to be too congested to allow, for instance, the efficient polymerization of α-olefins [4].

In order to alleviate this steric constraint of the metallocenes, one could utilize, in place of two cyclopentadienyl ligands, one cyclopentadienyl ligand that contains an additional coordinating site X or L tethered to the periphery of the five-membered ring *via* a bridge Z, where X is a one-electron, L is a two-electron ligand (using the neutral ligand formalism) [5], and Z is a covalent bridge of appropriate length (Fig. 7-1). Such bidentate ligands may form chelate complexes in which the cyclopentadienyl group and the additional donor group X or L are both interacting with one metal center [6].

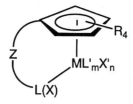

Figure 7-1. General formula for a metal complex containing a bifunctional cyclopentadienyl ligand. X and X′ denote one-electron, L and L′ two-electron ligands, following the neutral ligand formalism.

The replacement of one cyclopentadienyl moiety in a bridged bis(cyclopentadienyl) ligand by an amido ligand NR′, connected *via* a bridge Z, results in ligand systems that form complexes differing from both *ansa*-metallocenes and the simple half-sandwich complexes without the link Z (Fig. 7-2). The amido group is a three-electron ligand of the LX-type (including π-donation from the sp²-hybridized nitrogen atom), in contrast to the a five-electron L_2X-type cyclopentadienyl ligand.

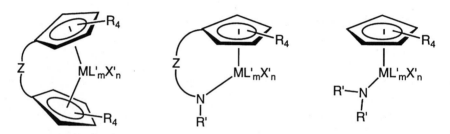

Figure 7-2. Relationship between *ansa*-metallocenes, linked amido–cyclopentadienyl, and half-sandwich amido complexes.

There are many other possibilities with which a cyclopentadienyl ligand may be replaced. An obviously close analog to the amido function would be a bridging alkoxo group (Z)O [7]. The imido group (Z)N=, a ligand isolobal to the cyclopentadienyl ligand, can also be expected to function as the second donor site, but so far only one example for such a bidentate ligand has been reported [8]. Numerous half-sandwich compounds containing neutral donor functions L, especially amine (NR'_2) and ether (OR′) groups at the tether (Z) have been described in the literature [9]. Recently, their potential use as volatile precursors for chemical vapor deposition and as catalysts for olefin polymerization has been noticed.

The cyclopentadienyl ligand is known to act mostly as an inert supporting ligand for a reactive transition metal center by not actively participating in a given substrate transformation. However, under certain conditions, for example during catalytic cycles, the cyclopentadienyl ligand may be involved in irreversible chemical reactions or may even dissociate from the metal. If a second donor (preferentially a multiply bonding ligand) tethered to the cyclopentadienyl ligand is bound as firmly as a cyclopentadienyl ligand, it will add to the stability of the entire ligand framework and prevent exchange or decomposition reactions.

By independently modifying the nature of each of the fragments C_5R_4, X, L, and Z in the bifunctional cyclopentadienyl ligand systems, a great potential for imparting novel properties to the resulting chelate complexes and for controlling the metal reactivity seems to emerge. The present review summarizes the coordination chemistry of linked amido–cyclopentadienyl ligands and the application of mainly group 3 and 4 metal complexes containing this type of ligand. Such complexes constitute a novel class of industrially extremely relevant homogeneous catalysts for olefin polymerization. They possess unique properties which differ significantly from the metallocene catalysts, and major attention is currently being paid to them [10]. The complexes containing a linked alkoxo–cyclopentadienyl ligand are similar to such complexes and are included in this review as well.

7.2 Complexes with Linked Amido–Cyclopentadienyl Ligands

In the late eighties Bercaw and Shapiro introduced the first complexes of bridged amido–cyclopentadienyl ligands [11] in the context of developing structurally well-characterized single-component olefin polymerization catalysts. As electronically more unsaturated and sterically more accessible analogs of *ansa*-scandocene complexes, precursor for complexes of the type $Sc(\eta^5:\eta^1\text{-}C_5Me_4SiMe_2NtBu)X$ (X = H, alkyl) were synthesized and in fact shown to exhibit much higher reactivity towards α-olefins [11, 12]. In order to explore sterically demanding derivatives of this dianionic ligand [13], iron and titanium complexes were synthesized shortly thereafter [14]. Based on these bridged amido–cyclopentadienyl ligands, a flurry of development occurred independently in the research laboratories of Dow Chemical and Exxon Chemical and culminated in their patent applications filed within two weeks of each other in 1990 [10]. A deluge of patents has appeared in the meantime and most of the data on this subject resides in the patent literature.

7.2.1 Synthesis

Following the traditional route to metallocene complexes, amido-bridged half-sandwich complexes are most commonly synthesized by first assembling the ligand $(C_5R_4H)Z(NHR')$ and then coordinating it at the metal center according to a suitable protocol. As the bridging function Z, mostly $SiMe_2$ is used, but carbon-based bridges $(CH_2)_n$ ($n = 2, 3$) or mixed group $SiMe_2CH_2$ have also appeared in the literature. For the introduction of the $SiMe_2$ link, metallated cyclopentadiene or any substituted cyclopentadiene $C_5R_4H_2$, usually as its lithium salt $Li(C_5R_4H)$, is reacted with dichlorodimethylsilane to give $(C_5R_4H)SiMe_2Cl$ (R = H, Me). The chlorodimethylsilyl substituted cyclopentadiene is further transformed into the ligand precursor $(C_5R_4H)SiMe_2NHR'$ using a variety of lithium amides $Li(NHR')$, or alternatively, using excess amine NH_2R' to remove the hydrogen chloride (Scheme 7-1). Thus the prototypical ligand $(C_5Me_4H)SiMe_2NHtBu$, containing the dimethylsilanediyl bridge, first described by Bercaw and Shapiro, is obtained as a moisture sensitive, distillable oil in overall yield of 76% based on $Li(C_5Me_4H)$ [12].

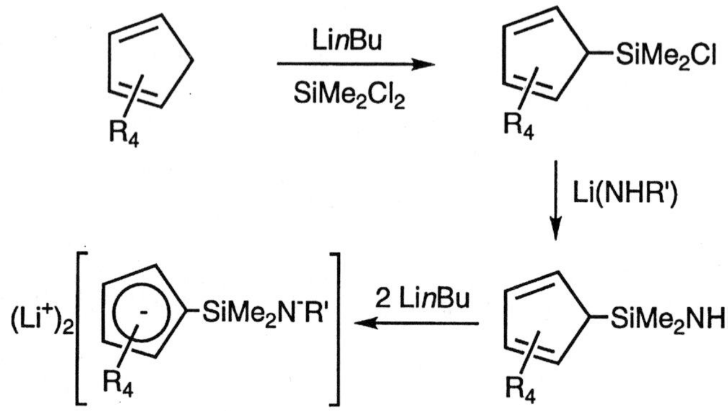

Scheme 7-1

There seems to be no limit as to the choice of the cyclopentadienyl moiety C_5R_4: it may also consist of annulated ring systems such as indenyl [15] or donor-substituted systems (see Sec. 7.3). In the case of the fluorenyl ligand $(C_{13}H_9)SiMe_2NHtBu$, the synthesis of $(C_{13}H_9)SiMe_2Cl$ is hampered by the substantial formation of $(C_{13}H_9)_2SiMe_2$ due to the high nucleophilicity of the fluorenyl anion [16].

A series of di- and trimethylene-linked ligands may be prepared by adding $[Br(CH_2)_nN^+H_2R']Br^-$ ($n = 2, 3$; R' = Me, iPr, tBu) to an excess of $Na(C_5H_5)$ or $Na(C_9H_7)$. These functionalized cyclopentadienes $C_5H_5(CH_2)_nNHR'$ exist as several double-bond isomers (Eq. 7-1) [17, 18].

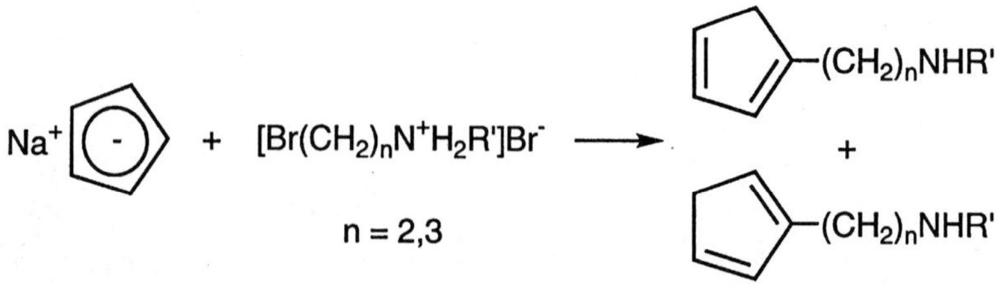

(7-1)

The double metalation is straightforwardly achieved with *n*-butyllithium and, as is commonly the case, used *in situ* without isolation. Occasionally isolation is possible: $Li_2(C_5Me_4SiMe_2NtBu)$ can be obtained as a tan powder. Based on what is known about both cyclopentadienyl and amido lithium complexes in solution, complicated structures can be presumed for such dianions. Temperature-dependent NMR spectra of $Li_2\{(C_{13}H_8)SiMe_2NtBu\}$ indicate an unsymmetrical structure devoid of a mirror plane [16]. When $ClCH_2SiMe_2NHtBu$ is reacted with two equivalents of fluorenyl lithium the structurally characterized monolithium derivative $Li(THF)\{(C_{13}H_8)CH_2SiMe_2NHtBu\}$ is isolated [19]. Interestingly, the reaction of potassium hydride with $(C_5Me_4H)SiMe_2NHtBu$ gives the mono(potassium) derivative $K(THF)(C_5Me_4SiMe_2NHtBu)$ that contains an intact amino function according to a single-crystal structure analysis [20]. The use of the magnesium derivative $\{MgCl(THF)\}_2(C_5Me_4SiMe_2NtBu)$ as a less reducing ligand transfer agent has been reported [10a, 33].

For the complexation of the linked amido–cyclopentadienyl ligand several different synthetic procedures have been developed. The metathetical reaction of the doubly metallated ligand precursor $(C_5R_4ZNR')^{2-}$ with appropriate metal halides appears to be the most common. The reaction of $Li_2(C_5Me_4SiMe_2NtBu)$ with $ScCl_3(THF)_3$ gives non-stoichiometric $Sc(\eta^5{:}\eta^1{-}C_5Me_4SiMe_2NtBu)Cl{\cdot}LiCl(THF)_x$ [12] (which probably contains an ate complex $[Sc(\eta^5{:}\eta^1{-}C_5Me_4SiMe_2NtBu)Cl_2]^-$) from which the THF can be removed by heating to 100°C under vacuum. After Soxhlet extraction, the chloro complex $Sc(\eta^5{:}\eta^1{-}C_5Me_4SiMe_2NtBu)Cl$ is obtained as a white powder in 51% yield (Eq. 7-2) [12].

$$\text{(7-2)}$$

The first example of a titanium complex containing a linked amido–cyclopentadienyl ligand, racemic $Ti\{\eta^5{:}\eta^1{-}(C_5H_3tBu)SiMe_2NtBu\}Cl_2$, was obtained from the reaction of $Li_2\{(C_5H_3tBu)SiMe_2NtBu\}$ with $TiCl_4(THF)_2$ as brown crystals in 35% yield and was shown by NMR spectroscopy to be a configurationally stable molecule [14]. The use of the tris(tetrahydrofuran) complex of titanium(III) chloride followed by oxidation of the titanium(III) intermediate instead of the more reduction-sensitive $TiCl_4(THF)_2$ leads to more reproducible and higher yields [21]. Apart from the conventional oxidants HCl, CCl_4 and AgCl, the best reagent for the chlorination of the titanium(III) intermediate proved to be lead dichloride, introduced by Teuben *et al.* [22]. Following this method, dichloro(titanium) complexes of the general type $Ti(\eta^5{:}\eta^1{-}C_5R_4SiMe_2NR')Cl_2$, with various cyclopentadienyl and indenyl moieties, as well as amido substituents R', including optically active ones, can be synthesized (Scheme 7-2) [15, 23–25].

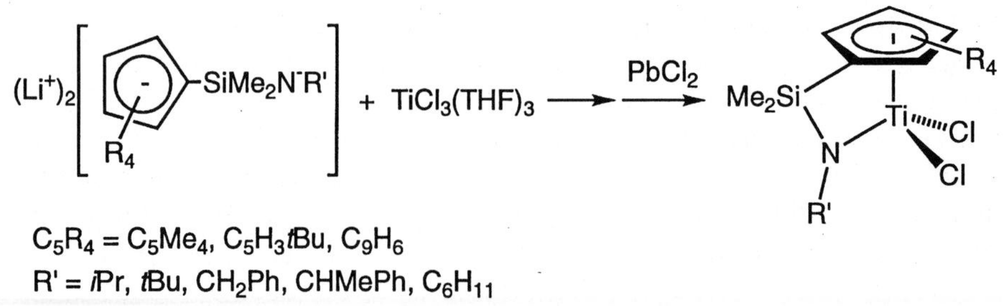

$(Li^+)_2 \left[\begin{array}{c} \bigcirc \\ R_4 \end{array} -SiMe_2N^-R' \right]$ + $TiCl_3(THF)_3$ $\xrightarrow{ PbCl_2 }$ Me$_2$Si $\begin{array}{c} \\ N \\ R' \end{array}$ Ti$^{\cdots}$Cl / Cl ... R$_4$

$C_5R_4 = C_5Me_4, C_5H_3tBu, C_9H_6$
$R' = iPr, tBu, CH_2Ph, CHMePh, C_6H_{11}$

Scheme 7-2

However, there are some disadvantages with this synthetic methodology: Occasionally only low yields are encountered during the synthesis of titanium complexes, e. g. when ligands containing an unsubstituted cyclopentadienyl ring is used (yield of $Ti(\eta^5:\eta^1\text{-}C_5H_4SiMe_2NtBu)Cl_2 < 5\%$) [26]. Not unexpectedly, reduction of the metal center is generally a synthetic obstacle, whenever strongly reducing (*i.e.* more ionic) cyclopentadienyl anions are employed [21]. An elegant solution for this problem is offered by the reaction of $C_5H_5(CH_2)_nNHR'$ with $TiCl_4$ in the presence of triethylamine to give complexes of the type $Ti\{\eta^5:\eta^1\text{-}C_5H_4(CH_2)_nNR'\}Cl_2$ in good yields [18]. In certain cases the preferred formation of bis(ligand) complexes of the type $Ti(\eta^5:\eta^1\text{-}C_5R_4ZNR')_2$ may account for the low yield (see Sec. 7.4). For example, irrespective of the molar ratio, the reaction between $Li_2(C_5H_4SiMe_2NtBu)$ and $ZrCl_4$ exclusively leads to the formation of $Zr(\eta^5:\eta^1\text{-}C_5H_4SiMe_2NtBu)_2$ [27]. According to another report an equimolar mixture of $Zr(\eta^5:\eta^1\text{-}C_5H_4SiMe_2NtBu)Cl_2$ and $[Li(THF)_n][Zr(\eta^5:\eta^1\text{-}C_5H_4SiMe_2NtBu)Cl_3]$ is obtained, whereas the reaction of $Li_2(C_5Me_4SiMe_2NtBu)$ with $ZrCl_4$ or $ZrCl_4(THF)_2$ gives the desired mono(ligand) complex $Zr(\eta^5:\eta^1\text{-}C_5Me_4SiMe_2\text{-}NtBu)Cl_2$ in 85% yields (Scheme 7-3) [26].

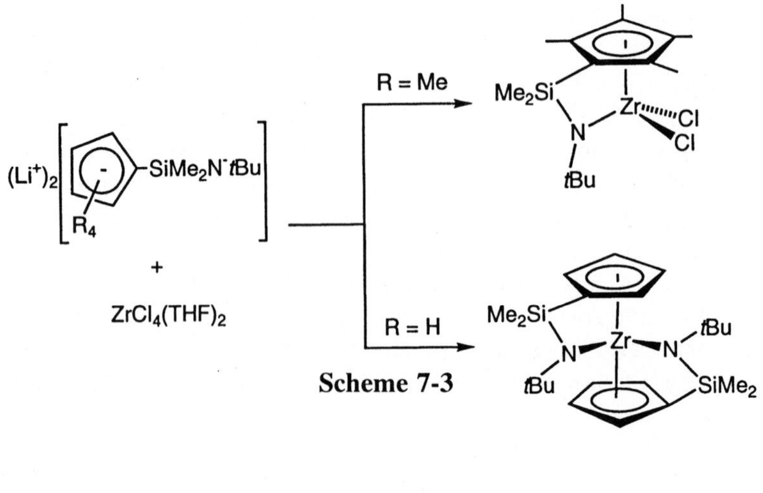

$(Li^+)_2 \left[\begin{array}{c} \bigcirc \\ R_4 \end{array} -SiMe_2N^-tBu \right]$

+

$ZrCl_4(THF)_2$

R = Me   Me$_2$Si $\begin{array}{c} \\ N \\ tBu \end{array}$ Zr$^{\cdots}$Cl / Cl

R = H   Me$_2$Si $\begin{array}{c} \\ N \\ tBu \end{array}$ Zr $\begin{array}{c} \\ N \\ SiMe_2 \end{array}$ tBu

Scheme 7-3

As mentioned above for the scandium complex $Sc(\eta^5:\eta^1\text{-}C_5Me_4SiMe_2NtBu)$-$Cl\bullet LiCl(THF)_x$, the expected higher Lewis acidity of the complexes of the linked amido–cyclopentadienyl ligand often results in the formation of solvent adducts [12]. Such solvent adducts may interfere with subsequent alkylation reactions or with activation procedures during olefin polymerization. In the case of fluorenyl-based systems, the resulting zirconium complexes are found to be only easily isolable as solvent adducts. Mono(tetrahydrofuran) or mono(diethyl ether) adducts $Zr(\eta^5:\eta^1\text{-}C_{13}H_8SiMe_2NtBu)Cl_2(L)$ (L = THF or Et_2O) are obtained by reacting $Li_2(C_{13}H_8SiMe_2NtBu)(THF)_n$ with $ZrCl_4(THF)_2$ or $ZrCl_4(Et_2O)_2$ [16]. Detailed NMR spectroscopic investigation revealed a labile coordination of one THF molecule in $Zr(\eta^5:\eta^1\text{-}C_{13}H_8SiMe_2NtBu)Cl_2(THF)$, which is irreversibly lost upon heating to give insoluble oligomers (Scheme 7-4). By changing the size of the bridging group Z from $SiMe_2$ to CH_2SiMe_2, the resulting fluorenyl complexes $M(\eta^5:\eta^1\text{-}C_{13}H_8CH_2SiMe_2NR')Cl_2$ (M = Ti, Zr, Hf; R' = tBu), synthesized from $Li_2(C_{13}H_8CH_2SiMe_2NtBu)(THF)_n$ and $TiCl_4(THF)_2$, $ZrCl_4$ and $HfCl_4$, are isolated without a solvent molecule [19]. The reaction of $Li_2(C_5Me_4SiMe_2NtBu)$ with chromium trichloride in THF affords the dark green chromium(III) complex also as its THF solvate $Cr(\eta^5:\eta^1\text{-}C_5Me_4SiMe_2NtBu)Cl(THF)$ [28]. In the absence of any Lewis base, some zirconium complexes tend to dimerize such as $\{Zr(\eta^5:\eta^1\text{-}C_5H_4SiMe_2NtBu)Cl\}_2(\mu\text{-}Cl)_2)$ [26].

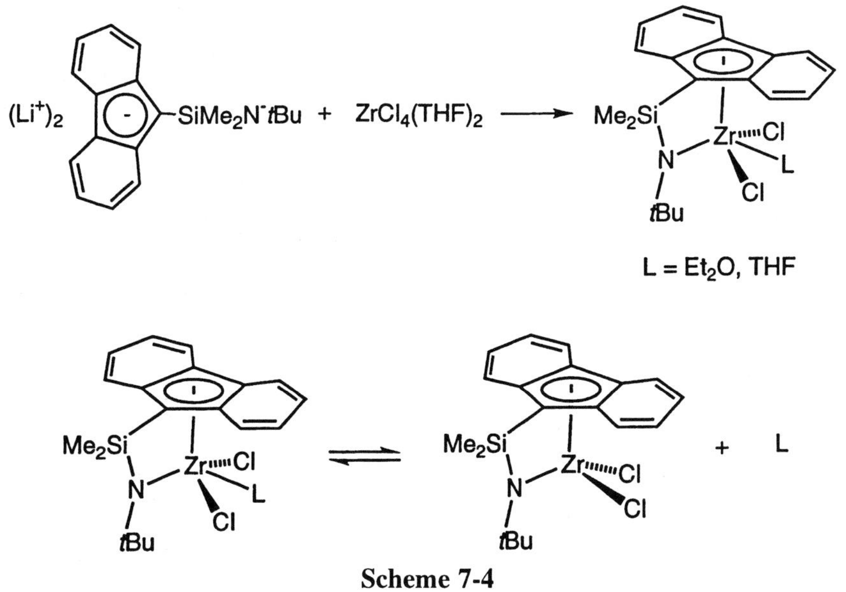

L = Et_2O, THF

Scheme 7-4

In order to avoid the problems associated with the metathesis method, the reaction of homoleptic metal amides $M(NR''_2)_n$ with the functionalized cyclopentadienes with amine elimination has been applied. This method introduced many years ago for the synthesis of metallocene amido complexes of the type $M(C_5R_5)_x(NR'')_y$, was recently expanded to the stereoselective synthesis of Brintzinger-type *ansa*-metallocenes [29]. It was first applied for the synthesis of complexes containing a linked amido–cyclopentadienyl ligand with a carbon bridge $C_5H_5(CH_2)_nNR'$ ($n = 2$, 3; R' = Me, *i*Pr, *t*Bu) [17, 18, 30]. The complexation of $M(NMe_2)_4$ (M = Ti, Zr, Hf) works well for these ligands and produces distillable $M\{\eta^5:\eta^1$-$C_5H_4(CH_2)_nNR'\}(NMe_2)_2$ or $M\{\eta^5:\eta^1$-$C_9H_6(CH_2)_nNR'\}(NMe_2)_2$ (M = Zr, Hf; $n = $ 2: R' = Me, *t*Bu; $n = $ 3: R' = Me) (Scheme 7-5). The reaction is more effective when less sterically hindered ligands, preferably containing two CH_2 units, are used. Whereas $C_5H_5(CH_2)_2NH$*t*Bu gives the desired product with $Ti(NMe_2)_4$ or $Zr(NMe_2)_4$, the double aminolysis starting from $C_5H_5(CH_2)_3NH$*t*Bu and $M(NMe_2)_4$ (M = Ti, Zr) does not occur even under forcing conditions and results in the formation of $M\{\eta^5$-$C_5H_4(CH_2)_3NH$*t*Bu$\}(NMe_2)_3$ (M = Ti, Zr). Decreased reactivity is observed using the titanium tetraamides $Ti(NMe_2)_4$, which do not react with indenyl substituted ligands.

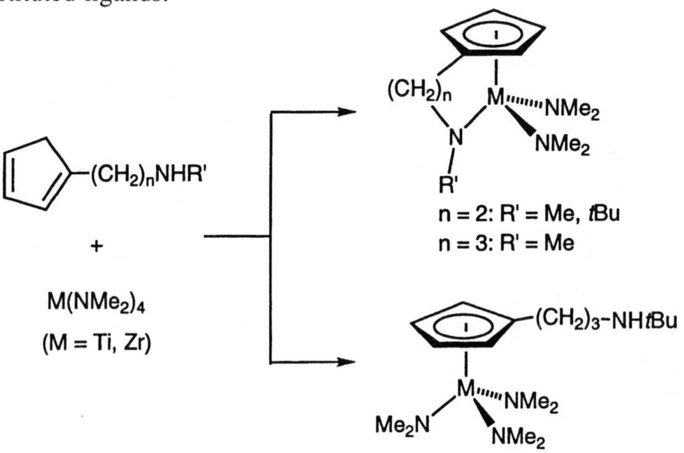

Scheme 7-5

Following analogous procedures, previously described dimethylsilanediyl-bridged amido–cyclopentadienyl ligands $C_5H_5SiMe_2NHR'$ (R' = *t*Bu, Ph) react with $Ti(NR'')_4$, $Zr(NR'')_4$, (R'' = Me, Et) and $Hf(NMe_2)_4$ to give the corresponding complexes of the general type $M(\eta^5:\eta^1$-$C_5H_4SiMe_2NR')(NR'')_2$ [26, 31]. At least in the case of zirconium, the analogous reaction appears to work for the less acidic $(C_5Me_4H)SiMe_2NH$*t*Bu' to give $Zr(\eta^5:\eta^1$-$C_5Me_4SiMe_2$*t*Bu$)(NMe_2)_2$. The driving force for the successful formation of these chelate complexes lies in the generation of the very volatile amines (bp for NMe_2H and NEt_2H: 7 and 55°C, respectively). Although often liquids, the products can be obtained in nearly quantitative yields and the reaction is accelerated with the more acidic ligands and with sterically less demanding cyclic systems [31]. Zirconium dichloro complexes $Zr[\eta^5:\eta^1$-$C_5H_4(CH_2)_2NR']Cl_2$ (R' = *t*Bu) are accessible by using $Zr(NMe_2)_2Cl_2(THF)_2$ and the ligand C_5H_5ZNHR'. The coordinated dimethylamine can be removed by subli-

mation (Eq. 7-3) [32]. Reaction of optically active indenyl ligand (–)-(S)-
$C_9H_6SiMe_2NCHMePh$ with $Ti(NMe_2)_4$ gives a 4:3 diastereomeric mixture of
$Ti(\eta^5:\eta^1-(-)-(S)-C_9H_6SiMe_2NCHMePh)(NMe_2)_2$ [33].

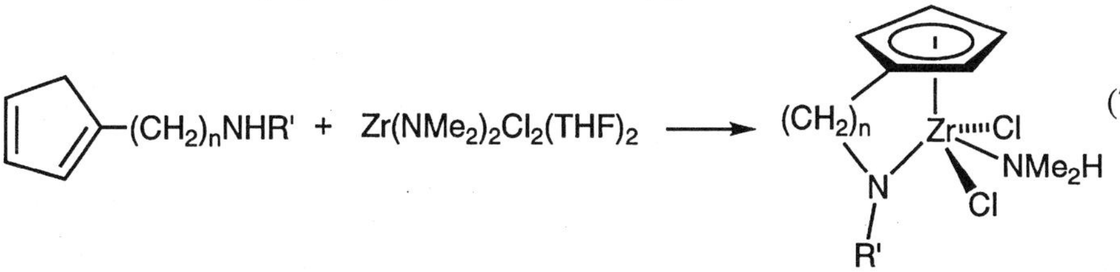

(7-3)

Analogously, niobium(V) and tantalum(V) complexes, as well as the
molybdenum(IV) complex of the type $M(\eta^5:\eta^1-C_5H_4SiMe_2NPh)(NMe_2)_n$ ($n = 3$:
Nb, Ta, $n = 2$: Mo), were obtained in moderate yields [34]. Similarly, using
$Y\{N(SiMe_3)_2\}_3$ the yttrium complex $Y(\eta^5:\eta^1-C_5Me_4SiMe_2NtBu)\{N(SiMe_3)_2\}$ was
synthesized in 40% yield by amine elimination [35].

An alternative approach to amido-bridged cyclopentadienyl complexes consists
of introducing the amido linkage within the preformed half-sandwich complex
[36]. The functionalized cyclopentadiene $C_5H_4(SiMe_2Cl)SiMe_3$, which may easily
be synthesized from $LiC_5H_5(SiMe_3)$ and $SiMe_2Cl_2$, is first reacted with $TiCl_4$ to
form $Ti(\eta^5-C_5H_4SiMe_2Cl)Cl_3$, with selective elimination of chlorotrimethylsilane.
The trichloro complex can be reacted with lithium amides LiNHR′ (R′ = tBu [36a],
iPr [25], CH_2Ph [25], (±)-, (–)-(S)-, (+)-(R)-CHMePh [24]) in the presence of a
base such as triethylamine to scavenge HCl. The resulting titanium complexes
$Ti(\eta^5:\eta^1-C_5H_4SiMe_2NR′)Cl_2$ are available in good yields. The mechanism for this
reaction implies that the amide anion first attacks titanium and then
intramolecularly bridges the silicon which is presumably less electrophilic than titanium (Scheme 7-6) [36].

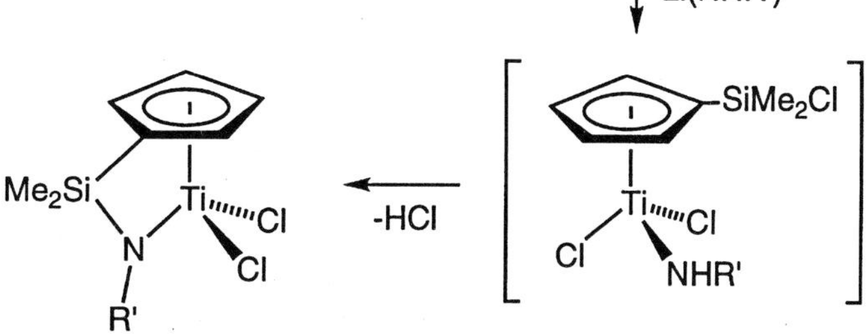

Scheme 7-6

However, this method is limited so far to the unsubstituted cyclopentadienyl system C_5H_4 and cannot be extended to substituted or annulated ring systems such as indenyl [15]. Furthermore, it does not apply to metal centers other than titanium. The reaction of $Zr(\eta^5\text{-}C_5H_4SiMe_2Cl)Cl_3$ with $LiNHtBu$ in the presence of NEt_3 leads to a mixture of two compounds: the desired $Zr(\eta^5:\eta^1\text{-}C_5H_4SiMe_2NtBu)Cl_2$ and the triethylamine adduct $Zr(\eta^5\text{-}C_5H_4SiMe_2NHtBu)Cl_3(NEt_3)$. When $Li(NHCHMePh)$ is reacted with $Zr(\eta^5\text{-}C_5H_4SiMe_2Cl)Cl_3$ only $Zr(\eta^5C_5H_4SiMe_2NHCHMePh)Cl_3(NEt_3)$ is isolated. The observed formation of complexes without the link indicates preferential attack of the amide NHR'^- at the silicon [36].

7.2.2 Hydrido and Alkyl Complexes

The synthesis of alkyl complexes from the chloro or amido derivatives are frequently studied in view of their importance as precursors for α-olefin polymerization catalysts. As suitable alkylating reagents, organolithium or Grignard reagents without β-hydrogens, including methyl, benzyl, trimethylsilylmethyl, neophyl and neopentyl, are employed. Generally, the bulkier alkyl groups provide more thermally and photochemically stable complexes, whereas some dimethyl titanium derivatives cannot be stored for prolonged periods of time at ambient temperatures, reminiscent of the highly sensitive half-sandwich analog $(\eta^5\text{-}C_5H_5)TiMe_3$ [37].

Remarkably, the only isolable monomeric scandium alkyl complex is $Sc(\eta^5:\eta^1\text{-}C_5Me_4SiMe_2NtBu)\{CH(SiMe_3)_2\}$, isolated from the reaction of the chloro complex $Sc(\eta^5:\eta^1\text{-}C_5Me_4SiMe_2NtBu)Cl$ with $LiCH(SiMe_3)_2$ [12]. Even the reaction with $LiCH_2SiMe_3$ does not afford any tractable products. $Sc(\eta^5:\eta^1\text{-}C_5Me_4SiMe_2\text{-}NtBu)\{CH(SiMe_3)_2\}$ can be hydrogenated in the presence of PMe_3 to form the dimeric hydride $\{Sc(\eta^5:\eta^1\text{-}C_5Me_4SiMe_2NtBu)(PMe_3)\}_2(\mu\text{-}H)_2$. It reacts with two equivalents of ethylene to give the unusual ethylene-bridged dimer $\{Sc(\eta^5:\eta^1\text{-}C_5Me_4SiMe_2NtBu)(PMe_3)\}_2(\mu\text{-}\eta^2:\eta^2\text{-}C_2H_4)$ and one equivalent of ethane, whereas the analogous reaction with propylene affords dimeric n-propyl complex $\{Sc(\eta^5:\eta^1\text{-}C_5Me_4SiMe_2NtBu)\}_2(\mu\text{-}nPr)_2$. The latter gives monomeric $Sc(\eta^5:\eta^1\text{-}C_5Me_4SiMe_2NtBu)(PMe_3)(nPr)$ upon addition of PMe_3 and undergoes rapid associative exchange *via* 16-electron $Sc(\eta^5:\eta^1\text{-}C_5Me_4SiMe_2NtBu)(PMe_3)_2(nPr)$ (Scheme 7-7).

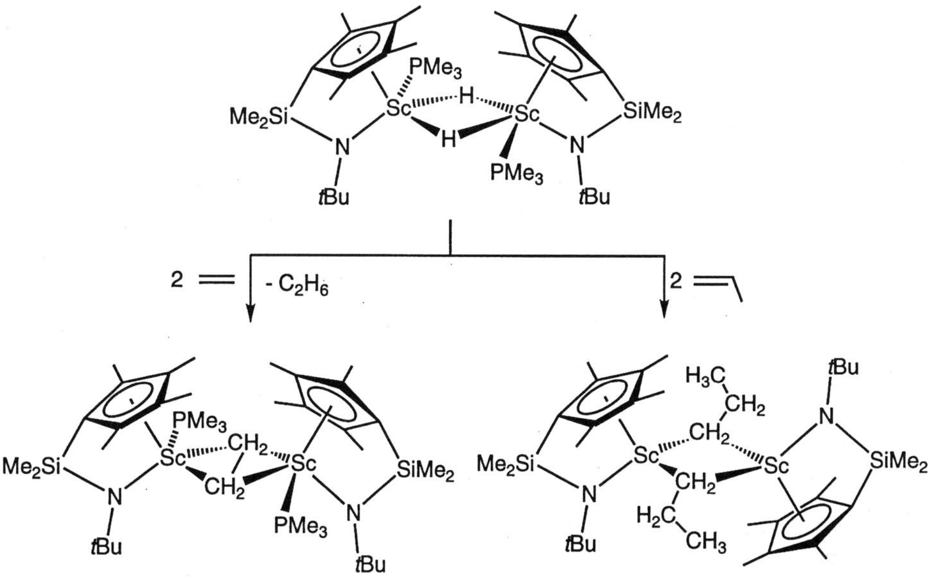

Scheme 7-7

A fairly extensive range of dialkyl complexes of group 4 metals, $M(\eta^5:\eta^1$-$C_5Me_4SiMe_2NtBu)R''_2$, have been reported. They are obtained by the reaction of the dichloro complexes with two equivalents of organolithium or Grignard reagents (Eq. 7-4) [19, 25, 36, 38, 39]. In the case of zirconium complexes containing carbon links, the amine adduct $Zr\{\eta^5:\eta^1$-$C_5H_4(CH_2)_3NMe\}Cl_2(NMe_2H)$ can be directly converted into the dialkyl [17]. Compared to their metallocene analogs, the more air- and moisture-sensitive dialkyl complexes clearly show a higher Lewis acidity (Eq. 7-5).

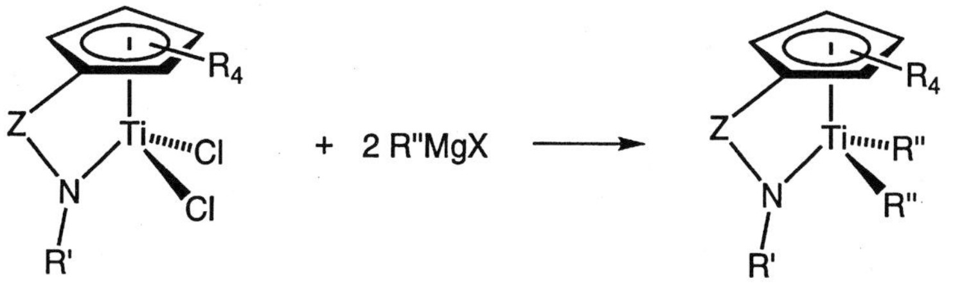

(7-4)

(7-5)

It appears that the retention of solvent molecules depends on the steric bulk of the alkyl groups. While zirconium dimethyl derivatives $Zr(\eta^5:\eta^1\text{-}C_{13}H_8\text{-}SiMe_2NtBu)$-$Me_2(THF)$ [16] or $Zr(\eta^5:\eta^1\text{-}C_5H_4(CH_2)_3NMe)Me_2(Et_2O)_{0.5}$ tend to retain solvents [17], the complexes with larger alkyls can be isolated without any coordinated solvent. Likewise, the chromium complex $Cr(\eta^5:\eta^1\text{-}C_5Me_4SiMe_2NtBu)R''$ retains a THF molecule for $R'' = Me$ and Ph, whereas for $R'' = CH_2SiMe_3$ crystal structure analysis shows a pseudotrigonal configuration (Eq. 7-6) [28].

$$\text{Me}_2\text{Si} \quad \begin{array}{c} \\ \text{Cr} \cdots \text{O} \\ \text{N} \quad \text{Cl} \\ \text{tBu} \end{array} \quad \xrightarrow{\text{LiCH}_2\text{SiMe}_3} \quad \text{Me}_2\text{Si} \quad \begin{array}{c} \\ \text{Cr} \\ \text{N} \quad \text{CH}_2\text{SiMe}_3 \\ \text{tBu} \end{array} \qquad (7\text{-}6)$$

In benzyl complexes of early transition metals the electronic unsaturation is relieved by a distorted coordination of the benzyl ligand. For example, the titanium compound $Ti(\eta^5:\eta^1\text{-}C_5Me_4SiMe_2NCH_2Ph)(CH_2Ph)_2$ exhibits an α-agostic bonding interaction of the CH_2-hydrogen atoms of one of the benzyl groups with the titanium (Fig. 7-3) [25]. In contrast to the other benzyl group the phenyl ring of one benzyl group is nearly coplanar to the cyclopentadienyl ring and the amido group's phenyl system. These different conformations of the benzyl ligands are related to the various metal^{en}–carbon distances (Ti–C1: 2.131(5) Å, Ti–C8: 2.157(5) Å) and in different angles at the benzylic carbon of the two benzyl groups. Compared to Ti–C8–C9 (110.6(3)°), the angle at the other carbon C1 of the other benzyl group is 131.3(3)°. The titanium–hydrogen distances for the former benzyl group are on average slightly shorter than those of the second benzyl group. The values are similar to those found in $Ti(\eta^5\text{-}C_5Me_5)(CH_2Ph)_3$ [40]. However, the unsymmetrical distortion cannot be detected by NMR data at room temperature, which reveal a C_s symmetry.

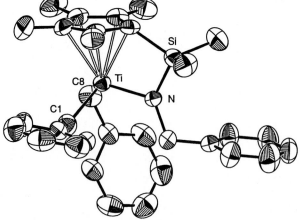

Figure 7-3. Molecular structure of $Ti(\eta^5:\eta^1\text{-}C_5Me_4SiMe_2NCH_2Ph)(CH_2Ph)_2$ [25].

In contrast to the titanium dibenzyl complex, $Ti(\eta^5:\eta^1\text{-}C_5Me_4SiMe_2NCH_2Ph)$-$(CH_2Ph)_2$, the NMR spectroscopic analysis of the zirconium dibenzyl $Zr\{\eta^5:\eta^1\text{-}C_5H_4(CH_2)_3NMe\}(CH_2Ph)_2$ [17] indicates some η^2-benzyl interaction [41], although the molecule possesses overall C_s symmetry. Since at $-80°C$ no change is observed, either both benzyl groups are distorted, or only one of them is distorted, the η^2-benzyl and η^1-benzyl ligand undergoing rapid exchange. The reaction of $Zr\{\eta^5:\eta^1\text{-}C_5H_4(CH_2)_3NMe\}Cl_2(NMe_2H)$ with one equivalent of $C_6H_5CH_2MgCl$ gives the dimeric mono(benzyl) complex $\{Zr(\eta^5:\eta^1\text{-}C_5H_4(CH_2)_3NMe)(CH_2Ph)(\mu\text{-}Cl)\}_2$. Instead of distorting the benzyl ligands, the electron deficiency is relieved by forming bridging chloro ligands. The crystal structure exhibits two different Zr–Cl distances due to the *trans* influence of the benzyl ligand [17]. The reaction of $Zr\{\eta^5:\eta^1\text{-}C_5H_4(CH_2)_3NMe\}Cl_2(NMe_2H)$ with excess of $LiBH_4$ gives the colorless bis(tetrahydroborate), $Zr\{\eta^5:\eta^1\text{-}C_5H_4(CH_2)_3NMe\}(\eta^3\text{-}BH_4)_2$, which contains rapidly exchanging terminal and bridging hydrido ligands.

Alkylidene complexes of early transition metals such as $Ti(\eta^5\text{-}C_5R_5)_2(=CR'_2)$ have played a pivotal role in the development of active initiators for olefin metathesis and ring{opening metathesis polymerization [42]. Thermolysis of $Ti\{\eta^5:\eta^1\text{-}C_5H_4(CH_2)_2NtBu\}(CH_2CMe_2R'')_2$ ($R'' = Me, Ph$) in the presence of PMe_3 affords the neopentylidene and neophylidene complexes $Ti\{\eta^5:\eta^1\text{-}C_5H_4(CH_2)_2NtBu\}$-$(=CHCMe_2R'')(PMe_3)$ in excellent yields, which exhibit characteristically low-field ^{13}C signals at δ 251 and 246 ppm, respectively (Eq. 7-7) [18, 43].

(7-7)

In contrast to the bis(cyclopentadienyl) metal carbene complexes, the Ti=C double bond in $Ti\{\eta^5:\eta^1\text{-}C_5H_4(CH_2)_2NtBu\}(=CHCMe_2R'')(PMe_3)$ ($R'' = Me, Ph$) are assumed to be rotationally fixed because two orientations of the alkylidene ligand are observed. While ^1H and ^{13}C NMR spectroscopic data indicate only one rotamer, NOESY experiments reveal an interaction between the α-proton and the *tert*-butyl group on the amido substituent, rendering the orientation of the large alkyl group towards the cyclopentadienyl ring most likely. Both complexes were reported to be inactive in ring-opening polymerization of strained cyclic olefins. On the other hand, the thermal decomposition of $Ti(\eta^5:\eta^1\text{-}C_5H_4SiMe_2NCH_2Ph)(CH_2SiMe_3)_2$ presumably gives an alkylidene complex that polymerizes norbornene with ring-opening [44].

Thermolysis of early transition metal dialkyl complexes that contain at least one phenyl group are known to give benzyne complexes [45]. The elimination of benzene to form benzyne intermediates at the zirconocene fragment allows the oxidative coupling with unsaturated substrates $R''_2C=X$ (X = CR''_2, NR'', O, S, P) and $R''C\equiv X$ (X = CR'', N) [45]. Electronically more unsaturated linked amido–cyclopentadienyl complexes behave similarly. The thermolysis of $Ti\{\eta^5:\eta^1\text{-}C_5H_4(CH_2)_2NtBu\}Ph_2$ in the presence of PMe_3 gives a phosphine-stabilized benzyne complex $Ti\{\eta^5:\eta^1\text{-}C_5H_4(CH_2)_2NtBu\}(\eta^2\text{-}C_6H_4)(PMe_3)$ that readily undergoes insertion reactions with alkynes, alkenes, nitriles or ketones (Scheme 7-8) [18, 46]. In a related β–hydride elimination reaction, the diethyl complex decomposes to give the ethylene complex $Ti\{\eta^5:\eta^1\text{-}C_5H_4(CH_2)_2NtBu\}(\eta^2\text{-}C_2H_4)(PMe_3)$ [18, 43].

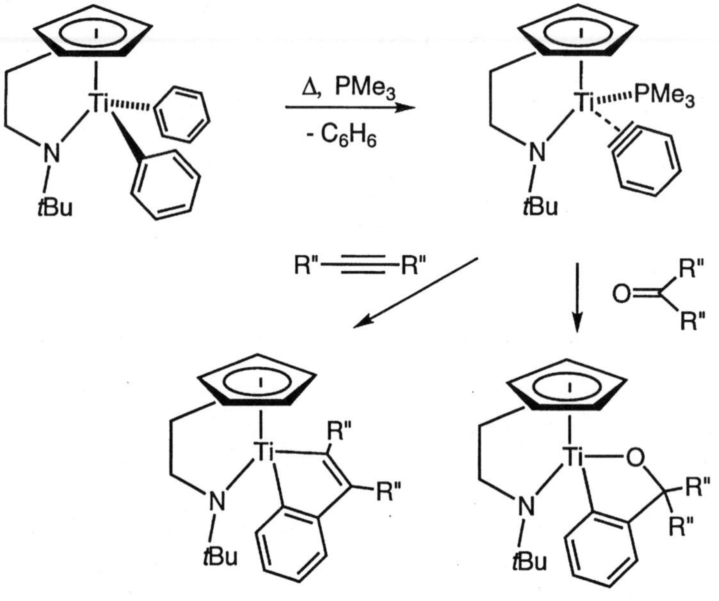

Scheme 7-8

When dichloro complexes $Ti(\eta^5:\eta^1\text{-}C_5Me_4SiMe_2NR')Cl_2$ (R' = Ph, tBu) are treated with two equivalents of *n*-butyllithium in the presence of 1,3-dienes such as piperylene and 2,4-hexadiene, diene complexes of the type $Ti(\eta^5:\eta^1\text{-}C_5Me_4\text{-}SiMe_2NR')(1,3\text{-diene})$ are formed [47]. The diene coordination mode prone-π at formally titanium(II) or supine-σ,π at formally titanium(IV) (titanacyclopentene) is highly sensitive to the identity of R'. While for R' = *t*Bu the former bonding situation is found, as authenticated by the crystal structure of $Ti(\eta^5:\eta^1\text{-}C_5Me_4SiMe_2NtBu)(\eta^4\text{-MeCH=CHCH=CHMe})$, for R' = Ph an inseparable mixture of both noninterconvertible isomers results (Scheme 7-9).

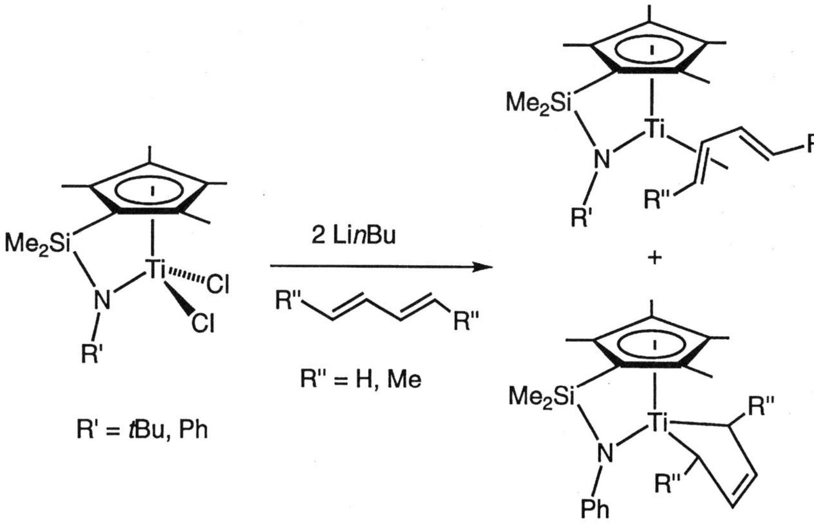

Scheme 7-9

7.2.3 Reactivity

Since the dichloro complexes $M(\eta^5:\eta^1\text{-}C_5R_4ZNR')Cl_2$ (M = Ti, Zr, Hf) are suitable precursors for alkylations, the conversion of the diamido complexes, $M(\eta^5:\eta^1\text{-}C_5R_4ZNR')(NMe_2)_2$, obtained by amine elimination, to the corresponding dichloro derivatives is quite important. Complete conversion, in good yields, can be achieved by treatment of the titanium complexes $Ti(\eta^5:\eta^1\text{-}C_5R_4ZNR')(NMe_2)_2$ with an excess of chlorotrimethylsilane or phosphorus pentachloride [26]. The reaction of the zirconium analog, $Zr(\eta^5:\eta^1\text{-}C_5R_4SiMe_2NtBu)(NMe_2)_2$, with two equivalents of chlorotrimethylsilane gives dimeric $\{Zr(\eta^5:\eta^1\text{-}C_5H_4SiMe_2NtBu)Cl(\mu\text{-}Cl)\}_2$ [26]. The reaction of two equivalents of protic reagents such as HCl or $(NEt_3H)Cl$ with $Ti(\eta^5:\eta^1\text{-}C_5H_4SiMe_2NtBu)(NMe_2)_2$ results in inseparable mixtures of $Ti(\eta^5:\eta^1\text{-}C_5H_4SiMe_2NtBu)Cl_2$ and $Ti(\eta^5\text{-}C_5H_4SiMe_2Cl)Cl_2(NMe_2)(NMe_2H)$. In contrast, the reaction with the corresponding zirconium complexes provides only one product, $Zr(\eta^5:\eta^1\text{-}C_5H_4SiMe_2NtBu)Cl_2(NMe_2H)$, with the NMe_2H ligand coordinated *trans* to the appended amido group [26]. Treatment of two equivalents of $(NMe_2H_2)Cl$ or $(NMe_2H_2)I$ with $M\{\eta^5:\eta^1\text{-}C_5H_4(CH_2)_3NMe\}(NMe_2)_2$ (M = Zr, Hf) gives $M\{\eta^5:\eta^1\text{-}C_5H_4(CH_2)_3NMe\}X_2(NMe_2H)$ (X = Cl, I) in good yields [17]. These results demonstrate the affinity of four-coordinate zirconium complexes for complexing additional L-type ligands (Eq. 7-8).

Noteworthy is the reactivity of the bridging Si–N bond, mainly in titanium complexes. The driving force is the stronger Si–Cl bond versus the Si–N bond [48], the electrophilicity of the silicon atom being enhanced by the ring strain of the chelate structure. Extreme moisture-sensitivity is documented in the electron-rich late metal complex $Fe\{(\eta^5:\eta^1\text{-}C_5H_3tBu)SiMe_2NtBu\}(CO)_2$ that gives the ferrocene derivative $Fe\{(\eta^5\text{-}C_5H_3tBu)SiMe_2NHtBu\}_2$ upon hydrolysis (Eq. 7-9) [14].

$$(7\text{-}8)$$

M = Zr, Hf

$$(7\text{-}9)$$

Reaction of carbon dioxide with complexes containing the $C_5R_4SiMe_2NR'$ ligand such as $Ti(\eta^5:\eta^1\text{-}C_5H_4SiMe_2NR')Cl_2$ (R' = tBu, CHMePh) results in the cleavage of the Si–N bridge with concomitant formation of isocyanate OCNR' and a dimeric siloxide complex with a silicon–oxygen link $\{Ti(\eta^5:\mu:\eta^1\text{-}C_5H_4SiMe_2O)Cl_2\}_2$. This unusual reaction implies the insertion of carbon dioxide into the titanium–nitrogen bond to form first a carbamato bridge which subsequently extrudes the isocyanate molecule OCNR' (Eq. 7-10) [36b].

The reaction of two equivalents of CO_2 with $Zr(\eta^5:\eta^1\text{-}C_5Me_4SiMe_2NtBu)Me_2$ affords $\{Zr(\eta^5:\eta^1\text{-}C_5Me_4SiMe_2NtBu)(\eta^2\text{-}O_2CMe)(\mu:\eta^2\text{-}O_2CMe)\}_2$ [38]. Using a 20-fold excess of CO_2 a sparingly soluble product is isolated and shown to be the dimeric bridged siloxide $\{Zr(\eta^5:\mu:\eta^1\text{-}C_5Me_4SiMe_2O)(\eta^2\text{-}O_2CMe)(\mu:\eta^2\text{-}O_2CMe)\}_2$. The conversion of the coordinated silylamido group into the bridged siloxide occurs again with concomitant elimination of OCNtBu (Eq. 7-11) [36].

$$(7\text{-}10)$$

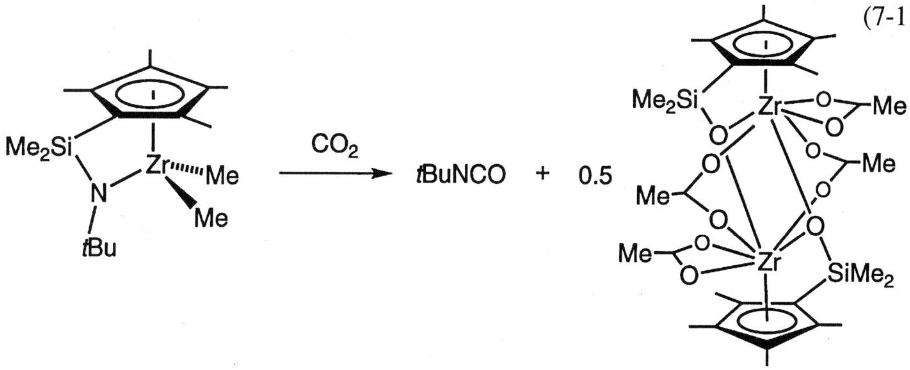

(7-11)

Irradiation of niobium and tantalum amides $M(\eta^5:\eta^1\text{-}C_5H_4SiMe_2NtBu)(NMe_2)_3$ causes Si–N bond cleavage followed by a peculiar rearrangement to give tert-butylimido complexes containing an $\eta^5\text{-}C_5H_4SiMe_2NMe_2$ ligand [34].

Whereas $Ti(\eta^5:\eta^1\text{-}C_5H_4SiMe_2NCH_2C_6H_3F_2\text{-}2,5)Cl_2$ is formed from $Ti(\eta^5:\eta^1\text{-}C_5H_4SiMe_2Cl)Cl_3$ and $Li(NHCH_2C_6H_3F_2\text{-}2,5)$ [25], the analogous reaction of the 2,6-difluorobenzyl derivative $Li(NHCH_2C_6H_3F_2\text{-}2,6)$ gives $Ti\{\eta^5:\eta^1\text{-}C_5H_4SiMe_2\text{-}NCH_2[C_6H_3F\text{-}6\text{-}(NHCH_2C_6H_3F_2\text{-}2,6)\text{-}2]\}Cl_2$ (Eq 7-12) [25]. The amido ligand is modified as a result of nucleophilic aromatic substitution at one *ortho*-fluorine atom by another molecule of 2,6-difluorobenzylamide. This substitution may be facilitated by a titanium–fluorine interaction. The lithiation of 2,5- or 2,6-difluorobenzylamime with *n*-butyllithium proceeds without complications.

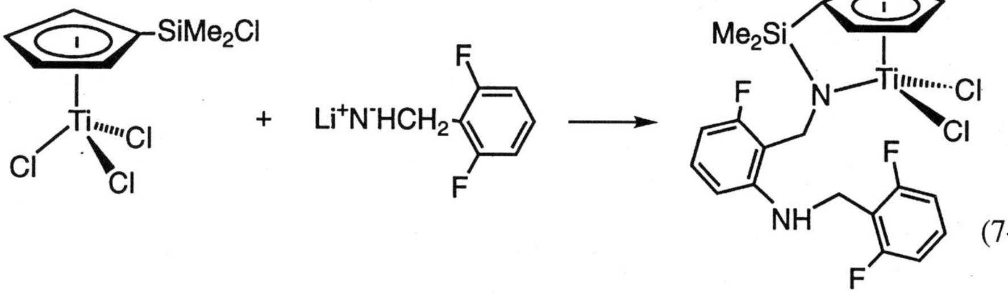

(7-12)

7.2.4   Structure

In the context of explaining the specific polymerization properties of group 4 metal complexes containing the linked amido–cyclopentadienyl ligand, single crystal structure analyses of numerous dichloro complexes of the type $M(\eta^5:\eta^1\text{-}C_5R_4ZNR')Cl_2$ have been performed. Some representative examples are compiled in Table 7-1. Fig. 7-4 shows the molecular structure of $Ti(\eta^5:\eta^1\text{-}C_5H_4SiMe_2NiPr)Cl_2$ [25] as a representative example.

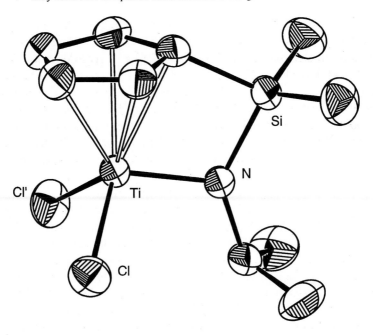

Figure 7-4. Molecular structure of Ti(η^5:η^1-C$_5$H$_4$SiMe$_2$NiPr)Cl$_2$

These complexes basically adopt a three-legged piano-stool or pseudotetrahedral geometry with the bifunctional amido–cyclopentadienyl ligand C$_5$R$_4$ZNR′ in addition to two terminal ligands X. The cyclopentadienyl ligand C$_5$R$_4$ is bonded in the usual η^5 fashion and the metal–ligand bond lengths are in the expected range. As a result of a certain constraint of the chelating ligand, the metal center is often unsymmetrically bound to the planar cyclopentadienyl ring. In the case of the dimethylsilanediyl or ethanediyl bridge (Z = SiMe$_2$ or CH$_2$CH$_2$), the bridging atom connected to the cyclopentadienyl periphery is displaced from the plane of the cyclic π system [49]. In the case of the longer bridge, Z = (CH$_2$)$_3$, this carbon atom is displaced away from the metal center and is located above the cyclopentadienyl ring [18, 25, 49].

The distance between the metal and the amido nitrogen is fairly sensitive to the nature of the metal fragment. For dichloro titanium(IV) complexes Ti(η^5:η^1-C$_5$R$_4$ZNR′)Cl$_2$, the bond lengths are around 1.9 Å. The metal–nitrogen distances are shorter than those observed for single bonds. In titanium derivatives containing titanium–nitrogen single bonds, the distance varies between 1.96 and 1.97 Å [50]. In bridged amido half-sandwich complexes of titanium, Ti–N distances shorter than 1.91 Å are sometimes observed. On the other hand these bond lengths are slightly larger than those found in nonbridged monocyclopentadienyl amido complexes. Thus, the Ti–N distance in Ti(η^5-C$_5$R$_5$)(NiPr$_2$)Cl$_2$ are 1.865(2) Å (R = H) [51] and 1.865(5) Å (R = Me) [51] compared to 1.901(3) Å (R = H) [26] and 1.908(6) Å (R

= Me) [52] in Ti(η^5:η^1-C$_5$R$_4$SiMe$_2$NtBu)Cl$_2$. An optimal overlap of the amido nitrogen p$_\pi$ orbital with the titanium d$_\pi$ orbital seems to be disturbed by the chelation. In the formally 18-electron diamido complex Ti(η^5:η^1-C$_5$H$_4$SiMe$_2$NtBu)(NMe$_2$)$_2$, a value of 1.972(4) Å is observed [26], reflecting a less pronounced double bond character based on a d$_\pi$–p$_\pi$ interaction. Also, in the titanium(II) diene complex Ti(η^5:η^1-C$_5$Me$_4$SiMe$_2$NtBu)(η^4-MeCH=CHCH=CHMe), a significantly longer metal–nitrogen bond of 2.007(4) Å is observed [47]. The sum of the bond angles around the appended amido nitrogen is practically always 360°. This is clearly due to a sp^2 hybridization caused by a d$_\pi$–p$_\pi$ interaction between the amido nitrogen and the metal center.

The Cp–M–N angles (Cp = centroid) in complexes M(η^5:η^1-C$_5$R$_4$ZNR′)X$_2$ are by 25–35° smaller than typical Cp–M–Cp angles in the corresponding 16-electron metallocene complexes M(η^5-C$_5$R$_5$)$_2$X$_2$, where angles vary between 125 and 135° [3]. Strain within the ligand system and an openness of the coordination sphere can be deduced from this finding. While there is a clear trend of increasing Cp–M–N angle with increasing length of the bridge Z, one should not overly emphasize a correlation of this geometric parameter with reactivity [32, 52]. The orbital interaction between the metal and a cyclopentadienyl ligand may not be directly comparable to that of an amido ligand. Chelation also results in a slight reduction of the metal–Cp distance as shown by the comparison of Ti(η^5:η^1-C$_5$H$_4$SiMe$_2$NtBu)Cl$_2$ (Ti–Cp: 2.019 Å) [26] and Ti(η^5-C$_5$H$_5$)(NHtBu)Cl$_2$ (Ti–Cp: 2.032 Å) [53].

Table 7-1. Some structural characteristics of half-sandwich complexes containing bridged amido–cyclopentadienyl ligands (Cp denotes the centroid of the cyclopentadienyl, N denotes the appended amido nitrogen)

Compound	Ref.	Cp–M–N (°)	M–Cp (Å)	M–N (Å)
Ti(η^5:η^1-C$_5$H$_4$SiMe$_2$NtBu)Cl$_2$	[26]	107.0	2.019	1.901(3)
Ti(η^5:η^1-C$_5$Me$_4$SiMe$_2$NtBu)Cl$_2$	[52]	107.6	2.030	1.908(6)
Ti{η^5:η^1-C$_5$Me$_4$(CH$_2$)$_2$NtBu}Cl$_2$	[52]	107.9	–	1.909(5)
Ti(η^5:η^1-C$_5$H$_4$SiMe$_2$NiPr)Cl$_2$	[25]	105.5	2.017	1.878(2)
Ti{η^5:η^1-C$_5$H$_4$(CH$_2$)$_2$NiPr}Cl$_2$	[18, 49]	104.4	2.008	1.864(2)
Ti{η^5:η^1-C$_5$H$_4$(CH$_2$)$_3$NiPr}Cl$_2$	[18, 49]	112.6	2.027	1.867(2)
Ti(η^5:η^1-C$_5$H$_4$SiMe$_2$NtBu)(NMe$_2$)$_2$	[26]	105.5	2.083	1.972(4)
Zr(η^5:η^1-C$_5$Me$_4$SiMe$_2$NtBu)Cl$_2$	[52]	102.0	2.163	2.056(6)
Zr(η^5:η^1-C$_5$Me$_4$SiMe$_2$NtBu)(NMe$_2$)$_2$	[26]	100.2	2.233	2.108(4)
Y(η^5:η^1-C$_5$Me$_4$SiMe$_2$NtBu){N(SiMe$_3$)$_2$}	[35]	97.7	2.300	2.184(7)

The most notable structural feature of the benzylamido derivative Ti(η^5:η^1-C$_5$H$_4$SiMe$_2$NCH$_2$C$_6$H$_3$F$_2$-2,5)Cl$_2$ in the solid state is the conformation of the benzylic group which is turned away from the metal center [25]. Single X-ray structural analysis of (−)-(S)-Ti(η^5:η^1-C$_5$R$_4$SiMe$_2$NCHMePh)Cl$_2$ (Fig. 7-5), the first optically active titanium complexes containing an amido–cyclopentadienyl ligand

[24], reveals a structure in which the phenyl group is arranged coplanarly to the C_5H_4 ring, turned away from the metal center. NOE measurements suggest restricted rotation about the nitrogen–methine-carbon bond based on a larger NOE between the methyl group of the stereogenic center and one of the two diastereotopic methyl groups of the silanediyl link.

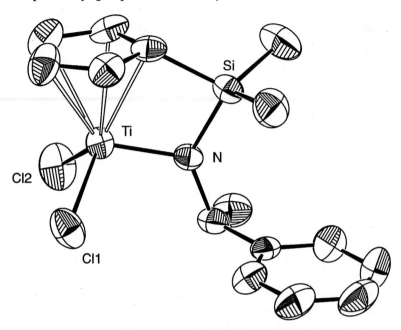

Figure 7-5. Molecular structure of $(-)$-(S)-Ti$(\eta^5{:}\eta^1$-$C_5R_4SiMe_2NCHMePh)Cl_2$ [24].

A similar characteristic feature for complexes containing a secondary amido substituent R′, is the tendency of the methine proton to orientate itself towards the metal center. This phenomenon is consistently found in *iso*-propylamido complexes such as Ti$(\eta^5{:}\eta^1$-$C_5H_4SiMe_2NiPr)Cl_2$ [25], Ti$\{\eta^5{:}\eta^1$-$C_5H_4(CH_2)_2NiPr\}Cl_2$ [18, 49] and Ti$\{\eta^5{:}\eta^1$-$C_5H_4(CH_2)_3NiPr\}Cl_2$ [18, 49].

Table 7-2. Structural characteristics of some complexes containing bridged amido–cyclopentadienyl ligands, where the amido substituent is a secondary alkyl group

Compound	Ref.	M–H$_{Methine}$ (Å)	M–N–C$_{Methine}$ (°)	δ(H$_{Methine}$) (ppm)
$(-)$-(S)-Ti$(\eta^5{:}\eta^1$-$C_5H_4SiMe_2NCHMePh)Cl_2$[24]		2.84	120.5(3)	6.54
Ti$(\eta^5{:}\eta^1$-$C_5H_4SiMe_2NiPr)Cl_2$	[25]	2.79	117.4(1)	5.69
Ti$\{\eta^5{:}\eta^1$-$C_5H_4(CH_2)_2NiPr\}Cl_2$	[18, 49]	2.67	114.2(2)	5.92
Ti$\{\eta^5{:}\eta^1$-$C_5H_4(CH_2)_3NiPr\}Cl_2$	[18, 49]	2.38	104.5(1)	6.57

Values for the M–N–$C_{Methine}$ angle and the M–$H_{Methine}$ distance are given in Table 7-2. A β-agostic interaction is not observed by ^{1}H NMR spectroscopy in complexes with Z = SiMe$_2$. However, it is detected in complexes containing less constrained, longer links (Z = (CH$_2$)$_3$). Here the Ti–N–$C_{Methine}$ angle and the Ti–$H_{Methine}$ distance for the β-agostic interaction are similar to those found in the unbridged cyclopentadienyl amido complex Ti(η^5-C$_5$H$_5$)(NiPr$_2$)Cl$_2$ [51], where one of the isopropyl groups is directed towards the metal center due to a β-agostic interaction of the methine proton, resulting in a Ti–$H_{Methine}$ distance of 2.25 Å and a Ti–N–$C_{Methine}$ angle of 101.4(2)°.

Finally, the upfield shifts for the C_{ipso} atom of the cyclopentadienyl group in the ^{13}C NMR spectra of complexes of the type Ti(η^5:η^1-C$_5$H$_4$SiMe$_2$NR′)Cl$_2$, compared to the values found for the nonlinked precursor Ti(η^5-C$_5$H$_4$SiMe$_2$Cl)Cl$_3$, are characteristic for the presence of a chelating amido group (Table 7-3).

Table 7-3. ^{13}C NMR chemical shifts for C_{ipso} atom in amido-bridged half-sandwich complexes and Ti(η^5-C$_5$H$_4$SiMe$_2$Cl)Cl$_3$

Complex	Ref.	δ(C_{ipso}) (ppm)
Ti(η^5-C$_5$H$_4$SiMe$_2$Cl)Cl$_3$	[36]	135.1
Ti(η^5:η^1-C$_5$H$_4$SiMe$_2$NtBu)Cl$_2$	[36]	110.0
Ti η^5:η^1-C$_5$H$_4$SiMe$_2$NiPr)Cl$_2$	[25]	108.1
Ti(η^5:η^1-C$_5$H$_4$SiMe$_2$NCH$_2$Ph)Cl$_2$	[25]	110.1
Ti(η^5:η^1-C$_5$H$_4$SiMe$_2$NCH$_2$C$_6$H$_3$F$_2$-2,5)Cl$_2$	[25]	109.7

7.3 Complexes with Polydentate Amido–Cyclopentadienyl Ligands

In order to attenuate the Lewis acidity of early transition metal centers, a new ligand system with a side chain incorporating an additional weak neutral donor site within the chelating amido-cyclopentadienyl ligand framework has been introduced [23]. Donor groups such as OMe or NMe$_2$ attached to the amido functionality offer new possibilities in tailoring the coordination sphere around a reactive transition metal center. The synthesis of group 4 metal complexes of such tridentate ligands can be achieved by following synthetic methodologies analogous to those for simpler amido substituents. Using the metathetical pathway, hexane-soluble tetramethylcyclopentadienyl titanium, zirconium and hafnium complexes M(η^5:η^1-C$_5$Me$_4$-SiMe$_2$NCH$_2$X)Cl$_2$ (M = Ti, Zr, Hf; X = CH=CH$_2$, CH$_2$OMe, CH$_2$NMe$_2$) [23, 55, 56] have been prepared (Scheme 7-10). The unsubstituted cyclopentadienyl-bridged titanium derivatives Ti(η^5:η^1-C$_5$H$_4$SiMe$_2$NCH$_2$X)Cl$_2$ (X = CH$_2$OMe, CH$_2$NMe$_2$) are accessible from Ti(η^5:η^1-C$_5$H$_4$SiMe$_2$Cl)Cl$_3$ [54].

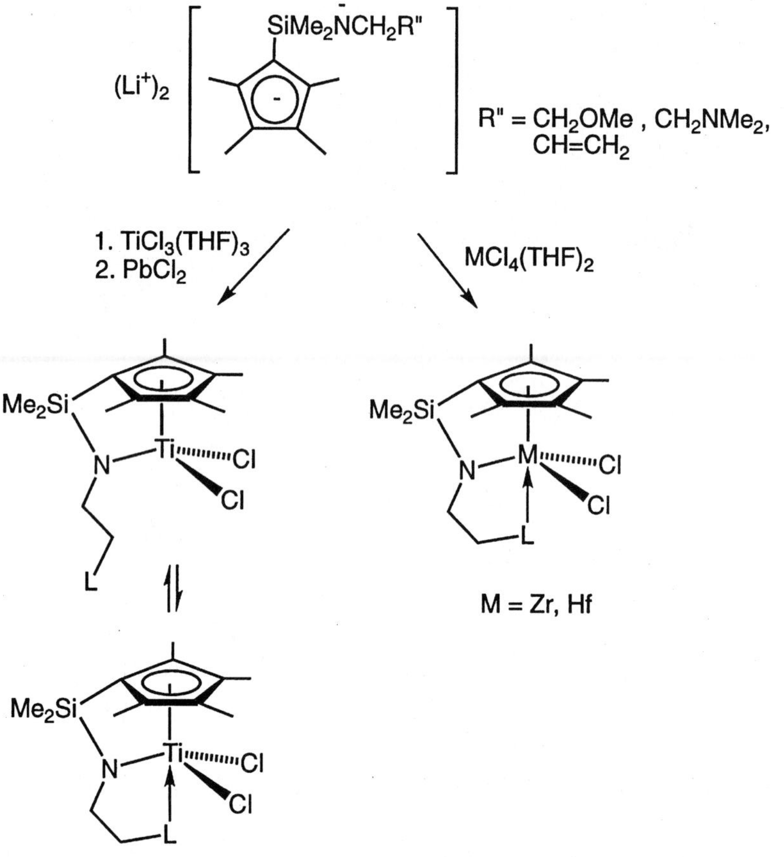

Scheme 7-10

Since the molecules contain a mirror plane irrespective of the coordination of the appended donor group in solution, the question whether the additional donor is rigidly bonded or in a fluxional manner cannot be decided by ^{1}H or ^{13}C NMR spectroscopy, including variable-temperature NMR spectra. The corresponding dimethyl complexes offer the possibility to record NOE measurements and to study the coordination mode. Whereas in titanium compounds Ti(η^5:η^1-C$_5$Me$_4$SiMe$_2$-NCH$_2$X)Me$_2$ a spatial relationship between the TiMe$_2$ signal and OMe and NMe$_2$ groups could not be established, homologous zirconium complexes Zr(η^5:η^1-C$_5$Me$_4$SiMe$_2$NCH$_2$X)Me$_2$ exhibit an NOE between the proton signals of the ZrMe$_2$ groups and the additional donor function OMe or NMe$_2$. Thus, the zirconium (and hafnium) complexes retain the intramolecular coordination in solution. This can be ascribed to the higher Lewis acidity of zirconium and hafnium compared to titanium. For example, the crystal structure of the complex Hf(η^5:η^1:η^1-C$_5$Me$_4$SiMe$_2$NCH$_2$CH$_2$OMe)Cl$_2$ shows a rather unusual trigonal bipyramidal configuration at the hafnium center with the OMe group occupying a site *trans* to the ring ligand (Fig. 7-6).

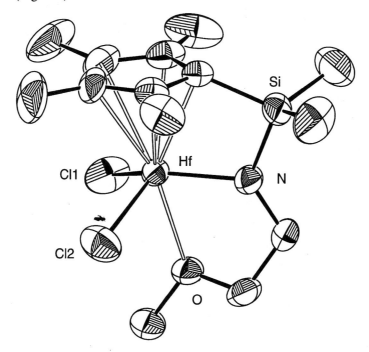

Figure 7-6. Molecular structure Hf(η^5:η^1:η^1-C$_5$Me$_4$SiMe$_2$NCH$_2$CH$_2$OMe)Cl$_2$ [55].

Dialkyl complexes M(η^5:η^1-C$_5$Me$_4$SiMe$_2$NCH$_2$CH$_2$OMe)R''$_2$ (M = Zr, Hf) can be synthesized with the alkyl ligands R'' including Me, Ph, CH$_2$SiMe$_3$ and CH$_2$Ph, but most remarkably, dialkyls with β-hydrogen atoms (R'' = Et, nPr, nBu) can be isolated in this series. The crystal structure of the thermally most stable (dec. > 90°C) complex Hf(η^5:η^1-C$_5$Me$_4$SiMe$_2$NCH$_2$CH$_2$OMe)(nBu)$_2$ shows two undistorted n-butyl groups [55].

NMR spectra of the planar chiral derivatives M{(η^5:η^1-C$_5$H$_3$$t$Bu)-SiMe$_2NCH_2CH_2NMe_2$}Cl$_2$(M = Ti, Zr) directly reveal the presence or absence of rigid intramolecular coordination of the NMe$_2$ function. A rigid coordination of the NMe$_2$ group prevents nitrogen inversion, giving rise to two nonequivalent methyl signals as in fact observed in the case of the zirconium complex Zr{(η^5:η^1-C$_5$H$_3$$t$Bu)SiMe$_2NCH_2CH_2NMe_2$}Cl$_2$. In the case of the titanium derivative, Ti{(η^5:η^1-C$_5$H$_3$$t$Bu)SiMe$_2NCH_2CH_2NMe_2$}Cl$_2$, only one signal for the NMe$_2$ group is observed in the temperature range of +80 to –80°C, suggesting that the coordination is not occurring or is fluxional on the NMR time scale [23]. A fluxional bonding mode (Eq. 7-13) is detected for the bulky dialkyl Zr{(η^5:η^1-C$_5$H$_3$$t$Bu)-SiMe$_2NCH_2CH_2NMe_2$}(CH$_2$SiMe$_3$)$_2$ with an activation energy of $\Delta G^{\ddagger}$(7°C) = 13.1 kcal mol^{-1} [56].

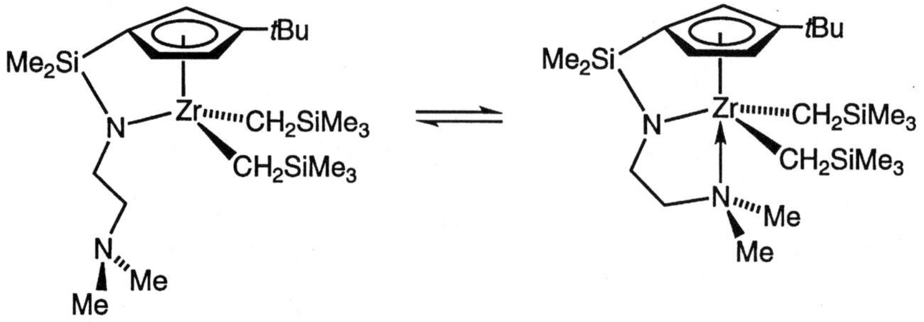

$$(7\text{-}13)$$

A tridendate ligand system with the additional donor attached directly to the cyclopentadienyl ring, $\{C_5H_4(CH_2CH_2NMe_2)\}SiMe_2NHtBu$, was prepared from 2-($N,N$-dimethylamino)ethylcyclopentadiene in a one-pot reaction (Scheme 7-11) [57]. Because of the 1,5-silyl migration two principal isomers are obtained in a 7 : 3 ratio for the 1,3- and 1,2-isomer. Treating $Zr(NMe_2)_4$ with the diprotonated ligand gives a mixture of the diamido complexes containing both the 1,2- and 1,3-substituted ligands (Scheme 7-11). Reaction with $(NMe_2H_2)Cl$ leads to the dichloro complex as the 1,3-isomer which can easily be methylated [58]. The complexation of this ligand with zirconium alkyls R_nZrCl_{4-n} ($R = CH_3$, $n = 3$; $R = CH_2SiMe_3$, $n = 2$) unexpectedly favors the formation of the 1,2 isomer (Eq. 7-14) [58]. Only one single resonance was observed for the NMe_2 group in the NMR spectra, suggesting fluxional coordination. The X-ray structure analysis of $Zr\{\eta^5:\eta^1:\eta^1\text{-}C_5H_3(CH_2CH_2NMe_2)SiMe_2NtBu\}ClMe$ reveals a weak donor bond between the zirconium and the amine ligand.

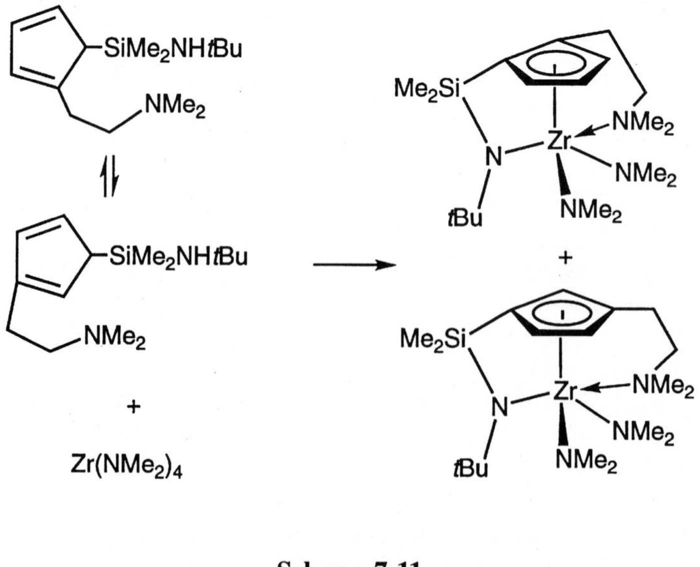

Scheme 7-11

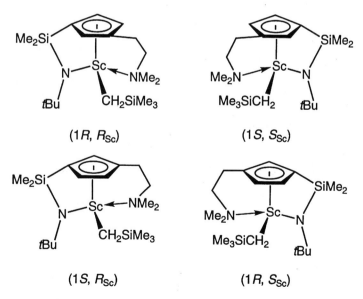

$$(7-14)$$

Reaction between $Sc(CH_2SiMe_3)_3(THF)_2$, obtained in situ from the reaction of $ScCl_3(THF)_3$ with $LiCH_2SiMe_3$, and the diprotonated ligand $\{C_5H_4(CH_2CH_2NMe_2)\}$-$SiMe_2NHtBu$, gives under alkane elimination the scandium complex $Sc\{\eta^5:\eta^1:\eta^1$-$C_5H_3(CH_2CH_2NMe_2)SiMe_2NtBu\}(CH_2SiMe_3)$. In this case the complex is produced as the 1,3-isomer [57]. Although two diastereomeric pairs of enantiomers are possible (Fig. 7-7), NMR data suggest the selective formation of only one. NOE measurements indicate the proximity of two ring protons to the methyl groups of the alkyl ligand, suggesting a molecule in which the alkyl ligand lies below the two adjacent ring protons, as in the enantiomers $(1S,R_{Sc})$ and $(1R,S_{Sc})$. The alkyl complex $Sc\{\eta^5:\eta^1:\eta^1$-$C_5H_3(CH_2CH_2NMe_2)SiMe_2NtBu\}(CH_2SiMe_3)$ undergoes hydrogenolysis to give dimeric hydride $\{Sc\{\eta^5:\eta^1:\eta^1$-$C_5H_3(CH_2CH_2NMe_2)SiMe_2NtBu\}(\mu$-$H)\}_2$ as a mixture of diastereomers. The alkyl loses $SiMe_4$ after heating to 70°C for 6 days, resulting in the formation of a mixture of dimeric products. For each, one diastereomer is characterized by X-ray crystallography [57].

$(1R, R_{Sc})$

$(1S, S_{Sc})$

$(1S, R_{Sc})$

$(1R, S_{Sc})$

Figure 7-7. Possible diastereomers for
$Sc\{\eta^5:\eta^1$-$C_5H_3(CH_2CH_2NMe_2)SiMe_2NtBu\}(CH_2SiMe_3)$ [57].

The hydroboration of the bridged amido half-sandwich complex containing a 2-pro-penyl group attached to the cyclopentadienyl ring, $Zr\{\eta^5:\eta^1\text{-}C_5H_3(C_3H_5)SiMe_2NtBu\}Cl_2$, with $[(C_6F_5)_2BH]_n$ [59], leads to a product of a highly regioselective 1,2 addition (Eq. 7-15). The ^{11}B NMR chemical shift is in agreement with data found for neutral, three-coordinate boron centers, suggesting the absence of any association between the boron center and the ligands on zirconium.

Reaction of the prolinol-functionalized ligand $(C_5Me_4H)SiMe_2OCH_2\{(-)\text{-}(S)\text{-}C_4H_7NH\}$ with $Zr(NMe_2)_4$ gives a linked amido complex that, upon treatment with $(NMe_2H_2)Cl$, forms a diastereomeric mixture of the N-protonated complexes (Eq. 7-16) [60].

$$\text{(7-15)}$$

$$\text{(7-16)}$$

7.4   Complexes with two Amido–Cyclopentadienyl Ligands

During the synthesis of half-sandwich complexes containing the linked amido–cyclopentadienyl ligands, it was occasionally noted that two of the dianionic ligands are coordinated at a metal center to give compounds of the composition $[M(\eta^5:\eta^1\text{-}C_5R_4ZNR')_2]$ [23, 27, 54, 55]. Evidently, the presence of excess ligand results in the double coordination of the linked amido–cyclopentadienyl ligands. Group 4 metals give neutral complexes with a characteristic C_2-symmetric metallocene structure, exemplified by the crystal structure of $Zr(\eta^5:\eta^1\text{-}C_5H_4\text{-}SiMe_2NPh)_2$ [31]. Group 3 metals give rise to anionic complexes of the type $[M(\eta^5:\eta^1\text{-}C_5R_4ZNR')_2]^-$. Such complexes are easily isolated when tridentate ligands of the type $C_5R_4SiMe_2NCH_2CH_2X$ (X = OMe, NMe_2) are used [61]. The reaction of anhydrous yttrium or lutetium trichloride with $Li_2(C_5R_4SiMe_2NCH_2CH_2X)$ produces, in high yields, hydrocarbon-soluble heterobimetallic complexes of the composition $Li[M(\eta^5:\eta^1-C_5R_4SiMe_2NCH_2CH_2X)_2]$ (M = Y, Lu) (Eq. 7-17). As shown by the crystal structure of $Li[Y(\eta^5:\eta^1-C_5Me_4SiMe_2NCH_2CH_2OMe)_2]$ (Fig. 7-8), the metallocene unit tightly coordinates the lithium ion with the ligand side chains. The amido nitrogen atoms exhibit a distorted tetrahedral geometry and a significantly longer yttrium–nitrogen bond than in $Y(\eta^5:\eta^1-C_5Me_4SiMe_2NtBu)-\{N(SiMe_3)_2\}$ [35]. When a *tert*-butylcyclopentadienyl ligand is used for C_5R_4, three diastereomeric pairs of enantiomers are expected (Fig. 7-9). Under kinetic control the sterically more constrained (*R,S*) isomer is preferentially formed, that converts into the thermodynamically more stable C_2-symmetric (*R,R*) isomer in donor solvents such as THF at room temperature (Eq. 7-18). The interconversion is first-order and activation parameters indicate a fairly ordered transition state. Intriguingly, these complexes are highly efficient polymerization catalysts for ε-caprolactone, producing poly(ε-caprolactone) of high molecular weight ($M_n > 30000$) and moderate polydispersity ($M_w/M_n < 2.0$) [62].

$$ (7\text{-}17) $$

X = OMe , NMe₂,

$$ (7\text{-}18) $$

(*P,R,S*)   (*P,R,R*)

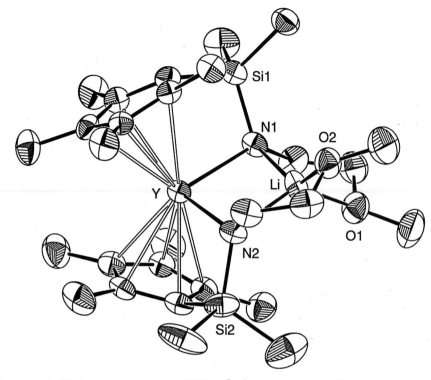

Figure 7-8. Molecular structure of Li[Y(η^5:η^1-C$_5$Me$_4$SiMe$_2$NCH$_2$CH$_2$OMe)$_2$] [61].

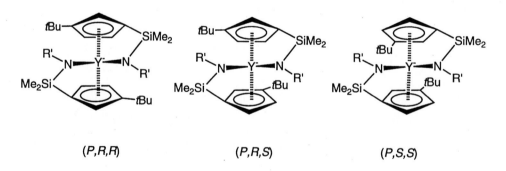

Figure 7-9. Possible diastereomers for Li{Y{(η^5:η^1-C$_5$H$_3$*t*Bu)SiMe$_2$NCH$_2$CH$_2$OMe}$_2$]. R′ denotes CH$_2$CH$_2$OMe coordinated to the lithium ion.

7.5 Complexes with Linked Alkoxo–Cyclopentadienyl Ligands

Compared to amido-bridged half-sandwich derivatives, complexes containing a cyclopentadienyl with an alkoxo-functionalized side chain (X = O in Fig. 7-1) are relatively scarce. The first example $Ti\{\eta^5:\eta^1-C_5Me_4(CH_2)_3O\}Cl_2$ was formed by thermolysis of the titanium ylide $Ti\{\eta^5-C_5Me_4(CH_2)_3OMe\}Cl_2(CHPPh_3)$ at 150°C (Eq 7-19) [63].

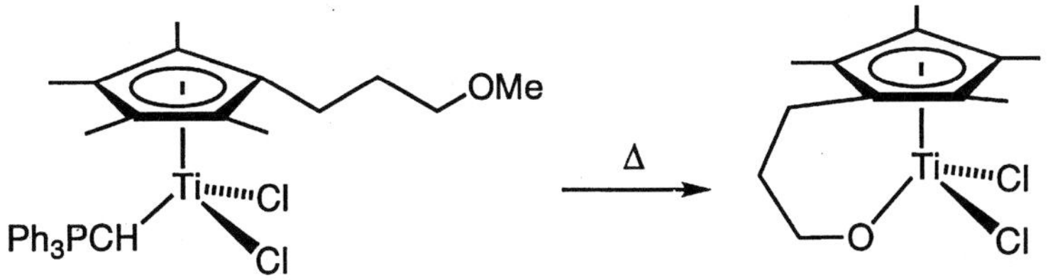

$$(7-19)$$

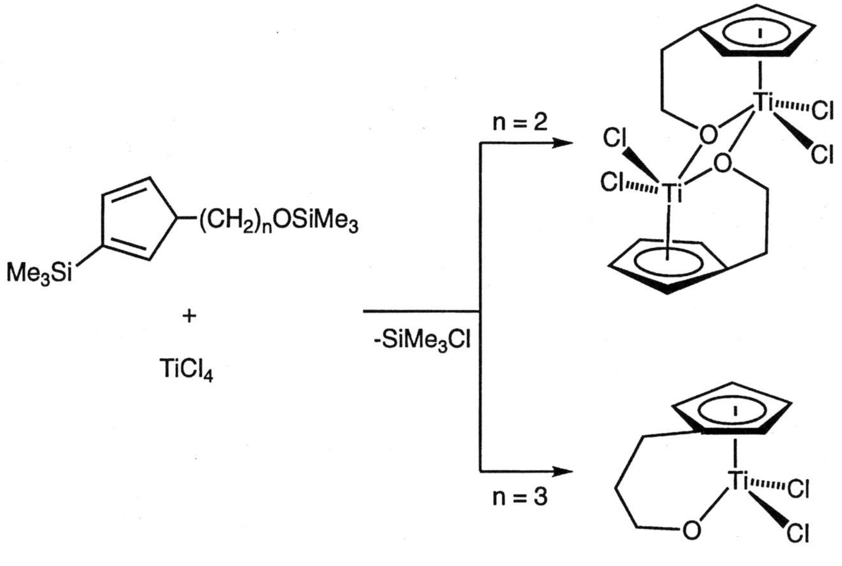

Scheme 7-12

More rationally, titanium complexes with a bridged alkoxo cyclopentadienyl ligand $Ti\{\eta^5:\eta^1-C_5Me_4(CH_2)_nO\}Cl_2$ are prepared in quantitative yields by reacting the ligands as trimethylsilyl ethers, $C_5H_4(SiMe_3)(CH_2)_nOSiMe_3$, with $TiCl_4$ (Scheme 7-12) [64]. The length of the bridge connecting the chelating alkoxo group and the cyclopentadienyl moiety influences the molecular structure dramatically: the $(CH_2)_3$ chain gives rise to a monomeric complex, whereas the shorter $(CH_2)_2$ chain affords a centrosymmetric dimer in which the alkoxo functions bridge the two titanium centers [64]. Alkoxo-functionalized fluorenyl complexes of zirconium of the type $Zr\{\eta^5:\eta^1\text{-}C_{13}H_8ZO\}Cl_2(THF)_2$ (Z = $(CH_2)_2$, $(CHR)_2$) have also been reported [65].

The monomeric titanium fulvene complex $Ti(\eta^5\text{-}C_5Me_5)\{\eta^7\text{-}C_5Me_3(CH_2)_2\}$, generated by thermolysis of paramagnetic titanium(III) complex $Ti(\eta^5\text{-}C_5Me_5)_2R$ (R = Me, Et, Pr), gives an alkoxo bridging titanium complex after reacting with acetophenone (Eq. 7-20) [66].

$$(7\text{-}20)$$

Likewise, the fulvene complex $Ti(C_5H_4Me)(Ph)(\eta^6\text{-}C_5H_4CH_2)$, generated by thermolysis of the diphenyl derivative $Ti(\eta^5\text{-}C_5H_4Me)_2Ph_2$, undergoes insertion with various aldehydes and ketones. NOE measurements indicate that acetaldehyde gives one diastereomer (> 95%) in which the methyl substituent and the titanium-bound phenyl ligand are disposed *cis* at the metallacyclic ring [67].

A remarkable formation of a chelate complex is observed during thermolysis of a titanium complex containing a tetramethylcyclopentadienyl ligand with a 2,6-dimethoxyphenyl group (Eq. 7-21). Similarly, an *ortho*-methoxybenzyl functionalized titanocene $Ti(\eta^5\text{-}C_5H_4CEt_2C_6H_4OMe)(\eta^5\text{-}C_5H_5)Cl_2$ is converted in the presence of LiBr into a benzyloxo bridged complex $Ti(\eta^5:\eta^1\text{-}C_5H_4CEt_2C_6H_4O)(\eta^5\text{-}C_5H_5)Cl$ following cleavage of the methoxy group (Eq. 7-22) [68].

$$(7\text{-}21)$$

$$(7\text{-}22)$$

The complexation of two deprotonated ligands $Li_2\{(C_5H_3tBu)CH_2CH_2O\}$ with $ZrCl_4$ may result in three enantiomeric pairs of diastereomeric $Zr\{(\eta^5:\eta^1\text{-}C_5H_3tBu)CH_2CH_2O\}_2$ due to the presence of two chiral planes and the metallocene helicity (*cf.* Sec. 7.4). However, only one enantiomeric pair of one diastereomer, *viz.* (*P,R,R*) and (*M,S,S*), is formed. The alkylation of $Zr\{(\eta^5:\eta^1\text{-}C_5H_3tBu)CH_2CH_2O\}_2$ with an excess of methyl triflate results in the formation of the unsymmetric complex $Zr\{\eta^5:\eta^1\text{-}C_5H_3tBu)CH_2CH_2O\}\{(\eta^5\text{-}C_5H_3tBu)CH_2CH_2OMe\}OTf$. The exchange of one alkoxy against a triflato ligand increases the electrophilicity of the zirconium center blocking further electrophilic attack of MeOTf at the remaining bridging alkoxo group. The stereospecific formation of *rac*-isomer $Zr\{(\eta^5\text{-}C_5H_3tBu)CH_2CH_2OSiMe_3\}_2Cl_2$ was achieved by reaction with chlorotrimethylsilane (Scheme 7-13) [69].

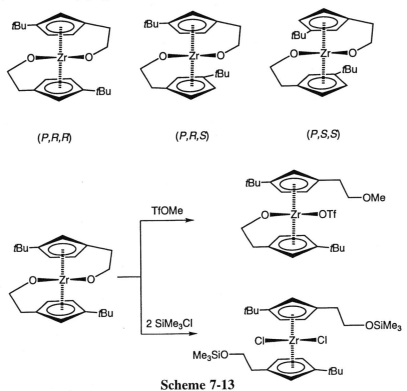

(*P,R,R*)   (*P,R,S*)   (*P,S,S*)

Scheme 7-13

7.6 Catalysis

7.6.1 Polymerization of Ethylene, α-Olefins and Dienes

As was mentioned in the introduction, the great interest in the group 4 metal complexes containing the linked amido–cyclopentadienyl ligand, sometimes referred to as 'constrained geometry catalysts' [52], stems from their great potential as a new generation of olefin polymerization catalysts. In particular, the possibility of producing polyolefins with new rheological properties and good processability at temperatures as high as 160°C, has stimulated frantic activity in synthesizing and testing such complexes.

The dimeric scandium hydride complex $\{Sc(\eta^5:\eta^1C_5Me_4SiMe_2NtBu)(PMe_3)\}_2(\mu$-H$)_2$ was found to be capable of catalyzing the aspecific oligomerization of the α-olefins propylene, 1-butene, and 1-pentene [12]. Although this presents an advantage over the scandocenes, the polymerization occurs slowly and to relatively low molecular weights (M_n = 4000 for poly(1-butene), 3000 for poly(1-pentene)). Addition of PMe$_3$ results in slower α-olefin polymerization rates. The dimeric Lewis base free *n*-alkyl complexes $\{Sc(\eta^5:\eta^1$-C$_5$Me$_4$SiMe$_2$NtBu$)\}_2(\mu$-CH$_2$CH$_2$R$)_2$ (R = Me, Et) are shown to be more active catalyst precursors than the hydrido complex, and polymers with higher molecular weights (M_n = 6000 for poly(1-pentene), 9600 for polypropylene) are obtained. Low-temperature ^{13}C NMR spectroscopic studies of the model complexes $\{Sc(\eta^5:\eta^1$-C$_5$Me$_4$SiMe$_2$NtBu)(PMe$_3$)(CH$_2$CHMeCH$_2$CH$_2$Me)$\}$ and $\{Sc(\eta^5:\eta^1$-C$_5$Me$_4$SiMe$_2$-NtBu)(PMe$_3$)$\{^{13}$CH$_2$CH(^{13}CH$_3$)$_2\}$ indicate that one PMe$_3$ adduct is in equilibrium with only one PMe$_3$-free species. By its symmetry it is concluded to be the monomeric 12-electron alkyl complex Sc($\eta^5:\eta^1$-C$_5$Me$_4$SiMe$_2$-NtBu)R″ (Scheme 7-14).

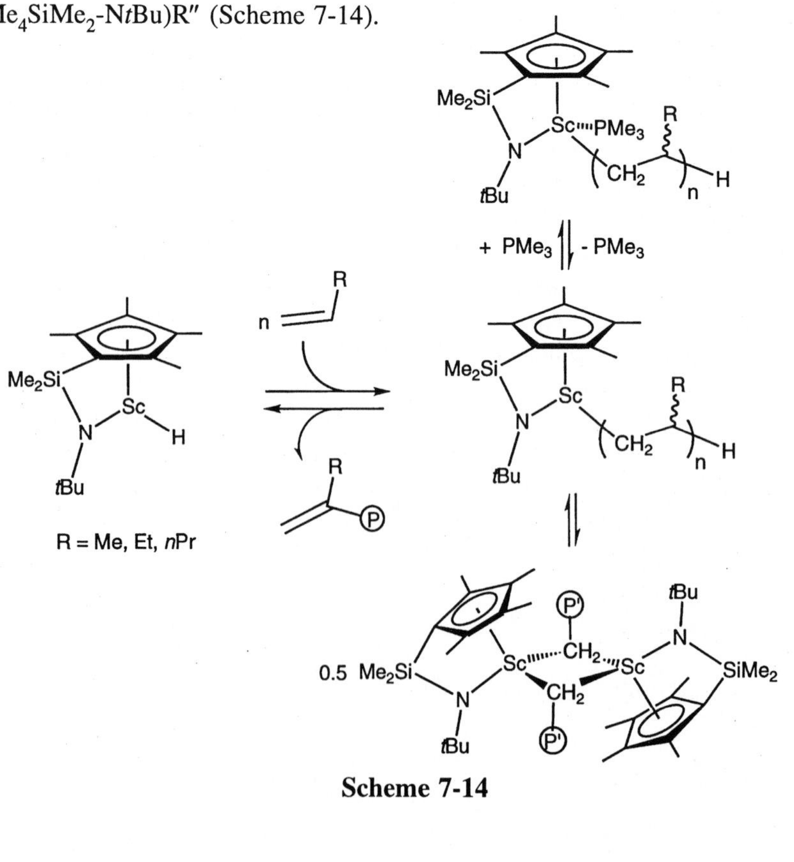

Scheme 7-14

Methylaluminoxane-activated titanium complexes $Ti(\eta^5:\eta^1\text{-}C_5Me_4ZNR')Cl_2$ appear to be commercially utilized catalysts for olefin polymerization. Preliminary activity–structure relationships [52] show that these catalysts form, depending on the nature of the ligand framework, high molecular weight polyethylene with long-chain branching, resulting from the incorporation of oligoethylene chains formed by β-hydride elimination. Also, superior properties as copolymerization catalysts were recognized, allowing efficient and uniform incorporation of higher α-olefin such as 1-octene to give low-density polyethylene with thermoelastic properties. This pronounced ability of these catalysts to incorporate higher olefins was ascribed to the more open coordination sphere, compared to conventional metallocene systems [52, 70]. Zirconium systems seem to be less active [52, 71], while the nature of the ligand substituents R in the C_5R_4 ring, R' of the amido substituent NR', and foremost the length of the bridge Z was found to influence the catalytic activity. Since catalyst precursors with the shorter bridge Z = $SiMe_2$ and CH_2CH_2 were found to give the best polymerization characteristics, the bite angle of chelating ligand (angle Cp–Ti–N, see Sec. 7.2.4) was proposed to be a crucial geometrical criterion for a catalyst to perform well [32]. It is obvious though that electronic properties imparted by R, R' and Z are important as well, since a peralkylated cyclopentadienyl ring also appears to be more preferable. Finally, the *tert*-butyl group as amido-substituent R' seems to be optimal [52].

In contrast to the 14-electron group 4 metallocenium polymerization catalysts $[M(\eta^5\text{-}C_5R_5)_2R'']^+$, but in analogy to the scandium catalysts described above, the 12-electron alkyl cation of the type $[M(\eta^5:\eta^1\text{-}C_5R_4ZNR')R'']^+$ is thought to be the active species [72]. It can be generated either by the reaction with methylaluminoxane, or by reacting a dialkyl complex with a Lewis acid such as $B(C_6F_5)_3$. Structurally characterized complexes include $[Zr(\eta^5:\eta^1\text{-}C_5Me_4\text{-}SiMe_2NtBu)Me]^+[FAl(C_{12}F_9)_3]^-$ [71b]. Variable-temperature NMR studies have revealed an activation barrier of 19.3(4) kcal mol^{-1} for the ion pair reorganization and symmetrization (Eq. 7-23), caused by the back-skip of the alkyl group in $[Zr(\eta^5:\eta^1\text{-}C_5Me_4SiMe_2NtBu)Me]^+$ $[MeB(C_6F_5)_3]^-$ [71]. The position of the equilibrium between solvent-separated ion pair and contact ion pair for the complexes $[Zr\{\eta^5:\eta^1\text{-}C_5H_4(CH_2)_nNR'\}(CH_2Ph)]^+[PhCH_2B(C_6F_5)_3]^-$ has been found to be strongly dependent on the bridge length n and the amido substituent R' [32]. Diene complexes $Ti(\eta^5:\eta^1\text{-}C_5Me_4SiMe_2NR')(1,3\text{-diene})$, described in Sec. 7.2, may serve as precursors for cationic species [47], since by adding $B(C_6F_5)_3$, a zwitterionic species form, as was also observed in a zirconocene catalyst generated from $Zr(\eta^5\text{-}C_5H_5)_2(\eta^4\text{-}1,3\text{-diene})$ and $B(C_6F_5)_3$ [73].

$$(7\text{-}23)$$

$$A^- = MeB^-(C_6F_5)_3,\ B^-(C_6F_4X)_4,\ FAl^-(C_{12}F_9)_3$$

Polymerization of propylene by methylaluminoxane-activated $Ti(\eta^5:\eta^1-C_5Me_4-SiMe_2NtBu)Cl_2$ results in atactic polypropylene with some slightly syndiotactic preference ([rrrr] = 20%) and 2–5% regioirregularity caused by 2,1-insertion. Chiral amido substituent R′ in one diastereomer of $Ti(\eta^5:\eta^1-C_9H_6SiMe_2-NCHMePh)Cl_2$ does not increase the isospecificity [33]. Quite intriguingly, the fluorenyl-based system $Zr(\eta^5:\eta^1-C_{13}H_8SiMe_2NtBu)Cl_2$ [16] is thought to function as a syndiospecific catalyst when activated with methylaluminoxane and as an isopecific catalyst when activated with a Lewis acid. It was suggested that a tight ion pair may force the back-skip of the growing polypropylene chain that takes place prior to the next insertion, a process which would eventually lead to isospecificity [74]. As is evident from the significant regioirregularity and low stereoselectivity, the replacement of one cyclopentadienyl ligand in an *ansa*-metallocene by an amido group opens up the reaction site considerably. Cyclopolymerization of 1,5-hexadiene is also achieved by methylaluminoxane-activated $Ti(\eta^5:\eta^1-C_5Me_4SiMe_2NtBu)Cl_2$. The polymer obtained contains randomly distributed *cis*- and *trans*-cyclopentane rings [75]. The chromium(III) complex $Cr(\eta^5:\eta^1-C_5Me_4-SiMe_2NtBu)CH_2SiMe_3$ is capable of polymerizing ethylene, but not 1-hexene, which is head-to-tail dimerized with a small amount of internal olefin formed [28].

The efficient copolymerization of ethylene with styrene became possible only with the use of metallocene catalysts. Conventional heterogeneous Ziegler catalysts normally induce homopolymerization of each of the monomers. Metallocene catalysts allowed poly(ethene-*co*-styrene) to be formed, but still suffer from low incorporation of styrene. One of the best copolymerization catalysts consists of the prototypical titanium complex $Ti(\eta^5:\eta^1-C_5Me_4SiMe_2NtBu)Cl_2$, that along with high activity results in the production of ethylene–styrene copolymer with up to 30 mol% styrene. By studying the influence of various substituents R and R′ in $Ti(\eta^5:\eta^1-C_5R_4SiMe_2NR')Cl_2$, it has been shown that both activity and the incorporation of styrene are sensitive to the nature of R and R′ [76]. Moreover, the maximum amount of styrene incorporated appears to be 66 mol% since, according to ^{13}C NMR spectroscopic microstructure analysis of the copolymer, no more than two styrene units coupled in a tail-to-tail manner are present in the chain [77]. This can be explained by the inability of styrene to be inserted in the polymer chain once a secondary insertion of the ultimate styrene has occurred, following the (regular) primary insertion of the penultimate unit. This mechanistic proposal is supported by the isolation of the scandium complex $\{Sc(\eta^5:\eta^1-C_5Me_4SiMe_2NtBu)-(PMe_3)\}(CHPhCH_2CH_2CHDPh)$, from $\{Sc(\eta^5:\eta^1-C_5Me_4SiMe_2NtBu)(PMe_3)\}_2(\mu-D)_2$, that cannot insert further styrene (Eq. 7-24) [12].

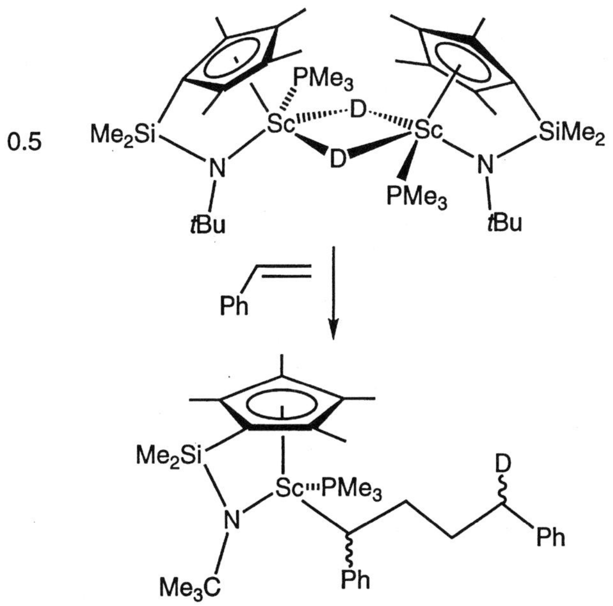

(7-24)

7.6.2 Hydrogenation and Hydroboration

Eclipsed by the overwhelming interest in utilizing the complexes of linked amido–cyclopentadienyl ligands in olefin polymerizations, only two other applications of this class have so far been reported in the open literature.

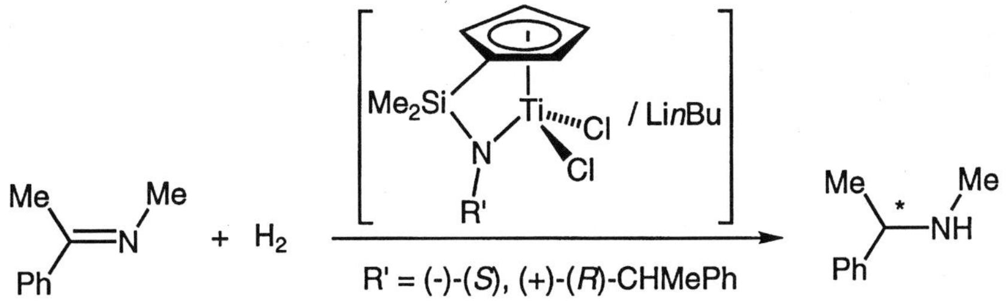

(7-25)

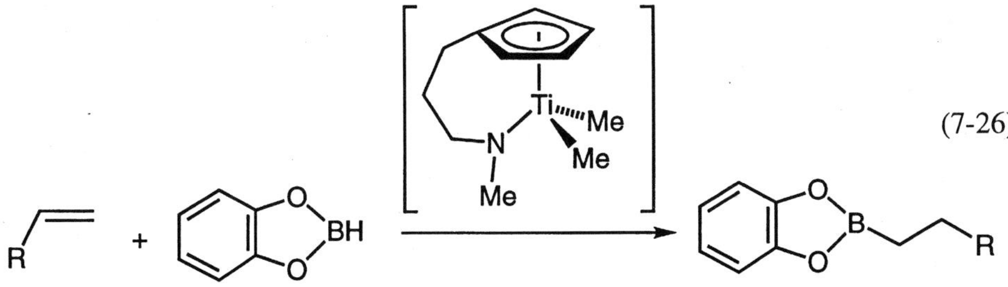

(7-26)

The enantioselective catalytic hydrogenation of substituted olefins and in particular of imines have been successfully achieved using reductively activated chiral Brintzinger-type titanocene derivatives [78]. When optically active titanium complexes of the type $Ti(\eta^5:\eta^1\text{-}C_5R_4SiMe_2NR')Cl_2$ (R = H, Me; R' = CHMePh) are treated with *n*-butyllithium, similarly active hydrogenation catalysts for imines are generated, albeit with low enantioselectivity (Eq. 7-25) [24]. Analogous to the titanocene system, the catalytically active species is presumed to be a titanium(III) hydrido species, possibly of the type $Ti(\eta^5:\eta^1\text{-}C_5R_4SiMe_2NR')H$. Hydroboration of alkenes with catecholborane is known to be catalyzed by metallocenes of group 3 metals [78]. The complexes $Ti\{\eta^5:\eta^1\text{-}C_5H_4(CH_2)_3NMe\}Me_2$ and $Zr\{\eta^5:\eta^1\text{-}C_5H_4\text{-}(CH_2)_3NMe\}X_2$ (X = BH_4, CH_2Ph) are moderately active as catalysts for the hydroboration of 1-hexene using catecholborane (Eq. 7-26) [79].

7.7 Conclusion

The introduction of the linked amido–cyclopentadienyl ligand as a replacement for the bridged bis(cyclopentadienyl) ligand in *ansa*-metallocenes has led to a new class of outstanding olefin polymerization catalysts. It is clear that the complexes of the type and $M(\eta^5:\eta^1\text{-}C_5R_4ZNR')L'_mX'_n$ exhibit some properties intermediate to those of *ansa*-metallocenes and half-sandwich complexes. However, many coordination chemical aspects have not been fully explored yet and it remains to be seen whether this ligand framework provides a metal template as versatile as the ubiquitous metallocene unit $M(\eta^5\text{-}C_5R_5)_2$. The linked amido–cyclopentadienyl complexes may be also considered as a hybrid between *ansa*–metallocenes and complexes containing chelating bis(amido) ligands derived from the diamines R'HN(Z)NHR'. The latter type of complexes has only just begun to emerge as yet another type of replacement for the bis(cyclopentadienyl) ligand and as homogeneous olefin catalyst with novel features complementary to both metallocenes and linked amido–cyclopentadienyl complexes [80].

References

[1]   J. Okuda, *Angew. Chem.* **1992**, *104*, 49; *Angew. Chem. Int. Ed. Engl.* **1992**, *31*, 47; J. Okuda, *Nachr. Chem. Tech. Lab.* **1993**, *41*, 8; R. Mülhaupt, *Nachr. Chem. Tech. Lab.* **1993**, *41*, 1341; A. M. Thayer, *Chem. Eng. News* **1995,** Sept. 11, 15; R. Mülhaupt, B. Rieger, *Chimia* **1996**, *50*, 10; W. Kaminsky, *Macromol. Chem. Phys.* **1996**, *197*, 3907.
[2]   H.-H. Brintzinger, D. Fischer, R. Mülhaupt, B. Rieger, R. Waymouth, *Angew. Chem.* **1995**, *107*, 1255, *Angew. Chem. Int. Ed. Engl.* **1995**, *34*, 1652; P. C. Möhring, N. J. Coville, *J. Organomet. Chem.* **1994**, *479*, 1.

[3] J. W. Lauher, R. Hoffmann, *J. Am. Chem. Soc.* **1976**, *98*, 1729.

[4] W. E. Piers, P. J. Shapiro, E. E. Bunel, J. E. Bercaw, *Synlett* **1990**, *2*, 74.

[5] M. L. H. Green, *J. Organomet. Chem.* **1995**, *500*, 127.

[6] J. Okuda, *Comments Inorg. Chem.* **1994**, *16*, 185.

[7] P. T. Wolczanski, *Polyhedron* **1995**, *14*, 3335.

[8] D. S. Williams, M. H. Schofield, R. R. Schrock, *Organometallics* **1993**, *12*, 4560; D. M. Antonelli, M. L. H. Green, P. Mountford, *J. Organomet. Chem.* **1992**, *438*, C4.

[9] J. C. Flores, J. C. W. Chien, M. D. Rausch, *Organometallics* **1994**, *13*, 4140; P. Jutzi, U. Siemeling, *J. Organomet. Chem.* **1995**, *500*, 175; W. A. Herrmann, M. J. Morawietz, T. Priermeier, K. Mashima, *J. Organomet. Chem.* **1995**, *486*, 291.

[10] (a) J. C. Stevens, F. J. Timmers, G. W. Rosen, G. W. Knight, S. Y. Lai (Dow Chemical Co.), *European Patent Application, EP 0 416 815 A2*, **1991** (filed August 30, 1990). (b) J. A. Canich (Exxon Chemical Co.), *European Patent Application, EP 0 420 436 A1*, **1991** (filed September 10, 1990).

[11] P. J. Shapiro, J. E. Bercaw, Abstracts of Papers, 195th Meeting of the American Chemical Society, Toronto, Canada, American Chemical Society, Washington, D. C., 1988, INOR 584; P. J. Shapiro, Ph. D. Dissertation, California Institute of Technology, 1990; P. J. Shapiro, E. E. Bunel, W. P. Schaefer, J. E. Bercaw, *Organometallics* **1990**, *9*, 867.

[12] P. J. Shapiro, W. D. Cotter, W. P. Schaefer, J. A. Labinger, J. E. Bercaw, *J. Am. Chem. Soc.* **1994**, *116*, 4623.

[13] J. Okuda, *Topics Current Chem.* **1991**, *160*, 97.

[14] J. Okuda, *Chem. Ber.* **1990**, *123*, 1649.

[15] F. Amor, J. Okuda, *J. Organomet. Chem.* **1996**, *520*, 245.

[16] J. Okuda, F. J. Schattenmann, S. Wocadlo, W. Massa, *Organometallics* **1995**, *14*, 789.

[17] A. K. Hughes, A. Meetsma, J. H. Teuben, *Organometallics* **1993**, *12*, 1936.

[18] P.-J. Sinnema, L. van der Veen, A. L. Spek, N. Veldman, J. H. Teuben, *Organometallics* **1997**, *16*, 4245.

[19] H. V. R. Dias, Z. Wang, S. G. Bott, *J. Organomet. Chem.* **1996**, *508*, 91.

[20] W. P. Schaefer, W. D. Cotter, J. E. Bercaw, *Acta Crystallogr.* **1993**, *C49*, 1489.

[21] This has long been known in titanocene chemistry: J. E. Bercaw, R. H. Marvich, L. G. Bell, H. H. Brintzinger, *J. Am. Chem. Soc.* **1972**, *94*, 1219; I. F. Urazowski, V. I. Ponomaryov, O. G. Ellert, I. E. Nifant'ev, D. A. Lemenovski, *J. Organomet. Chem.* **1988**, *356*, 181.

[22] G. A. Luinstra, J. H. Teuben, *J. Chem. Soc., Chem. Commun.* **1990**, 1470.

[23] K. E. du Plooy, U. Moll, S. Wocadlo, W. Massa, J. Okuda, *Organometallics* **1995**, *14*, 3129.

[24] J. Okuda, S. Verch, T. P. Spaniol, R. Stürmer, *Chem. Ber.* **1996**, *129*, 1429.

[25] J. Okuda, T. Eberle, T. P. Spaniol, *Chem. Ber.* **1997**, *130*, 209.

[26] D. W. Carpenetti, L. Kloppenburg, J. T. Kupec, J. L. Petersen, *Organometallics* **1996**, *15*, 1572.

[27] U. Böhme, K. H. Thiele, *J. Organomet. Chem.* **1994**, *472*, 39.

[28] Y. Liang, G. P. A. Yap, A. L. Rheingold, K. H. Theopold, *Organometallics* **1996**, *15*, 5284.

[29] G. Chandra, M. F. Lappert, *J. Chem. Soc. A* **1968**, 1940; M. F. Lappert, P. B. Power, A. R. Sanger, R. C. Srivastava, Metal and Metalloid Amides; Ellis Horwood: Chicester, West Sussex, U. K. 1980; G. M. Diamond, S. Rodewald, R. F. Jordan, *Organometallics* **1995**, *14*, 5.

[30] A. K. Hughes, S. M. B. Marsh, J. A. K. Howard, P. S. Ford, *J. Organomet. Chem.* **1997**, *528*, 195.

[31] W. A. Herrmann, M. J. A. Morawietz, *J. Organomet. Chem.* **1994**, *482*, 169; W. A. Herrmann, M. J. A. Morawietz, T. Priermeier, *Angew. Chem.* **1994**, *106*, 2025; *Angew. Chem., Int. Ed. Engl.* **1994**, *33*, 1946; M. J. A. Morawietz, Dissertation, Technische Universität München, 1995.

[32] P.-J. Sinnema, K. Liekelema, O. K. B. Stall, B. Hessen, J. H. Teuben, *J. Mol. Cat.*, in press.

[33] A. L. McKnight, M. A. Masood, R. M. Waymouth, D. A. Straus, *Organometallics* **1997**, *16*, 2879.

[34] W. A. Herrmann, W. Baratta, M. J. A. Morawietz, *J. Organomet. Chem.* **1995**, *497*, C4; W. A. Herrmann, W. Baratta, *J. Organomet. Chem.* **1996**, *506*, 357.

[35] Y. Mu, W. E. Piers, M.-A. MacDonald, M. J. Zaworotko, *Can. J. Chem.* **1995**, *73*, 2233.

[36] (a) S. Ciruelos, T. Cuenca, P. Gomez-Sal, A. Manzanero, P. Royo, *Organometallics* **1995**, *14*, 177; (b) S. Ciruelos, T. Cuenca, R. Gomez, P. Gomez-Sal, A. Manzanero, P. Royo, *Organometallics* **1996**, *15*, 5577.

[37] M. L. H. Green, C. R. Lucas, *J. Organomet. Chem.* **1974**, *73*, 259.

[38] L. Kloppenburg, J. L. Petersen, *Organometallics* **1996**, *15*, 7.

[39] P.-J. Sinnema, Organometaalchemie en Homogene Katalyse, Verslagen Werkgroep, Rijksuniversiteit Groningen, **1995**, *7*, 23.

[40] M. Mena, M. A. Pellinghelli, P. Royo, R. Serrano, A. Tiripicchio, *J. Chem. Soc., Chem. Commun.* **1986**, 1118.

[41] D. J. Crowther, R. F. Jordan, N. C. Baenziger, A. Verma, *Organometallics* **1990**, *9*, 2574; N. H. Dryden, P. Legzdins, J. Trotter, V. C. Yee, *Organometallics* **1991**, *10*, 2857.

[42] R. Beckhaus, *Angew. Chem.* **1997**, *109*, 694, *Angew. Chem. Int. Ed. Engl.* **1997**, *36*, 686.

[43] P.-J. Sinnema, Organometaalchemie en Homogene Katalyse, Verslagen Werkgroep, Rijksuniversiteit Groningen, **1995**, *8*, 93.

[44] J. Okuda, T. Eberle, unpublished results.

[45] S. L. Buchwald, R. B. Nielsen, *Chem. Rev.* **1988**, *88*, 1047.

[46] L. van der Veen, Organometaalchemie en Homogene Katalyse, Verslagen Werkgroep, Rijksuniversiteit Groningen, **1996**, *9*, 87.

[47] D. D. Devore, F. J. Timmers, D. L. Hasha, R. K. Rosen, T. J. Marks, P. A. Deck, C. L. Stern, *Organometallics* **1995**, *14*, 3132.

[48] R. Walsh, *Acc. Chem. Res.* **1981**, *14*, 246.

[49] P.-J. Sinnema, Organometaalchemie en Homogene Katalyse, Verslagen Werkgroep, Rijksuniversiteit Groningen, **1996**, *9*, 37.

[50] H. Bürger, K. Wiegel, U. Thewalt, D. Schomburg, *J. Organomet. Chem.* **1975**, *87*, 301; J. Feldman, J. C. Calabrese, *J. Chem. Soc., Chem. Commun.* **1991**, 1042.

[51] R. M. Pulpi, J. N. Coalter, J. L. Petersen, *J. Organomet. Chem.* **1995**, *497*, 17.

[52] (a) J. C. Stevens, *Metcon 93*, Houston, 26-28 May, **1993**, p. 157. (b) J. C. Stevens, *Stud. Surface Sci. Cat.* **1994**, *89*, 277.

[53] D. M. Giolando, K. Kirschbaum, L. J. Graves, U. Bolle, *Inorg. Chem.* **1992**, *31*, 3887.

[54] J. Okuda, K. E. du Plooy, W. Massa, H.-C. Kang, U. Rose, *Chem. Ber.* **1996**, *129*, 275.

[55] F. Amor, T. P. Spaniol, J. Okuda, *Organometallics* **1997**, *16*, 4765.

[56] F. Amor, K. E. du Plooy, T. P. Spaniol, J. Okuda, unpublished results.

[57] Y. Mu, W. E. Piers, D. C. MacQuarrie, M. J. Zaworotko, V. G. Young, Jr., *Organometallics* **1996**, *15*, 2720.

[58] Y. Mu, W. E. Piers, D. C. MacQuarrie, M. J. Zaworotko, *Can. J. Chem.* **1996**, *74*, 1696.

[59] R. E. v. H. Spence, W. E. Piers, *Organometallics* **1995**, *14*, 4617.

[60] Y. Mu, W. E. Piers, L. R. MacGillivray, M. J. Zaworotko, *Polyhedron* **1995**, *14*, 1.

[61] K. C. Hultzsch, T. P. Spaniol, J. Okuda, *Organometallics* **1997**, *16*, 4845.

[62] K. C. Hultzsch, J. Okuda, *Macromol. Rapid Commun.,* **1997**, *18*, 809.

[63] R. Fandos, A. Meetsma, J. H. Teuben, *Organometallics* **1991**, *10*, 59.

[64] G. Trouvé, D. A. Laske, A. Meetsma, J. H. Teuben, *J. Organomet. Chem.* **1996**, *511*, 255.

[65] B. Rieger, *J. Organomet. Chem.* **1991**, *420*, C17.

[66] J. W. Pattiasina, C. E. Hissink, J. L. de Boer, A. Meetsma, J. H. Teuben, *J. Am. Chem. Soc.* **1985**, *107*, 7758.

[67] G. Erker, U. Korek, *Z. Naturforsch.* **1989**, *44b*, 1593.

[68] Y. Qian, J. Huang, X. Chen, G. Li, W. Chen, B. Li, X. Jin, Q. Yang, *Polyhedron* **1994**, *13*, 1105.

[69] J. Christoffers, R. G. Bergman, *Angew. Chem.* **1995**, *107*, 2423; *Angew. Chem. Int. Ed. Engl.* **1995**, *34*, 2266.

[70] K. Soga, T. Uozumi, S. Nakamura, T. Toneri, T. Teranishi, T. Sano, T. Arai, *Macromol. Chem. Phys.* **1996**, *197*, 4237.

[71] (a) Y.-X. Chen, C. L. Stern, S. Yang, T. J. Marks, *J. Am. Chem. Soc.* **1996**, *118*, 12451. (b) L. Jia, X. Yang, C. L. Stern, T. J. Marks, *Organometallics* **1997**, *16*, 842.

[72] (a) T. K. Woo, L. Fan, T. Ziegler, *Organometallics* **1994**, *13*, 2252. (b) T. K. Woo, P. M. Margl, J. C. W. Lohrenz, P. E. Blöchl, T. Ziegler, *J. Am. Chem. Soc.* **1996**, *118*, 13021.

[73] B. Temme, G. Erker, J. Karl, H. Luftmann, R. Fröhlich, S. Kotila, *Angew. Chem.* **1995**, *107*, 1867, *Angew. Chem. Int. Ed. Engl.* **1994**, *34*, 1755.

[74] (a) T. Shiomura, T. Asanuma, N. Inoue, *Macromol. Rapid Commun.* **1996**, *17*, 9. (b) T. Shiomura, T. Asanuma, T. Sunaga, *Macromol. Rapid Commun.* **1997**, *18*, 169.

[75] F. G. Sernetz, R. Mülhaupt, R. M. Waymouth, *Polymer Bulletin* **1997**, *38*, 141.

[76] F. G. Sernetz, R. Mülhaupt, F. Amor, T. Eberle, J. Okuda, *J. Polym. Sci., Part A* **1997**, *35*, 1571.

[77] F. G. Sernetz, R. Mülhaupt, R. M. Waymouth, *Macromol. Chem. Phys.* **1996**, *197*, 1071; Y. Thomann, F. G. Sernetz, R. Thomann, J. Kressler, R. Mülhaupt, *Macromol. Chem. Phys.* **1997**, *198*, 739.

[78] A. H. Hoveyda, J. P. Morken, *Angew. Chem.* **1996**, *108*, 1378; *Angew. Chem. Int. Ed. Engl.* **1996**, *35*, 1262.

[79] E. A. Bijpost, R. Duchateau, J. H. Teuben, *J. Mol. Cat. A.* **1995**, *95*, 121.

[80] J. D. Scollard, D. H. McConville, *J. Am. Chem. Soc.* **1996**, *118*, 10008; R. Baumann, W. D. Davis, R. R. Schrock, *J. Am. Chem. Soc.* **1997**, *119*, 3830; S. Tinkler, R. J. Deeth, D. J. Duncalf, A. McCamley, *J. Chem. Soc., Chem. Comm.* **1996**, 2623.

8 Synthesis of Chiral Titanocene and Zirconocene Dichlorides

Ronald L. Halterman

8.1 Introduction

The synthesis of Group 4 metallocene dichloride complexes has progressed enormously from the preparation of simple unbridged complexes with two identical unsubstituted or singly substituted cyclopentadienyl or indenyl ligands in the 1950s and 60s. In the 1970s the additional aspect of chirality was introduced by incorporating either unsymmetrically disubstituted cyclopentadienyl ligands (planar chirality) or by incorporating a chiral substituent on the cyclopentadienyl rings. The first application of chiral metallocenes as asymmetric catalysts, the enantioselective hydrogenation of alkenes, appeared in the late 1970s. At the end of the 1970s and beginning of the 1980s bridged ligands were introduced to form *ansa*-metallocene dichlorides containing either two cyclopentadienyl or two indenyl ligands. The 1980s saw a slowly expanding use of chiral metallocenes as catalysts in organic synthesis, but the event that most energized the field was the discovery that C_2-symmetric ethylene-bridged *ansa*-bis(tetrahydroindenyl)zirconium dichloride formed an effective catalyst for the stereoregular isotactic polymerization of propene. New syntheses of substituted bridged bis(indenyl) and bis(tetrahydroindenyl)metal complexes and additional chiral bis(cyclopentadienyl)metal complexes for application as polymerization catalysts rapidly followed. The development of new metallocenes in the 1990s has continued to expand to produce better catalysts for an increased variety of polymerization reactions and a rapidly increasing number of catalytic reactions of usefulness in asymmetric organic synthesis. To meet these demands, new bridged cyclopentadienyl, indenyl, and now fluorenyl complexes, as well as new chiral substituted cyclopentadienyl and indenyl complexes are being produced by an increasingly sophisticated variety of synthetic methodologies.

The following chapter is intended to give an overview of the synthetic methodologies used over the past forty years in the synthesis of substituted ligands (unbridged and bridged) and the methods used to metalate these ligands to form titanocene or zirconocene dichlorides. By focusing on the synthetic methods used, the information should better guide practitioners to suitable adaptations for the synthesis of the next generations of desired metallocenes—whatever form they may take. The compounds are grouped according to ligand—the major division being unbridged or bridged ligands, the subdivisions being which of the three 'electronic' types of ligands (cyclopentadienyl, indenyl, or fluorenyl) are incorporated. Although they are properly considered to contain substituted cyclopentadienyl ligands, tetrahydroindenyl complexes are grouped with the indenyl ligands because of the strong historical co-development of these ligands. The preparation of most chiral or prochiral ligands is covered alongside the achiral versions since the synthetic chemistry is often based on the same chemistry. The preparation of ligands having stereogenic bridging groups is treated separately. The ramifications of ligand chirality on the formation of stereoisomeric metal complexes will be discussed in detail in the next section and will be tied into discussions of the metalation of particular ligands.

In deciding on the material to be covered, limitations had to be set. This chapter covers the preparation of the titanocene or zirconocene dichloride complexes. The metallocene dichlorides are defined as complexes have two η^5-carbocyclic π-ligands (cyclopentadienyl, indenyl or fluorenyl). These complexes are the most widely found and enable a reasonable comparison to be made of the synthetic methods used in their preparation. Heteroatom substituted ligands such as phosphacyclopentadienyl are not covered. Although many different metallocene complexes—such as metal dialkyls—are known, these complexes can usually be prepared from the metal dichloride complexes through fairly standard methods. The preparation of the less common hafnocene complexes is also not explicitly covered. Many papers dealing with the formation of zirconocene dichlorides also cover the corresponding hafnium complexes.

8.2 Ligand Chirality and Stereoisomeric Metal Complexes

The metalation of cyclopentadienyl, indenyl or fluorenyl ligands can produce achiral or variety of stereoisomeric metallocenes depending on the symmetry of the ligands being metalated [1]. In general, chiral Group 4 metallocenes can have chirality arising in three ways: 1) the chirality may be based on a stereogenic metal atom with four different ligands as in **1**; [2] 2) the chirality may be introduced by the coordination of the metal to chiral or prochiral ligands as in **2** [3], **3**[4] or **4** [5]; or 3) it may be due to both of these elements. Since this review covers the formation of metallocene dichlorides having nonstereogenic metal centers, only chirality due to the coordination of chiral or prochiral ligands appears.

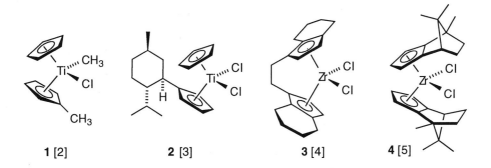

1 [2]   **2 [3]**   **3 [4]**   **4 [5]**

A key issue in determining what type of metal complexes (achiral, enantiomeric or diastereomeric) can form from particular ligands is how the two π-faces of a ligand are related to each other. Depending on the substituents on the ligands, the π-faces of the ligands can be related to one another in three ways: 1) the faces are equivalent or homotopic if the faces can interconvert by ligand rotation or if rotation about a C_2 axis can interconvert the two faces; 2) the faces are enantiotopic if only reflection through a mirror plane can interconvert the two faces: or 3) the faces are diastereotopic if a rotation or symmetry operation cannot interconvert the faces. Metalation of ligands with homotopic or equivalent faces will produce only a single isomeric complex while the metalation of ligands with enantiotopic or diastereotopic π-faces can produce enantiomeric and diastereomeric mixtures of metallocene dichlorides.

Ligands with Equivalent or Homotopic Faces. Two methods for determining facial equivalence are referred to here. If the π-faces of the ligand can interconvert by a ligand rotation about one of the single bonds between the ligand and a substituent (without the need for invoking a C_2-symmetry operation), the faces are termed *equivalent*. If the ligands faces can *only* be interconverted by a C_2-symmetry operation, the faces are described by a more restricted use of the term *homotopic*. In naming complexes derived from ligands with equivalent or homotopic π-faces, no stereochemical descriptor can be used for the π-coordination of the ligand since no new stereochemical element is introduced upon metalation.

Metalation of ligands with equivalent faces will produce only a single isomeric metallocene dichloride complex. The simplest example of equivalent cyclopentadienyl faces is seen in the metalation of unsubstituted cyclopentadienyl to form the single, achiral titanocene dichloride complex **5** [6] (Scheme 8-1). With a single substituent on the cyclopentadienyl anion—whether achiral as in methylcyclopentadienyl (**6**) [7] or chiral and enantiomerically pure as in menthylcyclopentadienyl (**7**) [3]—bis(cyclopentadienyl)metal dichlorides **8** and **9** can be formed only as a single isomer since the ligand π-faces are equivalent by rotation about the cyclopentadienyl–methyl or cyclopentadienyl–menthyl single bond. Polysubstituted cyclopentadienyl ligands will have equivalent faces if each of the substituents is equivalent as in, for example, 1,3-dimethylcyclopentadienyl (**10**) which can give only the single metallocene dichloride **11** [8].

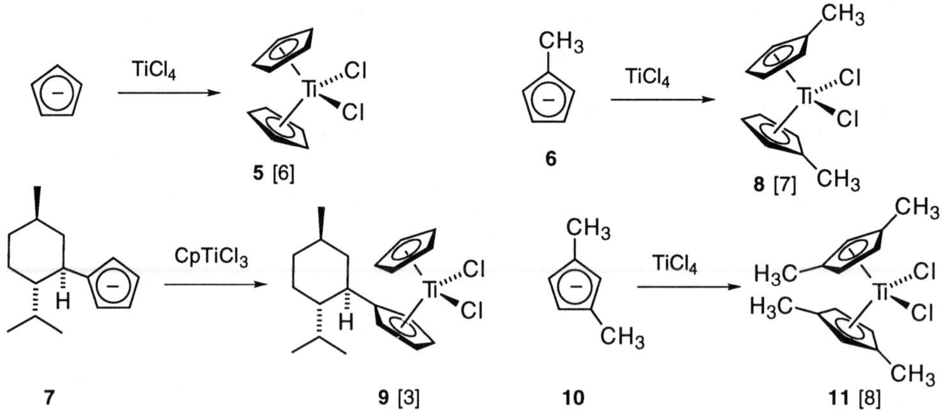

Figure 8-1. Metalation of cyclopentadienyl ligands with equivalent faces give only a single metallocene.

If a racemic mixture of chiral mono-substituted cyclopentadienyl ligands is metalated, a mixture of stereomeric complexes can result even though the ligand faces are equivalent. As illustrated in Scheme 8-2, addition of one equivalent of the racemic ligand **12** to the metal produces a racemic mixture of energetically equivalent enantiomeric cyclopentadienylmetal trichlorides **13**. These now chiral metal complexes can choose to react further with either of the enantiomeric ligands in the solution to form a mixture of *meso* and *dl* **14** [9]. If both ligands on the metal are the same enantiomer, the chiral complex will have C_2-symmetry. Since equal portions of the enantiomeric compounds **12** and **13** are present, the enantiomeric complexes will be formed in equal amounts and can be referred to as the racemic *dl* pair. If the two ligands on one metal are opposite enantiomers, this achiral complex will be the *meso* isomer. Since the *meso* isomer is a diastereomer of the *dl* pair, it can be energetically more or less stable and can be formed in greater or lower amounts than the *dl* pair.

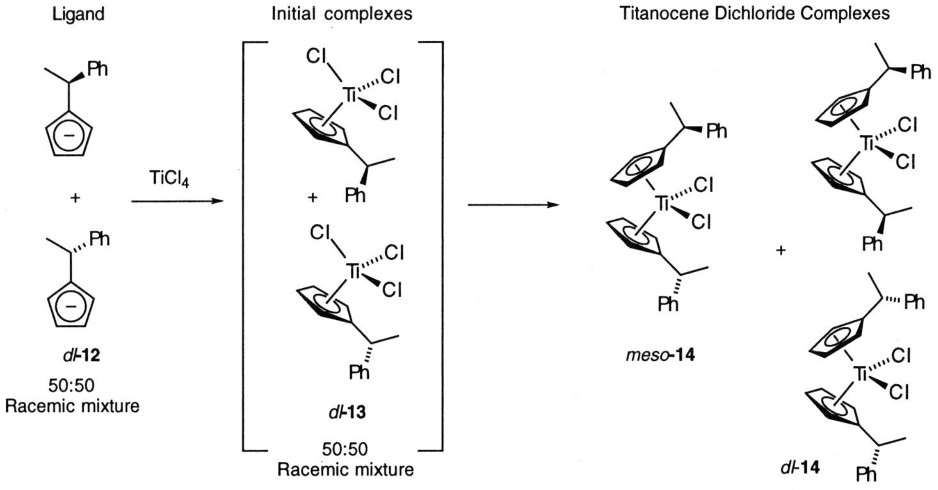

Scheme 8-2. Metalation of a racemic substituted chiral cyclopentadienyl ligand [9].

Bridged bis(cyclopentadienyl) ligands will have equivalent faces when they are un-substituted as in the tetramethylethylene-bridged bis(cyclopentadienyl) **15** [10] (Scheme 8-3). The introduction of chirality in the bridge as in **16** [11] does not ef-fect the equivalence of the cyclopentadienyl π-faces and the metalation of **15** or **16** can produce only the single *ansa*-metallocene dichlorides **17** or **18**, respectively. Di- or tetrasubstituted cyclopentadienyls can still have facial equivalence when the substituents are symmetrically placed about the bridging position as in **19**, whose metalation can produce only a single *ansa*-metallocene as in **20** [12].

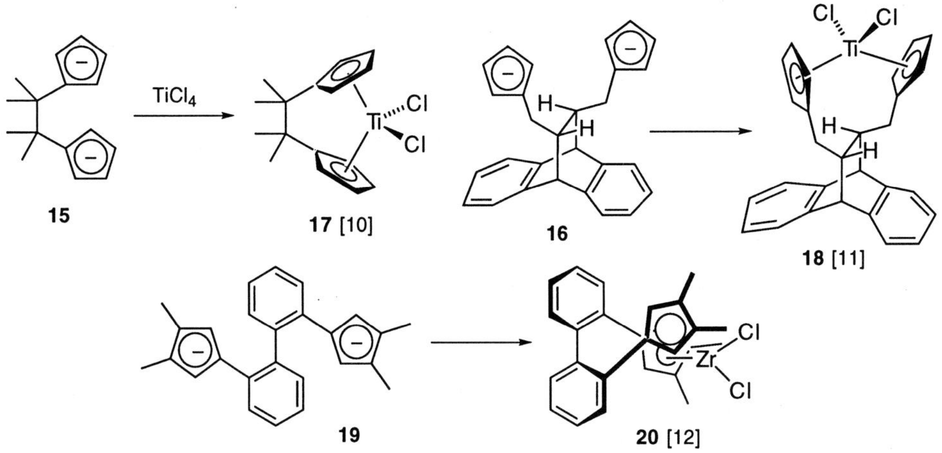

Scheme 8-3. Metalation of bridged, symmetrically substituted bis(cyclopentadienyl) ligands.

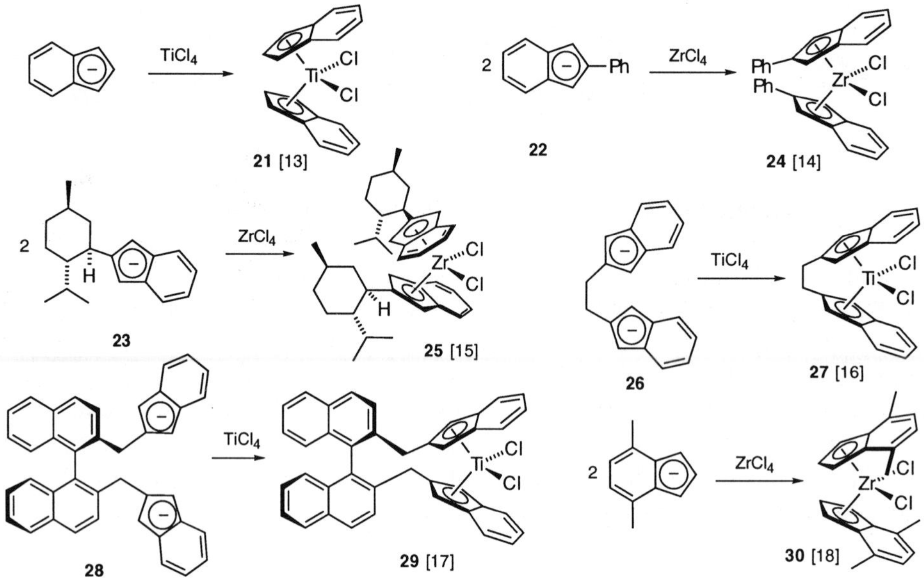

Scheme 8-4. Metalation of indenyl ligands with equivalent π-faces.

The simplest example of equivalent indenyl faces is seen in the metalation of the unsubstituted indenyl ligand to form a single, achiral complex **21** [13] (Scheme 8-4). Monosubstituted indenyl ligands will only have equivalent faces when the substituent, either achiral as in the phenyl-substituted **22** [14] or chiral as in the menthyl-substituted **23**, [15] is placed at the 2-position of the indenyl ligand. These 2-substituted indenyl ligands will give only the single metallocene dichlorides **24** and **25**. Metalation of indenyl ligands bridged at the 2-position with either an achiral tether as **26**, [16] or a chiral tether as in **28** [17] will lead to the single metallocene dichlorides **27** and **29**. More highly substituted indenyls will still have equivalent faces as long as the substituents are symmetrically placed about the 2-position as in the formation of **30** [18].

Ligands with Homotopic Faces. Chiral cyclopentadienes with homotopic faces (faces which can only interconvert by only a C_2-symmetry operation) such as **31** can inherently lead to only a single isomeric product **32** [19] due to the symmetry equivalence of the two ligand π-faces (Scheme 8-5). If the cyclopentadienyl ligand is bridged along the axis of symmetry, the resulting bridged ligand still has C_2-symmetry and can produce only a single product as with **33** [20]. Chiral C_2-symmetrical indenyl ligands would be possible by symmetrical placement of chiral auxiliaries at the 1,3- 4,7- or 5,6-positions—but such ligands have not been reported.

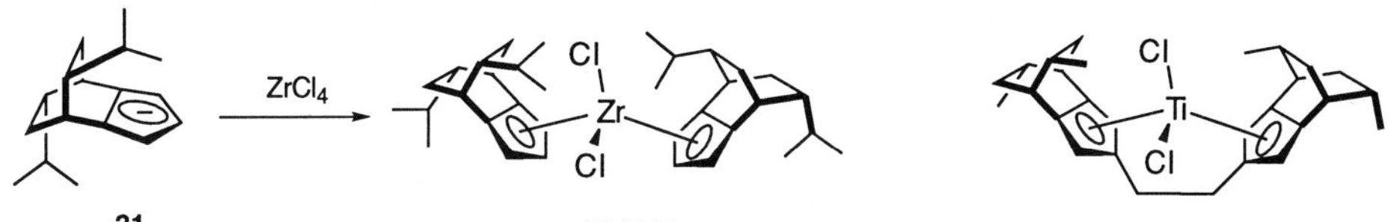

Scheme 8-5. Cyclopentadienyl ligands with homotopic faces.

Ligands with Enantiotopic Faces. Cyclopentadienyl ligands with at least two different achiral substituents will have enantiotopic faces as in 1-isopropyl-3-methyl-cyclopentadienyl **34** [21]. The approach of an achiral metal moiety (*e.g.*, CpTiCl$_3$) to either face of the prochiral ligand **34** will be energetically equivalent (enantiomeric transition states) and will result in the production of two enantiomeric complexes **35** in equal amounts as shown in Scheme 8-6. The chirality in complexes of this type has been described as 'planar' chirality and the π-coordination of ligand carries an *R* or *S* stereochemical descriptor [22].

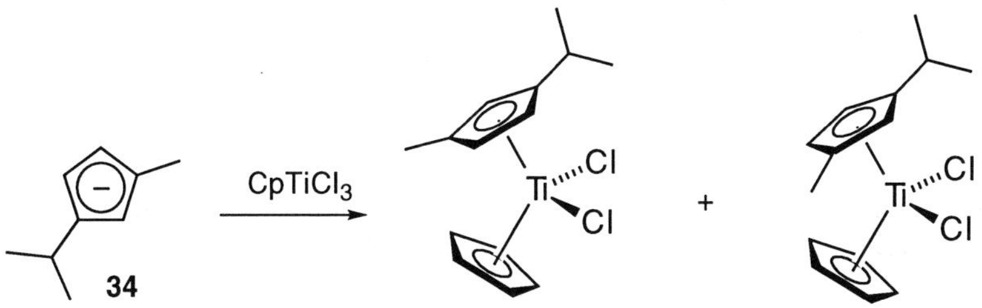

(p-*R*)-**35**   (50:50)   (p-*S*)-**35**

Scheme 8-6. Metalation of a cyclopentadienyl ligand having enantiotopic faces [21].

Note on Nomenclature. A note of definition is needed here on the *R/S* nomenclature used in describing metallocenes. According to rules set forth by Schlögl, the metal is treated as being individually bonded to all five of the π-atoms of the ligand and the configuration of one π-atom is used to describe the entire ligand [22]. The choice of the atom to be assigned has not been uniform in the literature. Schlögl has stated that the highest priority π-atom in unbridged ligands be assigned [22] whereas for *ansa*-metallocenes Brintzinger has used the site of the bridging group—the 1-indenyl position in (EBTHI)TiCl$_2$—to assign the configuration [23]. The Brintzinger convention has been widely adopted although it is often at odds with a rigorous application of Schögl's rule.

Misconceptions can be avoided if authors make clear which atom is being used for the assignment. The following definition is unambiguous. *"The planar chirality in cyclopentadienylmetal or indenylmetal complexes is assigned R or S based on the Cahn–Ingold–Prelog configuration of the 1-position of the ligand where the metal is treated as being individually bonded to all five of the π-atoms. The chiral-*

ity can be described as (1R) or (1S) to avoid ambiguity or as (p-R) or (p-S) to emphasize that this is a description of the planar chirality based on the 1-position." The 1-position is often the highest priority atom in unbridged metallocenes and has been commonly used in *ansa*-metallocenes. The 1-position is readily identified and is always stereogenic in planar chiral metallocenes. Even in cases where the bridging group in *ansa*-metallocenes is not at the 1-position, the 1-position can still be unambiguously used.

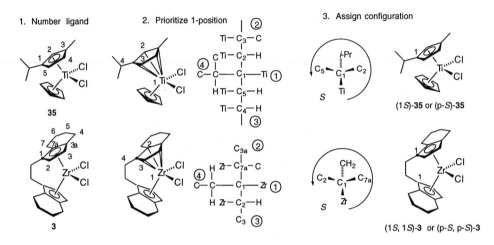

Scheme 8-7. Assignment of configuration in metallocenes having an element of planar chirality.

This method is illustrated in Scheme 8-7 for titanocene dichloride **35** and ethylenebis(tetrahydroindenyl)-zirconium dichloride **3**. In each case the 1-position is identified in Step 1. The priorities on C-1 are assigned in Step 2. The highest priority atom attached to C-1 is the metal. Of the three remaining carbon atoms attached to C-1 in **35**, the exocyclic isopropyl group is not further attached to the metal and receives the lowest priority. Of the remaining atoms, C-2 has the priority over C-5 since C-3 is more highly substituted than C-4. In Step 3 the lowest priority group is placed in back and the decreasing order of priorities indicate that **35** has the *S*-configuration at C-1. Thus, **35** can be assigned the *S*-configuration [(1*S*) or (p-*S*)]. Applying this same method to the enantiomer of *ansa*-metallocene **3** depicted in Scheme 8-7 results in the (*S*,*S*)-assignment.

Complexes with Two Enantiotopic Ligands. The situation where two equivalents of a prochiral cyclopentadienyl ligand add to one metal becomes somewhat more complicated. In this case, two diastereomeric bis(cyclopentadienyl)metal dichlorides (*meso* and *dl*) can be produced in different amounts (Scheme 8-8). The approaches of the achiral metal to either π-face of the ligand **34** are energetically equivalent and a racemic mixture of enantiomeric cyclopentadienylmetal trichlorides **36** is produced. When these now chiral metal complexes approach either prochiral face of the second equivalent of the cyclopentadienyl ligand **34**, the pathways are now diastereotopic and one approach may be energetically favored to give a potentially unequal mixture of the *meso* **37** and *dl* **38** isomers. In this particular case, no selectivity was seen in the metalation [21].

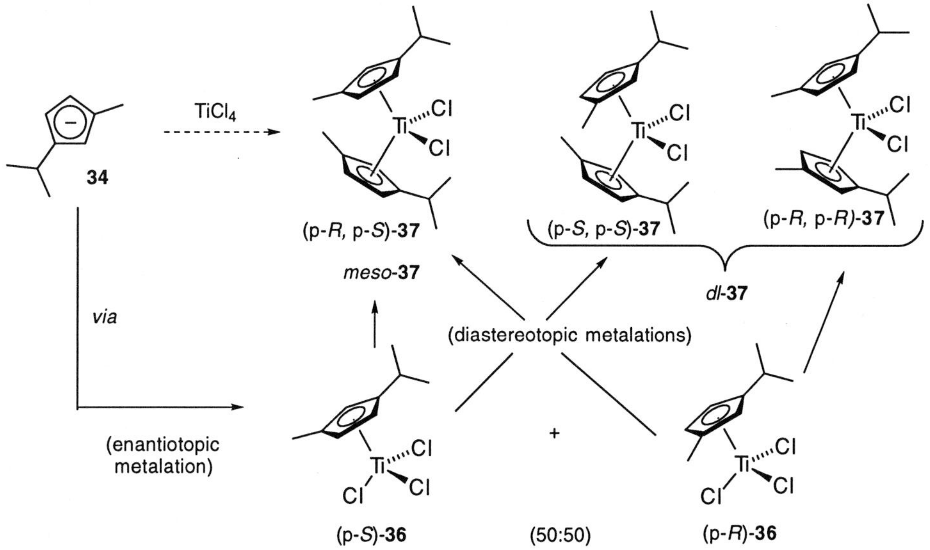

Scheme 8-8. Metalation of cyclopentadienyl ligands having enantiotopic faces [21].

Indenyl ligands with an achiral substituent at the 1-position also have enantiotopic ligand faces and their bis(indenyl)metal dichloride complexes also can exist as *dl* and *meso* isomers as seen in Scheme 8-9 for the metalation of 1-cyclohexylindenyl **38** to a mixture of *dl*-**39** and *meso*-**39**. In this case, the formation of the *dl* isomer is favored [24].

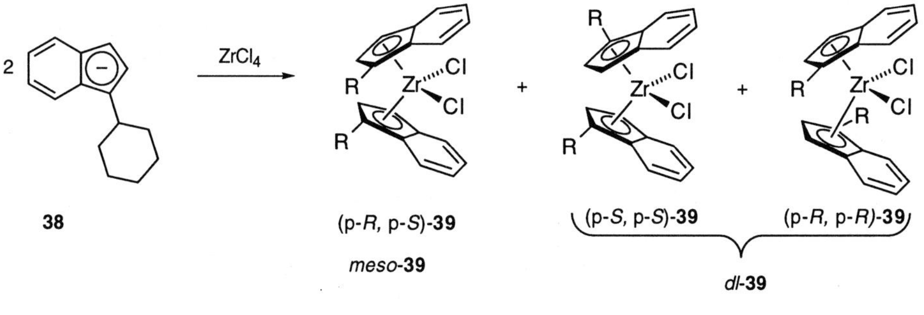

Scheme 8-9. Metalation of indenyl ligands having enantiotopic faces [24].

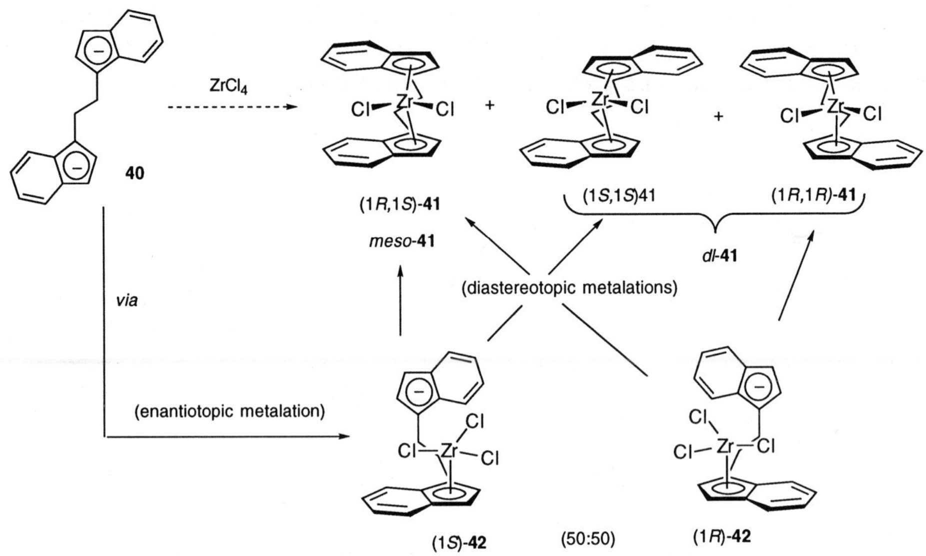

Scheme 8-10. Metalation of bridged bis(indenyl) having enantiotopic π-faces [4].

The metalation of Brintzinger's ethylene-bridged bis(indenyl) **40** can give *dl*- and *meso*-bis(indenyl)metal dichlorides **41** since the π-faces of each 1-substituted indenyl are enantiotopic [4] (Scheme 8-10). The initial metalation of **40** leads to formation of racemic enantiomeric indenylzirconium trichloride **42**. The approach of the now chiral indenylmetal fragment to either face of the second indenyl moiety is a diastereomeric process which in this case has a lower energy barrier for the selective formation of the *dl* isomers. Both of the coordinated π-ligands can be described by a stereochemical descriptor to give the *dl* pair as (1*R*,1*R*)- and (1*S*,1*S*)-**41** and the *meso* isomer as (1*R*,1*S*)-**41**.

Ligands with Diastereotopic Faces. Ligand π-faces are diastereotopic if a ligand rotation or symmetry operation (C_2-rotation or reflection through a mirror plane) cannot interconvert the π-faces. This situation arises when an unsymmetrically polysubstituted ligand also contains a chiral substituent. Metal complexes of such ligands will carry at least one *R*- or *S*-stereochemical descriptor for the ligand substituent(s) and one p-*R*- or p-*S*-stereochemical descriptor for the π-coordination of the ligand. When an enantiomerically pure ligand such as **43** (with one chiral substituent having an *R* configuration) is metalated with cyclopentadienyltitanium trichloride, one new stereochemical element is introduced and a mixture of two diastereomeric complexes (*R*, p-*R*)-**44** and (*R*, p-*S*)-**44** can be produced in unequal amounts (Scheme 8-11). When two equivalents of an enantiomerically pure ligand are combined on one metal, a mixture of three diastereomeric bis(cyclopentadienyl)metal dichlorides can be produced, two C_2-symmetric isomers [(*R*, p-*R*)- (*R*, p-*R*)]-**45**, [(*R*, p-*S*)- (*R*, p-*S*)]-**45**, and one C_1-symmetric isomer [(*R*, p-*R*)- (*R*, p-*S*)]-**45**. Complexes such as the C_1-symmetric isomer of **45** are often referred to as '*meso*-like' in reference to the *meso*-relationship between the elements of planar chirality without regard for any chiral substituents.

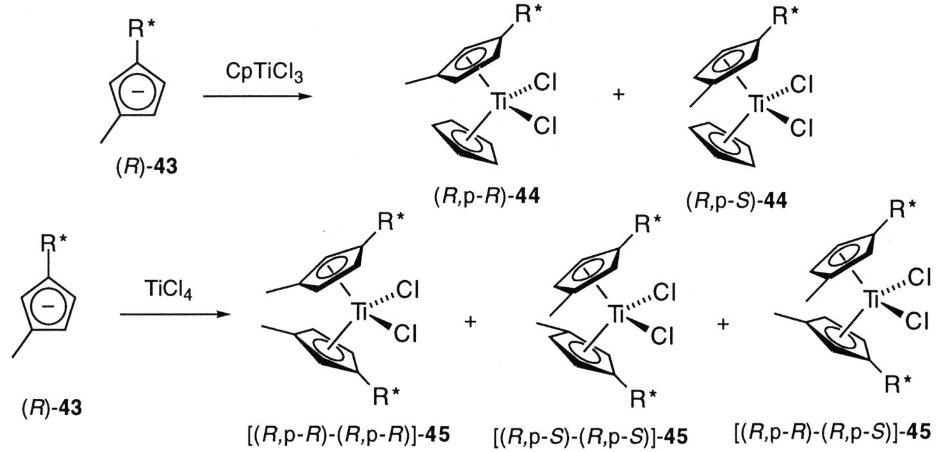

Scheme 8-11. Metalation of enantiomerically pure ligands
having diastereotopic faces.

If two equivalents of a racemic ligand having diasterotopic π-faces are added to one metal, the potential stereoisomeric mixture is now potentially horrendous—eleven stereoisomeric products, seven diastereomers and enantiomers of four these complexes can be produced. Unless the ligand exerts strong π-facial selectivity, a bewildering array of complexes will be formed in the metalation of diastereotopic ligands.

The most common examples of cyclopentadienyl ligands with diasteromeric faces are annelated cyclopentadienes derived from the chiral pool. For example, enantiomerically pure camphor-cyclopentadienyl **46** can be metalated with cyclo-pentadienyltitanium trichloride to form a mixture of the two expected diastereomeric complexes of **47** [25]. When two equivalents of **46** are metalated three diastereomeric complexes can be formed. In this case, a strong π-facial preference is exerted and one of the C_2-symmetrical isomers is formed as the major product (Scheme 8-12) [5, 25, 26].

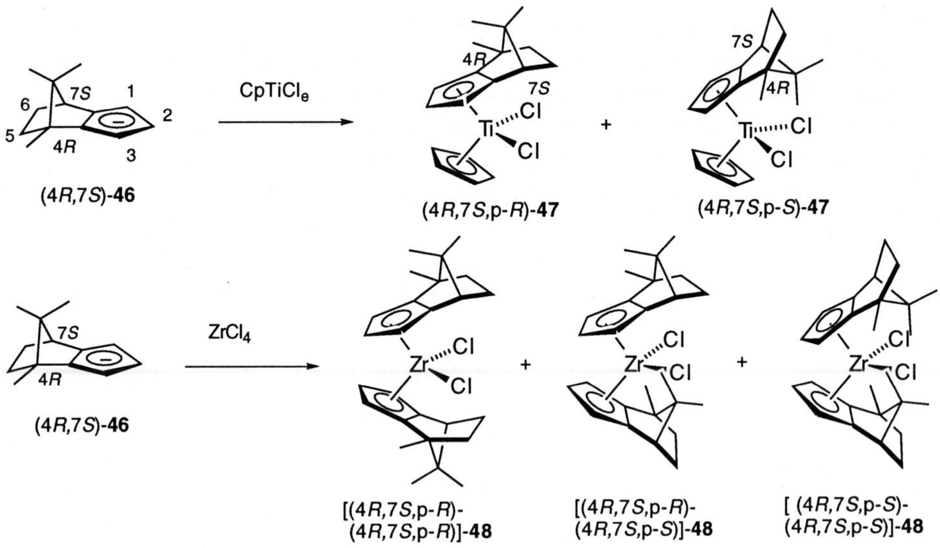

Scheme 8-12. Metalation of camphor-cyclopentadienyl [5, 25, 26].

Interestingly, not all ligands with diastereotopic π-faces will produce chiral complexes. The well-studied norbornadienyl ligand **49** has diastereotopic π-faces and a mirror plane—but one which does not interconvert the π-faces of the ligand. The ligand fits the definition of diastereotopic faces since it is unsymmetrically substituted (it has an *R*-carbon at C-3a and *S*-substituent at C-7a, indenyl numbering) and contains chiratopic atoms (Scheme 8-13). The diastereomeric metal complexes **50** can be described as *endo,endo* , *exo,exo*, or *endo,exo*, or using the stereochemical descriptors, $[(R,S,\text{p-}R)\text{-}\mathbf{49}]_2\text{TiCl}_2$, $[(R,S, \text{p-}S)\text{-}\mathbf{49}]_2\text{TiCl}_2$, or $[(R,S, \text{p-}R)\text{-}\mathbf{49}, (R,S, \text{p-}S)\text{-}\mathbf{49}]\text{TiCl}_2$ [27]. Each of these complexes is achiral and *meso*.

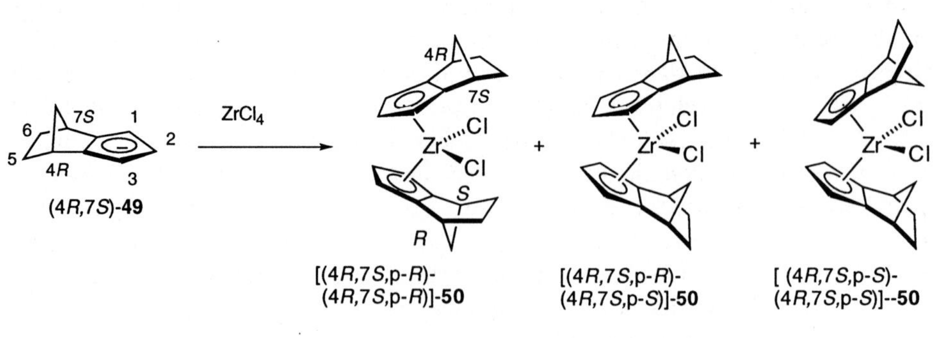

Scheme 8-13. Metalation of norbornacyclopentadienyl
having diastereotopic faces [27].

An example of an indenyl ligand having diastereotopic faces is the 1-(neo-menthyl)indenyl **51**. Metalation of this ligand with cyclopentadienylzirconium trichloride produces the expected mixture of two diastereomers of **52** [28] (Scheme 8-14). In this case about a 6:1 ratio of the diastereomers is realized. When two equivalents of **51** are added to ZrCl$_4$, a mixture of three diastereomers ensues—one C_1-symmetric and two, C_2-symmetric complexes—is formed [29].

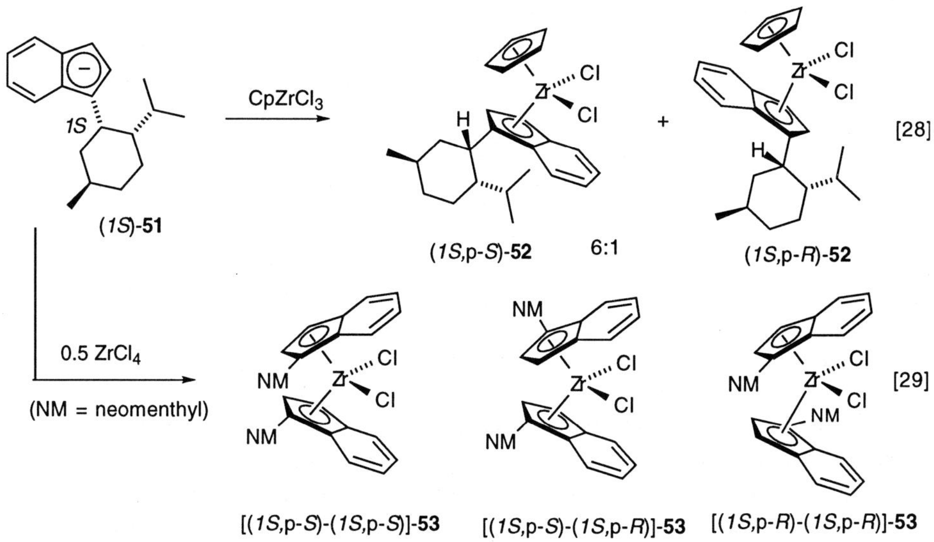

Scheme 8-14. Metalation of an enantiomerically
pure indenyl ligand having diastereotopic π-faces.

Bridged ligands containing a chiral auxiliary either as a substituent on the ligand or in the bridging group will have diastereotopic π-faces. An example of a bis(cyclopentadienyl) ligand with an achiral bridge and a chiral auxiliary on only one of the cyclopentadienyl ligands is the dimethyl(menthylCp)-(tetramethylCp)silane **54** (Scheme 8-15). Metalation of **54** should and does produce two diastereomeric complexes (3*S*,p-*R*)-**55** and (3*S*,p-*S*)-**55** [30]. If the racemic ligand **54** had been used, only two diastereomers would have been produced along with their enantiomers. An example of ligand in which both cyclopentadienyl groups have a chiral substituent and are bridged by an achiral group is the ethyl-ene-bridged bis(bicyclooctylcyclopentadienyl) **56**. Metalation of this ligand could give three diastereomers of **57**, but only a single diastereomeric titanocene dichloride was formed due to a high degree of facial selectivity [20].

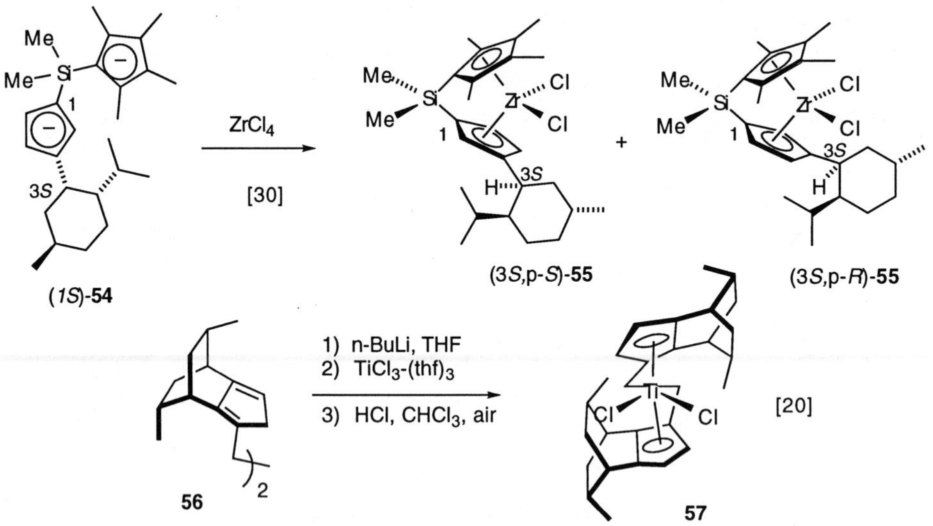

Scheme 8-15. Metalation of bridged ligands having diastereotopic faces.

One of the few examples of a bis(cyclopentadienyl) ligand having diastereotopic faces due to a chiral bridging group is the (*S,S*)-butane-2,3-diyl-bridged bis(3-methylcyclopentadienyl) **58** (Scheme 8-16) [31]. The metalation of such ligands with two diastereotopic faces can produce three diastereomeric complexes. If the racemic ligand is used, the diastereoselectivity is unchanged, but each diastereomer will occur with its enantiomer. In the metalation of **58** a 4.6:2.8:1 ratio of the *meso*-like (*S,S,*p-*R*, p-*S*)-**59** and the *dl*-like (*S,S,* p-*R*, p-*R*)-**59** and (*S,S,* p-*S*, p-*S*)-**59** is produced. An early example of a bis(indenyl) ligand having diastereotopic faces due to the presence of a chiral bridge at the 1-indenyl positions is the chiracene ligand **60** [32]. Metalation and reduction of the initial bis(indenyl)metals gave a 4.2:2.5:1 mixture of the three expected diastereomeric *ansa*-bis(tetrahydroindenyl)titanium dichlorides **61**. Metalation of racemic ligands of this type will produce the same diastereoselectivity along with enantiomers of the diastereomers.

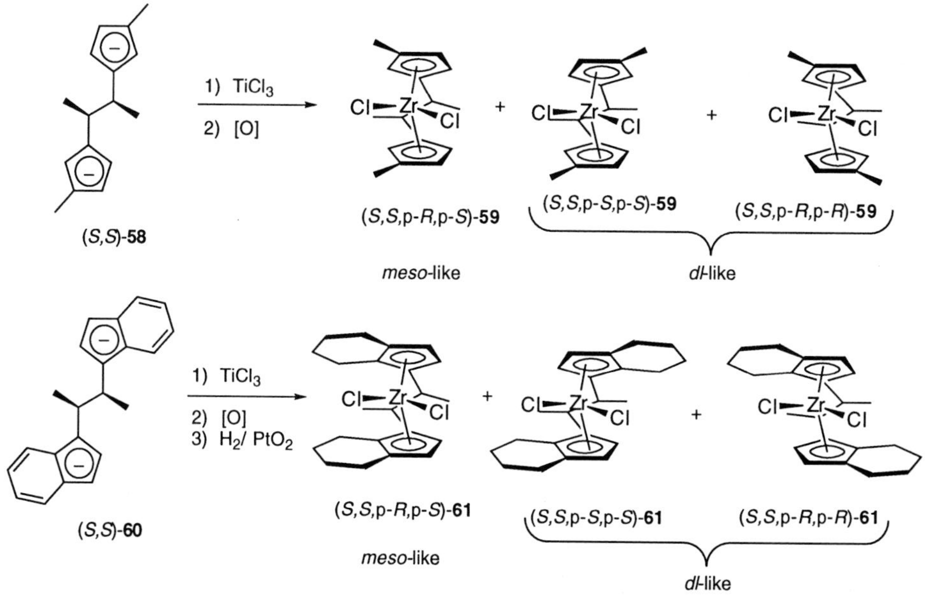

Scheme 8-16. Metalation of chiral bridged bis(cyclopentadienyl)
having diastereotopic π-faces.

In summary, the symmetry of the ligands play a major role in determining whether
one or up to eleven stereoisomeric metal complexes may be formed.

8.3   Unbridged Ligands

8.3.1   Cyclopentadienes

The preparation of substituted cyclopentadienes will be discussed according to the
number of substituents on the cyclopentadiene. The preparation of any chiral derivatives is included as part of each section. In cyclopentadiene syntheses a mixture
of double bond isomers is usually obtained. For simplicity only one of the double-
bond isomers will usually be depicted and the numbering will be according to the
relationship between the substituents without regard for the particular double bond
positioning.

Monosubstituted Achiral Cyclopentadienes. Two general, established routes have
been used most often to prepare monosubstituted cyclopentadienes, the alkylation
of a cyclopentadienyl anion or the formation of a pentafulvene followed by reduc-

tion or organometal addition. Both of these methods were already established in the 1930s [33] with more general applications appearing by the early 1960s [34, 35]. As shown in Scheme 8-17, a variety of different cyclopentadienylmetals in various solvents have reacted with a range of alkyl halides to give the monosubstituted cyclopentadienes [35–38]. The use of the sodium salt in THF or NH_3 is the most common method.

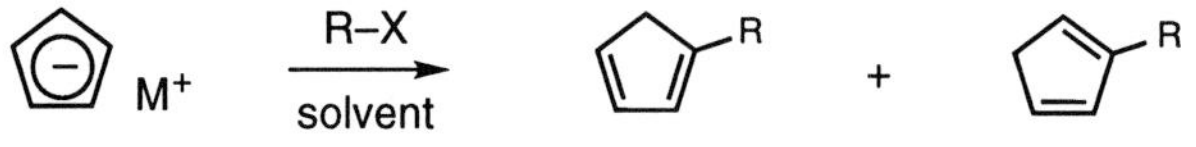

M = K,   PhH, R-X = MeI, EtBr, *n*-PrBr, *n*-BuBr (typical yields 50 to 80%) [36]

M = MgBr, Et$_2$O, R-X = allylBr, *tert*-BuBr   (44 to 51% yield) [35]

M = Na,   NH$_3$, R-X = MeBr, EtBr, *i*-PrBr, *tert*-BuBr, PhCH$_2$Cl   (typical yields 80 to 85%) [35,36]

M = Na,   THF, R-X = *n*-C$_n$H$_{2n+1}$Br(Cl), n = 5, 6, 7, 8, 10, 12, 14, 16, 18   (typical yields 40 to 80%) [37]

M = Na,   THF, R-X = 3-Bromopentane   (50% yield) [38]

Scheme 8-17. Preparation of monosubstituted cyclopentadienes by alkylation.

The facile condensation of aldehydes and ketones with cyclopentadiene to form a variety of fulvenes is the basis for the second common route to form monosubstituted cyclopentadienes. The cyclopentadienyl ligands are obtained by the reduction or deprotonation of the fulvenes or organometal addition to the fulvenes (Scheme 8-18). For example, 6,6-dimethylfulvene **62** could be reduced with LiAlH$_4$ to give isopropylcyclopentadienyl, reacted with phenyllithium or methyllithium to give (α,α-dimethylbenzyl)cyclopentadienyl **63** or *tert*-butylcyclopentadienyl, or it could be deprotonated with sodium amide to give isopropenylcyclopentadienyl [39]. Each of these cyclopentadienyllithium or sodium complexes could be directly metalated to form group 4 metallocene dichloride complexes. This method has been extended to the conversion of diphenylfulvene **64** to (diphenylmethyl)cyclopentadienyl **65** or (triphenylmethyl)cyclopentadienyl. A number of other dialkyl fulvenes such as fulvenes **66** or **67** have been used to generate *tert*-alkyl substituted cyclopentadienyls [41,42]. Chiral but racemic ligands are readily prepared by reducing the unsymmetrical fulvene **68** by LiAlH$_4$ [9]. A number of vinyl substituted cyclopentadienyls have been prepared by deprotonating pentamethylenefulvene **67** or the unsymmetrically substituted fulvenes **69** [43].

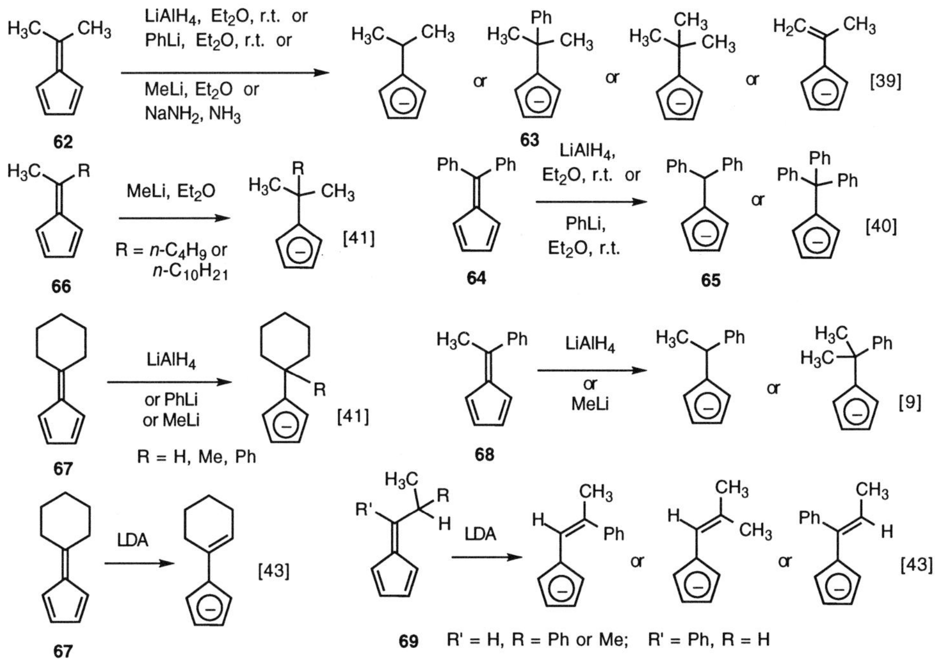

Scheme 8-18. Preparation of monosubstituted cyclopentadienes
by the fulvene route.

An additional, less general method for the preparation of monosubstituted cyclopentadienes is the dehydration of cyclopentenols. Monosubstituted cyclopentenols
have been generated by the addition of phenyllithium to cyclopentenone or by the
reduction of 3-methyl- or 3-phenylcyclopentenones [44, 45].

Chiral Monosubstituted Cyclopentadienes. Chiral, nonracemic monosubstituted
cyclopentadienes have most generally been formed by either the addition of cyclopentadienylmetals to chiral alkylating agents or by the enantioselective reduction
of fulvenes. The earliest synthesis of a chiral, nonracemic cyclopentadiene was Leblanc's 1976 asymmetric reduction of fulvene **68** with a quinine-modified LiAlH$_4$
reagent to give the cyclopentadienyl ligand **12** (Scheme 8-19). The enantioselectivity of the reduction was a modest 17.3% e.e. [46]. Additional examples of fulvenes
being used to prepare chiral monosubsituted cyclopentadienes utilized steroidal
precursors. In one synthesis steroidal ketone **70** was used to form fulvene **71** which
was further converted to the cyclopentadiene **72** [47]. In this case, a 3:1 mixture of
epimers were obtained due to incomplete selectivity in the addition of methyllithium to fulvene **71**. In a second synthesis 5α-cholestan-3-one **73** was similarly converted to a fulvene which underwent selective methyllithium addition to give only

the single diastereomer **74** [48]. In both cases, the double bonds in the cyclopentadiene existed as a mixture of regioisomers.

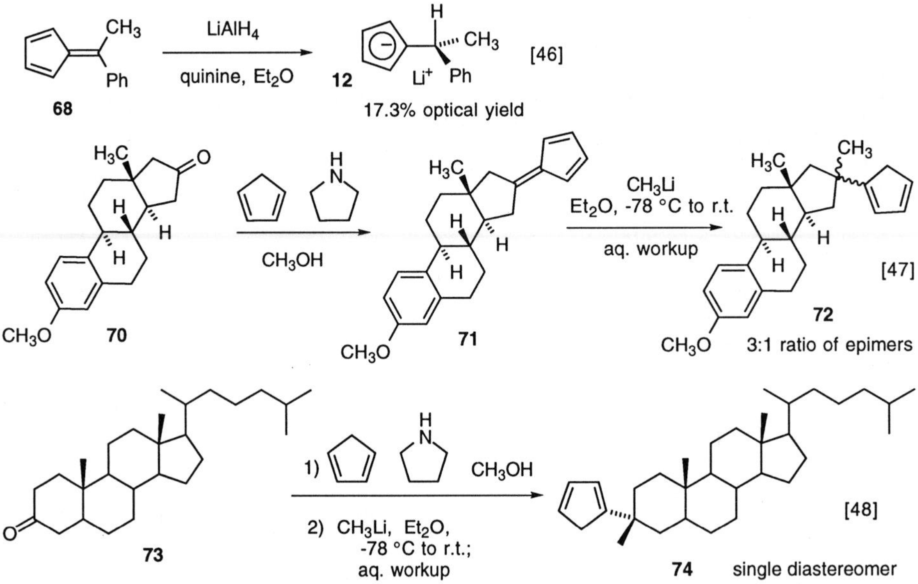

Scheme 8-19. Preparation of nonracemic monosubstituted
cyclopentadienes from fulvenes.

In 1978, Kagan and Cesarotti reported the first example of alkylating a cyclopentadienyl anion by a chiral nonracemic alkylating agent (Scheme 8-20). They employed the tosylates of menthol and neomenthyl to give natural product-derived neomenthylcyclopentadiene **75** and menthylcyclopentadiene **76** [3]. This menthyl framework was later extended to the (8-phenylmenthyl)cyclopentadiene **77** [5b]. The (2-methylbutyl)cyclopentadiene **78** and (2-phenylbutyl)cyclopentadiene **79** were also prepared using alkylating agents derived from the chiral pool [49]. Erker has demonstrated the use of steroid-derived alkylating agents to form chiral cyclopentadienes **80** and **81** [48].

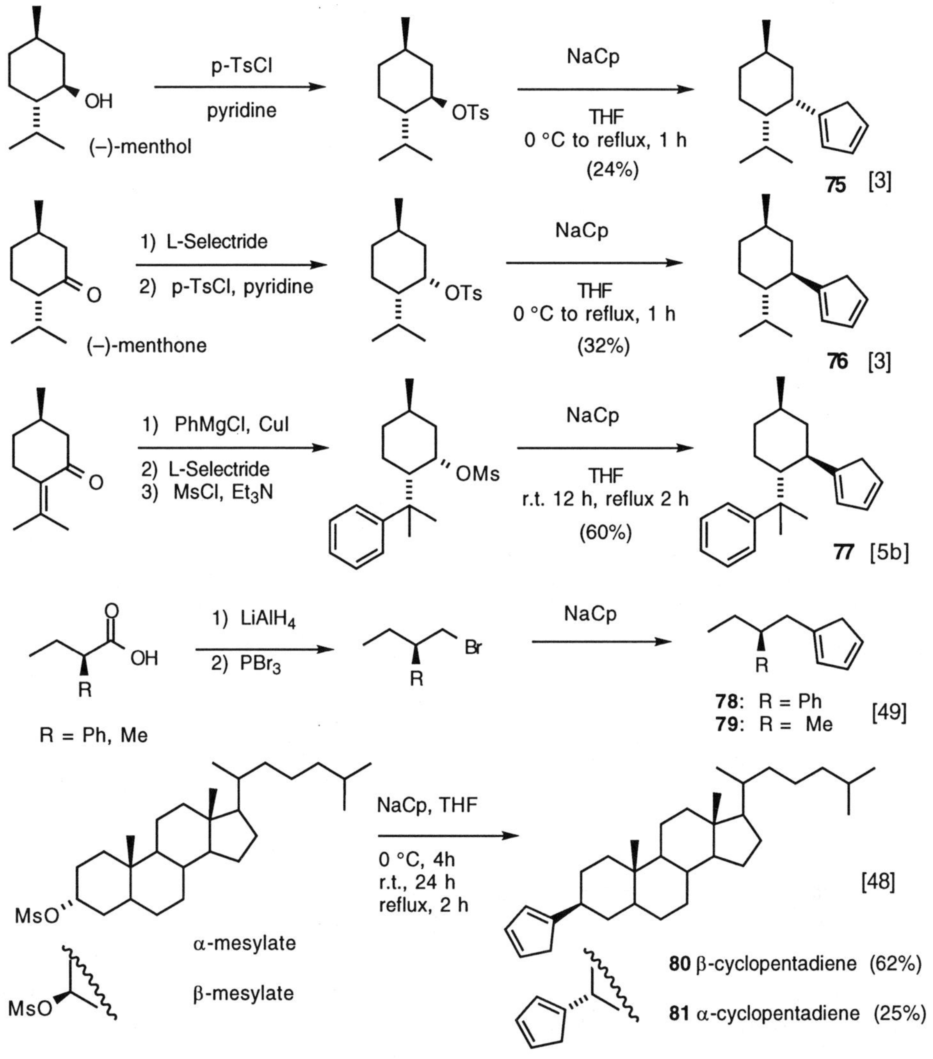

Scheme 8-20. Preparation of chiral cyclopentadienes by alkylation.

Palladium-catalyzed alkylations of chiral allyl acetates by cyclopentadienylsodium have been used to prepare the nonracemic cyclohexenylcyclopentadiene **82** as well as the terpene-derived cyclopentadienes **83–85** (Scheme 8-21) [50].

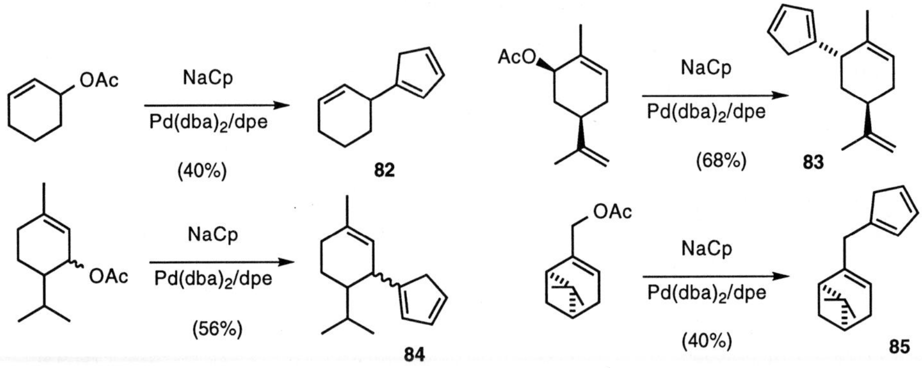

Scheme 8-21. Chiral cyclopentadienes by palladium-catalyzed alkylation [50].

Chiral Monosubstituted Cyclopentadienes Containing Donor Atoms. With the widespread occurrence of oxygen and nitrogen-containing chiral natural products, the possibilities for incorporating derivatives having donor atoms in chiral cyclopentadienes are numerous. Examples of several donor substituted cyclopentadienes have been prepared and are depicted in Scheme 8-22 [51–54].

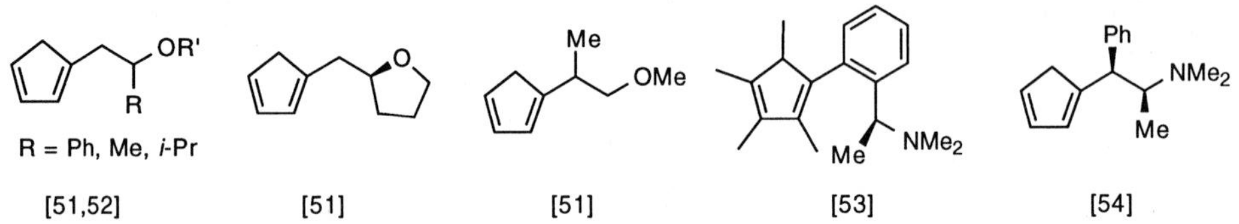

Scheme 8-22. Chiral donor-substituted cyclopentadienes.

Disubstituted Cyclopentadienes. The major problem in synthesizing disubstituted cyclopentadienes is controlling the selectivity for 1,2-disubstituted versus 1,3-disubstituted cyclopentadienes. In the dialkylation of cyclopentadiene, it is apparent that the 1,2-product is favored by electronic factors and the 1,3-product is favored by steric factors. With small alkyl groups, the 1,2-isomer is favored while with larger groups, the 1,3-product is favored (Scheme 8-23). For example, the dimethylation of cyclopentadiene favors the 1,2-dimethylcyclopentadiene over the 1,3-isomer in a 3.5:1 ratio [55]. The reaction of methylcyclopentadienylsodium with 1,2-dibromoethane still favors the 1,2-product **86** [55]. Diethylation moderately favors the 1,2-isomer in a 57:43 ratio [56]. Only with the dialkylation with isopropyl groups does the steric interaction predominate giving the 1,3-isomer in a 93:7 ratio [56]. The alkylation of methylcyclopentadiene with long chain *n*-alkyl groups or with secondary alkyl bromides gave approximately 1:1 mixtures of the 1,2- and 1,3-isomers [37]. The alkylation of *tert*-butylcyclopentadienylmagnesium complexes with *tert*-butylhalides give the 1,3-di-*tert*-butylcyclopentadiene as the only regioisomer [35, 57]. The *tert*-butyl group can also direct 1,3-alkylation with smaller reagents as seen in the formation of the ethanol derivatives **87** and **88** [58].

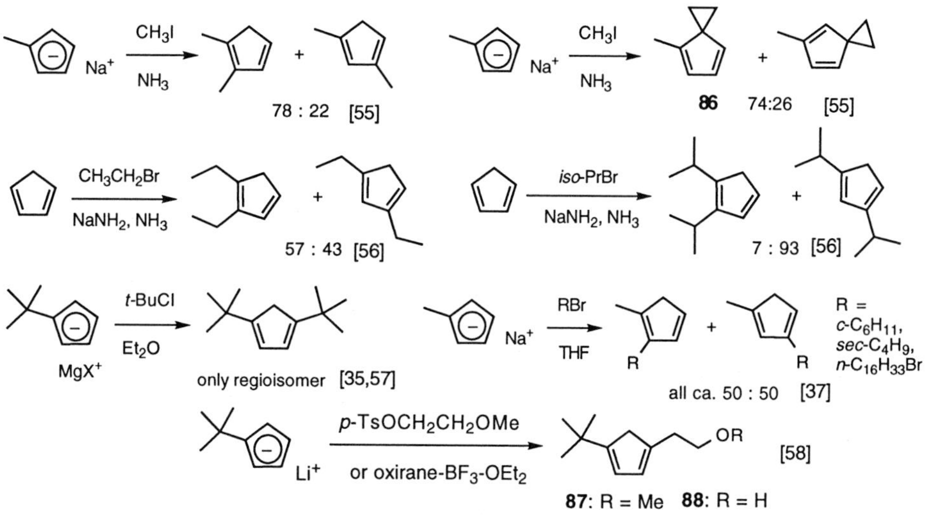

CH₃I / NH₃ , Na⁺   78 : 22 [55]

CH₃I / NH₃ , Na⁺   **86** 74:26 [55]

CH₃CH₂Br / NaNH₂, NH₃   57 : 43 [56]

iso-PrBr / NaNH₂, NH₃   7 : 93 [56]

t-BuCl / Et₂O , MgX⁺   only regioisomer [35,57]

RBr / THF , Na⁺   R = c-C₆H₁₁, sec-C₄H₉, n-C₁₆H₃₃Br   all ca. 50 : 50 [37]

p-TsOCH₂CH₂OMe or oxirane-BF₃-OEt₂ , Li⁺   OR [58]

87: R = Me **88**: R = H

Scheme 8-23. Preparation of disubstituted cyclopentadienes by dialkylation.

The condensation of ketones with monosubstituted cyclopentadienes can lead to very selective formation of fulvenes (Scheme 8-24). In an earlier procedure wherein sodium ethoxide is used as a base for the condensation of acetone with methylcyclopentadiene, an 80:20 ratio of 3-methyl and 2-methyl fulvenes **89** and **90** was formed [21]. Stone and Little have established a mild, high yield procedure for the formation of fulvenes using pyridine as the base [59]. This method has been used to selectively generate the 1,3-di-*tert*-butylcyclopentadiene **92** [60, 61]. Collins has applied this method to give exclusively the bis(3-substituted fulvene) **93** [62].

The addition of organometals to 3-substituted cyclopentenones has been used to selectively prepare 1,3-diphenylcyclopentadiene **94** [63] as well as 1-ethyl-3-methyl-, 1,3-dimethyl- and 3-methyl-1-phenylcyclopentadienes [8, 64, 65] (Scheme 8-25). 2-Methylcyclopent-2-enone has been converted by a similar method to 1,2-dimethylcyclopentadiene **98** [66]. This same product **98** is also available through a Friedel–Crafts cyclization method [67]. The novel synthesis of 1,2-di-*tert*-butylcyclopentadiene **100** utilizes an intramolecular pinacol coupling of diketone **99** followed by a double dehydration [68].

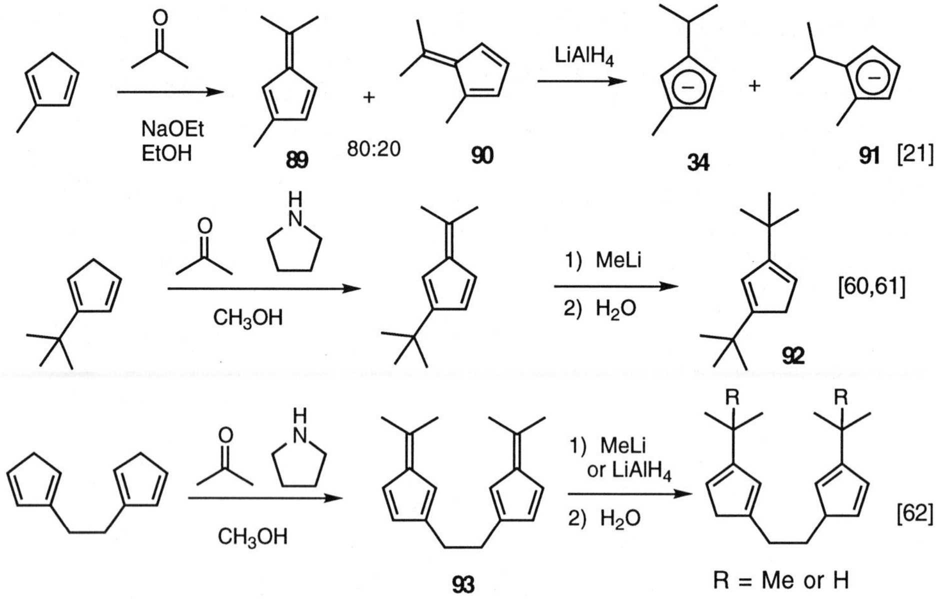

Scheme 8-24. Fulvene route to disubstituted cyclopentadienes.

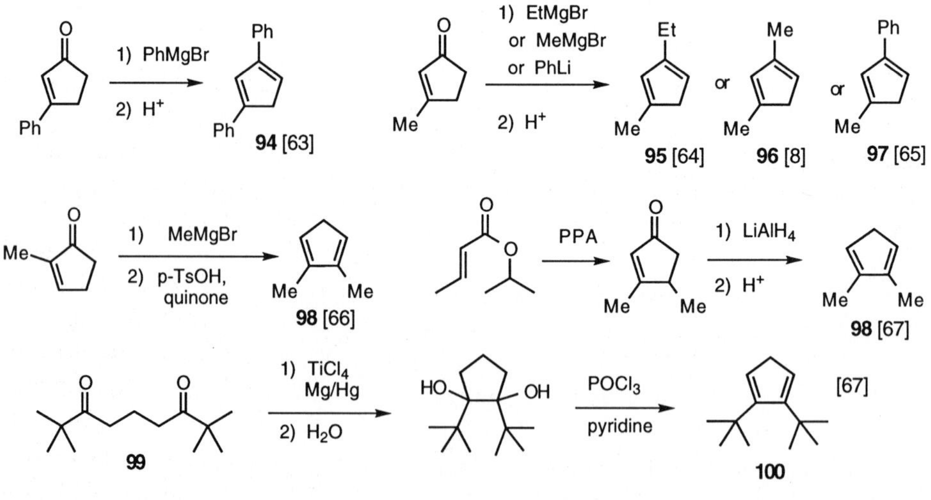

Scheme 8-25. Selective formation of 1,2- or 1,3-disubstituted cyclopentadienes.

Annelated Achiral Disubstituted Cyclopentadienes. The direct formation of annelated cyclopentadienes through the reaction with dihaloalkanes generally fails due to the kinetically more favorable generation of spiro products. Thus, the alkylation of cyclopentadienylsodium with 1,4-dibromobutane leads to the spiro product **101** [69] (Scheme 8-26). Fortunately, **101** can undergo a sigmatropic rearrangement to give tetrahydroindene **102** in good yield when the conditions are carefully controlled to prevent the thermodynamically favored rearrangement to **103** which ultimately leads to indane [69, 70]. The Skattebøl rearrangement from **104** has been used in an alternate synthesis of tetrahydroindene **102** [71]. Paquette has nicely extended this method of using dibromocyclopropanes to form intermediate carbenes which can rearrange to cyclopentadienes [72]. Another strategy for the synthesis of annelated cyclopentadienes has been to modify existing ring systems. For example, the Diels–Alder adducts **106** from cyclopentadiene and **109** from cyclopentadiene and cyclohexadiene, can be selectively hydrogenated and oxidized to the allylic alcohols **107** and **110**. Dehydration of **107** and **110** leads to the 4,7-methanotetrahydroindene **108** and the 4,7-ethanotetrahydroindene **111** [38].

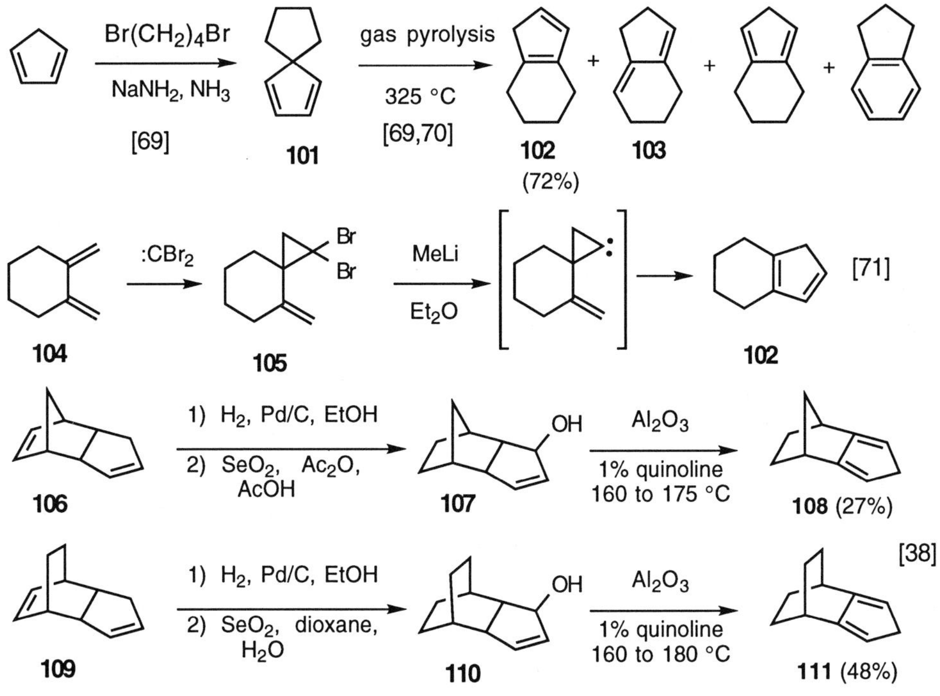

Scheme 8-26. Preparation of annelated achiral cyclopentadienes.

Annelated Chiral Disubstituted Cyclopentadienes. Reacting chiral dialkylating agents with cyclopentadienes can lead to chiral annelated cyclopentadienes (Scheme 8-27). In one case using the tartrate-derived ditosylate **112** the double alkylation lead directly to the ring fused product **113** due to the strain in the possible spiro product **114** [5]. A second C_2-symmetric cyclopentadiene **117** was generated from the racemic ditosylate **115**—going through the thermal rearrangement of spiro intermediate **116**. The enantiomerically enriched ligand could be made by replacing the toluenesulfonate groups with camphorsulfonates and resolving the sulfonates [73]. Related nonracemic bicyclooctane-annelated cyclopentadienes **120** and **121** bearing methyl or isopropyl groups were prepared from diols **118** which could be made in high enantiomeric purity through an asymmetric hydroboration reaction [19]. 2,2'-Di(bromomethyl)-1,1'-binaphthyl **122** could be converted the spiro-annelated cyclopentadiene **123** to the C_2-symmetric fused cyclopentadiene **124** [74].

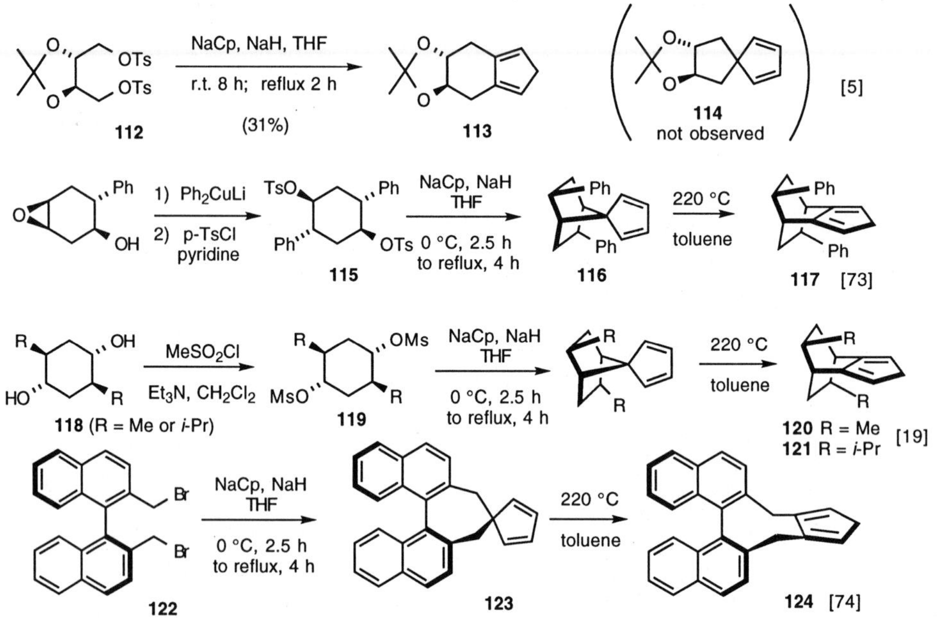

Scheme 8-27. Double alkylation route to chiral
annelated disubstituted cyclopentadienes.

A number of related syntheses have appeared in which the cyclopentadienyl ring is generated through a cyclopentenone synthesis (Scheme 8-28). In the first example, camphor (**125**) was converted to phosphonate ester **126** which underwent an intramolecular Waadsworth–Horner–Emmons reaction to form cyclopentenone **127**. Reduction and dehydration of **127** gave cam-CpH **128** [5]. A modified version of this synthesis has also appeared [25, 26]. A similar strategy was used to convert verbenone through cyclopentenone **129** to cyclopentadiene **130** [75]

Paquette has used the Skattebøl rearrangement to produce cam-CpH **128** and pin-CpH **134** from camphor [25] or the nopol-derived tosylate **132** [72] (Scheme 8-29). In the first example, the needed diene **131** was made by a coupling reaction and in the second case, **133** was produced by an elimination. The cyclopropana-

tion, carbene formation and rearrangement gave the cyclopentadienes in good overall yield.

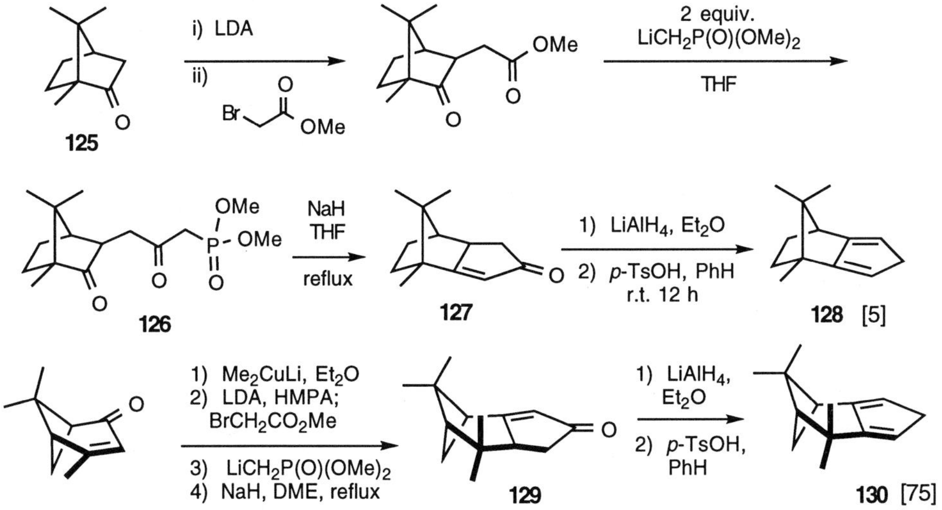

Scheme 8-28. Cyclopentenone route to annelated disubstituted
chiral cyclopentadienes.

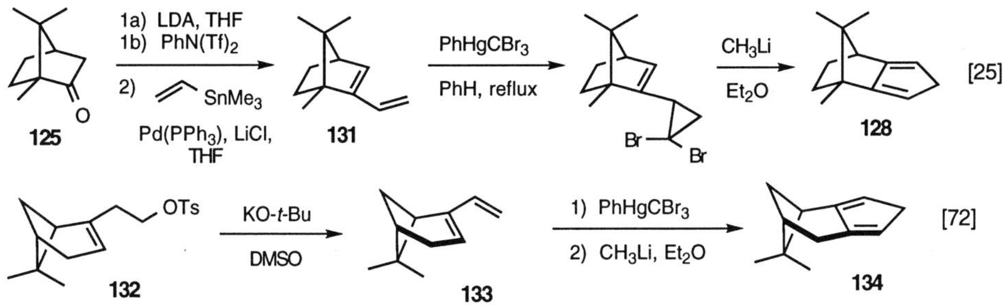

Scheme 8-29. Skattebøl route to chiral disubstituted annelated cyclopentadienes.

Achiral Trisubstituted Cyclopentadienes. The synthesis of trisubstituted cyclopentadienes through the polyalkylation of cyclopentadienes generally suffers from
poor regioselectivity (Scheme 8-30). The polymethylation of methylcyclopentadiene gave a mixture of 1,2,3-, 1,2,4-, 1,1,2-timethyl isomers in addition to mono-,
di-, tetra-, penta- and hexamethyl derivatives [55]. The alkylation of a (presumably
93:7) mixture of diisopropylcyclopentadienes surprisingly gave a 22:78 mixture of
the 1,2,3- and 1,2,4-isomers [76]. Evidently the more hindered site between the
isopropyl groups must be unusually reactive for this ratio to form. The more hindered 1,3-di-*tert*-butylcyclopentadiene reacts with *tert*-butyl chloride to give only

the sterically less hindered 1,2,4- isomer [77]. Similar results are obtained with two trimethylsilyl groups and one *tert*-butyl group **135** [78] or with three trimethylsilyl groups **136** [79].

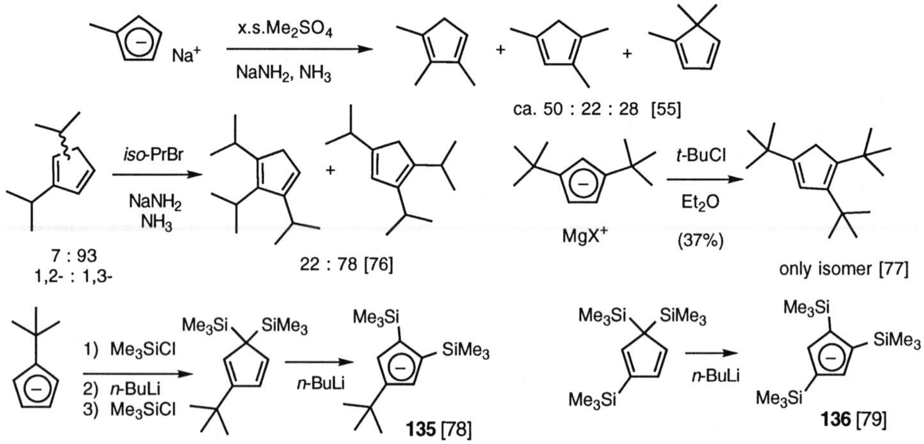

Scheme 8-30. Synthesis of trisubstituted cyclopentadienes through alkylations.

1,2,4-Trimethylcyclopentadiene **138** can be selectively prepared through a Grignard addition to enone **137** followed by dehydration (Scheme 8-31) [67]. The 1,2,3-trimethyl derivative **140** can be selectively prepared through reduction and dehydration of the readily available cyclopentenone **139** [80].

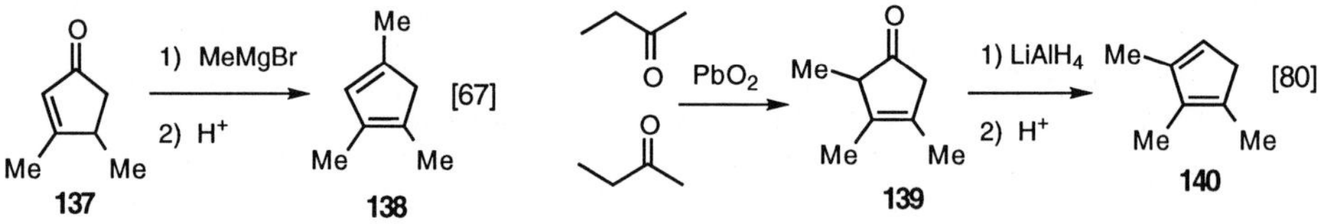

Scheme 8-31. Selective syntheses of trisubstituted cyclopentadienes.

The Friedel–Crafts cyclization of **141** gives cyclopentenone **142** [81] which can be reduced and dehydrated to give 2-methyltetrahydroindene **143** (Scheme 8-32) [82]. Aryl Grignard reagents can add to cyclopentenone **144** to give after dehydration 2-aryltetrahydroindene **145** [83]. An apparent exception to the usual lack of selectivity in the alkylation of sterically unhindered cyclopentadienes is the selective alkylation of norborna-CpH **146** to give the 1,2,4-substition pattern seen in **147** and **148** [84].

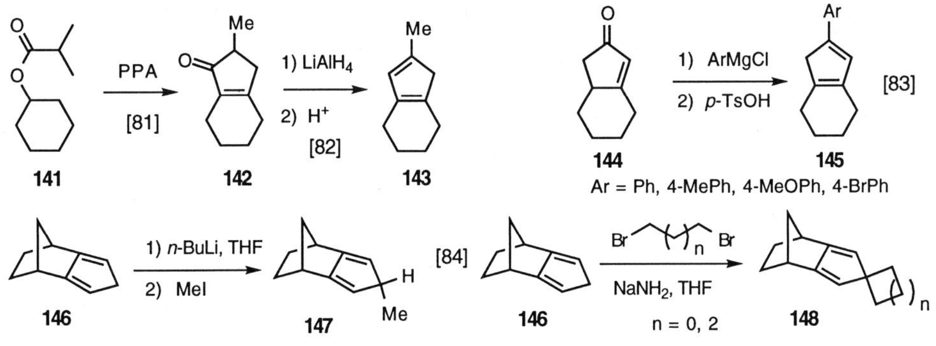

Scheme 8-32. Achiral annelated trisubstituted cyclopentadienes.

Chiral Trisubstituted Cyclopentadienes. Alkylation of BCOCp **120** with dibromoethane gave a mixture of 1,2,3- and 1,2,4- isomers **149** and **150** whereas fulvene formation was selective for the 1,2,4- isomer **151** [20] (Scheme 8-33). Camphor-derived trisubstituted cyclopentadienes **154** were regioselectively generated using a Nazarov cyclization of bis(allylic)alcohols **153**. These alcohols were readily prepared from the camphor-derived vinyllithium complex **152** [85].

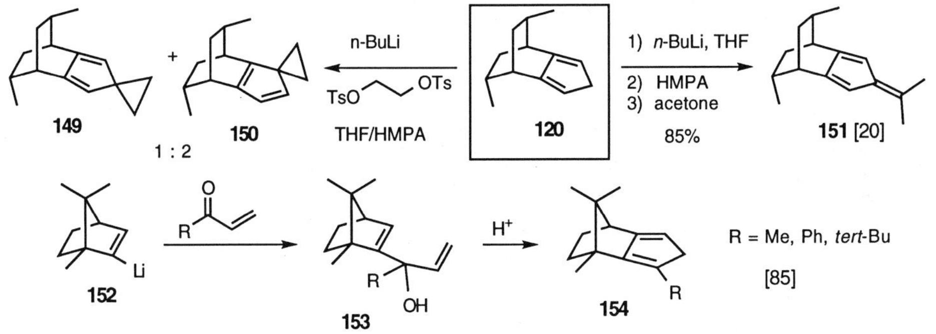

Scheme 8-33. Chiral trisubstituted annelated cyclopentadienes.

Tetrasubstituted Cyclopentadienes. A thorough review of very hindered cyclopentadienes has been recently published [1c] and only representative cyclopentadienes will be covered here. Tetramethylcyclopentadiene has been generated by a Grignard addition to trimethylcyclopentenone **139** followed by dehydration or by reduction and dehydration of tetramethylcyclopentadiene **156** [80] (Scheme 8-34). The alkylation of an isomeric mixture of triisopropylcyclopentadienes by isopropyl bromide gave the 1,2,3,4-tetrasubstituted cyclopentadiene **157** in good yield [76]. The pinacol coupling of tetraphenyl diketone **158** followed by double dehydration gave the tetraphenylcyclopentadiene **159** [86].

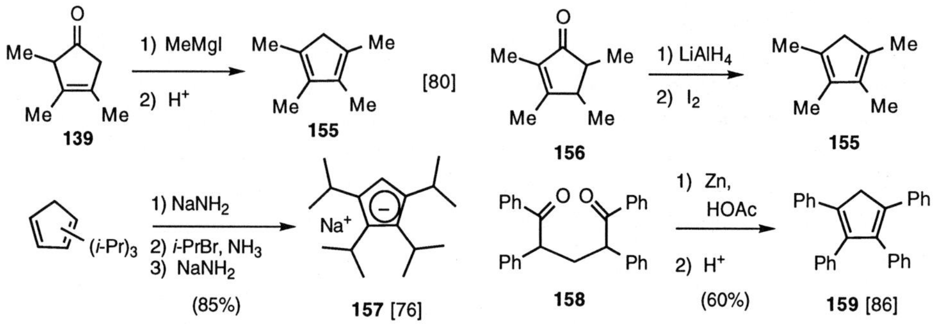

Scheme 8-34. Achiral tetrasubstituted cyclopentadienes.

Chiral Tetrasubstituted Cyclopentadienes. Two examples of tetrasubstituted chiral cyclopentadienes have been reported (Scheme 8-35). In each case a Nazarov ring closure approach was used in their synthesis. Addition of camphor-derived or verbenone-derived vinyl lithium complexes **152** or **160** to ethyl formate generated bis(allylic)alcohols which could be cyclized under acidic conditions to give good yields of the cyclopentadienes **161** [87] or **162** [88].

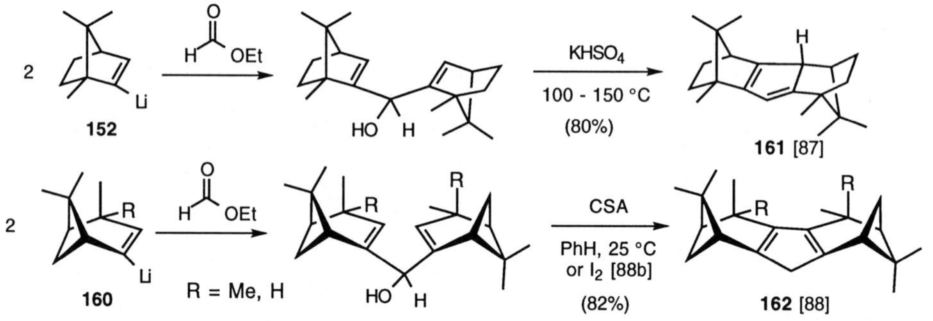

Scheme 8-35. Chiral tetrasubstituted cyclopentadienes.

Pentasubstituted Cyclopentadienes. Since the original publication of a pentamethylcyclopentadiene synthesis by de Vries in 1960 [89], several modified or new routes have appeared (Scheme 8-36). In most cases tetramethylcyclopentenone **156** is prepared and converted through the addition of methyllithium or methyl Grignard followed by dehydration to pentamethylcyclopentadiene **163** [89, 90]. Two related syntheses convert 3-pentanone to pyrone **164** through sequential aldol reactions. Pyrone **164** can then be cyclized under acidic conditions to cyclopentenone **156** [90]. Alternatively, tiglic acid can be esterified to 2-butyl tiglate **165** which is similarly cyclized to give **156** [91]. A variant of this route can also be used to make ethyltetramethylcyclopentadiene [91]. A novel approach was taken by Bercaw in which two equivalents of 2-buten-2-yllithium are added to an ester to give divinylcarbinol **166** which can undergo a cationic π-cyclization to give directly pentaalkylcyclopentadiene **163** [92]. This route has the distinct advantage of readily incorporating various substituents by starting with the appropriate ester. In this way the ethyl-, *n*-propyl-, *n*-butyl- and phenyltetramethylcyclopentadienes were prepared. Pentaethylcyclopentadiene has also been prepared from 4-heptanone going through a cyclopentenone route [93].

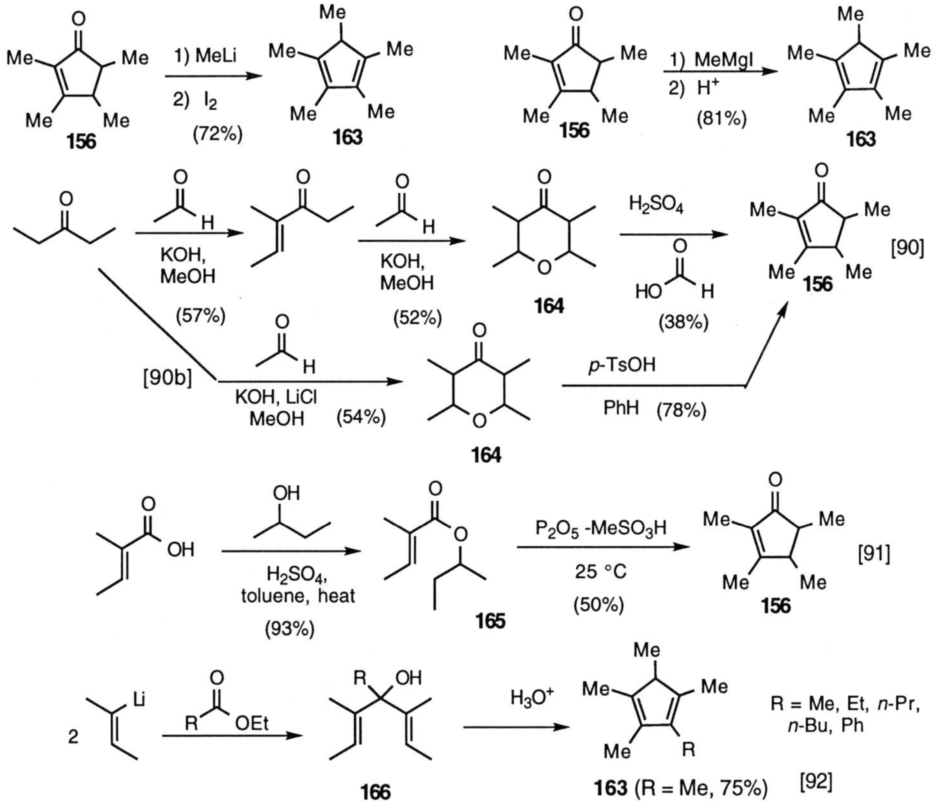

Scheme 8-36. Syntheses of pentamethylcyclopentadiene.

Chiral Pentasubstituted Cyclopentadienes. Bercaw's method of converting various esters to pentasubstituted cyclopentadienes [92] has been used to incorporate a chiral substituent into a pentasubstituted cyclopentadiene (Scheme 8-37). Addition of 2-buten-2-yllithium to the chiral ester **167** followed by cyclization of the intermediate divinylcarbinol gave (1-phenylpropyl)tetramethylcyclopentadiene **168** in high overall yield [94]. 1,2,3-Trimethylcyclopentadiene has been alkylated with the binaphthyl-based chiral dibromide **122** to give the chiral annelated pentasubstituted cyclopentadiene **169** [95].

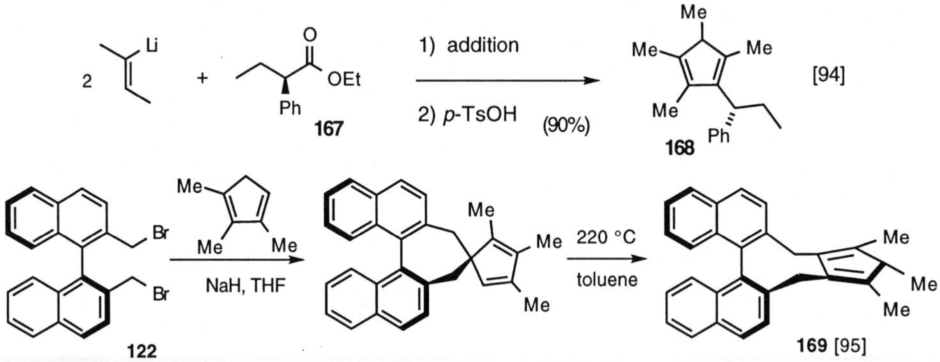

Scheme 8-37. Chiral pentasubstituted cyclopentadienes.

8.3.2 Indenes

1-Substituted Indenes. As with the generation and alkylation of cyclopentadienyl anions, the facile deprotonation of indene (pKa = 22) [96] enables its anion to re-act with alkylating agents to form a variety of alkylated indenes. Unlike the nonre-gioselective alkylation of most substituted cyclopentadienyls, the alkylation of the indenyl anion is electronically strongly preferred at the 1-position since alkylation at the 2-position would lead to the electronically less stable isoindene [97]. While the initial product is a 1-substituted indene, typical alkylation conditions generally promote the base-catalyzed double bond migration to form an equilibrium mixture of the 1- and 3-substituted indenes (Scheme 8-38) [97]. With most substituents such as alkyl or aryl groups, the 3-substituted indene is preferred. Since our main interest in substituted indenes is as the anionic ligand, it does not much matter which double bond isomer is present in the indene—the 1-substituted indenyl com-plex will be formed either from the 1- or 3-substituted indene.

The preparation of a number of 1- (or 3-) substituted indenes is summarized in Scheme 8-38. The alkylation yields are generally quite good, with only the alkyla-tion of indenyl anion by *tert*-butylbromide giving lower yields [97]. Through the careful control of reaction conditions, most notably the use of several equivalents of alkylating reagents, it is also possible to prepare unequilibrated 1-substituted in-denes if desired [98]. The introduction of a trimethylsilyl group on the 1-position is readily accomplished in high yield. While the trimethylsilyl group prefers to re-main on the allylic 1-position, it undergoes a 1,3-migration at room temperature [99, 100]. 3-(Methylthiol)indene can be prepared in good yield by the reaction of indenyllithium with dimethyldisulfide [101].

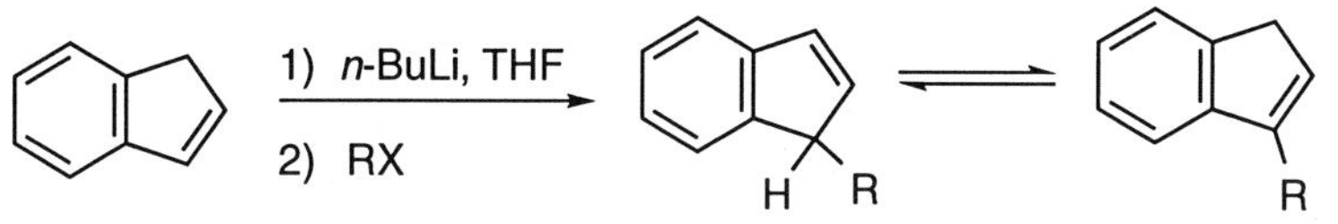

Mel (-20 °C, 18 h, 93%), EtBr (5 °C, 18 h, 74%), *i*-PrBr (25 °C, 11 h, 69%),
t-BuBr (35 °C, 8 h, 45%), PhCH$_2$Cl (25 °C, 4 h, 84%), Ph$_2$CHCl (25 °C, 4 h, 73%)

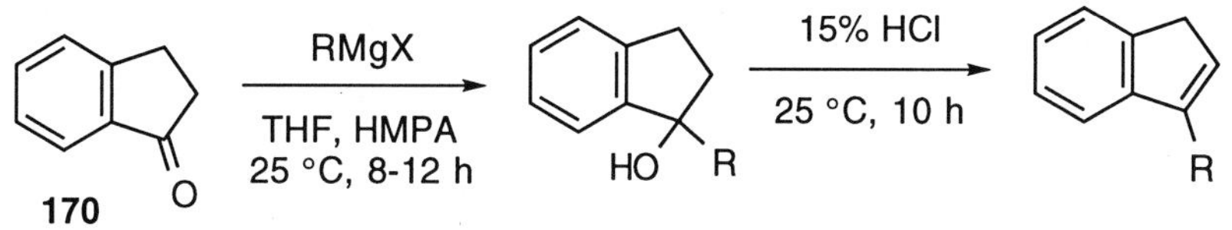

[R = Me (94%), Et (85%), *n*-Pr (90%), *n*-Bu (81%)]

Scheme 8-38. Preparation of 1- (or 3-)substituted indenes [97, 98].

A second general method for preparing 1-substituted indenes is through the addition of alkylmetals to 1-indanone **170** followed by acid-promoted dehydration. In this manner, simple alkyl groups could be introduced in good yields [102]. The addition of *para* substituted phenyl Grignard reagents to 1-indanone followed by dehydration was used to prepare a number of 3-arylindenes [96]

2-Substituted Indenes. Several methods for the preparation of 2-substituted indenes have been reported (Scheme 8-39). The alkylation of the enolate of 1-indanone (**170**) with alkyl halides followed by ketone reduction and dehydration is the least general with only the introduction of methyl being satisfactory [102]. A related method involves reaction of the aldol product **171** with either sodium borohydride or methylmagnesium iodide/Cu$_2$Cl$_2$ to give after ketone reduction and dehydration the 2-isopropylidene or 2-*tert*-butylidene [103]. The addition of *n*-alkyl or aryl Grignard reagents to 2-indanone (**172**) followed by dehydration can provide some 2-substituted indenes in good yields [14, 96, 104]. The addition of secondary Grignard reagents to 2-indanone, however, does not proceed well due to competing deprotonation or reduction of the ketone [15]. Perhaps the most general route to 2-substituted indenes is the nickel-catalyzed cross-coupling between a wide variety of alkyl or aryl Grignard reagents with 2-bromoindene (**173**) [102, 105]. This method even allows for the coupling of secondary and tertiary Grignard reagents [105].

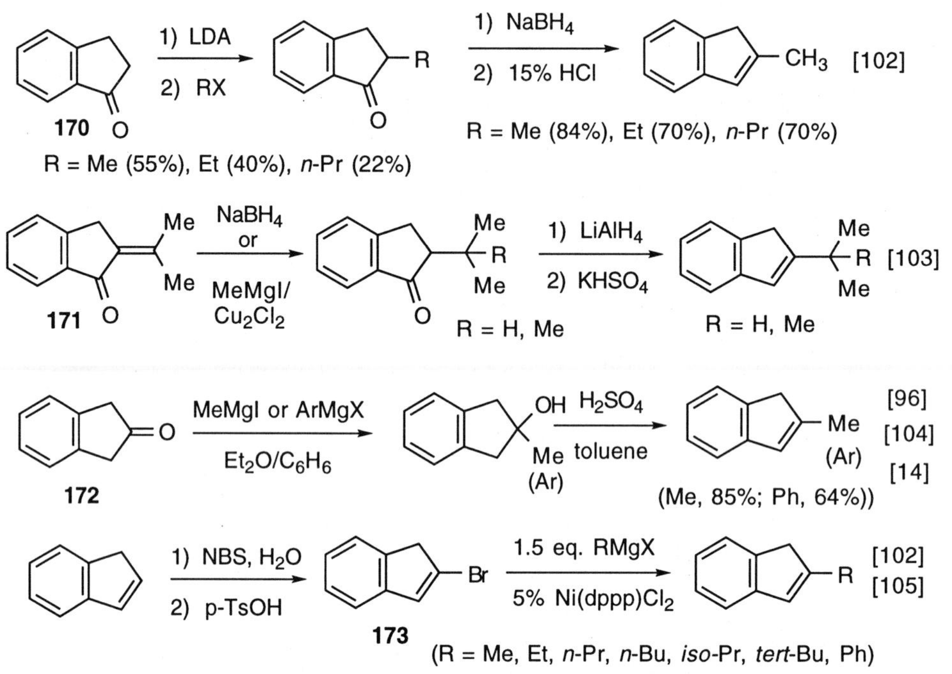

Scheme 8-39. Preparation of 2-substituted indenes.

Indenes with Substitution on the 6-Membered Ring. Most preparations of indenes bearing substituents on the 4-, 5-, 6- or 7-position utilize Friedel–Crafts chemistry to generate an indanone which can be reduced and dehydrated to the substituted indene (Scheme 8-40). When only one substituent is desired at the 4- (or 7-) position, the appropriate 3-(2-substituted phenyl)propionic acid is required. 4-Phenyl substituted indene **177a** can be built up through homologation of 2-phenylbenzyl bromide **174** to the propionyl chloride **175** which can be cyclized to the 4-phenylindanone **176** [106]. Reduction and dehydration gave 4-phenylindene **177a**. This method has also been applied in the preparation of related 4-substituted indenes. Through an analogous reaction sequence using 3-aryl-2-methylpropionic acid, the 2-methyl-4-phenylindene **177b** could be made [106].

The base-catalyzed condensation of ketoaldehyde **178** with cyclopentadiene gives 4-isopropylindene **179** directly [107]. Conversion of **179** to 2-indanone **180** enables the incorporation of an additional substituent at the 2-position through the addition of a Grignard reagent followed by dehydration. For example, 2-methyl-4-phenylindene **181** was formed in this manner [107]. Another general route to 4-substituted indenes utilizes a nickel-catalyzed cross-coupling reaction between 4-bromoindene **183** and a variety of Grignard reagents [102]. The main drawback of this route is the multistep Friedel–Crafts based synthesis of 4-bromoindene from 2-bromobenzyl bromide **182**.

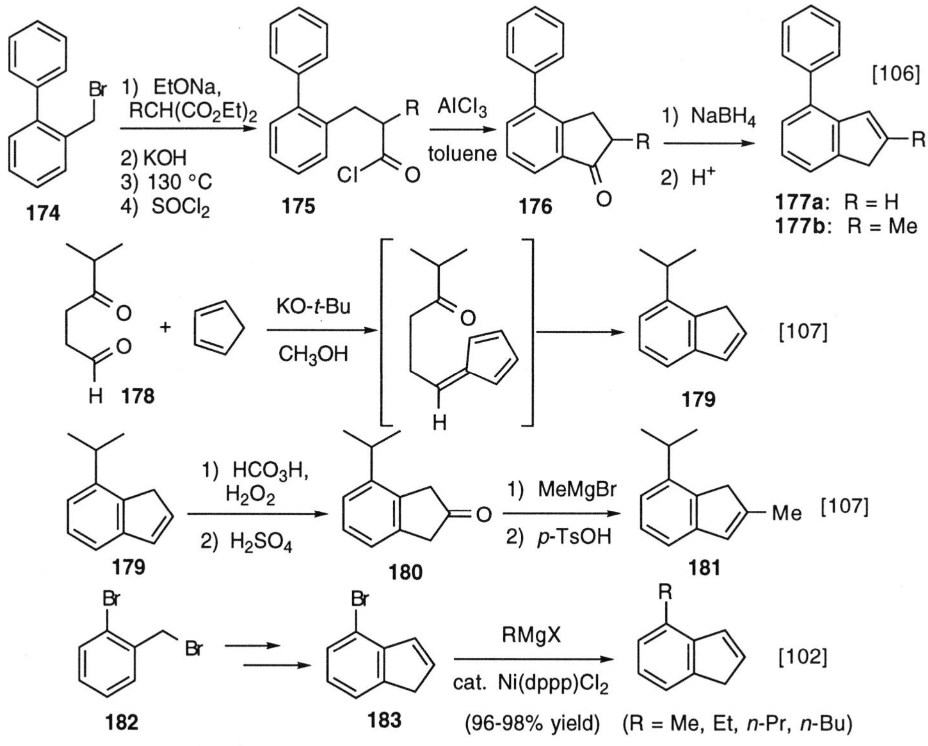

Scheme 8-40. Preparation of 4-substituted indenes.

As shown in Scheme 8-41, 4,7-disubstituted 1-indanones **185** are formed in good yield through the Friedel–Crafts acylation of electron-rich *para* substituted benzenes **184** with 3-chloropropionyl chloride followed by acid-promoted cyclization. Reduction and dehydration give the 4,7-disubstituted indenes **186**. Electron poor indenes, such as the 4,7-difluoroindene **186c** can be formed through conversion of the benzaldehyde **187** to the phenylpropionyl chloride **188** which can undergo a Friedel–Crafts acylation. Reduction and dehydration give the 4,7-difluoroindene **186c**. Indenes bearing substituents in the 5,6-positions have also been prepared through Friedel–Crafts cyclization of the appropriate 3-(3,4-disubstiuted phenyl)propionic acids [18b]. These syntheses are less efficient since more than one regioisomer can form. The 5,6-dimethoxy-, 5,6-dimethyl- and 5,6-dichloroindenes have been reported [18b, 108]. A very facile synthesis of 4,7-dimethylindene **186a** is available through the condensation of 2,5-hexanedione with cyclopentadiene [109].

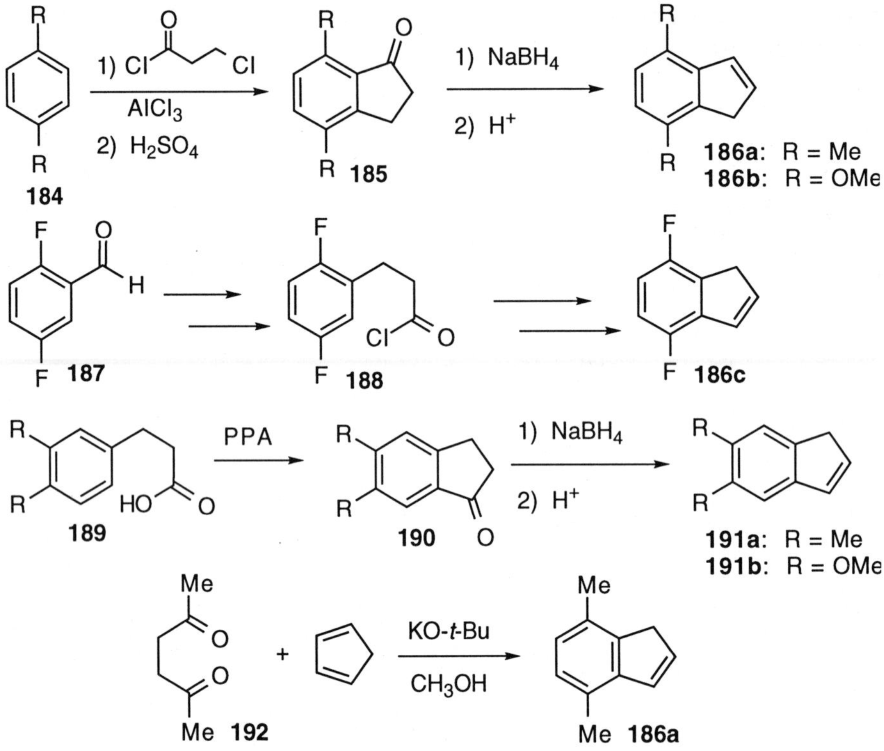

Scheme 8-41. Friedel–Crafts routes to 4,7- or 5,6-disubstituted indenes.

Polymethylated Indenes. More highly substituted indenes have also been prepared using Friedel–Crafts methodology (Scheme 8-42). The PPA-promoted acylation of *para*-xylene with methylacylic acid gave indanone **193** directly in fair yield. Reduction and dehydration gave 2,4,7-trimethylindene **194** [110a]. A related reaction sequence using tiglic acid chloride and *para*-xylene produced 2,3,4,7-tetramethylindene **196** [110a]. Similarly, 1,2,3,4-tetramethylbenzene and tigloyl chloride could be converted to hexamethylindanone **197** which upon addition of methyllithium followed by dehydration gave 1,2,3,4,5,6,7-heptamethylindene **198** [110b].

Benzannelated indenes have been prepared through two related synthetic approaches (Scheme 8-43). 2-(Bromomethyl)naphthalene could be alkylated with malonate diester to give after decarboxylation and activation the acid chloride **200**. Intramolecular Friedel–Crafts acylation of **200** followed by reduction and dehydration gave benz[*e*]indene **201a** [106]. Following the same reaction sequence but using methyl malonate diester, 2-methylbenz[*e*]indene **201b** could be formed. In each case both double bond isomers were present. The second approach utilizes the intermolecular acylation of naphthalene followed by cyclization to give either the methylbenzindanone derivative **202** or the benzindanone **203**. Regioisomers of both of these indanones were also formed. Reduction and dehydration led to the benz[*e*]indenes **201b** and **201a** [111].

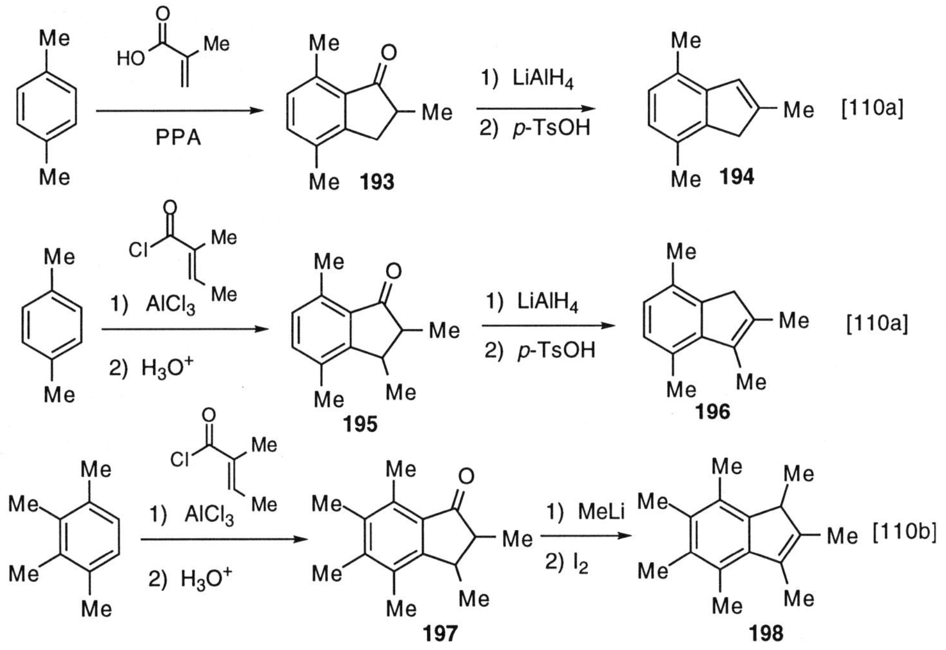

Scheme 8-42. Synthesis of polymethylated indenes [110].

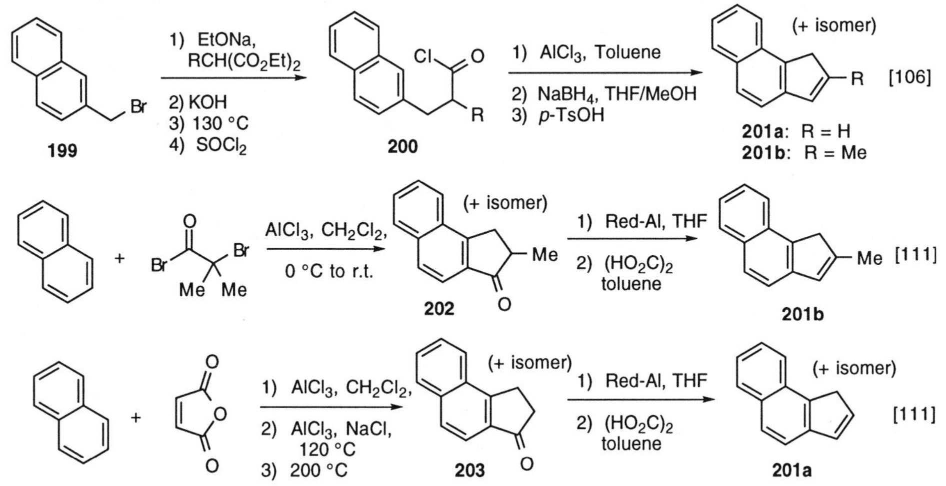

Scheme 8-43. Preparation of benzannelated indenes.

Indenes with Chiral Substituents. A limited number of indenes bearing chiral substituents have been prepared (Scheme 8-44). The most common method has been to treat indenyllithium with a chiral alkylsulfonate. In this way, 3-(neoisopinocamphyl)indene **205** was prepared from isopinocampheol **204** [112], and a series of menthyl **208**, neomenthyl **206**, neoisomenthyl **207** and isomenthyl **209** derivatives of indene [29]. Palladium-catalyzed cross-coupling reactions between menthylmagnesium chloride **211** and 3-indenyl triflate **210** or 2-bromoindene **173** provide another facile route to 3-menthylindene **208** and 2-menthylindene **212** [15, 114]. An example of an annelated chiral indene is the bicyclooctane-annelated indene **214** which was prepared by the bis(alkylation) of indene by chiral dimesylate **119** followed by a thermally-induced sigmatropic rearrangement of spiroindene **213** [115].

((Author - I cannot find reference 113 cited in the text))

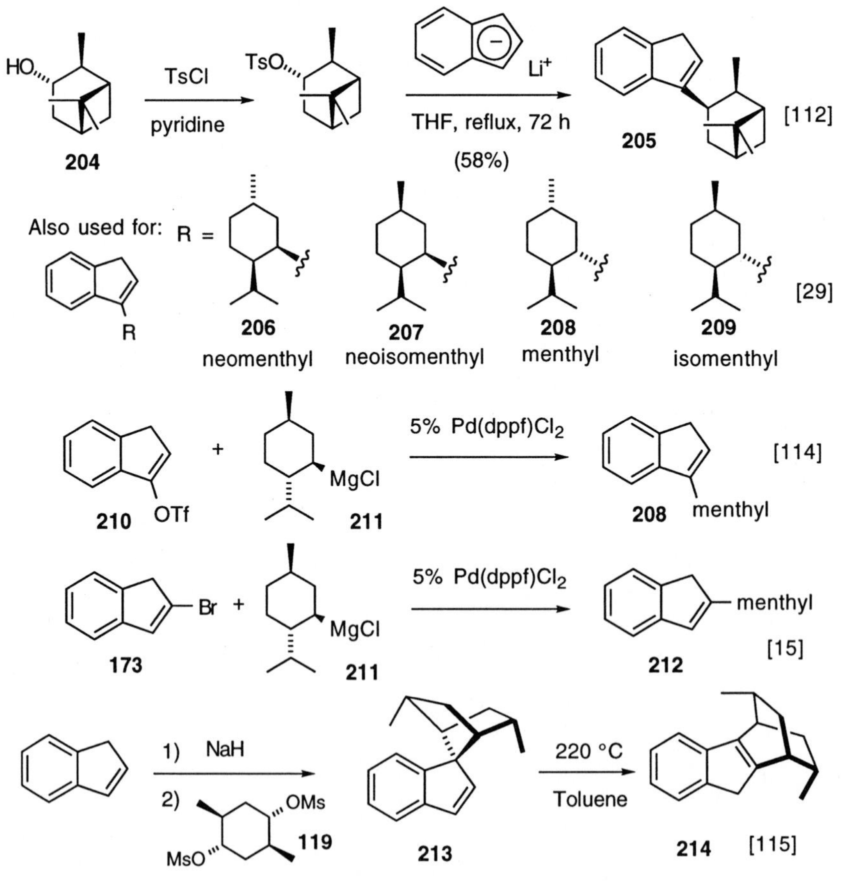

Scheme 8-44. Preparation of chiral indenes.

8.4 Bridged Ligands with Nonstereogenic Bridging Groups

8.4.1 Bis(cyclopentadienes)

The coverage of bis(cyclopentadienes) will be grouped according to an increasing level of complexity in the ligands. The major divisions are between nonprochiral ligands, prochiral ligands and ligands with chiral substituents. Ligands with stereogenic bridges will be covered in Sec. 8.5.

Unsubstituted Cyclopentadienes. Cyclopentadienes linked by simple alkyl tethers can be prepared by reacting the appropriate alkyl dihalide with cyclopentadienyl anion (Scheme 8-45). In this way, methylene chloride reacted to give the methylene-bridged bis(cyclopentadiene) **215** (only one of the double bond isomers shown) which could be deprotonated to the more stable dilithio salt **216** [116]. 1,2-Dibromoethane could be converted under improved conditions to ethylene-bridged bis(cyclopentadiene) **217** in good yield [62, 117]. The use of the magnesium salt of cyclopentadiene has proven to be crucial in minimizing the spiroalkylation of cyclopentadiene. 1,3-Propanobis(cyclopentadienyl)disodium **218** was similarly prepared [118]. The isopropylidene-bridged bis(cyclopentadiene) **219** could be prepared in a one-pot reaction between cyclopentadiene and acetone [119]. Presumably the intermediate 6,6-dimethylfulvene is formed and attacked by cyclopentadienyl anion. 6,6-Dimethylfulvene **62** can be

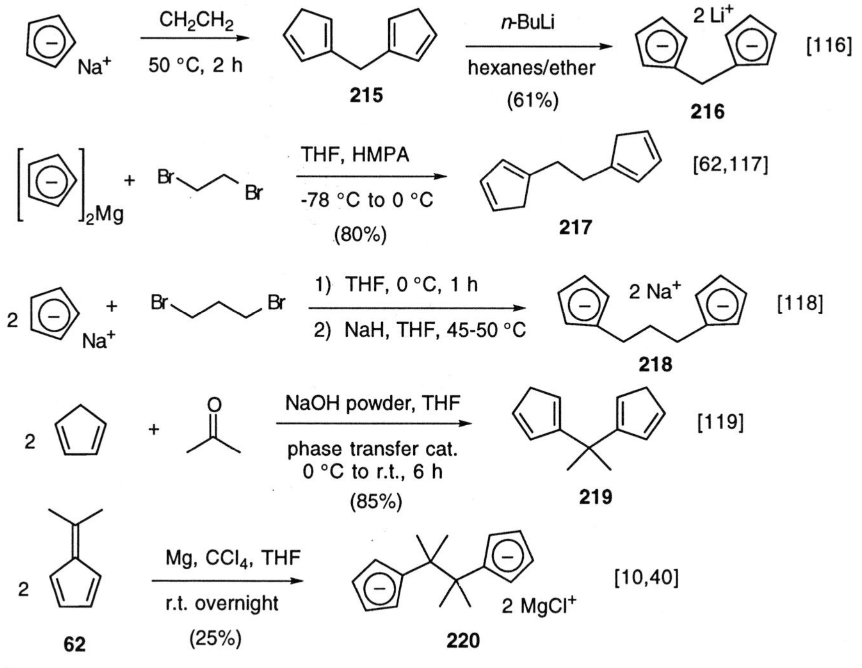

Scheme 8-45. Nonprochiral bis(cyclopentadienes) with simple alkyl tethers.

reductively coupled to give tetramethylethylene-bridged bis(cyclopentadienyl) salt **220** [10]. This coupling was observed as a side product in the dissolving metal reduction of 6,6-dimethylfulvene [40].

Silicon-containing bridges have become common owing to the facility with which chlorosilanes can react with cyclopentadienyl anions. The simple dimethylsilyl-bridged bis(cyclopentadienyl) ligand **221** (isolated as the more stable salt) was prepared from dichlorodimethylsilane (Scheme 8-46) [120]. Similarly, the disilyl-bridged complex **222** was produced [121]. The achiral *cis*-cyclopentane-1,3-diyl dimesylate **223** could be reacted with cyclopentadienylmagnesium bromide to give the achiral cyclopentane-1,3-diyl-bridged

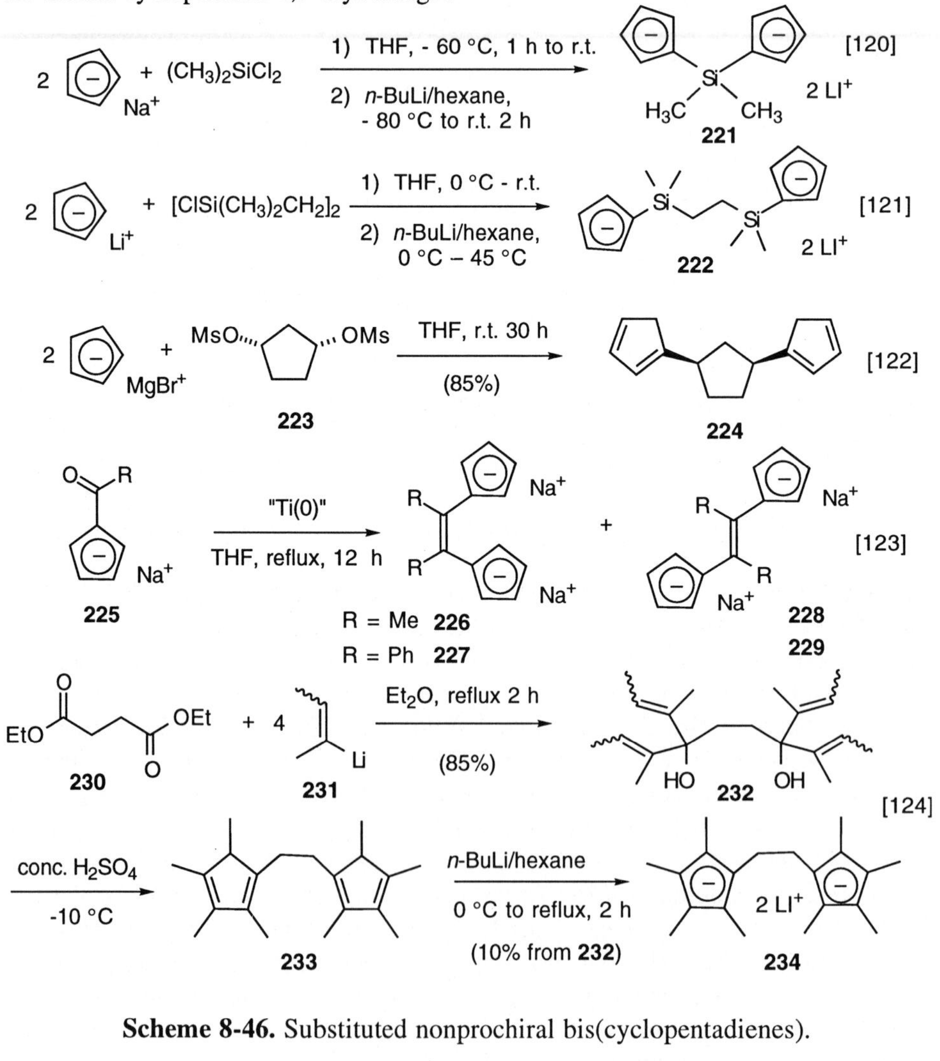

Scheme 8-46. Substituted nonprochiral bis(cyclopentadienes).

bis(cyclopentadiene) **224** [122]. The McMurry coupling of cyclopentadienyl ketone **225** gave a stereoisomeric mixture of the dimethyl- or diphenyl-etheno-bridged bis(cyclopentadienyls) **226–229** [123]. Brintzinger has also converted diester **230** to bis(divinylcarbinol) **232** which can undergo cyclization in acid to form the ethano-bis(pentamethylcyclopentadienyl) **234** [124].

Two syntheses of doubly bridged achiral bis(cyclopentadienes) have appeared (Scheme 8-47). The trimethylsilyl groups in **235** prefer the less hindered 1,3-orientation to the ethano-bridge and can help direct the formation of the novel bis(fulvene) **237**, albeit in low yield. Reduction of the fulvenes and desilylation gave the bis(cyclopentadiene) **238** [125]. Brintzinger has used a double Skattebøl approach to convert tetramethylenecyclooctane **239** to the doubly bridged dilithio salt **241** in good yield [126].

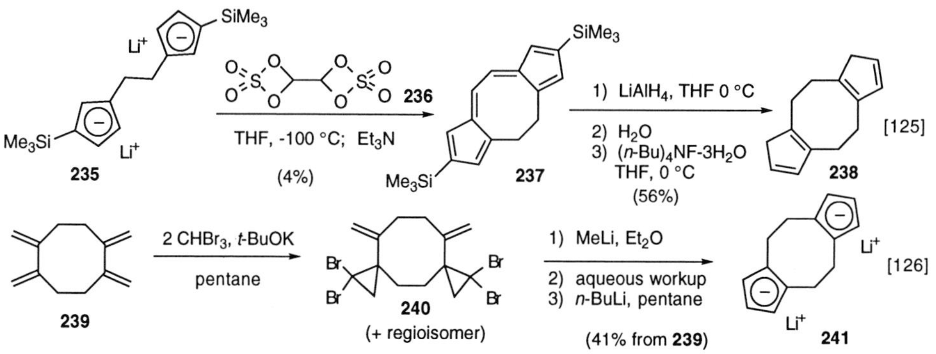

Scheme 8-47. Preparation of nonprochiral doubly bridged bis(cyclopentadienes).

Prochiral Bridged Bis(cyclopentadienes). The preparation of prochiral disubstituted bis(cyclopentadienes) suffer from the same regioselectivity problems seen in the synthesis of disubstituted cyclopentadienes. By concentrating on the more selective synthetic methods developed for the unbridged ligands—the use of hindered *tert*-butyl or trimethylsilyl groups or the use of fulvenes—good, simple syntheses have been developed (Scheme 8-48). *tert*-Butylcyclopentadienylsodium reacts with 1,2-dibromoethane at the less hindered 3-position to form selectively the ethano-bis(cyclopentadiene) **245a**. This reaction goes through a spiroalkylation followed by ring opening of the cyclopentane by *tert* butylcyclopentadienylsodium under forcing conditions [127]. The propensity of silyl groups to adopt the less hindered position was utilized in the bridging of substituted cyclopentadienyl derivatives to form **246**. The silicon atom in such cyclopentadienes undergoes [1,5]-shifts, but prefers to remain at the allylic carbon in a 1,3-relationship to the other substituent [128]. A number of related, silyl-bridged polysubstituted cyclopentadienes have been prepared [129]. 3-Substituted 6,6-dimethylfulvenes **244** undergo reductive coupling to give prochiral tetramethylethylene-bridged cyclopentadienes

247 [61]. Bis(dimethylfulvenes) can be regioselectively formed from bridged cyclopentadienes and reduced or alkylated to form bis(3-*tert*-butylcyclopentadienes) or bis(3-isopropylcyclopentadienes). In this way, ethylene-bridged, tetramethylethylene-bridged and cyclopentane-1,3-diyl-bridged bis(cyclopentadienes) **245** [62], **249** [130], and **250** [122] where formed from **217**, **248** and **224**, respectively.

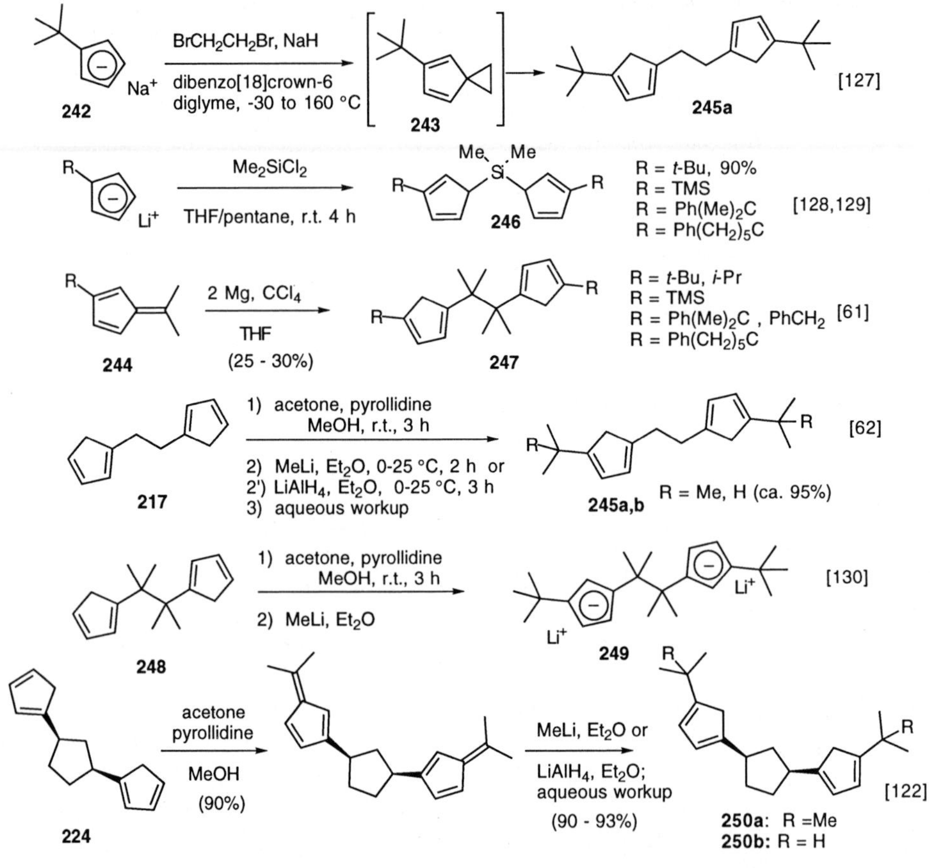

Scheme 8-48. Preparation of prochiral bis(cyclopentadienes).

Collins has published routes for the selective synthesis of prochiral ethanobis(cyclopentadienes) in those cases where the above procedures would not be adequate (Scheme 8-49). Each of the methods relies on the selective generation of bis(cyclopentenones) and their conversion to the desired ligands. Hydroboration of **217** followed by oxidation and double bond migration gave bis(enone) **251**. Organometal addition to the carbonyl followed by dehydration gave the methyl and ethyl derivatives **245c** and **245d** [62]. Bis(enone) **252** could be methylated and dehydrated to give bis(2-methylcyclopentadiene) **253**. Fulvene formation at the remote position followed by the usual reduction or methyllithium addition gave bis(2-methyl-4-alkylcyclopentadienes) **254a,b** [131]. The dimethyl derivative **254c** could be formed through sequential methyl additions to bis(enones) **252** and **255** [131].

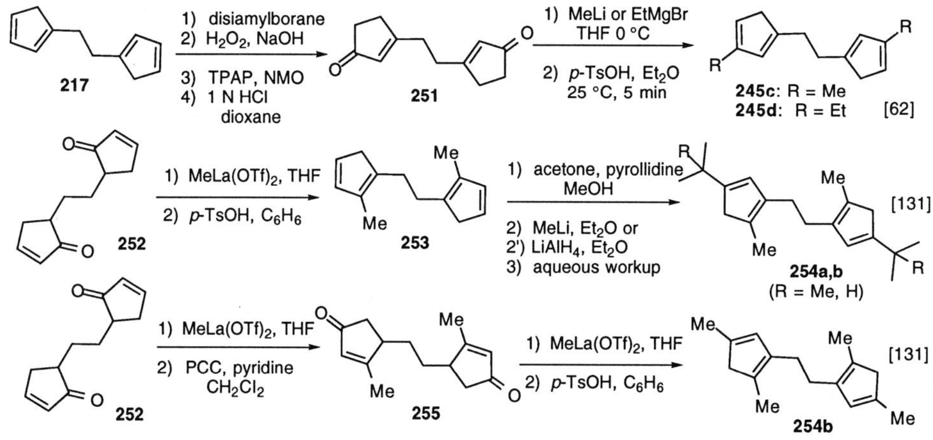

Scheme 8-49. Controlled synthesis of prochiral bis(cyclopentadienes).

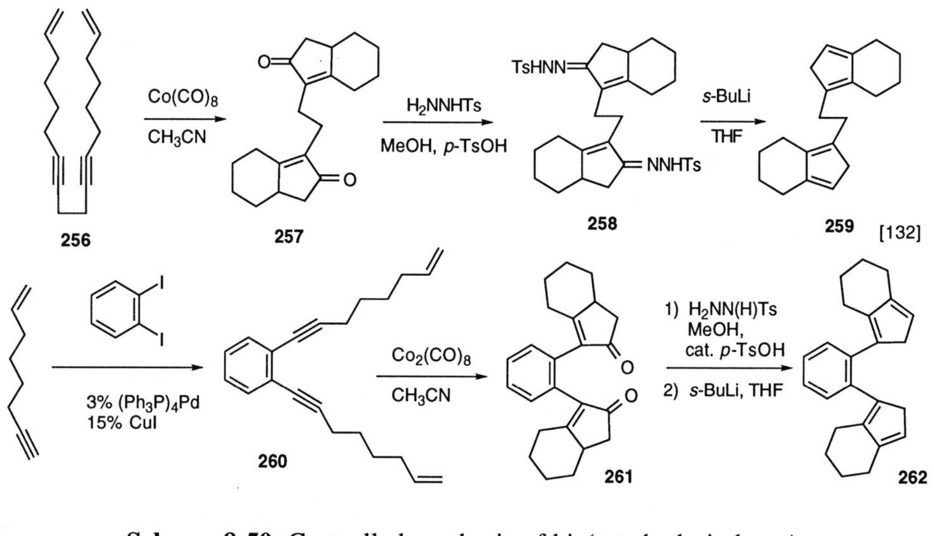

Scheme 8-50. Controlled synthesis of bis(tetrahydroindenes).

Halterman has used a double Pauson–Khand approach to selectively form bis(tetrahydroindenes) (Scheme 8-50). Bis(enyne) **256** underwent the cobalt-promoted cyclization with carbon monoxide to give the bis(cyclopentenone) **257** selectively bridged at the 1-indenyl position [132]. The usual reduction and dehydration of **257** did not produce the desired bis(tetrahydroindene) **259**, so an alternative dehydration involving the Shapiro elimination of bis(tosylhydrazone) **258** was successfully brought into play. This synthetic approach was readily extended to the cyclization of bis(enyne) **260** to bis(enone) **261**. Dehydration of **261** gave 1,2-phenyl-bridged bis(tetrahydroindene) **262** [132].

One report of a prochiral doubly bridged bis(cyclopentadiene has appeared (Scheme 8-51) [133]. Starting from either 1,2-dimethylcyclopentadiene or tetrahydroindene, the intermediate silyl-bridged bis(cyclopentadienyls) **263** could undergo a second reaction with dichlorodimethylsilane to give the doubly bridged complexes **264**.

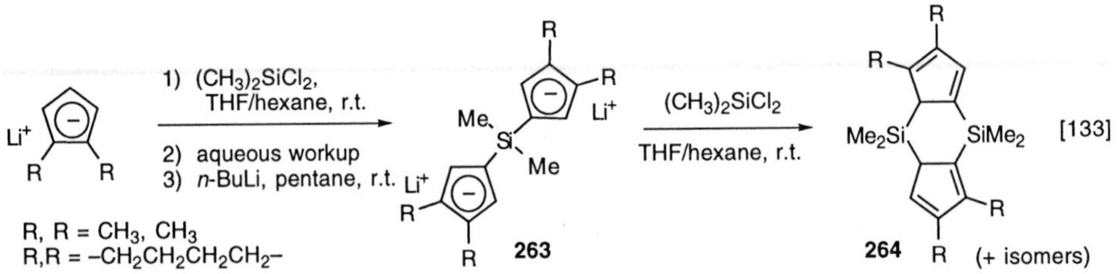

Scheme 8-51. Synthesis of prochiral doubly bridged bis(cyclopentadienes).

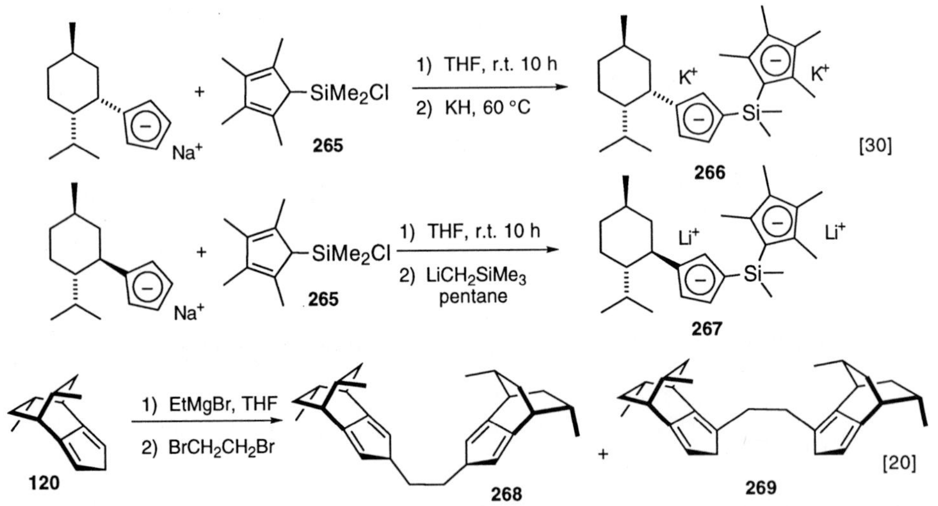

Scheme 8-52. Synthesis of bis(cyclopentadienes) having chiral substituents.

Chiral Bridged Bis(cyclopentadienes). Only a couple reports have appeared dealing with the bridging of chiral cyclopentadienes with achiral bridging groups (Scheme 8-52). As in the earlier preparation of substituted cyclopentadienes, overcoming regioselectivity problems is a concern here. Marks has nicely used the propensity of silyl substituents to adopt the less hindered position to achieve regioselectivity. Chlorodimethyl(tetramethylcyclopentadienyl)silane **265** reacted with neomenthylcyclopentadienylsodium or with menthylcyclopentadienylsodium to give after further deprotonation the silyl bridged chiral bis(cyclopentadienyls) **266** and **267** [30]. The bridging of bicyclooctane-annelated cyclopentadiene **120** with 1,2-dibromoethane, on the other hand, gave a regioisomeric mixture of bis(cyclopentadienes) **268** and **269** [20].

One selective synthesis of a bis(cyclopentadiene) has been undertaken (Scheme 8-53). In this sequence camphor is refunctionalized to give the cyanoaldehyde **270** which can undergo a bis(Wittig) reaction to give the tethered **271**. Conversion of the aldehyde groups in **271** to alkynes enabled an double Pauson–Khand reaction to convert bis(enyne) **272** to bis(cyclopentenone) **273**. Reduction and dehydration gave bis(cyclopentadiene) **274** having the bridging group selectively incorporated at only one position [134].

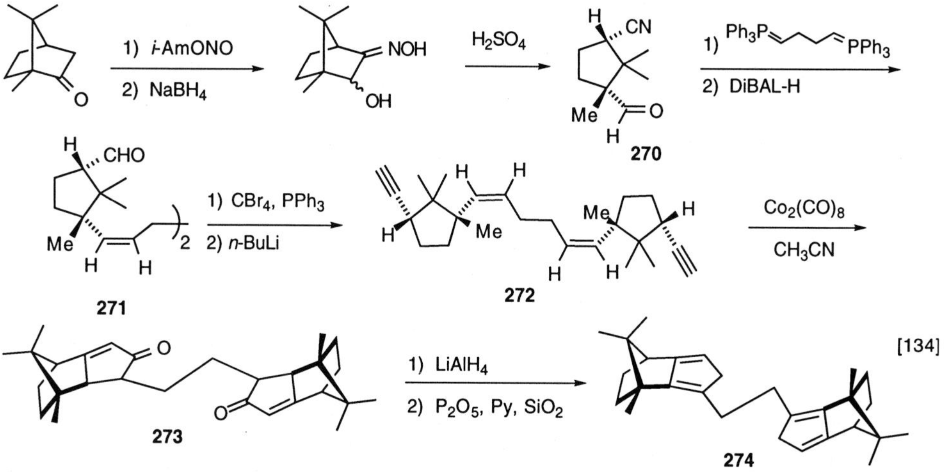

Scheme 8-53. Controlled synthesis of chiral bis(cyclopentadiene).

8.4.2 Bis(indenes)

The importance of Brintzinger's use of the ethanobis(1-indenyl) ligand in metal-locene chemistry cannot be overstated. Many derivatives of bis(indenes) have been prepared and will be discussed in the order of unsubstituted indenes followed by substituted bis(1-indenes) and finally bridging at other indenyl positions. Indenes bridged by stereogenic groups will be discussed in Sec. 8.5.

In Scheme 8-54, the preparation of unsubstituted bis(indenes) bridged by simple alkyl tethers is depicted. The original preparation of ethylenebis(indene) **275** was published in 1967 [135]. Improved conditions for the preparation of **275** include the convenient use of indenyllithium in THF [136, 137]. Bis(indenes) **276** [135] and **277** [138] having longer chain butyl- and dodecyl-tethers have been prepared. *cis*-Dimesylate **232** has been used to form bis(indene) **278** [122].

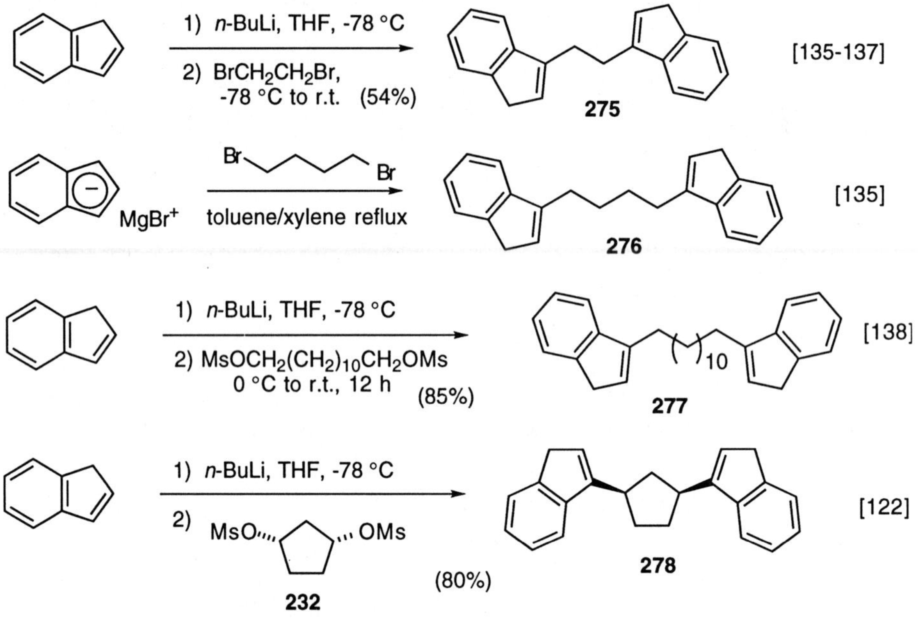

Scheme 8-54. Unsubstituted indenes with alkyl bridges.

Silyl-bridged bis(indenes) have also become common (Scheme 8-55). In these ligands, the silicon atom is fluctional and prefers the sp^3-1-indenyl position—resulting the generation of *meso* and *dl* isomers of the bis(indenes). No stereochemistry will be indicated here and it is inconsequential for metallocene chemistry since the stereocenter is no longer present in the deprotonated ligand. Unsubstituted indene is readily bridged to give the dimethylsilylene-bridged bis(indene) **279** [139, 140] or the disilane **280** [139]. Several substituted indenes whose preparation was described in Sec. 8.3.2 have been bridged by the dimethylsilylene group. Importantly, the silyl groups orient themselves away from the substituted 4-indenyl position as seen in **281** [106, 107] and **282** [106, 111]. The silyl group can still bridge even the more hindered 2,4,7-trimethylindenyl group in **283** [110a].

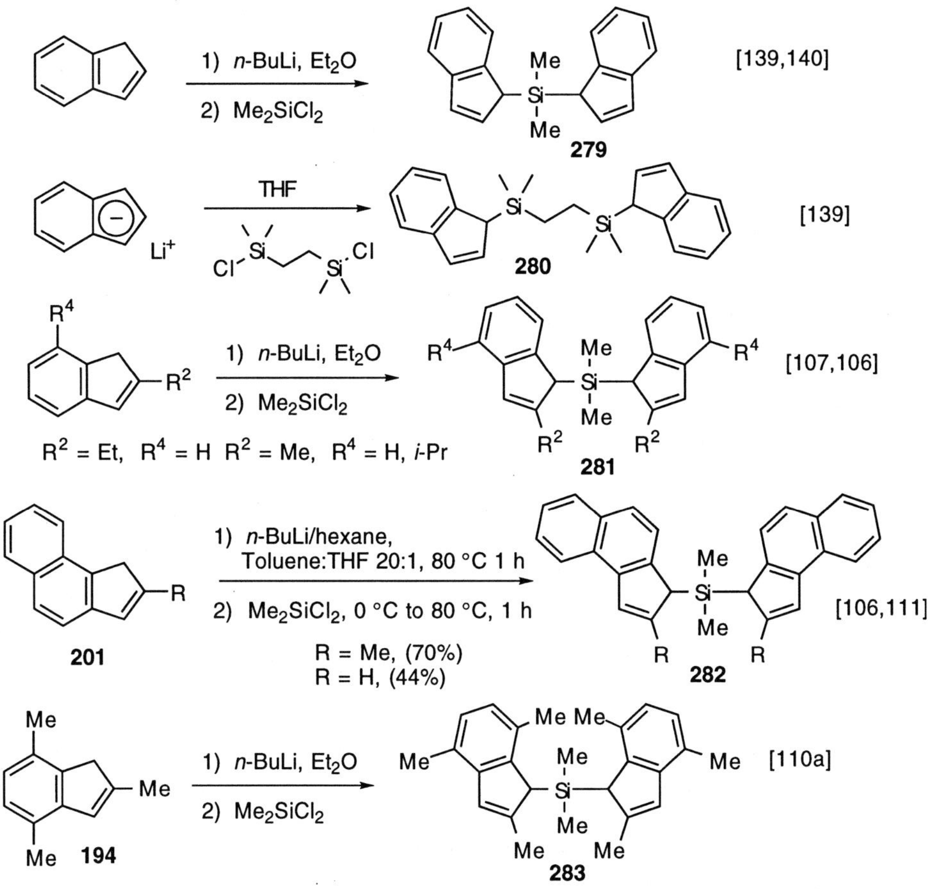

Scheme 8-55. Indenes with silicon bridges.

Dibromoethane has been used to tether a number of substituted indenes (Scheme 8-56). 4,7-Dimethylindene **186a** and 5,6-disubstituted indenes **191** gave bis(indenes) **284** and **285** [18b]. Two equivalents of 2,4,7-trimethylindene can be directly converted to bis(trimethylindene) **286** or one equivalent to give (bromoethyl)indene **287**. This monoindene could then be reacted with a different indene such as the 1,2,4,7-tetramethyl derivative to give unsymmetrically substituted bis(indene) **288** [110a].

Two novel bridging groups have been developed (Scheme 8-57). Brintzinger has reacted a titanium complex of dilithioferrocene with 1-indanone to give bis(indanol) **289** Dehydration of **289** gave ferrocene-bridged bis(indene) **290** [141]. Halterman has used the palladium-catalyzed coupling of 1,2-diiodobenzene with a zinc chloride complex of various indenyllithium complexes to give a series of 1,2-phenyl-bridged bis(indenes) **291** [134].

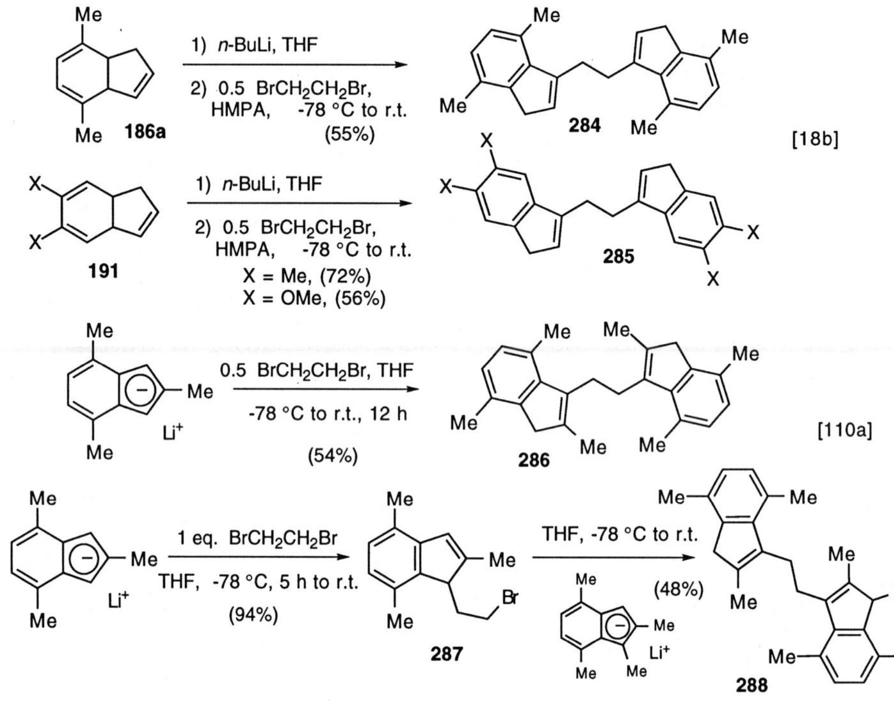

Scheme 8-56. Substituted indenes with ethano-bridge.

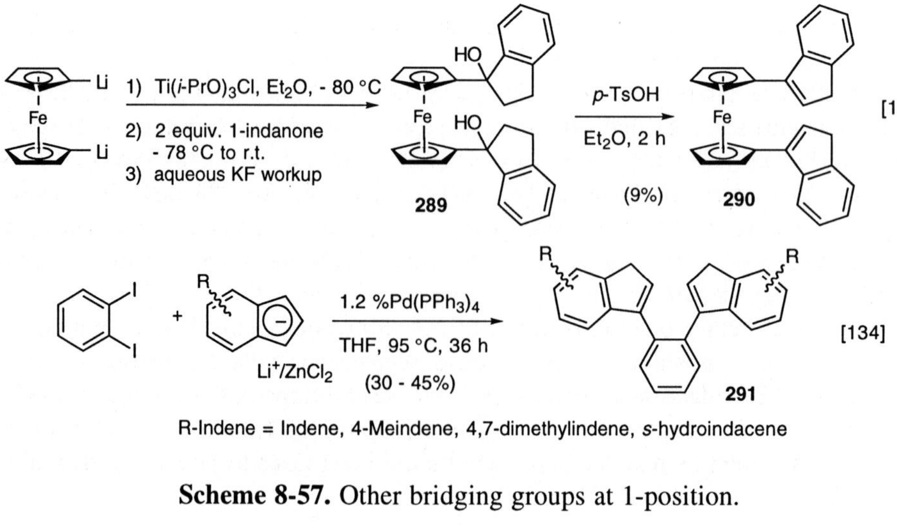

R-Indene = Indene, 4-Meindene, 4,7-dimethylindene, s-hydroindacene

Scheme 8-57. Other bridging groups at 1-position.

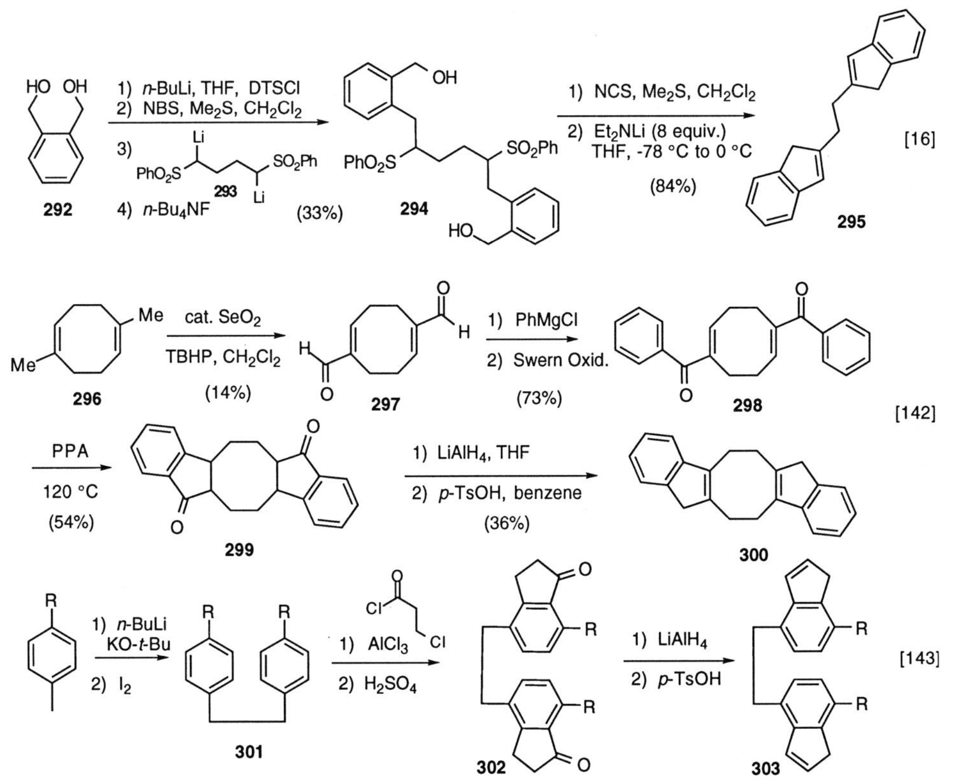

Scheme 8-58. Bridging at other positions, 2,2′ and 2,1′, 1,2′-doubly bridged.

While the common site for bridging indenes has been the 1-position, other bridging sites have been developed (Scheme 8-58). Nantz has used the dilithiated disulfone **293** to alkylate a benzyl bromide derivative to give **294**. Activation of the benzyl alcohol groups and a second sulfone alkylation followed by elimination gave ethylenebis(2-indenyl) **295** [16]. A doubly bridged bis(indene) has been prepared from 1,5-dimethylcyclooctadiene **296** [142] Allylic oxidation of **296** gave bis(formyl) **297** which could be reacted with phenyl Grignard to give after oxidation the bis(phenyl ketone) **298**. Treatment of **298** with polyphosphoric acid induced a Nazarov type ring closure to give bis(indanone) **299**. Reduction and dehydration of **299** gave the doubly bridged bis(indene) **300** containing one bridged between the 1- and 2′-indenyl positions and the second between the 2- and 1′-positions [142]. The novel 7,7-bridged bis(indenes) **303** has been prepared from *para*-xylene or cymene through coupling to the biphenyls **301**, Friedel–Crafts reactions to bis(indanones) **302** (+ regioisomeric ketones) and finally reduction and dehydration [143].

8.4.3 Bis(fluorenes)

A limited number of bridged bis(fluorenes) have been prepared with achiral bridges, mostly by Alt (Scheme 8-59). The simple bridging of fluorenyllithium with 1,2-dibromoethane or dichlorodimethylsilane gave the ethylene-bridged or the dimethylsilylene-bridged bis(fluorene) 304 [144] or 305 [145, 146]. Reaction of di-*tert*-butylfluorene 306 with dichlorodimethylsilane gave monofluorene 307 which could further react with fluorenyllithium to give the unsymmetrical bis(fluorene) 308 [147]. The benzannelated fluorene 309 could be bridged to give the ethanobis(fluorene) 310 [148]. The biphenyl-bridged bis(fluorenes) 312 could be prepared from 2,2′-dihalo-1,1′-biphenyls 311, but these bis(fluorenes) could not be deprotonated [148].

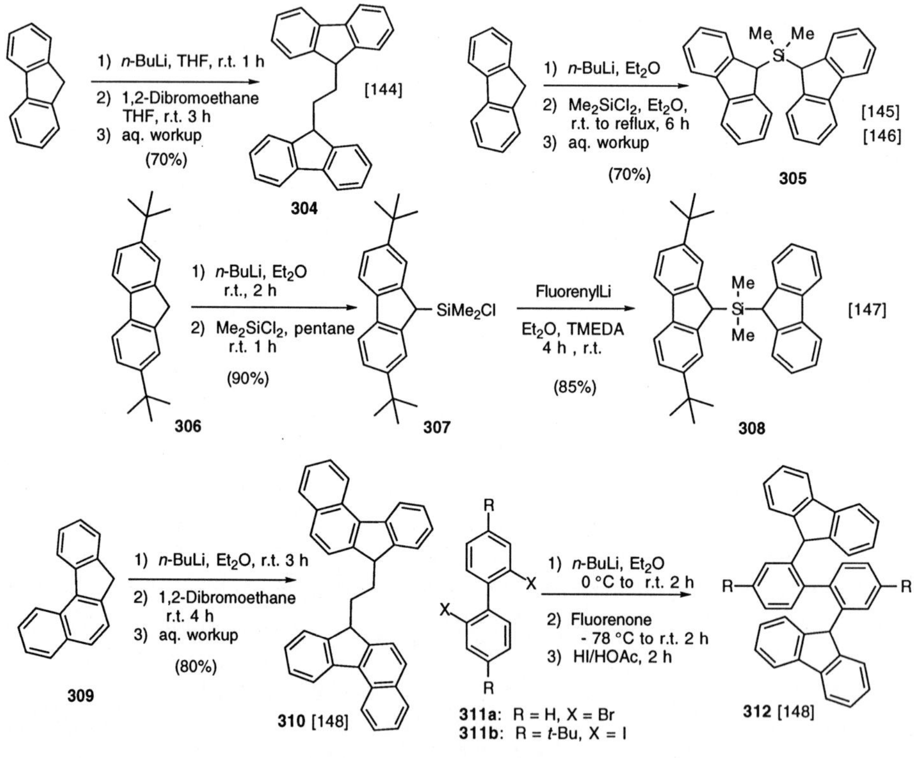

Scheme 8-59. Bis(fluorenes)

8.4.4 Mixed Ligands

This section will cover bridged ligands between two different ligand types, starting with cyclopentadienyl^indenyl ligands, then covering the more developed cyclopentadienyl^fluorenyl ligands and ending with indenyl^fluorenyl ligands. Bridged complexes with chiral bridging groups will be covered in Sec. 8.5.

Cyclopentadienyl^Indenyl. Indenyl anions add to fulvenes much more readily than do cyclopentadienyl anions and this method been used in the preparation of a number of cyclopentadienyl, indenyl compounds (Scheme 8-60). Green first applied this method with 6,6-dimethylfulvene and pentamethylenefulvene to give the mixed ligand **314** [149, 150]. The method has been extended to 6,6-diphenylfulvene [151]. A series of prochiral ligands has been prepared from the addition of 3-substituted indenes to 3-*tert*-butyl-6,6-dimethylfulvene to give ligands **315** [152]. Through sequential addition of one equivalent of indenyllithium and cyclopentadienylsodium to dichlorodimethylsilane, cyclopentadienylindenylsilane **316** has been prepared [140]. The substituted (*t*-butylcyclopentadienyl)(*t*-butylindenyl)silane **317** has been similarly prepared [152].

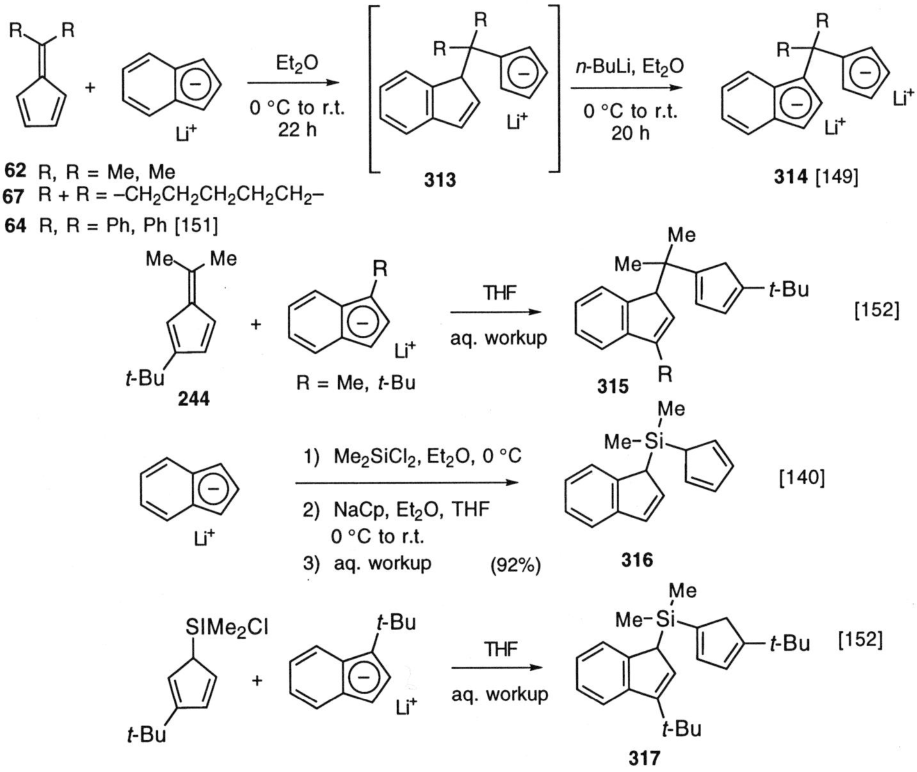

62 R, R = Me, Me
67 R + R = –CH₂CH₂CH₂CH₂CH₂–
64 R, R = Ph, Ph [151]

Scheme 8-60. Bridged cyclopentadiene, indenes.

Cyclopentadienyl^Fluorenyl. The most common mixed ligand is the combination of cyclopentadienes and fluorenes. The synthetic methods have been based on either the addition of fluorenyl anions to fulvenes or on the use of silyl bridges. The synthesis of the unsubstituted ligands are shown in Scheme 8-61. A variety of fulvenes have been applied in the fluorenyl addition reaction to yield the one-carbon-bridged ligands **318** [153–156]. The silyl-bridged ligand **320** has been best prepared from the initial addition of fluorenyllithium to dichlorodimethylsilane, followed by the addition of cyclopentadienyllithium [145].

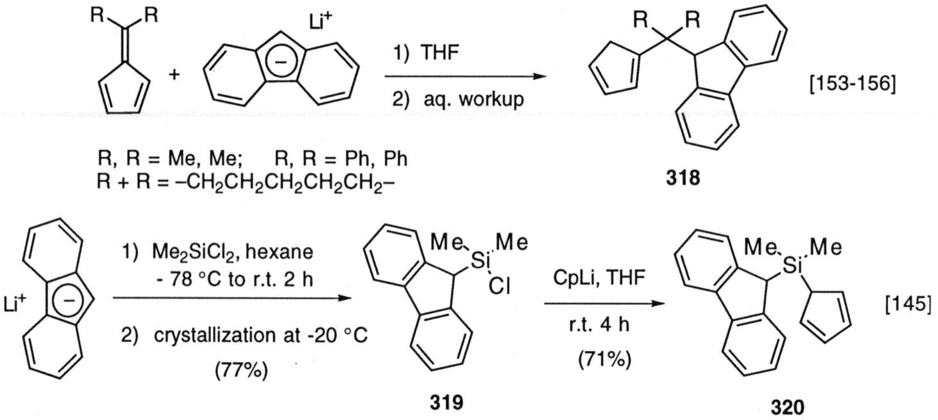

R, R = Me, Me;   R, R = Ph, Ph
R + R = –CH₂CH₂CH₂CH₂CH₂–

Scheme 8-61. Bridged unsubstituted cyclopentadiene, fluorenes.

A number of ligands having substituents either on the cyclopentadienyl or fluorenyl moieties has been prepared (Scheme 8-62). Isopropylidene-bridged 3-substituted cyclopentadiene, fluorene compounds **321** and **322** have been prepared by addition of fluorenyllithium to 3-substituted 6,6-dimethylfulvenes [110a, 157]. A number of substituted fluorenyl anions have been added to fulvenes to produce the substituted compounds **323** [156, 158]. The preparation of a number of methylene-bridged cyclopentadienyl, fluorenyl compounds **324** has been accomplished through the addition of the substituted fluorenyl anions to 6-dimethylaminofulvene followed by reduction of the incipient fulvene [159]. A substituted fluorenyl, cyclopentadienyl silyl compound **325** has prepared from intermediate **307** [147].

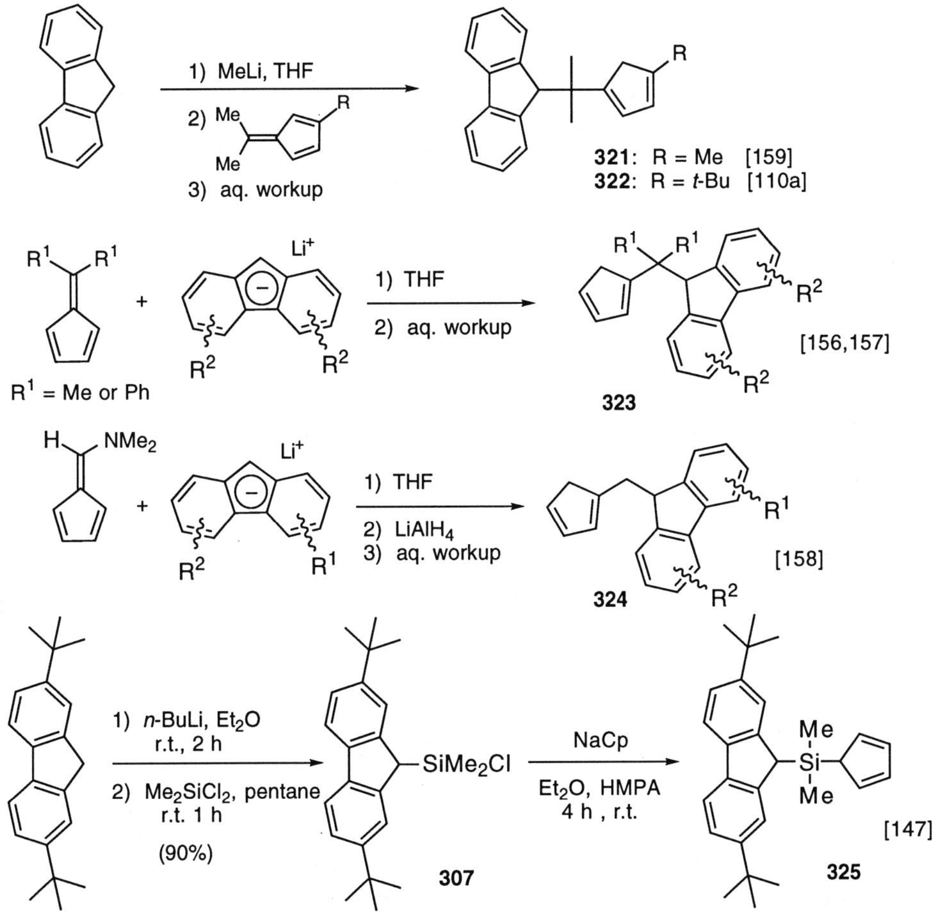

Scheme 8-62. Bridged substituted cyclopentadiene, fluorenes.

Indenyl^Fluorenyl. The addition of indenyllithium to fluorenylsilyl chloride **319** has been used to prepare the fluorenyl, indenyl mixed ligand **326** (Scheme 8-63) [145].

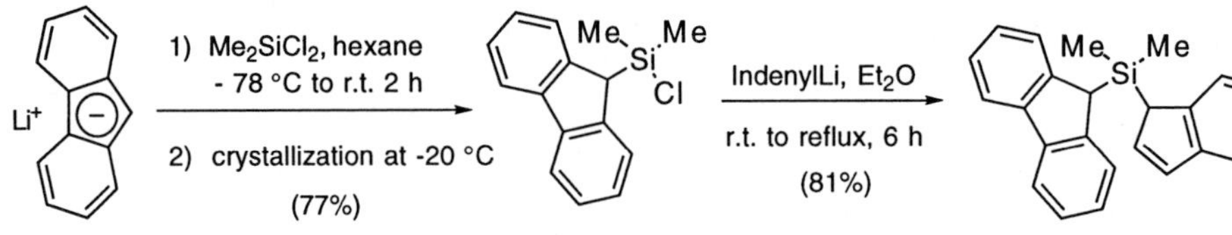

Scheme 8-63. Bridged indene, fluorene.

8.5 Bridged Ligands with Stereogenic Bridging Groups

The first examples of bridged ligands contained achiral bridges at the 1-position of the indenyl ligands. The resulting π-faces of the ligands were rendered enantiotopic and metalation would necessarily lead to an equal mixture of *d* and *l* enantiomers and possibly also the *meso* diastereomer. By incorporating chirality in the bridging group at the 1-position of indene, metalation of the resulting diastereotopic faces could now lead to three diastereomers in differing amounts. A hope was by judicious selection of the bridging group the formation of one of the diastereomeric metallocenes would be energetically favored. Ligands of this type will be discussed first in this section. A second approach to introducing chiral bridges was its placement at the symmetrical 2-position of indene. Metalation would lead inherently to only one stereoisomeric metallocene.

The historical development of ligands bridged by a stereogenic group started with an unsymmetrical one carbon bridge by Rausch and Chien. The bridging of bis(indenes) at the 1-position was initiated in 1991 in separate reports by Halterman and Green and has been the most developed type. Two examples of using chiral groups to bridge bis(cyclopentadienes) have been reported. A number of examples of biaryl bridging groups at the 2-position of indene and related ligands have appeared. Most recently one-atom chiral silylene bridges and chiral doubly bridged ligands have been introduced.

One-Carbon Chiral Bridge. The first example of a chiral bridging group was Rausch and Chien's 1990 report of the monocyclopentadienyl monoindenyl ligand *dl*-**327** [160] (Scheme 8-64). The ligand was formed by the addition of indenyllithium to pentamethylpentafulvene. Since the chirality is generated only upon formation of the bridging group from achiral starting materials, the ligand is racemic. Alt has reported two similar syntheses of chiral cyclopentadienyl, fluorenyl ligands. In the first, substituted indenyllithium is added to 6-dimethylaminofulvene and the intermediate fulvene reacted with methyllithium to give the racemic compound **328** [158]. The addition of a fluorenyllithium complex to unsymmetrically substituted fulvenes **329** led to the racemic ligands **330** [156]. In a unique synthesis, methyllithium or an allyl Grignard reagent adds to one fulvene in bis(fulvene) **331** and the resulting cyclopentadienyllithium adds to the second fulvene to form (after deprotonation) bis(cyclopentadienyl) dianions **332** [161]. As in Rausch and Chien's synthesis, the chirality is introduced only upon the formation of the final bridge and bis(cyclopentadienyls) **332** are therefore also racemic.

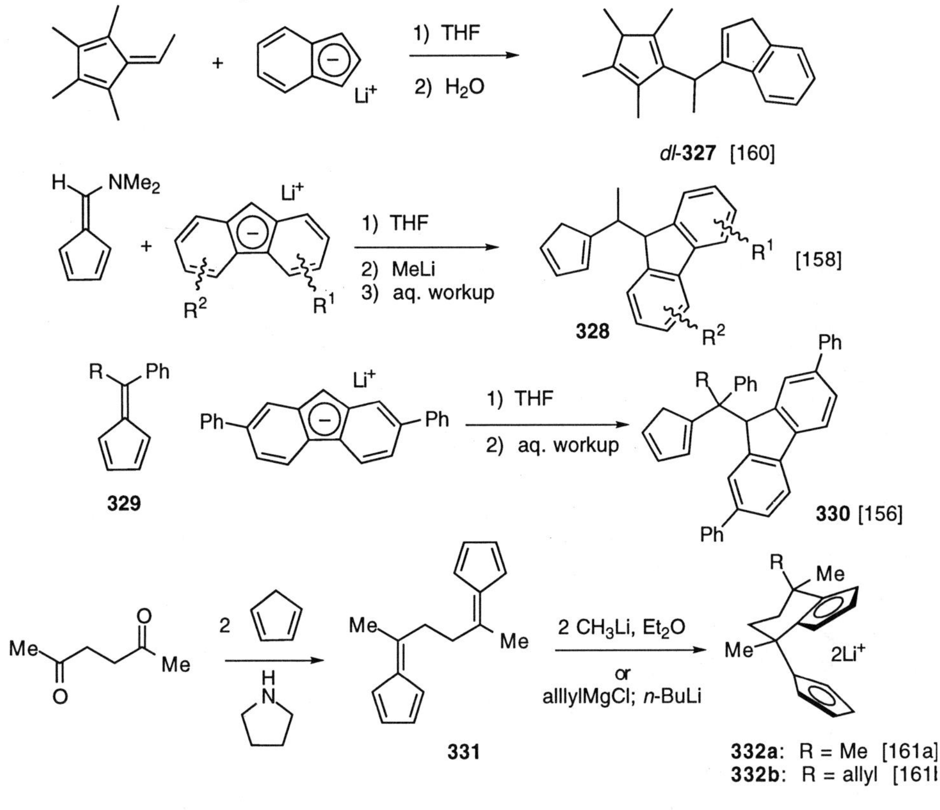

Scheme 8-64. Racemic one-carbon chiral bridged ligands.

Bis(indenes) Bridged at the 1-Position. The first reports of nonracemic chiral groups being used to bridge indenes were Halterman's 1991 report using the 1,1'-binaphthyl-2,2'-dimethyl bridge [162] and Green's report using a tartrate-derived 1,4-butanediyl bridge [163] (Scheme 8-65). Unlike the previous 1-carbon bridges, these ligands were prepared in nonracemic form. The known 1,1'-binaphthyl-2,2'-di(bromomethyl) **122** had been previously used to form a chiral cyclopentadiene. Dibromide **122** could be reacted with indenyllithium to form the bis(indene) in 71% yield. This method was later extended to include to bridged bis(4,7-dimethylindene) [164]. 4,7-Diisopropylindene could not be bridged by dibromide **122**. The tartrate-derived ditosylate **112** had also been previously used in a chiral cyclopenta-diene synthesis. Green was able to react ditosylate with indenylmagnesium bromide to give bis(indene) **334** in 76% yield [163].

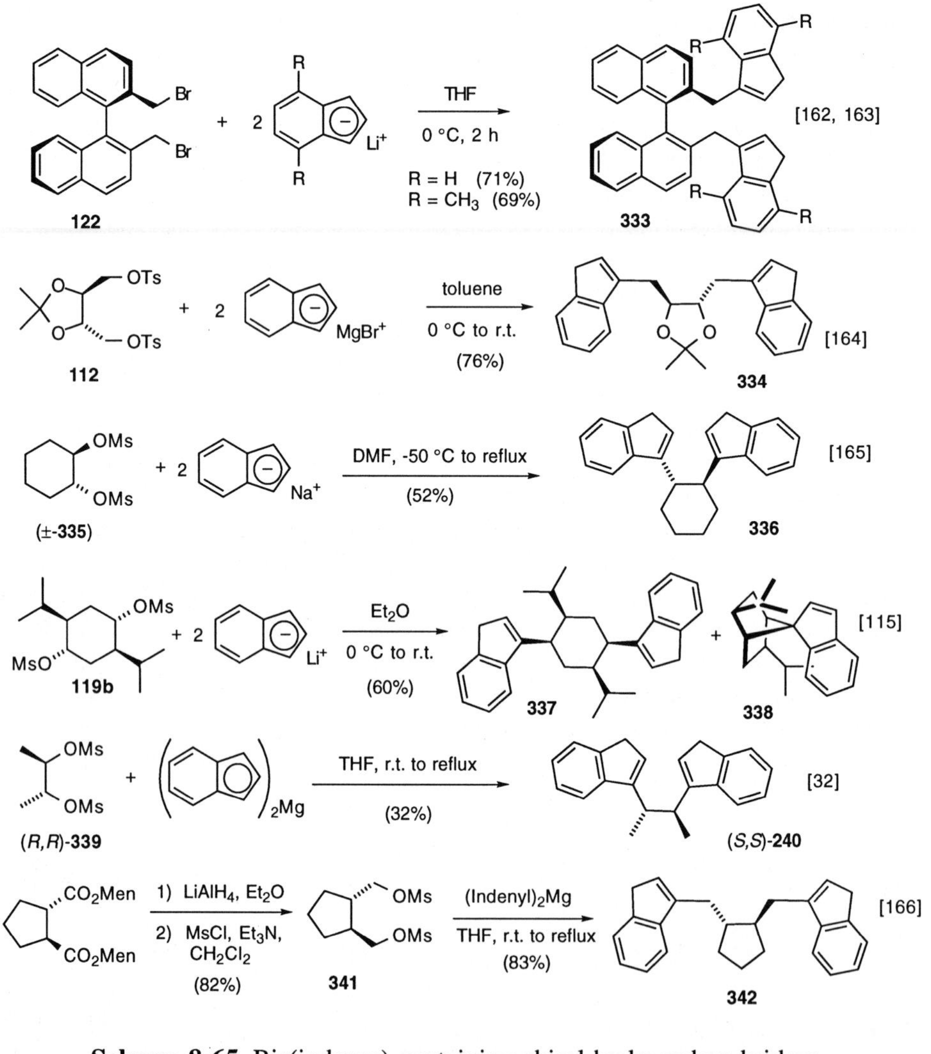

Scheme 8-65. Bis(indenes) containing chiral hydrocarbon bridges.

The first 2-carbon chiral bridging group was Rieger's use of racemic *trans*-cyclo-hexane-1,2-dimesylate **335**. The displacement of the unactivated secondary mesy-lates required rather forcing conditions—indenylsodium in refluxing DMF—but pro-duced *trans*-1,2-bis(indenyl)cyclohexane **336** in good yield [165]. The synthesis of diisopropylcyclohexane-1,4-dimesylate **119b** had been developed for a chiral cy-clopentadiene synthesis. Dimesylate **119b** could also be applied to the preparation of bridged bis(indene) **337** [115]. Interestingly, the alkylation of cyclopentadiene with the larger tethers tended to give spiro-annelated cyclopentadienes, whereas the alkylation of indenes generally led to bis(indenyl) products. Presumably the needed proton transfer for the formation of the spiro dialkylated ligand is slower in the indenes than the cyclopentadienes. In the case of the cyclohexane-1,4-dimesylate **119b**, the ratio of bis(indene) **337** to spiro-indene **338** was dependent of the conditions. The best conditions for forming the bis(indene) were to add indenyllithium to the dimesylate in ether which gave bis(indene) **337** in 60% yield. The use of HMPA/THF led completely to the spiro-annelated product **338**. Dimesylates **339** and **341** had been previously applied to the synthesis of chiral 1,2-diphosphines and 1,4-diphosphines. Bosnich has extended their use to include the bridging of indenes. Displacement of the secondary mesylates in **339** with bis(indenyl)magnesium gave a low yield of the bis(indene) **240** [32], but the primary mesylates in **341** cleanly displaced to give the 1,4-bridged bis(indene) **342** [166].

Bis(Cyclopentadienes) Bridged by Stereogenic Group. Two nonracemic chiral bridging groups for bis(cyclopentadienes) have been reported (Scheme 8-66). The first was by Nantz who employed a double-Skattebøl reaction of tetraene **345** to give the 2,3-butanediyl-bridged bis(3-methylcyclopentadiene) **346** in good yield [31]. The needed tetraene was efficiently produced by the alkylation of sulfonate **344** followed by elimination. The starting 2,3-dimethyl-1,4-butanediol is available in nonracemic form. The 1,4-bridged bis(cyclopentadiene) **349** was prepared in nonracemic form by Helmchen starting from the resolved Diels–Alder adduct **347** [11]. In this case triflates were used as the leaving groups in **348** for a cyclopentadienylmagnesium bromide displacement to give **349**.

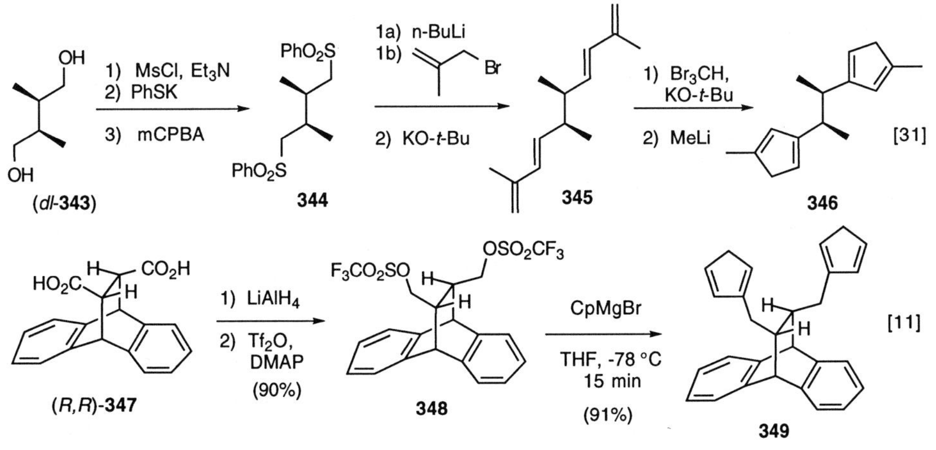

Scheme 8-66. Bis(cyclopentadienes bridged by chiral groups.

Biaryl-Bridging Groups for Bis(2-indenes) and Related Ligands. Brintzinger was the first of three groups to report the use of biaryl groups to bridge ligands in such a manner as to render the ligand π-faces equivalent (Scheme 8-67). The racemic 2,2′-diiodo-1,1′-biphenyl **350** was lithiated and reacted with dimethylcyclopentenone **137** to give (after dehydration and deprotonation) the bis(cyclopentadienyl) ligand **19** [12]. The π-faces of the cyclopentadienyl groups are equivalent by rotation about the bond to the biphenyl group. Since related 6,6′-dimethylbiphenyls have been resolved, the currently racemic dimethyl ligand **19** could possibly be prepared in nonracemic form. The initial addition produced a monocyclopentenol side product which was carried forward for separation at the metallocene stage.

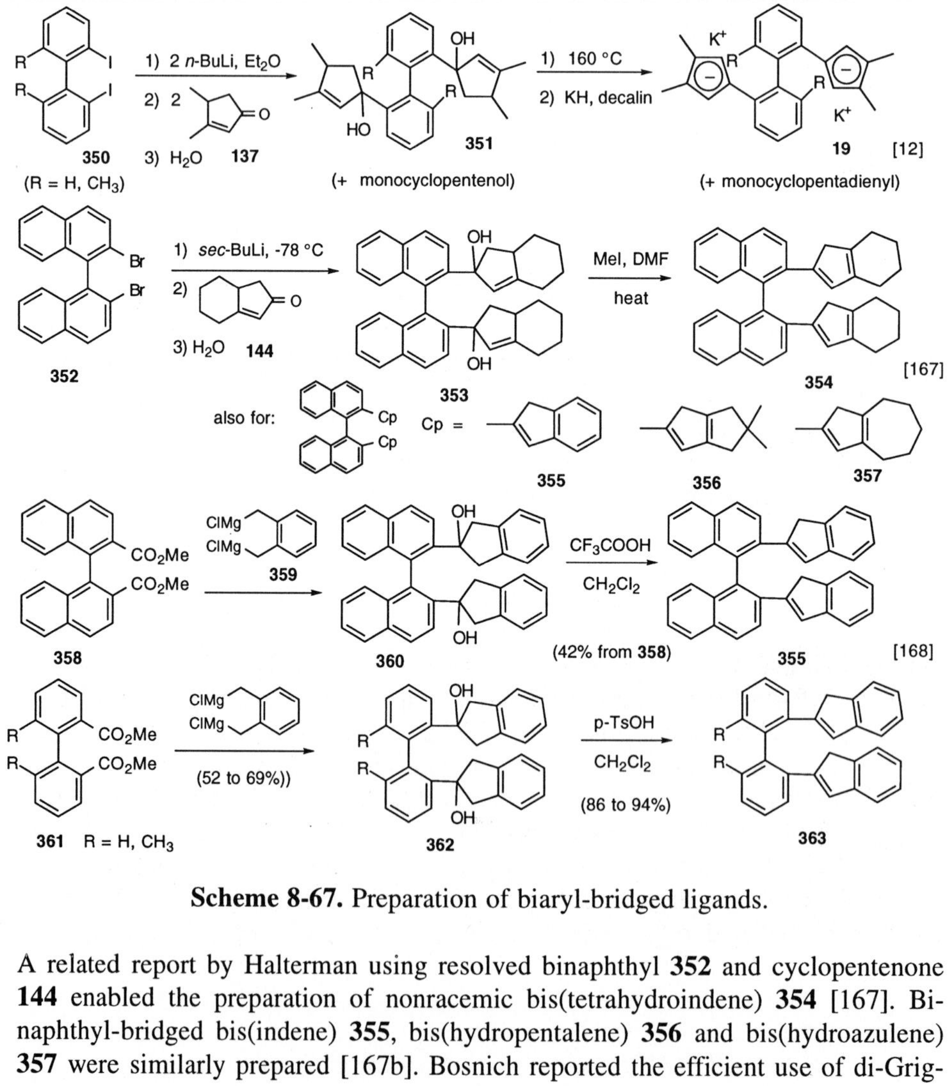

Scheme 8-67. Preparation of biaryl-bridged ligands.

A related report by Halterman using resolved binaphthyl **352** and cyclopentenone **144** enabled the preparation of nonracemic bis(tetrahydroindene) **354** [167]. Binaphthyl-bridged bis(indene) **355**, bis(hydropentalene) **356** and bis(hydroazulene) **357** were similarly prepared [167b]. Bosnich reported the efficient use of di-Grignard **359** to convert biaryl diesters **358** and **361** to indenols **360** and **362** which could be dehydrated to give the binaphthyl-bridged bis(indene) **355** and the biphenyl-bridged ligand **363** [168].

Chiral Silicon-Bridged Ligands. Two reports have appeared in which chiral silyl groups have been used to bridge indenes or cyclopentadienes (Scheme 8-68). Two equivalents of menthol can be used to derivatize tetrachlorosilane to the dichlorodimentholsilane **364**. Displacement of the two remaining chlorides with indenyllithium gave the chiral bridged bis(indene) **365** [169]. Binaphthol could react with the bis(cyclopentadienyl)silicon dichloride **366** to give the chiral bridged bis(cyclopentadiene) **367**. This ligand could also be further silylated and deprotonated to form a single isomer of **368** [170].

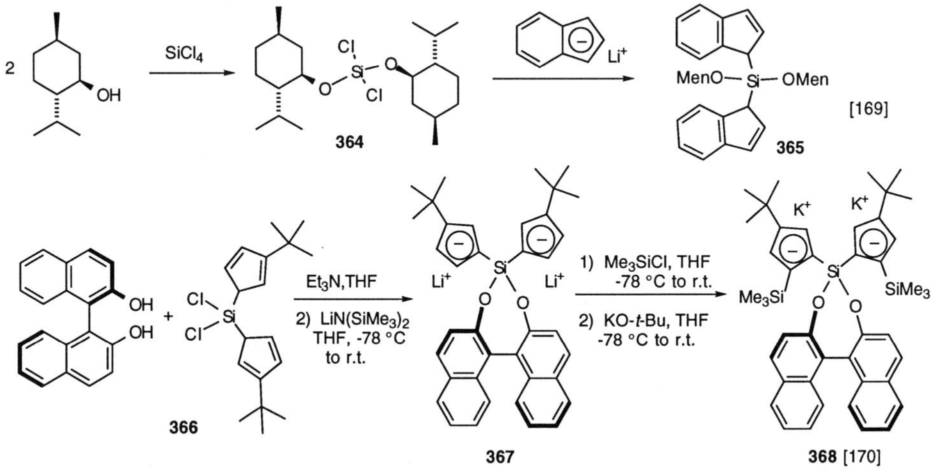

Scheme 8-68. Chiral silyl-bridged ligands.

Doubly Bridged Chiral Ligands. Two reports describe the use of Meerwein's diketone (**369**) to form chiral doubly bridged ligands (Scheme 8-69). Addition of 2-propenyl Grignard to **369** followed by dehydration gave tetraene **370**. This tetraene was cyclopropanated to enable a Skattebøl rearrangement to the doubly bridged bis(cyclopentadienyl) ligand **371** [171]. A different synthetic strategy was used to prepare bis(indene) **376**. In this synthesis, dinitrile **372** was reduced, reacted with phenyl Grignard to give after oxidation diketone **374**. Diketone **374** underwent a Nazarov type ring closure to give bis(indanone) **375** which could be reduced and dehydrated to the doubly bridged bis(indene) **376** [142].

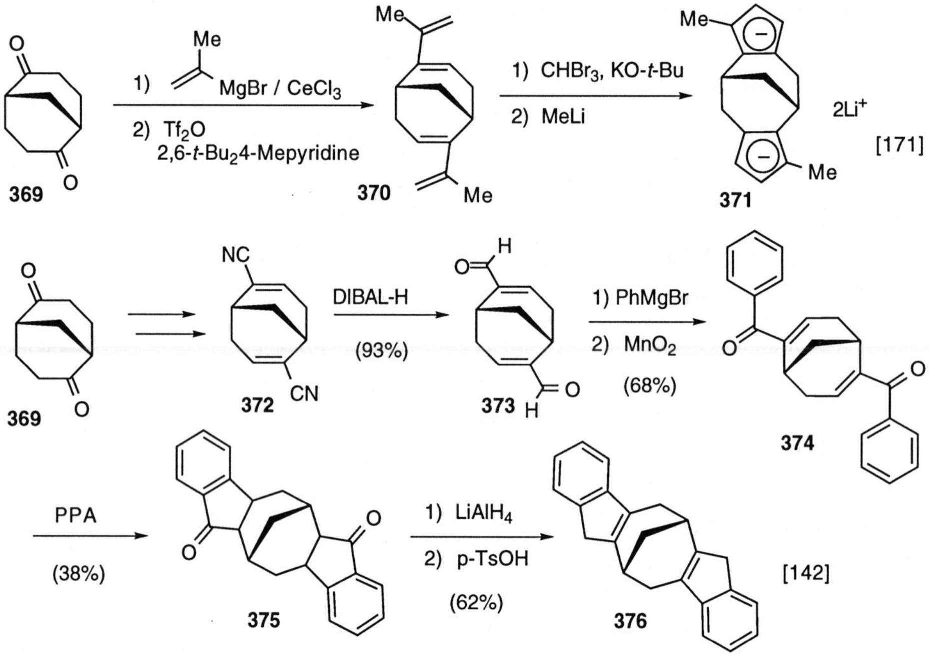

Scheme 8-69. Chiral doubly bridged ligands.

Chiral Bis(Fluorene) and Fluorene, Indene Ligands. Nonracemic styrene oxide has been used the preparation of nonracemic ligands containing fluorene and indene (Scheme 8-70). Ring-opening of styrene oxide by fluorenyllithium gave a mixture of regioisomer alcohols **377** and **378**. Activating the purified isomer **378** as the triflate enabled the displacement by indenyllithium or fluorenyllithium to give the chiral ligands **379** and **380** [172].

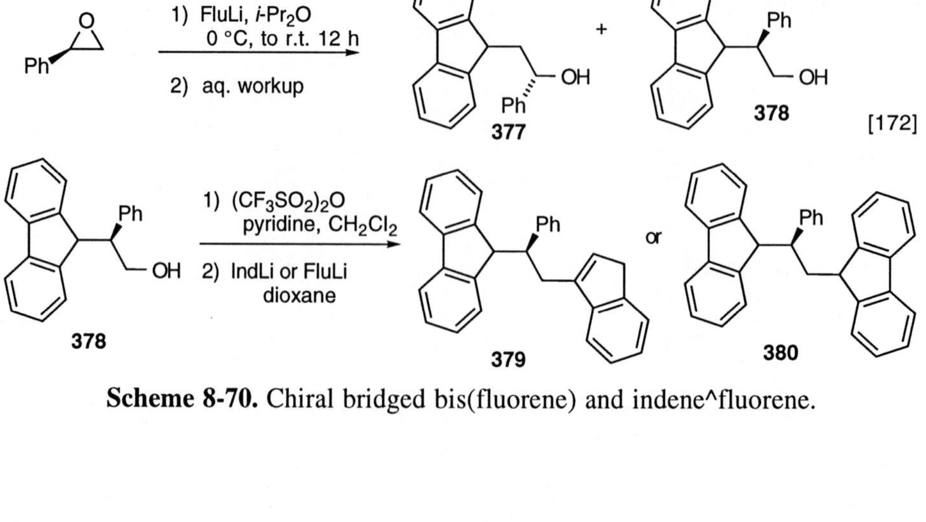

Scheme 8-70. Chiral bridged bis(fluorene) and indene^fluorene.

8.6 Unbridged Metal Complexes

8.6.1 Bis(cyclopentadienyl)metals

Far too many group 4 metallocene complexes have been prepared to cover completely in this section which will concentrate on general methods and the stereoselectivity of metallocene synthesis. Good reviews of simple metallocenes [173, 1b] and sterically hindered metallocenes [1c] are available and the references for the preparation of cyclopentadienyl ligands in Sec. 8.3.1 generally include the formation of metallocene complexes.

Achiral Bis(cyclopentadienyl)metal Complexes. Examples of common methods for the preparation of metallocene dihalides are shown in Scheme 8-71. The first [6] and most common method for preparing simple titanocene dihalides has been to react the cyclopentadienyl anion with titanium tetrahalide. Early examples of substituted complexes being prepared in this way are $(MeCp)_2TiCl_2$ [7], (i-PrCp)$_2$TiCl$_2$ and (t-BuCp)$_2$TiCl$_2$ [174]. NMR characterization of $(RCp)_2TiCl_2$ R = H, Me, Et, n-Pr, n-Bu, 1,2-Me$_2$, 1,3-Et$_2$, and 1-Et-3-Me have appeared [175] in addition to more complete reviews [173]. Early examples of the analogous preparation of $(RCp)_2ZrCl_2$ complexes include R = Me, Et, i-Pr, t-Bu, TMS [176]. The early use of (tetraamido)titanium or zirconium complexes to form metallocene complexes was described in 1968 [177]. Irradiation of solutions of two difference bis(cyclopentadienyl)titanium dichloride complexes was shown to interconvert the cyclopentadienyl ligands [178]. An early report of using titanium trichloride to form titanocene dichlorides was Brintzinger's addition of pentamethylcyclopentadienylsodium to TiCl$_3$, followed by oxidation using conc. HCl to form bis(pentamethylcyclopentadienyl)titanium dichloride in 40% yield [179]. The use of TiCl$_4$ in this reaction was less satisfactory. The benefits of using TiCl$_3$ presumably arise from improved steric interactions and the elimination of the oxidation of Me$_5$CpNa by the more oxidizing TiCl$_4$. Improved yields using TiCl$_3$ instead of TiCl$_4$ are now fairly commonly observed.

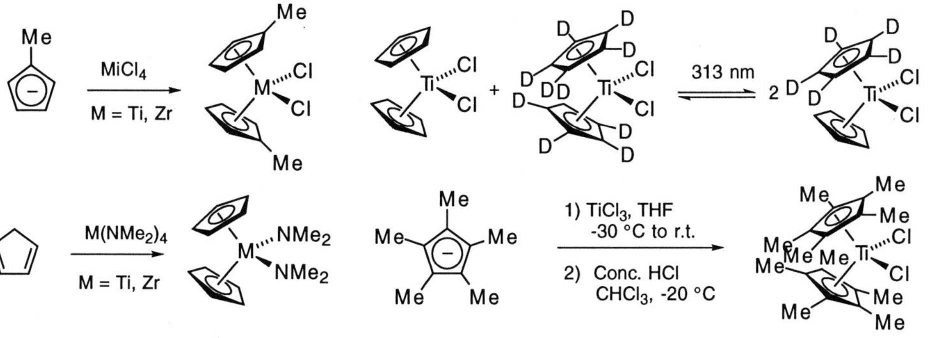

Scheme 8-71. Examples of bis(cyclopentadienyl)metal complex formation.

Achiral (Cp)(Cp′)titanium Complexes. Through the controlled use of stoichiometry [180] or insertion into the trimethylsilyl–cyclopentadienyl bond [27c], $(RCp)TiCl_3$ complexes can be prepared and reacted with a second cyclopentadienyl anion to form mixed bis(cyclopentadienyl)titanium dichloride complexes. Some examples are shown in Scheme 8-72. The addition of substituted cyclopentadienyl anions to $CpTiCl_3$ generates $(RCp)(Cp)TiCl_2$ **381** [173, 181] whereas the $(Me_5Cp)(Cp)TiCl_2$ complex was generated by the reverse sequence of adding CpLi to $Cp*TiCl_3$ [181].

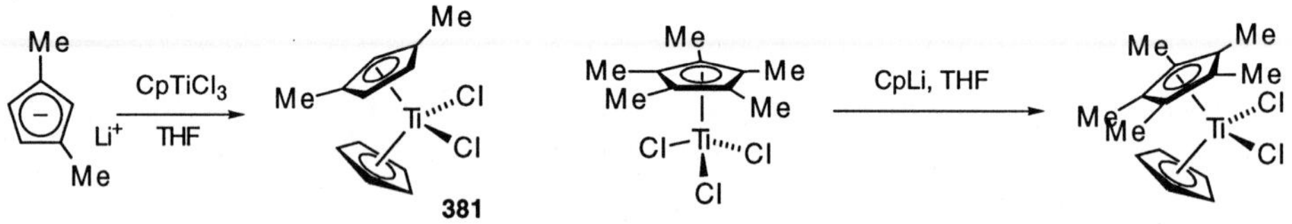

Scheme 8-72. Mixed bis(cyclopentadienyl)titanium dichlorides [181].

Chiral Bis(cyclopentadienyl)metal Complexes having Equivalent π-Faces. The introduction of a single chiral substituent on the cyclopentadienyl ligand leaves the faces of the ligand still equivalent. If the ligand is enantiomerically pure as in menthylcyclopentadiene **76**, only one isomeric $(R*Cp)_2$metal complex **382** can form [2]. Several chiral monosubsituted $(R*Cp)_2MCl_2$ complexes have been prepared containing the natural-product derived auxiliaries: neomenthyl **383** [3], 8-phenylmenthyl **384** [5b], carboxylic acid derived **385** [49], an estrone derivative **386** [47], cholestanol **387** [48] and cholestanone **388** [48] (Scheme 8-73). If both enantiomers of the ligand are present as in (1-phenylethyl)cyclopentadienyl **12**, then a mixture of *dl* and *meso* complexes **14** can form. Although the ratio of these stereoisomers in this particular case is 1:1 for both the titanium and zirconium complexes, selectivity for one isomer could have been possible [9, 182, 183]. The related (1-cyclohexylethyl)cyclopentadienyl ligand has been incorporated into an equimolar mixture of *dl* and *meso* isomers of zirconium complex **389** [183]. The *dl* isomers of both complexes **14** and **389** could be obtained in pure form through repeated recrystallizations of the isomeric mixture [182, 183].

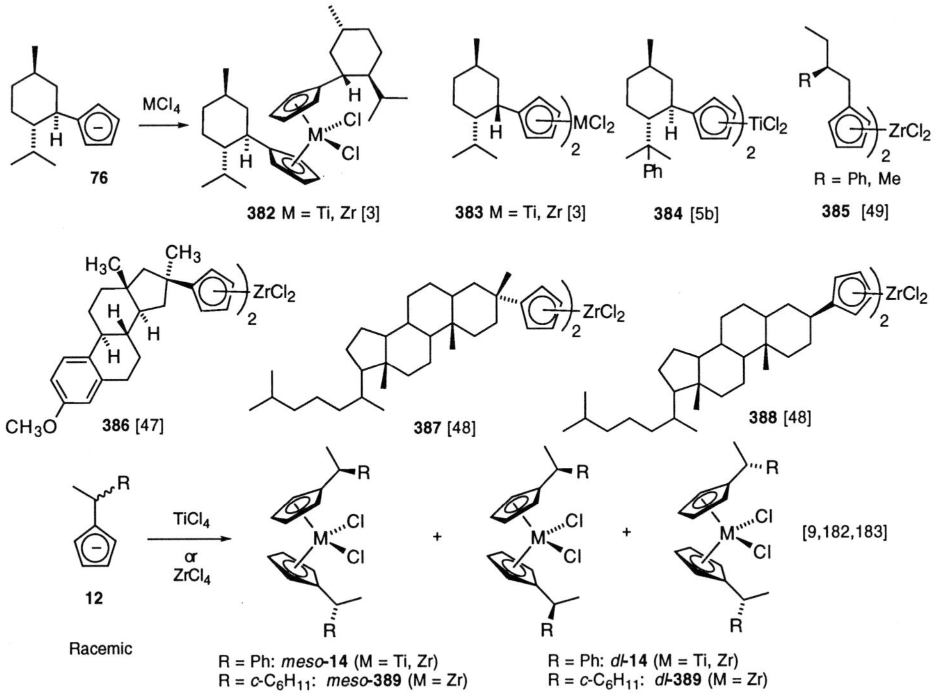

Scheme 8-73. Chiral bis(cyclopentadienyl)metal dichlorides.

The metalation of one enantiomer of C_2-symmetric cyclopentadienes can also inherently lead only to the formation of a single stereoisomeric bis(cyclopentadienyl)metal complex (Scheme 8-74). Examples of in the bis(cyclopentadienyl)metal complexes containing disubstituted, annelated cyclopentadienyl ligands are the bicyclooctane-annelated complexes **390** [73], **391** and **32** [19] and the binaphthyldimethylene-annelated complexes **392** [74]. C_2-symmetric tetrasubstituted, diannelated cyclopentadienes have been used to form monocyclocyclopentadienylmetal trichlorides **393** [87, 88] and the mixed metallocene complex **394** [88]. Apparently steric hindrance prevents two equivalents of the tetrasubstituted cyclopentadiene from being incorporated into a metallocene dichloride.

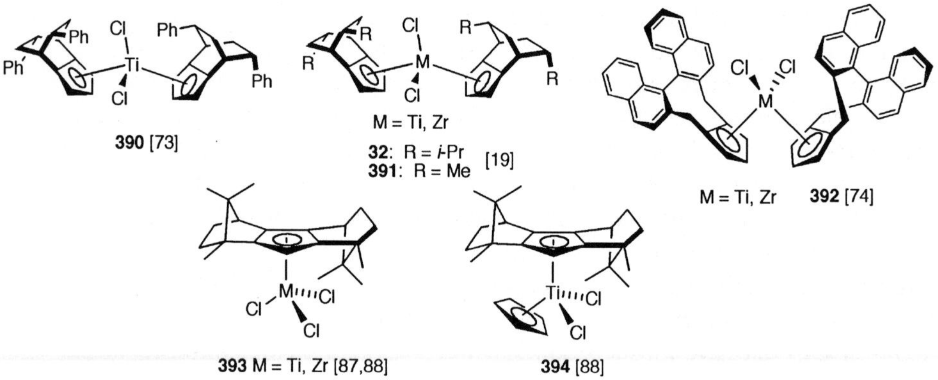

Scheme 8-74. Metal complexes of C$_2$-symmetric cyclopentadienes.

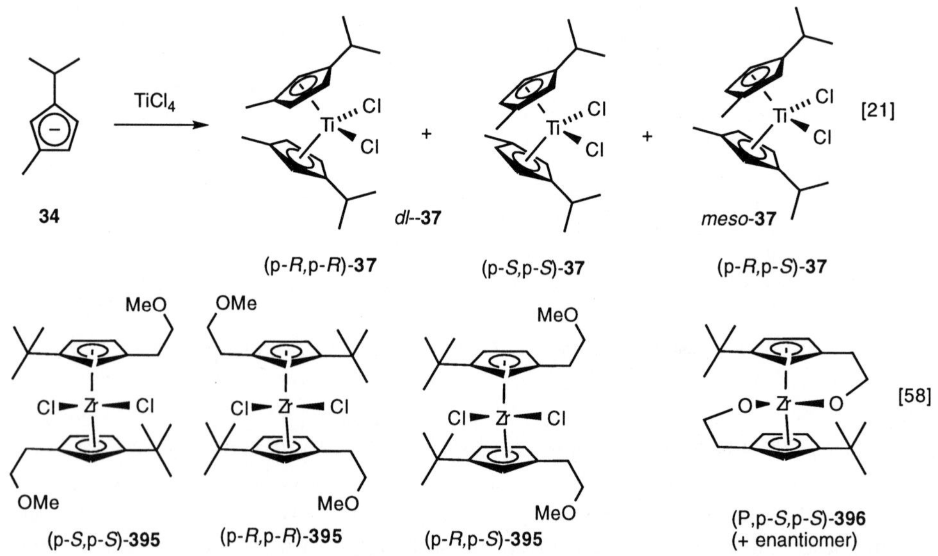

Scheme 8-75. Metalation of cyclopentadiene having enantiotopic faces.

Bis(cyclopentadienyl)metal Complexes having Enantiotopic Ligands. The preparation of metallocene complexes from cyclopentadienes having enantiotopic ligands is relatively undeveloped, perhaps due to the formation of stereoisomeric mixtures of the complexes (Scheme 8-75). An example is the metalation of the 1-isopropyl-3-methylcyclopentadienyl ligand **34** which leads to a mixture of *dl* and *meso* isomers of **37** [21]. Bergman has used chelating enantiotopic ligands to form a 3:1 mixture of *dl* and *meso* isomers of **395** and in a completely stereoselective metalation, the *dl* isomer **396** [58].

Bis(cyclopentadienyl)metal Complexes having Diastereotopic Faces. The metalation of cyclopentadienes containing diastereotopic faces can still lead to isomeric mixtures, but in many cases sufficient differentiation exists between the faces to enable selective metalations (Scheme 8-76). In the first example, the metalation of camphor-derived cyclopentadiene **128** was highly selective for one C_2-symmetric isomer **48**, with the C_1-symmetric isomer forming in trace amounts [5, 25, 26]. Metalation of the trisubstituted camphor-based cyclopentadiene **398** was also selective for the C_2-symmetric isomers **399** and **400** [85]. Metalations of verbenone-derived cyclopentadiene **130** [75] and pinanyl-cyclopentadiene **134** [72] were selective for a single isomer of **401** and **402**, respectively. The racemic phenylpentalene **403** was also selective for the single metallocene dichloride **404** [184].

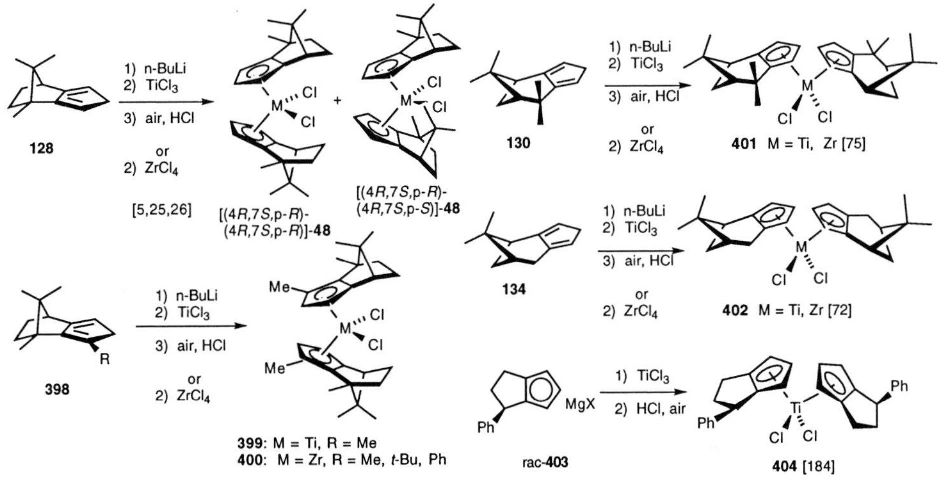

Scheme 8-76. Chiral bis(cyclopentadienyl)metal dichlorides.

A remarkable ability to control the facial selectivity in the metalation of norbornenylcyclopentadienyl ligand **49** has been demonstrated (Scheme 8-77) [27c]. Reaction of **49** with titanium trichloride at low temperature leads to the *endo,endo* isomer **405** whereas the reaction at room temperature leads to the *exo,exo* isomer **406**. The mixed titanocenes having the *endo* or *exo* configuration **407** or **408** could likewise be prepared through the reaction of **49** with CpTiCl$_3$.

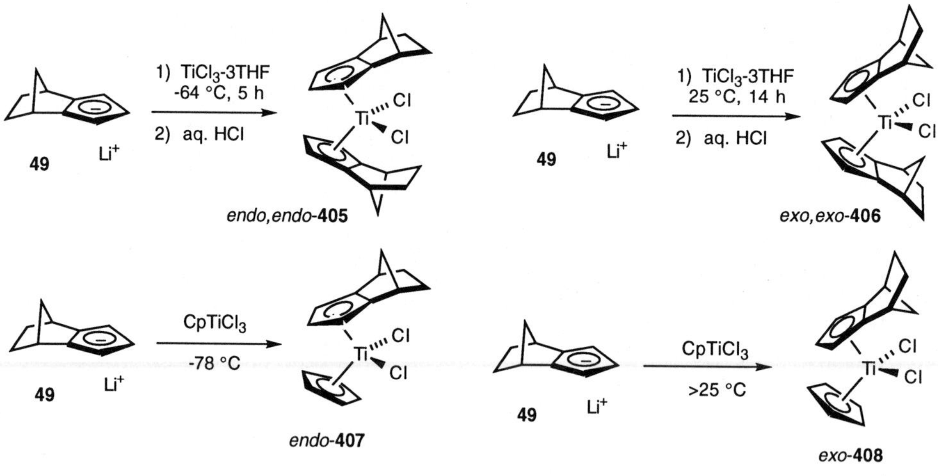

Scheme 8-77. Control of stereochemistry in isodicyclopentadienyltitanium complexes [27c].

8.6.2 Bis(indenes)

Achiral Complexes. The preparation of bis(η^5-indenyl)zirconium and titanium dichlorides is analogous to the preparation of cyclopentadienylmetal complexes. Two equivalents of the indenyl salt (originally the sodium salt [13a]) react with the metal tetrachloride to provide the metal complexes **409** (Scheme 8-78), whose solid state structures were obtained [13b]. A number of substituted bis(indenyl)metal complexes have been prepared from symmetrically substituted indenes. The lithium salt of several 2-arylindenes have been metalated by Waymouth to give complexes **410** in good yields [14]. 5,6-Disubstituted and 4,7-disubstituted indenes have been used to generate electronically varied bis(indenyl)zirconium dichloride complexes **411** and **412** [18]. These metalations were carried out using $ZrCl_4$ and the lithium salt of the indenes to give 30 to 65% yields of the metallocene dichlorides. Bis(2-aryltetrahydroindenyl)metal dichlorides **413** have been prepared directly by the metalation of the tetrahydroindene ligand **145** [83, 185]. The permethylated indene **198** has been converted into the very air sensitive bis(indenyl)zirconium dichloride **414** [186, 187].

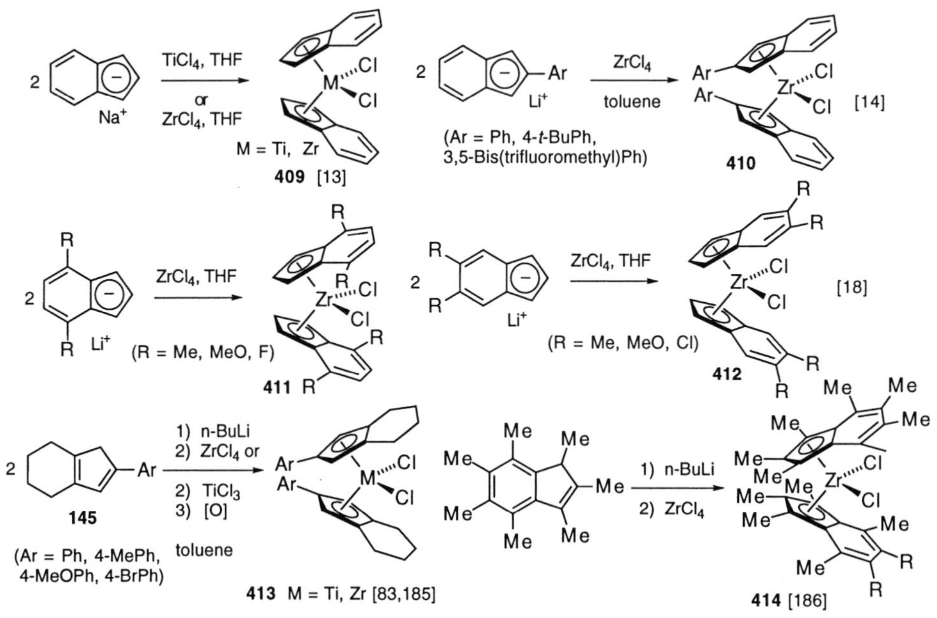

Scheme 8-78. Preparation of achiral bis(indenyl)metal dichloride complexes.

Chiral Complexes. A number of chiral unbridged bis(indenyl)zirconium dichloride complexes have been reported by Erker. The metalation of 1-cyclohexylindenyl-lithium with $ZrCl_4$ gave a 1:1 mixture of *rac-* and *meso*-bis(indenyl)zirconium dichloride complexes (Scheme 8-79). Catalytic hydrogenation of these complexes yielded the bis(tetrahydroindenyl)zirconium dichloride complexes. Both the indenyl and tetrahydroindenyl complexes were separable by fractional crystallization [24]. Indenes bearing the chiral substituents: neoisopinocamphyl, menthyl, neomenthyl, isomenthyl or neoisomenthyl have been metalated to give the bis(indenyl)zirconium dichloride complexes as varying mixtures of diastereomers [29, 112]. In none of these cases was the stereoselectivity between the two 'racemic-like' and the 'meso-like' diastereomers very high. The metalation of 4,7-dimethyl-1-neomenthylindene, however, did lead to a very high stereoselectivity for the formation of one 'racemic-like' bis(indenyl)zirconium complex a 98:1:1 ratio [10]. The chiral bis(indenyl)titanium dichloride complex **25** which contains equivalent ligand faces has been prepared from 2-menthylindene **415** [15].

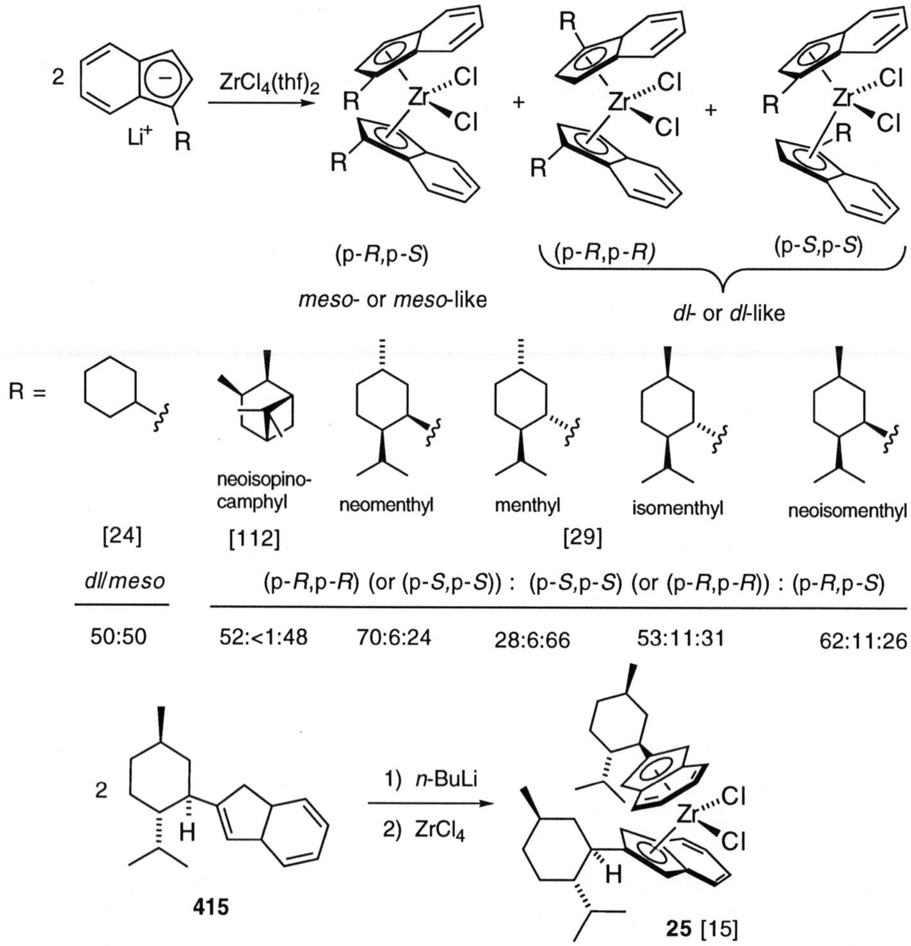
2
Li+ R
ZrCl4(thf)2
Zr
Cl
R
+
(p-R,p-S)
(p-R,p-R)
(p-S,p-S)
meso- or meso-like
dl- or dl-like
R =
neoisopino-
camphyl
neomenthyl
menthyl
isomenthyl
neoisomenthyl
[24]
[112]
[29]
dl/meso
(p-R,p-R) (or (p-S,p-S)) : (p-S,p-S) (or (p-R,p-R)) : (p-R,p-S)
50:50
52:<1:48
70:6:24
28:6:66
53:11:31
62:11:26
H
415
1) n-BuLi
2) ZrCl4
25 [15]

Scheme 8-79. Preparation of chiral unbridged bis(indenyl)zirconium dichlorides.

8.6.3 Bis(fluorenes)

Unbridged bis(fluorenyl)zirconium dichlorides can be prepared by the reaction of fluorenyl anions with zirconium tetrachloride (Scheme 8-80). The first example, the unsubstituted bis(fluorenyl)zirconium dichloride **409** was reported by Setton and Samuel in 1965 [13]. The methyl-substituted complex **410** was isolated by crystallization as only the *dl* isomer [188]. Alt has prepared a number of addition unbridged bis(fluorenyl)zirconium dichlorides such as the 9-substituted complexes **411** and **412** and the isomeric mixture of **413** [189]. Fluorenyl complexes are often difficult to purify and characterize since they tend to be much less soluble than corresponding cyclopentadienyl or indenyl complexes.

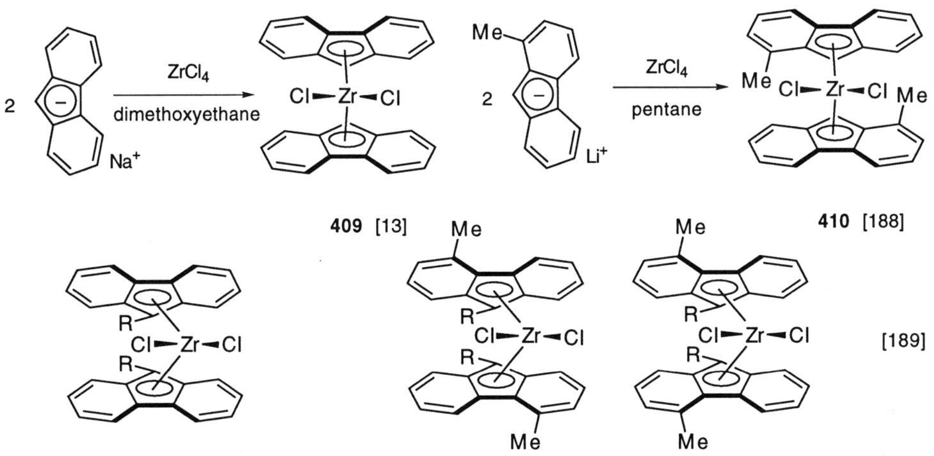

411: R = C_6H_{11}
412: R = Ph

dl-**413**: R = Ph

meso-**413**: R = Ph

Scheme 8-80. Bis(fluorenyl)zirconium dichlorides.

8.6.4 Mixed Ligands

Unbridged metallocene dichloride complexes containing mixed ring ligands are fairly undeveloped. Methods for the preparation of some of the few examples are given in this section.

Cyclopentadienyl, Indenyl. The mixed metallocene dichloride **414** has been prepared either by addition of indenyllithium to $CpTiCl_3$ or $CpZrCl_3$ [181,190] or by the addition of cyclopentadienyllithium to indenyltitanium trichloride **415** [191] (Scheme 8-81). Indenyltitanium trichloride **415** has been selectively formed from $TiCl_4$ and (indenyl)tributyltin. Complex **414** has also been prepared earlier using a thallium complex [192].

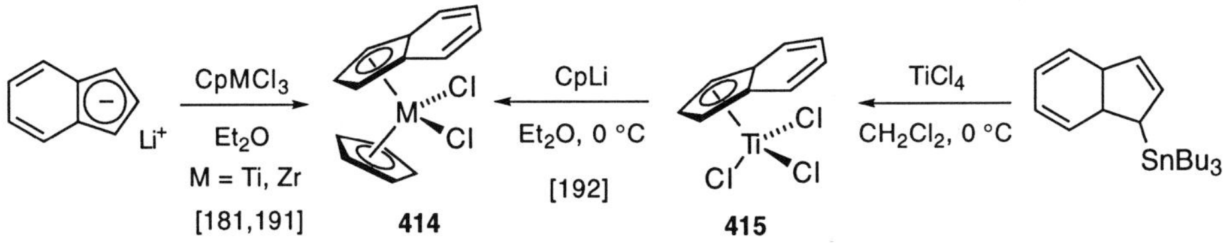

Scheme 8-81. Unbridged (cyclopentadienyl)(indenyl)metal dichlorides.

Cyclopentadienyl, Fluorenyl. Mixed complexes having cyclopentadienyl and fluorenyl ligands have been prepared by the addition of substituted fluorenyllithium complexes to either $CpZrCl_3$ or $(Me_5Cp)ZrCl_3$ to give complexes **416** [193] and **417** [194] (Scheme 8-82).

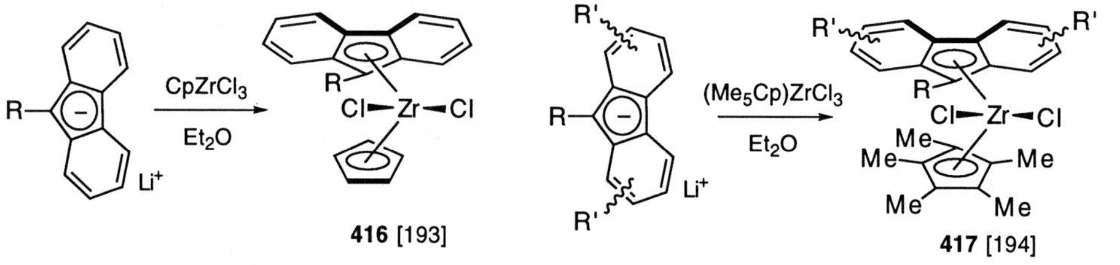

Scheme 8-82. Unbridged (cyclopentadienyl)(fluorenyl)metal dichlorides.

8.7   Metal Complexes with Nonstereogenic Bridging Groups

8.7.1 Bis(cyclopentadienyl)metals

Achiral Complexes. A good number of *ansa*-metallocene dichloride complexes which contain a nonstereogenic bridging group between two achiral cyclopentadienyl ligands have been prepared. With no stereochemical issues to overcome, the major synthetic hurdle is the differentiation between the formation of *ansa*-metallocenes versus oligomeric complexes. Examples shown in Scheme 8-83 are: one carbon bridged complexes **418** [116] and **419** [119, 195]; two carbon bridged **420** [117], **421** [10], **422** [124] and **423** [123]; propano-bridged complex **424** [118]; cyclopentane-1,3-diyl-bridged **425** [122]; silyl-bridged complexes **426** [120] and **427** [121] and the doubly bridged complex **428** [126,127].

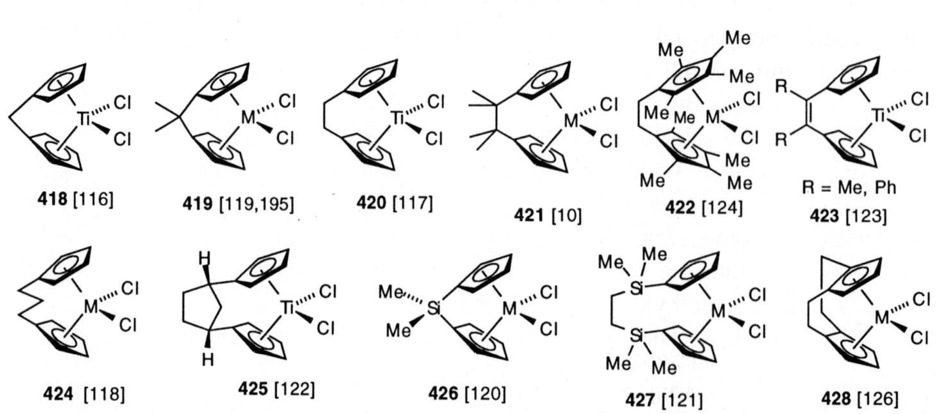

Scheme 8-83. Achiral bridged bis(cyclopentadienyl)metal complexes (M = Ti, Zr).

Chiral Complexes from Prochiral Ligands. Brintzinger reported the first chiral *ansa*-metallocene dichloride, *dl*-propanobis(3-*t*-butylcyclopentadienyl)titanium dichloride (**429**), in 1979 [196] (Scheme 8-84). The prochiral ligand **430** was metalated and the crude product purified by silica gel HPLC to give the *dl* pair **429**. The (1*S*,1′*S*) isomer could be resolved by reacting **429** with the sodium salt of (*S*)-binaphthol and purifying the ensuing mixture by HPLC. The single enantiomeric adduct was characterized by X-ray crystallography.

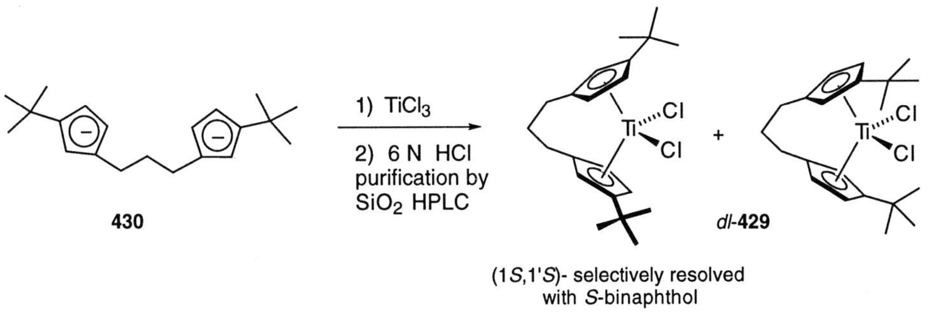

(1*S*,1′*S*)- selectively resolved
with *S*-binaphthol

Scheme 8-84. Propanobis(3-t-butylcyclopentadienyl)titanium dichloride.

Several other examples of *ansa*-bis(3-substituted cyclopentadienyl)metal dichlorides have been subsequently prepared having a dimethylsilylene-, an ethylene-, a tetramethylethylene- or a cyclopentane-1,3-diyl-bridging group. In most cases a mixture of three stereoisomers was formed—the *dl* pair and the *meso* complex. These metalations are summarized in Scheme 8-85. Only one of the *dl* complexes is depicted along with the ratios observed for the metalation. The *dl*/*meso* selectivity in the silyl-bridged zirconium complexes **431** was uniformly 1:1 [128]. Several other related silyl-bridged complexes have also been reported [129]. The *dl* isomers of the tetramethylethylene-bridged complexes **432** (M = Ti) and **433** (M = Zr) were moderately preferred over the *meso* isomers. In most of these cases the isomers could be separated. An additional study of the bis(*t*-butylcyclopentadienyl)titanium complex has been published which includes the use of Cp_2TiCl_2 as the metal source through ligand exchange [130]. Ethylene-bridged complexes **434** (M = Ti) were selective for the *meso* isomer [62]. The isopropyl- and *t*-butyl-substituted complexes could be separated by fractional crystallization. The analogous zirconium complexes **435** were moderately selective for the *dl* pair [197]. The metalation of cyclopentane-1,3-diyl-bridged bis(cyclopentadienes) **250** could give four isomers of **436**, the *dl* pair and two *meso* isomers having the substituents either both over the methylene group or the ethylene group of the cyclopentane. In each case, some selectivity was observed for the *dl* pair and both the *dl* pair and the first of the *meso* isomers could be purified by crystallization [122].

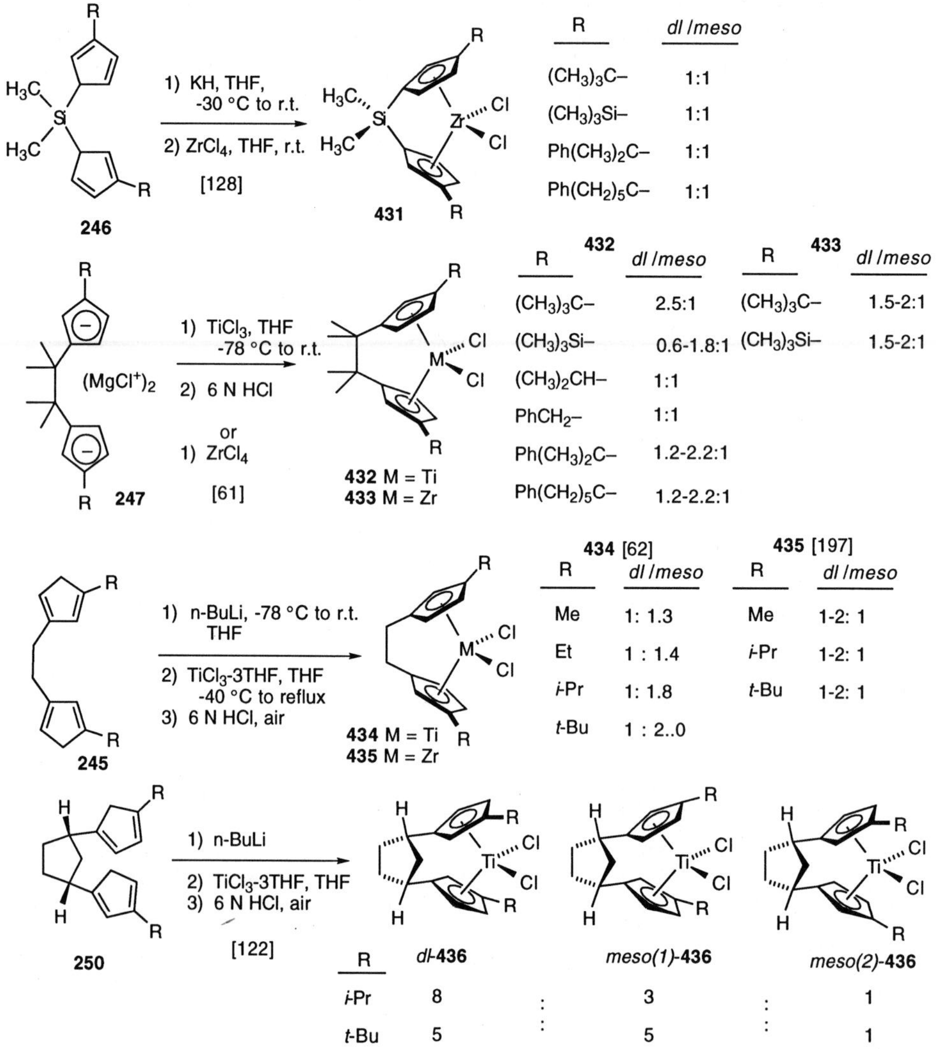
1) KH, THF, -30 °C to r.t.
2) ZrCl4, THF, r.t.
[128]
246
431
R dl/meso
(CH3)3C– 1:1
(CH3)3Si– 1:1
Ph(CH3)2C– 1:1
Ph(CH2)5C– 1:1
(MgCl+)2
1) TiCl3, THF -78 °C to r.t.
2) 6 N HCl
or
1) ZrCl4
[61]
247
432 M = Ti
433 M = Zr
432
R dl/meso
(CH3)3C– 2.5:1
(CH3)3Si– 0.6-1.8:1
(CH3)2CH– 1:1
PhCH2– 1:1
Ph(CH3)2C– 1.2-2.2:1
Ph(CH2)5C– 1.2-2.2:1
433
R dl/meso
(CH3)3C– 1.5-2:1
(CH3)3Si– 1.5-2:1
1) n-BuLi, -78 °C to r.t. THF
2) TiCl3-3THF, THF -40 °C to reflux
3) 6 N HCl, air
245
434 M = Ti
435 M = Zr
434 [62]
R dl/meso
Me 1: 1.3
Et 1 : 1.4
i-Pr 1: 1.8
t-Bu 1 : 2..0
435 [197]
R dl/meso
Me 1-2: 1
i-Pr 1-2: 1
t-Bu 1-2: 1
1) n-BuLi
2) TiCl3-3THF, THF
3) 6 N HCl, air
[122]
250
R dl-436 : meso(1)-436 : meso(2)-436
i-Pr 8 : 3 : 1
t-Bu 5 : 5 : 1

Scheme 8-85. ansa-bis(3-RCp)MCl_2 complexes.

Several examples of complexes having an additional substituent adjacent to the bridging group are summarized in Scheme 8-86. The ethylene-bridged titanium complexes **437** were initially only moderately selective for the *dl* pair, but the product mixture could be photochemically equilibrated to give a significant enhancement of the *dl* isomers [131]. The silyl-bridged zirconium complexes **438** were selective for the *dl* pair [128]. Several additional examples have been prepared and enhanced in the proportion of the *dl* isomers by crystallizations [129]. Another trisubstituted bis(cyclopentadienyl) system is the 1,2-phenylene-bridged bis(tetrahydroindenyl)metal dichloride complexes **439** and **440**. These tetrahydroindenyl complexes were formed directly from the hydroindene ligand without the usual need for hydrogenation of an initial bis(indenyl)metal complex. Both of these complexes were somewhat selective for the *dl* isomer and the *zirconium* complex could be photochemically enhanced in the *dl* isomer [132].

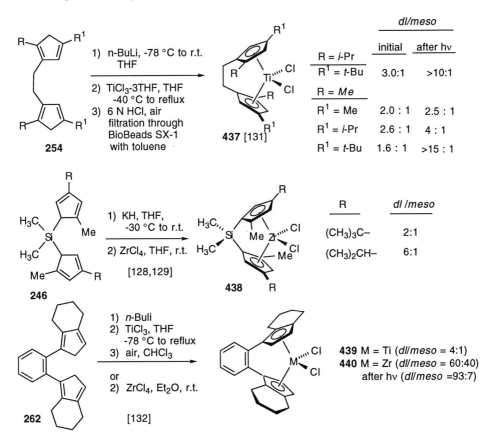

Scheme 8-86. ansa-Metallocene dichlorides of trisubstituted cyclopentadienes.

Two related complexes **441** and **442** having prochiral cyclopentadienyl ligands bridged by two silylene groups have been prepared by Brintzinger [133] (Scheme 8-87).

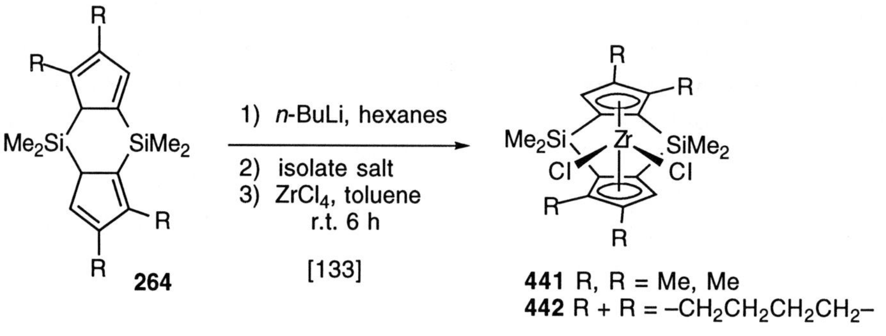

Scheme 8-87. Doubly bridged bis(cyclopentadienyl)zirconium dichlorides.

Bridged Complexes Containing Chiral Cyclopentadienyl Ligands. Marks has reported the metalation of the neomenthyl- and menthyl-containing bis(cyclopentadienyls) **266** and **267** (Scheme 8-88).

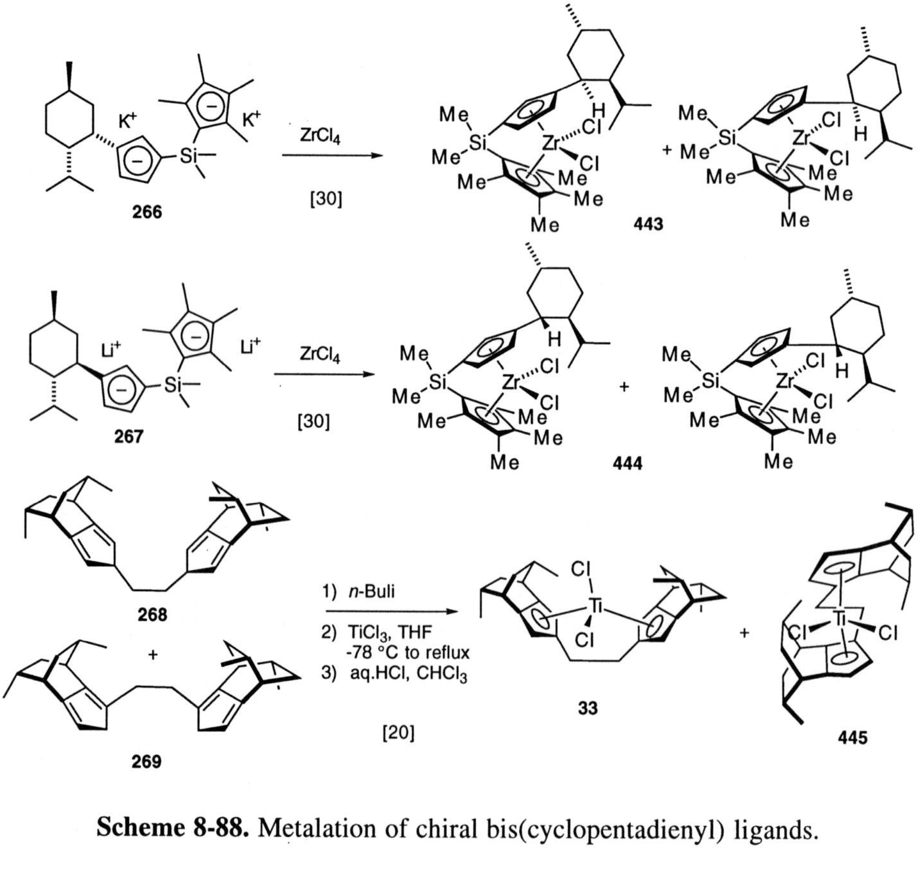

Scheme 8-88. Metalation of chiral bis(cyclopentadienyl) ligands.

These ligands contain one ligand with diastereotopic faces. Both diastereomeric complexes of **443** and **444** were formed and separated by fractional crystallization [30]. Halterman has reported the metalation of a mixture of ethanobis(BCOCpH) ligands **268** and **269**. Bis(cyclopentadiene) **268** is bridged along its axis of symmetry and the metalation of its homotopic faces could only lead to the single isomer **33**. Bis(cyclopentadiene) **269** contains diastereotopic faces and while three stereoisomeric complexes could have been formed, only C_2-symmetric complex **445** was observed [20].

8.7.2 Bis(indenes)

The first example of an *ansa*-bis(indenyl)metal complex was Brintzinger's 1982 report on an ethylene-bridged titanium complex (Scheme 8-89). The initial bis(indenyl)titanium dichloride complexes **446** were formed in *dl* to *meso* ratios of 2–10:1, depending on reaction conditions. These isomers could only be

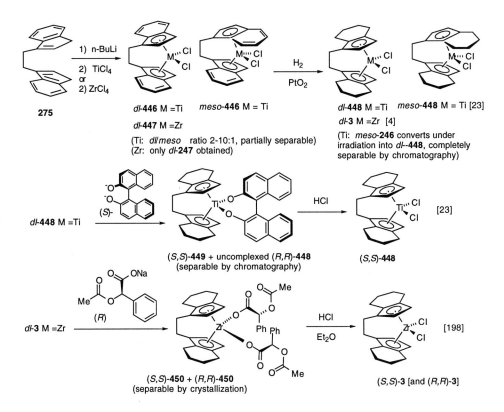

Scheme 8-89. Ethanobis(indenyl)metal dichlorides.

partially separated by chromatography. Hydrogenation of the 6-membered rings was possible and the resulting bis(tetrahydroindenyl)titanium dichloride complexes **448** could be separated. It was also found that irradiation converted the *meso* isomer into the *dl* pair [23]. An analogous reaction with zirconium gave only the *dl*-bis(indenyl)zirconium complex **447** which could also be hydrogenated to its tetrahydroindeny derivative **3** [4, 198]. The addition of hydrogen was shown to occur on the ligand face opposite the metal [199]. A number of synthetic modifications of the conditions for the formation of these complexes have appeared [136, 137]. The use of tetraamidozirconium and the bis(indene) **275** allows a facile preparation of *dl*-**447** under equilibrating conditions [200]. The *dl* isomers of both metals can be resolved. With titanium complex **448**, only the (*S,S*) enantiomer coordinates with (*S*)-binaphtholate to give complex **449** which can be purified by chromatography [23]. The zirconium complex **3** could be resolved *via* the mandelate complexes **450**. In this case both enantiomers form complexes, but these are separable by crystallization [201]. The dichlorides can be recovered by treatment of the alkoxide complexes with HCl.

A number of related bis(indenyl)metal dichlorides containing unsubstituted indenyl ligands have appeared and are shown in Scheme 8-90. The propano complex **251** and the cyclopentane complex **453** were selective for the *dl* isomer [122, 201]. The dodecyl-tethered complex **452** was selective for the *dl* isomer when the metalation was carried out in THF, but for the *meso* isomer when carried out in toluene [138].

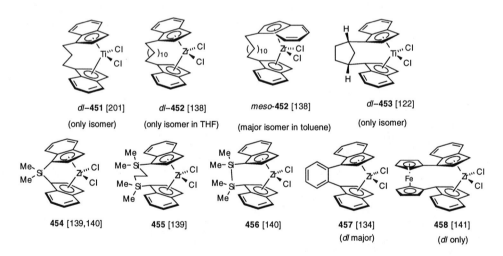

dl–**451** [201]	*dl*–**452** [138]	*meso*-**452** [138]	*dl*–**453** [122]
(only isomer)	(only isomer in THF)	(major isomer in toluene)	(only isomer)

454 [139,140]   **455** [139]   **456** [140]   **457** [134]   **458** [141]
(*dl* major)   (*dl* only)

Scheme 8-90. *ansa*-bis(Indenes) containing unsubstituted indenyl ligands.

Silyl-containing bridges are seen in complexes **454** [139, 140], **455** [139] and **456** [140]. The 1,2-phenylene-bridged complex **457** was selective for the *dl* isomer [134] while only the *dl* pair was observed with the ferrocene-bridged complex **458** [141].

More highly substituted bis(indenyl)zirconium dichloride complexes have also been prepared (Scheme 8-91). Included in this group are: ethylene-bridged complexes **459** and **460** [18b, 110a]; 3-substituted silyl-bridged complexes **461** [140]; substituted silyl-bridged complexes **462** [106, 107, 110a] and the benzannelated complexes **463** [106, 111].

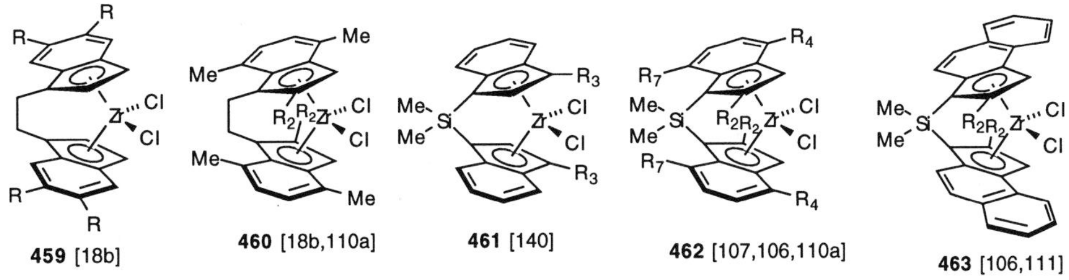

Scheme 8-91. Substituted bis(indenyl)zirconium dichloride complexes.

Several bis(indenyl)metal complexes having bridging groups other than at the 1-indenyl position have been prepared and are shown in Scheme 8-92. Only a single metal complex **464** could form in the metalation of C_2-symmetric bis(2-indenyl)ethane **295** [16] and in the metalation of the doubly bridged bis(indene) **300** to **465** and **466** [142]. Two sets of complexes having the indenes tethered through the 7-positions have been prepared. Both the titanium and zirconium complexes **467** and **468** formed as a mixture of stereoisomers in the metalation of **303**, but only the titanium complex **470** could be isolated in the metalation of bis(indene) **469** [143].

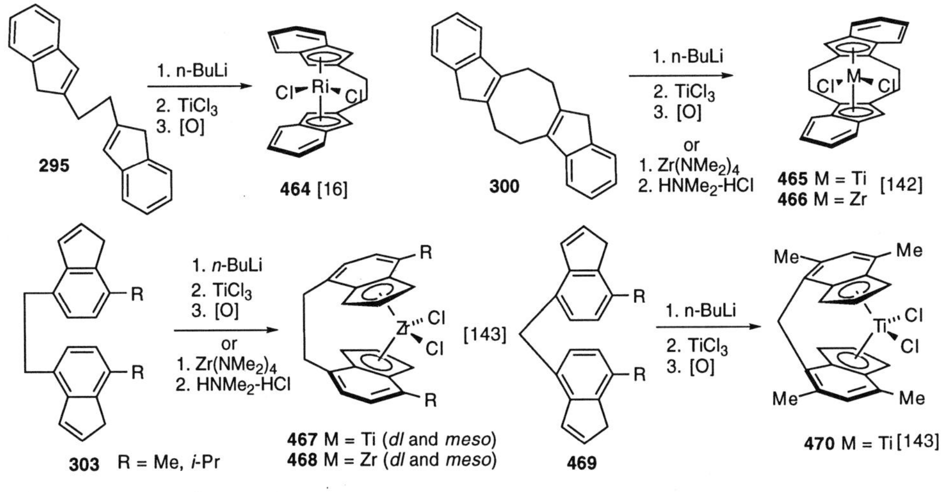

Scheme 8-92. *ansa*-bis(Indenyl)metal dichlorides bridged through 2- or 7-position.

8.7.3 Bis(fluorenes)

A limited number of *ansa*-bis(fluorenyl)zirconium dichloride complexes have been
prepared (Scheme 8-93). An ethylene-bridge and a dimethylsilylene-bridge are
found in the unsubstituted achiral bis(fluorenyl)zirconium complexes **471** [144]
and **472** [146]. Both the *meso* and *dl* forms of ethylene-bridged complex **473** were
formed [148]. One example of a complex containing two different fluorenyl
ligands is **474** [147].

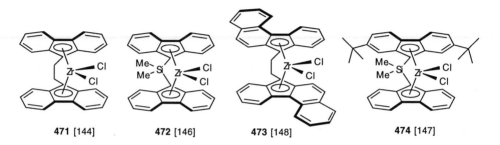

471 [144]   **472** [146]   **473** [148]   **474** [147]

Scheme 8-93. *ansa*-bis(Fluorenyl)zirconium dichlorides.

8.7.4 Mixed Ligands

Several examples of *ansa*-metallocenes containing one indenyl and one cyclopentadienyl moiety bridged by nonstereogenic bridging groups are given in Scheme 8-
94. Green was the first to report the isopropylidene and cyclohexylidene-bridged
zirconium complexes **475** [149]. Subsequently the titanium complex **475** was reported [150] as well as the diphenyl-substituted zirconium complex **475** [151]. The
parent silyl-bridged complex **476** is also known [140]. Two sets of substituted cyclopentadienyl, indenyl complexes have been prepared having either an isopropylidene bridge in **477** or a silyl bridge in **478** [152].

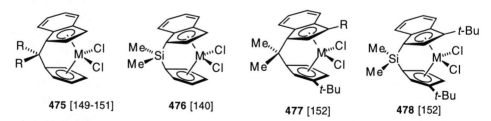

475 [149-151]   **476** [140]   **477** [152]   **478** [152]

Scheme 8-94. *ansa*-Metallocenes containing indenyl with cyclopentadienyl ligands.

Several mixed metallocenes containing a fluorenyl ligand tethered through a nonstereogenic group to a cyclopentadienyl ligand have been prepared (Scheme 8-95).

The first example was the isopropylidene-bridged complex **479** [153–156]. Additional examples are the silyl-bridged complex **480** [145], the di-(*t*-butyl) derivative **483** [147] and the unsymmetrically substituted complexes **481** [159, 110a] and **482** [156–158]. An interesting member of this last group is the methylene bridged complex **482** (R = H) [158]. The final example of ligands bridged by nonstereogenic groups is the fluorenyl, indenyl complex **484** [145].

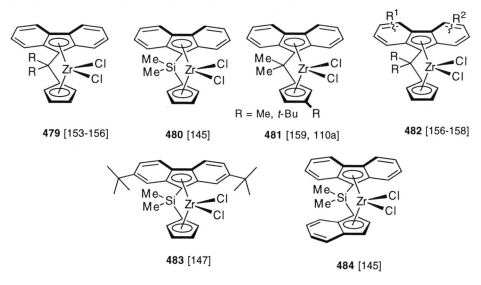

479 [153-156]   **480** [145]   **481** [159, 110a]   **482** [156-158]

R = Me, *t*-Bu   R

483 [147]   **484** [145]

Scheme 8-95. *ansa*-Metallocenes containing fluorenyl with cyclopentadienyl or indenyl ligands.

8.8 Metal Complexes with Stereogenic Bridging Groups

One-Carbon Bridged Complexes. In Rausch and Chien's bridged ligand **327**, only the indenyl ligand has diastereotopic π-faces and its metalation could produce two diastereomeric *ansa*-metallocene dichloride complexes. Only one isomer of the titanium or zirconium complexes **485** was isolated in each case and characterized by X-ray crystallography [160] (Scheme 8-96). Apparently, the methyl group in the bridging group directs the indenyl ligand to the observed less hindered orientation. Based on the low selectivity observed in the metalation of related chiral 1-substituted indenes, it is unlikely that the metal coordinates initially to the indene as that should lead to a diastereomeric mixture. Initial coordination to the equivalent faces of the cyclopentadienyl moiety would then allow the indenyl ligand to adopt the favorable conformation for the subsequent *ansa*-metalation. Several example of one-carbon bridged complexes in which the bridging carbon is part of a tetrahydroindenyl ring have been reported. Erker has prepared the titanium and zirconium complexes **486** and **487** [161]. The di-(*t*-butyl) derivatives **489** and **490** have also

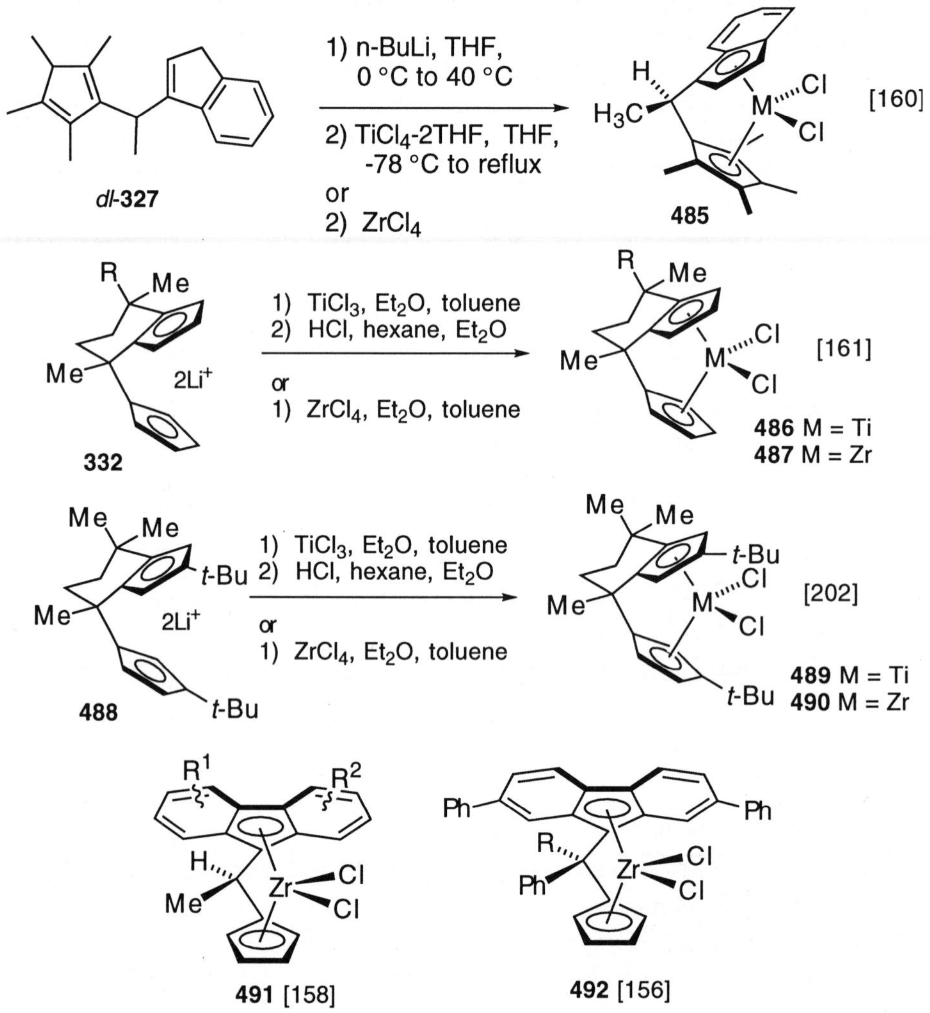

Scheme 8-96. One-carbon stereogenic bridging group.

been reported [202]. A number of *ansa*-metallocenes such as **491** and **492** having substituted fluorenyl groups tethered by a stereogenic group to cyclopentadienyl have been recently reported [158, 156].

Bis(indenes) Bridged by Stereogenic Groups at the 1-Position. Both indenyl ligands arising from deprotonation the dimethylbinaphthyl-bridged bis(indene) **333**

have diastereomeric faces and three diastereomeric *ansa*-metallocene complexes could be formed (Scheme 8-97). Treatment of the indenyllithium complexes of **333** with TiCl₃ followed by oxidation, or by ZrCl₄ produced in each case a single isolable diastereomeric *ansa*-metallocene complex **493** or **494** in low yield along with what was likely oligomeric metallocene complexes. Initial complexation of the metals to the first indenyl ligand was likely not selective. Due to interactions with the chiral bridge, only one diastereomeric initial complex could cyclize to an *ansa*-metallocene complex. The other initial diastereomeric complex presumably underwent intermolecular metalation to form oligomeric material. In the solid state **494a** adopts a C₁-symmetric

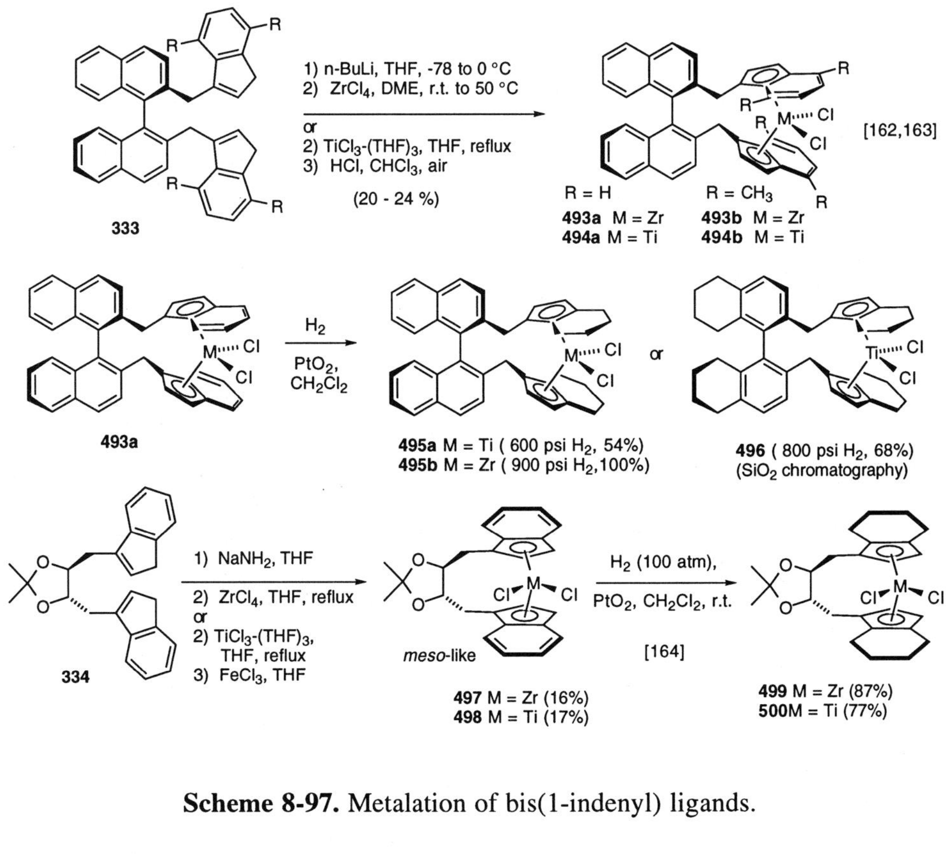

Scheme 8-97. Metalation of bis(1-indenyl) ligands.

conformation. In solution the flipping between two equivalent C₁-symmetric conformations of the titanium complex was found to have an energy barrier of about 13 kcal mol⁻¹. The zirconium complex **493a** was more mobile [162, 164]. The bis(indenyl)metal complexes **493a** and **494a** could be hydrogenated to give either the dimethylbinaphthyl-bridged bis(tetrahydroindenyl)titanium and -zirconium complexes **495** or the bitetralin-bridged bis(tetrahydroindenyl)titanium complex

496. The titanium complexes could be purified by silica gel chromatography. The bis(tetrahydroindenyl) complex **495a** was conformationally more mobile than the corresponding bis(indenyl)titanium complex **494a**. Metalation of Green's threitol-bridged bis(indene) **334** could also give three diastereomeric *ansa*-metallocenes. Bis(indene) **334** was deprotonated by sodium amide and metalated with either $ZrCl_4$ or with $TiCl_3$ followed by $FeCl_3$ oxidation to give low yields of the *meso*-like metallocene dichlorides **497** and **498**. Catalytic hydrogenation led to the *meso*-like bis(tetrahydroindenyl)metal complexes **499** and **500**. The complexes were purified by crystallization from hot toluene [163].

The dilithio salt of Rieger's 1,2-cyclohexane-bridged bis(indene) **336** was metalated with $ZrCl_4$, filtered through silica gel with methylene chloride and concentrated to give an undefined mixture of the three diastereomeric complexes **501**. These isomers could not be separated [165]. Metalation of the 1,4-cyclohexane-bridged bis(indene) **337** with $TiCl_3$ followed by oxidation gave only the one diastereomeric titanocene complex **502** having the indenyl groups oriented away from the isopropyl groups [115]. Metalation of the lithium salt of **337** with $ZrCl_4$ failed, but the authors did not attempt metalation under equilibrating conditions with tetrakis(dimethylamido)zirconium. The solid state structure of **502** indicates approximately C_2-symmetry. Bosnich's bis(indene) **240** was deprotonated with *n*-BuLi and metalated with either $ZrCl_4$-2THF or with $TiCl_4$-2THF [32]. The resulting crude bis(indenyl)metal complexes were hydrogenated to give the crude bis(tetrahydroindenyl)metal complexes **503** and **504**. Both these 'chiracene' zirconium and titanium complexes were purified by chromatography on silanized silica gel. In each case a mixture of two C_2-symmetric and one *meso*-like diastereomeric metal complex were formed. The *meso*-like complexes were the major isomers. Through recrystallization, one C_2-symmetric isomer of the zirconium and titanium complexes were obtained and their structures confirmed through X-ray crystallography. Both the initial mixtures of the bis(tetrahydroindenyl)titanium and zirconium complexes could be photolyzed with a 450-W mercury lamp to a photostationary state favoring one of the C_2-symmetric isomers. Metalation of the dilithio salt of bis(indene) **342** with $TiCl_4$-2$TiCl_4$ followed by catalytic hydrogenation gave only one C_2-symmetric stereoisomeric 'cyclacene' titanocene dichloride **505** (Scheme 8-98). This titanium complex was purified by silanized silica gel chromatography and recrystallization. Its solid state structure was determined [166].

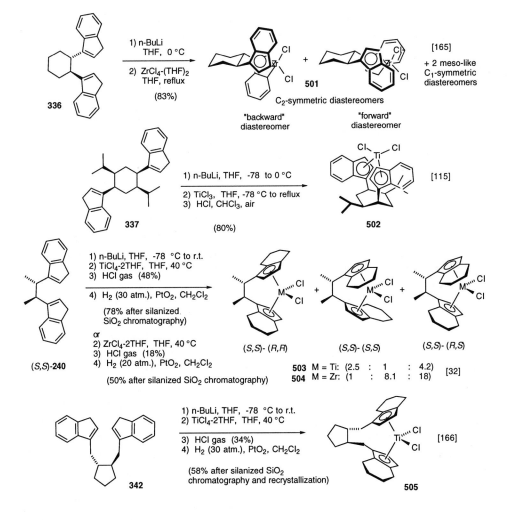

Scheme 8-98. Metalation of bis(1-indenyl) ligands.

Biaryl-Bridged Complexes. Brintzinger's biphenyl-bridged bis(cyclopentadienyl) ligand **19** has equivalent π-ligand faces and its metalation with either ZrCl₄ or TiCl₃ followed by oxidation gave the expected single isomeric metal complexes **506** and **507** (Scheme 8-99). The titanocene complexes were purified by silica gel chromatography, the zirconium complex by recrystallization. Since no synthetic intermediates were

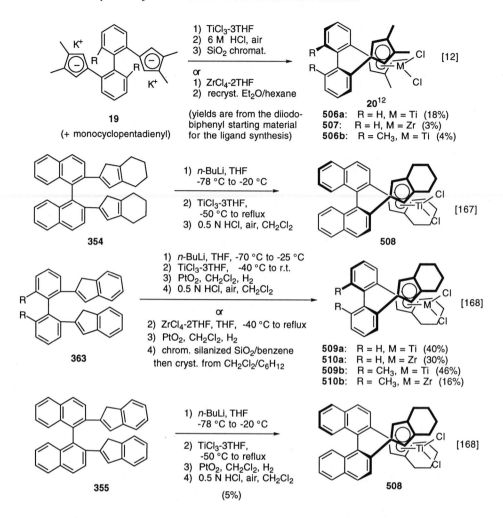

Scheme 8-99. Biaryl-bridged bis(cyclopentadienes) metal complexes.

isolated in this sequence, the reported yields are for several steps used to convert the 2,2′-diiodo-1,1′-biphenyls to the final metallocene dichlorides. The structures of all three complexes were determined by X-ray crystallography. Halterman's binaphthyl-bridged bis(tetrahydroindene) **354** was deprotonated by *n*-BuLi and metalated with TiCl$_3$ followed by oxidation to give the titanocene dichloride **508** in 68% yield or with ZrCl$_4$ to give the zirconium complex in good yield [167]. Although a nice series of ligands were mad, only the bis(hydroazulene) **357** was cleanly converted to a titanocene complex. The bis(indene) **355** and the bis(hydropentalene) **356** did not form *ansa*-metallocene complexes under the conditions examined. Bosnich was able to prepare the biphenyl bridged bis(tetrahydroindenyl)metal complexes **509** and **510** by hydrogenation of initially formed bis(indenyl)metal complexes [168]. Binaphthyl-bridged bis(tetrahy-droindene) **508** was similarly prepared from bis(indene) **355** [168]. In all of these cases, only a single stereoisomeric product could form due to the homotopicity of the ligand faces.

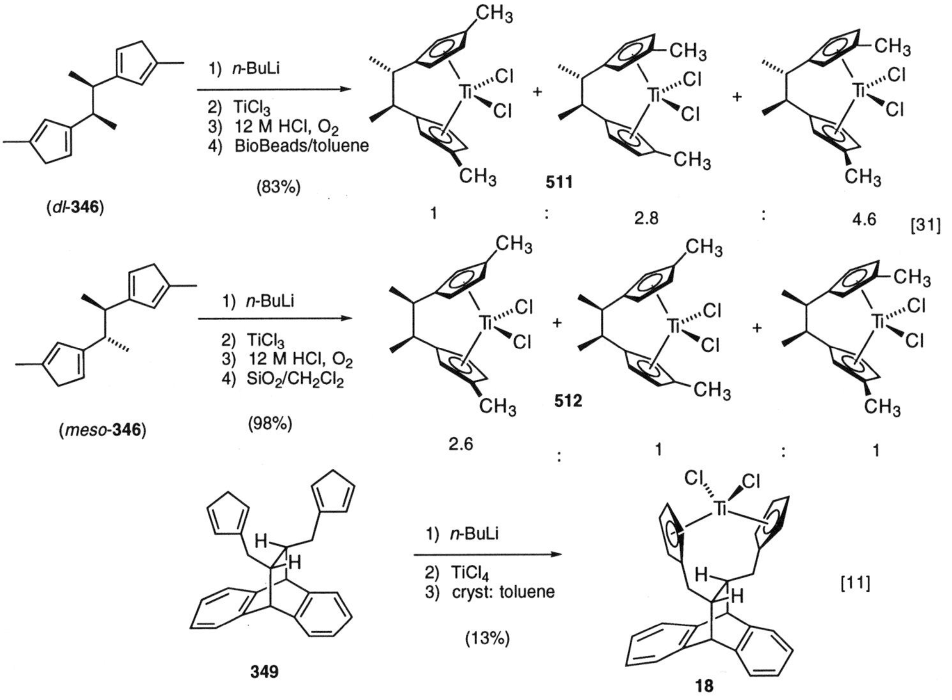

Scheme 8-100. Chiral *ansa*-bis(cyclopentadienyl)titanium dichloride complexes.

Bis(cyclopentadienyl)metal Complexes. Nantz has examined the ability of the 2,3-butane-diyl bridging group to exert its influence on the orientation of 3-methyl-cyclopentadienyl groups in titanocene complexes **511** and **512** (Scheme 8-100) [31]. In the metalation of C_2-symmetric bis(cyclopentadiene) *dl*-**346**, the *meso*-like isomer of **511** was favored, whereas in the metalation of *meso*-**346**, the formation of one of the two *rac*-like isomers of **512** was favored. In no case was the selectivity very pronounced and it is apparent that insufficient directing effects are at work here. The ligand faces in bis(cyclopentadienyl) **349** are equivalent and only the single titanocene **18** was formed upon its metalation [11].

Chiral Silyl-Bridged Complexes. Metalation of the di(menthoxy)silyl-bridged bis(indene) **365** leads to three stereoisomeric bis(indenyl)titanium dichlorides **513** with a small preference for the *meso*-like isomer (Scheme 8-101) [169].

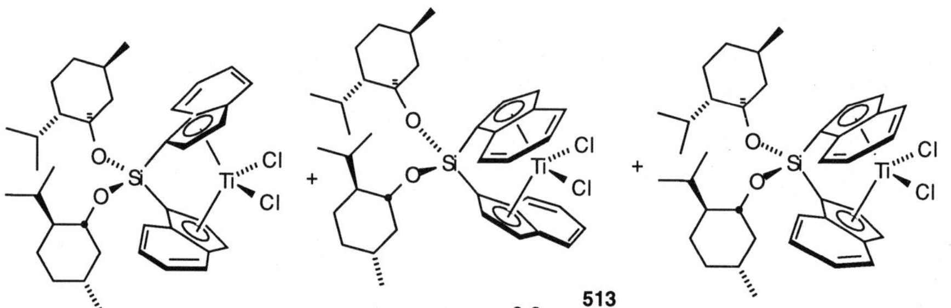

Scheme 8-101. Di(menthoxy)silyl-bridged bis(indenyl)titanium dichlorides.

Chiral Doubly Bridged Bis(cyclopentadienyl). One example has been published by Buchwald in which two cyclopentadienyl ligands are doubly bridged by a non-stereogenic group (Scheme 8-102). In this case the Meerwein's diketone-derived ligand **371** was metalated to give the titanocene complex **514**. The zirconium complex could not be isolated from an analogous attempted metalation [171].

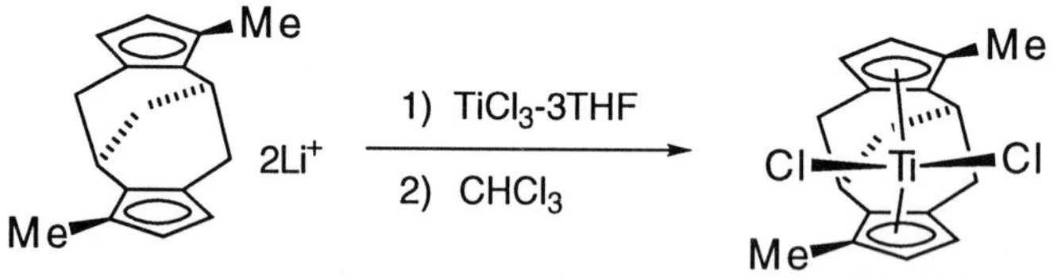

Scheme 8-102. Chiral doubly bridged bis(cyclopentadienyl)titanium complex.

Fluorenyl-Fluorenyl and Fluorenyl-Indenyl Complexes. Rieger has metalated the fluorenyl-indenyl ligand **379** to give the nonracemic metal complexes **515** and **516** in a 3:1 ratio (Scheme 8-103). These isomers could be separated or they could be hydrogenated then separated. In a related nonracemic bis(fluorenyl)zirconium complex **512** formed from bisfluorene) **380**, hydrogenation also led to the perhydrogenated complex **518** (including hydrogenation of the phenyl group to a cyclohexyl group) [172].

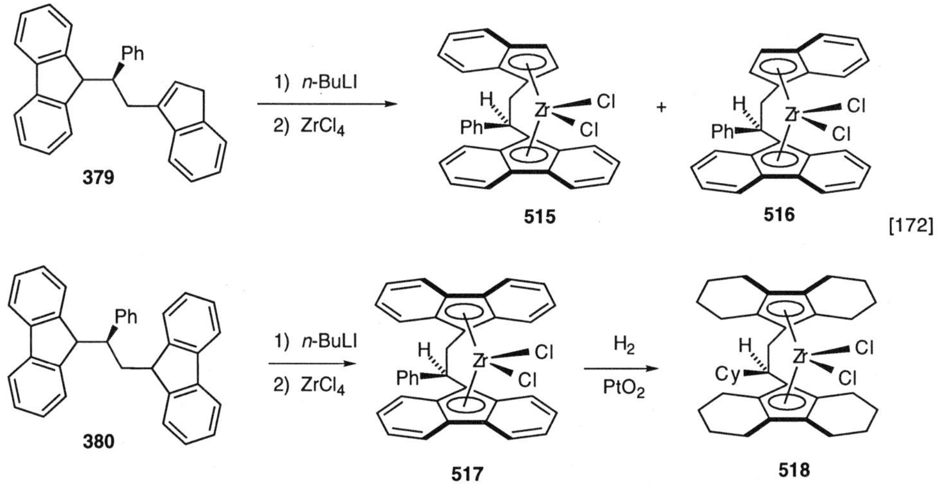

Scheme 8-103. Fluorenyl-containing metallocenes.

References

[1] a) Halterman, R. L. *Chem.Rev.* **1992**, *92*, 965. b) Okuda, J. *Topics Curr. Chem.* **1991**, *160*, 99. c) Janiak, C.; Schumann, H. *Adv. Organomet. Chem.* **1991**, *33*, 291.

[2] Moise, C.; Leblank, J.-C.; Tirouflet, J. *Tetrahedron Lett.* **1974**, 1723.

[3] Cesarotti, E.; Kagan, H. B.; Goddard, R.; Krüger, C.*J. Organomet. Chem.* **1978**, *162*, 297.

[4] Wild, F. R. P.; Wasiucionek, M.; Huttner, G.; Brintzinger, H. H. *J. Organomet. Chem.* **1985**, *288*, 63.

[5] a) Halterman, R. L.; Vollhardt, K. P. C. *Tetrahedron Lett.* **1986**, *27*, 1461. b) Halterman, R. L.; Vollhardt, K. P. C. *Organometallics* **1988**, *7*, 883.

[6] a) Wilkinson, G.; Pauson, P. L.; Birmingham, J. M.; Cotton, F. A. *J. Am. Chem. Soc.* **1953**, *75*, 1011.

  b) Wilkinson, G.; Birmingham, J. M. *J. Am. Chem. Soc.* **1954**, *76*, 4281-4284.

[7] Reynolds, L. T.; Wilkinson, G. *J. Inorg. Nucl. Chem.* **1959**, *9*, 86.

[8] Varga, V.; Mach, K.; Schmid, G.; Thewalt, U. *J. Organomet. Chem.* **1994**, *475*, 127.

[9] Leblanc, J. C.; Moise, C. *J. Organomet. Chem.* **1977**, *131*, 35.

[10] Schwemlein, H.; Brintzinger, H. H. *J. Organomet. Chem.* **1983**, *254*, 69.

[11] Gibis, K.-L.; Helmchen, G.; Huttner, G.; Zsolnai, L. *J. Organomet. Chem.* **1993**, *445*, 181.

[12] Huttenloch, M. E.; Diebold, J.; Rief, U.; Brintzinger, H. H.; Gilbert, A. M.; Katz, T. J. *Organometallics* **1992**, *11*, 3600.

[13] a) Samuel, E.; Setton, R.; *J. Organomet. Chem.* **1965**, *4*, 156. b) Atwood, J. L.; Hunter, W.E.; Hrncir, D. C.; Samuel, E.; Alt, H.; Rausch, M. D. *Inorg. Chem.* **1975**, *14*, 1757.

[14] a) Hauptman, E.; Waymouth, R. M.; Ziller, J. W. *J. Am. Chem. Soc.* **1995**, *117*, 11586. b) Waymouth, R. M.; Coates, G. W.; Hauptman, E. M. United States Patent 5,594,080, Jan. 14, 1997

[15] Shipman, J. *M.S. Thesis* University of Oklahoma, 1997.

[16] Nantz, M. H.; Hitchcock, S. R.; Sutton, S. C.; Smith, M.D. *Organometallics* **1993**, *12*, 5012.

[17]   Lotman, M. *M.S. Dissertation*, University of Oklahoma, 1992.
[18]   a) Piccolrovazzi, N.; Pino, P.; Consiglio, G.; Sironi, A.; Moret, M. *Organometallics* **1990**, *9*, 3098. b) Lee, I.-M.; Gauthier, W. J.; Ball, J.M.; Iyengar, B.; Collins, S. *Organometallics* **1992**, *11*, 2115.
[19]   a) Chen, Z.; Halterman, R. L. *Synlett* **1990**, 103. b) Chen, Z.; Eriks, K.; Halterman, R. L. *Organometallics* **1991**, *10*, 3449.
[20]   Halterman, R. L.; Chen, Z.; Khan, M. A. *Organometallics* **1996**, *15*, 3957.
[21]   Besançon, J.; Tirouflet, J.; Top, S.; Ea, B. H. *J. Organomet. Chem.* **1977**, *133*, 37.
[22]   a) Schlögl, K.; *Fortschr. Chem.Forsch.* **1966**, *6*, 479. b) Schlögl, K. *Top. Stereochem.* **1967**, *1*, 39.
[23]   Wild, F. R. W. P.; Zsolnai L.; Huttner, G.; Brintzinger, H. H. *J. Organomet. Chem.* **1982**, *232*, 233.
[24]   Krüger, C.; Lutz, F.; Nolte, M.; Erker, G.; Aulbach, M. *J.Organomet. Chem.* **1993**, *452*, 79.
[25]   Paquette, L. A.; Moriarty, K. J.; McKinney, J. A.; Rogers, R. D. *Organometallics* **1989**, *8*, 1707
[26]   McLaughlin, M. L.; McKinney, J. A.; Paquette, L. A. *Tetrahedron Lett.* **1986**, *27*, 5595.
[27]   a) Gallucci, J. C.; Gautheron, B.; Gugelchuk, M.; Meunier, P.; Paquette, L. A. *Organometallics* **1987**, *6*, 15. b) Paquette, L. A.; Moriarty, K. J.; Meunier, P.; Gautheron, B.; Crocq, V. *Organometallics* **1988**, *7*, 1873. c) Sornay, C.; Meunier, P.; Gautheron, B.; O'Doherty, G. A.; Paquette, L. A. *Organometallics* **1991**, *10*, 2082
[28]   Bell, L.; Whitby, R. J.; Jones, R. V. H.; Standen, M. C. H. *Tetrahedron Lett.* **1996**,*37*, 7139.
[29]   a) Knickmeier, M.; Erker, G.; Fox, T. *J. Am. Chem. Soc.* **1996**, *118*, 9623. b) Erker, G.; Aulbach, M. ; Knickmeier, M.; Wingermühle, D.; Krüger, C.; Nolte, M.; Werner, S. *J. Am. Chem. Soc.* **1993**, *115*, 4590.
[30]   a) Giardello, M. A.; Eisen, M. S.; Stern, C. L.; Marks, T. J. *J. Am. Chem. Soc.* **1993**, *115*, 3326. b) Giardello, M. A.; Eisen, M. S.; Stern, C. L.; Marks, T. J. *J. Am. Chem. Soc.* **1995**, *117*, 12114.
[31]   Sutton, S. C.; Nantz, M. H.; Parkin, S. R. *Organometallics* **1993**, *12*, 2248.
[32]   Rheingold, A. L.; Robinson, N. P.; Whelan, J.; Bosnich, B. *Organometallics* **1992**, *11*, 1869.
[33]   a) Ziegler, K.; Schafer, W. *Justus Liebigs Ann. Chem.* **1934**, *511*, 101. b) Alder, K.; Holzrichter, H.*Justus Liebigs Ann. Chem.* **1936**, *524*, 145.
[34]   Hafner, K.; Häfner, K. H.; König, C.; Kreuder, M.; Ploss, G.; Schulz, G.; Surm, E.; Vöpel, K. H. *Angew. Chem., Int. Ed. Engl.* **1963**, *2*, 123 and references therein.
[35]   Riemschneider, R. *Z. Naturforsch.* B**1963**, *18*, 641 and references therein.
[36]   Alder, K.; Ache, H.-J. *Chem. Ber.* **1962**, *95*, 503.
[37]   Hoch, M.; Rehder, D. *Chem. Ber.* **1988**, *121*, 1541.
[38]   Scroggins, W. T.; Rettig, M. F.; Wing, R. M. *Inorg. Chem.* **1976**, *15*, 1381.
[39]   a) Knox, G. R.; Munro, J. D.; Pauson, P. L.; Smith, G.H.; Watts, W. E. *J. Chem. Soc.* **1961**, 4619. b) Knox, G. R.; Pauson, P. L. *J. Chem. Soc.* **1961**, 4610.
[40]   Little, W. F.; Koestler, R. C. *J. Org. Chem.* **1961**, *26*, 3247.
[41]   Erker, G.; Nolte, R.; Krüger, C.; Schlund, R.; Benn, R.; Grondey, H.; Mynott, R. *J. Organomet. Chem.* **1989**, *364*, 119
[42]   Collins, S.; Hong, Y.; Kataoka, M.; Nguyen, T. *J. Org. Chem.***1990**, *55*, 3395.
[43]   Erker, G.; Aul, R. *Chem. Ber.* **1991**, *124*, 1301.
[44]   Pauson, P. L. *J. Am. Chem. Soc.* **1954**, *76*, 2187.
[45]   a) Riemschneider, R.; Nerin, R. *Monatsh. Chem.* **1960**, *91*, 829. b) Mironov, V. A.; Sobolev, E. V.; Elizarova, A. N. *Tetrahecron* **1963**, *19*, 1939.
[46]   Leblanc, J. C.; Moise, C. *J. Organomet. Chem.* **1976**, *120*, 65.
[47]   Erker, G.; Mollenkopf, C.; Grehl, M.; Schönecker, B. *Chem. Ber.* **1994**, *127*, 2341.
[48]   Erker, G.; Mollenkopf, C. *J. Organomet. Chem.* **1994**, *483*, 173.
[49]   Courtier, S.; Tainturier,G.; Gautheron, B. *J. Organomet. Chem.* **1980**, *195*, 291.
[50]   Fiaud, J. C.; Malleron, J. L. *Tetrahedron Lett.* **1980**, *21*, 4437.
[51]   a) Huang, Q.; Qian, Y. *Synthesis* **1987**, 910. b) Qian, Y.; Li, G.; Chen, W.; Li, B.; Jin, X. *J. Organomet. Chem.* **1989**, *373*, 185.

[52] Van de Weghe, P.; Bied, C.; Collin, J.; Marcalo, J.; Santos, I. *J. Organomet. Chem.* **1994**, *475*, 121.

[53] Adams, H.; Bailey, N. A.; Colley, M.; Schofield, P. A.; White, C. *J. Chem. Soc., Dalton Trans.* **1994**, 1445.

[54] van der Zeijden, A.A.H. *Tetrahedron Asymm.* **1995**, *6*, 913.

[55] a) McLean, S.; Haynes, P. *Tetrahedron* **1965**, *21*, 2343. b) McLean, S.; Haynes, P. *Tetrahedron* **1965**, *21*, 2313.

[56] Alder, K.; Ache, H.-J. *Chem. Ber.* **1962**, *95*, 503.

[57] Knothe, L.; Prinzbach, H.; Hadicke, E.; *Chem.Ber.* **1981**, *114*, 1656.

[58] Christoffers, J.; Bergman, R. G. *Angew. Chem., Int. Ed. Engl.* **1995**, *34*, 2266.

[59] Stone, K. J.; Little, R. D. *J. Org. Chem.* **1984**, *49*, 1849

[60] Okuda, J. *J. Organomet. Chem.* **1990**, *385*, C39

[61] Gutmann, S.; Burger, P.; Hund, H.-U.; Hofmann, J.; Brintzinger, H. H. *J. Organomet. Chem.* **1989**, *369*, 343.

[62] Collins, S.; Hong, Y.; Taylor, N. J. *Organometallics* **1990**, *9*, 2695.

[63] Drake, N. L.; Adams, J. R., Jr. *J. Chem. Soc* **1939**, *61*, 1326.

[64] Taylor, D. A. H. *J. Chem. Soc.* **1958**, 4779.

[65] Attig, T. G.; Wojcicki, A. *J. Organomet.Chem.***1974**, *82*, 397.

[66] Bursten, B. E.; Callstrom, M. R.; Jolly, C. A.; Paquette, L. A.; Sivik, M. R.; Tucker, R. S.; Wartchow, C. A. *Organometallics* **1994**, *13*, 127.

[67] Davies, A. G.; Lusztyk E.; Lusztyk, J. *J. Chem. Soc. Perkin Trans. II* **1982**, 729.

[68] Hughes, R. P.; Kowalski, A. S.; Lomprey, J. R.; Rheingold, A. L. *Organometallics* **1994**, *13*, 2691.

[69] Mironov, V. A.; Ivanov, A. P.; Kimelfled, Ya. M.; Petrovskaya, L. I.; Akhrem, A. A. *Tetrahedron Lett.* **1969**, 3347.

[70] Yang, Q.; Jensen, M. D. *Synlett* **1996**, 563.

[71] Reinarz, R. B.; Fonken, G. J. *Tetrahedron Lett.* **1973**, 4591.

[72] Paquette, L. A.; Gugelchuk, M.; McLaughlin, M. L. *J. Org. Chem.* **1987**, *52*, 4732.

[73] Halterman, R. L.; Vollhardt, K. P. C.; Welker, M. E.; Bläser, D.; Boese, R. *J. Am. Chem. Soc.* **1987**, *109*, 8105.

[74] a) Colletti, S. L.; Halterman, R. L. *Tetrahedron Lett.* **1990**, *30*, 3513. b) Colletti, S. L.; Halterman, R. L. *Organometallics* **1991**, *10*, 3438.

[75] Moriarty, K. J.; Rogers, R. D.; Paquette, L. A. *Organometallics* **1989**, *8*, 1512.

[76] Sitzman, H. *J. Organomet. Chem.* **1988**, *354*, 203.

[77] a) Vernier, C. G.; Casserly, E. W. *J. Am. Chem. Soc.* **1990**, *112*, 2808. b) Sitzmann, H.; Zhou, P.; Wolmershäuser, G. *Chem. Ber.* **1994**, *127*, 3.

[78] Okuda, J. *Chem. Ber.* **1989**, *122*, 1075.

[79] Okuda, J.; Herdtweck, E. *Chem. Ber.* **1988**, *121*, 1899.

[80] Mach, K.; Varga, V.; Antropiusová, H.; Polácek, J. *J. Organomet. Chem.* **1987**, *333*, 205.

[81] Conia, J.-M.; Leriverend, M.-L. *Bull. Chim. Soc. Fr.* **1970**, 2981.

[82] Dübner, F. Diplomarbeit, TU Berlin, 1997.

[83] Halterman, R. L.; Ramsey, T. M. *J. Organomet. Chem.* **1994**, *465*, 175.

[84] Paquette, L. A.; Charumilind, P.; Kravetz,T. M.; Böhm, M. C.; Gleiter, R. *J. Am. Chem. Soc.* **1983**, *105*, 3126.

[85] Halterman, R. L.; Tretyakov, A. *Tetrahedron (Symposium-in-Print)* **1995**, *51*, 4371.

[86] Schumann, H.; Janiak, C.; Zuckerman, J.J. *Chem. Ber.* **1988**, *121*, 207.

[87] a) Erker, G.; Schamberger, J.; van der Zeijden, A.A.H.; Denicke, S.; Krüger, C.; Goddard, R.; Nolte, M. *J. Organomet. Chem.* **1993**, *459*, 5832. b) Erker, G.; van der Zeijden, A. A. H. *Angew. Chem., Int. Ed. Engl.* **1990**, *29*, 512.

[88] a) Paquette, L. A.; Bzowej, E. I.; Kreuzholz, R. *Organometallics* **1996**, *15*, 4857. b) Garner, C. M.; Prince, M. E. *Tetrahedron Lett.* **1994**, *35*, 2463.

[89] de Vries, L. *J. Org. Chem.* **1960**, *25*, 1838.

[90] a) Burger, U.; Delay, A.; Mazenod, F. *Helv. Chim. Acta* **1974**, *57*, 2106. b) Kohl, F. X.; Jutzi, P. *J. Organomet. Chem.***1983**, *243*, 119.

542   8 *Synthesis of Chiral Titanocene and Zirconocene Dichlorides*

[91] Fietler, D.; Whitesides, G. M. *Inorg. Chem.* **1976**, *15*, 466.
[92] Threlkel, R. S.; Bercaw, J. E. *J. Organomet. Chem.* **1977**, *136*, 1.
[93] Stein, D.; Sitzmann, H. *J. Organomet. Chem.* **1991**, *402*, 249.
[94] Dormond, A.; Bouadili, A. El.; Moise, C. *Tetrahedron Lett.* **1983**, *24*, 3087.
[95] Colletti, S. L. Ph.D. Dissertation, Boston University, 1993.
[96] Greifenstein, L. G.; Lambert, J. B.; Nienhuis, R. J.; Fried, H. E.; Pagani, G. A. *J. Am. Chem. Soc.* **1981**, *46*, 5125
[97] Meurling, L. *Acta Chem. Scand.B* **1974**, *28*, 295.
[98] Cedheim, L.; Eberson, L. *Synthesis* **1973**, 159.
[99] a) Rakita, P. E.; Taylor, G. A. *Inorg. Chem.* **1972**, *11*, 2136. b) Andrews, M. N.; Rakita, P. E.; Taylor, G. A. *Tetrahedron Lett.* **1973**, 1851.
[100] a) Rigby, S. S.; Gupta, H. K.; Werstiuk, N. H.; Bain, A. D.; McGlinchey, M. J. *Polyhedron* **1995**, *14*, 2787. b) Luzikov, Yu. N.; Gergeyev, N. M.; Ustynyuk, Yu. A. *J. Organomet. Chem.* **1974**, *65*, 303.
[101] Ready, T. E.; Chien, J. C. W.; Rausch, M. D. *J. Organomet. Chem.* **1996**, *519*, 21.
[102] Adamczyk, M.; Watt , D. S.; Netzel, D. A. *J. Org. Chem.* **1984**, *49*, 4226.
[103] Edlund, U. *Organ. Mag. Reson.* **1978**, *11*, 516.
[104] Ribakove, E. C.; Kerber, R. C. *Organometallics* **1990**, *9*, 531.
[105] Seljeseth, C. *B.S. Thesis*, University of Oklahoma, 1997.
[106] Spaleck, W.; Küber, F.; Winter, A.; Rohrmann, J.; Bachmann, B.; Antberg, M.; Dolle, V.; Paulus, E. F. *Organometallics* **1994**, *13*, 954.
[107] Spaleck, W.; Antberg, M.; Rohrmann, J.; Winter, A.; Bachmann, B.; Kiprof, P.; Behm, J.; Herrmann, W. A. *Angew. Chem., Int. Ed. Engl.* **1992**, *31*, 1347.
[108] a) Cedheim, L. Eberson, L. *Acta. Chem. Scand. B* **1976**, *30*, 527. b) Normant-Chefnay, C. *Bull. Soc. Chim. Fr.* **1971**, 1351.
[109] Erker, G.; Psiorz, Fröhlich, R.; Grehl, M.; Krüger, C.; Noe, R.; Nolte, M. *Tetrahedron* **1995**, *51*, 4347.
[110] a) Kaminsky, W.; Rabe, O.; Schauwienold, A.-M.; Schupfner, G. U.; Hanss, J.; Kopf, J. *J. Organomet. Chem.* **1995**, *497*, 181-193. b) O'Hare, D.; Green, J. C.; Marder, T.; Collins, S.; Stringer, G.; Kakkar, A. K.; Kaltsoyannis, N.; Kuhn, A.; Lewis, R.; Mehnert, C.; Scott, P.; Kurmoo, M.; Pugh, S. *Organometallics* **1992**, *11*, 48.
[111] Stehling, U.; Diebold, J.; Kirsten, R.; Röll, W.; Jüngling, S.; Mülhaupt, R.; Langhauser, F. *Organometallics* **1994**, *13*, 964
[112] Erker, G.; Aulbach, M.; Krüger, C.; Werner, S. *J. Organomet. Chem.* **1993**, *450*, 1.
[113] a) Erker, G.; Aulbach, M.; Knickmeier, M.; Wingbermühle, D.; Krüger, C.; Nolte, M.; Werner, S. *J. Am. Chem. Soc.* **1993**, *115*, 4590. b) Knickmeier, M.; Erker, G.; Fox, T. *J. Am. Chem. Soc.* **1996**, *118*, 9623.
[114] Stenzel, O.; Schumann, H.; Halterman, R. L., unpublished results.
[115] Chen, Z.; Halterman, R. L. *J. Am. Chem. Soc.* **1992**, *114*, 2276.
[116] a) Katz, T. J.; Acton, N. *Tetrahedron Lett.* **1970**, 2497. b) Katz, T. J.; Acton, N.; Martin, G. *J. Am. Chem. Soc.* **1973**, *95*, 2934.
[117] Smith, J. A.; Brintzinger, H. H. *J. Organomet. Chem.* **1981**, *218*, 159.
[118] Hillman, M.; Weiss, A. J. *J. Organomet. Chem.* **1972**, *42*, 123.
[119] Nifant'ev, I.E.; Butakov, K. A.; Aliev, Z. G.; Urazovskii, I. F. *Organomet. Chem. USSR* **1991**, *4*, 622.
[120] Bajgur, C. S.; Tikkanen, W. R.; Petersen, J. L. *Inorg. Chem.* **1985**, *24*, 2539.
[121] Lang, H.; Seyferth, D. *Organometallics* **1991**, *10*, 347.
[122] Chen, Z.; Halterman, R. L. *Organometallics* **1994**, *13*, 3932.
[123] Burger, P.; Brintzinger, H. H. *J. Organomet. Chem.* **1991**, *407*, 207.
[124] Wochner, F.; Zsolnai, Huttner, G.; Brintzinger, H. H. *J. Organomet. Chem.* **1985**, *288*, 69.
[125] Mink, C.; Hafner, K. *Tetrahedron Lett.* **1994**, *35*, 4087.
[126] Dorer, B.; Prosenc, M.-H.; Rief, U.; Brintzinger, H. H. *Organometallics* **1994**, *13*, 3868.
[127] Hafner, K.; Thiele, G. F.; Mink, C. *Angew. Chem., Int. Ed. Engl.* **1988**, *27*, 1191.

[128] Wiesenfeldt, H.; Reinmuth, A.; Barsties, E.; Evertz, K.; Brintzinger, H. H. *J. Organomet. Chem.* **1989**, *369*, 359.

[129] Mise, T.; Miya, S.; Yamazaki, H. *Chem. Lett.* **1989**, 1853.

[130] Erickson, M. S.; Fronczek, F. R.; McLaughlin, M. L. *J. Organomet. Chem.* **1991**, *415*, 75.

[131] Collins, S.; Hong, Y.; Ramachandran, R.; Taylor, N. J. *Organometallics* **1991**, *10*, 2349.

[132] Halterman, R. L.; Ramsey, T. M.; Pailes, N. A.; Khan, M. A. *J. Organomet. Chem.* **1995** *497*, 43.

[133] Hüttenhofer, M.; Prosenc, M.-H.; Rief, U.; Schaper, F.; Brintzinger, H.-H. *Organometallics* **1996**, *15*, 4816.

[134] Tretyakov, A. Ph.D. Dissertation, University of Oklahoma, 1995.

[135] Maréchal, E.; Lepert, A. *Bull. Soc. Chim. Fr.* **1967**, 2954.

[136] Collins, S.; Kuntz, B. A.; Taylor, N. J.; Ward, D. G. *J. Organomet. Chem.* **1988**, *342*, 21.

[137] Grossman, R. B.; Doyle, R. A.; Buchwald, S. L. *Organometallics* **1991**, *10*, 1501.

[138] Erker, G.; Mollenkopf, C.; Grehl, M.; Fröhlich, R.; Krüger, C.; Noe, R.; Riedel, M. *Organometallics* **1994**, *13*, 1950.

[139] Herrmann, W. A.; Rohrmann, J.; Herdtweck, E.; Spaleck, W.; Winter, A. *Angew. Chem., Int. Ed. Engl.* **1989**, *28*, 1511.

[140] Spaleck, W.; Antberg, M.; Dolle, V.; Klein, R.; Rohrmann, J.; Winter, A. *New J. Chem.* **1990**, *14*, 499.

[141] Scott, P.; Rief, U.; Diebold, J.; Brintzinger, H. H. *Organometallics* **1993**, *12*, 3094.

[142] Halterman, R. L.; Tretyakov, A.; Combs, D.; Chang, J.; Khan, M. *Organometallics* **1997**, 6.

[143] Halterman, R. L.; Combs, D.; Khan, M. A., *Organometallics,* in press.

[144] Alt, H. G.; Milius, W.; Palackal, S. J. *J. Organomet. Chem.* **1994**, *472*, 113.

[145] Chen, Y.-X.; Rausch, M. D.; Chien, J. C. W. *J. Organomet. Chem.* **1995**, *497*, 1.

[146] Resconi, L.; Jones, R. L.; Rheingold, A. L.; Yap, G. P. A. *Organometallics* **1996**, *15*, 998.

[147] Patsidis, K.; Alt, H. G.; Hilius, W.; Palackal, S. J. *J. Organomet. Chem.* **1996**, *509*, 63.

[148] Alt, H. G.; Zenk, R. *J. Organomet. Chem.* **1996**, *512*, 51.

[149] Green, M. L. H.; Ishihara, N. *J. Chem. Soc., Dalton Trnas.* **1994**, 657.

[150] Willoughby, C. A.; Davis, W. M.; Buchwald, S. L. *J. Organomet. Chem.* **1995**, *497*, 11.

[151] Kaminsky, W.; Engehausen, R.; Kopf, J. *Angew. Chem., Int. Ed. Engl.* **1995**, *34*, 2273.

[152] Miyake, S.; Okumura, Y.; Inazawa, S. *Macromolecules* **1995**, *28*, 3074.

[153] Ewen, J. A.; Jones, R. L.; Razavi, A.; Ferrara, J. D. *J. Am. Chem. Soc.* **1988**, *110*, 6255.

[154] Razavi, A.; Ferrara, J. *J. Organomet. Chem.* **1992**, *435*, 299.

[155] Razavi, A.; Atwood, J. L. *J. Organomet. Chem.* **1993**, *459*, 117.

[156] Alt, H. G.; Zenk, R. *J. Organomet. Chem.* **1996**, *518*, 7.

[157] Razavi, A.; Atwood, J. L. *J. Organomet. Chem.* **1995**, *497*, 105.

[158] Alt, H. G.; Zenk, R.; Milius, W. *J. Organomet. Chem.* **1996**, *514*, 257.

[159] Alt, H. G.; Zenk, R. *J. Organomet. Chem.* **1996**, *526*, 295.

[160] a) Mallin, D. T.; Rausch, M. D.; Lin. Y.-L.; Dong, S.; Chien, J. C. W. *J. Am. Chem. Soc.* **1990**, *112*, 2030. b) Llinas, G. H.; Day, R. O.; Rausch, M. D.; Chien, J. C. W. *Organometallics* **1993**, *12*, 1283.

[161] a) Erker, G.; Psiorz, C.; Krüger, C.; Nolte, M. *Chem. Ber.* **1994**, *127*, 1551. b) Psiorz, C.; Erker, G.; Fröhlich, R.; Grehl, M. *Chem. Ber.* **1995**, *128*, 357.

[162] a) Burk, M. J.; Colletti, S. L., Halterman, R. L. *Organometallics* **1991**, *10*, 2998. b) Colletti, S. L.; Halterman, R. L. *J. Organomet. Chem.* **1993**, *455*, 99.

[163] Bandy, J. A.; Green, M. L. H.; Gardiner, I. M.; Prout, K. *J. Chem. Soc., Dalton Trans.* **1991**, 2207.

[164] Halterman, R. L.; Combs, D.;. Kihega, J. G.; Khan, M. A. *J. Organomet. Chem.* **1996**, *520*, 163.

[165] Rieger, B. *J. Organomet. Chem.* **1992**, *428*, C33.

[166] Hollis, T. K.; Rheingold, A. L.; Robinson, N. P.; Whelan, J.; Bosnich, B. *Organometallics* **1992**, *11*, 2812.

[167] a) Halterman, R. L.; Ramsey, T. M. *Organometallics* **1993**, *12*, 2879. b) Halterman, R. L.;

Ramsey, T. M. *J. Organomet. Chem.* **1997**, *530*, 225.

[168] Ellis, W. W.; Hollis, T. K.; Odenkirk, W.; Whelan, J.; Ostrander, R.; Rheingold, A. L.; Bosnich, B. *Organometallics* **1993**, *12*, 4391.

[169] Chen, Y.-X.; Rausch, M. D.; Chien, J. C. W. *J. Polym. Sci. A: Polym. Chem.* **1995**, *33*, 2093.

[170] Mitchell, J. P.; Hajela, S.; Brookhart, S. K.; Hardcastle, K. I.; Henling, L. M.; Bercaw, J. E. *J. Am. Chem. Soc.* **1996**, *118*, 1045.

[171] Grossman, R. B.; Tsai, J.-C.; Davis, W. M.; Gutiérrez, Buchwald, S.L. *Organometallics* **1994**, *13*, 3892.

[172] a) Jany, G.; Gawzi, R.; Steimann, M.; Rieger, B. *Organometallics* **1997**, *16*, 544. b) Rieger, B.; Jany, G.; Fawzi, R.; Steimann, M. *Organometallics* **1994**, *13*, 647. c) Rieger, B.; Jany, G. *Chem. Ber.* **1994**, *127*, 2417.

[173] a) Möhring, P. C.; Coville, N. J. *J. Mol. Catal.* **1992**, *77*, 41. b) Möhring, P. C.; Vlachakis, N.; Grimmer, N. E.; Coville, N. J. *J. Organomet. Chem.* **1994**, *483*, 159. c) Polymerization review: Möhring, P. C.; Coville, N. J. *J. Organomet. Chem.* **1994**, *279*, 1.

[174] Sullivan, M. F.; Little, W. F. *J. Organomet. Chem.* **1967**, *8*, 277.

[175] Davis, J. H.; Sun, H.; Redfield, D.; Stucky, G. D. *J. Magn. Reson.* **1980**, *37*, 441.

[176] Lappert, M. F.; Pickett, C. J.; Riley, P. I.; Yarrow, P. I. W. *J. Chem. Soc., Dalton Trans.* **1981**, 805.

[177] Chandra, G.; Lappert, M. F. *J. Chem. Soc. (A)* **1968**, 1940.

[178] Vitz, E.; Brubaker, C. H. Jr. *J. Organomet. Chem.* **1974**, *84*, C16.

[179] Bercaw, J. E.; Marvich, R. H.; Bell, L. G.; Brintzinger, H. H. *J. Am. Chem. Soc.* **1972**, *94*, 1219.

[180] a) Ti: Gorsich, R. D.; *J. Am. Chem. Soc.* **1960**, *82*, 4211. b) Zr: Erker, G.; Gerg, K.; Treschanke, L.; Engel, K. *Inorg. Chem.* **1982**, *21*, 1277.

[181] Ott, K. C.; deBoer, E. J. M.; Grubbs, R. H. *Organometallics* **1984**, *3*, 223.

[182] Erker, G.; Nolte, R.; Tsay, Y.-H.; Krüger, C. *Angew. Chem., Int. Ed. Engl.* **1989**, *28*, 628.

[183] Erker, G.; Nolte, R.; Aul, Wilker, S.; Krüger, C.; Noe, R. *J. Am. Chem. Soc.* **1991**, *113*, 7594.

[184] Burger, P.; Hund, H.-U.; Evertz, K.; Brintzinger, H.H. *J. Organomet. Chem.* **1989**, *378*, 153.

[185] Ramsey, T. M. *Ph.D. Dissertation*, Univeristy of Oklahoma, 1994.

[186] O'Hare, D.; Murphy, V.; Diamond, G. M.; Arnold, P.; Mountford, P. *Organometallics* **1994**, *13*, 4689.

[187] Kowala, C.; Wailes, P. C.; Weigold, H.; Wunderlich, J. A. *J. Chen. Soc., Chem. Commun.* **1974**, 993.

[188] Razavi, A.; Atwood, J. L. *J. Am. Chem. Soc.* **1993**, *115*, 7529.

[189] Patsidis, K.; Alt, H. G. *J. Organomet. Chem.* **1995**, *501*, 31.

[190] Schmid, M. A.; Alt, H. G.; Milius, W. *J. Organomet. Chem.* **1996**, *514*, 45.

[191] Hart, S. L.; Duncalf, D. J.; Hastings, J. J.; McCamley, A.; Taylor, P. C. *J. Chem. Soc., Dalton Trans.* **1996**, 2843.

[192] Sodhi, G. S.; Sharma, A. K.; Kaushik, N. K. *J. Organomet. Chem.* **1982**, *238*, 177.

[193] Schmid, M. A.; Alt, H. G.; Milius, W. *J. Organomet. Chem.* **1996**, *525*, 15.

[194] Schmid, M. A.; Alt, H. G.; Milius, W. *J. Organomet. Chem.* **1996**, *525*, 9.

[195] Nifant'ev, I. E.; Churakov, A. V.; Urazowski, I. F.; Mkoyan, Sh.G.; Atovmyan, L. O. *J. Organomet. Chem.* **1992**, *435*, 37.

[196] Schnutenhaus, H.; Brintzinger, H. H. *Angew. Chem., Int. Ed. Engl.* **1979**, *18*, 777.

[197] Collins, S.; Gauthier, W. J.; Holden, D. A.; Kuntz, B. A.; Taylor, N. J.; Ward, D. G. *Organometallics* **1991**, *10*, 2061.

[198] Schäfer, A.; Karl, E.; Zsolnai, L.; Huttner, G.; Brintzinger, H.-H. *J. Organomet. Chem.* **1987**, *328*, 87.

[199] Waymouth, R. M.; Bangerter, F.; Pino, P. *Inorg. Chem.* **1988**, *27*, 758.

[200] a) Diamond, G. M.; Rodewald, S.; Jordan, R. F. *Organometallics* **1995**, *14*, 5. b) Diamond, G. M.; Jordan, R. F.; Petersen, J. L. *Organometallics* **1996**, *15*, 4030.

[201] Röll, W.; Zsolnai, L.; Huttner, G.; Brintzinger, H. H. *J. Organomet. Chem.* **1987**, *322*, 65.

[202] Atovmyan, L.; Mkoyan, S.; Urazowski, I.; Broussier, R.; Ninoreille, S.; Perron, P.; Gautheron, B. *Organometallics* **1995**, *14*, 2601.